René L. Schilling, Lothar Partzsch
Brownian Motion
De Gruyter Graduate

Also of Interest

Stochastic Finance, 3rd Ed.
Hans Föllmer, Alexander Schied, 2011
ISBN 978-3-11-021804-6, e-ISBN 978-3-11-021805-3

Bernstein Functions, 2nd Ed.
René L. Schilling, Renming Song, Zoran Vondraček, 2012
ISBN 978-3-11-025229-3, e-ISBN 978-3-11-026933-8,
Set-ISBN 978-3-11-026900-0

Stochastic Calculus of Variations for Jump Processes
Yasushi Ishikawa, 2013
ISBN 978-3-11-028180-4, e-ISBN 978-3-11-028200-9,
Set-ISBN 978-3-11-028201-6

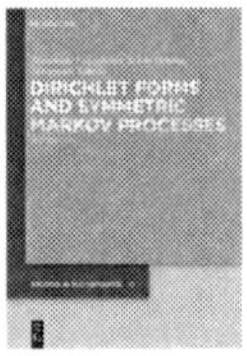

Dirichlet Forms and Symmetric Markov Processes, 2nd Ed.
Masatoshi Fukushima, Yoichi Oshima, Masayoshi Takeda, 2010
ISBN 978-3-11-021808-4, e-ISBN 978-3-11-021809-1,
Set-ISBN 978-3-11-173321-0

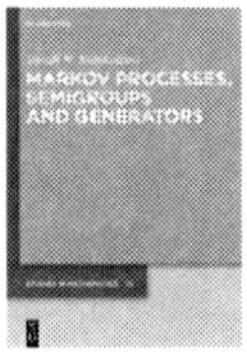

Markov Processes, Semigroups and Generators
Vassili N. Kolokoltsov, 2011
ISBN 978-3-11-025010-7, e-ISBN 978-3-11-025011-4,
Set-ISBN 978-3-11-916510-5

Stochastic Models for Fractional Calculus
Mark M. Meerschaert, Alla Sikorskii, 2011
ISBN 978-3-11-025869-1, e-ISBN 978-3-11-025816-5,
Set-ISBN 978-3-11-220472-6

René L. Schilling, Lothar Partzsch

Brownian Motion

An Introduction to Stochastic Processes

With a Chapter on Simulation by Björn Böttcher

2nd Edition

DE GRUYTER

Mathematics Subject Classification 2010
Primary: 60-01, 60J65; Secondary: 60H05, 60H10, 60J35, 60G46, 60J60, 60J25.

Authors

Prof. Dr. René L. Schilling
Technische Universität Dresden
Institut für Mathematische Stochastik
D-01062 Dresden
Germany

rene.schilling@tu-dresden.de
www.math.tu-dresden.de/sto/schilling

Dr. Lothar Partzsch
Technische Universität Dresden
Institut für Mathematische Stochastik
D-01062 Dresden
Germany

lothar.partzsch@tu-dresden.de

Online Resources

www.motapa.de/brownian_motion

ISBN 978-3-11-030729-0
e-ISBN 978-3-11-030730-6

Library of Congress Cataloging-in-Publication Data
A CIP catalog record for this book has been applied for at the Library of Congress.

Bibliographic information published by the Deutsche Nationalbibliothek
The Deutsche Nationalbibliothek lists this publication in the Deutsche Nationalbibliografie; detailed bibliographic data are available in the Internet at http://dnb.dnb.de.

Printing and binding: CPI buch bücher.de GmbH, Birkach
♾Printed on acid-free paper
Printed in Germany

MIX
Papier aus verantwortungsvollen Quellen
FSC® C003147

www.degruyter.com

Preface to the second edition

This second edition contains a good number of additions scattered throughout the text as well as numerous voluntary (and involuntary) changes. Most notably we have added Section 7.5 on the *Potential operator* of a Brownian motion, Chapter 11 *Brownian motion as a random fractal* and Chapter 20 on *Stratonovich's stochastic calculus*. The last addition prompted us to rewrite large portions of our presentation of the stochastic calculus; Chapter 17 on Itô's formula contains now full proofs also of the multivariate and time-dependent cases, and the SDE-chapters (Chapter 19 and 20) put more emphasis on solution strategies for concrete stochastic differential equations. Throughout the book we added further exercises which come with a detailed solution manual on the internet http://www.motapa.de/brownian_motion/. Despite of all of these changes we did not try to make our book into an encyclopedia, but we wanted to keep the easily accessible, introductory character of the exposition, and there are still plenty of topics deliberately left out.

After the publication of the first edition, we got many positive responses from colleagues and students alike, and quite a few led to improvements in the new edition. Thanks are due to our readers for telling us about errors and obscurities in the first edition. Very special thanks go to our student Franziska Kühn (who ploughed through the whole text and knows it inside out) and to Björn Böttcher and Julian Hollender for their valuable contributions. The de Gruyter editorial staff around Ms. Friederike Dittberner did a splendid job for this project and deserves our appreciation.

Without the patience – we tried it hard enough in all those moments when our thoughts were again diffusing in Wiener space – and support of our families this endeavour would not have been possible. Thank you!

Dresden, February 2014

René L. Schilling
Lothar Partzsch

Preface

Brownian motion is arguably the single most important stochastic process. Historically it was the first stochastic process in continuous time and with a continuous state space, and thus it influenced the study of Gaussian processes, martingales, Markov processes, diffusions and random fractals. Its central position within mathematics is matched by numerous applications in science, engineering and mathematical finance.

The present book grew out of several courses which we taught at the University of Marburg and TU Dresden, and it draws on the lecture notes [172] by one of us. Many students are interested in applications of probability theory and it is important to teach Brownian motion and stochastic calculus at an early stage of the curriculum. Such a course is very likely the first encounter with stochastic processes in continuous time, following directly on an introductory course on rigorous (i. e. measure-theoretic) probability theory. Typically, students would be familiar with the classical limit theorems of probability theory and basic discrete-time martingales, as it is treated, for example, by Jacod & Protter *Probability Essentials* [108], Williams *Probability with Martingales* [232], or in the more voluminous textbooks by Billingsley [15] and Durrett [61].

General textbooks on probability theory cover however, if at all, Brownian motion only briefly. On the other hand, there is a quite substantial gap to more specialized texts on Brownian motion which is not so easy to overcome for the novice. Our aim was to write a book which can be used in the classroom as an introduction to Brownian motion and stochastic calculus, and as a first course in continuous-time and continuous-state Markov processes. We also wanted to have a text which would be both a readily accessible mathematical back-up for contemporary applications (such as mathematical finance) and a foundation to get easy access to advanced monographs, e. g. Karatzas & Shreve [119], Revuz & Yor [189] or Rogers & Williams [195] (for stochastic calculus), Marcus & Rosen [154] (for Gaussian processes), Peres & Mörters [162] (for random fractals), Chung [31] or Port & Stone [182] (for potential theory) or Blumenthal & Getoor [18] (for Markov processes) to name but a few.

Things the readers are expected to know: Our presentation is basically self-contained, starting from 'scratch' with continuous-time stochastic processes. We do, however, assume some basic measure theory (as in [204]) and a first course on probability theory and discrete-time martingales (as in [108] or [232]). Some 'remedial' material is collected in the appendix, but this is really intended as a back-up.

How to read this book: Of course, nothing prevents you from reading it linearly. But there is more material here than one could cover in a one-semester course. Depending on your needs and likings, there are at least three possible selections: *BM and Itô calculus*, *BM and its sample paths* and *BM as a Markov process*. The diagram on page xiii will give you some ideas how things depend on each other and how to construct your own 'Brownian sample path' through this book.

Whenever special attention is needed and to point out traps & pitfalls, we have ! used the ! sign in the margin. Also in the margin, there are cross-references to exercises at the end of each chapter which we think fit (and are sometimes needed) at Ex. n.m that point.[1] They are not just drill problems but contain variants, excursions from and extensions of the material presented in the text. The proofs of the core material do not seriously depend on any of the problems.

Writing an introductory text also meant that we had to omit many beautiful topics. Often we had to stop at a point where we, hopefully, got you really interested... Therefore, we close every chapter with a brief outlook on possible texts for further reading.

Many people contributed towards the completion of this project: First of all the students who attended our courses and helped – often unwittingly – to shape the presentation of the material. We profited a lot from comments by Niels Jacob (Swansea) and Panki Kim (Seoul National University) who used an early draft of the manuscript in one of his courses. Special thanks go to our colleagues and students Björn Böttcher, Katharina Fischer, Julian Hollender, Felix Lindner and Michael Schwarzenberger who read substantial parts of the text, often several times and at various stages. They found countless misprints, inconsistencies and errors which we would never have spotted. Björn helped out with many illustrations and, more importantly, contributed Chapter 22 on simulation. Finally we thank our colleagues and friends at TU Dresden and our families who contributed to this work in many uncredited ways. We hope that they approve of the result.

Dresden, February 2012

René L. Schilling
Lothar Partzsch

1 For the readers' convenience there is a web page where additional material and solutions are available. The URL is http://www.motapa.de/brownian_motion/.

Contents

Dependence chart

As we have already mentioned in the preface, there are at least three paths through this book which highlight different aspects of Brownian motion: Brownian motion and Itô calculus, Brownian motion as a Markov process, and Brownian motion and its sample paths. Below we suggest some fast tracks “C”, “M” and “S” for each route, and we indicate how the other topics covered in this book depend on these fast tracks. This should help you to find your own personal sample path. Starred sections (in the grey ovals) contain results which can be used without proof and without compromising too much on rigour.

Getting started

For all three fast tracks you need to read Chapters 1 and 2 first. If you are not too much in a hurry, you should choose *one* construction of Brownian motion from Chapter 3. For the beginner we recommend either Sections 3.1, 3.2 or Section 3.4.

Basic stochastic calculus (C)

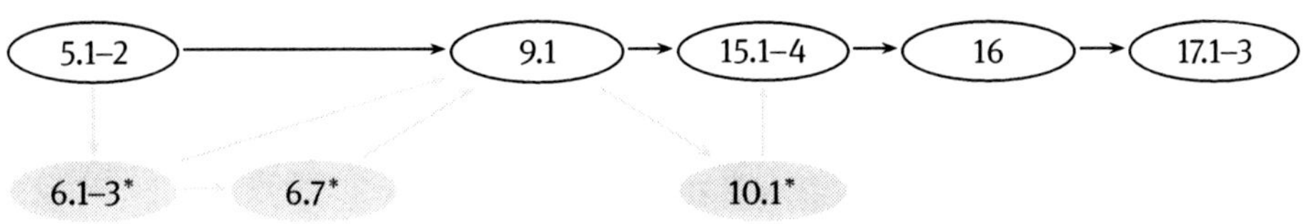

Basic Markov processes (M)

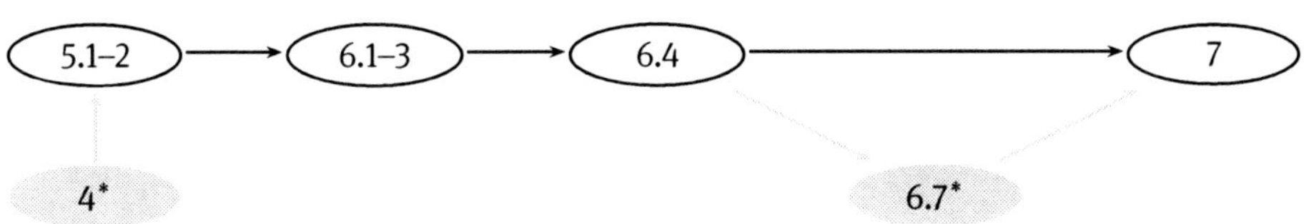

Basic sample path properties (S)

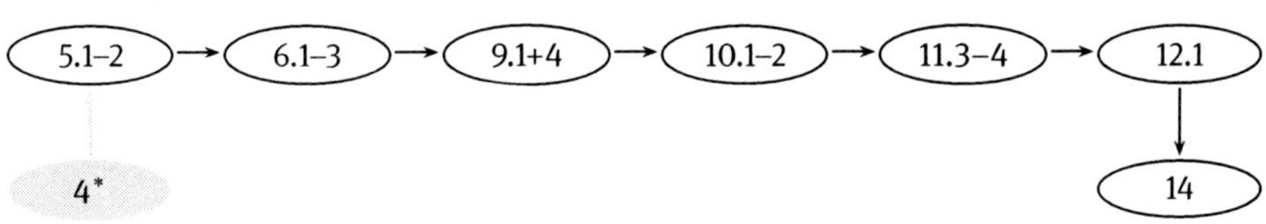

Dependence to the sections 5.1–21.2

The following list shows which prerequisites are needed for each section. A star as in 4* or 6.7* indicates that some result(s) from Chapter 4 or Section 6.7 are used which may be used without proof and without compromising too much on rigour. Starred sections are mentioned only where they are actually needed, while other prerequisites are repeated to indicate the full line of dependence. For example,

6.6: M *or* S *or* C, 6.1–3

indicates that the prerequisites for Section 6.6 are covered by either "M" or "S" or "C if you add 6.1, 6.2, 6.3". Since we do not refer to later sections with higher numbers, you will only need those sections in "M", "S", or "C and 6.1, 6.2, 6.3" with section numbers below 6.6. Likewise,

18.1: C, 17.4–5, 15.6*

means that 18.1 requires "C" plus the Sections 17.4 and 17.5. Some results from 15.6 are used, but they can be quoted without proof.

5.1: C *or* M *or* S
5.2: C *or* M *or* S
5.3: C *or* M *or* S
6.1: M *or* S *or* C
6.2: M *or* S *or* C, 6.1
6.3: M *or* S *or* C, 6.1–2
6.4: M *or* S *or* C, 6.1–3
6.5: M *or* S *or* C, 6.1–3
6.6: M *or* S *or* C, 6.1–3
6.7: M *or* S *or* C, 6.1–3
7.1: M, 4.2**or* C, 6.1, 4.2*
7.2: M *or* C, 6.1, 7.1
7.3: M *or* C, 6.1, 7.1–2
7.4: M *or* C, 6.1, 7.1–3
7.5: M *or* C, 6.1, 7.1–4
7.6: M *or* C, 6.1, 7.1–5
8.1: M *or* C, 6.1, 7.1–3
8.2: M, 8.1 *or* C, 6.1, 7.1–3, 8.1
8.3: M, 8.1–2 *or* C, 6.1, 7.1–4, 8.1–2
8.4: M, 6.7*, 8.1–3* *or* C, 6.1–4, 7, 6.7*, 8.1–3*
9.1: S *or* C *or* M
9.2: S *or* C *or* M, 9.1
9.3: S *or* C *or* M, 9.1
9.4: S *or* C *or* M, 9.1
10.1: S *or* C *or* M
10.2: S *or* C *or* M
10.3: S *or* C *or* M
11.1: S *or* M *or* C
11.2: S, 11.1, 10.3* *or* M, 11.1, 10.3* *or* C, 11.1, 10.3*
11.3: S *or* M *or* C, 6.1–3
11.4: S *or* M *or* C, 6.1–3
11.5: S, 11.3–4, 11.2* *or* M, 11.3–4, 11.2* *or* C, 6.1–3, 11.3–4, 11.2*
12.1: S, 10.3**or* C, 10.3**or* M, 10.3*
12.2: S *or* C, 12.1 *or* M, 12.1
13.1: S
13.2: S, 13.1
13.3: S, 13.1–2, 4*
14: S *or* C *or* M
15.1: C *or* M
15.2: C, 6.7**or* M, 15.1, 6.7*
15.3: C *or* M, 15.1–2
15.4: C *or* M, 15.1–3, 9.1*
15.5: C, 6.7*
15.6: C, 15.5
16: C *or* M, 15.1–4
17.1: C
17.2: C
17.3: C
17.4: C
17.5: C, 17.4
17.6: C, 17.4–5
17.7: C
18.1: C, 17.4–5, 15.6*
18.2: C, 17.4–5
18.3: C, 17.4–5, 18.1
18.4: C, 17.4–5
18.5: C, 15.5–6, 17.4–5, 18.2
18.6: C, 17.4–5, 18.2
18.7: C, 17.4–5
19.1: C
19.2: C, 19.1
19.3: C, 19.1
19.4: C, 19.1–3
19.5: C, 17.4–5, 19.1
19.6: C, 17.4, 17.7, 18.2–3, 19.1–5
19.7: C, 6.1, 17.4–5, 19.1, 19.5
19.8: C, 17.4–5, 19.1, 19.5–6
19.9: C, 17.4–5, 19.1, 19.5–6, 10.1*, 18.7*
20.1: C, 17.6
20.2: C, 17.6, 19.1, 19.5, 20.1
21.1: M *or* C, 6.1, 7
21.2: C, 6.1, 7, 17.4–5, 19, 21.1

Index of notation

This index is intended to aid cross-referencing, so notation that is specific to a single section is generally not listed. Some symbols are used locally, without ambiguity, in senses other than those given below; numbers following an entry are page numbers.

Unless otherwise stated, functions are real-valued and binary operations between functions such as $f \pm g$, $f \cdot g$, $f \wedge g$, $f \vee g$, comparisons $f \leqslant g$, $f < g$ or limiting relations $f_j \xrightarrow{j\to\infty} f$, $\lim_j f_j$, $\underline{\lim}_j f_j$, $\overline{\lim}_j f_j$, $\sup_j f_j$ or $\inf_j f_j$ are understood pointwise. 'Positive' and 'negative' always means '$\geqslant 0$' and '$\leqslant 0$'.

General notation: analysis

$\inf\emptyset$	$\inf\emptyset = +\infty$
$a \vee b$	maximum of a and b
$a \wedge b$	minimum of a and b
a^+	$a \vee 0$
a^-	$-(a \wedge 0)$
$\lfloor x \rfloor$	largest integer $n \leqslant x$
$\lvert x\rvert$	Euclidean norm in $\mathbb{R}^d$, $\lvert x\rvert^2 = x_1^2 + \cdots + x_d^2$
$\langle x, y\rangle$	scalar product in $\mathbb{R}^d$, $\sum_{j=1}^d x_j y_j$
I_d	unit matrix in $\mathbb{R}^{d\times d}$
$\mathbb{1}_A$	$\mathbb{1}_A(x) = \begin{cases} 1, & x \in A \\ 0, & x \notin A \end{cases}$
$\langle f, g\rangle_{L^2(\mu)}$	scalar product $\int fg\,d\mu$
Leb	Lebesgue measure
λ_T	Leb. measure on $[0, T]$
δ_x	point mass at x
$\mathfrak{D}$	domain
$\mathfrak{R}$	range
Δ	Laplace operator
∂_j	partial derivative $\frac{\partial}{\partial x_j}$
∇, ∇_x	gradient $(\frac{\partial}{\partial x_1}, \dots, \frac{\partial}{\partial x_d})^\top$

General notation: probability

$\sim$	"is distributed as"
$\overset{s}{\sim}$	"is sample of", 345
$\perp\!\!\!\perp$	"is stochastically independent"
$\xrightarrow{d}$	convergence in law
$\xrightarrow{\mathbb{P}}$	convergence in probab.
$\xrightarrow{L^p}$	convergence in $L^p(\mathbb{P})$
a. s.	almost surely (w. r. t. $\mathbb{P}$)
iid	independent and identically distributed
LIL	law of iterated logarithm
$\mathbb{P}$, $\mathbb{E}$	probability, expectation
$\mathbb{V}$, Cov	variance, covariance
$\mathsf{N}(\mu, \sigma^2)$	normal law in $\mathbb{R}$, mean μ, variance σ^2
$\mathsf{N}(m, \Sigma)$	normal law in $\mathbb{R}^d$, mean $m \in \mathbb{R}^d$, covariance matrix $\Sigma \in \mathbb{R}^{d\times d}$
BM	Brownian motion, 4
BM^1, BM^2	1-, 2-dimensional BM, 4
BM^d	d-dimensional BM, 4
(B0)–(B4)	4
(B3′)	6
(QB3)	13

Sets and σ-algebras

A^c	complement of the set A
$\overline{A}$	closure of the set A
$\mathbb{B}(x,r)$	open ball, centre x, radius r
$\overline{\mathbb{B}}(x,r)$	closed ball, centre x, radius r
$\operatorname{supp} f$	support, $\overline{\{f \neq 0\}}$
$\mathscr{B}(E)$	Borel sets of E
$\mathscr{F}_t^X$	$\sigma(X_s : s \leqslant t)$
$\mathscr{F}_{t+}$	$\bigcap_{u>t} \mathscr{F}_u$
$\overline{\mathscr{F}}_t$	completion of $\mathscr{F}_t$ with all subsets of $\mathbb{P}$ null sets
$\mathscr{F}_\infty$	$\sigma\left(\bigcup_{t\geqslant 0} \mathscr{F}_t\right)$
$\mathscr{F}_\tau, \mathscr{F}_{\tau+}$	52, 371
$\mathscr{P}$	progressive σ-algebra, 234

Stochastic processes

$(X_t, \mathscr{F}_t)_{t\geqslant 0}$	adapted process, 46
$\mathbb{P}^x, \mathbb{E}^x$	law of BM, starting at x, 60 law of Feller process, starting at x, 84–85
σ, τ	stopping times: $\{\sigma \leqslant t\} \in \mathscr{F}_t, t \geqslant 0$
τ_D, τ_D°	first hitting/entry time, 50
X_t^τ	stopped process $X_{t\wedge\tau}$
$\langle X\rangle_t$	quadratic variation, 220, 228, 381
$\langle X, Y\rangle_t$	quadratic covariation, 222
$\operatorname{var}_p(f;t)$	p-variation on $[0,t]$, 136
$\mathcal{S}_T$	simple processes, 224
$\mathcal{L}_T^2$	closure of $\mathcal{S}_T$, 228
$\mathcal{L}_{T,\mathrm{loc}}^2$	242
$L_{\mathscr{P}}^2$	$f \in L^2$ with $\mathscr{P}$ mble. representative, 234
$L_{\mathscr{P}}^2 = \mathcal{L}_T^2$	236
$\mathcal{M}^2, \mathcal{M}_T^2$	L^2 martingales, 220, 224
$\mathcal{M}_T^{2,c}$	continuous L^2 martingales, 224

Spaces of functions

$\mathcal{B}(E)$	Borel functions on E
$\mathcal{B}_b(E)$	– –, bounded
$\mathcal{C}(E)$	continuous functions on E
$\mathcal{C}_b(E)$	– –, bounded
$\mathcal{C}_\infty(E)$	– –, $\lim_{\lvert x\rvert\to\infty} f(x) = 0$
$\mathcal{C}_c(E)$	– –, compact support
$\mathcal{C}_{(0)}(E)$	– –, $f(0) = 0$,
$\mathcal{C}^k(E)$	k times continuously diff'ble functions on E
$\mathcal{C}_b^k(E)$	– –, bounded (with all derivatives)
$\mathcal{C}_\infty^k(E)$	– –, 0 at infinity (with all derivatives)
$\mathcal{C}_c^k(E)$	– –, compact support
$\mathcal{C}^{1,2}(I\times E)$	$f(\cdot,x) \in \mathcal{C}^1(I)$ and $f(t,\cdot) \in \mathcal{C}^2(E)$
$\mathcal{H}^1$	Cameron–Martin space, 192
$L^p(E,\mu), L^p(\mu), L^p(E)$	L^p space w. r. t. the measure space $(E, \mathscr{A}, \mu)$

1 Robert Brown's new thing

'I have some sea-mice – five specimens – in spirits. And I will throw in Robert Brown's new thing – "Microscopic Observations on the Pollen of Plants" – if you don't happen to have it already.'

George Eliot: Middlemarch
Book II, Chapter xvii

If you observe plant pollen in a drop of water through a microscope, you will see an incessant, irregular movement of the particles. The Scottish Botanist Robert Brown was not the first to describe this phenomenon – he refers to W. F. Gleichen-Rußwurm as *the discoverer of the motions of the Particles of the Pollen* [21, p. 164] – but his 1828 and 1829 papers [20; 21] are the first scientific publications investigating 'Brownian motion'. Brown points out that

- the motion is very irregular, composed of translations and rotations;
- the particles appear to move independently of each other;
- the motion is more active the smaller the particles;
- the composition and density of the particles have no effect;
- the motion is more active the less viscous the fluid;
- the motion never ceases;
- the motion is not caused by flows in the liquid or by evaporation;
- the particles are not animated.

Let us briefly sketch how the story of Brownian motion evolved.

Brownian motion and physics. Following Brown's observations, several theories emerged, but it was Einstein's 1905 paper [68] which gave the correct explanation: The atoms of the fluid perform a temperature-dependent movement and bombard the (in this scale) macroscopic particles suspended in the fluid. These collisions happen frequently and they do not depend on position nor time. In the introduction Einstein remarks: *It is possible, that the movements to be discussed here are identical with the so-called "Brownian molecular motion"; [...] If the movement discussed here can actually be observed (together with the laws relating to it that one would expect to find), then classical thermodynamics can no longer be looked upon as applicable with precision to bodies even of dimensions distinguishable in a microscope: An exact determination of actual atomic dimensions is then possible.*[1] [69, pp. 1–2]. And between the lines: This would settle the then ongoing discussion on the existence of atoms. It was Jean Perrin

1 Es ist möglich, daß die hier zu behandelnden Bewegungen mit der sogenannten "Brownschen Molekularbewegung" identisch sind; [...] Wenn sich die hier zu behandelnde Bewegung samt den für sie zu erwartenden Gesetzmäßigkeiten wirklich beobachten läßt, so ist die klassische Thermodynamik schon für mikroskopisch unterscheidbare Räume nicht mehr als genau gültig anzusehen und es ist dann eine exakte Bestimmung der wahren Atomgröße möglich [68, p. 549].

who combined in 1909 Einstein's theory and experimental observations of Brownian motion to prove the existence and determine the size of atoms, cf. [177; 178]. Independently of Einstein, M. von Smoluchowski arrived at an equivalent interpretation of Brownian motion, cf. [208].

Brownian motion and mathematics. As a mathematical object, Brownian motion can be traced back to the not completely rigorous definition of Bachelier [6] who makes no connection to Brown or Brownian motion. Bachelier's work was only rediscovered by economists in the 1960s, cf. [39]. The first rigorous mathematical construction of Brownian motion is due to Wiener [227] who introduces the *Wiener measure* on the space $\mathcal{C}[0,1]$ (which he calls *differential-space*) building on Einstein's and von Smoluchowski's work. Further constructions of Brownian motion were subsequently given by Wiener [228] (Fourier-Wiener series), Kolmogorov [128; 129] (giving a rigorous justification of Bachelier [6]), Lévy [144; 145, pp. 492–494, 17–20] (interpolation argument), Ciesielski [34] (Haar representation) and Donsker [49] (limit of random walks, invariance principle), see Chapter 3.

Let us start with Brown's observations to build a mathematical model of Brownian motion. To keep things simple, we consider a one-dimensional setting where each particle performs a random walk. We assume that each particle

- starts at the origin $x = 0$,
- changes its position only at discrete times $k\Delta t$ where $\Delta t > 0$ is fixed and for all $k = 1, 2, \dots$;
- moves Δx units to the left or to the right with equal probability;
- and Δx does not depend on any past positions nor the current position x nor on time $t = k\Delta t$.

Letting $\Delta t \to 0$ and $\Delta x \to 0$ in an appropriate way should give a random motion which is continuous in time and space.

Let us denote by X_t the random position of the particle at time $t \in [0, T]$. During the time $[0, T]$, the particle has changed its position $N = \lfloor T/\Delta t \rfloor$ times. Since the decision to move left or right is random, we will model it by independent, identically distributed Bernoulli random variables, ϵ_k, $k \geqslant 1$, where

$$\mathbb{P}(\epsilon_1 = 1) = \mathbb{P}(\epsilon_1 = 0) = \frac{1}{2}$$

so that

$$S_N = \epsilon_1 + \dots + \epsilon_N \quad \text{and} \quad N - S_N$$

denote the number of right and left moves, respectively. Thus

$$X_T = S_N \Delta x - (N - S_N)\Delta x = (2S_N - N)\Delta x = \sum_{k=1}^{N} (2\epsilon_k - 1)\Delta x$$

is the position of the particle at time $T = N\Delta t$. Since $X_0 = 0$ we find for any two times $t = n\Delta t$ and $T = N\Delta t$ that

$$X_T = (X_T - X_t) + (X_t - X_0) = \sum_{k=n+1}^{N} (2\epsilon_k - 1)\Delta x + \sum_{k=1}^{n} (2\epsilon_k - 1)\Delta x.$$

Since the ϵ_k are iid random variables, the two increments $X_T - X_t$ and $X_t - X_0$ are independent and

$$X_T - X_t \sim X_{T-t} - X_0$$

('$\sim$' indicates that the random variables have the same probability distribution). We write $\sigma^2(t) := \mathbb{V} X_t$. By Bienaymé's identity we get

$$\mathbb{V} X_T = \mathbb{V}(X_T - X_t) + \mathbb{V}(X_t - X_0) = \sigma^2(T - t) + \sigma^2(t)$$

which means that $t \mapsto \sigma^2(t)$ is linear:

$$\mathbb{V} X_T = \sigma^2(T) = \sigma^2 T,$$

where $\sigma > 0$ is the so-called *diffusion coefficient*. On the other hand, since $\mathbb{E}\, \epsilon_1 = \frac{1}{2}$ and $\mathbb{V}\, \epsilon_1 = \frac{1}{4}$ we get by a direct calculation that

$$\mathbb{V} X_T = N(\Delta x)^2 = \frac{T}{\Delta t} (\Delta x)^2$$

which reveals that

$$\frac{(\Delta x)^2}{\Delta t} = \sigma^2 = \text{const.}$$

The particle's position X_T at time $T = N\Delta t$ is the sum of N iid random variables,

$$X_T = \sum_{k=1}^{N} (2\epsilon_k - 1)\Delta x = (2S_N - N)\Delta x = S_N^* \sqrt{T}\sigma,$$

where

$$S_N^* = \frac{2S_N - N}{\sqrt{N}} = \frac{S_N - \mathbb{E}\, S_N}{\sqrt{\mathbb{V}\, S_N}}$$

is the normalization – i. e. mean 0, variance 1 – of the random variable S_N. A simple application of the central limit theorem now shows that in distribution

$$X_T = \sqrt{T}\sigma S_N^* \xrightarrow[\text{(i. e. } \Delta x, \Delta t \to 0)]{N \to \infty} \sqrt{T}\sigma G$$

where $G \sim \mathsf{N}(0, 1)$ is a standard normal distributed random variable. This means that, in the limit, the particle's position $B_T = \lim_{\Delta x, \Delta t \to 0} X_T$ is normally distributed with law $\mathsf{N}(0, T\sigma^2)$.

This approximation procedure yields for each $t \in [0, T]$ some random variable $B_t \sim \mathsf{N}(0, t\sigma^2)$. More generally

1.1 Definition. Let $(\Omega, \mathscr{A}, \mathbb{P})$ be a probability space. A *d-dimensional stochastic process* indexed by $I \subset [0, \infty)$ is a family of random variables $X_t : \Omega \to \mathbb{R}^d$, $t \in I$. We write $X = (X_t)_{t\in I}$. I is called the *index set* and $\mathbb{R}^d$ the *state space*.

The only requirement of Definition 1.1 is that the X_t, $t \in I$, are $\mathscr{A}/\mathscr{B}(\mathbb{R}^d)$ measurable. This definition is, however, too general to be mathematically useful; more information is needed on $(t, \omega) \mapsto X_t(\omega)$ as a function of two variables. Although the family $(B_t)_{t\in[0,T]}$ satisfies the condition of Definition 1.1, a realistic model of Brownian motion should have at least *continuous trajectories*: The *sample path* $[0, T] \ni t \mapsto B_t(\omega)$ should be a continuous function for all ω.

1.2 Definition. A *d-dimensional Brownian motion* $B = (B_t)_{t\geqslant 0}$ is a stochastic process indexed by $[0, \infty)$ taking values in $\mathbb{R}^d$ such that

$$B_0(\omega) = 0 \quad \text{for } \mathbb{P} \text{ almost all } \omega; \tag{B0}$$

$$\begin{gathered} B_{t_n} - B_{t_{n-1}}, \dots, B_{t_1} - B_{t_0} \text{ are independent} \\ \text{for all } n \geqslant 1,\ 0 = t_0 \leqslant t_1 < t_2 < \dots < t_n < \infty; \end{gathered} \tag{B1}$$

$$B_t - B_s \sim B_{t+h} - B_{s+h} \quad \text{for all } 0 \leqslant s < t,\ h \geqslant -s; \tag{B2}$$

$$B_t - B_s \sim \mathsf{N}(0, t-s)^{\otimes d}, \quad \mathsf{N}(0,t)(dx) = \frac{1}{\sqrt{2\pi t}} \exp\left(-\frac{x^2}{2t}\right) dx; \tag{B3}$$

$$t \mapsto B_t(\omega) \quad \text{is continuous for all } \omega. \tag{B4}$$

We use BM^d as shorthand for d-dimensional Brownian motion.

We will also speak of a Brownian motion if the index set is an interval of the form $[0, T)$ or $[0, T]$. We say that $(B_t + x)_{t\geqslant 0}$, $x \in \mathbb{R}^d$, is a d-dimensional Brownian motion *started at* x. Frequently we write $B(t, \omega)$ and $B(t)$ instead of $B_t(\omega)$ and B_t; this should cause no confusion.

By definition, Brownian motion is an $\mathbb{R}^d$-valued stochastic process starting at the origin (B0) with *independent increments* (B1), *stationary increments* (B2) and continuous paths (B4).

We will see later that this definition is redundant: (B0)–(B3) entail (B4) at least for almost all ω. On the other hand, (B0)–(B2) and (B4) automatically imply that the increment $B(t) - B(s)$ has a (possibly degenerate) Gaussian distribution (this is a consequence of the central limit theorem).

Before we discuss such details, we should settle the question, if there exists a process satisfying the requirements of Definition 1.2.

---

i **1.3 Further reading.** A good general survey on the history of continuous-time stochastic processes is the paper [37]. The role of Brownian motion in mathematical finance is explained in [8] and [39]. Good references for Brownian motion in physics are [158] and [164], for applications in modelling and engineering [149].

[8] Bachelier, Davis (ed.), Etheridge (ed.): *Louis Bachelier's Theory of Speculation: The Origins of Modern Finance.*
[37] Cohen: The history of noise.
[39] Cootner (ed.): *The Random Character of Stock Market Prices.*
[149] MacDonald: *Noise and Fluctuations.*
[158] Mazo: *Brownian Motion.*
[164] Nelson: *Dynamical Theories of Brownian Motion.*

Problems

A sequence of random variables $X_n : \Omega \to \mathbb{R}^d$ converges *weakly* (also: *in distribution* or *in law*) to a random variable X, $X_n \xrightarrow{d} X$ if, and only if, $\lim_{n\to\infty} \mathbb{E} f(X_n) = \mathbb{E} f(X)$ for all bounded and continuous functions $f \in \mathcal{C}_b(\mathbb{R}^d)$. This is equivalent to the convergence of the characteristic functions $\lim_{n\to\infty} \mathbb{E}\, e^{i\langle \xi, X_n\rangle} = \mathbb{E}\, e^{i\langle \xi, X\rangle}$ for all $\xi \in \mathbb{R}^d$.

1. Let $X, Y, X_n, Y_n : \Omega \to \mathbb{R}$, $n \geqslant 1$, be random variables.
   (a) If, for all $n \geqslant 1$, $X_n \perp\!\!\!\perp Y_n$ and if $(X_n, Y_n) \xrightarrow{d} (X, Y)$, then $X \perp\!\!\!\perp Y$.
   (b) Let $X \perp\!\!\!\perp Y$ such that $X, Y \sim \beta_{1/2} := \frac{1}{2}(\delta_0 + \delta_1)$ are Bernoulli random variables. We set $X_n := X + \frac{1}{n}$ and $Y_n := 1 - X_n$. Then $X_n \xrightarrow{d} X$, $Y_n \xrightarrow{d} Y$, $X_n + Y_n \xrightarrow{d} 1$ but (X_n, Y_n) does not converge weakly to (X, Y).
   (c) Assume that $X_n \xrightarrow{d} X$ and $Y_n \xrightarrow{d} Y$. Is it true that $X_n + Y_n \xrightarrow{d} X + Y$?
2. (Slutsky's Theorem) Let $X_n, Y_n : \Omega \to \mathbb{R}^d$, $n \geqslant 1$, be two sequences of random variables such that $X_n \xrightarrow{d} X$ and $X_n - Y_n \xrightarrow{\mathbb{P}} 0$. Then $Y_n \xrightarrow{d} X$.
3. (Slutsky's Theorem 2) Let $X_n, Y_n : \Omega \to \mathbb{R}$, $n \geqslant 1$, be two sequences of random variables.
   (a) If $X_n \xrightarrow{d} X$ and $Y_n \xrightarrow{\mathbb{P}} c$, then $X_n Y_n \xrightarrow{d} cX$. Is this still true if $Y_n \xrightarrow{d} c$?
   (b) If $X_n \xrightarrow{d} X$ and $Y_n \xrightarrow{\mathbb{P}} 0$, then $X_n + Y_n \xrightarrow{d} X$. Is this still true if $Y_n \xrightarrow{d} 0$?
4. Let $X_n, X, Y : \Omega \to \mathbb{R}$, $n \geqslant 1$, be random variables. If
   $$\lim_{n\to\infty} \mathbb{E}(f(X_n)g(Y)) = \mathbb{E}(f(X)g(Y)) \quad \text{for all } f \in \mathcal{C}_b(\mathbb{R}),\ g \in \mathcal{B}_b(\mathbb{R}),$$
   then $(X_n, Y) \xrightarrow{d} (X, Y)$. If $X = \phi(Y)$ for some $\phi \in \mathcal{B}(\mathbb{R})$, then $X_n \xrightarrow{\mathbb{P}} X$.
5. Let δ_j, $j \geqslant 1$, be iid Bernoulli random variables with $\mathbb{P}(\delta_j = \pm 1) = 1/2$. We set
   $$S_0 := 0, \quad S_n := \delta_1 + \dots + \delta_n \quad \text{and} \quad X_t^n := \frac{1}{\sqrt{n}} S_{\lfloor nt \rfloor}.$$
   A one-dimensional random variable G is Gaussian if it has the characteristic function $\mathbb{E}\, e^{i\xi G} = \exp(im\xi - \frac{1}{2}\sigma^2\xi^2)$ with $m \in \mathbb{R}$ and $\sigma \geqslant 0$. Prove that

(a) $X_t^n \xrightarrow{d} G_t$ where $t > 0$ and G_t is a Gaussian random variable.

(b) $X_t^n - X_s^n \xrightarrow{d} G_{t-s}$ where $t \geqslant s \geqslant 0$ and G_u is a Gaussian random variable. Do we have $G_{t-s} = G_t - G_s$?

(c) Let $0 \leqslant t_1 \leqslant \cdots \leqslant t_m$, $m \geqslant 1$. Determine the limit as $n \to \infty$ of the random vector $(X_{t_m}^n - X_{t_{m-1}}^n, \ldots, X_{t_2}^n - X_{t_1}^n, X_{t_1}^n)$.

6. Consider the condition

$$\begin{gathered}\text{for all } s < t \text{ the random variables } \frac{B(t) - B(s)}{\sqrt{t-s}} \text{ are} \\ \text{identically distributed, centered and have variance 1.}\end{gathered} \tag{B3'}$$

Show that (B0), (B1), (B2), (B3) and (B0), (B1), (B2), (B3′) are equivalent.
Hint: If $X \sim Y$, $X \perp\!\!\!\perp Y$ and $X \sim \frac{1}{\sqrt{2}}(X+Y)$, then $X \sim N(0,1)$, cf. Rényi [186, VI.5 Theorem 2].

7. Let G and G' be two independent real $\mathsf{N}(0,1)$-random variables. Show that

$$\Gamma^{\pm} := \frac{G \pm G'}{\sqrt{2}} \sim \mathsf{N}(0,1) \quad \text{and} \quad \Gamma^{-} \perp\!\!\!\perp \Gamma^{+}.$$

2 Brownian motion as a Gaussian process

Recall that a one-dimensional random variable Γ is *Gaussian* if it has the characteristic function

$$\mathbb{E}\, e^{i\xi\Gamma} = e^{im\xi - \frac{1}{2}\sigma^2\xi^2} \tag{2.1}$$

for some real numbers $m \in \mathbb{R}$ and $\sigma \geqslant 0$. If we differentiate (2.1) two times with respect to ξ and set $\xi = 0$, we see that

$$m = \mathbb{E}\,\Gamma \quad \text{and} \quad \sigma^2 = \mathbb{V}\,\Gamma. \tag{2.2}$$

A random vector $\Gamma = (\Gamma_1, \dots, \Gamma_n) \in \mathbb{R}^n$ is *Gaussian*, if $\langle \ell, \Gamma \rangle$ is for every $\ell \in \mathbb{R}^n$ a one-dimensional Gaussian random variable. This is the same as to say that

$$\mathbb{E}\, e^{i\langle \xi, \Gamma\rangle} = e^{i\,\mathbb{E}\langle \xi, \Gamma\rangle - \frac{1}{2}\,\mathbb{V}\langle \xi, \Gamma\rangle}. \tag{2.3}$$

Setting $m = (m_1, \dots, m_n) \in \mathbb{R}^n$ and $\Sigma = (\sigma_{jk})_{j,k=1\dots,n} \in \mathbb{R}^{n\times n}$ where

$$m_j := \mathbb{E}\,\Gamma_j \quad \text{and} \quad \sigma_{jk} := \mathbb{E}(\Gamma_j - m_j)(\Gamma_k - m_k) = \operatorname{Cov}(\Gamma_j, \Gamma_k),$$

we can rewrite (2.3) in the following form

$$\mathbb{E}\, e^{i\langle \xi, \Gamma\rangle} = e^{i\langle \xi, m\rangle - \frac{1}{2}\langle \xi, \Sigma\xi\rangle}. \tag{2.4}$$

We call m the *mean vector* and Σ the *covariance matrix* of Γ.

2.1 The finite dimensional distributions

Let us quickly establish some first consequences of the definition of Brownian motion. To keep things simple, we assume throughout this section that $(B_t)_{t\geqslant 0}$ is a one-dimensional Brownian motion.

2.1 Proposition. *Let $(B_t)_{t\geqslant 0}$ be a one-dimensional Brownian motion. Then B_t, $t \geqslant 0$, are Gaussian random variables with mean* 0 *and variance t:* Ex. 2.1

$$\mathbb{E}\, e^{i\xi B_t} = e^{-t\xi^2/2} \quad \textit{for all } t \geqslant 0,\ \xi \in \mathbb{R}. \tag{2.5}$$

Proof. Set $\phi_t(\xi) = \mathbb{E}\, e^{i\xi B_t}$. If we differentiate ϕ_t with respect to ξ, and use integration by parts we get

$$\begin{aligned}
\phi_t'(\xi) = \mathbb{E}\left(iB_t e^{i\xi B_t}\right) &\overset{\text{(B3)}}{=} \frac{1}{\sqrt{2\pi t}} \int_{\mathbb{R}} e^{ix\xi}\,(ix) e^{-x^2/(2t)}\,dx \\
&= \frac{1}{\sqrt{2\pi t}} \int_{\mathbb{R}} e^{ix\xi}\,(-it)\frac{d}{dx}\,e^{-x^2/(2t)}\,dx \\
&\overset{\text{parts}}{=} -t\xi\,\frac{1}{\sqrt{2\pi t}} \int_{\mathbb{R}} e^{ix\xi}\,e^{-x^2/(2t)}\,dx \\
&= -t\xi\,\phi_t(\xi).
\end{aligned}$$

Since $\phi_t(0) = 1$, (2.5) is the unique solution of the differential equation

$$\frac{\phi_t'(\xi)}{\phi_t(\xi)} = -t\xi \qquad \square$$

From the elementary inequality $1 \leqslant \exp([\frac{y}{2} - c]^2)$ we see that $e^{cy} \leqslant e^{c^2} e^{y^2/4}$ for all $c, y \in \mathbb{R}$. Therefore, $e^{cy} e^{-y^2/2} \leqslant e^{c^2} e^{-y^2/4}$ is integrable. Considering real and imaginary parts separately, it follows that the integrals in (2.5) converge for all $\xi \in \mathbb{C}$ and define an analytic function.

2.2 Corollary. *A one-dimensional Brownian motion $(B_t)_{t\geqslant 0}$ has exponential moments of all orders, i. e.*

$$\mathbb{E}\, e^{\zeta B_t} = e^{t\zeta^2/2} \quad \textit{for all } \zeta \in \mathbb{C}. \tag{2.6}$$

2.3 Moments. Note that for $k = 0, 1, 2, \dots$

$$\mathbb{E}(B_t^{2k+1}) = \frac{1}{\sqrt{2\pi t}} \int_{\mathbb{R}} x^{2k+1}\, e^{-x^2/(2t)}\, dx = 0 \tag{2.7}$$

and

$$\begin{aligned}
\mathbb{E}(B_t^{2k}) &= \frac{1}{\sqrt{2\pi t}} \int_{\mathbb{R}} x^{2k}\, e^{-x^2/(2t)}\, dx \\
&\overset{x=\sqrt{2ty}}{=} \frac{2}{\sqrt{2\pi t}} \int_0^\infty (2ty)^k\, e^{-y}\, \frac{2t\, dy}{2\sqrt{2ty}} \\
&= \frac{2^k t^k}{\sqrt{\pi}} \int_0^\infty y^{k-1/2}\, e^{-y}\, dy \\
&= t^k\, \frac{2^k\, \Gamma(k+1/2)}{\sqrt{\pi}}
\end{aligned} \tag{2.8}$$

where $\Gamma(\cdot)$ denotes Euler's Gamma function. In particular,

$$\mathbb{E}\, B_t = \mathbb{E}\, B_t^3 = 0, \quad \mathbb{V}\, B_t = \mathbb{E}\, B_t^2 = t \quad \text{and} \quad \mathbb{E}\, B_t^4 = 3t^2.$$

2.4 Covariance. For $s, t \geqslant 0$ we have

$$\operatorname{Cov}(B_s, B_t) = \mathbb{E}\, B_s B_t = s \wedge t.$$

Indeed, if $s \leqslant t$,

$$\mathbb{E}\, B_s B_t = \mathbb{E}\left(B_s(B_t - B_s)\right) + \mathbb{E}(B_s^2) \underset{2.3}{\overset{\text{(B1)}}{=}} s = s \wedge t.$$

2.5 Definition. A one-dimensional stochastic process $(X_t)_{t\geqslant 0}$ is called a *Gaussian process* if all vectors $\Gamma = (X_{t_1}, \dots, X_{t_n})$, $n \geqslant 1$, $0 \leqslant t_1 < t_2 < \dots < t_n$ are (possibly degenerate) Gaussian random vectors.

Let us show that a Brownian motion is a Gaussian process.

2.6 Theorem. *A one-dimensional Brownian motion $(B_t)_{t\geqslant 0}$ is a Gaussian process.*

For $t_0 := 0 < t_1 < \cdots < t_n$, $n \geqslant 1$, the vector $\Gamma := (B_{t_1}, \dots, B_{t_n})^\top$ is a Gaussian random variable with a strictly positive definite, symmetric covariance matrix $C = (t_j \wedge t_k)_{j,k=1\dots,n}$ and mean vector $m = 0 \in \mathbb{R}^n$: Ex. 2.2

$$\mathbb{E}\, e^{i\langle\xi,\Gamma\rangle} = e^{-\frac{1}{2}\langle\xi, C\xi\rangle}. \tag{2.9}$$

Moreover, the probability distribution of Γ is given by

$$\mathbb{P}(\Gamma \in dx) = \frac{1}{(2\pi)^{n/2}} \frac{1}{\sqrt{\det C}} \exp\left(-\frac{1}{2}\langle x, C^{-1}x\rangle\right) dx \tag{2.10a}$$

$$= \frac{1}{(2\pi)^{n/2}\sqrt{\prod_{j=1}^n (t_j - t_{j-1})}} \exp\left(-\frac{1}{2}\sum_{j=1}^n \frac{(x_j - x_{j-1})^2}{t_j - t_{j-1}}\right) dx \tag{2.10b}$$

with $x := (x_1, \dots, x_n) \in \mathbb{R}^n$ and $x_0 := 0$.

Proof. Set $\Delta := (B_{t_1} - B_{t_0}, B_{t_2} - B_{t_1}, \dots, B_{t_n} - B_{t_{n-1}})^\top$ and observe that we can write $B(t_k) - B(t_0) = \sum_{j=1}^k (B_{t_j} - B_{t_{j-1}})$. Thus,

$$\Gamma = \begin{pmatrix} 1 & & 0 \\ \vdots & \ddots & \\ 1 & \cdots & 1 \end{pmatrix} \Delta = M\Delta$$

where $M \in \mathbb{R}^{n\times n}$ is a lower triangular matrix with entries 1 on and below the diagonal. Ex. 2.3
Therefore,

$$\begin{aligned} \mathbb{E}\left(\exp\left[i\langle\xi, \Gamma\rangle\right]\right) &= \mathbb{E}\left(\exp\left[i\langle M^\top\xi, \Delta\rangle\right]\right) \\ &\overset{\text{(B1)}}{=} \prod_{j=1}^n \mathbb{E}\left(\exp\left[i(B_{t_j} - B_{t_{j-1}})(\xi_j + \cdots + \xi_n)\right]\right) \\ &\underset{(2.5)}{\overset{\text{(B2)}}{=}} \prod_{j=1}^n \exp\left[-\frac{1}{2}(t_j - t_{j-1})(\xi_j + \cdots + \xi_n)^2\right]. \end{aligned} \tag{2.11}$$

Observe that

$$\begin{aligned} &\sum_{j=1}^n t_j(\xi_j + \cdots + \xi_n)^2 - \sum_{j=1}^n t_{j-1}(\xi_j + \cdots + \xi_n)^2 \\ &\quad = t_n\xi_n^2 + \sum_{j=1}^{n-1} t_j\left((\xi_j + \cdots + \xi_n)^2 - (\xi_{j+1} + \cdots + \xi_n)^2\right) \\ &\quad = t_n\xi_n^2 + \sum_{j=1}^{n-1} t_j\xi_j(\xi_j + 2\xi_{j+1} + \cdots + 2\xi_n) \\ &\quad = \sum_{j=1}^n\sum_{k=1}^n (t_j \wedge t_k)\,\xi_j\xi_k. \end{aligned} \tag{2.12}$$

This proves (2.9). Since C is strictly positive definite and symmetric, the inverse C^{-1} exists and is again positive definite; both C and C^{-1} have unique positive definite, symmetric square roots. Using the uniqueness of the Fourier transform, the following calculation proves (2.10a):

$$\begin{aligned}\left(\frac{1}{2\pi}\right)^{n/2} \frac{1}{\sqrt{\det C}} \int_{\mathbb{R}^n} e^{i\langle x,\xi\rangle} e^{-\frac{1}{2}\langle x, C^{-1}x\rangle}\,dx &\overset{y=C^{-1/2}x}{=} \left(\frac{1}{2\pi}\right)^{n/2} \int_{\mathbb{R}^n} e^{i\langle (C^{1/2}y),\xi\rangle} e^{-\frac{1}{2}|y|^2}\,dy\\ &= \left(\frac{1}{2\pi}\right)^{n/2} \int_{\mathbb{R}^n} e^{i\langle y, C^{1/2}\xi\rangle} e^{-\frac{1}{2}|y|^2}\,dy\\ &= e^{-\frac{1}{2}|C^{1/2}\xi|^2} = e^{-\frac{1}{2}\langle\xi, C\xi\rangle}.\end{aligned}$$

Let us finally determine $\langle x, C^{-1}x\rangle$ and $\det C$. Since the entries of Δ are independent $\mathsf{N}(0, t_j - t_{j-1})$ distributed random variables we get

$$\mathbb{E}\, e^{i\langle\xi,\Delta\rangle} \overset{(2.5)}{=} \exp\left(-\frac{1}{2}\sum_{j=1}^n (t_j - t_{j-1})\xi_j^2\right) = \exp\left(-\frac{1}{2}\langle\xi, D\xi\rangle\right)$$

where $D \in \mathbb{R}^{n\times n}$ is a diagonal matrix with entries $(t_1 - t_0, \dots, t_n - t_{n-1})$. On the other hand, we have

$$e^{-\frac{1}{2}\langle\xi, C\xi\rangle} = \mathbb{E}\, e^{i\langle\xi,\Gamma\rangle} = \mathbb{E}\, e^{i\langle\xi, M\Delta\rangle} = \mathbb{E}\, e^{i\langle M^\top\xi,\Delta\rangle} = e^{-\frac{1}{2}\langle M^\top\xi, DM^\top\xi\rangle}.$$

Ex. 2.4 Thus, $C = MDM^\top$ and, therefore $C^{-1} = (M^\top)^{-1}D^{-1}M^{-1}$. Since M^{-1} is a two-band matrix with entries 1 on the diagonal and −1 on the first sub-diagonal below the diagonal, we see

$$\langle x, C^{-1}x\rangle = \langle M^{-1}x, D^{-1}M^{-1}x\rangle = \sum_{j=1}^n \frac{(x_j - x_{j-1})^2}{t_j - t_{j-1}}$$

as well as $\det C = \det(MDM^\top) = \det D = \prod_{j=1}^n (t_j - t_{j-1})$. This shows (2.10b). □

The proof of Theorem 2.6 actually characterizes Brownian motion among all Gaussian processes.

2.7 Corollary. *Let $(X_t)_{t\geqslant 0}$ be a one-dimensional Gaussian process such that the vector $\Gamma = (X_{t_1}, \dots, X_{t_n})^\top$ is a Gaussian random variable with mean 0 and covariance matrix $C = (t_j \wedge t_k)_{j,k=1,\dots,n}$. If $(X_t)_{t\geqslant 0}$ has continuous sample paths, then $(X_t)_{t\geqslant 0}$ is a one-dimensional Brownian motion.*

Proof. The properties (B4) and (B0) follow directly from the assumptions; note that $X_0 \sim \mathsf{N}(0,0) = \delta_0$. Set $\Delta = (X_{t_1} - X_{t_0}, \dots, X_{t_n} - X_{t_{n-1}})^\top$ and let $M \in \mathbb{R}^{n\times n}$ be the lower
Ex. 2.3 triangular matrix with entries 1 on and below the diagonal. Then, as in Theorem 2.6,

$\Gamma = M\Delta$ or $\Delta = M^{-1}\Gamma$ where M^{-1} is a two-band matrix with entries 1 on the diagonal and -1 on the first sub-diagonal below the diagonal. Since Γ is Gaussian, we see

$$\begin{aligned}\mathbb{E}\, e^{i\langle\xi,\Delta\rangle} = \mathbb{E}\, e^{i\langle\xi, M^{-1}\Gamma\rangle} = \mathbb{E}\, e^{i\langle (M^{-1})^\top\xi,\Gamma\rangle} &\overset{(2.9)}{=} e^{-\frac{1}{2}\langle (M^{-1})^\top\xi, C(M^{-1})^\top\xi\rangle}\\ &= e^{-\frac{1}{2}\langle \xi, M^{-1}C(M^{-1})^\top\xi\rangle}.\end{aligned}$$

A straightforward calculation shows that $M^{-1}C(M^{-1})^\top$ is just

$$\begin{pmatrix} 1 & & & \\ -1 & \ddots & & \\ & \ddots & \ddots & \\ & & -1 & 1\end{pmatrix}\begin{pmatrix} t_1 & t_1 & \cdots & t_1\\ t_1 & t_2 & \cdots & t_2\\ \vdots & \vdots & \ddots & \vdots\\ t_1 & t_2 & \cdots & t_n\end{pmatrix}\begin{pmatrix} 1 & -1 & & \\ & \ddots & \ddots & \\ & & \ddots & -1\\ & & & 1\end{pmatrix} = \begin{pmatrix} t_1 - t_0 & & & \\ & t_2 - t_1 & & \\ & & \ddots & \\ & & & t_n - t_{n-1}\end{pmatrix}.$$

Thus, Δ is a Gaussian random vector with uncorrelated, hence independent, components which are $\mathsf{N}(0, t_j - t_{j-1})$ distributed. This proves (B1), (B3) and (B2). □

2.2 Brownian motion in $\mathbb{R}^d$

We will now show that $B_t = (B_t^1, \dots, B_t^d)$ is a BM^d if, and only if, its coordinate processes B_t^j are independent one-dimensional Brownian motions. We call two stochastic processes $(X_t)_{t\geqslant 0}$ and $(Y_t)_{t\geqslant 0}$ (defined on the same probability space) *independent*, if the σ-algebras generated by these processes are independent: Ex. 2.11

$$\mathscr{F}_\infty^X \perp\!\!\!\perp \mathscr{F}_\infty^Y \tag{2.13}$$

where

$$\mathscr{F}_\infty^X := \sigma\left(\bigcup_{n\geqslant 1}\ \bigcup_{0\leqslant t_1<\dots<t_n<\infty} \sigma(X(t_1),\dots,X(t_n))\right). \tag{2.14}$$

Note that the family of sets $\bigcup_n \bigcup_{t_1,\dots,t_n} \sigma(X(t_1),\dots,X(t_n))$ is stable under finite intersections. Therefore, (2.13) follows already if Ex. 2.9 Ex. 2.10

$$(X(s_1),\dots,X(s_m)) \perp\!\!\!\perp (Y(t_1),\dots,Y(t_n))$$

for all $m, n \geqslant 1$, $s_1 < \dots < s_m$ and $t_1 < \dots < t_n$. Without loss of generality we can even assume that $m = n$ and $s_j = t_j$ for all j. This follows easily if we take the common refinement of the s_j and t_j.

The following simple characterization of d-dimensional Brownian motion will be very useful for our purposes.

2.8 Lemma. *Let $(X_t)_{t\geqslant 0}$ be a d-dimensional stochastic process. X satisfies* (B0)–(B3) *if, and only if, for all $n \geqslant 0$, $0 = t_0 < t_1 < \dots < t_n$, and $\xi_0, \dots, \xi_n \in \mathbb{R}^d$* Ex. 2.12

$$\mathbb{E}\left[\exp\left(i\sum_{j=1}^n \left\langle \xi_j, X_{t_j} - X_{t_{j-1}}\right\rangle + i\left\langle \xi_0, X_{t_0}\right\rangle\right)\right] = \exp\left(-\frac{1}{2}\sum_{j=1}^n |\xi_j|^2(t_j - t_{j-1})\right) \tag{2.15}$$

holds. If X has continuous sample paths, it is a BM^d.

Proof. Assume that X satisfies (B0)–(B3). Since the characteristic function of a Gaussian $\mathsf{N}(0,(t-s)I_d)$-random variable is $\exp\left(-\frac{1}{2}(t-s)|\xi|^2\right)$, we get

$$\begin{aligned}\mathbb{E}\left[\exp\left(i\sum_{j=1}^{n}\left\langle\xi_j, X_{t_j}-X_{t_{j-1}}\right\rangle + i\left\langle\xi_0, X_{t_0}\right\rangle\right)\right] &\overset{(B1)}{\underset{(B0)}{=}} \prod_{j=1}^{n}\mathbb{E}\left[\exp\left(i\left\langle\xi_j, X_{t_j}-X_{t_{j-1}}\right\rangle\right)\right]\\ &\overset{(B2)}{=} \prod_{j=1}^{n}\mathbb{E}\left[\exp\left(i\left\langle\xi_j, X_{t_j-t_{j-1}}\right\rangle\right)\right]\\ &\overset{(B3)}{=} \prod_{j=1}^{n}\exp\left(-\frac{1}{2}(t_j-t_{j-1})|\xi_j|^2\right).\end{aligned}$$

Conversely, assume that (2.15) holds. Fix $k\in\{0,1,\dots,n\}$ and pick $\xi_j=0$ for all $j\neq k$. Then

$$\begin{aligned}\mathbb{E}\left[\exp\left(i\left\langle\xi_k, X_{t_k}-X_{t_{k-1}}\right\rangle\right)\right] &= \exp\left(-\frac{1}{2}|\xi_k|^2(t_k-t_{k-1})\right)\\ &= \mathbb{E}\left[\exp\left(i\left\langle\xi_k, X_{t_k-t_{k-1}}\right\rangle\right)\right].\end{aligned}$$

This proves (B2), (B3) and, if we take $n=k=0$, also (B0).

Since $X_{t_j}-X_{t_{j-1}}\sim\mathsf{N}(0,(t_j-t_{j-1})I_d)$, (2.15) shows the independence of the increments $X_{t_1}-X_{t_0},\dots,X_{t_n}-X_{t_{n-1}}$, and (B1) follows. □

2.9 Corollary. *Let B be a d-dimensional Brownian motion. Then the coordinate processes B^j, $j=1,\dots,d$, are independent* BM^1.

Proof. Fix $n\geqslant 1$ and $t_0=0<t_1<\dots<t_n<\infty$. Since B is a BM^d, it satisfies (2.15) for all $\xi_1,\dots,\xi_n\in\mathbb{R}^d$. If we take $\xi_j=z_je_k$ where e_k is the kth unit vector of $\mathbb{R}^d$, we see that B^k satisfies (2.15) for all $z_1,\dots,z_n\in\mathbb{R}$. Since B^k inherits the continuity of its sample paths from B, Lemma 2.8 shows that B^k is a one-dimensional Brownian motion.

In order to see the independence of the coordinate processes, we have to show that the σ-algebras $(\mathscr{F}_t^{B^1})_{t\geqslant 0},\dots,(\mathscr{F}_t^{B^d})_{t\geqslant 0}$ are independent. As we have seen at the beginning of the section – for two processes, but the argument stays the same for finitely many processes – it is enough to verify that the random vectors

$$\left(B^k_{t_1},\dots,B^k_{t_n}\right),\quad k=1,2,\dots,d,$$

are independent for all choices of $n\geqslant 1$ and $t_0=0\leqslant t_1<\dots<t_n$. Since each $B^k_{t_\ell}$, $\ell=1,\dots,n$, can be written as $B^k_{t_\ell}=\sum_{j=1}^{\ell}(B^k_{t_j}-B^k_{t_{j-1}})$, it is enough to show that all increments $(B^k_{t_j}-B^k_{t_{j-1}})$, $j=1,\dots,n$ and $k=1,\dots,d$ are independent. This follows again with (2.15), since for all $\xi_1,\dots,\xi_n\in\mathbb{R}^d$

$$\mathbb{E}\left[\exp\left(i\sum_{j=1}^{n}\sum_{k=1}^{d}\xi_j^k(B^k_{t_j}-B^k_{t_{j-1}})\right)\right]=\mathbb{E}\left[\exp\left(i\sum_{j=1}^{n}\left\langle\xi_j, B_{t_j}-B_{t_{j-1}}\right\rangle\right)\right]$$

$$= \exp\left(-\frac{1}{2}\sum_{j=1}^{n}|\xi_j|^2(t_j - t_{j-1})\right) = \exp\left(-\frac{1}{2}\sum_{j=1}^{n}\sum_{k=1}^{d}(\xi_j^k)^2(t_j - t_{j-1})\right)$$

$$= \prod_{j=1}^{n}\prod_{k=1}^{d}\mathbb{E}\exp\left(i\xi_j^k(B_{t_j}^k - B_{t_{j-1}}^k)\right). \qquad \square$$

The converse of Corollary 2.9 is also true:

2.10 Theorem. *B is a* BM^d *if, and only if, the coordinate processes $B^1,\dots,B^d$ are independent* BM^1.

Proof. Because of Corollary 2.9 it is enough to check that $(B_t^1,\dots,B_t^d)$ is a BM^d if the B^j are independent BM^1.

Fix $n \geqslant 1$, $t_0 = 0 < t_1 < \cdots < t_n < \infty$ and $\xi_j^k \in \mathbb{R}$ where $j = 1,\dots,n$ and $k = 1,\dots d$. By assumption, each B^j satisfies (2.15), i. e. for all $k = 1,\dots,d$

$$\mathbb{E}\left[\exp\left(i\sum_{j=1}^{n}\xi_j^k(B_{t_j}^k - B_{t_{j-1}}^k)\right)\right] = \exp\left(-\frac{1}{2}\sum_{j=1}^{n}(\xi_j^k)^2(t_j - t_{j-1})\right).$$

Multiply the d resulting equalities; since $B^1,\dots,B^d$ are independent, we get

$$\mathbb{E}\left[\prod_{k=1}^{d}\exp\left(i\sum_{j=1}^{n}\xi_j^k(B_{t_j}^k - B_{t_{j-1}}^k)\right)\right] = \exp\left(-\frac{1}{2}\sum_{k=1}^{d}\sum_{j=1}^{n}(\xi_j^k)^2(t_j - t_{j-1})\right)$$

which is just (2.15) for $B = (B^1,\dots,B^d)$ and $\xi_j = (\xi_j^1,\dots,\xi_j^d) \in \mathbb{R}^d$ (note that we can neglect the t_0-term as $B(t_0) = 0$). The claim follows from Lemma 2.8. $\square$

Since a d-dimensional Brownian motion is a vector of independent one-dimensional Brownian motions, many properties of BM^1 also hold for BM^d, and often we need only consider the one-dimensional setting. Things are slightly different, if the coordinate processes are mixed:

2.11 Definition (Q-Brownian motion). Let $Q \in \mathbb{R}^{d\times d}$ be a symmetric, positive semi-definite $d{\times}d$ matrix. A *Q-Brownian motion* is a d-dimensional process $(X_t)_{t\geqslant 0}$ satisfying (B0)–(B2), (B4) and Ex. 2.17

$$X_t - X_s \sim \mathsf{N}(0,(t-s)Q) \quad \text{for all } s < t. \tag{QB3}$$

Clearly, BM^d is an I_d-BM. If $(B_t)_{t\geqslant 0}$ is a BM^d and if $\Sigma \in \mathbb{R}^{d\times d}$, then $X_t := \Sigma B_t$ is a Q-BM with $Q = \Sigma\Sigma^\top$. This follows immediately from

$$\mathbb{E}\, e^{i\langle \xi, X_t\rangle} = \mathbb{E}\, e^{i\langle \Sigma^\top\xi, B_t\rangle} = e^{-\frac{1}{2}t\,|\Sigma^\top\xi|^2} = e^{-\frac{1}{2}t\,\langle \xi, \Sigma\Sigma^\top\xi\rangle},$$

since the map $x \mapsto \Sigma x$ does not destroy the properties (B0)–(B2) and (B4).

The same calculation shows that any Q-BM $(X_t)_{t\geqslant 0}$ with a non-degenerate (i. e. strictly positive definite) Q is of the form ΣB_t where Σ is the unique positive definite square root of Q and $(B_t)_{t\geqslant 0}$ is some BM^d.

Since for Gaussian random vectors 'independent' and 'not correlated' coincide, it is easy to see that a Q-BM has independent coordinates if, and only if, Q is a diagonal matrix.

Fig. 2.1. Renewal at time a. The process $W_t := B_{t+a} - B_a$, $t \geqslant 0$, is again a BM^d.

2.3 Invariance properties of Brownian motion

The fact that a stochastic process is a Brownian motion is preserved under various operations at the level of the sample paths. Throughout this section $(B_t)_{t\geqslant 0}$ denotes a d-dimensional Brownian motion.

Ex. 2.19 **2.12 Reflection.** If $(B_t)_{t\geqslant 0}$ is a BM^d, so is $(-B_t)_{t\geqslant 0}$.

2.13 Renewal. Let $(B(t))_{t\geqslant 0}$ be a Brownian motion and fix some time $a > 0$. Then $(W(t))_{t\geqslant 0}$, $W(t) := B(t+a) - B(a)$, is again a BM^d. The properties (B0) and (B4) are obvious for $W(t)$. For all $s \leqslant t$

$$\begin{aligned} W(t) - W(s) &= B(t+a) - B(a) - (B(s+a) - B(a)) \\ &= B(t+a) - B(s+a) \\ &\overset{\text{(B3)}}{\sim} \mathsf{N}(0, (t-s)I_d) \end{aligned}$$

which proves (B3) and (B2) for the process W. Finally, if $t_0 = 0 < t_1 < \cdots < t_n$, then

$$W(t_j) - W(t_{j-1}) = B(t_j + a) - B(t_{j-1} + a) \quad \text{for all } j = 1, \ldots, n$$

i. e. the independence of the W-increments follows from (B1) for B at the times $t_j + a$, $j = 1, \ldots, n$.

A consequence of the independent increments property is that a Brownian motion has no memory. This is the essence of the next lemma.

Ex. 2.18 **2.14 Lemma** (Markov property of BM). *Let $(B(t))_{t\geqslant 0}$ be a d-dimensional Brownian motion and denote by $W(t) := B(t+a) - B(a)$ the shifted Brownian motion constructed in Paragraph 2.13. Then $(B(t))_{0\leqslant t\leqslant a}$ and $(W(t))_{t\geqslant 0}$ are independent, i. e. the σ-algebras generated by these processes are independent:*

$$\sigma\big(B(t) : 0 \leqslant t \leqslant a\big) =: \mathscr{F}_a^B \perp\!\!\!\perp \mathscr{F}_\infty^W := \sigma\big(W(t) : 0 \leqslant t < \infty\big). \tag{2.16}$$

In particular, $B(t) - B(s) \perp\!\!\!\perp \mathscr{F}_s^B$ for all $0 \leqslant s < t$.

Proof. Let $X_0, X_1, \dots, X_n$ be d-dimensional random variables. Then

$$\sigma\left(X_j \,:\, j = 0, \dots, n\right) = \sigma\left(X_0, X_j - X_{j-1} \,:\, j = 1, \dots, n\right). \tag{2.17}$$

Since X_0 and $X_j - X_{j-1}$ are $\sigma(X_j \,:\, j = 0, \dots, n)$ measurable, we see the inclusion '$\supset$'. For the converse we observe that $X_k = \sum_{j=1}^{k}(X_j - X_{j-1}) + X_0$, $k = 0, \dots, n$.

Let $0 = s_0 < s_1 < \dots < s_m = a = t_0 < t_1 < \dots < t_n$. By (B1) the random variables

$$B(s_1) - B(s_0), \;\dots, B(s_m) - B(s_{m-1}),\; B(t_1) - B(t_0),\; \dots, B(t_n) - B(t_{n-1})$$

are independent, thus

$$\sigma\big(B(s_j) - B(s_{j-1}) \,:\, j = 1, \dots, m\big) \perp\!\!\!\perp \sigma\big(B(t_k) - B(t_{k-1}) \,:\, k = 1, \dots, n\big).$$

Using $W(t_k - t_0) - W(t_{k-1} - t_0) = B(t_k) - B(t_{k-1})$ and $B(0) = W(0) = 0$, we can apply (2.17) to get

$$\sigma\big(B(s_j) \,:\, j = 1, \dots, m\big) \perp\!\!\!\perp \sigma\big(W(t_k - t_0) \,:\, k = 1, \dots, n\big)$$

and

$$\bigcup_{\substack{0<s_1<\dots<s_m\leqslant a\\ m\geqslant 1}} \sigma\big(B(s_j) \,:\, j = 1, \dots, m\big) \perp\!\!\!\perp \bigcup_{\substack{0<u_1<\dots<u_n\\ n\geqslant 1}} \sigma\big(W(u_k) \,:\, k = 1, \dots, n\big).$$

The families on the left and right-hand side are $\cap$-stable generators of $\mathscr{F}_a^B$ and $\mathscr{F}_\infty^W$, respectively, thus $\mathscr{F}_a^B \perp\!\!\!\perp \mathscr{F}_\infty^W$.

Finally, taking $a = s$, we see that $B(t) - B(s) = W(t - s)$ which is $\mathscr{F}_\infty^W$ measurable and therefore independent of $\mathscr{F}_s^B$. □

2.15 Time inversion. Let $(B_t)_{t\geqslant 0}$ be a Brownian motion and fix some time $a > 0$. Then $W_t := B_{a-t} - B_a$, $t \in [0, a]$, is again a BM^d. This follows as in 2.13 (see Fig. 2.2). Ex. 2.23

2.16 Scaling. Let $(B_t)_{t\geqslant 0}$ be a d-dimensional Brownian motion. For all $c > 0$ and $t > 0$ we have $B_{ct} \sim c^{1/2}\, B_t$. In particular, $(c^{-1/2} B_{ct})_{t\geqslant 0}$ is again a BM^d. Ex. 2.19

Denote by $\mathsf{N}(0, t)$ the normal law with mean 0 and variance t. The first assertion follows easily from (B0) and (B3) as $\mathsf{N}(0, ct) = c^{1/2}\,\mathsf{N}(0, t)$, i. e.

$$B_{ct} \sim \mathsf{N}(0, ct)^{\otimes d} = c^{1/2}\,\mathsf{N}(0, t)^{\otimes d} \sim c^{1/2}\, B_t.$$

The second claim is now obvious since scaling does not change the independence of the increments or the continuity of the sample paths.

2.17 Projective reflection at $t = \infty$. Let $(B_t)_{t\geqslant 0}$ be a BM^1. Then Ex. 2.24 Ex. 2.21

$$W(t) := \begin{cases} t\, B(\frac{1}{t}), & t > 0; \\ 0, & t = 0 \end{cases}$$

is again a BM^1.

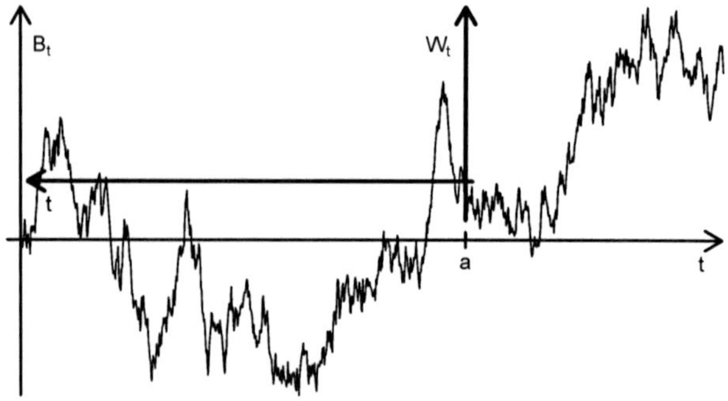

Fig. 2.2. Time inversion. The process $W_t := B_{a-t} - B_t$, $t \in [0, a]$, is again a BM^d.

It is clear that $(W(t_1), \dots, W(t_n))$ is for $0 < t_1 < \dots < t_n$ a Gaussian random vector. The mean is 0, and the covariance is given by

$$\operatorname{Cov}(W(t_j), W(t_k)) = \operatorname{Cov}\left(t_j B(\tfrac{1}{t_j}), t_k B(\tfrac{1}{t_k})\right) = t_j t_k \left(\tfrac{1}{t_j} \wedge \tfrac{1}{t_k}\right) = t_j \wedge t_k.$$

As $t \mapsto W_t$, $t > 0$, is continuous, Corollary 2.7 shows that $(W_t)_{t>0}$ satisfies (B1)–(B4) on $(0, \infty)$. All that remains to be shown is $\lim_{t\downarrow 0} W(t) = W(0) = 0$, i. e. the continuity of the sample paths at $t = 0$.

Note that the limit $\lim_{t\to 0} W(t, \omega) = 0$ if, and only if,

$$\forall n \geqslant 1 \quad \exists m \geqslant 1 \quad \forall r \in \mathbb{Q} \cap \left(0, \tfrac{1}{m}\right] \;:\; \left|rB(\tfrac{1}{r})\right| \leqslant \tfrac{1}{n}.$$

Thus,

$$\Omega^W := \left\{\lim_{t\to 0} W(t) = 0\right\} = \bigcap_{n\geqslant 1} \bigcup_{m\geqslant 1} \bigcap_{r\in\mathbb{Q}\cap(0,1/m]} \left\{|W(r)| \leqslant \tfrac{1}{n}\right\}.$$

We know already that $(W_t)_{t>0}$ and $(B_t)_{t>0}$ have the same finite dimensional distributions. Since Ω^W and the analogously defined set Ω^B are determined by countably many sets of the form $\left\{|W(r)| \leqslant \frac{1}{n}\right\}$ and $\left\{|B(r)| \leqslant \frac{1}{n}\right\}$, we conclude that $\mathbb{P}(\Omega^W) = \mathbb{P}(\Omega^B)$. Consequently,

$$\mathbb{P}(\Omega^W) = \mathbb{P}(\Omega^B) \overset{\text{(B4)}}{=} \mathbb{P}(\Omega) = 1.$$

This shows that $(W_t)_{t\geqslant 0}$ is, on the smaller probability space $(\Omega^W, \Omega^W \cap \mathscr{A}, \mathbb{P})$ equipped with the trace σ-algebra $\Omega^W \cap \mathscr{A}$, a a Brownian motion.

The d-dimensional analogue follows easily from the observation that the coordinate processes of a BM^d are independent BM^1, and vice versa, cf. Theorem 2.10.

---

i **2.18 Further reading.** The literature on Gaussian processes is quite specialized and technical. Among the most accessible monographs are [154] (with a focus on the Dynkin isomorphism theorem and local times) and [147]. The seminal, and still highly readable, paper [55] is one of the most influential original contributions in the field.

[55] Dudley, R. M.: Sample functions of the Gaussian process.
[147] Lifshits, M. A.: *Gaussian Random Functions.*
[154] Marcus, Rosen: *Markov Processes, Gaussian Processes, and Local Times.*

Problems

1. Show that there exist a random vector (U, V) such that U and V are one-dimensional Gaussian random variables but (U, V) is not Gaussian. **!**
Hint: Try $f(u, v) = g(u)g(v)(1 - \sin u \sin v)$ where $g(u) = (2\pi)^{-1/2} e^{-u^2/2}$.

2. Show that the covariance matrix $C = (t_j \wedge t_k)_{j,k=1,\dots,n}$ appearing in Theorem 2.6 is positive definite.

3. Verify that the matrix M in the proof of Theorem 2.6 and Corollary 2.7 is a lower triangular matrix with entries 1 on and below the diagonal. Show that the inverse matrix M^{-1} is a lower triangular matrix with entries 1 on the diagonal and -1 directly below the diagonal.

4. Let $A, B \in \mathbb{R}^{n\times n}$ be symmetric matrices. If $\langle Ax, x\rangle = \langle Bx, x\rangle$ for all $x \in \mathbb{R}^n$, then $A = B$.

5. Let $(B_t)_{t\geqslant 0}$ be a BM^1. Decide which of the following processes are Brownian motions: (a) $X_t := 2B_{t/4}$; (b) $Y_t := B_{2t} - B_t$; (c) $Z_t := \sqrt{t}\, B_1$ $(t \geqslant 0)$.

6. Let $(B(t))_{t\geqslant 0}$ be a BM^1.
   (a) Find the density of the random vector $(B(s), B(t))$ where $0 < s < t < \infty$.

   (b) (Brownian bridge) Find the conditional density of the vector $(B(s), B(t))$, $0 < s < t < 1$ under the condition $B(1) = 0$, and use this to determine $\mathbb{E}(B(s)B(t) \mid B(1) = 0)$.

   (c) Let $0 < t_1 < t_2 < t_3 < t_4 < \infty$. Determine the conditional density of the bivariate random variable $(B(t_2), B(t_3))$ given that $B(t_1) = B(t_4) = 0$. What is the conditional correlation of $B(t_2)$ and $B(t_3)$?

7. Find the correlation function $C(s, t) := \mathbb{E}(X_s X_t)$, $s, t \geqslant 0$, of the stochastic process $X_t := B_t^2 - t$, $t \geqslant 0$, where $(B_t)_{t\geqslant 0}$ is a BM^1.

8. (Ornstein–Uhlenbeck process) Let $(B_t)_{t\geqslant 0}$ be a BM^1, $\alpha > 0$ and consider the stochastic process $X_t := e^{-\alpha t/2} B_{e^{\alpha t}}$, $t \geqslant 0$.
   (a) Determine $m(t) = \mathbb{E}\, X_t$ and $C(s, t) = \mathbb{E}(X_s X_t)$, $s, t \geqslant 0$.

   (b) Find the probability density of $(X_{t_1}, \dots, X_{t_n})$ where $0 \leqslant t_1 < \dots < t_n < \infty$.

9. Let $(B_t)_{t\geqslant 0}$ be a BM^1. Show that $\mathscr{F}_\infty^B = \bigcup_{\substack{J\subset[0,\infty)\\ J \text{ countable}}} \sigma(B(t_j) : t \in J)$.

10. Let $X(t), Y(t)$ be any two stochastic processes. Show that

$$(X(u_1), \dots, X(u_p)) \perp\!\!\!\perp (Y(u_1), \dots, Y(u_p)) \quad \text{for all } u_1 < \dots < u_p,\ p \geqslant 1$$

implies

$$(X(s_1), \dots, X(s_n)) \perp\!\!\!\perp (Y(t_1), \dots, Y(t_m))$$

$$\text{for all} \quad s_1 < \dots < s_m,\ t_1 < \dots < t_n,\ m, n \geqslant 1.$$

11. Let $(\mathscr{F}_t)_{t\geqslant 0}$ and $(\mathscr{G}_t)_{t\geqslant 0}$ be any two filtrations and define $\mathscr{F}_\infty = \sigma(\bigcup_{t\geqslant 0}\mathscr{F}_t)$ and $\mathscr{G}_\infty = \sigma(\bigcup_{t\geqslant 0}\mathscr{G}_t)$. Show that $\mathscr{F}_t \perp\!\!\!\perp \mathscr{G}_t$ $(\forall\, t \geqslant 0)$ if, and only if, $\mathscr{F}_\infty \perp\!\!\!\perp \mathscr{G}_\infty$.

12. Use Lemma 2.8 to show that (all finite dimensional distributions of) a BM^d is invariant under rotations.

13. Let $B_t = (b_t, \beta_t)$, $t \geqslant 0$, be a two-dimensional Brownian motion.
   (a) Show that $W_t := \frac{1}{\sqrt{2}}(b_t + \beta_t)$ is a BM^1.
   (b) Are $X_t := (W_t, \beta_t)$ and $Y_t := \frac{1}{\sqrt{2}}(b_t+\beta_t, b_t-\beta_t)$, $t \geqslant 0$, two-dimensional Brownian motions?

14. Let $B_t = (b_t, \beta_t)$, $t \geqslant 0$, be a two-dimensional Brownian motion. For which values of $\lambda, \mu \in \mathbb{R}$ is the process $X_t := \lambda b_t + \mu\beta_t$ a BM^1?

15. Let $B_t = (b_t, \beta_t)$, $t \geqslant 0$, be a two-dimensional Brownian motion. Decide whether for $s > 0$ the process $X_t := (b_t, \beta_{s-t} - \beta_t)$, $0 \leqslant t \leqslant s$ is a two-dimensional Brownian motion.

16. Let $B_t = (b_t, \beta_t)$, $t \geqslant 0$, be a two-dimensional Brownian motion and $\alpha \in [0, 2\pi)$. Show that $W_t = (b_t \cdot \cos\alpha + \beta_t \cdot \sin\alpha, \beta_t \cdot \cos\alpha - b_t \cdot \sin\alpha)^\top$ is a two-dimensional Brownian motion. Find a suitable d-dimensional generalization of this observation.

17. Let X be a Q-BM. Determine $\operatorname{Cov}(X_t^j, X_s^k)$, the characteristic function of $X(t)$ and the transition probability (density) of $X(t)$.

18. Show that (B1) is equivalent to $B_t - B_s \perp\!\!\!\perp \mathscr{F}_s^B$ for all $0 \leqslant s < t$.

19. Let $(B_t)_{t\geqslant 0}$ be a BM^1 and set $\tau_a = \inf\{s \geqslant 0 : B_s = a\}$ where $a \in \mathbb{R}$. Show that $\tau_a \sim \tau_{-a}$ and $\tau_a \sim a^2\tau_1$.

20. A one-dimensional stochastic process $(X_t)_{t\geqslant 0}$ such that $\mathbb{E}(X_t^2) < \infty$ is called *stationary (in the wide sense)* if $m(t) = \mathbb{E}\,X_t \equiv$ const. and $C(s,t) = \mathbb{E}(X_s X_t) = g(t-s)$, $0 \leqslant s \leqslant t < \infty$ for some even function $g : \mathbb{R} \to \mathbb{R}$. Which of the following processes is stationary?

   (a) $W_t = B_t^2 - t$; (b) $X_t = e^{-\alpha t/2} B_{e^{\alpha t}}$;

   (c) $Y_t = B_{t+h} - B_t$; (d) $Z_t = B_{e^t}$?

21. Let $(B_t)_{t\in[0,1]}$ and $(\beta_t)_{t\in[0,1]}$ be independent one-dimensional Brownian motions. Show that the following process is again a Brownian motion:
$$W_t := \begin{cases} B_t, & \text{if } t \in [0,1], \\ B_1 + t\beta_{1/t} - \beta_1, & \text{if } t \in (1,\infty). \end{cases}$$

22. Find out whether the processes
$$X(t) := B(e^t) \quad\text{and}\quad X(t) := e^{-t/2}B(e^t) \qquad (t \geqslant 0)$$

have the *no-memory property*, i. e. $\sigma(X(t) : t \leqslant a) \perp\!\!\!\perp \sigma(X(t+a) - X(a) : t \geqslant 0)$ for $a > 0$.

23. Prove the time inversion property from Paragraph 2.15.

24. (Strong law of large numbers) Let $(B_t)_{t \geqslant 0}$ be a BM^1. Use Paragraph 2.17 to show that $\lim_{t \to \infty} B_t / t = 0$ a. s. and in mean square sense.

3 Constructions of Brownian motion

There are several different ways to construct Brownian motion and we will present a few of them here. Since a d-dimensional Brownian motion can be obtained from d independent one-dimensional Brownian motions, see Section 2.2, we restrict our attention to the one-dimensional setting. If this is your first encounter with Brownian motion, you should restrict your attention to Sections 3.1 and 3.2 or Section 3.4 which contain the most intuitive constructions.

3.1 A random orthogonal series

Some of the earliest mathematical constructions of Brownian motion are based on random orthogonal series. The idea is to write the paths $[0,1] \ni t \mapsto B_t(\omega)$ for (almost) every ω as a random series with respect to a complete orthonormal system (ONS) in the Hilbert space $L^2(dt) = L^2([0,1], dt)$ with scalar product $\langle f, g\rangle_{L^2} = \int_0^1 f(t)g(t)\,dt$. Assume that $(\phi_n)_{n\geqslant 0}$ is any complete ONS and let $(G_n)_{n\geqslant 0}$ be a sequence of real-valued iid Gaussian $\mathsf{N}(0,1)$-random variables on the probability space $(\Omega, \mathscr{A}, \mathbb{P})$. Set

$$W_N(t) := \sum_{n=0}^{N-1} G_n \langle \mathbb{1}_{[0,t)}, \phi_n\rangle_{L^2} = \sum_{n=0}^{N-1} G_n \int_0^t \phi_n(s)\,ds.$$

We want to show that $\lim_{N\to\infty} W_N(t)$ defines a Brownian motion on $[0,1]$.

Ex. 3.1 **3.1 Lemma.** *The limit $W(t) := \lim_{N\to\infty} W_N(t)$ exists for every $t \in [0,1]$ in $L^2(\mathbb{P})$ and the*
Ex. 3.2 *process $W(t)$ satisfies* (B0)–(B3).

Proof. Using the independence of the $G_n \sim \mathsf{N}(0,1)$ and Parseval's identity we get for every $t \in [0,1]$

$$\begin{aligned}\mathbb{E}(W_N(t)^2) &= \mathbb{E}\left[\sum_{m,n=0}^{N-1} G_n G_m \langle \mathbb{1}_{[0,t)}, \phi_m\rangle_{L^2} \langle \mathbb{1}_{[0,t)}, \phi_n\rangle_{L^2}\right] \\ &= \sum_{m,n=0}^{N-1} \underbrace{\mathbb{E}(G_n G_m)}_{=0\ (n\neq m),\ \text{or } =1\ (n=m)} \langle \mathbb{1}_{[0,t)}, \phi_m\rangle_{L^2} \langle \mathbb{1}_{[0,t)}, \phi_n\rangle_{L^2} \\ &= \sum_{n=0}^{N-1} \langle \mathbb{1}_{[0,t)}, \phi_n\rangle_{L^2}^2 \xrightarrow[N\to\infty]{} \langle \mathbb{1}_{[0,t)}, \mathbb{1}_{[0,t)}\rangle_{L^2} = t.\end{aligned}$$

This shows that $W(t) = L^2\text{-}\lim_{N\to\infty} W_N(t)$ exists. An analogous calculation yields for $s < t$ and $u < v$

$$\begin{aligned}\mathbb{E}(W(t) - W(s))(W(v) - W(u)) &= \sum_{n=0}^{\infty} \langle \mathbb{1}_{[0,t)} - \mathbb{1}_{[0,s)}, \phi_n\rangle_{L^2} \langle \mathbb{1}_{[0,v)} - \mathbb{1}_{[0,u)}, \phi_n\rangle_{L^2} \\ &= \langle \mathbb{1}_{[s,t)}, \mathbb{1}_{[u,v)}\rangle_{L^2},\end{aligned}$$

and now it is easy to see that

$$\mathbb{E}(W(t)-W(s))(W(v)-W(u)) = \begin{cases} t-s, & \text{if } [s,t)=[u,v); \\ 0, & \text{if } [s,t)\cap[u,v)=\emptyset; \\ (v\wedge t - u\vee s)^+, & \text{in general.} \end{cases}$$

With this calculation we find for all $0 \leqslant s < t \leqslant u < v$ and $\xi, \eta \in \mathbb{R}$

$$\begin{aligned}
&\mathbb{E}\left[\exp\left(i\xi(W(t)-W(s)) + i\eta(W(v)-W(u))\right)\right] \\
&= \lim_{N\to\infty} \mathbb{E}\left[\exp\left(i\sum_{n=0}^{N-1}\left(\xi\langle\mathbb{1}_{[s,t)},\phi_n\rangle + \eta\langle\mathbb{1}_{[u,v)},\phi_n\rangle\right)G_n\right)\right] \\
&\overset{\text{iid}}{=} \lim_{N\to\infty}\prod_{n=0}^{N-1}\mathbb{E}\left[\exp\left(\left(i\xi\langle\mathbb{1}_{[s,t)},\phi_n\rangle + i\eta\langle\mathbb{1}_{[u,v)},\phi_n\rangle\right)G_n\right)\right] \\
&= \lim_{N\to\infty}\prod_{n=0}^{N-1}\exp\left(-\frac{1}{2}\left|\xi\langle\mathbb{1}_{[s,t)},\phi_n\rangle + \eta\langle\mathbb{1}_{[u,v)},\phi_n\rangle\right|^2\right)
\end{aligned}$$

(since $\mathbb{E}\, e^{i\zeta G_n} = \exp(-\frac{1}{2}|\zeta|^2)$)

$$\begin{aligned}
&= \lim_{N\to\infty}\exp\left(-\frac{1}{2}\sum_{n=0}^{N-1}\left[\xi^2\langle\mathbb{1}_{[s,t)},\phi_n\rangle^2 + \eta^2\langle\mathbb{1}_{[u,v)},\phi_n\rangle^2\right]\right) \\
&\qquad\times \lim_{N\to\infty}\exp\Bigg(\underbrace{-\sum_{n=0}^{N-1}\xi\eta\langle\mathbb{1}_{[s,t)},\phi_n\rangle\langle\mathbb{1}_{[u,v)},\phi_n\rangle}_{\longrightarrow 0}\Bigg) \\
&= \exp\left(-\frac{1}{2}\xi^2(t-s)\right)\exp\left(-\frac{1}{2}\eta^2(v-u)\right).
\end{aligned}$$

This calculation shows

- $W(t)-W(s) \sim \mathsf{N}(0,t-s)$, if we take $\eta = 0$;
- $W(t-s) \sim \mathsf{N}(0,t-s)$, if we take $\eta = 0$, $s = 0$ and replace t by $t-s$;
- $W(t)-W(s) \perp\!\!\!\perp W(v)-W(u)$, since ξ, η are arbitrary.

The independence of finitely many increments can be seen in the same way. Since $W(0) = 0$ is obvious, we are done. □

The series defining $W(t)$ is a series of independent random variables which converges in $L^2(\mathbb{P})$, hence in probability. From classical probability theory we know that $\lim_{N\to\infty} W_N(t) = W(t)$ almost surely. But this is not enough to ensure that the path $t \mapsto W(t,\omega)$ is ($\mathbb{P}$ almost surely) continuous. The general theorem due to Itô–Nisio [106] would do the trick, but we prefer to give a direct proof using a special complete orthonormal system.

and they are a basis of $L^2([0,1], ds)$, i. e. a complete orthonormal system, see Theorem A.48 in the appendix.

3.2 Theorem (Lévy 1940; Ciesielski 1959). *There is a probability space $(\Omega, \mathscr{A}, \mathbb{P})$ and a sequence of iid standard normal random variables $(G_n)_{n\geqslant 0}$ such that* Ex. 3.3

$$W(t,\omega) = \sum_{n=0}^{\infty} G_n(\omega)\langle \mathbb{1}_{[0,t)}, H_n\rangle_{L^2}, \quad t \in [0,1], \tag{3.1}$$

is a Brownian motion.

Proof. Let $W(t)$ be as in Lemma 3.1 where we take the Haar functions H_n as orthonormal system in $L^2(dt)$. As probability space $(\Omega, \mathscr{A}, \mathbb{P})$ we use the probability space which supports the (countably many) independent Gaussian random variables $(G_n)_{n\geqslant 0}$ from the construction of $W(t)$. It is enough to prove that the sample paths $t \mapsto W(t,\omega)$ are continuous. By definition, the partial sums

$$t \mapsto W_N(t,\omega) = \sum_{n=0}^{N-1} G_n(\omega)\langle \mathbb{1}_{[0,t)}, H_n\rangle_{L^2} = \sum_{n=0}^{N-1} G_n(\omega) S_n(t). \tag{3.2}$$

are continuous for all $N \geqslant 1$, and it is sufficient to prove that (a subsequence of) $(W_N(t))_{N\geqslant 0}$ converges uniformly to $W(t)$.

The next step of the proof is similar to the proof of the Riesz-Fischer theorem on the completeness of L^p spaces, see e. g. [204, Theorem 12.7]. Consider the random variable

$$\Delta_j(t) := W_{2^{j+1}}(t) - W_{2^j}(t) = \sum_{k=0}^{2^j-1} G_{2^j+k} S_{2^j+k}(t).$$

If $k \neq \ell$, then $S_{2^j+k} S_{2^j+\ell} = 0$, and we see

$$\begin{aligned}
|\Delta_j(t)|^4 &= \sum_{k,\ell,p,q=0}^{2^j-1} G_{2^j+k} G_{2^j+\ell} G_{2^j+p} G_{2^j+q}\, S_{2^j+k}(t) S_{2^j+\ell}(t) S_{2^j+p}(t) S_{2^j+q}(t) \\
&= \sum_{k=0}^{2^j-1} G_{2^j+k}^4 |S_{2^j+k}(t)|^4 \\
&\leqslant \sum_{k=0}^{2^j-1} G_{2^j+k}^4\, 2^{-2j}.
\end{aligned}$$

Since the right-hand side does not depend on $t \in [0,1]$ and since $\mathbb{E}\, G_n^4 = 3$ (use 2.3 for $G_n \sim \mathsf{N}(0,1)$) we get

$$\mathbb{E}\Big[\sup_{t\in[0,1]} |\Delta_j(t)|^4\Big] \leqslant 3 \sum_{k=0}^{2^j-1} 2^{-2j} = 3 \cdot 2^{-j} \quad \text{for all } j \geqslant 1.$$

For $n < N$ we find using Minkowski's inequality in $L^4(\mathbb{P})$

$$\begin{aligned}\left\| \sup_{N>n} \sup_{t\in[0,1]} \left|W_{2^N}(t) - W_{2^n}(t)\right| \right\|_{L^4} &\leqslant \left\| \sum_{j=n+1}^{\infty} \sup_{t\in[0,1]} \underbrace{\left|W_{2^j}(t) - W_{2^{j-1}}(t)\right|}_{=|\Delta_{j-1}(t)|} \right\|_{L^4} \\ &\leqslant \sum_{j=n+1}^{\infty} \left\| \sup_{t\in[0,1]} \left|\Delta_{j-1}(t)\right| \right\|_{L^4} \\ &\leqslant 3^{1/4} \sum_{j=n+1}^{\infty} 2^{-(j-1)/4} \xrightarrow[n\to\infty]{} 0.\end{aligned}$$

By monotone convergence we get

$$\left\| \lim_{n\to\infty} \sup_{N>n} \sup_{t\in[0,1]} \left|W_{2^N}(t) - W_{2^n}(t)\right| \right\|_{L^4} = \lim_{n\to\infty} \left\| \sup_{N>n} \sup_{t\in[0,1]} \left|W_{2^N}(t) - W_{2^n}(t)\right| \right\|_{L^4} = 0.$$

! This shows that there exists a subset $\Omega_0 \subset \Omega$ with $\mathbb{P}(\Omega_0) = 1$ such that

Ex. 3.4
$$\lim_{n\to\infty} \sup_{N>n} \sup_{t\in[0,1]} \left|W_{2^N}(t,\omega) - W_{2^n}(t,\omega)\right| = 0 \quad \text{for all } \omega \in \Omega_0.$$

By the completeness of the space of continuous functions, there is a subsequence of $(W_{2^j}(t,\omega))_{j\geqslant1}$ which converges uniformly in $t \in [0,1]$. Since we know that $W(t,\omega)$ is the limiting function, we conclude that $W(t,\omega)$ inherits the continuity of the $W_{2^j}(t,\omega)$ for all $\omega \in \Omega_0$.

It remains to define Brownian motion everywhere on Ω. We set

$$\tilde{W}(t,\omega) := \begin{cases} W(t,\omega), & \omega \in \Omega_0 \\ 0, & \omega \notin \Omega_0. \end{cases}$$

As $\mathbb{P}(\Omega \setminus \Omega_0) = 0$, Lemma 3.1 remains valid for $\tilde{W}$ which means that $\tilde{W}(t,\omega)$ is a one-dimensional Brownian motion indexed by $[0,1]$. □

Strictly speaking, W is not a Brownian motion but has a restriction $\tilde{W}$ which is a Brownian motion.

3.3 Definition. Let $(X_t)_{t\in I}$ and $(Y_t)_{t\in I}$ be two $\mathbb{R}^d$-valued stochastic processes with the same index set. We call X and Y

Ex. 3.6 *indistinguishable* if they are defined on the same probability space and if

$$\mathbb{P}(X_t = Y_t \quad \forall t \in I) = 1;$$

Ex. 3.5 *modifications* if they are defined on the same probability space and

$$\mathbb{P}(X_t = Y_t) = 1 \quad \text{for all } t \in I;$$

equivalent if they have the same *finite dimensional distributions*, i. e.

$$(X_{t_1},\dots,X_{t_n}) \sim (Y_{t_1},\dots,Y_{t_n}) \quad \text{for all } t_1,\dots,t_n \in I,\ n \geqslant 1$$

(X, Y need not be defined on the same probability space).

Now it is easy to construct a real-valued Brownian motion for all $t \geqslant 0$.

3.4 Corollary. *Let $(W^k(t))_{t\in[0,1]}$, $k \geqslant 0$, be independent real-valued Brownian motions on the same probability space. Then*

$$B(t) := \begin{cases} W^0(t), & t \in [0,1); \\ W^0(1) + W^1(t-1), & t \in [1,2); \\ \sum_{j=0}^{k-1} W^j(1) + W^k(t-k), & t \in [k,k+1),\ k \geqslant 2. \end{cases} \tag{3.3}$$

is a BM^1 *indexed by* $t \in [0,\infty)$.

Proof. Let B^k be a copy of the process W from Theorem 3.2 and denote the corresponding probability space by $(\Omega_k, \mathscr{A}_k, \mathbb{P}_k)$.

On the product space $(\Omega, \mathscr{A}, \mathbb{P}) = \bigotimes_{k\geqslant 1}(\Omega_k, \mathscr{A}_k, \mathbb{P}_k)$, $\omega = (\omega_1, \omega_2, \ldots)$, we define processes $W^k(\omega) := B^k(\omega_k)$; by construction, these are independent Brownian motions on $(\Omega, \mathscr{A}, \mathbb{P})$.

Let us check (B0)–(B4) for the process $(B(t))_{t\geqslant 0}$ defined by (3.3). The properties (B0) and (B4) are obvious.

Let $s < t$ and assume that $s \in [\ell, \ell+1)$, $t \in [m, m+1)$ where $\ell \leqslant m$. Then

$$\begin{aligned} B(t) - B(s) &= \sum_{j=0}^{m-1} W^j(1) + W^m(t-m) - \sum_{j=0}^{\ell-1} W^j(1) - W^\ell(s-\ell) \\ &= \sum_{j=\ell}^{m-1} W^j(1) + W^m(t-m) - W^\ell(s-\ell) \\ &= \begin{cases} W^m(t-m) - W^m(s-m) \sim \mathsf{N}(0, t-s), & \ell = m \\ \underbrace{(W^\ell(1) - W^\ell(s-\ell)) + \sum_{j=\ell+1}^{m-1} W^j(1) + W^m(t-m)}_{\overset{\text{indep}}{\sim} \mathsf{N}(0,1-s+\ell)\star\mathsf{N}(0,1)^{\star m-\ell-1}\star\mathsf{N}(0,t-m) = \mathsf{N}(0,t-s)}, & \ell < m. \end{cases} \end{aligned}$$

This proves (B2). We show (B1) only for two increments $B(u) - B(t)$ and $B(t) - B(s)$ where $s < t < u$. As before, $s \in [\ell, \ell+1)$, $t \in [m, m+1)$ and $u \in [n, n+1)$ where $\ell \leqslant m \leqslant n$. Then

$$B(u) - B(t) = \begin{cases} W^m(u-m) - W^m(t-m), & m = n \\ (W^m(1) - W^m(t-m)) + \sum_{j=m+1}^{n-1} W^j(1) + W^n(u-n), & m < n. \end{cases}$$

By assumption, the random variables appearing in the representation of the increments $B(u) - B(t)$ and $B(t) - B(s)$ are independent which means that the increments are independent. The case of finitely many, not necessarily adjacent, increments is similar. □

3.3 Wiener's construction

This is also a series approach, but Wiener used the trigonometric functions $(e^{in\pi t})_{n\in\mathbb{Z}}$ as orthonormal basis for $L^2[0,1]$. In this case we obtain Brownian motion on $[0,1]$ as a Wiener-Fourier series

$$W(t,\omega) := tG_0(\omega) + \sqrt{2}\sum_{n=1}^{\infty}\frac{\sin(n\pi t)}{n\pi}G_n(\omega), \tag{3.4}$$

where $(G_n)_{n\geqslant 0}$ are iid standard normal random variables. Lemma 3.1 remains valid for (3.4) and shows that the series converges in L^2 and that the limit satisfies (B0)–(B3).

3.5 Theorem (Wiener 1923, 1924). *The random process $(W_t)_{t\in[0,1]}$ defined by* (3.4) *is a one-dimensional Brownian motion.*

Using Corollary 3.4 we can extend the construction of Theorem 3.5 to get a Brownian motion with time set $[0,\infty)$.

Proof of Theorem 3.5. It is enough to prove that (3.4) defines a continuous random process. Let

$$W_N(t,\omega) := tG_0(\omega) + \sqrt{2}\sum_{n=1}^{N}\frac{\sin(n\pi t)}{n\pi}G_n(\omega).$$

We will show that $(W_{2^n})_{n\geqslant 1}$ is a Cauchy sequence in $L^2(\mathbb{P})$ uniformly for all $t\in[0,1]$. Set

$$\Delta_j(t) := W_{2^{j+1}}(t) - W_{2^j}(t).$$

Using $|\operatorname{Im} z|\leqslant |z|$ for $z\in\mathbb{C}$, we see

$$|\Delta_j(t)|^2 = \left(\sum_{k=2^j+1}^{2^{j+1}}\frac{\sin(k\pi t)}{k}G_k\right)^2 \leqslant \left|\sum_{k=2^j+1}^{2^{j+1}}\frac{e^{ik\pi t}}{k}G_k\right|^2,$$

and since $|z|^2 = z\bar{z}$ we get

$$\begin{aligned}
|\Delta_j(t)|^2 &\leqslant \sum_{k=2^j+1}^{2^{j+1}}\sum_{\ell=2^j+1}^{2^{j+1}}\frac{e^{ik\pi t}e^{-i\ell\pi t}}{k\ell}G_kG_\ell\\
&= \sum_{k=2^j+1}^{2^{j+1}}\frac{G_k^2}{k^2} + 2\sum_{k=2^j+1}^{2^{j+1}}\sum_{\ell=2^j+1}^{k-1}\frac{e^{ik\pi t}e^{-i\ell\pi t}}{k\ell}G_kG_\ell\\
&\overset{m=k-\ell}{=} \sum_{k=2^j+1}^{2^{j+1}}\frac{G_k^2}{k^2} + 2\sum_{m=1}^{2^j-1}\sum_{\ell=2^j+1}^{2^{j+1}-m}\frac{e^{im\pi t}}{\ell(\ell+m)}G_\ell G_{\ell+m}\\
&\leqslant \sum_{k=2^j+1}^{2^{j+1}}\frac{G_k^2}{k^2} + 2\sum_{m=1}^{2^j-1}\left|\sum_{\ell=2^j+1}^{2^{j+1}-m}\frac{G_\ell G_{\ell+m}}{\ell(\ell+m)}\right|.
\end{aligned}$$

Note that the right-hand side is independent of t. On both sides we can therefore take the supremum over all $t \in [0,1]$ and then the mean value $\mathbb{E}(\cdots)$. From Jensen's inequality, $\mathbb{E}|Z| \leqslant \sqrt{\mathbb{E}|Z|^2}$, we conclude

$$\mathbb{E}\left(\sup_{t\in[0,1]} |\Delta_j(t)|^2\right) \leqslant \sum_{k=2^j+1}^{2^{j+1}} \frac{\mathbb{E}\,G_k^2}{k^2} + 2\sum_{m=1}^{2^j-1} \mathbb{E}\left|\sum_{\ell=2^j+1}^{2^{j+1}-m} \frac{G_\ell G_{\ell+m}}{\ell(\ell+m)}\right|$$
$$\leqslant \sum_{k=2^j+1}^{2^{j+1}} \frac{\mathbb{E}\,G_k^2}{k^2} + 2\sum_{m=1}^{2^j-1} \sqrt{\mathbb{E}\left(\sum_{\ell=2^j+1}^{2^{j+1}-m} \frac{G_\ell G_{\ell+m}}{\ell(\ell+m)}\right)^2}.$$

If we expand the square we get expressions of the form $\mathbb{E}(G_\ell G_{\ell+m} G_{\ell'} G_{\ell'+m})$; these expectations are zero whenever $\ell \neq \ell'$. Thus,

$$\mathbb{E}\left(\sup_{t\in[0,1]} |\Delta_j(t)|^2\right) = \sum_{k=2^j+1}^{2^{j+1}} \frac{1}{k^2} + 2\sum_{m=1}^{2^j-1} \sqrt{\sum_{\ell=2^j+1}^{2^{j+1}-m} \mathbb{E}\left(\frac{G_\ell G_{\ell+m}}{\ell(\ell+m)}\right)^2}.$$
$$= \sum_{k=2^j+1}^{2^{j+1}} \frac{1}{k^2} + 2\sum_{m=1}^{2^j-1} \sqrt{\sum_{\ell=2^j+1}^{2^{j+1}-m} \frac{1}{\ell^2(\ell+m)^2}}$$
$$\leqslant \sum_{k=2^j+1}^{2^{j+1}} \frac{1}{(2^j)^2} + 2\sum_{m=1}^{2^j-1} \sqrt{\sum_{\ell=2^j+1}^{2^{j+1}-m} \frac{1}{(2^j)^2(2^j)^2}}$$
$$\leqslant 2^j \cdot 2^{-2j} + 2\cdot 2^j \cdot \sqrt{2^j \cdot 2^{-4j}}$$
$$\leqslant 3\cdot 2^{-j/2}.$$

From this point onwards we can follow the proof of Theorem 3.2: For $n < N$ we find using Minkowski's inequality in $L^2(\mathbb{P})$

$$\left\|\sup_{N>n}\sup_{t\in[0,1]} \left|W_{2^N}(t) - W_{2^n}(t)\right|\right\|_{L^2} \leqslant \left\|\sum_{j=n+1}^{\infty} \sup_{t\in[0,1]} \underbrace{\left|W_{2^j}(t) - W_{2^{j-1}}(t)\right|}_{=|\Delta_{j-1}(t)|}\right\|_{L^2}$$
$$\leqslant \sum_{j=n+1}^{\infty} \left\|\sup_{t\in[0,1]} \left|\Delta_{j-1}(t)\right|\right\|_{L^2}$$
$$\leqslant 3^{1/4} \sum_{j=n+1}^{\infty} 2^{-(j-1)/4} \xrightarrow[n\to\infty]{} 0.$$

By monotone convergence we get

$$\left\|\lim_{n\to\infty}\sup_{N>n}\sup_{t\in[0,1]} \left|W_{2^N}(t) - W_{2^n}(t)\right|\right\|_{L^2} = \lim_{n\to\infty}\left\|\sup_{N>n}\sup_{t\in[0,1]} \left|W_{2^N}(t) - W_{2^n}(t)\right|\right\|_{L^2} = 0.$$

This shows that there exists a subset $\Omega_0 \subset \Omega$ with $\mathbb{P}(\Omega_0) = 1$ such that

$$\lim_{n\to\infty}\sup_{N>n}\sup_{t\in[0,1]} \left|W_{2^N}(t,\omega) - W_{2^n}(t,\omega)\right| = 0 \quad \text{for all } \omega \in \Omega_0.$$

! Ex. 3.4

By the completeness of the space of continuous functions, there is a subsequence of $(W_{2^j}(t,\omega))_{j\geqslant 1}$ which converges uniformly in $t \in [0,1]$. Since we know that $W(t,\omega)$ is the limiting function, we conclude that $W(t,\omega)$ inherits the continuity of the $W_{2^j}(t,\omega)$ for all $\omega \in \Omega_0$.

It remains to define the Brownian motion everywhere on Ω. We set

$$\tilde{W}(t,\omega) := \begin{cases} W(t,\omega), & \omega \in \Omega_0 \\ 0, & \omega \notin \Omega_0. \end{cases}$$

Since $\mathbb{P}(\Omega \setminus \Omega_0) = 0$, Lemma 3.1 remains valid for $\tilde{W}$ which means that $\tilde{W}(t,\omega)$ is a one-dimensional Brownian motion indexed by $[0,1]$. □

3.4 Lévy's original argument

Lévy's original argument is based on interpolation. Let $t_0 < t < t_1$ and assume that $(W_t)_{t\in[0,1]}$ is a Brownian motion. Then

$$G' := W(t) - W(t_0) \quad \text{and} \quad G'' := W(t_1) - W(t)$$

are independent Gaussian random variables with mean 0 and variances $t - t_0$ and $t_1 - t$, respectively. In order to simulate the positions $W(t_0), W(t), W(t_1)$ we could either determine $W(t_0)$ and then G' and G'' independently or, equivalently, simulate first $W(t_0)$ and $W(t_1) - W(t_0)$, and obtain $W(t)$ by interpolation. Let Γ be a further standard normal random variable which is independent of $W(t_0)$ and $W(t_1) - W(t_0)$. Then

$$\begin{aligned} &\frac{(t_1 - t)W(t_0) + (t - t_0)W(t_1)}{t_1 - t_0} + \sqrt{\frac{(t - t_0)(t_1 - t)}{t_1 - t_0}}\,\Gamma \\ &= W(t_0) + \frac{t - t_0}{t_1 - t_0}(W(t_1) - W(t_0)) + \sqrt{\frac{(t - t_0)(t_1 - t)}{t_1 - t_0}}\,\Gamma \end{aligned}$$

is – because of the independence of $W(t_0)$, $W(t_1) - W(t_0)$ and Γ – a Gaussian random variable with mean zero and variance t. Thus, we get a random variable with the same distribution as $W(t)$ and so

Ex. 3.8

$$\begin{gathered} \mathbb{P}(W(t) \in \cdot \mid W(t_0) = x_0, W(t_1) = x_1) = \mathsf{N}(m_t, \sigma_t^2) \\ m_t = \frac{(t_1 - t)x_0 + (t - t_0)x_1}{t_1 - t_0} \quad \text{and} \quad \sigma_t^2 = \frac{(t - t_0)(t_1 - t)}{t_1 - t_0}. \end{gathered} \tag{3.5}$$

If t is the midpoint of $[t_0, t_1]$, Lévy's method leads to the following prescription:

3.6 Algorithm. *Set $W(0) = 0$ and let $W(1)$ be an $\mathsf{N}(0,1)$ distributed random variable. Let $n \geqslant 1$ and assume that the random variables $W(k2^{-n})$, $k = 1, \dots, 2^n - 1$ have already*

$\Delta_n(t)$ is the distance between the polygonal arcs $W_{2^{n+1}}(t)$ and $W_{2^n}(t)$; the maximum is attained at one of the midpoints of the intervals $[(k-1)2^{-n}, k2^{-n}]$, $k = 1, \dots, 2^n$, see Figure 3.3. Thus

$$\begin{aligned}\sup_{0\leqslant t\leqslant 1} \left|W_{2^{n+1}}(t,\omega) - W_{2^n}(t,\omega)\right| &\leqslant \max_{1\leqslant k\leqslant 2^n} \left|\Delta_n((2k-1)2^{-n-1},\omega)\right| \\ &\leqslant c\sqrt{\frac{n\log 2}{2^{n+1}}} \quad \text{for all } n \geqslant N(\omega),\end{aligned}$$

which shows that the limit

$$W(t,\omega) := \lim_{N\to\infty} W_{2^N}(t,\omega) = \sum_{n=0}^{\infty} \left(W_{2^{n+1}}(t,\omega) - W_{2^n}(t,\omega)\right) + W_1(t,\omega)$$

exists for all $\omega \in \Omega_0$ uniformly in $t \in [0,1]$. Therefore, $t \mapsto W(t,\omega)$, $\omega \in \Omega_0$, inherits the continuity of the polygonal arcs $t \mapsto W_{2^n}(t,\omega)$. Set

$$\tilde W(t,\omega) := \begin{cases} W(t,\omega), & \omega \in \Omega_0, \\ 0, & \omega \notin \Omega_0. \end{cases}$$

By construction, we find for all $0 \leqslant j \leqslant k \leqslant 2^n$

$$\begin{aligned}\tilde W(k2^{-n}) - \tilde W(j2^{-n}) &= W_{2^n}(k2^{-n}) - W_{2^n}(j2^{-n}) \\ &= \sum_{\ell=j+1}^{k} \left(W_{2^n}(\ell 2^{-n}) - W_{2^n}((\ell-1)2^{-n})\right) \\ &\overset{\text{iid}}{\sim} \mathsf{N}(0,(k-j)2^{-n}).\end{aligned}$$

Since $t \mapsto \tilde W(t)$ is continuous and since the dyadic numbers are dense in $[0,t]$, we conclude that the increments $\tilde W(t_j) - \tilde W(t_{j-1})$, $0 = t_0 < t_1 < \dots < t_N \leqslant 1$ are independent $\mathsf{N}(0, t_j - t_{j-1})$ distributed random variables. This shows that $(\tilde W(t))_{t\in[0,1]}$ is a Brownian motion. □

3.8 Rigorous proof of (3.5). We close this section with a rigorous proof of Lévy's formula (3.5). Observe that for all $0 \leqslant s < t$ and $\xi \in \mathbb{R}$ Ex. 3.8

$$\begin{aligned}\mathbb{E}\left(e^{i\xi W(t)} \mid W(s)\right) &= e^{i\xi W(s)}\, \mathbb{E}\left(e^{i\xi(W(t)-W(s))} \mid W(s)\right) \\ &\overset{\text{(B1)}}{=} e^{i\xi W(s)}\, \mathbb{E}\left(e^{i\xi(W(t)-W(s))}\right) \\ &\underset{(2.5)}{\overset{\text{(B2)}}{=}} e^{-\frac{1}{2}(t-s)\xi^2}\, e^{i\xi W(s)}.\end{aligned}$$

If we apply this equality to the projectively reflected (cf. 2.17) Brownian motion $(tW(1/t))_{t>0}$, we get

$$\mathbb{E}\left(e^{i\xi tW(1/t)} \mid W(1/s)\right) = \mathbb{E}\left(e^{i\xi tW(1/t)} \mid sW(1/s)\right) = e^{-\frac{1}{2}(t-s)\xi^2}\, e^{i\xi sW(1/s)}.$$

Now set $b := \frac{1}{s} > \frac{1}{t} =: a$ and $\eta := \xi/a = \xi t$. Then

$$\begin{aligned}\mathbb{E}\left(e^{i\eta W(a)} \mid W(b)\right) &= \exp\left[-\frac{1}{2}a^2\left(\frac{1}{a}-\frac{1}{b}\right)\eta^2\right]\exp\left[i\eta\frac{a}{b}W(b)\right]\\ &= \exp\left[-\frac{1}{2}\frac{a}{b}(b-a)\eta^2\right]\exp\left[i\eta\frac{a}{b}W(b)\right].\end{aligned} \tag{3.6}$$

This is (3.5) for the case where $(t_0, t, t_1) = (0, a, b)$ and $x_0 = 0$.

Now we fix $0 \leqslant t_0 < t < t_1$ and observe that $W(s+t_0) - W(t_0)$, $s \geqslant 0$, is again a Brownian motion. Applying (3.6) yields

$$\mathbb{E}\left(e^{i\eta(W(t)-W(t_0))} \mid W(t_1) - W(t_0)\right) = e^{-\frac{1}{2}\frac{t-t_0}{t_1-t_0}(t_1-t)\eta^2}\, e^{i\eta\frac{t-t_0}{t_1-t_0}(W(t_1)-W(t_0))}.$$

On the other hand, $W(t_0) \perp\!\!\!\perp (W(t) - W(t_0), W(t_1) - W(t_0))$ and

$$\begin{aligned}&\mathbb{E}\left(e^{i\eta(W(t)-W(t_0))} \;\middle|\; W(t_1) - W(t_0)\right)\\ &= \mathbb{E}\left(e^{i\eta(W(t)-W(t_0))} \;\middle|\; W(t_1) - W(t_0), W(t_0)\right)\\ &= \mathbb{E}\left(e^{i\eta(W(t)-W(t_0))} \;\middle|\; W(t_1), W(t_0)\right)\\ &= e^{-i\eta W(t_0)}\,\mathbb{E}\left(e^{i\eta W(t)} \;\middle|\; W(t_1), W(t_0)\right).\end{aligned}$$

This proves

$$\mathbb{E}\left(e^{i\eta W(t)} \mid W(t_1), W(t_0)\right) = e^{-\frac{1}{2}\frac{t-t_0}{t_1-t_0}(t_1-t)\eta^2}\, e^{i\eta\frac{t-t_0}{t_1-t_0}(W(t_1)-W(t_0))}\, e^{i\eta W(t_0)},$$

and (3.5) follows.

3.9 Remark. The polygonal arc $W_{2^n}(t,\omega)$ can be expressed by Schauder functions, cf. Figure 3.1. From (3.2) and the proof of Theorem 3.2 we know that

$$W(t,\omega) = \underbrace{G_0(\omega)S_0(t)}_{=W_1(t,\omega)} + \sum_{n=0}^{\infty}\underbrace{\sum_{k=0}^{2^n-1} G_{2^n+k}(\omega)\,S_{2^n+k}(t)}_{=W_{2^{n+1}}(t,\omega)-W_{2^n}(t,\omega)},$$

where $S_n(t)$ are the Schauder functions. This shows that Lévy's construction and the Lévy–Ciesielski constructions are essentially the same.

For dyadic $t = k/2^n$, the expansion of W_t is finite. Observe that

1° — $W(1,\omega) = G_0(\omega)S_0(1)$

2° — $W(\frac{1}{2},\omega) = \underbrace{G_0(\omega)S_0(\tfrac{1}{2})}_{\text{interpolation}} + \underbrace{G_1(\omega)S_1(\tfrac{1}{2})}_{=\Gamma_1 \sim \mathsf{N}(0,1/4)}$

3° — $W(\frac{1}{4},\omega) = \underbrace{G_0(\omega)S_0(\tfrac{1}{4}) + G_1(\omega)S_1(\tfrac{1}{4})}_{\text{interpolation}} + \underbrace{G_2(\omega)S_2(\tfrac{1}{4})}_{=\Gamma_2 \sim \mathsf{N}(0,1/8)}$

$W(\frac{3}{4},\omega) = \overbrace{G_0(\omega)S_0(\tfrac{3}{4}) + G_1(\omega)S_1(\tfrac{3}{4})} + G_2(\omega)\underbrace{S_2(\tfrac{3}{4})}_{=0} + \overbrace{G_3(\omega)S_3(\tfrac{3}{4})}^{=\Gamma_3 \sim \mathsf{N}(0,1/8)}$

4° — ...

The 2^nth partial sum, $W_{2^n}(t,\omega)$, is therefore a piecewise linear interpolation of the Brownian path $W_t(\omega)$ at the points $(k2^{-n}, W(k2^{-n},\omega))$, $k = 0, 1, \ldots, 2^n$.

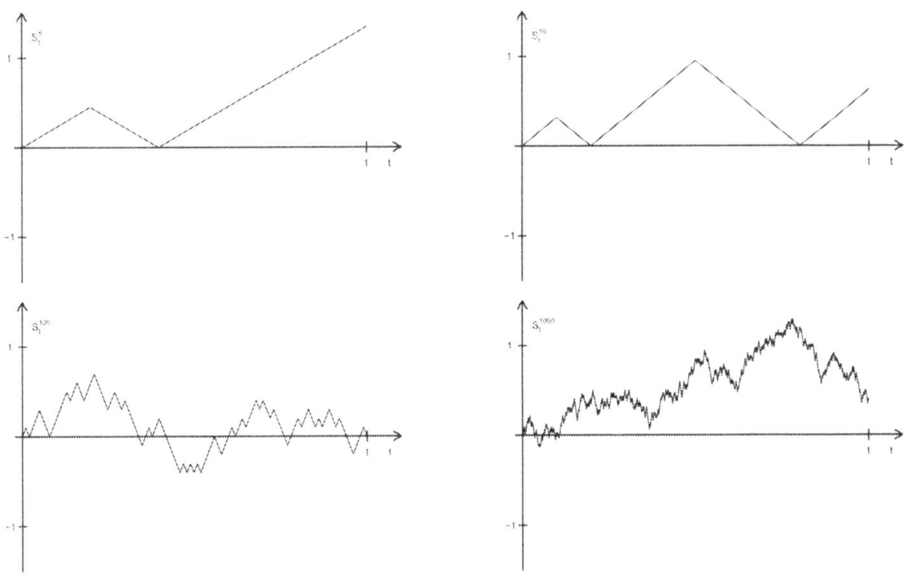

Fig. 3.4. The first few approximation steps in Donsker's construction: $n = 5, 10, 100$ and 1000.

3.5 Donsker's construction

Donsker's invariance theorem shows that Brownian motion is a limit of linearly interpolated random walks – pretty much in the way we have started the discussion in Chapter 1. As before, the difficult point is to prove the sample continuity of the limiting process.

On a probability space $(\Omega, \mathscr{A}, \mathbb{P})$ let ϵ_n, $n \geqslant 1$, be iid Bernoulli random variables such that $\mathbb{P}(\epsilon_1 = 1) = \mathbb{P}(\epsilon_1 = -1) = \frac{1}{2}$. Then

$$S_n := \epsilon_1 + \cdots + \epsilon_n$$

is a simple random walk. Interpolate linearly and apply Gaussian scaling

$$S^n(t) := \frac{1}{\sqrt{n}} \left(S_{\lfloor nt \rfloor} - (nt - \lfloor nt \rfloor)\epsilon_{\lfloor nt \rfloor + 1}\right), \quad t \in [0, 1],$$

see Figure 3.4 illustrating this for $n = 5, 10, 100$ and 1000 steps.

In particular, $S^n(j/n) = S_j/\sqrt{n}$. If $j = j(n)$ and $j/n = s = \text{const.}$, the central limit theorem shows that $S^n(j/n) = \sqrt{s}\, S_j/\sqrt{j} \xrightarrow{d} \sqrt{s}\, G$ as $n \to \infty$ where G is a standard normal random variable. Moreover, for $s = j/n$ and $t = k/n$, $0 \leqslant j < k \leqslant n$, the increment $S^n(t) - S^n(s) = (S_k - S_j)/\sqrt{n}$ is independent of $\epsilon_1, \ldots, \epsilon_j$, and therefore of all earlier increments of the same form. Moreover,

$$\mathbb{E}\left(S^n(t) - S^n(s)\right) = 0 \quad \text{and} \quad \mathbb{V}\left(S^n(t) - S^n(s)\right) = \tfrac{k-j}{n} = t - s;$$

in the limit we get a Gaussian increment with mean zero and variance $t - s$. Since independence and stationarity of the increments are distributional properties, they are inherited by the limiting process – which we will denote by $(B_t)_{t\in[0,1]}$. We have seen that $(B_q)_{q\in[0,1]\cap\mathbb{Q}}$ would have the properties (B0)–(B3) and it qualifies as a candidate for Brownian motion. If it had continuous sample paths, (B0)–(B3) would hold not only for rational times but for all $t \geqslant 0$. That the limit exists and is uniform in t is the essence of Donsker's invariance principle.

3.10 Theorem (Donsker 1951). *Let $(B(t))_{t\in[0,1]}$ be a one-dimensional Brownian motion, $(S^n(t))_{t\in[0,1]}$, $n \geqslant 1$, be as above and $\Phi : \mathcal{C}([0,1],\mathbb{R}) \to \mathbb{R}$ a uniformly continuous bounded functional. Then*

$$\lim_{n\to\infty} \mathbb{E}\,\Phi(S^n(\cdot)) = \mathbb{E}\,\Phi(B(\cdot)),$$

i. e. $S^n(\cdot)$ converges weakly to $B(\cdot)$.

We will give a proof Donsker's theorem in Chapter 14, Theorem 14.5. Since our proof relies on the existence of a Brownian motion, we cannot use it to construct $(B_t)_{t\geqslant 0}$. Nevertheless, it is an important result for the intuitive understanding of a Brownian motion as limit of random walks as well as for (simple) proofs of various limit theorems.

3.6 The Bachelier–Kolmogorov point of view

The starting point of this construction is the observation that the finite dimensional marginal distributions of a Brownian motion are Gaussian random variables. More precisely, for any number of times $t_0 = 0 < t_1 < \cdots < t_n$, $t_j \in I$, and all Borel sets $A_1, \ldots, A_n \in \mathscr{B}(\mathbb{R})$ the finite dimensional distributions

$$p_{t_1,\ldots,t_n}(A_1 \times \cdots \times A_n) = \mathbb{P}(B_{t_1} \in A_1, \ldots, B_{t_n} \in A_n)$$

are mean-zero normal laws with covariance matrix $C = (t_j \wedge t_k)_{j,k=1,\ldots,n}$. From Theorem 2.6 we know that they are given by

$$\begin{aligned} p_{t_1,\ldots,t_n}&(A_1 \times \cdots \times A_n) \\ &= \frac{1}{(2\pi)^{n/2}}\frac{1}{\sqrt{\det C}} \int\limits_{A_1\times\cdots\times A_n} \exp\left(-\tfrac{1}{2}\left\langle x, C^{-1}x\right\rangle\right) dx \\ &= \frac{1}{(2\pi)^{n/2}\sqrt{\prod_{j=1}^n (t_j - t_{j-1})}} \int\limits_{A_1\times\cdots\times A_n} \exp\left(-\frac{1}{2}\sum_{j=1}^n \frac{(x_j - x_{j-1})^2}{t_j - t_{j-1}}\right) dx. \end{aligned}$$

We will characterize a stochastic process in terms of its finite dimensional distributions – and we will discuss this approach in Chapter 4 below. The idea is to identify a stochastic process with an infinite dimensional measure $\mathbb{P}$ on the space of all sample

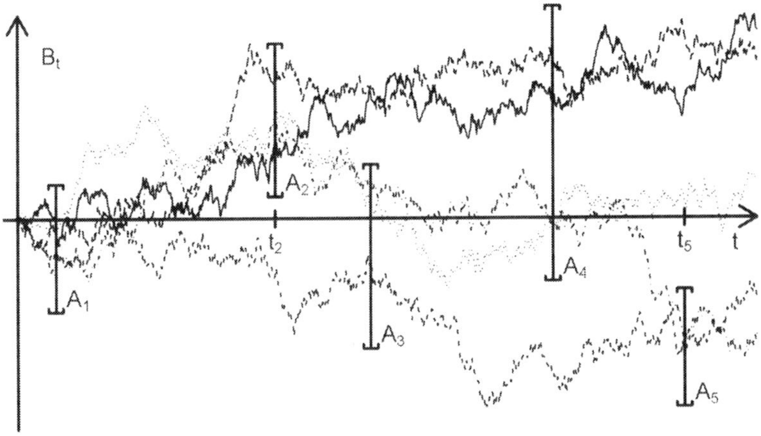

Fig. 3.5. Brownian paths meeting (and missing) the sets $A_1, \ldots, A_5$ at times $t_1, \ldots, t_5$

paths Ω such that the finite dimensional projections of $\mathbb{P}$ are exactly the finite dimensional distributions.

3.11 Further reading. Yet another construction of Brownian motion, using interpolation arguments and Bernstein polynomials, can be found in the well-written paper [131]. Donsker's theorem, without assuming the existence of BM, is e. g. proved in the classic monograph [14]; in a more general context it is contained in [56], the presentation in [119] is probably the easiest to read. Ciesielski's construction became popular through his very influential Aarhus lecture notes [35], one of the first books on Brownian motion after Wiener [230] and Lévy [145]. Good sources on random Fourier series and wavelet expansions are [115] and [179]. Constructions of Brownian motion in infinite dimensions, e. g. cylindrical Brownian motions etc., can be found in [183] and [43], but this needs a good knowledge of functional analysis.

[14] Billingsley: *Convergence of Probability Measures.*
[35] Ciesielski: *Lectures on Brownian Motion, Heat Conduction and Potential Theory.*
[43] DaPrato, Zabczyk: *Stochastic Equations in Infinite Dimensions.*
[56] Dudley: *Uniform Central Limit Theorems.*
[115] Kahane: *Some Random Series of Functions.*
[119] Karatzas, Shreve: *Brownian Motion and Stochastic Calculus.*
[131] Kowalski: Bernstein Polynomials and Brownian Motion.
[179] Pinsky: *Introduction to Fourier Analysis and Wavelets.*
[183] Prevôt, Röckner: *A Concise Course on Stochastic Partial Differential Equations.*

Problems

1. Use the Lévy–Ciesielski representation $B(t) = \sum_{n=0}^{\infty} G_n S_n(t)$, $t \in [0, 1]$, to obtain a series representation for $X := \int_0^1 B(t)\,dt$ and find the distribution of X.

2. (Random orthogonal measures) Let $(G_n)_{n\geqslant 0}$ be a sequence of iid Gaussian $\mathsf{N}(0,1)$ random variables, $(\phi_n)_{n\geqslant 0}$ be a complete ONS in $L^2([0,1],dt)$, $B \in \mathscr{B}[0,1]$ be a measurable set and define $W_N(B) := \sum_{n=0}^{N-1} G_n \langle \mathbb{1}_B, \phi_n\rangle_{L^2}$.
   (a) Show that the limit $W(B) := \lim_{N\to\infty} W_N(B)$ exists in $L^2(\mathbb{P})$.
   (b) Find $\mathbb{E}\, W(A)W(B)$ for sets $A, B \in \mathscr{B}[0,1]$.
   (c) Show that $B \mapsto \mu(B) := \mathbb{E}\left(|W(B)|^2\right)$ is a measure. Is $B \mapsto W(B)$ a measure?
   (d) Let $\mathcal{S}$ denote all step functions of the form $f(t) := \sum_{j<\infty} c_j \mathbb{1}_{B_j}(t)$ where $c_j \in \mathbb{R}$ and $B_j \in \mathscr{B}[0,1]$. Show that the operator $I(f) := \sum_j c_j W(B_j)$ is well-defined (i. e. independent of the particular representation of the step function).
   (e) Show that $\mathbb{E}(|I(f)|^2) = \int |f|^2\,d\mu$ and use this isometry between $L^2([0,1],\mu)$ and $L^2(\mathbb{P})$ to extend I to all $f \in L^2([0,1],\mu)$.

   Remark: If we take $B = [0,t]$, then $W_t := W([0,t])$, $t \in [0,1]$, is a Brownian motion.

3. Let $\phi_n = H_n$, $n = 0,1,2,\dots$, be the Haar functions. The following steps show that the *full* sequence $W_N(t,\omega) := \sum_{n=0}^{N-1} G_n(\omega)S_n(t)$, $N = 1,2,\dots$, converges uniformly for $\mathbb{P}$-almost all $\omega \in \Omega$.
   (a) Let $(a_n)_{n\geqslant 0} \subset \mathbb{R}$ be a sequence satisfying $a_n = O(n^\epsilon)$ for some $\epsilon \in (0,1/2)$. Show that $\sum_{n=0}^\infty a_n S_n(t)$ converges absolutely and uniformly in $t \in [0,1]$.
   (b) Let $(G_n)_{n\geqslant 0}$ be a sequence of iid $\mathsf{N}(0,1)$ random variables. Show that almost surely $|G_n| = O(\sqrt{\log n})$.

   Hint: Use $\mathbb{P}(|G_1| > x) \leqslant 2(2\pi)^{-1/2}x^{-1}e^{-x^2/2}$, $x > 0$, and the Borel–Cantelli lemma.

4. Let (S,d) be a complete metric space equipped with the σ-algebra $\mathscr{B}(S)$ of its Borel sets. Assume that $(X_n)_{n\geqslant 1}$ is a sequence of S-valued random variables such that

$$\lim_{n\to\infty} \sup_{m\geqslant n} \mathbb{E}\left(d(X_n, X_m)^p\right) = 0$$

   for some $p \in [1,\infty)$. Show that there is a subsequence $(n_k)_{k\geqslant 1}$ and a random variable X such that $\lim_{k\to\infty} X_{n_k} = X$ almost surely.

5. Let $(X_t)_{t\geqslant 0}$ and $(Y_t)_{t\geqslant 0}$ be two stochastic processes which are modifications of each other. Show that they have the same finite dimensional distributions.

6. Let $(X_t)_{t\in I}$ and $(Y_t)_{t\in I}$ be two processes with the same index set $I \subset [0,\infty)$ and state space. Show that

$$X, Y \text{ indistinguishable} \implies X, Y \text{ modifications} \implies X, Y \text{ equivalent}.$$

   Assume that the processes are defined on the same probability space and $t \mapsto X_t$ and $t \mapsto Y_t$ are right-continuous (or that I is countable). Do the reverse implications hold, too?

7. Let $(B_t)_{t\geqslant 0}$ be a real-valued stochastic process with exclusively continuous sample paths. Assume that $(B_q)_{q\in\mathbb{Q}\cap[0,\infty)}$ satisfies (B0)–(B3). Show that $(B_t)_{t\geqslant 0}$ is a BM.

8. Give a direct proof of the formula (3.5) using the joint probability distribution $(W(t_0), W(t), W(t_1))$ of the Brownian motion $W(t)$.

4 The canonical model

Ex. 4.1 Often we encounter the statement '*Let X be a random variable with law μ*' where μ is an *a priori* given probability distribution. There is no reference to the underlying probability space $(\Omega, \mathscr{A}, \mathbb{P})$, and actually the nature of this space is not important: While X is defined by the probability distribution, there is considerable freedom in the choice of our model $(\Omega, \mathscr{A}, \mathbb{P})$. The same situation is true for stochastic processes: Brownian motion is defined by distributional properties and our construction of BM, see e. g. Theorem 3.2, not only furnished us with a process but also a suitable probability space.

Ex. 4.2 By definition, a d-dimensional stochastic process $(X(t))_{t\in I}$ is a family of $\mathbb{R}^d$-valued random variables on the space $(\Omega, \mathscr{A}, \mathbb{P})$. Alternatively, we may understand a process as a map $(t,\omega) \mapsto X(t,\omega)$ from $I \times \Omega$ to $\mathbb{R}^d$ or as a map $\omega \mapsto \{t \mapsto X(t,\omega)\}$ from Ω into the space $(\mathbb{R}^d)^I = \{w,\ w : I \to \mathbb{R}^d\}$. If we go one step further and identify Ω with (a subset of) $(\mathbb{R}^d)^I$, we get the so-called *canonical model*. Of course, it is a major task to identify the correct σ-algebra and measure in $(\mathbb{R}^d)^I$. For Brownian motion it is enough to consider the subspace $\mathcal{C}_{(0)}[0,\infty) \subset (\mathbb{R}^d)^{[0,\infty)}$,

$$\mathcal{C}_{(0)} := \mathcal{C}_{(0)}[0,\infty) := \left\{w : [0,\infty) \to \mathbb{R}^d \ :\ w \text{ is continuous and } w(0) = 0\right\}.$$

In the first part of this chapter we will see how we can obtain a canonical model for BM defined on the space of continuous functions; in the second part we discuss Kolmogorov's construction of stochastic processes with given finite dimensional distributions.

4.1 Wiener measure

Let $I = [0,\infty)$ and denote by $\pi_t : (\mathbb{R}^d)^I \to \mathbb{R}^d$, $w \mapsto w(t)$, the canonical projection onto the tth coordinate. The natural σ-algebra on the infinite product $(\mathbb{R}^d)^I$ is the product σ-algebra

$$\mathscr{B}^I(\mathbb{R}^d) = \sigma\left\{\pi_t^{-1}(B) \ :\ B \in \mathscr{B}(\mathbb{R}^d),\ t \in I\right\} = \sigma\{\pi_t \ :\ t \in I\};$$

this shows that $\mathscr{B}^I(\mathbb{R}^d)$ is the smallest σ-algebra which makes all projections π_t measurable. Since Brownian motion has exclusively continuous sample paths, it is natural to replace $(\mathbb{R}^d)^I$ by $\mathcal{C}_{(0)}$. Unfortunately, cf. Corollary 4.6, $\mathcal{C}_{(0)}$ is not contained in $\mathscr{B}^I(\mathbb{R}^d)$; therefore, we have to consider the trace σ-algebra

$$\mathcal{C}_{(0)} \cap \mathscr{B}^I(\mathbb{R}^d) = \sigma\left(\pi_t|_{\mathcal{C}_{(0)}} \ :\ t \in I\right).$$

If we equip $\mathcal{C}_{(0)}$ with the metric of locally uniform convergence,

$$\rho(w,v) = \sum_{n=1}^{\infty}\left(1 \wedge \sup_{0\leqslant t\leqslant n}|w(t) - v(t)|\right)2^{-n},$$

$\mathcal{C}_{(o)}$ becomes a complete separable metric space. Denote by $\mathscr{O}_\rho$ the topology induced by ρ and consider the Borel σ-algebra $\mathscr{B}(\mathcal{C}_{(o)}) := \sigma(\mathscr{O}_\rho)$ on $\mathcal{C}_{(o)}$.

4.1 Lemma. *We have* $\mathcal{C}_{(o)} \cap \mathscr{B}^I(\mathbb{R}^d) = \mathscr{B}(\mathcal{C}_{(o)})$. Ex. 4.4

Proof. Under the metric ρ the map $\mathcal{C}_{(o)} \ni w \mapsto \pi_t(w)$ is (Lipschitz) continuous for every t, hence measurable with respect to the topological σ-algebra $\mathscr{B}(\mathcal{C}_{(o)})$. This means that

$$\sigma\left(\pi_t|_{\mathcal{C}_{(o)}}\right) \subset \mathscr{B}(\mathcal{C}_{(o)}) \;\; \forall t \in I, \quad \text{and so,} \quad \sigma\left(\pi_t|_{\mathcal{C}_{(o)}} \,:\, t \in I\right) \subset \mathscr{B}(\mathcal{C}_{(o)}).$$

Conversely, we have for $v, w \in \mathcal{C}_{(o)}$

$$\begin{aligned} \rho(v,w) &= \sum_{n=1}^{\infty}\Big(1 \wedge \sup_{t\in[0,n]\cap\mathbb{Q}} |w(t)-v(t)|\Big)2^{-n} \\ &= \sum_{n=1}^{\infty}\Big(1 \wedge \sup_{t\in[0,n]\cap\mathbb{Q}} |\pi_t(w)-\pi_t(v)|\Big)2^{-n}. \end{aligned} \tag{4.1}$$

Thus, $v \mapsto \rho(v,w)$ is measurable with respect to $\sigma\left(\pi_t|_{\mathcal{C}_{(o)}} \,:\, t \in \mathbb{Q}^+\right)$ and, consequently, with respect to $\mathcal{C}_{(o)} \cap \mathscr{B}^I(\mathbb{R}^d)$. Since $(\mathcal{C}_{(o)}, \rho)$ is separable, there exists a countable dense set $\mathcal{D} \subset \mathcal{C}_{(o)}$ and it is easy to see that every $U \in \mathscr{O}_\rho$ can be written as a countable union of the form

$$U = \bigcup_{\substack{\mathbb{B}^{(\rho)}(w,r)\subset U \\ w\in\mathcal{D},\, r\in\mathbb{Q}^+}} \mathbb{B}^{(\rho)}(w,r)$$

where $\mathbb{B}^{(\rho)}(w,r) = \left\{v \in \mathcal{C}_{(o)} \,:\, \rho(v,w) < r\right\}$ is an open ball in the metric ρ. Since $\mathbb{B}^{(\rho)}(w,r) \in \mathcal{C}_{(o)} \cap \mathscr{B}^I(\mathbb{R}^d)$, we get

$$\mathscr{O}_\rho \subset \mathcal{C}_{(o)} \cap \mathscr{B}^I(\mathbb{R}^d) \quad \text{and therefore} \quad \mathscr{B}(\mathcal{C}_{(o)}) = \sigma(\mathscr{O}_\rho) \subset \mathcal{C}_{(o)} \cap \mathscr{B}^I(\mathbb{R}^d).$$ □

Let $(B_t)_{t\geqslant 0}$ be a Brownian motion on some probability space $(\Omega, \mathscr{A}, \mathbb{P})$ and consider the map

$$\Psi : \Omega \to \mathcal{C}_{(o)}, \quad \omega \mapsto w := \Psi(\omega) := B(\cdot,\omega). \tag{4.2}$$

The metric balls $\mathbb{B}^{(\rho)}(v,r) \subset \mathcal{C}_{(o)}$ generate $\mathscr{B}(\mathcal{C}_{(o)})$. From (4.1) we conclude that the map $\omega \mapsto \rho(B(\cdot,\omega), v)$ is $\mathscr{A}$ measurable; therefore

$$\{\omega \,:\, \Psi(\omega) \in \mathbb{B}^{(\rho)}(v,r)\} = \{\omega \,:\, \rho(B(\cdot,\omega),v) < r\}$$

shows that Ψ is $\mathscr{A}/\mathscr{B}(\mathcal{C}_{(o)})$ measurable. The image measure

$$\mu(\Gamma) := \mathbb{P}\left(\Psi^{-1}(\Gamma)\right) = \mathbb{P}(\Psi \in \Gamma), \quad \Gamma \in \mathscr{B}(\mathcal{C}_{(o)}), \tag{4.3}$$

is a measure on $\mathcal{C}_{(o)}$. Assume that Γ is a *cylinder set*, i. e. a set of the form

$$\begin{aligned} \Gamma &= \left\{w \in \mathcal{C}_{(o)} \,:\, w(t_1) \in C_1, \dots, w(t_n) \in C_n\right\} \\ &= \left\{w \in \mathcal{C}_{(o)} \,:\, \pi_{t_1}(w) \in C_1, \dots, \pi_{t_n}(w) \in C_n\right\} \end{aligned}$$

where $n \geqslant 1$, $t_1 < t_2 < \cdots < t_n$ and $C_1, \ldots, C_n \in \mathscr{B}(\mathbb{R}^d)$. Because of $w = \Psi(\omega)$ and $\pi_t(w) = w(t) = B(t, \omega)$ we see

$$\mu(\Gamma) = \mu\left(\pi_{t_1} \in C_1, \ldots, \pi_{t_n} \in C_n\right) = \mathbb{P}\left(B(t_1) \in C_1, \ldots, B(t_n) \in C_n\right). \tag{4.4}$$

The family of all cylinder sets is a $\cap$-stable generator of the σ-algebra $\mathcal{C}_{(0)} \cap \mathscr{B}^I(\mathbb{R}^d)$. Therefore, the finite dimensional distributions (4.4) uniquely determine the measure μ (and $\mathbb{P}$). This proves the following theorem.

4.2 Theorem. *On the probability space* $(\mathcal{C}_{(0)}, \mathscr{B}(\mathcal{C}_{(0)}), \mu)$ *the family of projections* $(\pi_t)_{t\geqslant 0}$ *satisfies* (B0)–(B4), *i. e.* $(\pi_t)_{t\geqslant 0}$ *is a Brownian motion.*

Theorem 4.2 says that *every* Brownian motion $(\Omega, \mathscr{A}, \mathbb{P}, B_t, t \geqslant 0)$ admits an equivalent Brownian motion $(W_t)_{t\geqslant 0}$ on the probability space $(\mathcal{C}_{(0)}, \mathscr{B}(\mathcal{C}_{(0)}), \mu)$. Since $W(t)$ is just the projection π_t, we can identify $W(t, w)$ with its sample path $w(t)$.

4.3 Definition. Let $(\mathcal{C}_{(0)}, \mathscr{B}(\mathcal{C}_{(0)}), \mu, \pi_t)$ be as in Theorem 4.2. The measure μ is called the *Wiener measure*, $(\mathcal{C}_{(0)}, \mathscr{B}(\mathcal{C}_{(0)}), \mu)$ is called *Wiener space* or *path space*, and $(\pi_t)_{t\geqslant 0}$ is the *canonical (model of the) Brownian motion* process.

We will write $(\mathcal{C}_{(0)}(I), \mathscr{B}(\mathcal{C}_{(0)}(I)), \mu)$ if we consider Brownian motion on an interval I other than $[0, \infty)$.

4.4 Remark. We have seen in Section 2.3 that a Brownian motion is invariant under reflection, scaling, renewal, time inversion and projective reflection at $t = \infty$. Since these operations leave the finite dimensional distributions unchanged, and since the Wiener measure μ is uniquely determined by the finite dimensional distributions, μ will also be invariant. This observation justifies arguments of the type

$$\mathbb{P}(B(\cdot) \in A) = \mathbb{P}(W(\cdot) \in A)$$

where $A \in \mathscr{A}$ and W is the (path-by-path) transformation of B under any of the above operations. Below is an overview on the transformations both at the level of the sample paths and the canonical space $\mathcal{C}_{(0)}$. Note that the maps $S : \mathcal{C}_{(0)} \to \mathcal{C}_{(0)}$ in Table 4.1 are bijective.

Let us finally show that the product σ-algebra $\mathscr{B}^I(\mathbb{R}^d)$ is fairly small in the sense that the set $\mathcal{C}_{(0)}$ is not measurable. We begin with an auxiliary result which shows that $\Gamma \in \mathscr{B}^I(\mathbb{R}^d)$ is uniquely determined by *countably many* indices.

Ex. 4.4 **4.5 Lemma.** *Let* $I = [0, \infty)$. *For every* $\Gamma \in \mathscr{B}^I(\mathbb{R}^d)$ *there is a countable set* $S = S_\Gamma \subset I$ *such that*

$$f \in (\mathbb{R}^d)^I,\ w \in \Gamma : f|_S = w|_S \implies f \in \Gamma. \tag{4.5}$$

Proof. Set $\Sigma := \left\{\Gamma \subset (\mathbb{R}^d)^I : (4.5) \text{ holds for some countable set } S = S_\Gamma\right\}$. We claim that Σ is a σ-algebra.

- Clearly, $\emptyset \in \Sigma$;

Table 4.1. Transformations of Brownian motion.

	time set I	paths $W(t,\omega)$	$w \in \mathcal{C}_{(0)}(I)$, $w \mapsto Sw(\cdot)$
Reflection	$[0,\infty)$	$-B(t,\omega)$	$-w(t)$
Renewal	$[0,\infty)$	$B(t+a,\omega)-B(a,\omega)$	$w(t+a)-w(a)$
Time inversion	$[0,a]$	$B(a-t,\omega)-B(a,\omega)$	$w(a-t)-w(a)$
Scaling	$[0,\infty)$	$\sqrt{c}\,B(\frac{t}{c},\omega)$	$\sqrt{c}\,w(\frac{t}{c})$
Proj. reflection	$[0,\infty)$	$t\,B(\frac{1}{t},\omega)$	$t\,w(\frac{1}{t})$

- Let $\Gamma \in \Sigma$ with $S = S_\Gamma$. Then we find for Γ^c and S that
$$f \in (\mathbb{R}^d)^I,\ w \in \Gamma^c,\ f|_S = w|_S \implies f \notin \Gamma$$
(otherwise, we would have $f \in \Gamma$ and then (4.5) would imply that $w \in \Gamma$). This means that (4.5) holds for Γ^c with $S = S_\Gamma = S_{\Gamma^c}$, i. e. $\Gamma^c \in \Sigma$.
- Let $\Gamma_1, \Gamma_2, \ldots \in \Sigma$ and set $S := \bigcup_{j\geqslant 1} S_{\Gamma_j}$. Obviously, S is countable and
$$\begin{aligned} f \in (\mathbb{R}^d)^I,\ w \in \bigcup_{j\geqslant 1}\Gamma_j,\ f|_S = w|_S &\implies \exists j_0 : w \in \Gamma_{j_0},\ f|_{S_{\Gamma_{j_0}}} = w|_{S_{\Gamma_{j_0}}} \\ &\overset{(4.5)}{\implies} f \in \Gamma_{j_0} \text{ hence } f \in \bigcup_{j\geqslant 1}\Gamma_j. \end{aligned}$$
This proves $\bigcup_{j\geqslant 1}\Gamma_j \in \Sigma$.

Since all sets of the form $\{\pi_t \in C\}$, $t \in I$, $C \in \mathscr{B}(\mathbb{R}^d)$ are contained in Σ, we find
$$\mathscr{B}^I(\mathbb{R}^d) = \sigma(\pi_t : t \in I) \subset \sigma(\Sigma) = \Sigma. \qquad \square$$

4.6 Corollary. *If $I = [0,\infty)$ then $\mathcal{C}_{(0)} \notin \mathscr{B}^I(\mathbb{R}^d)$.*

Proof. Assume that $\mathcal{C}_{(0)} \in \mathscr{B}^I(\mathbb{R}^d)$. By Lemma 4.5 there exists a countable set $S \subset I$ such that
$$f \in (\mathbb{R}^d)^I,\ w \in \mathcal{C}_{(0)} : f|_S = w|_S \implies f \in \mathcal{C}_{(0)}.$$
Fix $w \in \mathcal{C}_{(0)}$, pick $t_0 \notin S$, $\gamma \in \mathbb{R}^d \setminus \{w(t_0)\}$ and define a new function by
$$f(t) := \begin{cases} w(t), & t \neq t_0 \\ \gamma, & t = t_0 \end{cases}.$$
Then $f|_S = w|_S$, but $f \notin \mathcal{C}_{(0)}$, contradicting our assumption. $\square$

4.2 Kolmogorov's construction

We have seen in the previous section that the finite dimensional distributions of a d-dimensional Brownian motion $(B_t)_{t\geqslant 0}$ uniquely determine a measure μ on the space of all sample paths, such that the projections $(\pi_t)_{t\geqslant 0}$ are, under μ, a Brownian motion. This measure μ is uniquely determined by the (finite dimensional) projections

$$\pi_{t_1,\dots,t_n}(w) := (w(t_1),\dots,w(t_n))$$

and the corresponding finite dimensional distributions

$$p_{t_1,\dots,t_n}(C_1\times\cdots\times C_n) = \mu(\pi_{t_1,\dots,t_n}\in C_1\times\cdots\times C_n)$$

where $t_1,\dots,t_n\in I$ and $C_1,\dots,C_n\in\mathscr{B}(\mathbb{R}^d)$. From (4.4) it is obvious that the conditions

$$\left.\begin{aligned} p_{t_1,\dots,t_n}(C_1\times\cdots\times C_n) &= p_{t_{\sigma(1)},\dots,t_{\sigma(n)}}(C_{\sigma(1)}\times\cdots\times C_{\sigma(n)})\\ \text{for all permutations}\quad &\sigma:\{1,2,\dots,n\}\to\{1,2,\dots,n\}\\ p_{t_1,\dots,t_{n-1},t_n}(C_1\times\cdots\times C_{n-1}\times\mathbb{R}^d) &= p_{t_1,\dots,t_{n-1}}(C_1\times\cdots\times C_{n-1})\end{aligned}\right\}\tag{4.6}$$

are necessary for a family $p_{t_1,\dots,t_n}$, $n\geqslant 1$, $t_1,\dots,t_n\in I$ to be finite dimensional distributions of a stochastic process.

4.7 Definition. Assume that for every $n\geqslant 1$ and all $t_1,\dots,t_n\in I$, $p_{t_1,\dots,t_n}$ is a probability measure on $((\mathbb{R}^d)^n,\mathscr{B}((\mathbb{R}^d)^n))$. If these measures satisfy the *consistency conditions* (4.6), we call them *consistent* or *projective.*

In fact, the consistency conditions (4.6) are even sufficient for $p_{t_1,\dots,t_n}$ to be finite dimensional distributions of some stochastic process. The following deep theorem is due to Kolmogorov. A proof is given in the Appendix A.1.

4.8 Theorem (Kolmogorov 1933). *Let $I\subset[0,\infty)$ and $p_{t_1,\dots,t_n}$ be probability measures defined on $((\mathbb{R}^d)^n,\mathscr{B}((\mathbb{R}^d)^n))$ for all $t_1,\dots,t_n\in I$, $n\geqslant 1$. If the family of measures is consistent, then there exists a probability measure μ on $((\mathbb{R}^d)^I,\mathscr{B}^I(\mathbb{R}^d))$ such that*

$$p_{t_1,\dots,t_n}(C) = \mu(\pi_{t_1,\dots,t_n}^{-1}(C))\quad\textit{for all } C\in\mathscr{B}((\mathbb{R}^d)^n).$$

Using Theorem 4.8 we can construct a stochastic process for any family of consistent finite dimensional probability distributions.

Ex. 4.2 **4.9 Corollary** (Canonical process). *Let $I\subset[0,\infty)$ and $p_{t_1,\dots,t_n}$ be probability measures*
Ex. 4.5 *defined on $((\mathbb{R}^d)^n,\mathscr{B}((\mathbb{R}^d)^n))$ for all $t_1,\dots,t_n\in I$, $n\geqslant 1$. If the family of measures is consistent, there exist a probability space $(\Omega,\mathscr{A},\mathbb{P})$ and a d-dimensional stochastic process $(X_t)_{t\geqslant 0}$ such that for all $C_1,\dots,C_n\in\mathscr{B}(\mathbb{R}^d)$*

$$\mathbb{P}(X(t_1)\in C_1,\dots,X(t_n)\in C_n) = p_{t_1,\dots,t_n}(C_1\times\cdots\times C_n).$$

Proof. We take $\Omega = (\mathbb{R}^d)^I$, the product σ-algebra $\mathscr{A} = \mathscr{B}^I(\mathbb{R}^d)$ and $\mathbb{P} = \mu$, where μ is the measure constructed in Theorem 4.8. Then $X_t := \pi_t$ defines a stochastic process with finite dimensional distributions $\mu\circ(\pi_{t_1},\dots,\pi_{t_n})^{-1} = p_{t_1,\dots,t_n}$. □

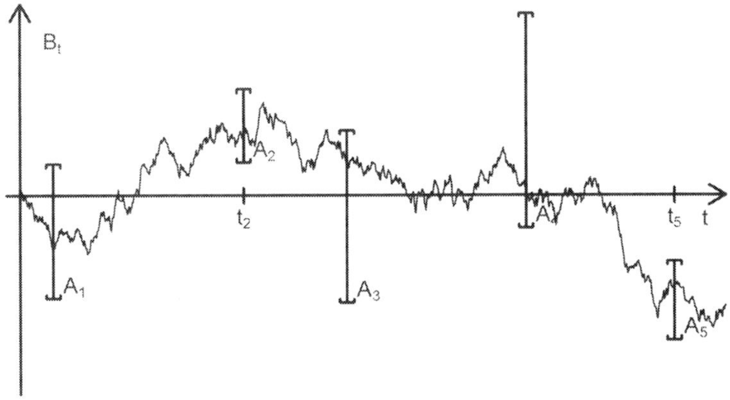

Fig. 4.1. A Brownian path meeting (and missing) the sets $A_1, \dots, A_5$ at times $t_1, \dots, t_5$

4.10 Kolmogorov's construction of BM. For $d = 1$ and $0 < t_1 < \cdots < t_n$ the n-variate Gaussian measures

$$p_{t_1,\dots,t_n}(A_1 \times \cdots \times A_n) = \left(\frac{1}{2\pi}\right)^{n/2} \frac{1}{\sqrt{\det C}} \int\limits_{A_1\times\cdots\times A_n} e^{-\frac{1}{2}\langle x, C^{-1}x\rangle}\, dx \tag{4.7}$$

with $C = (t_j \wedge t_k)_{j,k=1,\dots,n}$, are probability measures. If $t_1 = 0$, we use $\delta_0 \otimes p_{t_2,\dots,t_n}$ instead. It is not difficult to check that the family (4.7) is consistent. Therefore, Corollary 4.9 proves that there exists an $\mathbb{R}$-valued stochastic process $(B_t)_{t\geqslant 0}$. By construction, (B0) and (B3) are satisfied. From (B3) we get immediately (B2); (B1) follows if we read (2.11) backwards: Let $C = (t_j \wedge t_k)_{j,k=1,\dots,n}$ where $0 < t_1 < \cdots < t_n < \infty$; then we have for all $\xi = (\xi_1, \dots, \xi_n)^\top \in \mathbb{R}^n$

$$\begin{aligned}
\mathbb{E}\left[\exp\left(i\sum_{j=1}^n \xi_j B_{t_j}\right)\right] &= \exp\left(-\frac{1}{2}\langle \xi, C\xi\rangle\right) \\
&\overset{(2.12)}{=} \prod_{j=1}^n \exp\left(-\frac{1}{2}(t_j - t_{j-1})(\xi_j + \cdots + \xi_n)^2\right) \\
&\underset{(2.5)}{\overset{(B2)}{=}} \prod_{j=1}^n \mathbb{E}\left[\exp\left(i(B_{t_j} - B_{t_{j-1}})(\xi_j + \cdots + \xi_n)\right)\right].
\end{aligned}$$

On the other hand, we can easily rewrite the expression on the left-hand side as

$$\mathbb{E}\left[\exp\left(i\sum_{j=1}^n \xi_j B_{t_j}\right)\right] = \mathbb{E}\left[\exp\left(i\sum_{j=1}^n (B_{t_j} - B_{t_{j-1}})(\xi_j + \cdots + \xi_n)\right)\right].$$

Using the substitution $\eta_j := \xi_j + \cdots + \xi_n$ we see that

$$\mathbb{E}\left[\exp\left(i\sum_{j=1}^n \eta_j (B_{t_j} - B_{t_{j-1}})\right)\right] = \prod_{j=1}^n \mathbb{E}\left[\exp\left(i\eta_j (B_{t_j} - B_{t_{j-1}})\right)\right]$$

for all $\eta_1, \dots, \eta_n \in \mathbb{R}$ which finally proves (B1).

This gives a new way to prove the existence of a Brownian motion. The problem is, as in all other constructions of Brownian motion, to check the continuity of the sample paths (B4). This follows from yet another theorem of Kolmogorov.

Ex. 10.1 **4.11 Theorem** (Kolmogorov 1934; Slutsky 1937; Chentsov 1956). *Denote by $(X_t)_{t\geqslant 0}$ a stochastic process on $(\Omega, \mathscr{A}, \mathbb{P})$ taking values in $\mathbb{R}^d$. If*

$$\mathbb{E}\left(|X(t) - X(s)|^{\alpha}\right) \leqslant c\,|t-s|^{1+\beta} \quad \textit{for all } s,t \geqslant 0 \tag{4.8}$$

holds for some constants $c > 0$ and $\alpha, \beta > 0$, then $(X_t)_{t\geqslant 0}$ has a modification $(X'_t)_{t\geqslant 0}$ which has only continuous sample paths.

We will give the proof in a different context in Chapter 10 below. For a process satisfying (B0)–(B3), we can take $\alpha = 4$ and $\beta = 1$ since, cf. (2.8),

$$\mathbb{E}\left(|B(t) - B(s)|^4\right) = \mathbb{E}\left(|B(t-s)|^4\right) = 3\,|t-s|^2.$$

---

i **4.12 Further reading.** The concept of Wiener space has been generalized in many directions. Here we only mention Gross' concept of abstract Wiener space [137], [237], [151] and the Malliavin calculus, i. e. stochastic calculus of variations on an (abstract) Wiener space [152]. The results of Cameron and Martin are nicely summed up in [230].

[137] Kuo: *Gaussian Measures in Banach Spaces.*
[151] Malliavin: *Integration and Probability.*
[152] Malliavin: *Stochastic Analysis.*
[230] Wiener et al.: *Differential Space, Quantum Systems, and Prediction.*
[237] Yamasaki: *Measures on infinite dimensional spaces.*

Problems

1. Let $F : \mathbb{R} \to [0,1]$ be a distribution function.
   (a) Show that there exists a probability space $(\Omega, \mathscr{A}, \mathbb{P})$ and a random variable X such that $F(x) = \mathbb{P}(X \leqslant x)$.
   (b) Show that there exists a probability space $(\Omega, \mathscr{A}, \mathbb{P})$ and an iid sequence of random variables X_n such that $F(x) = \mathbb{P}(X_n \leqslant x)$.
   (c) State and prove the corresponding assertions for the d-dimensional case.
2. Show that there exists a stochastic process $(X_t)_{t\geqslant 0}$ such that the random variables X_t are independent $\mathsf{N}(0,t)$ random variables. This process also satisfies $\mathbb{P}(\lim_{s\to t} X_s \text{ exists}) = 0$ for every $t > 0$.
3. Let S be any set and $\mathscr{E}$ be a family of subsets of S. Show that

$$\sigma(\mathscr{E}) = \bigcup \{\sigma(\mathscr{C}) : \mathscr{C} \subset \mathscr{E},\ \mathscr{C} \text{ is countable}\}.$$

4. Let $\emptyset \neq E \in \mathscr{B}(\mathbb{R}^d)$ contain at least two elements, and $\emptyset \neq I \subset [0, \infty)$ be any index set. We denote by E^I the product space $E^I = \{f : f : I \to E\}$ which we equip with its natural product topology (i. e. $f_n \to f$ if, and only if, $f_n(t) \to f(t)$ for every $t \in I$). Then we can define the Borel σ-algebra $\mathscr{B}(E^I)$ and the product σ-algebra $\mathscr{B}^I(E) := \sigma(\pi_t : t \in I)$ where $\pi_t : E^I \to E$, $\pi_t(f) := f(t)$ is the canonical projection onto the tth coordinate.
   (a) Show that for every $B \in \mathscr{B}^I(E)$ there exists a countable set $J \subset I$ and a set $A \in \mathscr{B}^J(E)$ such that $B = \pi_J^{-1}(A)$. (π_J is the coordinate projection $\pi_J : E^I \to E^J$.)
   Hint: Problem 4.3.
   (b) If I is not countable, then each $\emptyset \neq B \in \mathscr{B}^I(E)$ is not countable.
   Hint: Part a)
   (c) Show that $\mathscr{B}^I(E) \subset \mathscr{B}(E^I)$.
   Hint: Part b)
   (d) If I is not countable, then $\mathscr{B}^I(E) \neq \mathscr{B}(E^I)$.
   (e) If I is countable, then $\mathscr{B}^I(E) = \mathscr{B}(E^I)$.
5. Let μ be a probability measure on $(\mathbb{R}^d, \mathscr{B}(\mathbb{R}^d))$, $v \in \mathbb{R}^d$ and consider the d-dimensional stochastic process $X_t(\omega) := \omega + tv$, $t \geqslant 0$, on the probability space $(\mathbb{R}^d, \mathscr{B}(\mathbb{R}^d), \mu)$. Give an interpretation of X_t if $\mu = \delta_x$ for some $x \in \mathbb{R}^d$ and determine the family of finite-dimensional distributions of $(X_t)_{t \geqslant 0}$.

5 Brownian motion as a martingale

Let I be some subset of $[0, \infty]$ and $(\mathscr{F}_t)_{t\in I}$ a filtration, i. e. an increasing family of sub-σ-algebras of $\mathscr{A}$. Recall that a *martingale* $(X_t, \mathscr{F}_t)_{t\in I}$ is a real or complex stochastic process $X_t : \Omega \to \mathbb{R}^d$ or $X_t : \Omega \to \mathbb{C}$ satisfying

a) $\mathbb{E}\,|X_t| < \infty$ for all $t \in I$;
b) X_t is $\mathscr{F}_t$ measurable for every $t \in I$;
c) $\mathbb{E}(X_t \mid \mathscr{F}_s) = X_s$ for all $s, t \in I, s \leqslant t$.

If X_t is real-valued and if we have in c) '$\geqslant$' or '$\leqslant$' instead of '=', we call $(X_t, \mathscr{F}_t)_{t\in I}$ a *sub-* or *supermartingale*, respectively.

We call a stochastic process $(X_t)_{t\in I}$ *adapted* to the filtration $(\mathscr{F}_t)_{t\in I}$ if X_t is for each $t \in I$ an $\mathscr{F}_t$ measurable random variable. We write briefly $(X_t, \mathscr{F}_t)_{t\geqslant 0}$ for an adapted process. Clearly, X is $\mathscr{F}_t$ adapted if, and only if, $\mathscr{F}_t^X \subset \mathscr{F}_t$ for all $t \in I$ where $\mathscr{F}_t^X$ is the natural filtration $\sigma(X_s \,:\, s \in I,\; s \leqslant t)$ of the process X.

5.1 Some 'Brownian' martingales

Brownian motion is a martingale with respect to its *natural filtration*, i. e. the family of σ-algebras $\mathscr{F}_t^B := \sigma(B_s \,:\, s \leqslant t)$. Recall that, by Lemma 2.14,

$$B_t - B_s \perp\!\!\!\perp \mathscr{F}_s^B \quad \text{for all } 0 \leqslant s \leqslant t. \tag{5.1}$$

It is often necessary to enlarge the canonical filtration $(\mathscr{F}_t^B)_{t\geqslant 0}$ of a Brownian motion $(B_t)_{t\geqslant 0}$. The property (5.1) is equivalent to the independent increments property (B1) of $(B_t)_{t\geqslant 0}$, see Lemma 2.14 and Problem 2.18, and it is this property which preserves the Brownian character of a filtration.

Ex. 5.1 **5.1 Definition.** Let $(B_t)_{t\geqslant 0}$ be a d-dimensional Brownian motion. A filtration $(\mathscr{F}_t)_{t\geqslant 0}$ is
Ex. 5.2 called *admissible*, if

a) $\mathscr{F}_t^B \subset \mathscr{F}_t$ for all $t \geqslant 0$;
b) $B_t - B_s \perp\!\!\!\perp \mathscr{F}_s$ for all $0 \leqslant s \leqslant t$.

If $\mathscr{F}_0$ contains all subsets of $\mathbb{P}$ null sets, $(\mathscr{F}_t)_{t\geqslant 0}$ is an *admissible complete* filtration.

The natural filtration $(\mathscr{F}_t^B)_{t\geqslant 0}$ is always admissible. We will discuss further examples of admissible filtrations in Lemma 6.20.

5.2 Example. Let $(B_t)_{t\geqslant 0}$, $B_t = (B_t^1, \dots, B_t^d)$, be a d-dimensional Brownian motion and $(\mathscr{F}_t)_{t\geqslant 0}$ an admissible filtration.

a) $(B_t)_{t\geqslant 0}$ is a martingale with respect to $\mathscr{F}_t$.
Indeed: Let $0 \leqslant s \leqslant t$. Using the conditions 5.1 a) and b) we get

$$\mathbb{E}(B_t \mid \mathscr{F}_s) = \mathbb{E}(B_t - B_s \mid \mathscr{F}_s) + \mathbb{E}(B_s \mid \mathscr{F}_s) = \mathbb{E}(B_t - B_s) + B_s = B_s.$$

b) $M(t) := |B_t|^2$ is a positive submartingale with respect to $\mathscr{F}_t$.
Indeed: For all $0 \leqslant s < t$ we can use (the conditional form of) Jensen's inequality to get

$$\mathbb{E}(M_t \mid \mathscr{F}_s) = \sum_{j=1}^d \mathbb{E}((B_t^j)^2 \mid \mathscr{F}_s) \geqslant \sum_{j=1}^d \mathbb{E}(B_t^j \mid \mathscr{F}_s)^2 = \sum_{j=1}^d (B_s^j)^2 = M_s.$$

c) $M_t := |B_t|^2 - d \cdot t$ is a martingale with respect to $\mathscr{F}_t$.
Indeed: Since $|B_t|^2 - d \cdot t = \sum_{j=1}^d \left((B_t^j)^2 - t\right)$ it is enough to consider $d = 1$. For $s < t$ we see

$$\begin{aligned}
\mathbb{E}\left[B_t^2 - t \,\middle|\, \mathscr{F}_s\right] &= \mathbb{E}\left[((B_t - B_s) + B_s)^2 - t \,\middle|\, \mathscr{F}_s\right] \\
&= \mathbb{E}\left[(B_t - B_s)^2 + 2B_s(B_t - B_s) + B_s^2 - t \,\middle|\, \mathscr{F}_s\right] \\
&\overset{B_t - B_s \perp\!\!\!\perp \mathscr{F}_s}{=} \underbrace{\mathbb{E}\left[(B_t - B_s)^2\right]}_{= t-s} + 2B_s \underbrace{\mathbb{E}\left[B_t - B_s\right]}_{=0} + B_s^2 - t \\
&= B_s^2 - s.
\end{aligned}$$

d) $M^\xi(t) := e^{i\langle \xi, B(t)\rangle + \frac{t}{2}|\xi|^2}$ is for all $\xi \in \mathbb{R}^d$ a complex-valued martingale with respect to $\mathscr{F}_t$. Ex. 5.5
Indeed: For $0 \leqslant s < t$ we have, cf. Example a),

$$\begin{aligned}
\mathbb{E}(M^\xi(t) \mid \mathscr{F}_s) &= e^{\frac{t}{2}|\xi|^2}\, \mathbb{E}(e^{i\langle \xi, B(t)-B(s)\rangle} \mid \mathscr{F}_s)\, e^{i\langle \xi, B(s)\rangle} \\
&\underset{B_t - B_s \sim B_{t-s}}{\overset{B_t - B_s \perp\!\!\!\perp \mathscr{F}_s}{=}} e^{\frac{t}{2}|\xi|^2}\, \mathbb{E}(e^{i\langle \xi, B(t-s)\rangle})\, e^{i\langle \xi, B(s)\rangle} \\
&= e^{\frac{t}{2}|\xi|^2}\, e^{-\frac{t-s}{2}|\xi|^2}\, e^{i\langle \xi, B(s)\rangle} \\
&= M^\xi(s).
\end{aligned}$$

e) $M^\zeta(t) := \exp\left(\sum_{j=1}^d (\zeta_j B^j(t) - \frac{t}{2}\zeta_j^2)\right)$ is for all $\zeta \in \mathbb{C}^d$ a complex-valued martingale with respect to $\mathscr{F}_t$. Ex. 5.6 Ex. 5.7
Indeed: Since $\mathbb{E}\, e^{|\zeta||B(t)|} < \infty$ for all $\zeta \in \mathbb{C}^d$, cf. (2.6), we can use the argument of Example d).

Since Brownian motion is a martingale we have all martingale tools at our disposal. The following maximal estimate will be particularly useful. It is a direct consequence of Doob's maximal inequality A.10 and the observation that for a process with continuous sample paths $\sup_{s\in D\cap[0,t]}|B_s| = \sup_{s\in[0,t]}|B_s|$ holds for any dense subset $D \subset [0, \infty)$.

5.3 Lemma. *Let $(B_t)_{t\geqslant 0}$ be a d-dimensional Brownian motion. Then we have for all $t \geqslant 0$ and $p > 1$*

$$\mathbb{E}\left[\sup_{s\leqslant t}|B_s|^p\right] \leqslant \left(\frac{p}{p-1}\right)^p \mathbb{E}[|B_t|^p].$$

The following characterization of a Brownian motion through conditional characteristic functions is quite useful.

5.4 Lemma. *Let $(X_t)_{t\geqslant 0}$ be a continuous $\mathbb{R}^d$-valued stochastic process with $X_0 = 0$ and which is adapted to the filtration $(\mathscr{F}_t)_{t\geqslant 0}$. Then $(X_t, \mathscr{F}_t)_{t\geqslant 0}$ is a BM^d if, and only if, for all $0 \leqslant s \leqslant t$ and $\xi \in \mathbb{R}^d$*

$$\mathbb{E}[e^{i\langle \xi, X_t - X_s\rangle} \mid \mathscr{F}_s] = e^{-\frac{1}{2}|\xi|^2(t-s)}. \tag{5.2}$$

Proof. We verify (2.15), i. e. for all $n \geqslant 0$, $0 = t_0 < t_1 < \cdots < t_n$, $\xi_0, \dots, \xi_n \in \mathbb{R}^d$ and $X_{t_{-1}} := 0$:

$$\mathbb{E}\left[\exp\left(i\sum_{j=0}^{n}\left\langle \xi_j, X_{t_j} - X_{t_{j-1}}\right\rangle\right)\right] = \exp\left[-\frac{1}{2}\sum_{j=1}^{n}|\xi_j|^2(t_j - t_{j-1})\right].$$

Using the tower property we find

$$\begin{aligned}
&\mathbb{E}\left[\exp\left(i\sum_{j=0}^{n}\left\langle \xi_j, X_{t_j} - X_{t_{j-1}}\right\rangle\right)\right]\\
&\quad = \mathbb{E}\left[\mathbb{E}\left(\exp\left(i\left\langle \xi_n, X_{t_n} - X_{t_{n-1}}\right\rangle\right) \,\Big|\, \mathscr{F}_{t_{n-1}}\right)\exp\left(i\sum_{j=0}^{n-1}\left\langle \xi_j, X_{t_j} - X_{t_{j-1}}\right\rangle\right)\right]\\
&\quad = \exp\left(-\frac{1}{2}|\xi_n|^2(t_n - t_{n-1})\right)\mathbb{E}\left[\exp\left(i\sum_{j=0}^{n-1}\left\langle \xi_j, X_{t_j} - X_{t_{j-1}}\right\rangle\right)\right].
\end{aligned}$$

Iterating this step we get (2.15), and the claim follows from Lemma 2.8. □

All examples in 5.2 were of the form '$f(B_t)$'. We can get many more examples of this type. For this we begin with a real-variable lemma.

Ex. 5.8 **5.5 Lemma.** *Let $p(t,x) := (2\pi t)^{-d/2}\exp(-|x|^2/2t)$ be the transition density of a d-dimensional Brownian motion. Then*

$$\frac{\partial}{\partial t}p(t,x) = \frac{1}{2}\Delta_x p(t,x) \quad \textit{for all } x \in \mathbb{R}^d,\ t > 0, \tag{5.3}$$

where $\Delta = \Delta_x = \sum_{j=1}^{d}\frac{\partial^2}{\partial x_j^2}$ is the d-dimensional Laplace operator. Moreover,

$$\begin{aligned}
\int p(t,x)\,\frac{1}{2}\Delta_x f(t,x)\,dx &= \int f(t,x)\,\frac{1}{2}\Delta_x p(t,x)\,dx\\
&= \int f(t,x)\,\frac{\partial}{\partial t}\,p(t,x)\,dx
\end{aligned} \tag{5.4}$$

for all functions $f \in \mathcal{C}^{1,2}\left((0,\infty)\times\mathbb{R}^d\right) \cap \mathcal{C}\left([0,\infty)\times\mathbb{R}^d\right)$ satisfying

$$|f(t,x)| + \left|\frac{\partial f(t,x)}{\partial t}\right| + \sum_{j=1}^{d}\left|\frac{\partial f(t,x)}{\partial x_j}\right| + \sum_{j,k=1}^{d}\left|\frac{\partial^2 f(t,x)}{\partial x_j \partial x_k}\right| \leqslant c(t)\,e^{C|x|} \tag{5.5}$$

for all $t > 0$, $x \in \mathbb{R}^d$, some $C > 0$ and a locally bounded function $c : (0,\infty) \to [0,\infty)$.

Proof. The identity (5.3) can be directly verified. We leave this (lengthy but simple) computation to the reader. The first equality in (5.4) follows if we integrate by parts (twice); note that the condition (5.5) imposed on $f(t,x)$ guarantees that the marginal terms vanish and that all integrals exist. Finally, the second equality in (5.4) follows if we plug (5.3) into the second integral. □

5.6 Theorem. *Let $(B_t, \mathscr{F}_t)_{t\geqslant 0}$ be a d-dimensional Brownian motion and assume that the function $f \in \mathcal{C}^{1,2}((0,\infty)\times\mathbb{R}^d,\mathbb{R}) \cap \mathcal{C}([0,\infty)\times\mathbb{R}^d,\mathbb{R})$ satisfies* (5.5). *Then* Ex. 5.10 Ex. 17.7

$$M_t^f := f(t,B_t) - f(0,B_0) - \int_0^t Lf(r,B_r)\,dr, \quad t \geqslant 0, \tag{5.6}$$

is an $\mathscr{F}_t$ martingale. Here $Lf(t,x) = \frac{\partial}{\partial t} f(t,x) + \frac{1}{2}\Delta_x f(t,x)$.

Proof. Observe that $M_t^f - M_s^f = f(t,B_t) - f(s,B_s) - \int_s^t Lf(r,B_r)\,dr$ for $0 \leqslant s < t$. We have to show $\mathbb{E}\left(M_t^f - M_s^f \mid \mathscr{F}_s\right) = 0$. Using $B_t - B_s \perp\!\!\!\perp \mathscr{F}_s$ we see from Lemma A.3 that

$$\begin{aligned}
&\mathbb{E}\left(M_t^f - M_s^f \mid \mathscr{F}_s\right)\\
&\quad= \mathbb{E}\left(f(t,B_t) - f(s,B_s) - \int_s^t Lf(r,B_r)\,dr \,\Bigg|\, \mathscr{F}_s\right)\\
&\quad= \mathbb{E}\left(f(t,(B_t-B_s)+B_s) - f(s,B_s) - \int_s^t Lf(r,(B_r-B_s)+B_s)\,dr \,\Bigg|\, \mathscr{F}_s\right)\\
&\quad\overset{\text{A.3}}{=} \mathbb{E}\left(f(t,(B_t-B_s)+z) - f(s,z) - \int_s^t Lf(r,(B_r-B_s)+z)\,dr\right)\Bigg|_{z=B_s}.
\end{aligned}$$

Set $\phi(t,x) = \phi_{s,z}(t,x) := f(t+s,x+z)$, $s>0$, and observe that $B_t - B_s \sim B_{t-s}$. Since $f \in \mathcal{C}^{1,2}((0,\infty)\times\mathbb{R}^d)$, the shifted function ϕ satisfies (5.5) on $[0,T]\times\mathbb{R}^d$ where we can take $C = c(t) \equiv \gamma$ and the constant γ depends only on s, T and z.

We have to show that for all $z \in \mathbb{R}$

$$\mathbb{E}\left(M_{t-s}^\phi - M_0^\phi\right) = \mathbb{E}\left(\phi(t-s,B_{t-s}) - \phi(0,B_0) - \int_0^{t-s} L\phi(r,B_r)\,dr\right) = 0.$$

Let $0 < \epsilon < u$. Then

$$\begin{aligned}
\mathbb{E}\left(M_u^\phi - M_\epsilon^\phi\right) &= \mathbb{E}\left(\phi(u,B_u) - \phi(\epsilon,B_\epsilon) - \int_\epsilon^u L\phi(r,B_r)\,dr\right)\\
&\overset{(*)}{=} \int \left(p(u,x)\phi(u,x) - p(\epsilon,x)\phi(\epsilon,x)\right)dx\\
&\qquad - \int_\epsilon^u \int p(r,x)\left(\frac{\partial\phi(r,x)}{\partial r} + \frac{1}{2}\Delta_x\phi(r,x)\right)dx\,dr
\end{aligned}$$

$$\begin{aligned}
&\overset{5.5}{=} \int \left(p(u,x)\phi(u,x) - p(\epsilon,x)\phi(\epsilon,x)\right) dx \\
&\qquad - \int_\epsilon^u \int \left(p(r,x)\frac{\partial \phi(r,x)}{\partial r} + \phi(r,x)\frac{\partial p(r,x)}{\partial r} \right) dx\, dr \\
&= \int \left(p(u,x)\phi(u,x) - p(\epsilon,x)\phi(\epsilon,x)\right) dx - \int_\epsilon^u \int \frac{\partial}{\partial r}\left(p(r,x)\phi(r,x)\right) dx\, dr \\
&\overset{(*)}{=} \int \left(p(u,x)\phi(u,x) - p(\epsilon,x)\phi(\epsilon,x)\right) dx - \int \int_\epsilon^u \frac{\partial}{\partial r}\left(p(r,x)\phi(r,x)\right) dr\, dx
\end{aligned}$$

and this equals zero. In the lines marked with an asterisk $(*)$ we used Fubini's theorem which is indeed applicable because of (5.5) and $\epsilon > 0$. Moreover,

$$\left|M_u^\phi - M_\epsilon^\phi\right| \leqslant \gamma e^{\gamma|B_u|} + \gamma e^{\gamma|B_\epsilon|} + (u-\epsilon)\sup_{\epsilon\leqslant r\leqslant u} \gamma e^{\gamma|B_r|} \leqslant (2+u)\gamma \sup_{0\leqslant r\leqslant u} e^{\gamma|B_r|}.$$

By Doob's maximal inequality for $p = \gamma$ – without loss of generality we can assume that $\gamma > 1$ – and for the submartingale $e^{|B_r|}$ we see that

$$\mathbb{E}\left(\sup_{0\leqslant r\leqslant u} e^{\gamma|B_r|}\right) \leqslant \left(\frac{\gamma}{\gamma-1}\right)^\gamma \mathbb{E}\left(e^{\gamma|B_u|}\right) < \infty.$$

Now we can use dominated convergence to get

$$\mathbb{E}\left(M_u^\phi - M_0^\phi\right) = \lim_{\epsilon\downarrow 0} \mathbb{E}\left(M_u^\phi - M_\epsilon^\phi\right) = 0. \qquad \square$$

5.2 Stopping and sampling

If we want to know when a Brownian motion $(B_t)_{t\geqslant 0}$

- leaves or enters a set for the first time,
- hits its running maximum,
- returns to zero,

we have to look at random times. A random time $\tau : \Omega \to [0,\infty]$ is a *stopping time (with respect to* $(\mathscr{F}_t)_{t\geqslant 0}$*)* if

$$\{\tau \leqslant t\} \in \mathscr{F}_t \quad \text{for all } t \geqslant 0. \tag{5.7}$$

Ex. 5.12 Typical examples of stopping times are entry and hitting times of a process $(X_t)_{t\geqslant 0}$ into a set $A \in \mathscr{B}(\mathbb{R}^d)$:

(first) entry time into A: $\quad \mathring{\tau}_A := \inf\{t \geqslant 0 : X_t \in A\}$,
(first) hitting time of A: $\quad \tau_A := \inf\{t > 0 : X_t \in A\}$, $\qquad (\inf\emptyset = \infty)$
(first) exit time from A: $\quad \tau_{A^c} := \inf\{t > 0 : X_t \notin A\}$,

Note that $\tau_A^\circ \leqslant \tau_A$. If $t \mapsto X_t$ and $(\mathscr{F}_t)_{t\geqslant 0}$ are sufficiently regular, τ_A°, τ_A are stopping times for every Borel set $A \in \mathscr{B}(\mathbb{R}^d)$. In this generality, the proof is very hard, cf. [18, Chapter I.10].

For our purposes it is enough to consider closed and open sets A. The natural filtration $\mathscr{F}_t^X := \sigma(X_s \,:\, s \leqslant t)$ of a stochastic process $(X_t)_{t\geqslant 0}$ is relatively small. For many interesting stopping times we have to consider the slightly larger filtration

$$\mathscr{F}_{t+}^X := \bigcap_{u>t} \mathscr{F}_u^X = \bigcap_{n\geqslant 1} \mathscr{F}_{t+\frac{1}{n}}^X. \tag{5.8}$$

5.7 Lemma. *Let $(X_t)_{t\geqslant 0}$ be a d-dimensional stochastic process with right-continuous sample paths and $U \subset \mathbb{R}^d$ an open set. The first hitting time τ_U satisfies*

$$\{\tau_U < t\} \in \mathscr{F}_t^X \quad \text{and} \quad \{\tau_U \leqslant t\} \in \mathscr{F}_{t+}^X \quad \text{for all } t \geqslant 0,$$

i. e. it is a stopping time with respect to $\mathscr{F}_{t+}^X$.

It is not difficult to see that, for open sets U, the stopping times τ_U° and τ_U coincide. Ex. 5.13

Proof. The second assertion follows immediately from the first as

$$\{\tau_U \leqslant t\} = \bigcap_{n\geqslant 1} \{\tau_U < t + \tfrac{1}{n}\} \in \bigcap_{n\geqslant 1} \mathscr{F}_{t+\frac{1}{n}}^X = \mathscr{F}_{t+}^X. \tag{5.9}$$

For the first assertion we claim that for all $t > 0$

$$\{\tau_U < t\} = \bigcup_{\mathbb{Q}^+ \ni r < t} \underbrace{\{X(r) \in U\}}_{\in \mathscr{F}_r^X} \in \mathscr{F}_t^X.$$

Indeed: '$\subset$' If $\tau_U(\omega) < t$, then there is some $s < t$ with $X(s,\omega) \in U$. Since the paths are continuous from the right and U is open, we find some $s < r < t$, $r \in \mathbb{Q}$, with $X(r,\omega) \in U$. Hence $\omega \in \bigcup_{\mathbb{Q}^+\ni r<t} \{X(r) \in U\}$.

'$\supset$' Conversely, let $\omega \in \{X(r) \in U\}$ for some $r \in (0,t) \cap \mathbb{Q}$. Since U is open and $t \mapsto X(t)$ right-continuous, we have $\tau_U(\omega) \leqslant r < t$. □

5.8 Lemma. *Let $(X_t)_{t\geqslant 0}$ be a d-dimensional stochastic process with continuous sample paths and $F \subset \mathbb{R}^d$ a closed set. Then τ_F° is an $\mathscr{F}_t^X$ stopping time whereas τ_F is an $\mathscr{F}_{t+}^X$ stopping time.*

Proof. Denote by $d(x,F) := \inf_{y\in F} |x - y|$ the distance of x and F. If we can show that Ex. 5.14

$$\{\tau_F^\circ \leqslant t\} = \left\{\omega \in \Omega \,:\, \inf_{r\in\mathbb{Q}\cap[0,t]} d(X_r(\omega), F) = 0\right\} =: \Omega_t,$$

then τ_F° is a stopping time since, obviously, $\Omega_t \in \mathscr{F}_t$.

Indeed: '$\supset$' Assume that $\omega \in \Omega_t$. Then there is a sequence $(r_j)_{j\geqslant 1} \subset \mathbb{Q}^+ \cap [0,t]$ and some $s \leqslant t$ such that

$$r_j \xrightarrow[j\to\infty]{} s, \quad d(X(r_j,\omega), F) \xrightarrow[j\to\infty]{} 0 \quad \text{and} \quad X(r_j,\omega) \xrightarrow[j\to\infty]{} X(s,\omega).$$

Since $d(x, F)$ is continuous, we get $d(X(s, \omega), F) = 0$, i. e. $X(s, \omega) \in F$, as F is closed. Thus, $\mathring{\tau}_F(\omega) \leqslant s \leqslant t$.

'$\subset$' Now we assume that $\omega \notin \Omega_t$. Then

$$\inf_{\mathbb{Q}^+ \ni r \leqslant t} d(X(r, \omega), F) \geqslant \delta > 0 \overset{\text{cts.}}{\underset{\text{paths}}{\Longrightarrow}} X(s, \omega) \notin F, \quad \forall s \leqslant t \implies \mathring{\tau}_F(\omega) > t.$$

For τ_F we set $\tau_n := \inf\{s \geqslant 0 : X_{1/n+s} \in F\}$. Then

$$\tau_F = \inf_{n \geqslant 1} \inf\{t \geqslant \tfrac{1}{n} : X_t \in F\} = \inf_{n \geqslant 1}\left(\tfrac{1}{n} + \tau_n\right)$$

and $\{\tau_F < t\} = \bigcup_{n \geqslant 1}\{\tau_n < t - \frac{1}{n}\} \in \mathscr{F}_t^X$. As in (5.9) we see that $\{\tau_F \leqslant t\} \in \mathscr{F}_{t+}^X$. □

5.9 Example (First passage time). For a BM[1] $(B_t)_{t \geqslant 0}$ the entry time $\tau_b := \mathring{\tau}_{\{b\}}$ into the closed set $\{b\}$ is often called *first passage time*. Observe that

$$\sup_{t \geqslant 0} B_t \geqslant \sup_{n \geqslant 1} B_n = \sup_{n \geqslant 1}(\xi_1 + \xi_2 + \cdots + \xi_n)$$

where the random variables $\xi_j = B_j - B_{j-1}$ are iid standard normal random variables. Then

$$S := \sup_{n \geqslant 1}(\xi_1 + \xi_2 + \cdots + \xi_n) = \xi_1 + \sup_{n \geqslant 2}(\xi_2 + \cdots + \xi_n) =: \xi_1 + S'.$$

By the iid property of the ξ_j we get that $S \sim S'$ and $\xi_1 \perp\!\!\!\perp S'$. Since $\xi_1 \sim \mathsf{N}(0, 1)$ is not trivial, we conclude that $S = \infty$ a. s. The same argument applies to the infimum $\inf_{t \geqslant 0} B_t$ and we get

$$\mathbb{P}\left(\sup_{t \geqslant 0} B_t = +\infty, \quad \inf_{t \geqslant 0} B_t = -\infty\right) = 1, \tag{5.10}$$

hence

$$\mathbb{P}\left(\tau_b < \infty\right) = 1 \quad \text{for all } b \in \mathbb{R}. \tag{5.11}$$

For an alternative martingale proof of this fact, see Theorem 5.13 below.

Because of the continuity of the sample paths, (5.10) immediately implies that the random set $\{t \geqslant 0 : B_t(\omega) = b\}$ is a. s. unbounded. In particular, a one-dimensional Brownian motion is point recurrent, i. e. it returns to each level $b \in \mathbb{R}$ time and again.

Ex. 5.15 As usual, we can associate σ-algebras with an $\mathscr{F}_t$ stopping time τ:
Ex. 5.16

$$\mathscr{F}_{\tau(+)} := \left\{A \in \mathscr{F}_\infty := \sigma\left(\bigcup_{t \geqslant 0} \mathscr{F}_t\right) : A \cap \{\tau \leqslant t\} \in \mathscr{F}_{t(+)} \ \forall t \geqslant 0\right\}.$$

! Note that, despite the slightly misleading notation, $\mathscr{F}_\tau$ is a family of sets and $\mathscr{F}_\tau$ does not depend on ω.

If we apply the optional stopping theorem, see A.18, we obtain rather surprising results.

Ex. 5.6 **5.10 Theorem** (Wald's identities. Wald 1944). *Let $(B_t, \mathscr{F}_t)_{t \geqslant 0}$ be a BM[1] and assume that*
Ex. 5.19 *τ is an $\mathscr{F}_t$ stopping time. If $\mathbb{E}\,\tau < \infty$, then $B_\tau \in L^2(\mathbb{P})$ and we have*

$$\mathbb{E}\, B_\tau = 0 \quad \text{and} \quad \mathbb{E}\, B_\tau^2 = \mathbb{E}\,\tau.$$

Proof. Since $(\tau \wedge t)_{t \geqslant 0}$ is a family of bounded stopping times, the optional stopping theorem (Theorem A.18) applies and shows that $(B_{\tau\wedge t}, \mathscr{F}_{\tau\wedge t})_{t\geqslant 0}$ is a martingale. In particular, $\mathbb{E}\, B_{\tau\wedge t} = \mathbb{E}\, B_0 = 0$.

Using Lemma 5.3 for $p = 2$ we see that

$$\mathbb{E}\left[B_{\tau\wedge t}^2\right] \leqslant \mathbb{E}\left[\sup_{s\leqslant t} B_s^2\right] \leqslant 4\,\mathbb{E}\left[B_t^2\right] = 4t,$$

i. e. $B_{\tau\wedge t} \in L^2(\mathbb{P})$. Again by optional stopping we conclude with Example 5.2 c) that $\left(M_t^2 := B_{\tau\wedge t}^2 - \tau\wedge t, \mathscr{F}_{\tau\wedge t}\right)_{t\geqslant 0}$ is a martingale, hence

$$\mathbb{E}\left[M_t^2\right] = \mathbb{E}\left[M_0^2\right] = 0 \quad \text{or} \quad \mathbb{E}\left[B_{\tau\wedge t}^2\right] = \mathbb{E}(\tau \wedge t).$$

By the martingale property we see $\mathbb{E}\left[B_{\tau\wedge t} B_{\tau\wedge s}\right] = \mathbb{E}\left[B_{\tau\wedge s}^2\right]$ for all $s \leqslant t$, i. e.

$$\mathbb{E}\left[(B_{\tau\wedge t} - B_{\tau\wedge s})^2\right] = \mathbb{E}\left[B_{\tau\wedge t}^2 - B_{\tau\wedge s}^2\right] = \mathbb{E}(\tau\wedge t - \tau\wedge s) \xrightarrow[s,t\to\infty]{\text{dom. conv.}} 0.$$

Thus, $(B_{\tau\wedge t})_{t\geqslant 0}$ is an L^2-Cauchy sequence; the fact that $\tau < \infty$ a. s. and the continuity of Brownian motion yield $L^2\text{-}\lim_{t\to\infty} B_{\tau\wedge t} = B_\tau$. In particular, we have $B_\tau \in L^2(\mathbb{P})$ and, by monotone convergence,

$$\mathbb{E}\left[B_\tau^2\right] = \lim_{t\to\infty} \mathbb{E}\left[B_{\tau\wedge t}^2\right] = \lim_{t\to\infty} \mathbb{E}[\tau\wedge t] = \mathbb{E}\,\tau.$$

In particular, we see that $\mathbb{E}\, B_\tau = \lim_{t\to\infty} \mathbb{E}\, B_{t\wedge\tau} = 0$, as L^2-convergence implies L^1-convergence. □

The following corollaries are typical applications of Wald's identities.

5.11 Corollary. *Let $(B_t)_{t\geqslant 0}$ be a BM[1] and $\tau = \inf\{t \geqslant 0 : B_t \notin (-a, b)\}$ be the first entry time into the set $(-a,b)^c$, $a, b \geqslant 0$. Then* Ex. 5.17

$$\mathbb{P}(B_\tau = -a) = \frac{b}{a+b}, \quad \mathbb{P}(B_\tau = b) = \frac{a}{a+b} \quad \textit{and} \quad \mathbb{E}\,\tau = ab. \tag{5.12}$$

Proof. Note that $t \wedge \tau$ is a bounded stopping time. Because of $B_{t\wedge\tau} \in [-a, b]$, we have $|B_{t\wedge\tau}| \leqslant a \vee b$, and so

$$\mathbb{E}(t\wedge\tau) \stackrel{5.10}{=} \mathbb{E}\left(B_{t\wedge\tau}^2\right) \leqslant a^2 \vee b^2.$$

By monotone convergence $\mathbb{E}\,\tau \leqslant a^2 \vee b^2 < \infty$. We can apply Theorem 5.10 and get

$$-a\,\mathbb{P}(B_\tau = -a) + b\,\mathbb{P}(B_\tau = b) = \mathbb{E}\, B_\tau = 0.$$

Since $\tau < \infty$ a. s., we also know

$$\mathbb{P}(B_\tau = -a) + \mathbb{P}(B_\tau = b) = 1.$$

Solving this system of equations yields

$$\mathbb{P}(B_\tau = -a) = \frac{b}{a+b} \quad \text{and} \quad \mathbb{P}(B_\tau = b) = \frac{a}{a+b}.$$

The second identity from Theorem 5.10 finally gives

$$\mathbb{E}\,\tau = \mathbb{E}\left(B_\tau^2\right) = a^2\,\mathbb{P}(B_\tau = -a) + b^2\,\mathbb{P}(B_\tau = -a) = ab.$$ □

! The next result is quite unexpected: A one-dimensional Brownian motion hits 1 infinitely often with probability one – but the average time the Brownian motion needs to get there is infinite.

5.12 Corollary. *Let $(B_t)_{t\geqslant 0}$ be a* BM^1 *and $\tau_1 = \inf\{t \geqslant 0 : B_t = 1\}$ be the first passage time of the level* 1. *Then* $\mathbb{E}\,\tau_1 = \infty$.

Proof. We know from Example 5.9 that $\tau_1 < \infty$ almost surely. Therefore, $B_{\tau_1} = 1$ and $\mathbb{E}\,B_{\tau_1} = \mathbb{E}\,B^2_{\tau_1} = 1$. In view of Wald's identities, Theorem 5.10, this is only possible if $\mathbb{E}\,\tau_1 = \infty$. □

5.3 The exponential Wald identity

Let $(B_t)_{t\geqslant 0}$ be a BM^1 and $\tau_b = \mathring{\tau}_{\{b\}}$ the first passage time of the level b. Recall from Example 5.2 d) that $M^\xi(t) := e^{\xi B(t) - \frac{1}{2}\xi^2 t}$, $t \geqslant 0$ and $\xi > 0$, is a martingale. Applying the optional stopping theorem (Theorem A.18) to the bounded stopping times $t \wedge \tau_b$ we see that

$$1 = \mathbb{E}\,M^\xi(0) = \mathbb{E}\,M^\xi(t\wedge\tau_b) = \mathbb{E}\left[e^{\xi B(t\wedge\tau_b) - \frac{1}{2}\xi^2(t\wedge\tau_b)}\right].$$

Since $B(t\wedge\tau_b) \leqslant b$ for $b > 0$, we have $0 \leqslant e^{\xi B(t\wedge\tau_b) - \frac{1}{2}\xi^2(t\wedge\tau_b)} \leqslant e^{\xi b}$. Using the fact that $e^{-\infty} = 0$ we get

$$\lim_{t\to\infty} e^{\xi B(t\wedge\tau_b) - \frac{1}{2}\xi^2(t\wedge\tau_b)} = \begin{cases} e^{\xi B(\tau_b) - \frac{1}{2}\xi^2\tau_b}, & \text{if} \quad \tau_b < \infty, \\ 0, & \text{if} \quad \tau_b = \infty. \end{cases}$$

Thus,

$$1 = \mathbb{E}\left[e^{\xi B(t\wedge\tau_b) - \frac{1}{2}\xi^2(t\wedge\tau_b)}\right] \xrightarrow[t\to\infty]{\text{dom. conv.}} e^{\xi b}\,\mathbb{E}\left[\mathbb{1}_{\{\tau_b<\infty\}}\,e^{-\frac{1}{2}\xi^2\tau_b}\right]$$

which shows

$$\mathbb{E}\left[\mathbb{1}_{\{\tau_b<\infty\}}\,e^{-\frac{1}{2}\xi^2\tau_b}\right] = e^{-\xi b}.$$

By monotone convergence we get

$$\mathbb{P}(\tau_b < \infty) = \lim_{\xi\downarrow 0}\mathbb{E}\left[\mathbb{1}_{\{\tau_b<\infty\}}\,e^{-\frac{1}{2}\xi^2\tau_b}\right] = 1.$$

Inserting this into the previous equality is (almost) the proof of

5.13 Theorem. *Let $(B_t)_{t\geqslant 0}$ be a* BM^1. *Then the first passage time $\tau_b = \mathring{\tau}_{\{b\}}$, $b \in \mathbb{R}$, is a. s. finite and its Laplace transform is given by*

$$\mathbb{E}\,e^{-\zeta\tau_b} = e^{-\sqrt{2\zeta}\,|b|}, \quad \zeta \geqslant 0.$$

Proof. For $b \geqslant 0$ there is nothing left to show. If $b < 0$ we apply the calculation preceding Theorem 5.13 with $-\xi$ instead of ξ. □

In the proof of Theorem 5.13 we use $\mathbb{E}\exp(\xi B(\tau_b) - \frac{1}{2}\xi^2\tau_b) = 1$ for $\xi > 0$. A similar stopping result holds for stopping times τ which satisfy $\mathbb{E}\, e^{\tau/2} < \infty$. This is a very special case of the Novikov condition, cf. Theorem 18.4 below or [119, Chapter 3.5.D] and [189, pp. 332-3].

5.14 Theorem (Exponential Wald identity). *Let $(B_t)_{t\geqslant 0}$ be a* BM^1 *and let τ be a $\mathscr{F}_t^B$ stopping time such that $\mathbb{E}\, e^{\tau/2} < \infty$. Then*

$$\mathbb{E}\, e^{B(\tau)-\frac{1}{2}\tau} = 1.$$

Proof. Fix some $c \in (0,1)$ and pick $p = p(c) > 1$ such that $p < \frac{1}{c(2-c)}$. From the optional stopping theorem we know that $M_t := e^{cB_{t\wedge\tau}-\frac{1}{2}c^2(t\wedge\tau)}$, $t \geqslant 0$, is a martingale. Using the Hölder inequality for the conjugate exponents $1/pc$ and $1/(1-pc)$ we find

$$\begin{aligned}\mathbb{E}\left[e^{pcB_{t\wedge\tau}-\frac{1}{2}pc^2(t\wedge\tau)}\right] &= \mathbb{E}\left[e^{pc(B_{t\wedge\tau}-\frac{1}{2}(t\wedge\tau))}e^{\frac{1}{2}pc(1-c)(t\wedge\tau)}\right]\\ &\leqslant \underbrace{\left[\mathbb{E}\, e^{B_{t\wedge\tau}-\frac{1}{2}(t\wedge\tau)}\right]^{pc}}_{=[\mathbb{E}\, M_t]^{pc}=[\mathbb{E}\, M_0]^{pc}=1}\left[\mathbb{E}\, e^{\frac{1}{2}\frac{pc(1-c)}{1-pc}\tau}\right]^{1-pc} \leqslant \left(\mathbb{E}\, e^{\tau/2}\right)^{1-pc}.\end{aligned}$$

In the last step we used that $pc(1-c)/(1-pc) \leqslant 1$. This shows that the pth moment of the martingale M_t is uniformly bounded for all $t > 0$, i. e. $(M_t)_{t\geqslant 0}$ is uniformly integrable.

Therefore we can let $t \to \infty$ and find, using uniform integrability and Hölder's inequality for the exponents $1/c$ and $1/(1-c)$,

$$1 = \mathbb{E}\, e^{cB_\tau-\frac{1}{2}c^2\tau} = \mathbb{E}\left[e^{cB_\tau-\frac{1}{2}c\tau}e^{\frac{1}{2}c(1-c)\tau}\right] \leqslant \left[\mathbb{E}\, e^{B_\tau-\frac{1}{2}\tau}\right]^c\left[\mathbb{E}\, e^{\frac{1}{2}c\tau}\right]^{1-c}.$$

Since the last factor is bounded by $\left(\mathbb{E}\, e^{\tau/2}\right)^{1-c} \xrightarrow[c\to 1]{} 1$, we find as $c \to 1$

$$1 \leqslant \mathbb{E}\, e^{B_\tau-\frac{1}{2}\tau} = \mathbb{E}\left[\lim_{t\to\infty} e^{B_{t\wedge\tau}-\frac{1}{2}(t\wedge\tau)}\right] \leqslant \varliminf_{t\to\infty}\mathbb{E}\left[e^{B_{t\wedge\tau}-\frac{1}{2}(t\wedge\tau)}\right] = 1,$$

and the assertion follows. □

5.15 Remark. A close inspection of the proof of Theorem 5.14 shows that we have actually shown $\mathbb{E}\, e^{cB(\tau)-\frac{1}{2}c^2\tau} = 1$ for all $0 \leqslant c \leqslant 1$ and all stopping times τ satisfying $\mathbb{E}\, e^{\tau/2} < \infty$. If we replace in this calculation B by $-B$, we get

$$\mathbb{E}\, e^{\xi B(\tau)-\frac{1}{2}\xi^2\tau} = 1 \quad \text{for all } -1 \leqslant \xi \leqslant 1.$$

5.16 Further reading. The monograph [189] is one of the best and most comprehensive sources on martingales in connection with Brownian motion. A completely 'martingale' view of the world and everything else can be found in (the second volume of) [195]. Applications of Brownian motion and martingales to analysis are in [60].

[60] Durrett: *Brownian Motion and Martingales in Analysis.*
[189] Revuz, Yor: *Continuous Martingales and Brownian Motion*
[195] Rogers, Williams: *Diffusions, Markov Processes and Martingales.*

Problems

1. Let $(B_t)_{t\geqslant 0}$ be a BM^d and assume that X is a d-dimensional random variable which is independent of $\mathscr{F}_\infty^B$.
   (a) Show that $\widetilde{\mathscr{F}}_t := \sigma(X, B_s : s \leqslant t)$ defines an admissible filtration for $(B_t)_{t\geqslant 0}$.
   (b) The completion $\overline{\mathscr{F}}_t^B$ of $\mathscr{F}_t^B$ is the smallest σ-algebra which contains $\mathscr{F}_t^B$ and all subsets of $\mathbb{P}$ null sets. Show that $(\overline{\mathscr{F}}_t^B)_{t\geqslant 0}$ is admissible.
2. Let $(\mathscr{F}_t)_{t\geqslant 0}$ be an admissible filtration for the Brownian motion $(B_t)_{t\geqslant 0}$. Mimic the proof of Lemma 2.14 and show that $\mathscr{F}_t \perp\!\!\!\perp \mathscr{F}_{[t,\infty)}^W := \sigma(B_u - B_t : u \geqslant t)$ for each $t > 0$.
3. Let $(X_t, \mathscr{F}_t)_{t\geqslant 0}$ be a martingale and denote by $\mathscr{F}_t^*$ be the completion of $\mathscr{F}_t$ (completion means to add all subsets of $\mathbb{P}$-null sets).
   (a) Show that $(X_t, \mathscr{F}_t^*)_{t\geqslant 0}$ is a martingale.
   (b) Let $(\widetilde{X}_t)_{t\geqslant 0}$ be a modification of $(X_t)_{t\geqslant 0}$. Show that $(\widetilde{X}_t, \mathscr{F}_t^*)_{t\geqslant 0}$ is a martingale.
4. Let $(X_t, \mathscr{F}_t)_{t\geqslant 0}$ be a sub-martingale with continuous paths and $\mathscr{F}_{t+} = \bigcap_{u>t} \mathscr{F}_u$. Show that $(X_t, \mathscr{F}_{t+})_{t\geqslant 0}$ is again a sub-martingale.
5. Let $(B_t)_{t\geqslant 0}$ be a BM^1. Find a polynomial $\pi(t,x)$ in x and t which is of order 4 in the variable x such that $\pi(t, B_t)$ is a martingale.
   Hint: One possibility is to use the exponential Wald identity $\mathbb{E} \exp\left(\xi B_\tau - \frac{1}{2}\xi^2\tau\right) = 1$, $-1 \leqslant \xi \leqslant 1$ for a suitable stopping time τ and use a power-series expansion of the left-hand side in ξ.
6. This exercise contains a recipe how to obtain 'polynomial' martingales with leading term B_t^n, where $(B_t, \mathscr{F}_t)_{t\geqslant 0}$ is a BM^1.
   (a) We know that $M_t^\xi := \exp\left(\xi B_t - \frac{1}{2}\xi^2 t\right)$, $t \geqslant 0$, $\xi \in \mathbb{R}$, is a martingale. Differentiate $\mathbb{E}(M_t^\xi \mathbb{1}_F) = \mathbb{E}(M_s^\xi \mathbb{1}_F)$, $F \in \mathscr{F}_s$, $s \leqslant t$, at $\xi = 0$ and show that
   $$B_t^2 - t, \quad B_t^3 - 3tB_t, \quad B_t^4 - 6tB_t^2 + 3t^2, \quad \ldots$$
   are martingales.
   (b) Find a general expression for the polynomial $b \mapsto P_n(b, \xi)$ of degree n such that $P_n(B_t, 0)$ is a martingale.
   (c) Let $\tau = \tau_{(-a,b)^c}$ be the first exit time of a Brownian motion $(B_t)_{t\geqslant 0}$ from the interval $(-a, b)$ with $-a < 0 < b$. Show that
   $$\mathbb{E}(\tau^2) \leqslant 4\,\mathbb{E}(B_\tau^4) \quad \text{and} \quad \mathbb{E}(B_\tau^4) \leqslant 36\,\mathbb{E}(\tau^2).$$
7. Let $(B_t)_{t\geqslant 0}$ be a BM^d. Find all $c \in \mathbb{R}$ such that $\mathbb{E}\, e^{c|B_t|}$ and $\mathbb{E}\, e^{c|B_t|^2}$ are finite.
8. Let $p(t,x) = (2\pi t)^{-d/2} \exp\left(-|x|^2/(2t)\right)$, $x \in \mathbb{R}^d$, $t > 0$, be the transition density of a d-dimensional Brownian motion.

(a) Show that $p(t,x)$ is a solution for the heat equation, i. e.

$$\frac{\partial}{\partial t}\,p(t,x) = \frac{1}{2}\,\Delta_x p(t,x) = \frac{1}{2}\sum_{j=1}^{d}\frac{\partial^2}{\partial x_j^2}\,p(t,x), \qquad x \in \mathbb{R}^d,\ t > 0.$$

(b) Let $f \in \mathcal{C}^{1,2}((0,\infty)\times\mathbb{R}^d) \cap \mathcal{C}([0,\infty)\times\mathbb{R}^d)$ be such that for some constants $c, c_t > 0$ and all $x \in \mathbb{R}^d,\ t > 0$

$$|f(t,x)| + \left|\frac{\partial f(t,x)}{\partial t}\right| + \sum_{j=1}^{d}\left|\frac{\partial f(t,x)}{\partial x_j}\right| + \sum_{j,k=1}^{d}\left|\frac{\partial^2 f(t,x)}{\partial x_j \partial x_k}\right| \leqslant c_t\, e^{c|x|}.$$

Show that

$$\int p(t,x)\frac{1}{2}\,\Delta_x f(t,x)\,dx = \int f(t,x)\frac{1}{2}\,\Delta_x p(t,x)\,dx.$$

9. Let $(B_t, \mathscr{F}_t)_{t\geqslant 0}$ be a BM^1. Show that $X_t = \exp(aB_t + bt)$, $t \geqslant 0$, is a martingale if, and only if, $a^2/2 + b = 0$.

10. Let $(B_t, \mathscr{F}_t)_{t\geqslant 0}$ be a one-dimensional Brownian motion. Which of the following processes are martingales?

(a) $U_t = e^{cB_t}$, $c \in \mathbb{R}$;
(b) $V_t = tB_t - \int_0^t B_s\,ds$;
(c) $W_t = B_t^3 - tB_t$;
(d) $X_t = B_t^3 - 3\int_0^t B_s\,ds$;
(e) $Y_t = \frac{1}{3}\,B_t^3 - tB_t$;
(f) $Z_t = e^{B_t - ct}$, $c \in \mathbb{R}$.

11. Let $(B_t, \mathscr{F}_t)_{t\geqslant 0}$ be a BM^d. Show that $X_t = \frac{1}{d}\,|B_t|^2 - t$, $t \geqslant 0$, is a martingale.

12. Let $(X_t, \mathscr{F}_t)_{t\geqslant 0}$ be a d-dimensional stochastic process and $A, A_n, C \in \mathscr{B}(\mathbb{R}^d)$, $n \geqslant 1$. Then
(a) $A \subset C$ implies $\mathring{\tau}_A \geqslant \mathring{\tau}_C$ and $\tau_A \geqslant \tau_C$;
(b) $\mathring{\tau}_{A\cup C} = \min\{\mathring{\tau}_A, \mathring{\tau}_C\}$ and $\tau_{A\cup C} = \min\{\tau_A, \tau_C\}$;
(c) $\mathring{\tau}_{A\cap C} \geqslant \max\{\mathring{\tau}_A, \mathring{\tau}_C\}$ and $\tau_{A\cap C} \geqslant \max\{\tau_A, \tau_C\}$;
(d) $A = \bigcup_{n\geqslant 1} A_n$, then $\mathring{\tau}_A = \inf_{n\geqslant 1}\mathring{\tau}_{A_n}$ and $\tau_A = \inf_{n\geqslant 1}\tau_{A_n}$;
(e) $\tau_A = \inf_{n\geqslant 1}\left(\frac{1}{n} + \mathring{\tau}_n\right)$ where $\mathring{\tau}_n = \inf\{s \geqslant 0 : X_{s+1/n} \in A\}$;
(f) find an example of a set A such that $\mathring{\tau}_A < \tau_A$.

13. Let $U \subset \mathbb{R}^d$ be an open set and assume that $(X_t)_{t\geqslant 0}$ is a stochastic process with continuous paths. Show that $\tau_U = \mathring{\tau}_U$.

14. Show that the function $d(x,A) := \inf_{y\in A}|x-y|$, $A \subset \mathbb{R}^d$, is continuous.

15. Let τ be a stopping time. Check that $\mathscr{F}_\tau$ and $\mathscr{F}_{\tau+}$ are σ-algebras.

16. Let τ be a stopping time for the filtration $(\mathscr{F}_t)_{t\geqslant 0}$. Show that
(a) $F \in \mathscr{F}_{\tau+} \iff \forall t \geqslant 0 : F \cap \{\tau < t\} \in \mathscr{F}_t$;

(b) $\{\tau \leqslant t\} \in \mathscr{F}_{\tau\wedge t}$ for all $t \geqslant 0$.

17. Let $\tau := \tau^{\circ}_{(-a,b)^c}$ be the first entry time of a BM^1 into the set $(-a,b)^c$.
 (a) Show that τ has finite moments $\mathbb{E}\,\tau^n$ of any order $n \geqslant 1$.
 Hint: Use Example 5.2 d) and show that $\mathbb{E}\,e^{c\tau} < \infty$ for some $c > 0$.

 (b) Evaluate $\mathbb{E}\int_0^\tau B_s\,ds$.
 Hint: Find a martingale which contains $\int_0^t B_s\,ds$.

18. Let $(B_t)_{t\geqslant 0}$ be a BM^d. Find $\mathbb{E}\,\tau_R$ where $\tau_R = \inf\{t \geqslant 0 : |B_t| = R\}$.

19. Let $(B_t)_{t\geqslant 0}$ be a BM^1 and σ, τ be two stopping times such that $\mathbb{E}\,\tau, \mathbb{E}\,\sigma < \infty$.
 (a) Show that $\sigma \wedge \tau$ is a stopping time.

 (b) Show that $\{\sigma \leqslant \tau\} \in \mathscr{F}_{\sigma\wedge\tau}$.

 (c) Show that $\mathbb{E}(B_\tau B_\sigma) = \mathbb{E}\,\tau \wedge \sigma$
 Hint: Consider $\mathbb{E}(B_\sigma B_\tau \mathbb{1}_{\{\sigma\leqslant\tau\}}) = \mathbb{E}(B_{\sigma\wedge\tau} B_\tau \mathbb{1}_{\{\sigma\leqslant\tau\}})$ and use optional stopping.

 (d) Deduce from c) that $\mathbb{E}(|B_\tau - B_\sigma|^2) = \mathbb{E}\,|\tau - \sigma|$.

6 Brownian motion as a Markov process

We have seen in 2.13 that for a d-dimensional Brownian motion $(B_t)_{t\geqslant 0}$ and any $s > 0$ the shifted process $W_t := B_{t+s} - B_s$, $t \geqslant 0$, is again a BM^d which is independent of $(B_t)_{0\leqslant t\leqslant s}$. Since $B_{t+s} = W_t + B_s$, we can interpret this as a renewal property: Rather than going from $0 = B_0$ straight away to $x = B_{t+s}$ in $(t + s)$ units of time, we stop after time s at $x' = B_s$ and move, using a further Brownian motion W_t for t units of time, from x' to $x = B_{t+s}$. This situation is shown in Figure 6.1:

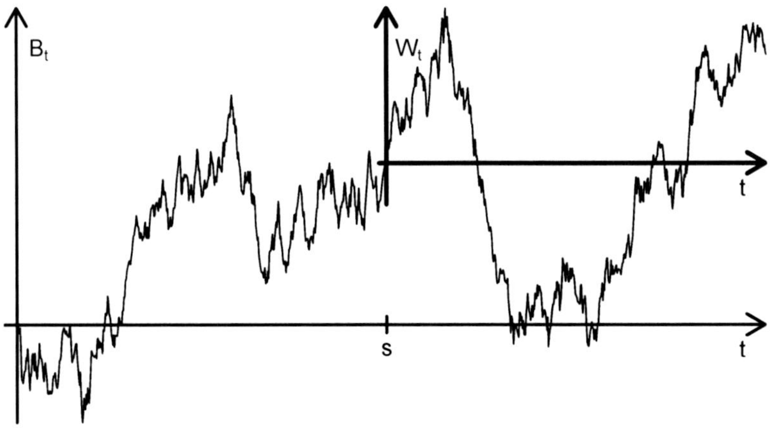

Fig. 6.1. $W_t := B_{t+s} - B_s$, $t \geqslant 0$, is a Brownian motion in the new coordinate system with origin (s, B_s).

6.1 The Markov property

In fact, we know from Lemma 2.14 that the paths up to time s and thereafter are stochastically independent, i. e. $\mathscr{F}_s^B \perp\!\!\!\perp \mathscr{F}_\infty^W := \sigma(B_{t+s} - B_s : t \geqslant 0)$. If we use any admissible filtration $(\mathscr{F}_t)_{t\geqslant 0}$ instead of the natural filtration $(\mathscr{F}_t^B)_{t\geqslant 0}$, the argument of Lemma 2.14 remains valid, and we get Ex. 5.2

6.1 Theorem (Markov property of BM). *Let $(B_t)_{t\geqslant 0}$ be a BM^d and $(\mathscr{F}_t)_{t\geqslant 0}$ some admissible filtration. For every $s > 0$, the process $W_t := B_{t+s} - B_s$, $t \geqslant 0$, is also a BM^d and $(W_t)_{t\geqslant 0}$ is independent of $\mathscr{F}_s$, i. e. $\mathscr{F}_\infty^W = \sigma(W_t, t \geqslant 0) \perp\!\!\!\perp \mathscr{F}_s$.* Ex. 6.1 Ex. 6.2

Theorem 6.1 justifies our intuition that we can split a Brownian path into two independent pieces

$$B(t + s) = B(t + s) - B(s) + B(s) = W(t) + y\,\big|_{y=B(s)}.$$

Observe that $W(t) + y$ is a Brownian motion started at $y \in \mathbb{R}^d$. Let us introduce the following notation

$$\mathbb{P}^x(B_{t_1} \in A_1, \dots, B_{t_n} \in A_n) := \mathbb{P}(B_{t_1} + x \in A_1, \dots, B_{t_n} + x \in A_n) \tag{6.1}$$

where $0 \leqslant t_1 < \cdots < t_n$ and $A_1, \dots, A_n \in \mathscr{B}(\mathbb{R}^d)$. We will write $\mathbb{E}^x$ for the corresponding mathematical expectation. Clearly, $\mathbb{P}^0 = \mathbb{P}$ and $\mathbb{E}^0 = \mathbb{E}$. This means that $\mathbb{P}^x(B_s \in A)$ denotes the probability that a Brownian particle starts at time $t = 0$ at the point x and travels in s units of time into the set A.

Since the finite dimensional distributions (6.1) determine $\mathbb{P}^x$ uniquely, cf. Theorem 4.8, $\mathbb{P}^x$ is a well-defined measure on $(\Omega, \mathscr{F}_\infty)$.[1]

Ex. 6.3 Moreover, $x \mapsto \mathbb{E}^x u(B_t) = \mathbb{E}^0 u(B_t + x)$ is for all $u \in \mathcal{B}_b(\mathbb{R}^d)$ a measurable function.

We will need the following formulae for conditional expectations, a proof of which can be found in the appendix, Lemma A.3. Assume that $\mathscr{X} \subset \mathscr{A}$ and $\mathscr{Y} \subset \mathscr{A}$ are independent σ-algebras. If X is an $\mathscr{X}/\mathscr{B}(\mathbb{R}^d)$ and Y a $\mathscr{Y}/\mathscr{B}(\mathbb{R}^d)$ measurable random variable, then

$$\mathbb{E}\left[\Phi(X,Y) \mid \mathscr{X}\right] = \mathbb{E}\,\Phi(x,Y)\big|_{x=X} = \mathbb{E}\left[\Phi(X,Y) \mid X\right] \tag{6.2}$$

for all bounded Borel measurable functions $\Phi : \mathbb{R}^d \times \mathbb{R}^d \to \mathbb{R}$.

If $\Psi : \mathbb{R}^d \times \Omega \to \mathbb{R}$ is bounded and $\mathscr{B}(\mathbb{R}^d) \otimes \mathscr{Y}/\mathscr{B}(\mathbb{R})$ measurable, then

$$\mathbb{E}\left[\Psi(X(\cdot),\cdot) \mid \mathscr{X}\right] = \mathbb{E}\,\Psi(x,\cdot)\big|_{x=X} = \mathbb{E}\left[\Psi(X(\cdot),\cdot) \mid X\right]. \tag{6.3}$$

6.2 Theorem (Markov property). *Let $(B_t)_{t\geqslant 0}$ be a BM^d with admissible filtration $(\mathscr{F}_t)_{t\geqslant 0}$. Then*

$$\mathbb{E}\left[u(B_{t+s}) \mid \mathscr{F}_s\right] = \mathbb{E}\left[u(B_t + x)\right]\Big|_{x=B_s} = \mathbb{E}^{B_s}\left[u(B_t)\right] \tag{6.4}$$

holds for all bounded measurable functions $u : \mathbb{R}^d \to \mathbb{R}$ and

$$\mathbb{E}\left[\Psi(B_{\bullet+s}) \mid \mathscr{F}_s\right] = \mathbb{E}\left[\Psi(B_\bullet + x)\right]\Big|_{x=B_s} = \mathbb{E}^{B_s}\left[\Psi(B_\bullet)\right] \tag{6.5}$$

holds for all bounded $\mathscr{B}(\mathcal{C})/\mathscr{B}(\mathbb{R})$ measurable functionals $\Psi : \mathcal{C}[0,\infty) \to \mathbb{R}$ which may depend on the whole Brownian path.

Proof. We write $u(B_{t+s}) = u(B_s + B_{t+s} - B_s) = u(B_s + (B_{t+s} - B_s))$. Then (6.4) follows from (6.2) and (B1) with $\Phi(x,y) = u(x+y)$, $X = B_s$, $Y = B_{t+s} - B_s$ and $\mathscr{X} = \mathscr{F}_s$.

In the same way we can derive (6.5) from (6.3) and 2.14 if we take $\mathscr{Y} = \mathscr{F}_\infty^W$ where $W_t := B_{t+s} - B_s$, $\mathscr{X} = \mathscr{F}_s$ and $X = B_s$. □

6.3 Remark. a) Theorem 6.1 really is a result for processes with stationary and independent increments, whereas Theorem 6.2 is not necessarily restricted to such

---

1 The corresponding canonical Wiener measure μ^x from Theorem 4.2 is, consequently, concentrated on the continuous functions $w : [0,\infty) \to \mathbb{R}^d$ satisfying $w(0) = x$.

processes. In general any d-dimensional, $\mathscr{F}_t$ adapted stochastic process $(X_t)_{t\geqslant 0}$ with (right-)continuous paths is called a *Markov process* if it satisfies

$$\mathbb{E}\left[u(X_{t+s}) \mid \mathscr{F}_s\right] = \mathbb{E}\left[u(X_{t+s}) \mid X_s\right] \quad \text{for all } s, t \geqslant 0,\ u \in \mathcal{B}_b(\mathbb{R}^d). \tag{6.4a}$$

Since we can express $\mathbb{E}\left[u(X_{t+s})|X_s\right]$ as an (essentially unique) function of X_s, say $g_{u,s,t+s}(X_s)$, we have

$$\mathbb{E}\left[u(X_{t+s}) \mid \mathscr{F}_s\right] = \mathbb{E}^{s,X_s}\left[u(X_{t+s})\right] \text{ with } \mathbb{E}^{s,x}\left[u(X_{t+s})\right] := g_{u,s,t+s}(x). \tag{6.4b}$$

Note that we can always select a Borel measurable version of $g_{u,s,t+s}(\cdot)$. If $g_{u,s,t+s}$ depends only on the difference $(t+s)-s=t$, we call the Markov process *homogeneous* and we write

$$\mathbb{E}\left[u(X_{t+s}) \mid \mathscr{F}_s\right] = \mathbb{E}^{X_s}\left[u(X_t)\right] \quad \text{with} \quad \mathbb{E}^x\left[u(X_t)\right] := g_{u,t}(x). \tag{6.4c}$$

Obviously (6.4) implies both (6.4a) and (6.4c), i. e. Brownian motion is also a Markov process in this more general sense. For homogeneous Markov processes we have

$$\mathbb{E}\left[\Psi(X_{\bullet+s}) \mid \mathscr{F}_s\right] = \mathbb{E}\left[\Psi(X_{\bullet+s})|X_s\right] = \mathbb{E}^{X_s}\left[\Psi(X_\bullet)\right] \tag{6.5a}$$

for bounded measurable functionals $\Psi : \mathcal{C}[0,\infty) \to \mathbb{R}$ of the sample paths. Although this relation seems to be more general than (6.4a), (6.4b), it is actually possible to derive (6.5a) directly from (6.4a), (6.4b) by standard measure-theoretic techniques.

b) It is useful to write (6.4) in integral form and with all ω: **!**

$$\begin{aligned}\mathbb{E}\left[u(B_{t+s}) \mid \mathscr{F}_s\right](\omega) &= \int_\Omega u(B_t(\omega') + B_s(\omega))\,\mathbb{P}(d\omega') \\ &= \int_\Omega u(B_t(\omega'))\,\mathbb{P}^{B_s(\omega)}(d\omega') \\ &= \mathbb{E}^{B_s(\omega)}\left[u(B_t)\right].\end{aligned}$$

Markov processes are often used in the literature. In fact, the Markov property (6.4a), (6.4b) can be seen as a means to obtain the finite dimensional distributions from one-dimensional distributions. We will show this for a Brownian motion $(B_t)_{t\geqslant 0}$ where we only use the property (6.4). The guiding idea is as follows: If a particle starts at x and is by time t in a set A and by time u in C, we can realize this as

$$x \xrightarrow[\text{Brownian motion}]{\text{time } t} A \quad \left\{\begin{array}{l}\text{stop there at } B_t = a,\\ \text{then restart from } a \in A\end{array}\right\} \quad \xrightarrow[\text{Brownian motion}]{\text{time } u-t} C.$$

Let us become a bit more rigorous. Let $0 \leqslant s < t < u$ and let $\phi, \psi, \chi : \mathbb{R}^d \to \mathbb{R}$ bounded measurable functions. Then

$$\begin{aligned}\mathbb{E}\left[\phi(B_s)\,\psi(B_t)\,\chi(B_u)\right] &= \mathbb{E}\left[\phi(B_s)\,\psi(B_t)\,\mathbb{E}\left(\chi(B_u)\mid \mathscr{F}_t\right)\right] \\ &\overset{(6.4)}{=} \mathbb{E}\left[\phi(B_s)\,\psi(B_t)\,\mathbb{E}^{B_t}\,\chi(B_{u-t})\right] \\ &\overset{(6.4)}{=} \mathbb{E}\Big[\phi(B_s)\,\underbrace{\mathbb{E}^{B_s}\left(\psi(B_{t-s})\,\mathbb{E}^{B_{t-s}}\,\chi(B_{u-t})\right)}_{=\int \psi(a)\int \chi(b)\,\mathbb{P}^a(B_{u-t}\in db)\,\mathbb{P}^{B_s}(B_{t-s}\in da)}\Big].\end{aligned}$$

If we rewrite all expectations as integrals we get

$$\begin{aligned}&\iiint \phi(x)\psi(y)\chi(z)\mathbb{P}(B_s \in dx, B_t \in dy, B_u \in dz) \\ &\quad = \int\left[\phi(c)\int\psi(a)\int\chi(b)\,\mathbb{P}^a(B_{u-t}\in db)\,\mathbb{P}^c(B_{t-s}\in da)\right]\mathbb{P}(B_s \in dc).\end{aligned}$$

This shows that we find the joint distribution of the vector (B_s, B_t, B_u) if we know the one-dimensional distributions of each B_t. For $s = 0$ and $\phi(0) = 1$ these relations are also known as *Chapman-Kolmogorov equations*. By iteration we get

6.4 Theorem. *Let $(B_t)_{t\geqslant 0}$ be a d-dimensional Brownian motion (or a general Markov process), $x_0 = 0$ and $t_0 = 0 < t_1 < \cdots < t_n$. Then*

$$\mathbb{P}^0(B_{t_1} \in dx_1, \ldots, B_{t_n} \in dx_n) = \prod_{j=1}^{n} \mathbb{P}^{x_{j-1}}(B_{t_j - t_{j-1}} \in dx_j). \tag{6.6}$$

6.2 The strong Markov property

Let us now show that (6.4) remains valid if we replace s by a stopping time $\sigma(\omega)$. Since stopping times can attain the value $+\infty$, we have to take care that B_σ is defined even in this case. If $(X_t)_{t\geqslant 0}$ is a stochastic process and σ a stopping time, we set

$$X_\sigma(\omega) := \begin{cases} X_{\sigma(\omega)}(\omega) & \text{if } \sigma(\omega) < \infty, \\ \lim_{t\to\infty} X_t(\omega) & \text{if } \sigma(\omega) = \infty \text{ and if the limit exists,} \\ 0 & \text{in all other cases.}\end{cases}$$

6.5 Theorem (Strong Markov property). *Let $(B_t)_{t\geqslant 0}$ be a BM^d with admissible filtration $(\mathscr{F}_t)_{t\geqslant 0}$ and $\sigma < \infty$ some a. s. finite stopping time. Then $(W_t)_{t\geqslant 0}$, where $W_t := B_{\sigma+t} - B_\sigma$, is again a BM^d, which is independent of $\mathscr{F}_{\sigma+}$.*

Proof. We know, cf. Lemma A.16, that $\sigma_j := (\lfloor 2^j\sigma\rfloor + 1)/2^j$ is a decreasing sequence of stopping times such that $\inf_{j\geqslant 1}\sigma_j = \sigma$. For all $0 \leqslant s < t$, $\xi \in \mathbb{R}^d$ and all $F \in \mathscr{F}_{\sigma+}$ we find by the continuity of the sample paths

$$
\begin{aligned}
\mathbb{E}\left[e^{i\langle\xi,\,B_{\sigma+t}-B_{\sigma+s}\rangle}\,\mathbb{1}_F\right] &= \lim_{j\to\infty}\mathbb{E}\left[e^{i\langle\xi,\,B_{\sigma_j+t}-B_{\sigma_j+s}\rangle}\,\mathbb{1}_F\right]\\
&= \lim_{j\to\infty}\sum_{k=1}^{\infty}\mathbb{E}\left[e^{i\langle\xi,\,B_{k2^{-j}+t}-B_{k2^{-j}+s}\rangle}\,\mathbb{1}_{\{\sigma_j=k2^{-j}\}}\cdot\mathbb{1}_F\right]\\
&= \lim_{j\to\infty}\sum_{k=1}^{\infty}\mathbb{E}\Big[\overbrace{e^{i\langle\xi,\,B_{k2^{-j}+t}-B_{k2^{-j}+s}\rangle}}^{\perp\!\!\!\perp\,\mathscr{F}_{k2^{-j}},\ \sim B_{t-s}\text{ by (B1)/5.1b), (B2)}}\,\underbrace{\mathbb{1}_{\{(k-1)2^{-j}\leqslant\sigma<k2^{-j}\}}\,\mathbb{1}_F}_{\in\mathscr{F}_{k2^{-j}}\text{ as }F\in\mathscr{F}_{\sigma+}}\Big]\\
&= \lim_{j\to\infty}\sum_{k=1}^{\infty}\mathbb{E}\left[e^{i\langle\xi,\,B_{t-s}\rangle}\right]\mathbb{P}\left(\{(k-1)2^{-j}\leqslant\sigma<k2^{-j}\}\cap F\right)\\
&= \mathbb{E}\left[e^{i\langle\xi,\,B_{t-s}\rangle}\right]\mathbb{P}(F).
\end{aligned}
$$

In the last equality we used $\dot{\bigcup}_{k=1}^{\infty}\{(k-1)2^{-j}\leqslant\sigma<k2^{-j}\}=\{\sigma<\infty\}$ for all $j\geqslant 1$.

The same calculation applies to finitely many increments. Let $t_0=0<t_1<\dots<t_n$ and $\xi_1,\dots,\xi_n\in\mathbb{R}^d$ and $F\in\mathscr{F}_{\sigma+}$. Then

$$
\mathbb{E}\left[e^{i\sum_{j=1}^{n}\langle\xi_j,\,B_{\sigma+t_j}-B_{\sigma+t_{j-1}}\rangle}\,\mathbb{1}_F\right]=\prod_{j=1}^{n}\mathbb{E}\left[e^{i\langle\xi_j,\,B_{t_j-t_{j-1}}\rangle}\right]\mathbb{P}(F).
$$

This shows that the increments $B_{\sigma+t_j}-B_{\sigma+t_{j-1}}$ are independent and distributed like $B_{t_j-t_{j-1}}$. Moreover, all increments are independent of $F\in\mathscr{F}_{\sigma+}$. Using the same argument as in the proof of Theorem 6.1 shows that all random vectors of the form $(B_{\sigma+t_1}-B_\sigma,\dots,B_{\sigma+t_n}-B_{\sigma+t_{n-1}})$ and, consequently, $\mathscr{F}_\infty^W:=\sigma(B_{\sigma+t}-B_\sigma,t\geqslant 0)$ are independent of $\mathscr{F}_{\sigma+}$. □

We can now use (6.2), (6.3) (or Lemma A.3) to deduce from Theorem 6.5 the following consequences of the strong Markov property.

6.6 Theorem. *Let $(B_t)_{t\geqslant0}$ be a BM^d with admissible filtration $(\mathscr{F}_t)_{t\geqslant0}$ and σ be a stopping time. Then we have for all $t\geqslant 0$, $u\in\mathcal{B}_b(\mathbb{R}^d)$ and $\mathbb{P}$ almost all $\omega\in\{\sigma<\infty\}$* Ex. 6.4

$$
\mathbb{E}\left[u(B_{t+\sigma})\mid\mathscr{F}_{\sigma+}\right](\omega)=\mathbb{E}\left[u(B_t+x)\right]\Big|_{x=B_\sigma(\omega)}=\mathbb{E}^{B_\sigma(\omega)}u(B_t). \tag{6.7}
$$

For all bounded $\mathscr{B}(\mathcal{C})/\mathscr{B}(\mathbb{R})$ measurable functionals $\Psi:\mathcal{C}[0,\infty)\to\mathbb{R}$ which may depend on a whole Brownian path and $\mathbb{P}$ almost all $\omega\in\{\sigma<\infty\}$ this becomes

$$
\mathbb{E}\left[\Psi(B_{\bullet+\sigma})\mid\mathscr{F}_{\sigma+}\right]=\mathbb{E}\left[\Psi(B_\bullet+x)\right]\Big|_{x=B_\sigma}=\mathbb{E}^{B_\sigma}\left[\Psi(B_\bullet)\right]. \tag{6.8}
$$

Let $\tau=\tau(B_\bullet)$ be a stopping time which can be expressed as a functional of a Brownian path, e. g. a first hitting time, and assume that σ is a further stopping time such that $\sigma\leqslant\tau$ a. s. Denote by $\tau'=\tau(B_{\bullet+\sigma})$ the stopping time τ for the shifted process $B_{\bullet+\sigma}$, which is the remaining time, counting from σ, until the event described by τ happens. Set $W_\bullet:=B_{\bullet+\sigma}-B_\sigma$; then $\tau'=\tau(W_\bullet+B_\sigma)$, and the functionals $u(B_\tau)$ and $u(W_{\tau'}+B_\sigma)$ have the same distribution. Thus, (6.8) implies the following result.

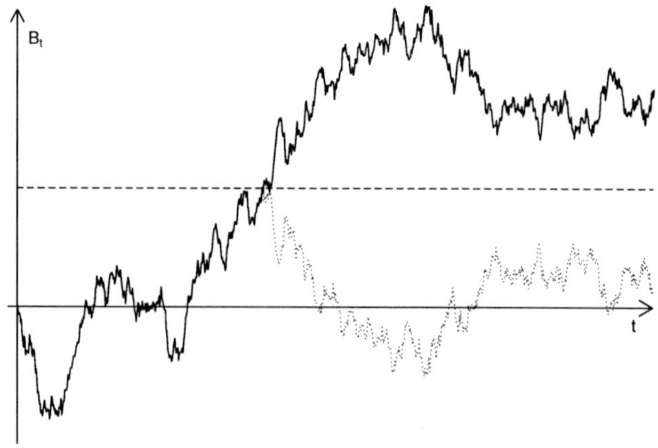

Fig. 6.2. Reflection upon reaching the level b for the first time.

6.7 Corollary. *Let $(B_t)_{t\geqslant 0}$ be a BM^d with admissible filtration $(\mathscr{F}_t)_{t\geqslant 0}$, $\tau = \tau(B_\bullet)$ a first hitting time and σ a stopping time such that $\sigma \leqslant \tau$ almost surely. Set $\tau' = \tau(B_{\bullet+\sigma})$ and $W_\bullet := B_{\bullet+\sigma} - B_\sigma$. Then*

$$\mathbb{E}\left[u(B_\tau) \mid \mathscr{F}_{\sigma+}\right](\omega) = \mathbb{E}\left[u(W_{\tau'} + x)\right]\big|_{x=B_\sigma(\omega)} = \mathbb{E}^{B_\sigma(\omega)} u(W_{\tau'}) \tag{6.9}$$

holds for all $u \in \mathcal{B}_b(\mathbb{R}^d)$ and $\mathbb{P}$ almost all $\omega \in \{\tau < \infty\}$.

Often one identifies the Brownian motions B and W appearing in (6.9), and writes only B.

6.8 Remark. Let $(X_t)_{t\geqslant 0}$ be an $\mathscr{F}_t$ adapted, d-dimensional stochastic process with (right-)continuous paths. If $(X_t)_{t\geqslant 0}$ satisfies for all $t \geqslant 0$ and $u \in \mathcal{B}_b(\mathbb{R}^d)$

$$\mathbb{E}\left[u(X_{\sigma+t}) \mid \mathscr{F}_{\sigma+}\right] = \mathbb{E}\left[u(X_{\sigma+t}) \mid X_\sigma\right] \quad \text{on} \quad \{\sigma < \infty\}, \tag{6.10}$$

it is called a *strong Markov process*. For a Brownian motion this follows from (6.7). Just as for the simple Markov property, the (general analogue of) property (6.8) seems only stronger than (6.7); as a matter of fact one can derive it directly from (6.7) using standard techniques.

6.3 Desiré André's reflection principle

Let us briefly discuss one of the most famous applications of the strong Markov property of a Brownian motion. Let $(B_t)_{t\geqslant 0}$ be a one-dimensional Brownian motion and $\tau_b = \inf\{t \geqslant 0 : B_t = b\}$ the first passage time for b. Assume that B has reached the level b for the first time and that $\tau_b < t$. Stop at τ_b and start anew from $B_{\tau_b} = b$. Since $W_s + b = B_{\tau_b+s} - B_{\tau_b} + b = B_{\tau_b+s}$ is, by Theorem 6.5, again a Brownian motion (started at b), the probability to be at time t above or below b is the same, cf. Figure 6.2. Since

$B_{\tau_b} = b$ and since τ_b is $\mathscr{F}^B_{\tau_b}$ measurable, cf. Lemma A.15, we have

$$\begin{aligned}\mathbb{P}(\tau_b \leqslant t, B_t < b) &= \mathbb{P}\big(\underbrace{\{\tau_b \leqslant t\}}_{\in \mathscr{F}^B_{\tau_b}} \cap \underbrace{\{B_{\tau_b+(t-\tau_b)} - B_{\tau_b} < 0\}}_{\in \mathscr{F}^W_\infty \perp\!\!\!\perp \mathscr{F}^B_{\tau_b},\ \sim W_{t-\tau_b}}\big)\\ &= \mathbb{P}(\{\tau_b \leqslant t\} \cap \{B_{\tau_b+(t-\tau_b)} - B_{\tau_b} > 0\})\\ &= \mathbb{P}(\tau_b \leqslant t, B_t > b)\end{aligned}$$

and so, for all $b > 0$,

$$\begin{aligned}\mathbb{P}(\tau_b \leqslant t) &= \mathbb{P}(\tau_b \leqslant t, B_t \geqslant b) + \mathbb{P}(\tau_b \leqslant t, B_t < b)\\ &= \mathbb{P}(\tau_b \leqslant t, B_t \geqslant b) + \mathbb{P}(\tau_b \leqslant t, B_t > b)\\ &= \mathbb{P}(B_t \geqslant b) + \mathbb{P}(B_t > b)\\ &= 2\mathbb{P}(B_t \geqslant b).\end{aligned}$$

Observe that $\{\tau_b \leqslant t\} = \{M_t \geqslant b\}$ where $M_t := \sup_{s\leqslant t} B_s$. Therefore,

$$\mathbb{P}(M_t \geqslant b) = \mathbb{P}(\tau_b \leqslant t) = 2\mathbb{P}(B_t \geqslant b) \quad \text{for all } b > 0. \tag{6.11}$$

From this we can calculate the probability distributions of various functionals of a Brownian motion.

6.9 Theorem (Lévy 1939). *Let $(B_t)_{t\geqslant 0}$ be a BM^1, $b \in \mathbb{R}$, and set* Ex. 6.5 Ex. 6.6 Ex. 6.7

$$\tau_b = \inf\{t \geqslant 0 : B_t = b\}, \quad M_t := \sup_{s\leqslant t} B_s, \quad \text{and} \quad m_t := \inf_{s\leqslant t} B_s.$$

Then, for all $x \geqslant 0$,

$$M_t \sim |B_t| \sim -m_t \sim M_t - B_t \sim B_t - m_t \sim \sqrt{\frac{2}{\pi t}}\, e^{-x^2/(2t)}\, dx \tag{6.12}$$

and for $t > 0$

$$\tau_b \sim \frac{|b|}{\sqrt{2\pi t^3}}\, e^{-b^2/(2t)}\, dt. \tag{6.13}$$

Proof. The equalities (6.11) immediately show that $M_t \sim |B_t|$. Using the symmetry 2.12 of a Brownian motion we get $m_t = \inf_{s\leqslant t} B_s = -\sup_{s\leqslant t}(-B_s) \sim -M_t$. The invariance of BM^1 under time reversals, cf. 2.15, gives

$$M_t - B_t = \sup_{s\leqslant t}(B_s - B_t) = \sup_{s\leqslant t}(B_{t-s} - B_t) \sim \sup_{s\leqslant t} B_s = M_t,$$

and if we combine this again with the symmetry 2.12 we get $M_t - B_t \sim B_t - m_t$.

The formula (6.13) follows for $b > 0$ from the fact that $\{\tau_b > t\} = \{M_t < b\}$ if we differentiate the relation

$$\mathbb{P}(\tau_b > t) = \mathbb{P}(M_t < b) = \sqrt{\frac{2}{\pi t}} \int_0^b e^{-x^2/(2t)}\, dx.$$

For $b < 0$ we use the symmetry of a Brownian motion. □

! **6.10 Remark.** (6.12) tells us that for each single instant of time the random variables, e. g. M_t and $|B_t|$, have the same distribution. This is no longer true if we consider the finite dimensional distributions, i. e. the laws of processes $(M_t)_{t\geqslant 0}$ and $(|B_t|)_{t\geqslant 0}$ are different. In our example this is obvious since $t \mapsto M_t$ is a. s. increasing while $t \mapsto |B_t|$ is positive but not necessarily monotone. On the other hand, the processes $-m$ and M are equivalent, i. e. they do have the same law.

In order to derive (6.11) we used the Markov property of a Brownian motion, i. e. the fact that $\mathscr{F}_s^B$ and $\mathscr{F}_\infty^W$, $W_t = B_{t+s} - B_s$ are independent. Often the proof is based on (6.7) but this leads to the following difficulty:

$$\begin{aligned}\mathbb{P}(\tau_b \leqslant t, B_t < b) &= \mathbb{E}\left(\mathbb{1}_{\{\tau_b\leqslant t\}}\,\mathbb{E}\left[\mathbb{1}_{(-\infty,0)}(B_t - b) \mid \mathscr{F}_{\tau_b}^B\right]\right)\\ &\overset{?!}{=} \mathbb{E}\left(\mathbb{1}_{\{\tau_b\leqslant t\}}\,\mathbb{E}^{B_{\tau_b}}\left[\mathbb{1}_{(-\infty,0)}(B_{t-\tau_b} - b)\right]\right)\\ &= \mathbb{E}\left(\mathbb{1}_{\{\tau_b\leqslant t\}}\,\mathbb{E}^{b}\left[\mathbb{1}_{(-\infty,0)}(B_{t-\tau_b} - b)\right]\right)\\ &= \mathbb{E}\left(\mathbb{1}_{\{\tau_b\leqslant t\}}\,\mathbb{E}\left[\mathbb{1}_{(-\infty,0)}(B_{t-\tau_b})\right]\right)\\ &= \mathbb{E}\left(\mathbb{1}_{\{\tau_b\leqslant t\}}\,\mathbb{E}\left[\mathbb{1}_{(0,\infty)}(B_{t-\tau_b})\right]\right).\end{aligned}$$

! Notice that in the step marked by '?!' there appears the expression $t - \tau_b$. As it is written, we do not know whether $t - \tau_b$ is positive – and if it were, our calculation would certainly not be covered by (6.7).

In order to make such arguments work we need the following stronger version of the strong Markov property. Mind the dependence of the random times on the fixed path ω in the expression below!

6.11 Theorem. *Let $(B_t)_{t\geqslant 0}$ be a BM^d, τ an $\mathscr{F}_t^B$ stopping time and $\eta \geqslant \tau$ where η is an $\mathscr{F}_{\tau+}^B$ measurable random time. Then we have for all $\omega \in \{\eta < \infty\}$ and all $u \in \mathcal{B}_b(\mathbb{R}^d)$*

$$\begin{aligned}\mathbb{E}\left[u(B_\eta) \mid \mathscr{F}_{\tau+}^B\right](\omega) &= \mathbb{E}^{B_\tau(\omega)}\left[u(B_{\eta(\omega)-\tau(\omega)}(\cdot))\right] \qquad (6.14)\\ &= \int u(B_{\eta(\omega)-\tau(\omega)}(\omega'))\,\mathbb{P}^{B_\tau(\omega)}(d\omega').\end{aligned}$$

The requirement that τ is a stopping time and η is a $\mathscr{F}_{\tau+}$ measurable random time with $\eta \geqslant \tau$ looks artificial. Typical examples of (τ,η) pairs are $(\tau, \tau + t)$, $(\tau, \tau \vee t)$ or $(\sigma \wedge t, t)$ where σ is a further stopping time.

Proof of Theorem 6.11. Replacing $u(B_\eta)$ by $u(B_\eta)\mathbb{1}_{\{\eta<\infty\}}$ we may assume that $\eta < \infty$. Set $\sigma_j := (\lfloor 2^j(\eta - \tau)\rfloor + 1)/2^j$. Then $\inf_{j\geqslant 1}\sigma_j = \eta - \tau$ and σ_j is clearly $\mathscr{F}_{\tau+}^B$ measurable, since η and τ are $\mathscr{F}_{\tau+}^B$ measurable.

For all $F \in \mathscr{F}^B_{\tau+}$ we know that $F \cap \{\sigma_j = k/2^j\} \in \mathscr{F}^B_{\tau+}$, i. e.

$$\begin{aligned}\int_F u(B_{\tau+\sigma_j})\,d\mathbb{P} &= \sum_{k=1}^{\infty} \int_{F\cap\{\sigma_j=k/2^j\}} u(B_{\tau+k/2^j})\,d\mathbb{P}\\ &= \sum_{k=1}^{\infty} \int_{F\cap\{\sigma_j=k/2^j\}} \mathbb{E}\left[u(B_{\tau+k/2^j}) \mid \mathscr{F}^B_{\tau+}\right] d\mathbb{P}\\ &\overset{(6.7)}{=} \sum_{k=1}^{\infty} \int_{F\cap\{\sigma_j=k/2^j\}} \mathbb{E}^{B_\tau(\omega)}\, u(B_{k/2^j})\,\mathbb{P}(d\omega)\\ &= \int_F \mathbb{E}^{B_\tau(\omega)}\, u(B_{\sigma_j(\omega)})\,\mathbb{P}(d\omega).\end{aligned}$$

Since $t \mapsto B_t$ is continuous, we have $\lim_{j\to\infty} B_{\sigma_j} = B_{\eta-\tau}$, and we get (6.14) for $u \in \mathcal{C}_b(\mathbb{R}^d)$ using dominated convergence.

For every compact set $K \subset \mathbb{R}^d$ we can approximate $\mathbb{1}_K$ from above by a sequence of continuous functions. Therefore, Ex. 6.10

$$\int_F \mathbb{1}_K(B_\eta)\,d\mathbb{P} = \int_F \mathbb{E}^{B_\tau(\omega)}\, \mathbb{1}_K(B_{\eta(\omega)-\tau(\omega)})\,\mathbb{P}(d\omega).$$

As the compact sets generate the Borel σ-algebra, we can use the uniqueness theorem for measures to extend the above equality to all functions of the form $\mathbb{1}_K$ where $K \in \mathscr{B}(\mathbb{R}^d)$; then we get (6.14) for all measurable step functions and a Beppo–Levi argument finally shows (6.14) for $\mathcal{B}_b(\mathbb{R}^d)$. □

The proof of the reflection principle uses only the strong Markov property and the symmetry of a Brownian motion. Therefore, we can rephrase the reflection principle in the following more general version. Let $(B(t), \mathscr{F}_t)_{t\geqslant 0}$ be a Brownian motion and τ an a. s. finite stopping time. We consider the process

$$W(t,\omega) := \begin{cases} B(t,\omega), & \text{if } 0 \leqslant t < \tau(\omega) \leqslant +\infty,\\ 2B(\tau(\omega),\omega) - B(t,\omega), & \text{if } \tau(\omega) \leqslant t < \infty, \end{cases} \tag{6.15}$$

see Figure 6.3.

Observe that the trajectories of B and W coincide in the interval $[0, \tau(\omega))$. On $[\tau(\omega), \infty)$ we have $W(t,\omega) = B(\tau(\omega),\omega) - \big(B(t,\omega) - B(\tau(\omega),\omega)\big)$, i. e. every trajectory of W is the reflection of the original trajectory with respect to the axis $y = B(\tau(\omega),\omega)$.

From Theorem 6.5 we know that $(B(\tau) - B(\tau+t))_{t\geqslant 0}$ is a Brownian motion which is independent of the stopped process $(B(t\wedge\tau))_{t\geqslant 0}$. Therefore it is plausible that the process W, which is a concatenation of these two processes, is again a Brownian motion.

6.12 Theorem. *Let $(B(t), \mathscr{F}_t)_{t\geqslant 0}$ be a BM^d and let $(W(t))_{t\geqslant 0}$ the process given by (6.15). Then $(W(t))_{t\geqslant 0}$ is again a Brownian motion.*

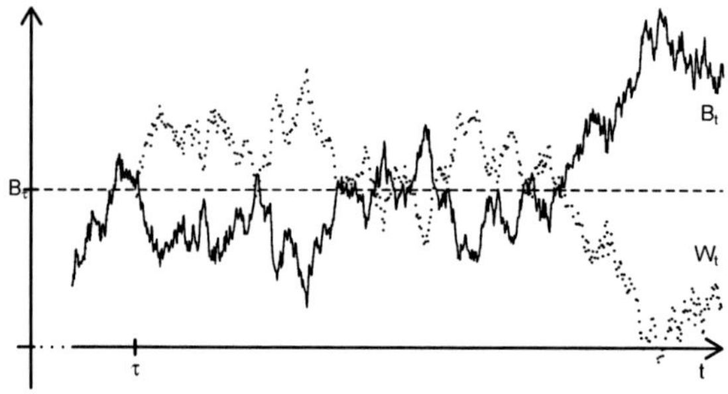

Fig. 6.3. Both B_t and the reflection W_t are Brownian motions.

Proof. Without restriction we may assume that $\tau(\omega) < \infty$ for all $\omega \in \Omega$. The mapping $\Phi : \mathcal{C}_{(0)} \times [0,\infty) \times \mathcal{C}_{(0)} \to \mathcal{C}_{(0)}$, where $\mathcal{C}_{(0)} = \{f \in \mathcal{C}[0,\infty) : f(0) = 0\}$, defined by

$$\Phi(f,t,g) := \begin{cases} f(s), & \text{if } 0 \leqslant s < t, \\ f(t) + g(s-t), & \text{if } t \leqslant s < \infty, \end{cases}$$

is continuous if we equip $\mathcal{C}_{(0)} \times [0,\infty) \times \mathcal{C}_{(0)}$ with the topology of locally uniform convergence (in $\mathcal{C}_{(0)}$) and pointwise convergence (in $[0,\infty)$); in particular, the mapping Φ is $\mathscr{B}(\mathcal{C}_{(0)}) \otimes \mathscr{B}[0,\infty) \otimes \mathscr{B}(\mathcal{C}_{(0)})/\mathscr{B}(\mathcal{C}_{(0)})$ measurable.

By Lemma 6.13, $\omega \mapsto (B(\cdot \wedge \tau(\omega), \omega), \tau(\omega))$ is $\mathscr{F}_{\tau+}/\mathscr{B}(\mathcal{C}_{(0)}) \otimes \mathscr{B}[0,\infty)$ measurable. Moreover, $\omega \mapsto B(\cdot + \tau(\omega), \omega) - B(\tau(\omega), \omega)$ is independent of $\mathscr{F}_{\tau+}$ and has the same probability distribution as $\omega \mapsto B(\cdot, \omega)$. Because of the symmetry of the Wiener measure, $B(\cdot + \tau) - B(\tau) \sim -B(\cdot + \tau) + B(\tau)$, and we see that

$$W = \Phi(B(\cdot \wedge \tau), \tau, B(\tau) - B(\cdot + \tau)) \quad \text{and} \quad B = \Phi(B(\cdot \wedge \tau), \tau, B(\cdot + \tau) - B(\tau))$$

have the same law. Since W has, by construction, continuous paths, the claim follows. □

Ex. 6.4 **6.13 Lemma** (cf. Corollary 6.25). *Let $(B(t), \mathscr{F}_t)_{t \geqslant 0}$ be a BM[1] and $\tau < \infty$ an a. s. finite stopping time. Then $B(t \wedge \tau)$ is $\mathscr{F}_{\tau+}$ measurable for every $t \geqslant 0$.*

Proof. We have to show that $\{B(t \wedge \tau) \in U\} \cap \{\tau < s\} \in \mathscr{F}_s$ for all Borel sets $U \subset \mathbb{R}$ and $s \geqslant 0$, cf. Definition A.14. It is clearly enough to consider only open sets U and dyadic numbers $s = k2^{-n}$. From Lemma A.16 we know that $\tau_j := (\lfloor 2^j \tau \rfloor + 1)/2^j$ is a decreasing sequence of stopping times such that $\inf_{j \geqslant 1} \tau_j = \tau$. Since $t \mapsto B(t, \omega)$ is continuous and U open, we see that for all $n \geqslant 1$

$$\{B(t \wedge \tau) \in U\} = \bigcup_{m \geqslant n} \bigcap_{l \geqslant m} \{B(t \wedge \tau_l) \in U\}.$$

By definition, $\{\tau < k2^{-n}\} = \{\tau_n \leqslant k2^{-n}\} = \{\tau_m \leqslant k2^{-n}\}$ for all $m \geqslant n$. Therefore,

$$\{B(t \wedge \tau) \in U\} \cap \{\tau < k2^{-n}\} = \bigcup_{m \geqslant n} \bigcap_{l \geqslant m} \Big(\underbrace{\big\{B(t \wedge \tau_l) \in U\big\} \cap \big\{\tau_l \leqslant k2^{-n}\big\}}_{=A(l,k,n)} \Big).$$

From $A(l,k,n) = \bigcup_{j:\,j2^{-l}\leqslant k2^{-n}} \left\{B(t\wedge j2^{-l}) \in U\right\} \cap \left\{\tau_l = j2^{-l}\right\} \in \mathscr{F}_{k2^{-n}}$, we conclude that $\{B(t\wedge\tau) \in U\} \cap \{\tau < k2^{-n}\} \in \mathscr{F}_{k2^{-n}}$. □

6.4 Transience and recurrence

In Section 5.2 we used Wald's identities to obtain the distribution of a BM^1 when exiting from the interval $(-a, b)$, cf. Corollary 5.11. Without much effort we can extend this to continuous martingales.

6.14 Corollary. *Let $(M_t^x, \mathscr{F}_t)_{t\geqslant 0}$ be a real-valued martingale with continuous paths and $M_0^x = x \in \mathbb{R}$ a. s. If the first exit time $\tau = \tau_{(r,R)^c}$ from the interval (r, R), $r < R$, is a. s. finite, then we have for all $x \in (r, R)$*

$$\mathbb{P}(M_\tau^x = r) = \frac{R-x}{R-r} \quad \text{and} \quad \mathbb{P}(M_\tau^x = R) = \frac{x-r}{R-r}.$$

If we combine Corollary 6.14 with Theorem 5.6 for a suitable function f we obtain the exit probabilities of a BM^d from an annulus $\mathbb{B}(0,R)\setminus\mathbb{B}(0,r)$.

6.15 Theorem. *Let $(B_t, \mathscr{F}_t)_{t\geqslant 0}$ be a Brownian motion with a right-continuous admissible filtration, $r < R$ and denote by $\tau_{\mathbb{B}(0,r)}$ and $\tau_{\mathbb{B}^c(0,R)}$ the first hitting times of the sets $\mathbb{B}(0,r)$ and $\mathbb{B}^c(0,R)$. Then, for all $r < |x| < R$,*

$$\mathbb{P}^x\left(\tau_{\mathbb{B}(0,r)} < \tau_{\mathbb{B}^c(0,R)}\right) = \begin{cases} \dfrac{R-|x|}{R-r}, & \text{if } d = 1,\\[2ex] \dfrac{\log R - \log|x|}{\log R - \log r}, & \text{if } d = 2,\\[2ex] \dfrac{R^{2-d} - |x|^{2-d}}{R^{2-d} - r^{2-d}}, & \text{if } d \geqslant 3. \end{cases} \tag{6.16}$$

Proof. Note that $\tau := \tau_{\mathbb{B}(0,r)}\wedge\tau_{\mathbb{B}^c(0,R)}$ is the first exit time from $\mathbb{B}(0,R)\setminus\mathbb{B}(0,r)$. Therefore, (6.16) follows, in dimension $d = 1$, directly from Corollary 6.14 where we set $M_t = B_t + x$.

If $d = 2$, we use Theorem 5.6 with $f(|x|) = -\log|x|$ and $0 < r \leqslant |x| \leqslant R$ (the attentive reader might first want to multiply f by a cut-off function $\chi_{r,R} \in \mathcal{C}_c(\mathbb{R}^2)$ such that $\chi_{r,R}|_{\mathbb{B}(0,R)\setminus\mathbb{B}(0,r)} \equiv 1$ and $\chi_{r,R}|_{\mathbb{B}(0,r/2)} \equiv 0$). Since

$$\frac{\partial f(|x|)}{\partial x_j} = \frac{-x_j}{|x|^2} \quad \text{and} \quad \frac{\partial^2 f(|x|)}{\partial x_j^2} = \frac{2x_j^2 - |x|^2}{|x|^4}, \qquad j = 1, 2,$$

we find $\Delta f(x) \equiv 0$ on $\mathbb{B}(0,R)\setminus\mathbb{B}(0,r)$. Therefore, $M_t = f(|B_t + x|)$ is a martingale with continuous paths. Moreover, $|B_t|^2 = (B_t^1)^2 + (B_t^2)^2 \geqslant (B_t^1)^2$ where the coordinate processes $(B_t^j)_{t\geqslant 0}$ are one-dimensional Brownian motions; since any one-dimensional

Brownian motion leaves every interval $[-R, R]$ with probability one, cf. Corollary 5.11, we see that $\tau < \infty$ a. s. We can now apply Corollary 6.14 and conclude that

$$\mathbb{P}^x\big(\tau_{\mathbb{B}(0,r)} < \tau_{\mathbb{B}^c(0,R)}\big) = \frac{f(R) - f(x)}{f(R) - f(r)} = \frac{\log R - \log|x|}{\log R - \log r}.$$

If $d = 3$, we can use the same argument with $f(|x|) = |x|^{2-d}$. Now $\Delta f(|x|) = 0$ since for $j = 1, 2, \dots, d$

$$\frac{\partial f(|x|)}{\partial x_j} = \frac{(2-d)x_j}{|x|^d} \quad \text{and} \quad \frac{\partial^2 f(|x|)}{\partial x_j^2} = \frac{(2-d)|x|^2 - d(2-d)x_j^2}{|x|^{d+2}}. \qquad \square$$

The results of Theorem 6.15 become especially interesting for degenerate annuli, i. e. for $r \to 0$ and $R \to \infty$. Letting first $r \to 0$ and then $R \to \infty$ yields

$$\mathbb{P}^x\big(\tau_{\mathbb{B}(0,r)} < \tau_{\mathbb{B}^c(0,R)}\big) \xrightarrow[r\to 0]{} \mathbb{P}^x\big(\tau_{\{0\}} \leqslant \tau_{\mathbb{B}^c(0,R)}\big) \xrightarrow[R\to\infty]{} \mathbb{P}^x\big(\tau_{\{0\}} < \infty\big).$$

If we consider only $R \to \infty$ we get

$$\mathbb{P}^x\big(\tau_{\mathbb{B}(0,r)} < \tau_{\mathbb{B}^c(0,R)}\big) \xrightarrow[R\to\infty]{} \mathbb{P}^x\big(\tau_{\mathbb{B}(0,r)} < \infty\big).$$

Since we know from Theorem 6.15 the values for the respective probabilities, we arrive at

6.16 Corollary. *Let $(B_t, \mathscr{F}_t)_{t\geqslant 0}$ be a BM^d with a right-continuous filtration. Then we have for all $|x| \geqslant r > 0$*

$$\mathbb{P}^x\big(\tau_{\{0\}} < \infty\big) = \begin{cases} 1, & \text{if } d = 1, \\ 0, & \text{if } d \geqslant 2, \end{cases} \tag{6.17}$$

$$\mathbb{P}^x\big(\tau_{\mathbb{B}(0,r)} < \infty\big) = \begin{cases} 1, & \text{if } d = 1, 2, \\ \left(\frac{|x|}{r}\right)^{2-d}, & \text{if } d \geqslant 3. \end{cases} \tag{6.18}$$

If $x = 0$, $\tau_{\{0\}}$ is the first return time to 0, and (6.17) is still true:

$$\{\tau_{\{0\}} < \infty\} = \{\exists t > 0 : B_t = 0\} = \bigcup_{n\geqslant 1}\{\exists t > 1/n : B_t = 0\}.$$

By the Markov property, and the fact that $\mathbb{P}^0(B_{1/n} = 0) = 0$, we have

$$\mathbb{P}^0(\exists t > 1/n : B_t = 0) = \mathbb{E}^0\, \mathbb{P}^{B_{1/n}}(\exists t > 0 : B_t = 0) = \mathbb{E}^0\, \underbrace{\mathbb{P}^{B_{1/n}}\big(\tau_{\{0\}} < \infty\big)}_{=0 \text{ or } 1 \text{ by } (6.17)}$$

which is 0 or 1 according to $d = 1$ or $d \geqslant 2$.

By the (strong) Markov property, a re-started Brownian motion is again a Brownian motion. This is the key to the following result.

6.17 Corollary. *Let $(B(t))_{t\geqslant 0}$ be a* BM^d *and* $x \in \mathbb{R}^d$.

a) *If* $d = 1$, *Brownian motion is* point recurrent, *i. e.*

$$\mathbb{P}^x(B(t) = 0 \;\; \textit{infinitely often}) = 1.$$

b) *If* $d = 2$, *Brownian motion is* neighbourhood recurrent, *i. e.*

$$\mathbb{P}^x(B(t) \in \mathbb{B}(0,r) \;\; \textit{infinitely often}) = 1 \quad \textit{for every } r > 0.$$

c) *If* $d \geqslant 3$, *Brownian motion is* transient, *i. e.* $\mathbb{P}^x(\lim_{t\to\infty} |B(t)| = \infty) = 1$.

Proof. Let $d = 1, 2$, $r > 0$ and denote by $\tau_{\mathbb{B}(0,r)}$ the first hitting time of $\mathbb{B}(0,r)$ for $(B(t))_{t\geqslant 0}$. We know from Corollary 6.16 that $\tau_{\mathbb{B}(0,r)}$ is a. s. finite. By the strong Markov property, Theorem 6.6, we find

$$\mathbb{P}^0(\exists t > 0 \,:\, B(t + \tau_{\mathbb{B}^c(0,r)}) \in \mathbb{B}(0,r)) = \mathbb{E}^0\left(\mathbb{P}^{B(\tau_{\mathbb{B}^c(0,r)})}(\tau_{\mathbb{B}(0,r)} < \infty)\right) \overset{(6.18)}{=} 1.$$

This means that $(B(t))_{t\geqslant 0}$ visits every neighbourhood $\mathbb{B}(0,r)$ time and again. Since $B(t) - y$, $y \in \mathbb{R}^d$, is also a Brownian motion, the same argument shows that $(B(t))_{t\geqslant 0}$ visits any ball $\mathbb{B}(y,r)$ infinitely often. In one dimension this is, because of the continuity of the paths of a Brownian motion, only possible if every *point* is visited infinitely often.

Let $d \geqslant 3$. By the strong Markov property, Theorem 6.6, and Corollary 6.16 we find for all $r < R$

$$\mathbb{P}^0(\exists t \geqslant \tau_{\mathbb{B}^c(0,R)} \,:\, B(t) \in \mathbb{B}(0,r)) = \mathbb{E}^0\left(\mathbb{P}^{B(\tau_{\mathbb{B}^c(0,R)})}(\tau_{\mathbb{B}(0,r)} < \infty)\right) = \left(\frac{r}{R}\right)^{d-2}.$$

Take $r = \sqrt{R}$ and let $R \to \infty$. This shows that the return probability of $(B(t))_{t\geqslant 0}$ to *any compact* neighbourhood of 0 is zero, i. e. $\lim_{t\to\infty} |B(t)| = \infty$. □

6.5 Lévy's triple law

In this section we show how we can apply the reflection principle repeatedly to obtain for a BM^1, $(B_t)_{t\geqslant 0}$, the joint distribution of B_t, the running minimum $m_t = \inf_{s\leqslant t} B_s$ and maximum $M_t = \sup_{s\leqslant t} B_s$. This is known as P. Lévy's *loi à trois variables*, [145, VI.42.6, 4°, p. 213]. In a different context this formula appears already in Bachelier [7, Nos. 413 and 504]. An interesting application of the triple law can be found in Section 12.2, footnote 1.

6.18 Theorem (Lévy 1948). *Let $(B_t)_{t\geqslant 0}$ be a* BM^1 *and denote by m_t and M_t its running minimum and maximum respectively. Then* Ex. 6.8

$$\mathbb{P}(m_t > a, M_t < b, B_t \in dx) = \frac{dx}{\sqrt{2\pi t}} \sum_{n=-\infty}^{\infty} \left(e^{-\frac{(x+2n(b-a))^2}{2t}} - e^{-\frac{(x-2a-2n(b-a))^2}{2t}}\right) \tag{6.19}$$

for all $t > 0$ and $a < 0 < b$.

Proof. Let $a < 0 < b$, denote by $\tau = \inf\{t \geqslant 0 : B_t \notin (a,b)\}$ the first exit time from the interval (a,b), and pick $I = [c,d] \subset (a,b)$. We know from Corollary 5.11 that τ is an a. s. finite stopping time.

Since $\{\tau > t\} = \{m_t > a, M_t < b\}$ we have

$$\begin{aligned}
&\mathbb{P}(m_t > a, M_t < b, B_t \in I)\\
&\quad= \mathbb{P}(\tau > t, B_t \in I)\\
&\quad= \mathbb{P}(B_t \in I) - \mathbb{P}(\tau \leqslant t, B_t \in I)\\
&\quad= \mathbb{P}(B_t \in I) - \mathbb{P}(B_\tau = a, \tau \leqslant t, B_t \in I) - \mathbb{P}(B_\tau = b, \tau \leqslant t, B_t \in I).
\end{aligned}$$

For the last step we used the fact that $\{B_\tau = a\}$ and $\{B_\tau = b\}$ are, up to the null set $\{\tau = \infty\}$, a disjoint partition of Ω.

We want to use the reflection principle repeatedly. Denote by $\mathrm{r}_a x := 2a - x$ and $\mathrm{r}_b x := 2b - x$ the reflection in the (x,t)-plane with respect to the line $x = a$ and $x = b$, respectively, cf. Figure 6.4. Using Theorem 6.11 we find

$$\mathbb{P}(B_\tau = b, \tau \leqslant t, B_t \in I \mid \mathscr{F}_\tau)(\omega) = \mathbb{1}_{\{B_\tau = b\}\cap\{\tau\leqslant t\}}(\omega)\mathbb{P}^{B_\tau(\omega)}(B_{t-\tau(\omega)} \in I).$$

Because of the symmetry of Brownian motion,

$$\begin{aligned}
\mathbb{P}^b(B_{t-\tau(\omega)} \in I) &= \mathbb{P}(B_{t-\tau(\omega)} \in I - b)\\
&= \mathbb{P}(B_{t-\tau(\omega)} \in b - I)\\
&= \mathbb{P}^b(B_{t-\tau(\omega)} \in \mathrm{r}_b I),
\end{aligned}$$

and, therefore,

$$\begin{aligned}
\mathbb{P}(B_\tau = b, \tau \leqslant t, B_t \in I \mid \mathscr{F}_\tau)(\omega) &= \mathbb{1}_{\{B_\tau = b\}\cap\{\tau\leqslant t\}}(\omega)\mathbb{P}^{B_\tau(\omega)}(B_{t-\tau(\omega)} \in \mathrm{r}_b I)\\
&= \mathbb{P}(B_\tau = b, \tau \leqslant t, B_t \in \mathrm{r}_b I \mid \mathscr{F}_\tau)(\omega).
\end{aligned}$$

This shows that

$$\mathbb{P}\left(B_\tau = b, \tau \leqslant t, B_t \in I\right) = \mathbb{P}\left(B_\tau = b, \tau \leqslant t, B_t \in \mathrm{r}_b I\right) = \mathbb{P}\left(B_\tau = b, B_t \in \mathrm{r}_b I\right).$$

Further applications of the reflection principle yield

$$\begin{aligned}
\mathbb{P}(B_\tau = b, B_t \in \mathrm{r}_b I) &= \mathbb{P}(B_t \in \mathrm{r}_b I) - \mathbb{P}(B_\tau = a, B_t \in \mathrm{r}_b I)\\
&= \mathbb{P}(B_t \in \mathrm{r}_b I) - \mathbb{P}(B_\tau = a, B_t \in \mathrm{r}_a\mathrm{r}_b I)\\
&= \mathbb{P}(B_t \in \mathrm{r}_b I) - \mathbb{P}(B_t \in \mathrm{r}_a\mathrm{r}_b I) + \mathbb{P}(B_\tau = b, B_t \in \mathrm{r}_a\mathrm{r}_b I)
\end{aligned}$$

and, finally,

$$\mathbb{P}(B_\tau = b, \tau \leqslant t, B_t \in I) = \mathbb{P}(B_t \in \mathrm{r}_b I) - \mathbb{P}(B_t \in \mathrm{r}_a\mathrm{r}_b I) + \mathbb{P}(B_t \in \mathrm{r}_b\mathrm{r}_a\mathrm{r}_b I) - + \cdots$$

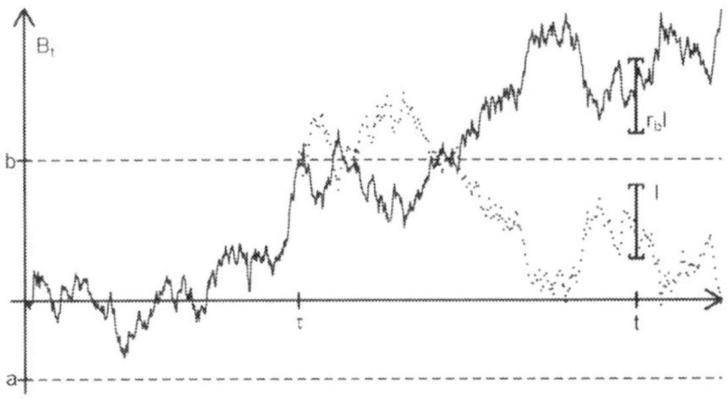

Fig. 6.4. A [reflected] Brownian path visiting the [reflected] interval at time t.

Since the quantities a and b play exchangeable roles in all calculations up to this point, we get

$$\begin{aligned}
&\mathbb{P}(m_t > a, M_t < b, B_t \in I)\\
&\quad= \mathbb{P}(B_t \in I) - \mathbb{P}(B_t \in \mathrm{r}_b I) + \mathbb{P}(B_t \in \mathrm{r}_a\mathrm{r}_b I) - \mathbb{P}(B_t \in \mathrm{r}_b\mathrm{r}_a\mathrm{r}_b I) + - \cdots\\
&\qquad\qquad - \mathbb{P}(B_t \in \mathrm{r}_a I) + \mathbb{P}(B_t \in \mathrm{r}_b\mathrm{r}_a I) - \mathbb{P}(B_t \in \mathrm{r}_a\mathrm{r}_b\mathrm{r}_a I) + - \cdots\\
&\quad= \mathbb{P}(B_t \in I) - \mathbb{P}(\mathrm{r}_a B_t \in \mathrm{r}_a\mathrm{r}_b I) + \mathbb{P}(B_t \in \mathrm{r}_a\mathrm{r}_b I) - \mathbb{P}(\mathrm{r}_a B_t \in (\mathrm{r}_a\mathrm{r}_b)^2 I) + - \cdots\\
&\qquad\qquad - \mathbb{P}(\mathrm{r}_a B_t \in I) + \mathbb{P}(B_t \in \mathrm{r}_b\mathrm{r}_a I) - \mathbb{P}(\mathrm{r}_a B_t \in \mathrm{r}_b\mathrm{r}_a I) + - \cdots
\end{aligned}$$

In the last step we used that $\mathrm{r}_a\mathrm{r}_a = \mathrm{id}$. All probabilities in these identities are probabilities of mutually disjoint events, i. e. the series converges absolutely (with a sum < 1); therefore we may rearrange the terms in the series. Since $\mathrm{r}_a I = 2a - I$ and $\mathrm{r}_b\mathrm{r}_a I = 2b - (2a - I) = 2(b-a) + I$, we get for all $n \geqslant 0$

$$(\mathrm{r}_b\mathrm{r}_a)^n I = 2n(b-a) + I \quad\text{and}\quad (\mathrm{r}_a\mathrm{r}_b)^n I = 2n(a-b) + I = 2(-n)(b-a) + I.$$

Therefore,

$$\begin{aligned}
&\mathbb{P}(m_t > a, M_t < b, B_t \in I)\\
&\quad= \sum_{n=-\infty}^{\infty} \mathbb{P}(B_t \in 2n(a-b) + I) - \sum_{n=-\infty}^{\infty} \mathbb{P}(2a - B_t \in 2n(a-b) + I)\\
&\quad= \sum_{n=-\infty}^{\infty} \int_I \left(e^{-\frac{(x-2n(a-b))^2}{2t}} - e^{-\frac{(2a-x-2n(a-b))^2}{2t}} \right) \frac{dx}{\sqrt{2\pi t}}
\end{aligned}$$

which is the same as (6.19). □

6.6 An arc-sine law

There are quite a few functionals of a Brownian path which follow an arc-sine distribution. Most of them are more or less connected with the zeros of $(B_t)_{t\geqslant 0}$. Consider, for example, the largest zero of B_s in the interval $[0, t]$:

$$\xi_t := \sup\{s \leqslant t \;:\; B_s = 0\}.$$

Obviously, ξ_t is not a stopping time since $\{\xi_t \leqslant r\}$, $r < t$, cannot be contained in $\mathscr{F}_r^B$; otherwise we could forecast on the basis of $[0, r] \ni s \mapsto B_s$ whether $B_s = 0$ for some future time $s \in (r, t)$ or not. Nevertheless we can determine the probability distribution of ξ_t.

Ex. 6.11 **6.19 Theorem.** *Let $(B_t)_{t\geqslant 0}$ be a* BM[1] *and write ξ_t for the largest zero of B_s in the interval*
Ex. 6.12 $[0, t]$. *Then*
Ex. 6.13

$$\mathbb{P}(\xi_t < s) = \frac{2}{\pi} \arcsin \sqrt{\frac{s}{t}} \quad \textit{for all } 0 \leqslant s \leqslant t. \tag{6.20}$$

Proof. Set $h(s) := \mathbb{P}(\xi_t < s)$. By the Markov property we get

$$\begin{aligned} h(s) &= \mathbb{P}(B_u \neq 0 \quad \text{for all } u \in [s, t]) \\ &= \int \mathbb{P}^{B_s(\omega)}(B_{u-s} \neq 0 \quad \text{for all } u \in [s, t]) \, \mathbb{P}(d\omega) \\ &= \int \mathbb{P}^b(B_{u-s} \neq 0 \quad \text{for all } u \in [s, t]) \, \mathbb{P}(B_s \in db) \\ &= \int \mathbb{P}^0(B_v \neq -b \quad \text{for all } v \in [0, t-s]) \, \mathbb{P}(B_s \in db) \\ &= \int \mathbb{P}^0(\tau_{-b} \geqslant t - s) \, \mathbb{P}(B_s \in db), \end{aligned}$$

where τ_{-b} is the first passage time for $-b$. Using

$$\mathbb{P}(B_s \in db) = \mathbb{P}(B_s \in -db) \quad \text{and} \quad \mathbb{P}(\tau_b \geqslant t - s) = \sqrt{\frac{2}{\pi(t-s)}} \int_0^b e^{-x^2/2(t-s)} \, dx,$$

cf. Theorem 6.9, we get

$$\begin{aligned} h(s) &= 2 \int_0^\infty \sqrt{\frac{2}{\pi(t-s)}} \int_0^b e^{-\frac{x^2}{2(t-s)}} \, dx \, \frac{1}{\sqrt{2\pi s}} \, e^{-\frac{b^2}{2s}} \, db \\ &= \frac{2}{\pi} \int_0^\infty \int_0^{\beta\sqrt{\frac{s}{t-s}}} e^{-\frac{\xi^2}{2}} \, d\xi \, e^{-\frac{\beta^2}{2}} \, d\beta. \end{aligned}$$

If we differentiate the last expression with respect to s we find

$$h'(s) = \frac{1}{\pi} \int_0^\infty e^{-\frac{\beta^2}{2}} e^{-\frac{\beta^2}{2}\frac{s}{t-s}} \frac{t}{t-s} \frac{\beta}{\sqrt{s(t-s)}} \, d\beta = \frac{1}{\pi} \frac{1}{\sqrt{s(t-s)}}, \; 0 < s < t.$$

Since $\frac{d}{ds} \arcsin \sqrt{\frac{s}{t}} = \frac{1}{2} \frac{1}{\sqrt{s(t-s)}} = \frac{\pi}{2} h'(s)$ and $0 = \arcsin 0 = h(0)$, we are done. □

6.7 Some measurability issues

Let $(B_t)_{t\geqslant 0}$ be a BM[1]. Recall from Definition 5.1 that a filtration $(\mathscr{G}_t)_{t\geqslant 0}$ is admissible if

$$\mathscr{F}_t^B \subset \mathscr{G}_t \quad \text{and} \quad B_t - B_s \perp\!\!\!\perp \mathscr{G}_s \quad \text{for all } s < t.$$

Typically, one considers the following types of enlargements:

$$\mathscr{G}_t := \sigma(\mathscr{F}_t^B, X) \quad \text{where } X \perp\!\!\!\perp \mathscr{F}_\infty^B \text{ is a further random variable,}$$

$$\mathscr{G}_t := \mathscr{F}_{t+}^B := \bigcap_{u>t} \mathscr{F}_u^B,$$

$$\mathscr{G}_t := \overline{\mathscr{F}}_t^B := \sigma(\mathscr{F}_t^B, \mathscr{N})$$

where $\mathscr{N} = \{M \subset \Omega : \exists N \in \mathscr{A}, M \subset N, \mathbb{P}(N) = 0\}$ is the system of all subsets of measurable $\mathbb{P}$ null sets; $\overline{\mathscr{F}}_t^B$ is the *completion* of $\mathscr{F}_t^B$.

6.20 Lemma. *Let $(B_t)_{t\geqslant 0}$ be a BM^d. Then $\mathscr{F}_{t+}^B$ and $\overline{\mathscr{F}}_t^B$ are admissible in the sense of Definition 5.1.* Ex. 5.1

Proof. **1°** $\mathscr{F}_{t+}^B$ is admissible: Let $0 \leqslant t \leqslant u$, $F \in \mathscr{F}_{t+}$, and $f \in \mathcal{C}_b(\mathbb{R}^d)$. Since $F \in \mathscr{F}_{t+\epsilon}$ for all $\epsilon > 0$, we find with (5.1)

$$\mathbb{E}\left[\mathbb{1}_F \cdot f(B(u+\epsilon) - B(t+\epsilon))\right] = \mathbb{E}\left[\mathbb{1}_F\right] \cdot \mathbb{E}\left[f(B(u+\epsilon) - B(t+\epsilon))\right].$$

Letting $\epsilon \to 0$ proves 5.1 b); condition 5.1 a) is trivially satisfied.

2° $\overline{\mathscr{F}}_t^B$ is admissible: Let $0 \leqslant t \leqslant u$. By the very definition of the completion we can find for every $\tilde{F} \in \overline{\mathscr{F}}_t^B$ some $F \in \mathscr{F}_t^B$ and $N \in \mathscr{N}$ such that the symmetric difference $(\tilde{F} \setminus F) \cup (F \setminus \tilde{F})$ is in $\mathscr{N}$. Therefore,

$$\begin{aligned}\mathbb{E}(\mathbb{1}_{\tilde{F}} \cdot (B_u - B_t)) = \mathbb{E}(\mathbb{1}_F \cdot (B_u - B_t)) &\overset{(5.1)}{=} \mathbb{P}(F)\,\mathbb{E}(B_u - B_t) \\ &= \mathbb{P}(\tilde{F})\,\mathbb{E}(B_u - B_t),\end{aligned}$$

which proves 5.1 b); condition 5.1 a) is clear. □

6.21 Theorem. *Let $(B_t)_{t\geqslant 0}$ be a BM^d. Then $(\overline{\mathscr{F}}_t^B)_{t\geqslant 0}$ is a right-continuous filtration, i. e. $\overline{\mathscr{F}}_{t+}^B = \overline{\mathscr{F}}_t^B$ holds for all $t \geqslant 0$.*

Proof. Let $F \in \overline{\mathscr{F}}_{t+}^B$. It is enough to show that $\mathbb{1}_F = \mathbb{E}\left[\mathbb{1}_F \mid \overline{\mathscr{F}}_t^B\right]$ almost surely. If this is true, $\mathbb{1}_F$ is, up to a set of measure zero, $\overline{\mathscr{F}}_t^B$ measurable. Since $\mathscr{N} \subset \overline{\mathscr{F}}_t^B \subset \overline{\mathscr{F}}_{t+}^B$, we get $\overline{\mathscr{F}}_{t+}^B \subset \overline{\mathscr{F}}_t^B$.

Fix $t < u$ and pick $0 \leqslant s_1 < \cdots < s_m \leqslant t < u = t_0 \leqslant t_1 < \cdots < t_n$ and $\eta_1, \ldots, \eta_m, \xi_1, \ldots, \xi_n \in \mathbb{R}^d$, $m, n \geqslant 1$. Then we have

$$\mathbb{E}\left[e^{i\sum_{j=1}^m \langle \eta_j, B(s_j)\rangle} \mid \overline{\mathscr{F}}_t^B\right] = e^{i\sum_{j=1}^m \langle \eta_j, B(s_j)\rangle} = \mathbb{E}\left[e^{i\sum_{j=1}^m \langle \eta_j, B(s_j)\rangle} \mid \overline{\mathscr{F}}_{t+}^B\right]. \tag{6.21}$$

As in the proof of Theorem 6.1 we get

$$\sum_{j=1}^{n} \langle \xi_j, B(t_j) \rangle = \sum_{k=1}^{n} \langle \zeta_k, B(t_k) - B(t_{k-1}) \rangle + \langle \zeta_1, B(u) \rangle, \quad \zeta_k = \xi_k + \cdots + \xi_n.$$

Therefore,

$$\begin{aligned}
\mathbb{E}\left[e^{i \sum_{j=1}^{n} \langle \xi_j, B(t_j) \rangle} \,\middle|\, \overline{\mathscr{F}}_u^B \right] &= \mathbb{E}\left[e^{i \sum_{k=1}^{n} \langle \zeta_k, B(t_k) - B(t_{k-1}) \rangle} \,\middle|\, \overline{\mathscr{F}}_u^B \right] e^{i \langle \zeta_1, B(u) \rangle} \\
&\overset{6.20}{\underset{(B1)}{=}} \prod_{k=1}^{n} \mathbb{E}\left[e^{i \langle \zeta_k, B(t_k) - B(t_{k-1}) \rangle} \right] e^{i \langle \zeta_1, B(u) \rangle} \\
&\overset{(2.15)}{=} \prod_{k=1}^{n} e^{-\frac{1}{2}(t_k - t_{k-1})|\zeta_k|^2} e^{i \langle \zeta_1, B(u) \rangle}.
\end{aligned}$$

Exactly the same calculation with u replaced by t and $t_0 = t$ tells us

$$\mathbb{E}\left[e^{i \sum_{j=1}^{n} \langle \xi_j, B(t_j) \rangle} \,\middle|\, \overline{\mathscr{F}}_t^B \right] = \prod_{k=1}^{n} e^{-\frac{1}{2}(t_k - t_{k-1})|\zeta_k|^2} e^{i \langle \zeta_1, B(t) \rangle}.$$

Now we can use the backwards martingale convergence Theorem A.8 and let $u \downarrow t$ (along a countable sequence) in the first calculation. This gives

$$\begin{aligned}
\mathbb{E}\left[e^{i \sum_{j=1}^{n} \langle \xi_j, B(t_j) \rangle} \,\middle|\, \overline{\mathscr{F}}_{t+}^B \right] &= \lim_{u \downarrow t} \mathbb{E}\left[e^{i \sum_{j=1}^{n} \langle \xi_j, B(t_j) \rangle} \,\middle|\, \overline{\mathscr{F}}_u^B \right] \\
&= \lim_{u \downarrow t} \prod_{k=1}^{n} e^{-\frac{1}{2}(t_k - t_{k-1})|\zeta_k|^2} e^{i \langle \zeta_1, B(u) \rangle} \\
&= \prod_{k=1}^{n} e^{-\frac{1}{2}(t_k - t_{k-1})|\zeta_k|^2} e^{i \langle \zeta_1, B(t) \rangle} \\
&= \mathbb{E}\left[e^{i \sum_{j=1}^{n} \langle \xi_j, B(t_j) \rangle} \,\middle|\, \overline{\mathscr{F}}_t^B \right].
\end{aligned} \tag{6.22}$$

Given $0 \leqslant u_1 < u_2 < \cdots < u_N$, $N \geqslant 1$, and $t \geqslant 0$, then t splits the times u_j into two subfamilies, $s_1 < \cdots < s_m \leqslant t < t_1 < \cdots < t_n$, where $m + n = N$ and $\{u_1, \ldots, u_N\} = \{s_1, \ldots, s_m\} \cup \{t_1, \ldots, t_n\}$. If we combine the equalities (6.21) and (6.22), we see for $\Gamma := (B(u_1), \ldots, B(u_N)) \in \mathbb{R}^{Nd}$ and all $\theta \in \mathbb{R}^{Nd}$

$$\mathbb{E}\left[e^{i\langle \theta, \Gamma \rangle} \,\middle|\, \overline{\mathscr{F}}_{t+}^B \right] = \mathbb{E}\left[e^{i\langle \theta, \Gamma \rangle} \,\middle|\, \overline{\mathscr{F}}_t^B \right].$$

Using the definition and then the L^2 symmetry of the conditional expectation we get for all $F \in \overline{\mathscr{F}}_{t+}^B$

$$\int \mathbb{1}_F \cdot e^{i\langle \theta, \Gamma \rangle} \, d\mathbb{P} = \int \mathbb{1}_F \cdot \mathbb{E}\left[e^{i\langle \theta, \Gamma \rangle} \,\middle|\, \overline{\mathscr{F}}_t^B \right] d\mathbb{P} = \int \mathbb{E}\left[\mathbb{1}_F \,\middle|\, \overline{\mathscr{F}}_t^B \right] \cdot e^{i\langle \theta, \Gamma \rangle} \, d\mathbb{P}.$$

This shows that the image measures $(\mathbb{1}_F \mathbb{P}) \circ \Gamma^{-1}$ and $\left(\mathbb{E}\left[\mathbb{1}_F \,\middle|\, \overline{\mathscr{F}}_t^B \right] \mathbb{P} \right) \circ \Gamma^{-1}$ have the same characteristic functions, i. e.

$$\int_A \mathbb{1}_F \, d\mathbb{P} = \int_A \mathbb{E}\left[\mathbb{1}_F \,\middle|\, \overline{\mathscr{F}}_t^B \right] d\mathbb{P} \tag{6.23}$$

holds for all $A \in \sigma(\Gamma) = \sigma(B(u_1), \dots, B(u_N))$ and all $0 \leqslant u_1 < \dots < u_N$ and $N \geqslant 1$. Since $\bigcup_{u_1<\dots<u_N, N\geqslant 1} \sigma(B(u_1), \dots, B(u_N))$ is a $\cap$-stable generator of $\mathscr{F}^B_\infty$, the uniqueness theorem for measures shows that (6.23) holds for all $A \in \mathscr{F}^B_\infty$ and, by completion of the measures, for all $A \in \overline{\mathscr{F}}^B_\infty$. Therefore, $\mathbb{1}_F = \mathbb{E}\left[\mathbb{1}_F \,\middle|\, \overline{\mathscr{F}}^B_t\right]$ almost surely. □

6.22 Corollary (Blumenthal's 0-1-law. Blumenthal 1957). *Let $(B(t))_{t\geqslant 0}$ be a BM^d. Then $\mathscr{F}^B_{0+}$ is trivial, i. e. if $F \in \mathscr{F}^B_{0+}$ then $\mathbb{P}(F) = 0$ or 1.*

Proof. By Theorem 6.21 we know that $\overline{\mathscr{F}}^B_0 = \overline{\mathscr{F}}^B_{0+} \supset \mathscr{F}^B_{0+}$. On the other hand we have $\overline{\mathscr{F}}^B_0 = \sigma(\mathscr{N}, \mathscr{F}^B_0) = \sigma(\mathscr{N})$ since $\mathscr{F}^B_0 \subset \sigma(\mathscr{N})$. This proves our claim.

Let us indicate the *classical proof* which does not rely on Theorem 6.21. Since $\mathscr{F}^B_{t+}$ is admissible, cf. Lemma 6.20, we find that $\mathscr{F}^B_{0+}$ is independent of $B_t - B_0 = B_t$ for all $t \geqslant 0$. This means that

$$\mathscr{F}^B_{0+} \perp\!\!\!\perp \bigcap_{t>0} \sigma\Bigg(\underbrace{\bigcup_{0\leqslant t_1<\dots<t_n\leqslant t, n\geqslant 1} \sigma(B_{t_1}, \dots, B_{t_n})}_{\cap\text{-stable generator}}\Bigg) = \mathscr{F}^B_{0+},$$

i. e. $\mathbb{P}(F) = \mathbb{P}(F \cap F) = \mathbb{P}(F) \cdot \mathbb{P}(F)$ so that $\mathbb{P}(F) = 0$ or 1. □

6.23 Remark (Usual conditions). Many authors assume that the underlying filtered probability space $(\Omega, \mathscr{A}, \mathbb{P}, \mathscr{F}_t, t \geqslant 0)$ satisfies the *usual conditions* or *usual hypotheses*, i. e.

a) $\mathscr{F}_0$ contains all subsets of $\mathbb{P}$ null sets of $\mathscr{A}$;

b) $(\mathscr{F}_t)_{t\geqslant 0}$ is right-continuous, i. e. $\mathscr{F}_t = \mathscr{F}_{t+}$ holds for all $t \geqslant 0$.

Theorem 6.21 ensures that a Brownian motion together with the completion of its canonical filtration satisfies the usual conditions.

We close this section with a result that exhibits the interaction of a Brownian motion with the underlying measurability structure.

6.24 Lemma (Progressive measurability). *Let $(B_t)_{t\geqslant 0}$ be a BM^d with an admissible filtration $(\mathscr{F}_t)_{t\geqslant 0}$. Then $(B_t)_{t\geqslant 0}$ is progressively measurable i. e.*

$$B(\cdot,\cdot) : ([0,t] \times \Omega, \mathscr{B}[0,t] \otimes \mathscr{F}_t) \to (\mathbb{R}^d, \mathscr{B}(\mathbb{R}^d))$$

is, for every $t \geqslant 0$, $\mathscr{B}[0,t] \otimes \mathscr{F}_t / \mathscr{B}(\mathbb{R}^d)$ measurable.

Proof. Set

$$B^j(s,\omega) := \begin{cases} B\left(\frac{k}{2^j}t, \omega\right), & \frac{k}{2^j}t < s \leqslant \frac{k+1}{2^j}t,\ k = 0, 1, \dots, 2^j - 1, \\ B(0,\omega), & s = 0. \end{cases}$$

Because of the (left-)continuity of the sample paths, we get

$$B(s,\omega) = \lim_{j\to\infty} B^j(s,\omega) \quad \text{for all } s \in [0,t],\ \omega \in \Omega.$$

For all $A \in \mathscr{B}(\mathbb{R}^d)$ we have that

$$\begin{aligned}\{(s,\omega) \in [0,t]\times\Omega \, : \, B^j(s,\omega) \in A\} \\ = \{0\} \times \{B(0,\cdot) \in A\} \cup \bigcup_{k=0}^{2^j-1} \left(\tfrac{k}{2^j}t, \tfrac{k+1}{2^j}t\right] \times \left\{B(\tfrac{k}{2^j}t,\cdot) \in A\right\}.\end{aligned}$$

This set is contained in $\mathscr{B}[0,t] \otimes \mathscr{F}_t$. Therefore, the processes $(B^j(t))_{t\geqslant 0}$, $j \geqslant 0$, and the limit $(B(t))_{t\geqslant 0}$ are progressively measurable. □

Ex. 6.4 **6.25 Corollary.** *Let $(B_t)_{t\geqslant 0}$ be a BM^d and τ be an $\mathscr{F}_t$ stopping time. Then $B_\tau \mathbb{1}_{\{\tau<\infty\}}$ is $\mathscr{F}_\tau$ measurable.*

Proof. Consider the maps

$$(\{\tau \leqslant t\}, \mathscr{F}_t) \xrightarrow{\omega\mapsto(\tau(\omega),\omega)} ([0,t]\times\Omega, \mathscr{B}[0,t]\otimes\mathscr{F}_t) \xrightarrow{(s,\omega)\mapsto B(s,\omega)} (\mathbb{R}^d, \mathscr{B}(\mathbb{R}^d)).$$

The second map is measurable, since $(B_t)_{t\geqslant 0}$ is progressively measurable. For the first map we observe that for all $r \leqslant t$ and $\Gamma \in \mathscr{F}_t$

$$\underbrace{\{\tau \leqslant r\}}_{\in \mathscr{F}_r \subset \mathscr{F}_t} \cap \Gamma \in \mathscr{F}_t.$$

Thus, $\{B(\tau) \in A\} \cap \{\tau \leqslant t\} = \{\omega \in \{\tau \leqslant t\} \, : \, B(\tau(\omega),\omega) \in A\} \in \mathscr{F}_t$, which proves that $\{B(\tau) \in A\} \cap \{\tau < \infty\} \in \mathscr{F}_\tau$, i.e. $B(\tau)\mathbb{1}_{\{\tau<\infty\}}$ is $\mathscr{F}_\tau$ measurable. □

6.26 Further reading. The (strong) Markov property is a central theme in most books on Brownian motion. The account in [31] excels in its clarity, elegance and precision. Shift operators and the Markov property on the path space are explored in [119]. For the general theory of Markov processes we refer to [72] and the classic book [18].

[18] Blumenthal, Getoor: *Markov Processes and Potential Theory.*
[31] Chung: *Lectures from Markov Processes to Brownian Motion*
[72] Ethier, Kurtz: *Markov Processes: Characterization and Convergence.*
[119] Karatzas, Shreve: *Brownian Motion and Stochastic Calculus.*

Problems

1. Let $(B_t, \mathscr{F}_t)_{t\geqslant 0}$ be a BM^1.
   (a) Show that $X_t := |B_t|$, $t \geqslant 0$, is also a Markov process for the filtration $(\mathscr{F}_t)_{t\geqslant 0}$, i.e. for all $s, t \geqslant 0$ and $u \in \mathcal{B}_b(\mathbb{R})$

   $$\mathbb{E}\left(u(X_{t+s}) \mid \mathscr{F}_s\right) = \mathbb{E}\left(u(X_{t+s}) \mid X_s\right).$$

   (b) Find in a) the function $g_{u,s,t+s}(x)$ with $\mathbb{E}(u(X_{t+s}) \mid X_s) = g_{u,s,t+s}(X_s)$.

(c) Solve a) and b) for the process $Y_t := \sup_{s\leqslant t} B_s - B_t$, $t \geqslant 0$.

(d) Compare the transition functions from b) and c) for X and Y, respectively.

2. Let $(B_t, \mathscr{F}_t)_{t\geqslant 0}$ be a BM^1 and set $M_t := \sup_{s\leqslant t} B_s$ and $I_t = \int_0^t B_s\,ds$.
   (a) Show that the two-dimensional process $(B_t, M_t)_{t\geqslant 0}$ is a Markov process for $(\mathscr{F}_t)_{t\geqslant 0}$.

   (b) Show that the two-dimensional process $(B_t, I_t)_{t\geqslant 0}$ is a Markov process for $(\mathscr{F}_t)_{t\geqslant 0}$.

   (c) Are $(M_t)_{t\geqslant 0}$ and $(I_t)_{t\geqslant 0}$ Markov processes for $(\mathscr{F}_t)_{t\geqslant 0}$?

3. Let $(B(t))_{t\geqslant 0}$ be a BM^d and let Z be a bounded $\mathscr{F}_\infty^B$ measurable random variable. Then $x \mapsto \mathbb{E}^x Z$ is in $\mathcal{B}_b(\mathbb{R}^d)$.
   Hint: $\mathscr{F}_\infty^B$ is generated by sets of the form $G := \bigcap_{j=1}^n \{B(t_j) \in A_j\}$. Then, $\mathbb{E}^x \mathbb{1}_G \in \mathcal{B}_b(\mathbb{R}^d)$. Consider the family $\Sigma := \{A : \mathbb{E}^x \mathbb{1}_A \in \mathscr{B}_b(\mathbb{R}^d)\}$; this is a Dynkin system containing all sets G, hence $\mathscr{F}_\infty^B$.

4. Let $(B_t, \mathscr{F}_t)_{t\geqslant 0}$ be a BM^1 and τ a stopping time for the filtration $(\mathscr{F}_t)_{t\geqslant 0}$. Show that B_τ is measurable with respect to $\mathscr{F}_{\tau+}$.
   Hint: Use in Theorem 6.6 $u(x) = u_n(x) \to x$, $u \in \mathcal{C}_b(\mathbb{R}^d)$; see also Corollary 6.25.

5. Let $(B_t)_{t\geqslant 0}$ be a BM^1. Show that $\mathbb{P}(\sup_{s\leqslant t} |B_s| \geqslant x) \leqslant 2\mathbb{P}(|B_t| \geqslant x)$, $x, t \geqslant 0$.

6. Let $(B_t)_{t\geqslant 0}$ be a BM^1 and denote by $\tau_b = \inf\{s \geqslant 0 : B_s = b\}$ the first time when B_t reaches $b \in \mathbb{R}$. Show that
   (a) $\tau_b \sim \tau_{-b}$;
   (b) $\tau_{cb} \sim c^2\tau_b$, $c \in \mathbb{R}$;
   (c) if $0 < a < b$, then $\tau_b - \tau_a \perp\!\!\!\perp \{\tau_\alpha : \alpha \in [0, a]\}$.

7. Let $\tau = \mathring{\tau}_{(a,b)^c}$ be the first exit time of a Brownian motion from the interval (a, b).
   (a) Find $\mathbb{E}^x e^{-\lambda\tau}$ for all $x \in (a, b)$ and $\lambda > 0$.

   (b) Find $\mathbb{E}^x \left(e^{-\lambda\tau} \mathbb{1}_{\{B_\tau = a\}}\right)$ for all $x \in (a, b)$ and $\lambda > 0$.

8. Let $(B_t)_{t\geqslant 0}$ be a BM^1 and set $M_t := \sup_{s\leqslant t} B_s$. Find the distribution of (M_t, B_t).

9. Let $(B_t)_{t\geqslant 0}$ be a BM^1 and let τ_0 be the first hitting time of 0. Find the 'density' of $\mathbb{P}^x(B_t \in dz, \tau_0 > t)$, i. e. find the function $f_{t,x}(z)$ such that

$$\mathbb{P}^x(B_t \in A, \tau_0 > t) = \int_A f_{t,x}(z)\,dz \quad \text{for all } A \in \mathscr{B}(\mathbb{R}).$$

10. Let $K \subset \mathbb{R}^d$ be a compact set. Show that there is a decreasing sequence of continuous functions $\phi_n(x)$ such that $\mathbb{1}_K = \inf_n \phi_n$.
    Hint: Let $U \supset K$ be an open set and $\phi(x) := d(x, U^c)/(d(x, K) + d(x, U^c))$.

11. Let $(B_t)_{t\geqslant 0}$ be a BM^1. Find the distribution of $\tilde{\xi}_t := \inf\{s \geqslant t : B_s = 0\}$.

12. Let $(B_t)_{t\geqslant 0}$ be a BM^1 and set $M_t = \sup_{s\leqslant t} B_s$. Denote by ξ_t the largest zero of B_s before time t and by η_t the largest zero of $Y_s = M_s - B_s$ before time t. Show that $\xi_t \sim \eta_t$.

13. Let $(B_t)_{t\geqslant 0}$ be a BM^1 and $0 < u < v < w < \infty$. Find the following probabilities:
    (a) $\mathbb{P}(B_t = 0 \text{ for some } t \in (u, v))$;
    (b) $\mathbb{P}(B_t \neq 0\ \forall t \in (u, w) \mid B_t \neq 0\ \forall t \in (u, v))$;
    (c) $\mathbb{P}(B_t \neq 0\ \forall t \in (0, w) \mid B_t \neq 0\ \forall t \in (0, v))$.

7 Brownian motion and transition semigroups

The Markov property allows us to study Brownian motion through an analytic approach. For this we associate with a Brownian motion $(B_t)_{t\geqslant 0}$ two families of linear operators, the *(transition) semigroup*

$$P_t u(x) = \mathbb{E}^x u(B_t), \qquad u \in \mathcal{B}_b(\mathbb{R}^d),\ x \in \mathbb{R}^d,\ t \geqslant 0, \tag{7.1}$$

and the *resolvent*

$$U_\alpha u(x) = \mathbb{E}^x \left[\int_0^\infty u(B_t) e^{-\alpha t}\, dt \right], \qquad u \in \mathcal{B}_b(\mathbb{R}^d),\ x \in \mathbb{R}^d,\ \alpha > 0. \tag{7.2}$$

Throughout this chapter we assume that $(B_t)_{t\geqslant 0}$ is a d-dimensional Brownian motion which is adapted to the filtration $(\mathscr{F}_t)_{t\geqslant 0}$. We will assume that $(\mathscr{F}_t)_{t\geqslant 0}$ is right-continuous, i. e. $\mathscr{F}_t = \mathscr{F}_{t+}$ for all $t \geqslant 0$, cf. Theorem 6.21 for a sufficient condition. Many results of this chapter will be true, with only minor changes in the proofs, for general Markov processes.

7.1 The semigroup

A(n operator) semigroup $(P_t)_{t\geqslant 0}$ on a Banach space $(\mathfrak{B}, \|\cdot\|)$ is a family of linear operators $P_t : \mathfrak{B} \to \mathfrak{B}$, $t \geqslant 0$, which satisfies

$$P_t P_s = P_{t+s} \quad \text{and} \quad P_0 = \mathrm{id}. \tag{7.3}$$

In this chapter we will consider the following Banach spaces:

$\mathcal{B}_b(\mathbb{R}^d)$ the family of all bounded Borel measurable functions $f : \mathbb{R}^d \to \mathbb{R}$ equipped with the uniform norm $\|\cdot\|_\infty$;

$\mathcal{C}_\infty(\mathbb{R}^d)$ the family of all continuous functions $f : \mathbb{R}^d \to \mathbb{R}$ vanishing at infinity, i. e. $\lim_{|x|\to\infty} u(x) = 0$, equipped with the uniform norm $\|\cdot\|_\infty$. Ex. 7.1

Unless it is too ambiguous, we write $\mathcal{B}_b$ and $\mathcal{C}_\infty$ instead of $\mathcal{B}_b(\mathbb{R}^d)$ and $\mathcal{C}_\infty(\mathbb{R}^d)$.

7.1 Lemma. *Let $(B_t)_{t\geqslant 0}$ be a d-dimensional Brownian motion with filtration $(\mathscr{F}_t)_{t\geqslant 0}$. Then* (7.1) *defines a semigroup of operators on $\mathcal{B}_b(\mathbb{R}^d)$.*

Proof. It is obvious that P_t is a linear operator. Since $x \mapsto e^{-|x-y|^2/2t}$ is measurable, we see that

$$P_t u(x) = \int_{\mathbb{R}^d} u(y)\, \mathbb{P}^x(B_t \in dy) = \frac{1}{(2\pi t)^{d/2}} \int_{\mathbb{R}^d} u(y)\, e^{-|x-y|^2/2t}\, dy$$

is again a Borel measurable function. (For a general Markov process we must use (6.4c) with a Borel measurable $g_{u,t}(\cdot)$.) The semigroup property (7.3) is now a direct conse- !

quence of the Markov property (6.4): For $s, t \geqslant 0$ and $u \in \mathcal{B}_b(\mathbb{R}^d)$

$$\begin{aligned} P_{t+s}u(x) = \mathbb{E}^x u(B_{t+s}) &\overset{\text{tower}}{=} \mathbb{E}^x \left[\mathbb{E}^x \left(u(B_{t+s}) \mid \mathscr{F}_s \right) \right] \\ &\overset{(6.4)}{=} \mathbb{E}^x \left[\mathbb{E}^{B_s}(u(B_t)) \right] \\ &= \mathbb{E}^x \left[P_t u(B_s) \right] \\ &= P_s P_t u(x) \end{aligned}$$

□

For many properties of the semigroup the Banach space $\mathcal{B}_b(\mathbb{R}^d)$ is too big, and we have to consider the smaller space $\mathcal{C}_\infty(\mathbb{R}^d)$.

7.2 Lemma. *A d-dimensional Brownian motion $(B_t)_{t \geqslant 0}$ is uniformly stochastically continuous, i. e.*

$$\lim_{t \to 0} \sup_{x \in \mathbb{R}^d} \mathbb{P}^x(|B_t - x| > \delta) = 0 \quad \textit{for all } \delta > 0. \tag{7.4}$$

Proof. This follows immediately from the translation invariance of a Brownian motion and Chebyshev's inequality

$$\mathbb{P}^x(|B_t - x| > \delta) = \mathbb{P}(|B_t| > \delta) \leqslant \frac{\mathbb{E}\,|B_t|^2}{\delta^2} = \frac{td}{\delta^2}.$$

□

After this preparation we can discuss the properties of the semigroup $(P_t)_{t \geqslant 0}$.

Ex. 7.2 **7.3 Proposition.** *Let $(P_t)_{t \geqslant 0}$ be the transition semigroup of a d-dimensional Brownian*
Ex. 7.3 *motion $(B_t)_{t \geqslant 0}$ and $\mathcal{C}_\infty = \mathcal{C}_\infty(\mathbb{R}^d)$, $\mathcal{C}_b = \mathcal{C}_b(\mathbb{R}^d)$ and $\mathcal{B}_b = \mathcal{B}_b(\mathbb{R}^d)$. For all $t \geqslant 0$*
Ex. 7.4 a) P_t is conservative*: $P_t 1 = 1$.*
Ex. 7.5 b) P_t *is a* contraction *on $\mathcal{B}_b$: $\|P_t u\|_\infty \leqslant \|u\|_\infty$ for all $u \in \mathcal{B}_b$.*
c) P_t is positivity preserving*: For $u \in \mathcal{B}_b$ we have $u \geqslant 0 \Rightarrow P_t u \geqslant 0$;*
d) P_t is sub-Markovian*: For $u \in \mathcal{B}_b$ we have $0 \leqslant u \leqslant 1 \Rightarrow 0 \leqslant P_t u \leqslant 1$;*
e) P_t *has the* Feller property*: If $u \in \mathcal{C}_\infty$ then $P_t u \in \mathcal{C}_\infty$;*
f) P_t is strongly continuous on $\mathcal{C}_\infty$*: $\lim_{t \to 0} \|P_t u - u\|_\infty = 0$ for all $u \in \mathcal{C}_\infty$;*
g) P_t *has the* strong Feller property*: If $u \in \mathcal{B}_b$, then $P_t u \in \mathcal{C}_b$.*

7.4 Remark. A semigroup of operators $(P_t)_{t \geqslant 0}$ on $\mathcal{B}_b(\mathbb{R}^d)$ which satisfies
a)–d) is called a *Markov semigroup*;
b)–d) is called a *sub-Markov semigroup*;
b)–f) is called a *Feller semigroup*;
b)–d), g) is called a *strong Feller semigroup*.

Proof of Proposition 7.3. a) We have $P_t 1(x) = \mathbb{E}^x \mathbb{1}_{\mathbb{R}^d}(B_t) = \mathbb{P}^x(B_t \in \mathbb{R}^d) = 1$ for all $x \in \mathbb{R}^d$.

b) We have $|P_t u(x)| = |\,\mathbb{E}^x u(B_t)| \leqslant \int |u(B_t)|\, d\mathbb{P}^x \leqslant \|u\|_\infty$ for all $x \in \mathbb{R}^d$.

c) Since $u \geqslant 0$ we see that $P_t u(x) = \mathbb{E}^x u(B_t) \geqslant 0$.

d) Positivity follows from c). The proof of b) shows that $P_t u \leqslant 1$ if $u \leqslant 1$.

e) Note that $P_t u(x) = \mathbb{E}^x u(B_t) = \mathbb{E}\, u(B_t + x)$. Since $|u(B_t + x)| \leqslant \|u\|_\infty$ is integrable, the dominated convergence theorem, can be used to show

$$\lim_{x\to y} \mathbb{E}\, u(B_t + x) = \mathbb{E}\, u(B_t + y) \quad \text{and} \quad \lim_{|x|\to\infty} \mathbb{E}\, u(B_t + x) = 0.$$

f) Fix $\epsilon > 0$. Since $u \in \mathcal{C}_\infty(\mathbb{R}^d)$ is uniformly continuous, there is some $\delta > 0$ such that

$$|u(x) - u(y)| < \epsilon \quad \text{for all } |x - y| < \delta.$$

Thus,

$$\begin{aligned}
\|P_t u - u\|_\infty &= \sup_{x\in\mathbb{R}^d} \left|\mathbb{E}^x u(B_t) - u(x)\right| \\
&\leqslant \sup_{x\in\mathbb{R}^d} \mathbb{E}^x \left|u(B_t) - u(x)\right| \\
&\leqslant \sup_{x\in\mathbb{R}^d} \left(\int_{|B_t - x|<\delta} \left|u(B_t) - u(x)\right| d\mathbb{P}^x + \int_{|B_t - x|\geqslant\delta} \left|u(B_t) - u(x)\right| d\mathbb{P}^x \right) \\
&\leqslant \epsilon \sup_{x\in\mathbb{R}^d} \mathbb{P}^x(|B_t - x| < \delta) + 2\,\|u\|_\infty \sup_{x\in\mathbb{R}^d} \mathbb{P}^x(|B_t - x| \geqslant \delta) \\
&\leqslant \epsilon + 2\,\|u\|_\infty\, \mathbb{P}(|B_t| \geqslant \delta).
\end{aligned}$$

Since $(B_t)_{t\geqslant 0}$ is stochastically continuous, we get $\overline{\lim}_{t\to 0} \|P_t u - u\|_\infty \leqslant \epsilon$, cf. (7.4), and the claim follows as $\epsilon \to 0$.

g) Since continuity is a local property, it is enough to consider $z \to x$ for all points $z, x \in \mathbb{B}(0, R)$ and any $R > 0$. Let $u \in \mathcal{B}_b(\mathbb{R}^d)$. Then $P_t u$ is given by

$$P_t u(z) = p_t \star u(z) = \frac{1}{(2\pi t)^{d/2}} \int u(y)\, e^{-|z-y|^2/2t}\, dy.$$

If $|y| \geqslant 2R$ we get $|z - y|^2 \geqslant (|y| - |z|)^2 \geqslant \frac{1}{4}|y|^2$ which shows that the integrand is bounded by

$$|u(y)|\, e^{-|z-y|^2/2t} \leqslant \|u\|_\infty \left(\mathbb{1}_{\mathbb{B}(0,2R)}(y) + e^{-|y|^2/8t}\, \mathbb{1}_{\mathbb{B}^c(0,2R)}(y) \right)$$

and the right-hand side is integrable. Therefore we can use dominated convergence to get $\lim_{z\to x} P_t u(z) = P_t u(x)$. This proves that $P_t u \in \mathcal{C}_b(\mathbb{R}^d)$. □

The semigroup notation offers a simple way to express the finite dimensional distributions of a Markov process. Let $(B_t)_{t\geqslant 0}$ be a d-dimensional Brownian motion, $s < t$ and $f, g \in \mathcal{B}_b(\mathbb{R}^d)$. Then

$$\begin{aligned}
\mathbb{E}^x \left[f(B_s) g(B_t)\right] &= \mathbb{E}^x \left[f(B_s)\, \mathbb{E}^{B_s} g(B_{t-s})\right] \\
&= \mathbb{E}^x \left[f(B_s) P_{t-s} g(B_s)\right] = P_s\left[f P_{t-s} g\right](x).
\end{aligned}$$

If we iterate this, we get the following counterpart of Theorem 6.4.

7.5 Theorem. *Let* $(B_t)_{t\geqslant 0}$ *be a* BM^d, $t_0 = 0 < t_1 < \cdots < t_n$, $C_1, \dots, C_n \in \mathscr{B}(\mathbb{R}^d)$ *and* $x \in \mathbb{R}^d$. *If* $(P_t)_{t\geqslant 0}$ *is the transition semigroup of* $(B_t)_{t\geqslant 0}$, *then*

$$\mathbb{P}^x(B_{t_1} \in C_1, \dots, B_{t_n} \in C_n) = P_{t_1}\big[\mathbb{1}_{C_1} P_{t_2-t_1}\big[\mathbb{1}_{C_2} \cdots P_{t_n-t_{n-1}} \mathbb{1}_{C_n}\big]\dots\big](x).$$

Ex. 7.6 **7.6 Remark (From semigroups to processes).** So far we have considered semigroups given by a Brownian motion. It is an interesting question whether we can construct a Markov process starting from a semigroup of operators. If $(T_t)_{t\geqslant 0}$ is a Markov semigroup where $p_t(x, C) := T_t \mathbb{1}_C(x)$ is an *everywhere defined* kernel such that

$$\begin{aligned} C &\mapsto p_t(x, C) \quad \text{is a probability measure for all } x \in \mathbb{R}^d, \\ (t, x) &\mapsto p_t(x, C) \quad \text{is measurable for all } C \in \mathscr{B}(\mathbb{R}^d), \end{aligned}$$

this is indeed possible. We can use the formula from Theorem 7.5 to define for every $x \in \mathbb{R}^d$

$$p^x_{t_1,\dots,t_n}(C_1 \times \dots \times C_n) := T_{t_1}\big[\mathbb{1}_{C_1} T_{t_2-t_1}\big[\mathbb{1}_{C_2} \cdots T_{t_n-t_{n-1}} \mathbb{1}_{C_n}\big]\dots\big](x)$$

where $0 \leqslant t_1 < t_2 < \cdots < t_n$ and $C_1, \dots, C_n \in \mathscr{B}(\mathbb{R}^d)$. For fixed $x \in \mathbb{R}^d$, this de-
Ex. 7.7 fines a family of finite dimensional measures on $(\mathbb{R}^{d\cdot n}, \mathscr{B}^n(\mathbb{R}^d))$ which satisfies the Kolmogorov consistency conditions (4.6). The permutation property is obvious, while the projectivity condition follows from the semigroup property:

$$\begin{aligned} &p^x_{t_1,\dots,t_k,\dots,t_n}(C_1 \times \dots \times \mathbb{R}^d \times \dots \times C_n) \\ &= T_{t_1}\Big[\mathbb{1}_{C_1} T_{t_2-t_1}\big[\mathbb{1}_{C_2} \cdots \mathbb{1}_{C_{k-1}} \underbrace{T_{t_k-t_{k-1}} \mathbb{1}_{\mathbb{R}^d} T_{t_{k+1}-t_k}}_{= T_{t_k-t_{k-1}} T_{t_{k+1}-t_k} = T_{t_{k+1}-t_{k-1}}} \mathbb{1}_{C_{k+1}} \cdots T_{t_n-t_{n-1}} \mathbb{1}_{C_n}\big]\dots\Big](x) \\ &= T_{t_1}\Big[\mathbb{1}_{C_1} T_{t_2-t_1}\big[\mathbb{1}_{C_2} \cdots \mathbb{1}_{C_{k-1}} T_{t_{k+1}-t_{k-1}} \mathbb{1}_{C_{k+1}} \cdots T_{t_n-t_{n-1}} \mathbb{1}_{C_n}\big]\dots\Big](x) \\ &= p^x_{t_1,\dots,t_{k-1},t_{k+1},\dots,t_n}(C_1 \times \dots \times C_{k-1} \times C_{k+1} \times \dots \times C_n). \end{aligned}$$

Now we can use Kolmogorov's construction, cf. Theorem 4.8 and Corollary 4.9, to get measures $\mathbb{P}^x$ and a canonical process $(X_t)_{t\geqslant 0}$ such that $\mathbb{P}^x(X_0 = x) = 1$ for every $x \in \mathbb{R}^d$.

We still have to check that the process $X = (X_t)_{t\geqslant 0}$ is a Markov process. Denote by $\mathscr{F}_t := \sigma(X_s : s \leqslant t)$ the canonical filtration of X. We will show that

$$p_{t-s}(X_s, C) = \mathbb{P}^x(X_t \in C \mid \mathscr{F}_s) \quad \text{for all } s < t \text{ and } C \in \mathscr{B}(\mathbb{R}^d). \tag{7.5}$$

A further application of the tower property gives (6.4a). Since $\mathscr{F}_s$ is generated by an $\cap$-stable generator $\mathscr{G}$ consisting of sets of the form

$$G = \bigcap_{j=1}^m \{X_{s_j} \in C_j\}, \quad m \geqslant 1,\ 0 \leqslant s_1 < \cdots < s_m = s,\ C_1, \dots, C_m \in \mathscr{B}(\mathbb{R}^d),$$

it is enough to show that for every $G \in \mathscr{G}$

$$\int_G p_{t-s}(X_s, C)\, d\mathbb{P}^x = \int_G \mathbb{1}_C(X_t)\, d\mathbb{P}^x.$$

Because of the form of G and since $s = s_m$, this is the same as

$$\int \prod_{j=1}^{m} \mathbb{1}_{C_j}(X_{s_j}) p_{t-s_m}(X_{s_m}, C)\, d\mathbb{P}^x = \int \prod_{j=1}^{m} \mathbb{1}_{C_j}(X_{s_j}) \mathbb{1}_C(X_t)\, d\mathbb{P}^x.$$

By the definition of the finite dimensional distributions we see

$$\begin{aligned}
&\int \prod_{j=1}^{m} \mathbb{1}_{C_j}(X_{s_j}) p_{t-s_m}(X_{s_m}, C)\, d\mathbb{P}^x \\
&\quad = T_{s_1}\Big[\mathbb{1}_{C_1} T_{s_2-s_1}\Big[\mathbb{1}_{C_2} \cdots \mathbb{1}_{C_{m-1}} T_{s_m-s_{m-1}}[\mathbb{1}_{C_m} p_{t-s_m}(\cdot, C)] \cdots \Big]\Big](x) \\
&\quad = T_{s_1}\Big[\mathbb{1}_{C_1} T_{s_2-s_1}\Big[\mathbb{1}_{C_2} \cdots \mathbb{1}_{C_{m-1}} T_{s_m-s_{m-1}}[\mathbb{1}_{C_m} T_{t-s_m} \mathbb{1}_C] \cdots \Big]\Big](x) \\
&\quad = p^x_{s_1,\dots,s_m,t}(C_1 \times \dots \times C_{m-1} \times C_m \times C) \\
&\quad = \int \prod_{j=1}^{m} \mathbb{1}_{C_j}(X_{s_j}) \mathbb{1}_C(X_t)\, d\mathbb{P}^x.
\end{aligned}$$

□

7.7 Remark. Feller processes. Denote by $(T_t)_{t\geqslant 0}$ a Feller semigroup on $\mathcal{C}_\infty(\mathbb{R}^d)$. Remark 7.6 allows us to construct a d-dimensional Markov process $(X_t)_{t\geqslant 0}$ whose transition semigroup is $(T_t)_{t\geqslant 0}$. The process $(X_t)_{t\geqslant 0}$ is called a *Feller process*. Let us briefly outline this construction: By the Riesz representation theorem, cf. e. g. Rudin [198, Theorem 2.14], there exists a unique (compact inner regular) kernel $p_t(x, \cdot)$ such that

$$T_t u(x) = \int u(y)\, p_t(x, dy), \quad u \in \mathcal{C}_\infty(\mathbb{R}^d). \tag{7.6}$$

Obviously, (7.6) extends T_t to $\mathcal{B}_b(\mathbb{R}^d)$, and $(T_t)_{t\geqslant 0}$ becomes a Feller semigroup in the sense of Remark 7.4.

Let $K \subset \mathbb{R}^d$ be compact. Then $U_n := \{x + y : x \in K, y \in \mathbb{B}(0, 1/n)\}$ is open and Ex. 7.8

$$u_n(x) := \frac{d(x, U_n^c)}{d(x, K) + d(x, U_n^c)}, \quad \text{where} \quad d(x, A) := \inf_{a\in A} |x - a|, \tag{7.7}$$

is a sequence of continuous functions such that $\inf_{n\geqslant 1} u_n = \mathbb{1}_K$. By monotone convergence, $p_t(x, K) = \inf_{n\geqslant 1} T_t u_n(x)$, and $(t, x) \mapsto p_t(x, K)$ is measurable. This follows from the joint continuity of $(t, x) \mapsto T_t u(x)$ for $u \in \mathcal{C}_\infty(\mathbb{R}^d)$, cf. Problem 7.9. Since the kernel Ex. 7.9
is inner regular, we have $p_t(x, C) = \sup\{p_t(x, K) : K \subset C,\ K \text{ compact}\}$ for all Borel sets $C \subset \mathbb{R}^d$ and, therefore, $(t, x) \mapsto p_t(x, C)$ is measurable.

Obviously, $p_t(x, \mathbb{R}^d) \leqslant 1$. If $p_t(x, \mathbb{R}^d) < 1$, the following trick can be used to make Remark 7.6 work: Consider the one-point compactification of $\mathbb{R}^d$ by adding the point ∂ and define

$$\begin{cases} p_t^\partial(x, \{\partial\}) := 1 - p_t(x, \mathbb{R}^d) & \text{for } x \in \mathbb{R}^d,\ t \geqslant 0, \\ p_t^\partial(\partial, C) := \delta_\partial(C) & \text{for } C \in \mathscr{B}(\mathbb{R}^d \cup \{\partial\}),\ t \geqslant 0 \\ p_t^\partial(x, C) := p_t(x, C) & \text{for } x \in \mathbb{R}^d,\ C \in \mathscr{B}(\mathbb{R}^d),\ t \geqslant 0. \end{cases} \tag{7.8}$$

With some effort we can always get a version of $(X_t)_{t\geqslant 0}$ with right-continuous sample paths. Moreover, a Feller process is a strong Markov process, cf. Section A.5 in the appendix.

7.2 The generator

Let $\phi : [0,\infty) \to \mathbb{R}$ be a continuous function which satisfies the functional equation $\phi(t)\phi(s) = \phi(t+s)$ and $\phi(0) = 1$. Then there is a unique $a \in \mathbb{R}$ such that $\phi(t) = e^{at}$. Obviously,

$$a = \frac{d^+}{dt}\phi(t)\Big|_{t=0} = \lim_{t\to 0}\frac{\phi(t)-1}{t}.$$

Since a strongly continuous semigroup $(P_t)_{t\geqslant 0}$ on a Banach space $(\mathfrak{B}, \|\cdot\|)$ also satisfies the functional equation $P_tP_su = P_{t+s}u$ it is not too wild a guess that – in an appropriate sense – $P_tu = e^{tA}u$ for some operator A in $\mathfrak{B}$. In order to keep things technically simple, we will only consider Feller semigroups.

7.8 Definition. Let $(P_t)_{t\geqslant 0}$ be a Feller semigroup on $\mathcal{C}_\infty(\mathbb{R}^d)$. Then

$$Au := \lim_{t\to 0}\frac{P_tu-u}{t} \quad \Big(\text{the limit is taken in } (\mathcal{C}_\infty(\mathbb{R}^d), \|\cdot\|_\infty)\Big) \tag{7.9a}$$

$$\mathfrak{D}(A) := \left\{u \in \mathcal{C}_\infty(\mathbb{R}^d) \;\Big|\; \exists g \in \mathcal{C}_\infty(\mathbb{R}^d) : \lim_{t\to 0}\left\|\frac{P_tu-u}{t} - g\right\|_\infty = 0\right\} \tag{7.9b}$$

is the *(infinitesimal) generator* of the semigroup $(P_t)_{t\geqslant 0}$.

Ex. 7.10 Clearly, (7.9) defines a linear operator $A : \mathfrak{D}(A) \to \mathcal{C}_\infty(\mathbb{R}^d)$.

7.9 Example. Let $(B_t)_{t\geqslant 0}$, $B_t = (B_t^1, \dots, B_t^d)$, be a d-dimensional Brownian motion and denote by $P_tu(x) = \mathbb{E}^x u(B_t)$ the transition semigroup and by $(A, \mathfrak{D}(A))$ the generator. Then $\mathcal{C}^2_\infty(\mathbb{R}^d) \subset \mathfrak{D}(A)$ where

$$A|_{\mathcal{C}^2_\infty(\mathbb{R}^d)} = \frac{1}{2}\Delta = \frac{1}{2}\sum_{j=1}^d \partial_j^2, \qquad \partial_j = \frac{\partial}{\partial x_j},$$

and $\mathcal{C}^2_\infty(\mathbb{R}^d) := \left\{u \in \mathcal{C}_\infty(\mathbb{R}^d) : \partial_j u,\ \partial_j\partial_k u \in \mathcal{C}_\infty(\mathbb{R}^d),\ j,k = 1,\dots,d\right\}$.

Let $u \in \mathcal{C}^2_\infty(\mathbb{R}^d)$. By the Taylor formula we get for some point $\xi_t = \xi(x,t,\omega)$ between x and $x + B_t(\omega)$

$$\begin{aligned}
&\left|\mathbb{E}\left[\frac{u(B_t+x)-u(x)}{t} - \frac{1}{2}\sum_{j=1}^d \partial_j^2 u(x)\right]\right| \\
&= \left|\mathbb{E}\left[\frac{1}{t}\sum_{j=1}^d \partial_j u(x)B_t^j + \frac{1}{2t}\sum_{j,k=1}^d \partial_j\partial_k u(\xi_t)B_t^jB_t^k - \frac{1}{2}\sum_{j=1}^d \partial_j^2 u(x)\right]\right| \\
&= \frac{1}{2}\left|\mathbb{E}\left[\sum_{j,k=1}^d \big(\partial_j\partial_k u(\xi_t) - \partial_j\partial_k u(x)\big)\frac{B_t^jB_t^k}{t}\right]\right|
\end{aligned}$$

In the last equality we used $\mathbb{E}\,B_t^j = 0$ and $\mathbb{E}\,B_t^jB_t^k = \delta_{jk}t$. From the integral representation of the remainder term in the Taylor formula, $\frac{1}{2}\partial_j\partial_k u(\xi_t) = \int_0^1 \partial_j\partial_k u(x+rB_t)(1-r)\,dr$, we see that $\omega \mapsto \partial_j\partial_k u(\xi_t(\omega))$ is measurable. Applying the Cauchy–Schwarz inequality two times, we find

$$\leqslant \frac{1}{2}\,\mathbb{E}\left[\left(\sum_{j,k=1}^{d} \left|\partial_j\partial_k u(\xi_t) - \partial_j\partial_k u(x)\right|^2\right)^{\frac{1}{2}} \underbrace{\left(\sum_{j,k=1}^{d} \frac{(B_t^jB_t^k)^2}{t^2}\right)^{\frac{1}{2}}}_{=\sum_{l=1}^d (B_t^l)^2/t=|B_t|^2/t}\right]$$

$$\leqslant \frac{1}{2}\left(\sum_{j,k=1}^{d} \mathbb{E}\left(\left|\partial_j\partial_k u(\xi_t) - \partial_j\partial_k u(x)\right|^2\right)\right)^{\frac{1}{2}} \left(\frac{\mathbb{E}\left(|B_t|^4\right)}{t^2}\right)^{\frac{1}{2}}.$$

By the scaling property 2.16, $\mathbb{E}(|B_t|^4) = t^2\,\mathbb{E}(|B_1|^4) < \infty$. Since

$$\left|\partial_j\partial_k u(\xi_t) - \partial_j\partial_k u(x)\right|^2 \leqslant 4\|\partial_j\partial_k u\|_\infty^2 \quad \text{for all } j,k = 1,\dots,d,$$

and since the expression on the left converges (uniformly in x) to 0 as $t \to 0$, we get by dominated convergence

$$\lim_{t\to 0}\sup_{x\in\mathbb{R}^d}\left|\mathbb{E}\left[\frac{u(B_t+x)-u(x)}{t} - \frac{1}{2}\sum_{j=1}^{d}\partial_j^2 u(x)\right]\right| = 0.$$

This shows that $\mathcal{C}_\infty^2(\mathbb{R}^d) \subset \mathfrak{D}(A)$ and $A = \frac{1}{2}\Delta$ on $\mathcal{C}_\infty^2(\mathbb{R}^d)$.

We will see in Example 7.16 below that $\mathfrak{D}(A) = \mathcal{C}_\infty^2(\mathbb{R})$ and $A = \frac{1}{2}\Delta$ if $d = 1$. **!**

For $d > 1$ we have $\mathfrak{D}(A) \subsetneq \mathcal{C}_\infty^2(\mathbb{R}^d)$, cf. [105, pp. 234–5] and Example 7.25. In general, $\mathfrak{D}(A)$ is the Sobolev-type space $\left\{u \in L^2(\mathbb{R}^d) : u,\ \Delta u \in \mathcal{C}_\infty(\mathbb{R}^d)\right\}$ where Δu is understood in the sense of distributions, cf. Example 7.17.

Since every Feller semigroup $(P_t)_{t\geqslant 0}$ is given by a family of measurable kernels, cf. Remark 7.7, we can define linear operators of the type $L = \int_0^t P_s\,ds$ in the following way: For $u \in \mathcal{B}_b(\mathbb{R}^d)$

$$Lu \text{ is a Borel function given by } x \mapsto Lu(x) := \int_0^t\!\!\int u(y)\,p_s(x,dy)\,ds.$$

The following lemma describes the fundamental relation between generators and semigroups.

7.10 Lemma. *Let $(P_t)_{t\geqslant 0}$ be a Feller semigroup with generator $(A,\mathfrak{D}(A))$.*

a) *For all $u \in \mathfrak{D}(A)$ and $t > 0$ we have $P_tu \in \mathfrak{D}(A)$. Moreover,*

$$\frac{d}{dt}P_tu = AP_tu = P_tAu \quad \textit{for all } u \in \mathfrak{D}(A),\ t > 0. \tag{7.10}$$

b) *For all $u \in \mathcal{C}_\infty(\mathbb{R}^d)$ we have $\int_0^t P_s u\,ds \in \mathfrak{D}(A)$ and*

$$P_t u - u = A \int_0^t P_s u\,ds, \qquad u \in \mathcal{C}_\infty(\mathbb{R}^d),\ t > 0 \tag{7.11a}$$

$$= \int_0^t AP_s u\,ds, \qquad u \in \mathfrak{D}(A),\ t > 0 \tag{7.11b}$$

$$= \int_0^t P_s Au\,ds, \qquad u \in \mathfrak{D}(A),\ t > 0. \tag{7.11c}$$

Proof. a) Let $0 < \epsilon < t$ and $u \in \mathfrak{D}(A)$. The semigroup and contraction properties give

$$\begin{aligned}\left\|\frac{P_\epsilon P_t u - P_t u}{\epsilon} - P_t Au\right\|_\infty &= \left\|\frac{P_{t+\epsilon} u - P_t u}{\epsilon} - P_t Au\right\|_\infty \\ &= \left\|P_t \frac{P_\epsilon u - u}{\epsilon} - P_t Au\right\|_\infty \\ &\leqslant \left\|\frac{P_\epsilon u - u}{\epsilon} - Au\right\|_\infty \xrightarrow[\epsilon\to 0]{} 0.\end{aligned}$$

By the definition of the generator, $P_t u \in \mathfrak{D}(A)$ and $\frac{d^+}{dt} P_t u = AP_t u = P_t Au$. Similarly,

$$\begin{aligned}&\left\|\frac{P_t u - P_{t-\epsilon} u}{\epsilon} - P_t Au\right\|_\infty \\ &\leqslant \left\|P_{t-\epsilon}\frac{P_\epsilon u - u}{\epsilon} - P_{t-\epsilon} Au\right\|_\infty + \left\|P_{t-\epsilon} Au - P_{t-\epsilon} P_\epsilon Au\right\|_\infty \\ &\leqslant \left\|\frac{P_\epsilon u - u}{\epsilon} - Au\right\|_\infty + \left\|Au - P_\epsilon Au\right\|_\infty \xrightarrow[\epsilon\to 0]{} 0.\end{aligned}$$

where we used the strong continuity in the last step.

This shows that we have $\frac{d^-}{dt} P_t u = AP_t u = P_t Au$.

b) Let $u \in \mathcal{B}_b(\mathbb{R}^d)$ and $t, \epsilon > 0$. By Fubini's theorem

$$\begin{aligned}P_\epsilon \int_0^t P_s u(x)\,ds &= \iint\int_0^t u(y) p_s(z, dy)\,ds\, p_\epsilon(x, dz) \\ &= \int_0^t\iint u(y) p_s(z, dy)\, p_\epsilon(x, dz)\,ds = \int_0^t P_\epsilon P_s u(x)\,ds,\end{aligned}$$

and so,

$$\begin{aligned}\frac{P_\epsilon - \mathrm{id}}{\epsilon}\int_0^t P_s u(x)\,ds &= \frac{1}{\epsilon}\int_0^t \left(P_{s+\epsilon} u(x) - P_s u(x)\right)ds \\ &= \frac{1}{\epsilon}\int_t^{t+\epsilon} P_s u(x)\,ds - \frac{1}{\epsilon}\int_0^\epsilon P_s u(x)\,ds.\end{aligned}$$

The contraction property and the strong continuity give for $r \geqslant 0$

$$\left| \frac{1}{\epsilon} \int_r^{r+\epsilon} P_s u(x)\, ds - P_r u(x) \right| \leqslant \frac{1}{\epsilon} \int_r^{r+\epsilon} \left| P_s u(x) - P_r u(x) \right| ds$$
$$\leqslant \sup_{r \leqslant s \leqslant r+\epsilon} \|P_s u - P_r u\|_\infty \tag{7.12}$$
$$\leqslant \sup_{s \leqslant \epsilon} \|P_s u - u\|_\infty \xrightarrow[\epsilon \to 0]{} 0.$$

This shows that $\int_0^t P_s u\, ds \in \mathfrak{D}(A)$ as well as (7.11a). If $u \in \mathfrak{D}(A)$, then

$$\int_0^t P_s A u(x)\, ds \overset{(7.10)}{=} \int_0^t A P_s u(x)\, ds \overset{(7.10)}{=} \int_0^t \frac{d}{ds} P_s u(x)\, ds = P_t u(x) - u(x)$$
$$\overset{(7.11a)}{=} A \int_0^t P_s u(x)\, ds.$$

This proves (7.11c) and (7.11b). □

7.11 Corollary. *Let $(P_t)_{t \geqslant 0}$ be a Feller semigroup with generator $(A, \mathfrak{D}(A))$.* Ex. 21.3

a) *$\mathfrak{D}(A)$ is a dense subset of $\mathcal{C}_\infty(\mathbb{R}^d)$.*

b) *$(A, \mathfrak{D}(A))$ is a* closed operator, *i. e.*

$$\left.\begin{array}{l} (u_n)_{n \geqslant 1} \subset \mathfrak{D}(A),\ \lim_{n\to\infty} u_n = u, \\ (Au_n)_{n \geqslant 1} \text{ is a Cauchy sequence} \end{array}\right\} \implies \left\{\begin{array}{l} u \in \mathfrak{D}(A) \text{ and} \\ Au = \lim_{n\to\infty} Au_n. \end{array}\right. \tag{7.13}$$

c) *If $(T_t)_{t \geqslant 0}$ is another Feller semigroup with generator $(A, \mathfrak{D}(A))$, then $P_t = T_t$.*

Proof. a) Let $u \in \mathcal{C}_\infty(\mathbb{R}^d)$. From Lemma 7.10 b) we see $u_\epsilon := \epsilon^{-1} \int_0^\epsilon P_s u\, ds \in \mathfrak{D}(A)$ and, by an argument similar to (7.12), we find $\lim_{\epsilon\to 0} \|u_\epsilon - u\|_\infty = 0$.

b) Let $(u_n)_{n \geqslant 1} \subset \mathfrak{D}(A)$ be such that $\lim_{n\to\infty} u_n = u$ and $\lim_{n\to\infty} Au_n = w$ uniformly. Therefore

$$P_t u(x) - u(x) = \lim_{n\to\infty} \left(P_t u_n(x) - u_n(x)\right)$$
$$\overset{(7.11c)}{=} \lim_{n\to\infty} \int_0^t P_s A u_n(x)\, ds = \int_0^t P_s w(x)\, ds.$$

As in (7.12) we find

$$\lim_{t\to 0} \frac{P_t u(x) - u(x)}{t} = \lim_{t\to 0} \frac{1}{t} \int_0^t P_s w(x)\, ds = w(x)$$

uniformly for all x, thus $u \in \mathfrak{D}(A)$ and $Au = w$.

c) Let $0 < s < t$ and assume that $0 < |h| < \min\{t-s, s\}$. Then we have for $u \in \mathfrak{D}(A)$

$$\frac{P_{t-s}T_s u - P_{t-s+h}T_{s-h}u}{h} = \frac{P_{t-s} - P_{t-s+h}}{h} T_s u + P_{t-s+h}\frac{T_s u - T_{s-h}u}{h}.$$

Ex. 7.11 Letting $h \to 0$, Lemma 7.10 shows that the function $s \mapsto P_{t-s}T_s u$, $u \in \mathfrak{D}(A)$, is differentiable, and

$$\frac{d}{ds}P_{t-s}T_s u = \left(\frac{d}{ds}P_{t-s}\right)T_s u + P_{t-s}\frac{d}{ds}T_s u = -P_{t-s}AT_s u + P_{t-s}AT_s u = 0.$$

Thus,

$$0 = \int_0^t \frac{d}{ds}P_{t-s}T_s u(x)\,ds = P_{t-s}T_s u(x)\Big|_{s=0}^{s=t} = T_t u(x) - P_t u(x).$$

Therefore, $T_t u = P_t u$ for all $t > 0$ and $u \in \mathfrak{D}(A)$. Since $\mathfrak{D}(A)$ is dense in $\mathcal{C}_\infty(\mathbb{R}^d)$, we can approximate $u \in \mathcal{C}_\infty(\mathbb{R}^d)$ by a sequence $(u_j)_{j\geqslant 1} \subset \mathfrak{D}(A)$, i. e.

$$T_t u = \lim_{j\to\infty} T_t u_j = \lim_{j\to\infty} P_t u_j = P_t u. \qquad \square$$

7.3 The resolvent

Let $(P_t)_{t\geqslant 0}$ be a Feller semigroup. Then the integral

$$U_\alpha u(x) := \int_0^\infty e^{-\alpha t}\, P_t u(x)\,dt, \quad \alpha > 0,\ x \in \mathbb{R}^d,\ u \in \mathcal{B}_b(\mathbb{R}^d), \tag{7.14}$$

exists and defines a linear operator $U_\alpha : \mathcal{B}_b(\mathbb{R}^d) \to \mathcal{B}_b(\mathbb{R}^d)$.

7.12 Definition. Let $(P_t)_{t\geqslant 0}$ be a Feller semigroup and $\alpha > 0$. The operator U_α given by (7.14) is the *α-potential operator*.

The α-potential operators share many properties of the semigroup. In particular, we have the following counterpart of Proposition 7.3.

Ex. 7.13 **7.13 Proposition.** *Let $(P_t)_{t\geqslant 0}$ be a Feller semigroup with α-potential operators $(U_\alpha)_{\alpha>0}$*
Ex. 7.14 *and generator $(A, \mathfrak{D}(A))$.*
Ex. 7.16
a) *αU_α is conservative whenever $(P_t)_{t\geqslant 0}$ is conservative: $\alpha U_\alpha 1 = 1$;*
b) *αU_α is a contraction on $\mathcal{B}_b$: $\|\alpha U_\alpha u\|_\infty \leqslant \|u\|_\infty$ for all $u \in \mathcal{B}_b$;*
c) *αU_α is positivity preserving: For $u \in \mathcal{B}_b$ we have $u \geqslant 0 \Rightarrow \alpha U_\alpha u \geqslant 0$;*
d) *αU_α has the Feller property: If $u \in \mathcal{C}_\infty$, then $\alpha U_\alpha u \in \mathcal{C}_\infty$;*
e) *αU_α is strongly continuous on $\mathcal{C}_\infty$: $\lim_{\alpha\to\infty} \|\alpha U_\alpha u - u\|_\infty = 0$;*
f) *$(U_\alpha)_{\alpha>0}$ is the resolvent: For all $\alpha > 0$, $\alpha\,\mathrm{id} - A$ is invertible with a bounded inverse $U_\alpha = (\alpha\,\mathrm{id} - A)^{-1}$;*

g) *Resolvent equation:* $U_\alpha u - U_\beta u = (\beta - \alpha)U_\beta U_\alpha u$ *for all* $\alpha, \beta > 0$ *and* $u \in \mathcal{B}_b$*. In particular,* $\lim_{\beta\to\alpha} \|U_\alpha u - U_\beta u\|_\infty = 0$ *for all* $u \in \mathcal{B}_b$*.*
h) *There is a one-to-one relationship between* $(P_t)_{t\geqslant 0}$ *and* $(U_\alpha)_{\alpha>0}$*.*
i) *The resolvent* $(U_\alpha)_{\alpha>0}$ *is sub-Markovian if, and only if,* $(P_t)_{t\geqslant 0}$ *is sub-Markovian: Let* $0 \leqslant u \leqslant 1$*. Then* $0 \leqslant P_t u \leqslant 1 \;\; \forall t \geqslant 0$ *if, and only if,* $0 \leqslant \alpha U_\alpha u \leqslant 1 \;\; \forall \alpha > 0$*.*

Proof. a)–c) follow immediately from the integral formula (7.14) for $U_\alpha u(x)$ and the corresponding properties of the Feller semigroup $(P_t)_{t\geqslant 0}$.

d) Since $P_t u \in \mathcal{C}_\infty$ and $|\alpha e^{-\alpha t} P_t u(x)| \leqslant \alpha e^{-\alpha t}\, \|u\|_\infty \in L^1(dt)$, this follows from dominated convergence.

e) By the integral representation (7.14) and the triangle inequality we find

$$\|\alpha U_\alpha u - u\|_\infty \leqslant \int_0^\infty \alpha e^{-\alpha t} \|P_t u - u\|_\infty \, dt = \int_0^\infty e^{-s} \|P_{s/\alpha} u - u\|_\infty \, ds.$$

As $\|P_{s/\alpha} u - u\|_\infty \leqslant 2\|u\|_\infty$ and $\lim_{\alpha\to\infty} \|P_{s/\alpha} u - u\|_\infty = 0$, the claim follows with dominated convergence.

f) For every $\epsilon > 0, \alpha > 0$ and $u \in \mathcal{C}_\infty(\mathbb{R}^d)$ we have

$$\begin{aligned}\frac{1}{\epsilon}\Big(P_\epsilon(U_\alpha u) - U_\alpha u\Big) &= \frac{1}{\epsilon}\int_0^\infty e^{-\alpha t}(P_{t+\epsilon} u - P_t u)\, dt \\ &= \frac{1}{\epsilon}\int_\epsilon^\infty e^{-\alpha(s-\epsilon)} P_s u\, ds - \frac{1}{\epsilon}\int_0^\infty e^{-\alpha s} P_s u\, ds \\ &= \underbrace{\frac{e^{\alpha\epsilon} - 1}{\epsilon}}_{\to\alpha,\ \epsilon\to 0} \int_0^\infty e^{-\alpha s} P_s u\, ds - \underbrace{\frac{e^{\alpha\epsilon}}{\epsilon}\int_0^\epsilon e^{-\alpha s} P_s u\, ds.}_{\to u,\ \epsilon\to 0 \text{ as in (7.12)}}\end{aligned}$$

This shows that we have in the sense of uniform convergence

$$\lim_{\epsilon\to 0} \frac{P_\epsilon U_\alpha u - U_\alpha u}{\epsilon} = \alpha U_\alpha u - u,$$

i. e. $U_\alpha u \in \mathfrak{D}(A)$ and $AU_\alpha u = \alpha U_\alpha u - u$. Hence, $(\alpha \operatorname{id} - A)U_\alpha u = u$ for all $u \in \mathcal{C}_\infty$.

Since for $u \in \mathfrak{D}(A)$

$$U_\alpha A u = \int_0^\infty e^{-\alpha t} P_t A u \, dt \overset{(7.11)}{=} A \int_0^\infty e^{-\alpha t} P_t u\, dt = AU_\alpha u,$$

it is easy to see that $u = (\alpha \operatorname{id} - A)U_\alpha u = U_\alpha(\alpha \operatorname{id} - A)u$.

This shows that U_α is the left- and right-inverse of $\alpha \operatorname{id} - A$. Since U_α is bounded, $\alpha \operatorname{id} - A$ has a bounded inverse, i. e. $\alpha > 0$ is in the resolvent set of A, and $(U_\alpha)_{\alpha>0}$ is the resolvent.

g) Let $\alpha, \beta > 0$ and $u \in \mathcal{B}_b$. Using the definition of U_α and Fubini's theorem we see that

$$U_\alpha U_\beta u(x) = \int_0^\infty\int_0^\infty e^{-\alpha t}e^{-\beta s}P_tP_su(x)\,ds\,dt$$
$$= \int_0^\infty\int_0^\infty e^{-\beta s}e^{-\alpha t}P_sP_tu(x)\,dt\,ds = U_\beta U_\alpha u(x).$$

Thus, by part f),

$$U_\alpha u - U_\beta u = (\beta\,\mathrm{id} - A)U_\beta U_\alpha u - (\alpha\,\mathrm{id} - A)U_\alpha U_\beta u = (\beta - \alpha)U_\beta U_\alpha u.$$

Since αU_α and βU_β are contractive, we find

$$\|U_\alpha u - U_\beta u\|_\infty = \frac{|\beta - \alpha|}{\beta\alpha}\|\alpha U_\alpha \beta U_\beta u\|_\infty \leqslant \left|\frac{1}{\alpha} - \frac{1}{\beta}\right| \|u\|_\infty \xrightarrow[\beta\to\alpha]{} 0.$$

h) Formula (7.14) shows that $(P_t)_{t\geqslant 0}$ defines $(U_\alpha)_{\alpha>0}$. On the other hand, (7.14) means that $\nu(\alpha) := U_\alpha u(x)$ is the Laplace transform of $\phi(t) := P_tu(x)$ for every $x \in \mathbb{R}^d$. Therefore, ν determines $\phi(t)$ for Lebesgue almost all $t \geqslant 0$, cf. [205, Proposition 1.2] for the uniqueness of the Laplace transform. Since $\phi(t) = P_tu(x)$ is continuous for $u \in \mathcal{C}_\infty$, this shows that $P_tu(x)$ is uniquely determined for all $u \in \mathcal{C}_\infty$, and so is $(P_t)_{t\geqslant 0}$, cf. Remark 7.7.

i) Assume that $u \in \mathcal{B}_b$ satisfies $0 \leqslant u \leqslant 1$. From (7.14) we get that $0 \leqslant P_tu \leqslant 1$, $t \geqslant 0$, implies $0 \leqslant \alpha U_\alpha u \leqslant 1$. Conversely, let $0 \leqslant u \leqslant 1$ and $0 \leqslant \alpha U_\alpha u \leqslant 1$ for all $\alpha > 0$. From the resolvent equation g) we see for all $u \in \mathcal{B}_b$ with $u \geqslant 0$

$$\frac{d}{d\alpha}U_\alpha u(x) = \lim_{\beta\to\alpha}\frac{U_\alpha u(x) - U_\beta u(x)}{\alpha - \beta} \overset{\text{g)}}{=} -\lim_{\beta\to\alpha}U_\beta U_\alpha u(x) = -U_\alpha^2 u(x) \leqslant 0.$$

Ex. 7.15 Iterating this shows that $(-1)^n\frac{d^n}{d\alpha^n}U_\alpha u(x) \geqslant 0$. This means that $\nu(\alpha) := U_\alpha u(x)$ is completely monotone, hence the Laplace transform of a *positive* measure, cf. Bernstein's theorem [205, Theorem 1.4]. As $U_\alpha u(x) = \int_0^\infty e^{-\alpha t}P_tu(x)\,dt$, we infer that $P_tu(x) \geqslant 0$ for Lebesgue almost all $t > 0$.

The formula for the derivative of U_α yields $\frac{d}{d\alpha}\alpha U_\alpha u(x) = (\mathrm{id} - \alpha U_\alpha)U_\alpha u(x)$. By iteration we get $(-1)^{n+1}\frac{d^n}{d\alpha^n}\alpha U_\alpha u(x) = n!(\mathrm{id} - \alpha U_\alpha)U_\alpha^n u(x)$. Plug in $u \equiv 1$; then it follows that $\Upsilon(\alpha) := 1 - \alpha U_\alpha 1(x) = \int_0^\infty \alpha e^{-\alpha t}(1 - P_t1(x))\,dt$ is completely monotone. Thus, $P_t1(x) \leqslant 1$ for Lebesgue almost all $t > 0$, and using the positivity of P_t for the positive function $1 - u$, we get $P_tu \leqslant P_t1 \leqslant 1$ for almost all $t > 0$.

Since $t \mapsto P_tu(x)$ is continuous if $u \in \mathcal{C}_\infty$, we see that $0 \leqslant P_tu(x) \leqslant 1$ for *all* $t > 0$ if $u \in \mathcal{C}_\infty$ and $0 \leqslant u \leqslant 1$, and we extend this to $u \in \mathcal{B}_b$ with the help of Remark 7.7. □

7.14 Example. Let $(B_t)_{t\geqslant 0}$ be a Brownian motion and denote by $(U_\alpha)_{\alpha>0}$ the resolvent. Then

$$U_\alpha u(x) = G_{\alpha,d} * u(x) = \int G_{\alpha,d}(y)u(x - y)\,dy \quad \text{for all } u \in \mathcal{B}_b(\mathbb{R}^d)$$

with the convolution kernel

$$G_{\alpha,d}(y) = \frac{1}{\pi^{d/2}} \left(\frac{\sqrt{2\alpha}}{2|y|} \right)^{\frac{d}{2}-1} K_{\frac{d}{2}-1}(\sqrt{2\alpha}\,|y|), \qquad d \geqslant 1, \tag{7.15}$$

where $K_\nu(x)$ is a Bessel function of the third kind, cf. [167, 10.32.10, p. 253 and 10.39.2 p. 254]. For $d = 1, 2$ (7.15) becomes

$$G_{\alpha,1}(y) = \frac{e^{-\sqrt{2\alpha}\,|y|}}{\sqrt{2\alpha}} \quad \text{and} \quad G_{\alpha,2}(y) = \frac{1}{\pi} K_0(\sqrt{2\alpha}\,|y|).$$

Indeed: Let $d = 1$ and denote by $(P_t)_{t\geqslant 0}$ the transition semigroup. Using Fubini's theorem we find for $\alpha > 0$ and $u \in \mathcal{B}_b(\mathbb{R})$

$$\begin{aligned} U_\alpha u(x) = \int_0^\infty e^{-\alpha t}\, P_t u(x)\, dt &= \int_0^\infty e^{-\alpha t}\, \mathbb{E}\, u(x + B_t)\, dt \\ &= \int\int_0^\infty e^{-\alpha t}\, u(x+y) \frac{1}{\sqrt{2\pi t}}\, e^{-y^2/2t}\, dt\, dy \\ &= \int r_\alpha(y^2)\, u(x-y)\, dy \end{aligned}$$

where

$$r_\alpha(\eta) = \int_0^\infty \frac{1}{\sqrt{2\pi t}}\, e^{-(\alpha t + \eta/2t)}\, dt = \frac{e^{-\sqrt{2\alpha\eta}}}{\sqrt{2\alpha}} \int_0^\infty \sqrt{\frac{\alpha}{\pi t}}\, e^{-\left(\sqrt{\alpha t} - \sqrt{\eta/2t}\right)^2} dt.$$

Changing variables according to $t = \eta/(2\alpha s)$ and $dt = -\eta/(2\alpha s^2)\, ds$ gives

$$r_\alpha(\eta) = \frac{e^{-\sqrt{2\alpha\eta}}}{\sqrt{2\alpha}} \int_0^\infty \sqrt{\frac{\eta}{2\pi s^3}}\, e^{-\left(\sqrt{\eta/2s} - \sqrt{\alpha s}\right)^2} ds.$$

If we add these two equalities and divide by 2, we get

$$r_\alpha(\eta) = \frac{e^{-\sqrt{2\alpha\eta}}}{\sqrt{2\alpha}} \frac{1}{\sqrt{\pi}} \int_0^\infty \frac{1}{2} \left(\sqrt{\frac{\alpha}{s}} + \sqrt{\frac{\eta}{2s^3}} \right) e^{-\left(\sqrt{\alpha s} - \sqrt{\eta/2s}\right)^2} ds.$$

A further change of variables, $y = \sqrt{\alpha s} - \sqrt{\eta/2s}$ and $dy = \frac{1}{2}\left(\sqrt{\frac{\alpha}{s}} + \sqrt{\frac{\eta}{2s^3}}\right) ds$, reveals that

$$r_\alpha(\eta) = \frac{e^{-\sqrt{2\alpha\eta}}}{\sqrt{2\alpha}} \frac{1}{\sqrt{\pi}} \int_{-\infty}^\infty e^{-y^2}\, dy = \frac{e^{-\sqrt{2\alpha\eta}}}{\sqrt{2\alpha}}.$$

For a d-dimensional Brownian motion, $d \geqslant 2$, a similar calculation leads to the kernel

$$r_\alpha(\eta) = \int_0^\infty \frac{1}{(2\pi t)^{d/2}}\, e^{-(\alpha t + \eta/2t)}\, dt = \frac{1}{\pi^{d/2}} \left(\frac{\alpha}{2\eta} \right)^{\frac{d}{4}-\frac{1}{2}} K_{\frac{d}{2}-1}(\sqrt{2\alpha\eta})$$

which cannot be expressed by elementary functions if $d \geqslant 2$.

The following result is often helpful if we want to determine the domain of the generator.

7.15 Theorem (Dynkin 1956; Reuter 1957). *Let $(A, \mathfrak{D}(A))$ be the generator of a Feller semigroup and assume that $(L, \mathfrak{D}(L))$, $\mathfrak{D}(L) \subset \mathcal{C}_\infty(\mathbb{R}^d)$, extends $(A, \mathfrak{D}(A))$, i. e.*

$$\mathfrak{D}(A) \subset \mathfrak{D}(L) \quad \text{and} \quad L|_{\mathfrak{D}(A)} = A. \tag{7.16}$$

If for any $u \in \mathfrak{D}(L)$

$$Lu = u \implies u = 0, \tag{7.17}$$

then $(A, \mathfrak{D}(A)) = (L, \mathfrak{D}(L))$.

Proof. Pick $u \in \mathfrak{D}(L)$ and set

$$g := u - Lu \quad \text{and} \quad h := (\operatorname{id} - A)^{-1} g \in \mathfrak{D}(A).$$

Then

$$h - Lh \overset{(7.16)}{=} h - Ah = (\operatorname{id} - A)(\operatorname{id} - A)^{-1} g = g \overset{\text{def}}{=} u - Lu.$$

Thus, $h - u = L(h - u)$ and, by (7.17), $h = u$. Since $h \in \mathfrak{D}(A)$ we get that $u \in \mathfrak{D}(A)$, and so $\mathfrak{D}(L) \subset \mathfrak{D}(A)$. □

7.16 Example. Let $(A, \mathfrak{D}(A))$ be the generator of the transition semigroup $(P_t)_{t \geqslant 0}$ of a BM^d. We know from Example 7.9 that $\mathcal{C}^2_\infty(\mathbb{R}^d) \subset \mathfrak{D}(A)$.

If $d = 1$, then $\mathfrak{D}(A) = \mathcal{C}^2_\infty(\mathbb{R}) = \{u \in \mathcal{C}_\infty(\mathbb{R}) : u, u', u'' \in \mathcal{C}_\infty(\mathbb{R})\}$.

Ex. 7.12 *Proof.* We know from Proposition 7.13 f) that U_α is the inverse of $\alpha \operatorname{id} - A$. In particular, $U_\alpha(\mathcal{C}_\infty(\mathbb{R})) = \mathfrak{D}(A)$. This means that any $u \in \mathfrak{D}(A)$ can be written as $u = U_\alpha f$ for some $f \in \mathcal{C}_\infty(\mathbb{R})$.

Using the formula (7.15) for U_α from Example 7.14, we see for all $f \in \mathcal{C}_\infty(\mathbb{R})$

$$U_\alpha f(x) = \frac{1}{\sqrt{2\alpha}} \int_{-\infty}^{\infty} e^{-\sqrt{2\alpha}\,|y-x|} f(y)\, dy.$$

By the differentiability lemma for parameter-dependent integrals, cf. [204, Theorem 11.5], we find

$$\begin{aligned}
\frac{d}{dx} U_\alpha f(x) &= \int_{-\infty}^{\infty} e^{-\sqrt{2\alpha}\,|y-x|} \operatorname{sgn}(y - x)\, f(y)\, dy \\
&= \int_{x}^{\infty} e^{-\sqrt{2\alpha}\,(y-x)} f(y)\, dy - \int_{-\infty}^{x} e^{-\sqrt{2\alpha}\,(x-y)} f(y)\, dy, \\
\frac{d^2}{dx^2} U_\alpha f(x) &= \sqrt{2\alpha} \int_{-\infty}^{\infty} e^{-\sqrt{2\alpha}|y-x|} f(y)\, dy - 2f(x).
\end{aligned}$$

Since $\left|\frac{d}{dx} U_\alpha f(x)\right| \leqslant 2\sqrt{2\alpha} U_\alpha |f|(x)$ and $\frac{d^2}{dx^2} U_\alpha f(x) = 2\alpha U_\alpha f(x) - 2f(x)$, Proposition 7.13 d) shows that $U_\alpha f \in \mathcal{C}_\infty(\mathbb{R})$, thus $\mathcal{C}^2_\infty(\mathbb{R}) \supset \mathfrak{D}(A)$. □

If $d > 1$, the arguments of Example 7.16 are no longer valid. !

7.17 Example. We have seen in Example 7.9 that the generator $(A, \mathfrak{D}(A))$ of a d-dimensional Brownian motion satisfies $\mathcal{C}^2_\infty(\mathbb{R}^d) \subset \mathfrak{D}(A)$. Using Theorem 7.15 we can describe the generator precisely: $(A, \mathfrak{D}(A)) = (\frac{1}{2}\nabla^2_x, \mathfrak{D}(\nabla^2_x))$ where

$$\mathfrak{D}(\nabla^2_x) = \big\{u \in \mathcal{C}_\infty(\mathbb{R}^d) \;:\; \nabla^2_x u \in \mathcal{D}'(\mathbb{R}^d) \cap \mathcal{C}_\infty(\mathbb{R}^d)\big\} \tag{7.18}$$

and ∇^2_x denotes the distributional (weak) derivative in the space $\mathcal{D}'(\mathbb{R}^d)$ of Schwartz distributions or generalized functions.

Recall that the Schwartz distributions $\mathcal{D}'(\mathbb{R}^d)$ are the topological dual of the test functions $\mathcal{D}(\mathbb{R}^d) = \mathcal{C}^\infty_c(\mathbb{R}^d)$, see for example [199, Chapter 6]. From Hölder's inequality ($p = 1, q = \infty$) it follows immediately that

$$\frac{1}{2}\nabla^2_x u[\phi] := \int u(x) \cdot \tfrac{1}{2}\Delta\phi(x)\,dx, \quad u \in \mathcal{C}_\infty(\mathbb{R}^d),\ \phi \in \mathcal{C}^\infty_c(\mathbb{R}^d) \tag{7.19}$$

defines a generalized function $\frac{1}{2}\nabla^2_x u \in \mathcal{D}'(\mathbb{R}^d)$. If $u \in \mathcal{C}^2_\infty(\mathbb{R}^d)$, then (7.19) is just the familiar integration by parts formula

$$\int \tfrac{1}{2}\Delta u(x) \cdot \phi(x)\,dx = \int u(x) \cdot \tfrac{1}{2}\Delta\phi(x)\,dx, \quad u \in \mathcal{C}^2_\infty(\mathbb{R}^d),\ \phi \in \mathcal{C}^\infty_c(\mathbb{R}^d). \tag{7.20}$$

Therefore, $\frac{1}{2}\nabla^2_x\big|_{\mathcal{C}_\infty(\mathbb{R}^d)}$ extends $\frac{1}{2}\Delta\big|_{\mathcal{C}^2_\infty(\mathbb{R}^d)} = A\big|_{\mathcal{C}^2_\infty(\mathbb{R}^d)}$; in fact, it even extends $(A, \mathfrak{D}(A))$.

To see this, we note that by Proposition 7.13 f) $\mathfrak{D}(A) = U_1(\mathcal{C}_\infty(\mathbb{R}^d))$ and every function $u \in \mathfrak{D}(A)$ can be represented as $u = U_1 f$ with $f = u - Au$. Since $A = \frac{1}{2}\Delta$ on $\mathcal{C}^\infty_c(\mathbb{R}^d)$, we get for $\phi \in \mathcal{C}^\infty_c(\mathbb{R}^d)$

$$\begin{aligned}
(u - \tfrac{1}{2}\nabla^2_x u)[\phi] &= \int u(x) \cdot \big(\phi(x) - \tfrac{1}{2}\Delta\phi(x)\big)\,dx \\
&= \int G_{1,d} * f(x) \cdot \big(\phi(x) - \tfrac{1}{2}\Delta\phi(x)\big)\,dx \\
&= \int f(x) \cdot G_{1,d} * \big(\phi(x) - \tfrac{1}{2}\Delta\phi(x)\big)\,dx \\
&= \int f(x) \cdot U_1(\operatorname{id} - \tfrac{1}{2}\Delta)\phi(x)\,dx \\
&= \int f(x) \cdot \phi(x)\,dx.
\end{aligned}$$

This shows that $u - \frac{1}{2}\nabla^2_x u \in \mathcal{D}'(\mathbb{R}^d)$ is represented by $f = u - Au$, i. e.

$$(\tfrac{1}{2}\nabla^2_x, \mathfrak{D}(\nabla^2_x)), \qquad \mathfrak{D}(\nabla^2_x) = \big\{u \in \mathcal{C}_\infty(\mathbb{R}^d) \;:\; \nabla^2_x u \in \mathcal{D}'(\mathbb{R}^d) \cap \mathcal{C}_\infty(\mathbb{R}^d)\big\}$$

extends $(A, \mathfrak{D}(A))$.

Now let $u \in \mathfrak{D}(\nabla^2_x)$ such that $u - \frac{1}{2}\nabla^2_x u = 0$. Therefore, we find for all $\phi \in \mathcal{C}^\infty_c(\mathbb{R}^d)$ and all $z \in \mathbb{R}^d$

$$(u - \tfrac{1}{2}\nabla^2_x u)[\tau_z\phi] = 0 \quad \text{with } \tau_z\phi(x) := \phi(x - z).$$

Since $\tau_z\phi$ has compact support, we can integrate this identity with respect to the kernel $G_{1,d}(z)$ of the convolution operator U_1, and get

$$\begin{aligned} 0 &= \iint \left(u(x) - \tfrac{1}{2}\nabla_x^2 u(x)\right)\phi(x-z)G_{1,d}(z)\,dz\,dx \\ &= \int u(x)U_1(\operatorname{id} - \tfrac{1}{2}\Delta)\phi(x)\,dx \\ &= \int u(x)\phi(x)\,dx. \end{aligned}$$

This shows that $u = 0$ (Lebesgue almost everywhere), and Theorem 7.15 proves that $\left(\frac{1}{2}\nabla_x^2, \mathfrak{D}(\nabla_x^2)\right) = (A, \mathfrak{D}(A))$.

7.18 Remark. If we combine Examples 7.16 and 7.17, we find

$$\mathcal{C}_\infty^2(\mathbb{R}) = \left\{u \in \mathcal{C}_\infty(\mathbb{R}) \;:\; \nabla_x^2 u \in \mathcal{D}'(\mathbb{R}) \cap \mathcal{C}_\infty(\mathbb{R})\right\}.$$

This is also a consequence of the (local) regularity of solutions of *Poisson's equation* $\Delta u = f$. For $d = 1$ the situation is easy since we deal with an ODE $u'' = f$. In higher dimensions the situation is entirely different. If u is a radial solution, i. e. depending only on $|x|$, then

$$\Delta u = f \in \mathcal{C}_\infty(\mathbb{R}^d), \quad u \text{ radial} \implies u \in \mathcal{C}_\infty^2(\mathbb{R}^d).$$

In general, this is wrong, see also the discussion in Example 7.25. As soon as we assume, however, that $f \in \mathcal{C}_\infty^\alpha(\mathbb{R}^d)$ is Hölder continuous of order $\alpha \in (0,1]$, then

$$\Delta u = f, \quad f \in \mathcal{C}_\infty^\alpha(\mathbb{R}^d) \implies u \in \mathcal{C}_\infty^{2+\alpha}(\mathbb{R}^d)$$

where $u \in \mathcal{C}_\infty^{2+\alpha}(\mathbb{R}^d)$ means that $u \in \mathcal{C}_\infty^2(\mathbb{R}^d)$ and $\partial_j\partial_k u \in \mathcal{C}_\infty^\alpha(\mathbb{R}^d)$ for $j,k = 1,\dots,d$. Proofs can be found in Dautray–Lions [42, Chapter II.2, pp. 283–296], a classical presentation is in Günter [91, §14, pp. 82–86].

7.4 The Hille-Yosida theorem and positivity

We will now discuss the structure of generators of strongly continuous contraction semigroups and Feller semigroups.

7.19 Theorem (Hille 1948; Yosida 1948; Lumer–Phillips 1961). *Let $(\mathfrak{B}, \|\cdot\|)$ be a Banach space. A linear operator $(A, \mathfrak{D}(A))$ is the infinitesimal generator of a strongly continuous contraction semigroup if, and only if,*

a) *$\mathfrak{D}(A)$ is a dense subset of $\mathfrak{B}$;*
b) *A is* dissipative, *i. e.* $\forall \lambda > 0,\ u \in \mathfrak{D}(A) \;:\; \|\lambda u - Au\| \geqslant \lambda\|u\|$;
c) *$\mathfrak{R}(\lambda \operatorname{id} - A) = \mathfrak{B}$ for some (or all) $\lambda > 0$.*

We do not give a proof of this theorem but refer the interested reader to Ethier–Kurtz [72] and Pazy [173]. In fact, Theorem 7.19 is too general for transition semigroups of stochastic processes. Let us now show which role is played by the positivity and sub-Markovian property of such semigroups. To keep things simple, we restrict ourselves to the Banach space $\mathfrak{B} = \mathcal{C}_\infty$ and Feller semigroups.

7.20 Lemma. *Let $(P_t)_{t\geqslant 0}$ be a Feller semigroup with generator $(A, \mathfrak{D}(A))$. Then A satisfies the* positive maximum principle (PMP)

$$u \in \mathfrak{D}(A),\ u(x_0) = \sup_{x\in\mathbb{R}^d} u(x) \geqslant 0 \implies Au(x_0) \leqslant 0. \tag{7.21}$$

For the Laplace operator (7.21) is quite familiar: At a (global) maximum the second derivative is negative.

Proof. Let $u \in \mathfrak{D}(A)$ and assume that $u(x_0) = \sup u \geqslant 0$. Since P_t preserves positivity, we get $u^+ - u \geqslant 0$ and $P_t(u^+ - u) \geqslant 0$ or $P_t u^+ \geqslant P_t u$. Since $u(x_0)$ is positive, $u(x_0) = u^+(x_0)$ and so

$$\frac{1}{t}(P_t u(x_0) - u(x_0)) \leqslant \frac{1}{t}(P_t u^+(x_0) - u^+(x_0)) \leqslant \frac{1}{t}(\|u^+\|_\infty - u^+(x_0)) = 0.$$

Letting $t \to 0$ therefore shows $Au(x_0) = \lim_{t\to 0} t^{-1}\,(P_t u(x_0) - u(x_0)) \leqslant 0$. ☐

In fact, the converse of Lemma 7.20 is also true.

7.21 Lemma. *Assume that the linear operator $(A, \mathfrak{D}(A))$ satisfies conditions a) and c) of Theorem 7.19 and the PMP (7.21). Then 7.19 b) holds, in particular $\lambda\,\mathrm{id} - A$ is invertible, and both the resolvent $U_\lambda = (\lambda\,\mathrm{id} - A)^{-1}$ and the semigroup generated by A are positivity preserving.*

Proof. Let $u \in \mathfrak{D}(A)$ and x_0 be such that $|u(x_0)| = \sup_{x\in\mathbb{R}^d} |u(x)|$. We may assume that $u(x_0) \geqslant 0$, otherwise we could consider $-u$. Then we find for all $\lambda > 0$

$$\|\lambda u - Au\|_\infty \geqslant \lambda u(x_0) - \underbrace{Au(x_0)}_{\leqslant 0 \text{ by PMP}} \geqslant \lambda u(x_0) = \lambda\|u\|_\infty.$$

This is 7.19 b). Because of Proposition 7.13 i) it is enough to show that U_λ preserves positivity. If we use the PMP for $-u$, we get Feller's minimum principle:

$$v \in \mathfrak{D}(A),\ v(x_0) = \inf_{x\in\mathbb{R}^d} v(x) \leqslant 0 \implies Av(x_0) \geqslant 0.$$

Now assume that $v \geqslant 0$. If $U_\lambda v$ does not attain its infimum, we see that $U_\lambda v \geqslant 0$ as $U_\lambda v$ is continuous and vanishes at infinity. Otherwise, let x_0 be a minimum point of $U_\lambda v$, then $U_\lambda v(x_0) \leqslant 0$ since $U_\lambda v$ vanishes at infinity. By Feller's minimum principle

$$\lambda U_\lambda v(x_0) - v(x_0) \overset{\text{7.13 f)}}{=} AU_\lambda v(x_0) \geqslant 0$$

and, therefore, $\lambda U_\lambda v(x) \geqslant \inf_x \lambda U_\lambda v(x) = \lambda U_\lambda v(x_0) \geqslant v(x_0) \geqslant 0$. ☐

The next result shows that positivity allows us to determine the domain of the generator using pointwise rather than uniform convergence.

7.22 Theorem. *Let $(P_t)_{t\geqslant 0}$ be a Feller semigroup generated by $(A, \mathfrak{D}(A))$. Then*

$$\mathfrak{D}(A) = \left\{ u \in \mathcal{C}_\infty(\mathbb{R}^d) \;\middle|\; \exists\, g \in \mathcal{C}_\infty(\mathbb{R}^d)\ \forall x : \lim_{t\to 0} \frac{P_t u(x) - u(x)}{t} = g(x) \right\}. \tag{7.22}$$

Proof. We define the *weak generator* of $(P_t)_{t\geqslant 0}$ as

$$A_w u(x) := \lim_{t\to 0} \frac{P_t u(x) - u(x)}{t} \quad \text{for all } u \in \mathfrak{D}(A_w) \text{ and } x \in \mathbb{R}^d,$$

where $\mathfrak{D}(A_w)$ is set on the right-hand side of (7.22). Clearly, $(A_w, \mathfrak{D}(A_w))$ extends $(A, \mathfrak{D}(A))$ and the calculation in the proof of Lemma 7.20 shows that $(A_w, \mathfrak{D}(A_w))$ also satisfies the positive maximum principle. Using the argument of (the first part of the proof of) Lemma 7.21 we infer that $(A_w, \mathfrak{D}(A_w))$ is dissipative. In particular, we find for all $u \in \mathfrak{D}(A_w)$

$$A_w u = u \implies 0 = \|u - A_w u\| \geqslant \|u\| \implies \|u\| = 0 \implies u = 0.$$

Therefore, Theorem 7.15 shows that $(A_w, \mathfrak{D}(A_w)) = (A, \mathfrak{D}(A))$. □

7.5 The potential operator

Let $(P_t)_{t\geqslant 0}$ be a Feller semigroup with generator $(A, \mathfrak{D}(A))$ and α-potential operators $(U_\alpha)_{\alpha>0}$. We have seen in Proposition 7.13 f) that the α-potential operator U_α gives a representation of the solution of the equation

$$\alpha u - Au = f \in \mathcal{C}_\infty(\mathbb{R}^d) \implies u = U_\alpha f \in \mathfrak{D}(A).$$

Now we are interested in the equation $Au = f$ where $\alpha = 0$. It is a natural guess that $u = -\lim_{\alpha\to 0} U_\alpha f$; this is indeed true, but we need a few preparations.

7.23 Definition (Yosida 1965). Let $(P_t)_{t\geqslant 0}$ be a Feller semigroup on $\mathcal{C}_\infty(\mathbb{R}^d)$ with α-potential operators $(U_\alpha)_{\alpha>0}$. The operator $(U_0, \mathfrak{D}(U_0))$ defined by

$$U_0 u := \lim_{\alpha\to 0} U_\alpha u \quad \big(\text{the limit is taken in } (\mathcal{C}_\infty(\mathbb{R}^d), \|\cdot\|_\infty)\big) \tag{7.23a}$$

$$\mathfrak{D}(U_0) := \left\{ u \in \mathcal{C}_\infty(\mathbb{R}^d) \;\middle|\; \exists\, g \in \mathcal{C}_\infty(\mathbb{R}^d) : \lim_{\alpha\to 0} \|U_\alpha u - g\|_\infty = 0 \right\} \tag{7.23b}$$

is called the *potential operator* or *zero-resolvent* of the semigroup $(P_t)_{t\geqslant 0}$.

Clearly, (7.23) defines a linear operator $U_0 : \mathfrak{D}(U_0) \to \mathcal{C}_\infty(\mathbb{R}^d)$. The potential operator U_0 behaves like the α-potential operators $(U_\alpha)_{\alpha>0}$.

7.24 Lemma. *Let $(P_t)_{t\geqslant 0}$ be a Feller semigroup with generator $(A, \mathfrak{D}(A))$ and potential operators $(U_\alpha)_{\alpha>0}$, $(U_0, \mathfrak{D}(U_0))$.*

a) $U_\alpha(\mathfrak{D}(U_0)) \subset \mathfrak{D}(U_0)$ *for all* $\alpha > 0$;
b) $U_0 u - U_\alpha u = \alpha U_0 U_\alpha u = \alpha U_\alpha U_0 u$ *for all* $u \in \mathfrak{D}(U_0)$ *and* $\alpha > 0$;
c) $U_0(\mathfrak{D}(U_0)) \subset \mathfrak{D}(A)$ *and* $AU_0 u = -u$ *for all* $u \in \mathfrak{D}(U_0)$;
d) $\{u \in \mathcal{C}_\infty(\mathbb{R}^d) : u \geqslant 0, \ \sup_{\alpha>0} U_\alpha u \in \mathcal{C}_\infty(\mathbb{R}^d)\} \subset \mathfrak{D}(U_0)$;
e) *If* $\lim_{\alpha\to 0} \|\alpha U_\alpha u\|_\infty = 0$ *for all* $u \in \mathcal{C}_\infty(\mathbb{R}^d)$, *then* $\mathfrak{D}(U_0)$ *is dense in* $\mathcal{C}_\infty(\mathbb{R}^d)$. *Moreover,* $\mathfrak{D}(U_0) = \mathfrak{R}(A)$ *and* $U_0 = -A^{-1}$.

Proof. a) & b) Let $u \in \mathfrak{D}(U_0)$. By the resolvent equation, Proposition 7.13 g), we get

$$\lim_{\beta\to 0} U_\beta U_\alpha u = \lim_{\beta\to 0} \frac{1}{\beta - \alpha}(U_\alpha u - U_\beta u) = -\frac{1}{\alpha}(U_\alpha u - U_0 u).$$

This proves that $U_\alpha u \in \mathfrak{D}(U_0)$. Moreover, since $\lim_{\beta\to 0} U_\beta U_\alpha u = U_0 U_\alpha u$, we get the first equality in b). The second equality follows from the same calculation if we interchange α and β, and use that U_α is a continuous operator, cf. Proposition 7.13 b).

c) Since $U_\alpha = (\alpha\,\mathrm{id} - A)^{-1}$, cf. Proposition 7.13 f), we have $AU_\alpha u = -u + \alpha U_\alpha u$. For each $u \in \mathfrak{D}(U_0)$, we get

$$\lim_{\alpha\to 0} U_\alpha u = U_0 u \quad \text{and} \quad \lim_{\alpha\to 0} AU_\alpha u = -u + \lim_{\alpha\to 0} \alpha U_\alpha u = -u.$$

As $(A, \mathfrak{D}(A))$ is a closed operator, see Corollary 7.11, we conclude that $U_0 u \in \mathfrak{D}(A)$ and $AU_0 u = -u$.

d) Since $u \geqslant 0$, $P_t u \geqslant 0$ and we can use monotone convergence to get a (pointwise) extension of U_0

$$\widetilde{U}_0 u(x) := \sup_{\alpha>0} \int_0^\infty e^{-\alpha t} P_t u(x)\,dt = \int_0^\infty P_t u(x)\,dt \quad \text{for all } x \in \mathbb{R}^d,\ u \in \mathcal{C}_\infty^+(\mathbb{R}^d).$$

If $\widetilde{U}_0 u \in \mathcal{C}_\infty(\mathbb{R}^d)$, (a variant of) Dini's theorem [197, Theorem 7.13, p. 150] shows that the limit $\lim_{\alpha\to 0} U_\alpha u = \widetilde{U}_0 u$ is uniform, which means that $u \in \mathfrak{D}(U_0)$ and $U_0 u = \widetilde{U}_0 u$. Ex. 7.16

e) Assume that $\lim_{\alpha\to 0} \alpha U_\alpha u = 0$ for all $u \in \mathcal{C}_\infty(\mathbb{R}^d)$. Since U_α is the resolvent, we get

$$AU_\alpha u = U_\alpha A u = -u + \alpha U_\alpha u \quad \text{for all } u \in \mathfrak{D}(A)$$

and so

$$\lim_{\alpha\to 0} U_\alpha A u = \lim_{\alpha\to 0}(-u + \alpha U_\alpha u) = -u.$$

This proves that $Au \in \mathfrak{D}(U_0)$ and $U_0 A u = -u$. Together with c) we see that U_0 and A are inverses of each other and $\mathfrak{D}(U_0) = \mathfrak{R}(A)$.

Let $u \in \mathfrak{D}(A)$. As $Au \in \mathfrak{D}(U_0)$, we have by part a) that $U_\alpha A u \in \mathfrak{D}(U_0)$, and our calculation shows $\lim_{\alpha\to 0} AU_\alpha u = -u$. Since $AU_\alpha u = U_\alpha A u \in \mathfrak{D}(U_0)$, we conclude that $\mathfrak{D}(A) \subset \overline{\mathfrak{D}(U_0)}$, hence $\overline{\mathfrak{D}(A)} = \overline{\mathfrak{D}(U_0)} = \mathcal{C}_\infty(\mathbb{R}^d)$. □

7.25 Example (Günter 1934, 1957). Let $(B_t)_{t\geqslant 0}$ be a BM^d with semigroup $(P_t)_{t\geqslant 0}$, generator $(A, \mathfrak{D}(A))$ and resolvent $(U_\alpha)_{\alpha>0}$. Then

$$\mathcal{C}_\infty^2(\mathbb{R}^d) \subsetneq \mathfrak{D}(A) \quad \text{for all } d \geqslant 3.$$

Indeed: Because of Lemma 7.24 c),d) it is enough to show that there are positive functions $u, v \in \mathcal{C}_\infty(\mathbb{R}^d)$ such that

$$\sup_{\alpha>0} U_\alpha u,\ \sup_{\alpha>0} U_\alpha v \in \mathcal{C}_\infty(\mathbb{R}^d) \quad \text{but} \quad U_0(u-v)(x) = \sup_{\alpha>0} U_\alpha u(x) - \sup_{\alpha>0} U_\alpha v(x) \notin \mathcal{C}^2(\mathbb{R}^d).$$

Using monotone convergence and Tonelli's theorem we get for any $u \in \mathcal{C}_\infty^+(\mathbb{R}^d)$ and $d \geqslant 3$

$$\begin{aligned}
\sup_{\alpha>0} U_\alpha u(x) &= \int_0^\infty P_t u(x)\,dt \\
&= \int_0^\infty \int_{\mathbb{R}^d} \frac{1}{(2\pi t)^{d/2}} e^{-|y|^2/2t} u(x-y)\,dy\,dt \\
&= \int_{\mathbb{R}^d} \int_0^\infty \frac{1}{(2\pi t)^{d/2}} e^{-|y|^2/2t}\,dt\, u(x-y)\,dy \\
&\overset{r=|y|^2/2t}{=} \int_{\mathbb{R}^d} \frac{1}{2\,\pi^{d/2}} \int_0^\infty r^{\frac{d}{2}-2} e^{-r}\,dr\, |y|^{2-d}\, u(x-y)\,dy \\
&= \int_{\mathbb{R}^d} \frac{\Gamma(\frac{d}{2}-1)}{2\,\pi^{d/2}} |y|^{2-d}\, u(x-y)\,dy \\
&= c_d \int_{\mathbb{R}^d} |y|^{2-d}\, u(x-y)\,dy
\end{aligned}$$

with $c_d = \pi^{-d/2}\Gamma(\frac{d}{2})/(d-2)$. Since $\int_{|y|\leqslant 1} |y|^{2-d}\,dy < \infty$, we see with dominated convergence that the function $U_0 u(x)$ is in $\mathcal{C}_\infty(\mathbb{R}^d)$ for all $u \in \mathcal{C}_\infty^+(\mathbb{R}^d) \cap L^1(dy)$.

Ex. 7.17 For the following calculation we assume $d = 3$. In Problem 7.17 we sketch how to deal with $d \geqslant 4$.

Pick any positive $f \in \mathcal{C}_c([0,1))$ such that $f(0) = 0$. We denote by (x, y, z) points in $\mathbb{R}^3$ and set $r^2 := x^2 + y^2 + z^2$. Then let

$$u(x,y,z) := \frac{3z^2}{r^2} f(r), \quad v(x,y,z) := f(r), \quad w(x,y,z) := u(x,y,z) - v(x,y,z).$$

We will show that there is some f such that $w \in \mathfrak{D}(U_0)$ and $U_0 w \notin \mathcal{C}^2(\mathbb{R}^3)$. The first assertion follows directly from Lemma 7.24 d). Using the integral formula for $U_0 u$ and $U_0 v$ and, introducing polar coordinates, we get for $z \in (0, 1/2)$

$$U_0 w(0,0,z) = \int_0^1 r^2 f(r) \left(\int_0^\pi \frac{(3\cos^2\theta - 1)\sin\theta}{\sqrt{z^2 + r^2 - 2rz\cos\theta}}\,d\theta \right) dr.$$

Changing variables according to $t = \sqrt{z^2 + r^2 - 2zr\cos\theta}$ gives for the inner integral

$$\int_0^\pi \frac{(3\cos^2\theta - 1)\sin\theta}{\sqrt{z^2 + r^2 - 2rz\cos\theta}}\,d\theta = \frac{1}{rz}\int_{|z-r|}^{z+r}\left[3\left(\frac{z^2 + r^2 - t^2}{2zr}\right)^2 - 1\right]dt$$
$$= \begin{cases} \frac{4}{5}\frac{r^2}{z^3} & \text{for } r < z, \\ \frac{4}{5}\frac{z^2}{r^3} & \text{for } r \geqslant z. \end{cases}$$

Thus,

$$U_0 w(0,0,z) = \frac{4}{5}\left(\frac{1}{z^3}\int_0^z r^4 f(r)\,dr + z^2\int_z^1 \frac{f(r)}{r}\,dr\right).$$

We may differentiate this function in z and get

$$\frac{d}{dz}U_0 w(0,0,z) = \frac{4}{5}\left(-\frac{3}{z^4}\int_0^z r^4 f(r)\,dr + 2z\int_z^1 \frac{f(r)}{r}\,dr\right).$$

It is not hard to see that $\lim_{z\to 0+}\frac{d}{dz}U_0 w(0,0,z) = 0$. In order to calculate the second derivative in z-direction at $z = 0$ we consider

$$\frac{\frac{d}{dz}U_0 w(0,0,z) - \frac{d}{dz}U_0 w(0,0,0)}{z} = \frac{4}{5}\left(-\frac{3}{z^5}\int_0^z r^4 f(r)\,dr + 2\int_z^1 \frac{f(r)}{r}\,dr\right).$$

Letting $z \to 0$ yields

$$\lim_{z\to 0+}\frac{\frac{d}{dz}U_0 w(0,0,z) - \frac{d}{dz}U_0 w(0,0,0)}{z} = \frac{4}{5}\left(-\frac{3}{5}f(0) + 2\lim_{z\to 0+}\int_z^1 \frac{f(r)}{r}\,dr\right).$$

All we have to do is to pick a positive $f \in \mathcal{C}_c([0,1))$ such that the integral $\int_{0+}^1 f(r)\,\frac{dr}{r}$ diverges; a canonical candidate is $f(r) = |\log r|^{-1}\chi(r)$ with a suitable cut-off function $\chi \in \mathcal{C}_c^+([0,1))$ and $\chi|_{[0,1/2]} \equiv 1$.

This shows that $U_0 w \notin \mathcal{C}^2(\mathbb{R}^3)$, but $U_0 w$ is in the domain of the generator of three-dimensional Brownian motion.

7.26 Lemma. *Let $(B_t)_{t\geqslant 0}$ be a BM^d, and denote the corresponding resolvent by $(U_\alpha)_{\alpha>0}$. Then the condition of Lemma 7.24 e) holds, i. e. $\lim_{\alpha\to 0}\|\alpha U_\alpha u\|_\infty = 0$ for all $u \in \mathcal{C}_\infty(\mathbb{R}^d)$.*

Proof. Fix $\epsilon > 0$. Since $\mathcal{C}_c(\mathbb{R}^d)$ is dense in $\mathcal{C}_\infty(\mathbb{R}^d)$, there is some $u_\epsilon \in \mathcal{C}_c(\mathbb{R}^d)$ such that $\|u_\epsilon - u\|_\infty < \epsilon$. By Lemma 7.13 b) we get

$$\|\alpha U_\alpha u\|_\infty \leqslant \|\alpha U_\alpha u_\epsilon\|_\infty + \underbrace{\|\alpha U_\alpha(u - u_\epsilon)\|_\infty}_{\leqslant \|u - u_\epsilon\|_\infty} \leqslant \|\alpha U_\alpha u_\epsilon\|_\infty + \epsilon.$$

Therefore, it is enough to prove the claim for $u \in \mathcal{C}_c(\mathbb{R}^d)$ with, say, $\operatorname{supp} u \subset \mathbb{B}(0,R)$. From Example 7.14 we know that $U_\alpha u(x) = G_{\alpha,d} * u(x)$ for all $u \in \mathcal{C}_\infty(\mathbb{R}^d)$. Using the explicit representation (7.15) and a simple change of variables yield

$$\alpha U_\alpha u(x) = \frac{1}{2\pi^{d/2}} \int_{\mathbb{R}^d} (2|z|)^{1-\frac{d}{2}} K_{\frac{d}{2}-1}(|z|)\, u(x - z/\sqrt{2\alpha})\, dz.$$

By Proposition 7.13 d) $\alpha U_\alpha \in \mathcal{C}_\infty(\mathbb{R}^d)$, and so there is some $S = S_\epsilon > 0$ such that

$$\|\alpha U_\alpha u\|_\infty \leqslant \sup_{|x|\leqslant S} |\alpha U_\alpha u(x)| + \epsilon.$$

Observe that

$$\left| x - \frac{z}{\sqrt{2\alpha}} \right| \geqslant \frac{1}{R\sqrt{2\alpha}} - S \quad \text{for all } |x| \leqslant S,\ |z| \geqslant \frac{1}{R}.$$

Therefore, we have for small $\alpha \leqslant c = c(R, S)$

$$\begin{aligned} \alpha U_\alpha u(x) &= \frac{1}{2\pi^{d/2}} \Bigg(\int_{|z|<1/R} + \int_{|z|\geqslant 1/R} \Bigg) (2|z|)^{1-\frac{d}{2}} K_{\frac{d}{2}-1}(|z|)\, u(x - \tfrac{z}{\sqrt{2\alpha}})\, dz \\ &\leqslant \frac{1}{2\pi^{d/2}} \|u\|_\infty \int_{|z|<1/R} (2|z|)^{1-\frac{d}{2}} K_{\frac{d}{2}-1}(|z|)\, dz. \end{aligned}$$

By the well-known asymptotic behaviour of the Bessel functions near the origin, cf. [167, 10.27.3, 10.30(i), pp. 251–252], we get

$$K_\nu(x) \approx \begin{cases} \frac{1}{2}\sqrt{\pi}\left(\frac{1}{2}x\right)^{-\frac{1}{2}}, & \text{if } \nu = -\frac{1}{2}, \\ -\log x, & \text{if } \nu = 0, \\ \frac{1}{2}\Gamma(\nu)\left(\frac{1}{2}x\right)^{-\nu}, & \text{if } \nu > 0, \end{cases} \qquad \text{as } x \to 0+,$$

Thus, if $R > R_\epsilon$ is large, $\sup_{|x|\leqslant S} |\alpha U_\alpha u(x)| \leqslant \epsilon$. This proves $\lim_{\alpha\to 0} \|\alpha U_\alpha u\|_\infty = 0$ for any $u \in \mathcal{C}_c(\mathbb{R}^d)$. □

7.27 Theorem (Yosida 1970). *Let $(B_t)_{t\geqslant 0}$ be a BM^d, and denote the corresponding resolvent by $(U_\alpha)_{\alpha>0}$. Then the potential operator is densely defined and it is the inverse of the generator of $(B_t)_{t\geqslant 0}$: $U_0 = -A^{-1}$. Moreover,*

$$U_0 u(x) = G_0 * u(x) = \int G_0(y) u(x - y)\, dy \tag{7.24}$$

for all $u \in \mathfrak{D}(U_0) \cap \mathcal{C}_c(\mathbb{R}^d)$ (with $\int u(x)\, dx = 0$ if $d = 1, 2$) where

$$G_0(y) = G_{0,d}(y) = \begin{cases} -|y| & \textit{if } d = 1 \quad \textit{(linear potential)}; \\ \frac{1}{\pi} \log \frac{1}{|y|} & \textit{if } d = 2 \quad \textit{(logarithmic potential)}; \\ \frac{\Gamma(d/2)}{\pi^{d/2}(d-2)} |y|^{2-d} & \textit{if } d \geqslant 3 \quad \textit{(Newton potential)}. \end{cases} \tag{7.25}$$

Proof. Lemmas 7.24 e) and 7.26 show that $\mathfrak{D}(U_0)$ is dense in $\mathcal{C}_\infty(\mathbb{R}^d)$ and $U_0 = -A^{-1}$.

The formula for $G_{0,d}(y)$ in dimension $d \geqslant 3$ follows from the calculation in Example 7.25: Just write $u \in \mathfrak{D}(U_0) \cap \mathcal{C}_c(\mathbb{R}^d)$ as $u = u^+ - u^-$.

Now we consider the case $d = 1$. Let $u \in \mathfrak{D}(U_0) \cap \mathcal{C}_c(\mathbb{R})$ with $\int u(y)\,dy = 0$. Then, by dominated convergence,

$$U_0 u(x) = \lim_{\alpha \to 0} \int_{\mathbb{R}} \frac{e^{-\sqrt{2\alpha}|x-y|}}{\sqrt{2\alpha}} u(y)\,dy$$

$$= \lim_{\alpha \to 0} \int_{\mathbb{R}} \frac{e^{-\sqrt{2\alpha}|x-y|} - 1}{\sqrt{2\alpha}} u(y)\,dy = \int_{\mathbb{R}} -|x-y| u(y)\,dy.$$

Finally, let $d = 2$ and $u \in \mathfrak{D}(U_0) \cap \mathcal{C}_c(\mathbb{R}^2)$ such that $\int u(y)\,dy = 0$. Observe that

$$G_{\alpha,2}(y) = \frac{1}{\pi} K_0(\sqrt{2\alpha}|y|) = -\frac{1}{\pi^2} \int_0^\pi e^{\sqrt{2\alpha}|y|\cos\theta} \left[\gamma + \log\left(2\sqrt{2\alpha}|y|\sin^2\theta\right)\right] d\theta$$

where $\gamma = 0.57721\ldots$ is Euler's constant, cf. [167, 10.32.5, p. 252]. Then,

$$\begin{aligned}
&U_0 u(x) \\
&= -\frac{1}{\pi^2} \lim_{\alpha \to 0} \int_{\mathbb{R}^2} \int_0^\pi e^{\sqrt{2\alpha}|x-y|\cos\theta} \left[\gamma + \log\left(2\sqrt{2\alpha}\sin^2\theta\right) + \log|x-y|\right] d\theta\, u(y)\,dy \\
&= -\frac{1}{\pi} \int_{\mathbb{R}^2} \log|x-y|\, u(y)\,dy \\
&\qquad - \frac{1}{\pi^2} \lim_{\alpha \to 0} \sqrt{2\alpha} \log\sqrt{2\alpha} \int_{\mathbb{R}^2} \left[\int_0^\pi \frac{e^{\sqrt{2\alpha}|x-y|\cos\theta} - 1}{\sqrt{2\alpha}}\, d\theta\right] u(y)\,dy \\
&= -\frac{1}{\pi} \int_{\mathbb{R}^2} \log|x-y|\, u(y)\,dy.
\end{aligned}$$

□

Using the positivity of the semigroup $(P_t)_{t \geqslant 0}$, there is a natural (pointwise) extension of U_0 to positive measurable functions $u \in \mathcal{B}_b^+(\mathbb{R}^d)$.

$$\begin{aligned}
U_0 u(x) = \sup_{\alpha > 0} \int_0^\infty e^{-\alpha t} P_t u(x)\,dt &= \sup_{\alpha > 0} \sup_{r > 0} \int_0^r e^{-\alpha t} P_t u(x)\,dt \\
&= \sup_{r > 0} \sup_{\alpha > 0} \int_0^r e^{-\alpha t} P_t u(x)\,dt \\
&= \sup_{r > 0} \int_0^r P_t u(x)\,dt \in [0, \infty].
\end{aligned}$$

The last expression yields an alternative way to define the potential operator.

7.28 Definition (Hunt 1957). Let $(P_t)_{t\geqslant 0}$ be a Feller semigroup on $\mathcal{C}_\infty(\mathbb{R}^d)$. The operator $(G, \mathfrak{D}(G))$ defined by

$$Gu := \lim_{r\to\infty} \int_0^r P_t u\,dt \quad \big(\text{the limit is taken in } (\mathcal{C}_\infty(\mathbb{R}^d), \|\cdot\|_\infty)\big) \tag{7.26a}$$

$$\mathfrak{D}(G) := \Big\{u \in \mathcal{C}_\infty(\mathbb{R}^d) \;\Big|\; \exists\, g \in \mathcal{C}_\infty(\mathbb{R}^d) : \lim_{r\to\infty} \Big\|\textstyle\int_0^r P_t u\,dt - g\Big\|_\infty = 0\Big\} \tag{7.26b}$$

is called *Hunt's potential operator* of the semigroup $(P_t)_{t\geqslant 0}$.

If we consider U_0 and G as operators in the Banach space $(\mathcal{C}_\infty(\mathbb{R}^d), \|\cdot\|_\infty)$, then Yosida's definition is more general than Hunt's.

7.29 Theorem (Berg 1974). *Let $(P_t)_{t\geqslant 0}$ be a Feller semigroup on $\mathcal{C}_\infty(\mathbb{R}^d)$ with generator $(A, \mathfrak{D}(A))$, potential operators $(U_\alpha)_{\alpha>0}$, $(U_0, \mathfrak{D}(U_0))$ and Hunt's potential operator $(G, \mathfrak{D}(G))$.*

a) *$(U_0, \mathfrak{D}(U_0))$ extends $(G, \mathfrak{D}(G))$, i. e. $\mathfrak{D}(G) \subset \mathfrak{D}(U_0)$ and $U_0|_{\mathfrak{D}(G)} = G$.*

b) *If $\lim_{t\to\infty} \|P_t u\|_\infty = 0$ for all $u \in \mathcal{C}_\infty(\mathbb{R}^d)$, then $G = U_0 = -A^{-1}$.*

Proof. a) Pick $u \in \mathfrak{D}(G)$ and set $G_r u := \int_0^r P_t u\,dt$ for $r > 0$. Integrating by parts yields

$$\int_0^r e^{-t\alpha} P_t u\,dt = \big[e^{-t\alpha} G_t u\big]_{t=0}^r + \int_0^r \alpha e^{-t\alpha} G_t u\,dt.$$

Letting $r \to \infty$ shows (all limits are taken in $\mathcal{C}_\infty(\mathbb{R}^d)$)

$$U_\alpha u = \int_0^\infty e^{-t\alpha} P_t u\,dt = \int_0^\infty \alpha e^{-t\alpha} G_t u\,dt \overset{s=t\alpha}{=} \int_0^\infty e^{-s} G_{s/\alpha} u\,ds.$$

By assumption, $G_{s/\alpha} u = \int_0^{s/\alpha} P_t u\,dt \to Gu$ uniformly as $\alpha \to 0$. So, $\lim_{\alpha\to 0} U_\alpha u = Gu$, and this proves $u \in \mathfrak{D}(U_0)$ and $U_0 u = Gu$.

b) Lemma 7.24 e) shows that $\mathfrak{D}(U_0) = \mathfrak{R}(A)$ and $U_0 = -A^{-1}$. Therefore, it is enough to show that $\mathfrak{D}(G) = \mathfrak{R}(A)$ and $G = -A^{-1}$. Let $G_t u := \int_0^t P_s u\,ds$. From (7.11c) we see

$$P_t u - u = \int_0^t P_s A u\,ds = G_t A u \quad \text{for all } u \in \mathfrak{D}(A)$$

and, if $t \to \infty$, we get $Au \in \mathfrak{D}(G)$ and $-u = GAu$. On the other hand, by (7.11a),

$$P_t u - u = A \int_0^t P_s u\,ds = AG_t u \quad \text{for all } u \in \mathfrak{D}(G) \subset \mathcal{C}_\infty(\mathbb{R}^d).$$

Letting $t \to \infty$ yields, because of the closedness of A, that $Gu = \lim_{t\to\infty} G_t u \in \mathfrak{D}(A)$ and $-u = AGu$. Thus, $\mathfrak{D}(G) = \mathfrak{R}(A)$ and $G = -A^{-1}$. ☐

7.6 Dynkin's characteristic operator

We will now give a probabilistic characterization of the infinitesimal generator. As so often, a simple martingale relation turns out to be extremely helpful. The following theorem should be compared with Theorem 5.6. Recall that a Feller process is a (strong[1]) Markov process with right-continuous trajectories whose transition semigroup $(P_t)_{t\geqslant 0}$ is a Feller semigroup. Of course, a Brownian motion is a Feller process.

7.30 Theorem. *Let $(X_t, \mathscr{F}_t)_{t\geqslant 0}$ be a Feller process on $\mathbb{R}^d$ with transition semigroup $(P_t)_{t\geqslant 0}$ and generator $(A, \mathfrak{D}(A))$. Then* Ex. 7.18

$$M_t^u := u(X_t) - u(x) - \int_0^t Au(X_r)\,dr \quad \textit{for all } u \in \mathfrak{D}(A)$$

is an $\mathscr{F}_t$ martingale.

Proof. Let $u \in \mathfrak{D}(A)$, $x \in \mathbb{R}^d$ and $s, t > 0$. By the Markov property (6.4c),

$$\begin{aligned}
\mathbb{E}^x\left(M_{s+t}^u \mid \mathscr{F}_t\right) &= \mathbb{E}^x\left(u(X_{s+t}) - u(x) - \int_0^{s+t} Au(X_r)\,dr \,\middle|\, \mathscr{F}_t\right)\\
&= \mathbb{E}^{X_t} u(X_s) - u(x) - \int_0^t Au(X_r)\,dr - \mathbb{E}^x\left(\int_t^{s+t} Au(X_r)\,dr \,\middle|\, \mathscr{F}_t\right)\\
&= P_s u(X_t) - u(x) - \int_0^t Au(X_r)\,dr - \mathbb{E}^{X_t}\left(\int_0^s Au(X_r)\,dr\right)\\
&= P_s u(X_t) - u(x) - \int_0^t Au(X_r)\,dr - \int_0^s \mathbb{E}^{X_t}\left(Au(X_r)\right)dr\\
&= P_s u(X_t) - u(x) - \int_0^t Au(X_r)\,dr - \underbrace{\int_0^s P_r Au(X_t)\,dr}_{=P_s u(X_t) - u(X_t) \text{ by (7.11c)}}\\
&= u(X_t) - u(x) - \int_0^t Au(X_r)\,dr\\
&= M_t^u.
\end{aligned}$$

□

The following result, due to Dynkin, could be obtained from Theorem 7.30 by optional stopping, but we prefer to give an independent proof. Ex. 7.19

7.31 Proposition (Dynkin 1956). *Let $(X_t, \mathscr{F}_t)_{t\geqslant 0}$ be a Feller process on $\mathbb{R}^d$ with transition semigroup $(P_t)_{t\geqslant 0}$, resolvent $(U_\alpha)_{\alpha>0}$ and generator $(A, \mathfrak{D}(A))$. If σ is an almost surely*

1 Feller processes have the strong Markov property, see Theorem A.25 in the appendix.

finite $\mathscr{F}_t$ stopping time, then

$$U_\alpha f(x) = \mathbb{E}^x\left(\int_0^\sigma e^{-\alpha t} f(X_t)\,dt\right) + \mathbb{E}^x\left(e^{-\alpha\sigma} U_\alpha f(X_\sigma)\right), \quad f \in \mathcal{B}_b(\mathbb{R}^d) \tag{7.27}$$

for all $x \in \mathbb{R}^d$. If $\mathbb{E}^x \sigma < \infty$, then Dynkin's formula *holds*

$$\mathbb{E}^x u(X_\sigma) - u(x) = \mathbb{E}^x\left(\int_0^\sigma Au(X_t)\,dt\right), \quad u \in \mathfrak{D}(A). \tag{7.28}$$

Proof. Since $\sigma < \infty$ a. s., we get

$$\begin{aligned} U_\alpha f(x) &= \mathbb{E}^x\left(\int_0^\infty e^{-\alpha t} f(X_t)\,dt\right) \\ &= \mathbb{E}^x\left(\int_0^\sigma e^{-\alpha t} f(X_t)\,dt\right) + \mathbb{E}^x\left(\int_\sigma^\infty e^{-\alpha t} f(X_t)\,dt\right). \end{aligned}$$

Using the strong Markov property (6.7), the second integral becomes

$$\begin{aligned} \mathbb{E}^x\left(\int_\sigma^\infty e^{-\alpha t} f(X_t)\,dt\right) &= \mathbb{E}^x\left(\int_0^\infty e^{-\alpha(t+\sigma)} f(X_{t+\sigma})\,dt\right) \\ &= \int_0^\infty \mathbb{E}^x\left(e^{-\alpha(t+\sigma)}\,\mathbb{E}^x\left[f(X_{t+\sigma}) \mid \mathscr{F}_{\sigma+}\right]\right) dt \\ &= \int_0^\infty e^{-\alpha t}\,\mathbb{E}^x\left(e^{-\alpha\sigma}\,\mathbb{E}^{X_\sigma} f(X_t)\right) dt \\ &= \mathbb{E}^x\left(e^{-\alpha\sigma}\int_0^\infty e^{-\alpha t} P_t f(X_\sigma)\,dt\right) \\ &= \mathbb{E}^x\left(e^{-\alpha\sigma} U_\alpha f(X_\sigma)\right). \end{aligned}$$

If $u \in \mathfrak{D}(A)$, there is some $f \in \mathcal{C}_\infty(\mathbb{R}^d)$ such that $u = U_\alpha f$, hence $f = \alpha u - Au$. Using the first identity, we find by dominated convergence – we use $\mathbb{E}^x \sigma < \infty$ in the step $(*)$ below to get an integrable majorant –:

$$\begin{aligned} u(x) &= \mathbb{E}^x\left(\int_0^\sigma e^{-\alpha t} f(X_t)\,dt\right) + \mathbb{E}^x\left(e^{-\alpha\sigma} u(X_\sigma)\right) \\ &\overset{(*)}{=} \mathbb{E}^x\left(\int_0^\sigma e^{-\alpha t}(\alpha u(X_t) - Au(X_t))\,dt\right) + \mathbb{E}^x\left(e^{-\alpha\sigma} u(X_\sigma)\right) \\ &\xrightarrow[\alpha\to 0]{} -\mathbb{E}^x\left(\int_0^\sigma Au(X_t)\,dt\right) + \mathbb{E}^x\left(u(X_\sigma)\right). \end{aligned}$$

□

Assume that the filtration $(\mathscr{F}_t)_{t\geqslant 0}$ is right-continuous. Then the first hitting time of the open set $\mathbb{R}^d \setminus \{x\}$, $\tau_x := \inf\{t > 0 : X_t \neq x\}$, is a stopping time, cf. Lemma 5.7. One can show that $\mathbb{P}^x(\tau_x \geqslant t) = \exp(-\lambda(x)t)$ for some exponent $\lambda(x) \in [0, \infty]$, see Theorem A.26 in the appendix. For a Brownian motion we have $\lambda(x) \equiv \infty$, since for $t > 0$

$$\mathbb{P}^x(\tau_x \geqslant t) = \mathbb{P}^x\Big(\sup_{s\leqslant t} |B_s - x| = 0\Big) \leqslant \mathbb{P}^x(|B_t - x| = 0) = \mathbb{P}(B_t = 0) = 0.$$

Under $\mathbb{P}^x$ the random time τ_x is the waiting time of the process at the starting point. Therefore, $\lambda(x)$ tells us how quickly a process leaves its starting point, and we can use it to give a classification of the starting points.

7.32 Definition. Let $(X_t)_{t\geqslant 0}$ be a Feller process. A point $x \in \mathbb{R}^d$ is said to be

a) an *exponential holding point* if $0 < \lambda(x) < \infty$,

b) an *instantaneous point* if $\lambda(x) = \infty$,

c) an *absorbing point* if $\lambda(x) = 0$.

7.33 Lemma. *Let $(X_t, \mathscr{F}_t)_{t\geqslant 0}$ be a Feller process with a right-continuous filtration, transition semigroup $(P_t)_{t\geqslant 0}$ and generator $(A, \mathfrak{D}(A))$. If x_0 is not absorbing, then*

a) *there exists some $u \in \mathfrak{D}(A)$ such that $Au(x_0) \neq 0$;*

b) *there is an open neighbourhood $U = U(x_0)$ of x_0 such that for $V := \mathbb{R}^d \setminus \overline{U}$ the first hitting time $\tau_V := \inf\{t > 0 : X_t \in V\}$ satisfies $\mathbb{E}^{x_0} \tau_V < \infty$.*

Proof. a) We show that $Aw(x_0) = 0$ for all $w \in \mathfrak{D}(A)$ implies that x_0 is an absorbing point. Let $u \in \mathfrak{D}(A)$. By Lemma 7.10 a) $w = P_t u \in \mathfrak{D}(A)$ and

$$\frac{d}{dt} P_t u(x_0) = AP_t u(x_0) = 0 \quad \text{for all } t \geqslant 0.$$

Thus, (7.11b) shows that $P_t u(x_0) = u(x_0)$. Since $\mathfrak{D}(A)$ is dense in $\mathcal{C}_\infty(\mathbb{R}^d)$ we get $\mathbb{E}^{x_0} u(X_t) = P_t u(x_0) = u(x_0)$ for all $u \in \mathcal{C}_\infty(\mathbb{R}^d)$. Using the sequence from (7.7), we can approximate $x \mapsto \mathbb{1}_{\{x_0\}}(x)$ from above by $(u_n)_{n\geqslant 0} \subset \mathcal{C}_\infty(\mathbb{R}^d)$. By monotone convergence,

$$\mathbb{P}^{x_0}(X_t = x_0) = \lim_{n\to\infty} \underbrace{\mathbb{E}^{x_0} u_n(X_t)}_{=u_n(x_0)} = \lim_{n\to\infty} u_n(x_0) = \mathbb{1}_{\{x_0\}}(x_0) = 1.$$

Since $t \mapsto X_t$ is a. s. right-continuous, we find Ex. 7.20

$$1 = \mathbb{P}^{x_0}\big(X_q = x_0 \ \forall q \in \mathbb{Q} \cap [0, \infty)\big) = \mathbb{P}^{x_0}\big(X_t = x_0 \ \forall t \in [0, \infty)\big)$$

which means that x_0 is an absorbing point.

b) If x_0 is not absorbing, then part a) shows that there is some $u \in \mathfrak{D}(A)$ such that $Au(x_0) > 0$. Since Au is continuous, there exists some $\epsilon > 0$ and an open neighbourhood $U = U(x_0, \epsilon)$ such that $Au|_U \geqslant \epsilon > 0$. Let τ_V be the first hitting time of the open set $V = \mathbb{R}^d \setminus \overline{U}$. From Dynkin's formula (7.28) with $\sigma = \tau_V \wedge n$, $n \geqslant 1$, we deduce

$$\epsilon\, \mathbb{E}^{x_0}(\tau_V \wedge n) \leqslant \mathbb{E}^{x_0}\left(\int_0^{\tau_V\wedge n} Au(X_s)\, ds\right) = \mathbb{E}^{x_0} u(X_{\tau_V\wedge n}) - u(x_0) \leqslant 2\|u\|_\infty.$$

Finally, monotone convergence shows that $\mathbb{E}^{x_0}\tau_V \leqslant 2\|u\|_\infty/\epsilon < \infty$. □

7.34 Definition (Dynkin 1956). Let $(X(t))_{t\geqslant 0}$ be a Feller process and $\tau_r^x := \tau_{\overline{\mathbb{B}}^c(x,r)}$ the first hitting time of the set $\overline{\mathbb{B}}^c(x,r)$. *Dynkin's characteristic operator* is the linear operator defined by

$$\mathfrak{A}u(x) := \begin{cases} \lim\limits_{r\to 0} \dfrac{\mathbb{E}^x u(X(\tau_r^x)) - u(x)}{\mathbb{E}^x \tau_r^x}, & \text{if } x \text{ is not absorbing,} \\ 0, & \text{if } x \text{ is absorbing,} \end{cases} \tag{7.29}$$

on the set $\mathfrak{D}(\mathfrak{A})$ consisting of all $u \in \mathcal{B}_b(\mathbb{R}^d)$ such that the limit in (7.29) exists for each non-absorbing point $x \in \mathbb{R}^d$.

Note that, by Lemma 7.33, $\mathbb{E}^x \tau_r^x < \infty$ for sufficiently small $r > 0$.

7.35 Theorem. *Let $(X_t)_{t\geqslant 0}$ be a Feller process with generator $(A, \mathfrak{D}(A))$ and characteristic operator $(\mathfrak{A}, \mathfrak{D}(\mathfrak{A}))$.*
a) *$\mathfrak{A}$ is an extension of A;*
b) *$(\mathfrak{A}, \mathfrak{D}) = (A, \mathfrak{D}(A))$ if $\mathfrak{D} = \{u \in \mathfrak{D}(\mathfrak{A}) : u, \mathfrak{A}u \in \mathcal{C}_\infty(\mathbb{R}^d)\}$.*

Proof. a) Let $u \in \mathfrak{D}(A)$ and assume that x is not an absorbing point. Since we have $Au \in \mathcal{C}_\infty(\mathbb{R}^d)$, there exists for every $\epsilon > 0$ an open neighbourhood $U_0 = U_0(\epsilon, x)$ of x such that

$$|Au(y) - Au(x)| < \epsilon \quad \text{for all } \ y \in U_0.$$

Since x is not absorbing, we find some open neighbourhood $U_1 \subset \overline{U}_1 \subset U_0$ of x such that $\mathbb{E}^x \tau_{\overline{U}_1^c} < \infty$. By Dynkin's formula (7.28) for the stopping time $\sigma = \tau_{\overline{\mathbb{B}}^c(x,r)}$ where $\mathbb{B}(x,r) \subset U_1 \subset U_0$ we see

$$\begin{aligned} &\left| \mathbb{E}^x u(X_{\tau_{\overline{\mathbb{B}}^c(x,r)}}) - u(x) - Au(x)\, \mathbb{E}\, \tau_{\overline{\mathbb{B}}^c(x,r)} \right| \\ &\leqslant \mathbb{E}^x \Bigg(\int\limits_{[0,\tau_{\overline{\mathbb{B}}^c(x,r)})} \underbrace{|Au(X_s) - Au(x)|}_{\leqslant\epsilon}\, ds \Bigg) \leqslant \epsilon\, \mathbb{E}^x \tau_{\overline{\mathbb{B}}^c(x,r)}. \end{aligned}$$

Thus, $\lim_{r\to 0} \Big(\mathbb{E}^x u(X_{\tau_{\overline{\mathbb{B}}^c(x,r)}}) - u(x) \Big) / \mathbb{E}^x \tau_{\overline{\mathbb{B}}^c(x,r)} = Au(x)$.

If x is an absorbing point, $Au(x) = 0$ by Lemma 7.33 a), hence $Au = \mathfrak{A}u$ for all $u \in \mathfrak{D}(A)$.

b) Since $(\mathfrak{A}, \mathfrak{D})$ satisfies the positive maximum principle (7.21), the claim follows like Theorem 7.22. □

Let us finally show that processes with continuous sample paths are generated by differential operators.

7.36 Definition. Let $L : \mathfrak{D}(L) \subset \mathcal{B}_b(\mathbb{R}^d) \to \mathcal{B}_b(\mathbb{R}^d)$ be a linear operator. It is called *local* if for every $x \in \mathbb{R}^d$ and $u, w \in \mathfrak{D}(L)$ satisfying $u|_{\mathbb{B}(x,\epsilon)} \equiv w|_{\mathbb{B}(x,\epsilon)}$ in some open neighbourhood $\mathbb{B}(x,\epsilon)$ of x we have $Lu(x) = Lw(x)$.

Typical local operators are differential operators and multiplication operators, e. g. $u \mapsto \sum_{i,j} a_{ij}(\cdot)\partial_i\partial_j u(\cdot) + \sum_j b_j(\cdot)\partial_j u(\cdot) + c(\cdot)u(\cdot)$. A typical non-local operator is an integral operator, e. g. $u \mapsto \int_{\mathbb{R}^d} k(x,y)\,u(y)\,dy$. A deep result by J. Peetre says that local operators are essentially differential operators:

7.37 Theorem (Peetre 1960). *Suppose that $L : \mathcal{C}_c^\infty(\mathbb{R}^d) \to \mathcal{C}_c^k(\mathbb{R}^d)$ is a linear operator where $k \geqslant 0$ is fixed. If $\operatorname{supp} Lu \subset \operatorname{supp} u$, then $Lu = \sum_{\alpha\in\mathbb{N}_0^d} a_\alpha(\cdot)\frac{\partial^\alpha}{\partial x^\alpha}u$ with finitely many, uniquely determined distributions $a_\alpha \in \mathcal{D}'(\mathbb{R}^d)$ (i. e. the topological dual of $\mathcal{C}_c^\infty(\mathbb{R}^d)$) which are locally represented by functions of class $\mathcal{C}^k$.*

Let L be as in Theorem 7.37 and assume that $L = A|_{\mathcal{C}_c^\infty(\mathbb{R}^d)}$ where $(A, \mathfrak{D}(A))$ is the generator of a Markov process. Then L has to satisfy the positive maximum principle (7.21). In particular, L is at most a *second order differential operator*. This follows from the fact that we can find test functions $\phi \in \mathcal{C}_c^\infty(\mathbb{R}^d)$ such that x_0 is a global maximum while $\partial_j\partial_k\partial_l\phi(x_0)$ has arbitrary sign. **!**

Every Feller process with continuous sample paths has a local generator.

7.38 Theorem. *Let $(X_t)_{t\geqslant 0}$ be a Feller process with continuous sample paths. Then the generator $(A, \mathfrak{D}(A))$ is a local operator.*

Proof. Because of Theorem 7.35 it is enough to show that $(\mathfrak{A}, \mathfrak{D}(\mathfrak{A}))$ is a local operator. If x is absorbing, this is clear. Assume, therefore, that x is not absorbing. If the functions $u, w \in \mathfrak{D}(\mathfrak{A})$ coincide in $\mathbb{B}(x,\epsilon)$ for some $\epsilon > 0$, we know for all $r < \epsilon$

$$u(X(\tau_{\overline{\mathbb{B}}^c(x,r)})) = w(X(\tau_{\overline{\mathbb{B}}^c(x,r)}))$$

where we used that $X(\tau_{\mathbb{B}^c(x,r)}) \in \overline{\mathbb{B}}(x,r) \subset \mathbb{B}(x,\epsilon)$ because of the continuity of the trajectories. Thus, $\mathfrak{A}u(x) = \mathfrak{A}w(x)$. □

If the domain $\mathfrak{D}(A)$ is sufficiently rich, e. g. if it contains the test functions $\mathcal{C}_c^\infty(\mathbb{R}^d)$, then the local property of A follows from the speed of the process moving away from its starting point.

7.39 Theorem. *Let $(X_t)_{t\geqslant 0}$ be a Feller process such that $\mathcal{C}_c^\infty(\mathbb{R}^d)$ is in the domain of the generator $(A, \mathfrak{D}(A))$. Then $(A, \mathcal{C}_c^\infty(\mathbb{R}^d))$ is a local operator if, and only if,*

$$\forall \epsilon > 0,\ r > 0,\ K \subset \mathbb{R}^d \text{ compact} : \ \lim_{h\to 0}\sup_{t\leqslant h}\sup_{x\in K}\frac{1}{h}\,\mathbb{P}^x(r > |X_t - x| > \epsilon) = 0. \tag{7.30}$$

Proof. Conversely, assume that $(A, \mathcal{C}_c^\infty(\mathbb{R}^d))$ is local. Fix $\epsilon > 0$, $r > 3\epsilon$, pick any compact set K and cover it by finitely many balls $\mathbb{B}(x_j, \epsilon)$, $j = 1, \dots, N$.

Since $\mathcal{C}_c^\infty(\mathbb{R}^d) \subset \mathfrak{D}(A)$, there exist functions $u_j \in \mathcal{C}_c^\infty(\mathbb{R}^d)$, $j = 1, \dots, N$, with Ex. 7.8

$$u_j \geqslant 0, \quad u_j|_{\mathbb{B}(x_j,\epsilon)} \equiv 0 \quad \text{and} \quad u_j|_{\overline{\mathbb{B}}(x_j, r+\epsilon)\setminus\mathbb{B}(x_j,2\epsilon)} \equiv 1$$

and

$$\lim_{t\to 0}\left\|\frac{1}{t}(P_t u_j - u_j) - Au_j\right\|_\infty = 0 \quad \text{for all } j = 1, \dots, N.$$

Therefore, there is for every $\delta > 0$ and all $j = 1, \dots, N$ some $h = h_\delta > 0$ such that for all $t < h$ and all $x \in \mathbb{R}^d$

$$\frac{1}{t} \left| \mathbb{E}^x u_j(X_t) - u_j(x) - tAu_j(x) \right| \leqslant \delta.$$

If $x \in \mathbb{B}(x_j, \epsilon)$, then $u_j(x) = 0$ and, by locality, $Au_j(x) = 0$. Thus

$$\frac{1}{t} \mathbb{E}^x u_j(X_t) \leqslant \delta;$$

for $x \in \mathbb{B}(x_j, \epsilon)$ and $r > |X_t - x| > 3\epsilon$ we have $r + \epsilon > |X_t - x_j| > 2\epsilon$, therefore

$$\frac{1}{t}\mathbb{P}^x(r > |X_t - x| > 3\epsilon) \leqslant \frac{1}{t}\mathbb{P}^x(r + \epsilon > |X_t - x_j| > 2\epsilon) \leqslant \frac{1}{t} P_t u_j(x) \leqslant \delta.$$

This proves

$$\begin{aligned}\overline{\lim_{h\to 0}} \frac{1}{h} \sup_{x\in K} \sup_{t\leqslant h} \mathbb{P}^x(r > |X_t - x| > 3\epsilon) \\ \leqslant \overline{\lim_{h\to 0}} \sup_{x\in K} \sup_{t\leqslant h} \frac{1}{t} \mathbb{P}^x(r > |X_t - x| > 3\epsilon) \leqslant \delta.\end{aligned}$$

Conversely, Fix $x \in \mathbb{R}^d$ and assume that $u, w \in \mathcal{C}_c^\infty(\mathbb{R}^d) \subset \mathfrak{D}(A)$ coincide on $\mathbb{B}(x, \epsilon)$ for some $\epsilon > 0$. Moreover, pick r in such a way that $\operatorname{supp} u \cup \operatorname{supp} w \subset \mathbb{B}(x, r)$. Then

$$\begin{aligned}|Au(x) - Aw(x)| &= \lim_{t\to 0} \left| \frac{\mathbb{E}^x u(X_t) - u(x)}{t} - \frac{\mathbb{E}^x w(X_t) - w(x)}{t} \right| \\ &= \lim_{t\to 0} \frac{1}{t} \left| \mathbb{E}^x \left(u(X_t) - w(X_t)\right) \right| \\ &\leqslant \lim_{t\to 0} \frac{1}{t} \int_{\mathbb{R}^d} \underbrace{|u(y) - w(y)|}_{=0 \text{ for } |y-x|<\epsilon \text{ or } |y-x|\geqslant r} \mathbb{P}^x(X_t \in dy) \\ &\leqslant \|u - w\|_\infty \lim_{t\to 0} \frac{1}{t} \mathbb{P}^x(r > |X_t - x| > \epsilon) \overset{(7.30)}{=} 0,\end{aligned}$$

proving that $(A, \mathcal{C}_c^\infty(\mathbb{R}^d))$ is a local operator. □

7.40 Further reading. Semigroups in probability theory are nicely treated in the first chapter of [72]; the presentation in (the first volume of) [65] is still up-to-date. Both sources approach things from the analysis side. The survey paper [102] gives a brief self-contained introduction from a probabilistic point of view. The Japanese school has a long tradition approaching stochastic processes through semigroups, see for example [103] and [104]. We did not touch the topic of quadratic forms induced by a generator, so-called Dirichlet forms. The classic text is [85]. The monograph by Port–Stone [182] contains a detailed discussion of potentials from Hunt's perspective.

[65] Dynkin: *Markov Processes* (volume 1).
[72] Ethier, Kurtz: *Markov Processes. Characterization and Convergence.*
[85] Fukushima, Oshima, Takeda: *Dirichlet Forms and Symmetric Markov Processes.*
[102] Itô: Semigroups in Probability Theory.
[103] Itô: *Stochastic Processes.*
[104] Itô: *Essentials of Stochastic Processes.*
[182] Port, Stone: *Brownian Motion and Classical Potential Theory.*

Problems

1. Show that $\mathcal{C}_\infty := \{u : \mathbb{R}^d \to \mathbb{R} : \text{continuous and } \lim_{|x|\to\infty} u(x) = 0\}$ equipped with the uniform topology is a Banach space. Show that in this topology $\mathcal{C}_\infty$ is the closure of $\mathcal{C}_c$, i. e. the family of continuous functions with compact support.

2. Let $(B_t)_{t\geqslant 0}$ be a BM^d with transition semigroup $(P_t)_{t\geqslant 0}$. Show, using arguments similar to those in the proof of Proposition 7.3 f), that $(t, x, u) \mapsto P_t u(x)$ is continuous. As usual, we equip $[0,\infty) \times \mathbb{R}^d \times \mathcal{C}_\infty(\mathbb{R}^d)$ with the norm $|t| + |x| + \|u\|_\infty$.

3. Let $(X_t)_{t\geqslant 0}$ be a d-dimensional Feller process and let $f, g \in \mathcal{C}_\infty(\mathbb{R}^d)$. Show that the function $x \mapsto \mathbb{E}^x(f(X_t)g(X_{t+s}))$ is also in $\mathcal{C}_\infty(\mathbb{R}^d)$.
   Hint: Markov property.

4. Let $(B_t)_{t\geqslant 0}$ be a BM^d and set $u(t,x) := P_t u(x)$. Adapt the proof of Proposition 7.3 g) and show that for $u \in \mathcal{B}_b(\mathbb{R}^d)$ the function $u(t,\cdot) \in \mathcal{C}^\infty$ and $u \in \mathcal{C}^{1,2}$ (i. e. once continuously differentiable in t, twice continuously differentiable in x).

5. Let $(P_t)_{t\geqslant 0}$ be the transition semigroup of a BM^d and denote by L^p, $1 \leqslant p < \infty$, the space of pth power integrable functions with respect to Lebesgue measure on $\mathbb{R}^d$. Set $u_n := (-n) \vee u \wedge n$ for every $u \in L^p$. Verify that
   (a) $u_n \in \mathcal{B}_b(\mathbb{R}^d) \cap L^p(\mathbb{R}^d)$ and $L^p\text{-}\lim_{n\to\infty} u_n = u$.
   (b) $\tilde{P}_t u := \lim_{n\to\infty} P_t u_n$ extends P_t and gives a contraction semigroup on L^p.
   Hint: Use Young's inequality for convolutions, see e. g. [204, Theorem 14.6].
   (c) $\tilde{P}_t$ is *sub-Markovian*, i. e. if $u \in L^p$ and $0 \leqslant u \leqslant 1$ a. e., then $0 \leqslant \tilde{P}_t u \leqslant 1$ a. e.
   (d) $\tilde{P}_t$ is *strongly continuous*, i. e. $\lim_{t\to 0} \|P_t u - u\|_{L^p} = 0$ for all $u \in L^p$.
   Hint: For $u \in L^p$, $y \mapsto \|u(\cdot + y) - u\|_{L^p}$ is continuous, cf. [204, Theorem 14.8].

6. Let $(T_t)_{t\geqslant 0}$ be a Markov semigroup given by $T_t u(x) = \int_{\mathbb{R}^d} u(y)\, p_t(x, dy)$ where $p_t(x, C)$ is a kernel in the sense of Remark 7.6. Show that the semigroup property of $(T_t)_{t\geqslant 0}$ entails the Chapman–Kolmogorov equations
$$p_{t+s}(x, C) = \int p_t(y, C)\, p_s(x, dy) \quad \text{for all} \quad x \in \mathbb{R}^d \quad \text{and} \quad C \in \mathscr{B}(\mathbb{R}^d).$$

7. Show that $p^x_{t_1,\dots,t_n}(C_1 \times \dots \times C_n)$ of Remark 7.6 define measures on $\mathscr{B}(\mathbb{R}^{d\cdot n})$.

8. (a) Let $d(x, A) := \inf_{a\in A} |x - a|$ be the distance between the point $x \in \mathbb{R}^d$ and the set $A \in \mathscr{B}(\mathbb{R}^d)$. Show that $x \mapsto d(x, A)$ is continuous.
   (b) Let u_n be as in (7.7). Show that $u_n \in \mathcal{C}_c(\mathbb{R}^d)$ and $\inf u_n(x) = \mathbb{1}_K$. Draw a picture if $d = 1$ and $K = [0,1]$.
   (c) Let χ_n be a sequence of type δ, i. e. $\chi_n \in \mathcal{C}_c^\infty(\mathbb{R}^d)$, $\operatorname{supp}\chi_n \subset \mathbb{B}(0, 1/n)$, $\chi_n \geqslant 0$ and $\int \chi_n(x)\,dx = 1$. Show that $v_n := \chi_n \star u_n \in \mathcal{C}_c^\infty(\mathbb{R}^d)$ are smooth functions such that $\lim_{n\to\infty} v_n = \mathbb{1}_K$.

9. Let $(T_t)_{t\geqslant 0}$ be a Feller semigroup, i. e. a strongly continuous, positivity preserving sub-Markovian semigroup on $(\mathcal{C}_\infty(\mathbb{R}^d), \|\cdot\|_\infty)$, cf. Remark 7.4.
   (a) Show that $(t,x) \mapsto T_t u(x)$ is jointly continuous for $t \geqslant 0, x \in \mathbb{R}^d$ and every $u \in \mathcal{C}_\infty(\mathbb{R}^d)$.
   (b) Conclude that for every compact set $K \subset \mathbb{R}^d$ the function $(t,x) \mapsto T_t \mathbb{1}_K(x)$ is measurable.
   (c) Show that $(t,x,u) \mapsto T_t u(x)$ is jointly continuous for all $t \geqslant 0, x \in \mathbb{R}^d$ and $u \in \mathcal{C}_\infty(\mathbb{R}^d)$.
10. Let $A, B \in \mathbb{R}^{d\times d}$ and set $P_t := \exp(tA) := \sum_{j=0}^\infty (tA)^j/j!$.
    (a) Show that P_t is a strongly continuous semigroup. Is it contractive?
    (b) Show that $\frac{d}{dt}e^{tA}$ exists and that $\frac{d}{dt}e^{tA} = Ae^{tA} = e^{tA}A$.
    (c) Show that $e^{tA}e^{tB} = e^{tB}e^{tA}\ \forall t \iff AB = BA$.
    Hint: Differentiate $e^{tA}e^{sB}$ at $t = 0$ and $s = 0$.
    (d) (Trotter product formula) Show that $e^{A+B} = \lim_{k\to\infty}(e^{A/k}e^{B/k})^k$.
11. Let $(P_t)_{t\geqslant 0}$ and $(T_t)_{t\geqslant 0}$ be two Feller semigroups with generators A, resp., B.
    (a) Show that $\frac{d}{ds}P_{t-s}T_s = -P_{t-s}AT_s + P_{t-s}BT_s$.
    (b) Is it possible that $U_t := T_t P_t$ is again a Feller semigroup?
    (c) (Trotter product formula) Assume that $U_t f := \lim_{n\to\infty}(T_{t/n}P_{t/n})^n f$ exists strongly and locally uniformly for t. Show that U_t is a Feller semigroup and determine its generator.
12. Complete the following alternative argument for Example 7.16. Let $(u_n)_{n\geqslant 1} \subset \mathcal{C}^2_\infty(\mathbb{R})$ such that $\frac{1}{2}u_n''$ is a Cauchy sequence in $\mathcal{C}_\infty$ and that u_n converges uniformly to u.
    (a) Show that $u_n(x) - u_n(0) - xu_n'(0) = \int_0^x\int_0^y u_n''(z)\,dz\,dy \to \int_0^x\int_0^y 2g(z)\,dz\,dy$ uniformly. Conclude that $c' := \lim_{n\to\infty} u_n'(0)$ exists.
    (b) Show that $u_n'(x) - u_n'(0) \to \int_0^x 2g(z)\,dz$ uniformly. Deduce from this and part a) that $u_n'(x)$ converges uniformly to $\int_0^x 2g(z)\,dz + c'$ where $c' = \int_{-\infty}^0 2g(z)\,dz$.
    (c) Use a) and b) to show that $u_n(x) - u_n(0) \to \int_0^x\int_{-\infty}^y 2g(z)\,dz$ uniformly and argue that $u_n(x)$ has the uniform limit $\int_{-\infty}^x\int_{-\infty}^y 2g(z)\,dz\,dy$.
13. Let U_α be the α-potential operator of a BM^d. Give a probabilistic interpretation of $U_\alpha\mathbb{1}_C$ and $\lim_{\alpha\to 0}U_\alpha\mathbb{1}_C$ if $C \subset \mathbb{R}^d$ is a Borel set.
14. Use the definition (7.14) of U_α and Fubini's theorem to prove the resolvent equation from Proposition 7.13 g) by a direct calculation without using Proposition 7.13 f).
    Hint: Start with $(\beta-\alpha)U_\alpha U_\beta$.

15. Let $(U_\alpha)_{\alpha>0}$ be the α-potential operator of a BM^d. Use the resolvent equation to prove the following formulae for $f \in \mathcal{B}_b$ and $x \in \mathbb{R}^d$:

$$\frac{d^n}{d\alpha^n} U_\alpha f(x) = n!(-1)^n U_\alpha^{n+1} f(x)$$

and

$$\frac{d^n}{d\alpha^n}(\alpha U_\alpha) f(x) = n!\,(-1)^{n+1}(\mathrm{id} - \alpha U_\alpha) U_\alpha^n f(x).$$

16. (Dini's theorem for $\mathcal{C}_\infty$) Let $(f_n)_{n\geqslant 1} \subset \mathcal{C}_\infty(\mathbb{R}^d)$ be a sequence of functions such that $0 \leqslant f_n \leqslant f_{n+1}$ and $f := \sup_n f_n \in \mathcal{C}_\infty(\mathbb{R}^d)$. Show that $\lim_{n\to\infty} \|f - f_n\|_\infty = 0$.

17. Let $(A, \mathfrak{D}(A))$ be the generator of a BM^d. Adapt the arguments of Example 7.25 and show that $\mathcal{C}^2_\infty(\mathbb{R}^d) \subsetneq \mathfrak{D}(A)$ for any dimension $d \geqslant 3$.
Instructions. Use in Example 7.25 d-dimensional polar coordinates

$$\begin{aligned}
y_d &= r\cos\theta_{d-2},\\
y_{d-1} &= r\cos\theta_{d-3}\cdot\sin\theta_{d-2},\\
y_{d-2} &= r\cos\theta_{d-4}\cdot\sin\theta_{d-3}\cdot\sin\theta_{d-2},\\
&\ \ \vdots\\
y_2 &= r\cos\phi\sin\theta_1\cdot\ldots\cdot\sin\theta_{d-2},\\
y_1 &= r\sin\phi\sin\theta_1\cdot\ldots\cdot\sin\theta_{d-2},
\end{aligned}$$

where $(r, \phi, \theta_1, \ldots, \theta_{d-2}) \in (0,\infty)\times(0,2\pi)\times(0,\pi)^{d-2}$. The functional determinant is

$$r^{d-1}(\sin\theta_1)^1(\sin\theta_2)^2\ldots(\sin\theta_{d-2})^{d-2}.$$

The calculations will lead to the following integrals ($|x| < 1$)

$$\int_0^\pi \frac{(\sin\theta)^{d-2}}{(1+x^2-2x\cos\theta)^{(d-2)/2}}\,d\theta \quad\text{and}\quad \int_0^\pi \frac{(\sin\theta)^{d}}{(1+x^2-2x\cos\theta)^{(d-2)/2}}\,d\theta \tag{$*$}$$

which have the values a_d and $b_d x^2 + c_d$, respectively, with constants $a_d, b_d, c_d \in \mathbb{R}$ depending on the dimension, cf. Gradshteyn–Ryzhik [89, 3.665.2, p. 384].
Remark. A simple evaluation of the integrals ($*$) is only possible in $d = 3$. For $d > 3$ these are most elegantly evaluated using Gegenbauer polynomials, cf. [89, 8.93, pp. 1029–1031] and the discussion on 'stack exchange' `http://math.stackexchange.com/questions/395336`.

18. Let $(B_t)_{t\geqslant 0}$ be a BM^1 and consider the two-dimensional process $X_t := (t, B_t)$, $t \geqslant 0$.
   (a) Show that $(X_t)_{t\geqslant 0}$ is a Feller process.
   (b) Determine the associated transition semigroup, resolvent and generator.
   (c) State and prove Theorem 7.30 for this process and compare the result with Theorem 5.6

19. Show that Dynkin's formula (7.28) in Proposition 7.31 follows from Theorem 7.30.

20. Let $t \mapsto X_t$ be a right-continuous stochastic process. Show that for closed sets F

$$\mathbb{P}(X_t \in F \ \ \forall t \in \mathbb{R}^+) = \mathbb{P}(X_q \in F \ \ \forall q \in \mathbb{Q}^+).$$

8 The PDE connection

We want to discuss some relations between partial differential equations (PDEs) and Brownian motion. For many classical PDE problems probability theory yields concrete representation formulae for the solutions in the form of expected values of a Brownian functional. These formulae can be used to get generalized solutions of PDEs (which require less smoothness of the initial/boundary data or the boundary itself) and they are amenable to Monte–Carlo simulations. Purely probabilistic existence proofs for classical PDE problems are, however, rare: Classical solutions require smoothness, which does usually not follow from martingale methods.[1] This explains the role of Proposition 7.3 g) and Proposition 8.10. Let us point out that $\frac{1}{2}\Delta$ has two meanings: It is the restriction of the generator $(A, \mathfrak{D}(A))$ of a Brownian motion, $A|_{\mathcal{C}^2_\infty} = \frac{1}{2}\Delta|_{\mathcal{C}^2_\infty}$, but it can also be seen as partial differential operator $L = \sum_{j,k=1}^d \frac{\partial^2}{\partial x_j \partial x_k}$ which acts on all $\mathcal{C}^2$ functions. Of course, on $\mathcal{C}^2_\infty$ both meanings coincide, and it is this observation which makes the method work. Throughout this chapter $(A, \mathfrak{D}(A))$ denotes the generator of a BM^d.

The origin of the probabilistic approach are the pioneering papers by Kakutani [116; 117], Kac [110; 111] and Doob [51]. As an illustration of the method we begin with the elementary (inhomogeneous) heat equation. In this context, classical PDE methods are certainly superior to the clumsy-looking Brownian motion machinery. Its elegance and effectiveness become obvious in the Feynman–Kac formula and the Dirichlet problem which we discuss in Sections 8.3 and 8.4. Moreover, since many second-order differential operators generate diffusion processes, see Chapter 21, only minor changes in our proofs yield similar representation formulae for the corresponding PDE problems.

A martingale relation is the key ingredient. Let $(B_t, \mathscr{F}_t)_{t\geqslant 0}$ be a BM^d. Recall from Theorem 5.6 that for $u \in \mathcal{C}^{1,2}((0,\infty) \times \mathbb{R}^d) \cap \mathcal{C}([0,\infty) \times \mathbb{R}^d)$ satisfying

$$|u(t,x)| + \left|\frac{\partial u(t,x)}{\partial t}\right| + \sum_{j=1}^d \left|\frac{\partial u(t,x)}{\partial x_j}\right| + \sum_{j,k=1}^d \left|\frac{\partial^2 u(t,x)}{\partial x_j \partial x_k}\right| \leqslant c(t)\, e^{C|x|} \tag{8.1}$$

for all $t > 0$ and $x \in \mathbb{R}^d$ with some constant $C > 0$ and a locally bounded function $c : (0,\infty) \to [0,\infty)$, the process

$$M^u_s = u(t-s, B_s) - u(t, B_0) + \int_0^s \left(\frac{\partial}{\partial t} - \frac{1}{2}\Delta_x\right) u(t-r, B_r)\, dr, \quad s \in [0,t), \tag{8.2}$$

is an $\mathscr{F}_s$ martingale for every measure $\mathbb{P}^x$, i. e. for every starting point $B_0 = x$ of a Brownian motion.

1 A notable exception is, of course, Malliavin's calculus, cf. [150; 152].

8.1 The heat equation

In Chapter 7 we have seen that the generator $(A, \mathfrak{D}(A))$ of a Brownian motion extends the Laplacian $(\frac{1}{2}\Delta, \mathcal{C}^2_\infty(\mathbb{R}^d))$. In fact, applying Lemma 7.10 to a Brownian motion yields

$$\frac{d}{dt}P_t f(x) = AP_t f(x) \quad \text{for all } f \in \mathfrak{D}(A).$$

Since $P_t f(x) = \mathbb{E}^x f(B_t)$ is smooth, cf. Problem 7.4, we get $\frac{1}{2}\Delta_x P_t u(x) = AP_t f(x)$. Setting $u(t,x) := P_t f(x) = \mathbb{E}^x f(B_t)$ this almost proves the following lemma. Ex. 7.4

8.1 Lemma. *Let $(B_t)_{t\geqslant 0}$ be a BM^d, $f \in \mathfrak{D}(A)$ and $u(t,x) := \mathbb{E}^x f(B_t)$. Then $u(t,x)$ is the unique bounded solution of the heat equation with initial value f,*

$$\frac{\partial}{\partial t}u(t,x) - \frac{1}{2}\Delta_x u(t,x) = 0, \tag{8.3a}$$

$$u(0,x) = f(x). \tag{8.3b}$$

Proof. It remains to show uniqueness. Let $u(t,x)$ be a bounded solution of (8.3). Its Laplace transform is given by

$$v_\lambda(x) = \int_0^\infty u(t,x)\, e^{-\lambda t}\, dt.$$

Since we may differentiate under the integral sign, we get

$$\begin{aligned}
\lambda v_\lambda(x) - \frac{1}{2}\Delta_x v_\lambda(x) &= \lambda v_\lambda(x) - \int_0^\infty \frac{1}{2}\Delta_x u(t,x)\, e^{-\lambda t}\, dt \\
&= \lambda v_\lambda(x) - \int_0^\infty \frac{\partial}{\partial t} u(t,x)\, e^{-\lambda t}\, dt.
\end{aligned}$$

Integration by parts and (8.3b) yield

$$\Big(\lambda\,\mathrm{id} - \frac{1}{2}\Delta_x\Big) v_\lambda(x) = \lambda v_\lambda(x) + \int_0^\infty u(t,x)\, \frac{\partial}{\partial t}\, e^{-\lambda t}\, dt - u(t,x)e^{-\lambda t}\Big|_{t=0}^{t=\infty} = f(x).$$

Since the equation $(\lambda\,\mathrm{id} - \frac{1}{2}\Delta_x)v_\lambda = f$ has for all $f \in \mathfrak{D}(A)$ and $\lambda > 0$ the unique solution $v_\lambda = U_\lambda f$ where U_λ is the resolvent operator, and since the Laplace transform is unique, cf. [205, Proposition 1.2], we conclude that $u(t,x)$ is unique. □

Lemma 8.1 is a PDE proof. Let us sketch a further, more probabilistic approach, which will be important if we consider more complicated PDEs. It is, admittedly, too complicated for the simple setting considered in Lemma 8.1, but it allows us to remove the restriction that $f \in \mathfrak{D}(A)$, see the discussion in Remark 8.4 below. Ex. 8.1

8.2 Probabilistic proof of Lemma 8.1. *Uniqueness:* Assume that $u(t,x)$ satisfies (8.1) and $|u(t,x)| \leqslant c\,e^{c|x|}$ on $[0,T]\times\mathbb{R}^d$ where $c = c_T$. By (8.2),

$$M_s^u := u(t-s,B_s) - u(t,B_0) + \int_0^s \underbrace{\left(\frac{\partial}{\partial t} - \frac{1}{2}\Delta_x\right) u(t-r,B_r)\,dr}_{=0 \text{ by (8.3a)}},$$

is a martingale w. r. t. $(\mathscr{F}_s)_{s\in[0,t)}$, and so is $N_s^u := u(t-s,B_s)$, $s\in[0,t)$. By assumption, u is exponentially bounded. Therefore, $(N_s^u)_{s\in[0,t)}$ is dominated by the integrable function $c\sup_{0\leqslant s\leqslant t}\exp(c\,|B_s|)$, hence uniformly integrable. By the martingale convergence theorem, Theorem A.6, the limit $\lim_{s\to t} N_s^u$ exists a. s. and in L^1. Thus,

$$u(t,x) = \mathbb{E}^x\, u(t,B_0) = \mathbb{E}^x\, N_0^u = \lim_{s\to t}\mathbb{E}^x\, N_s^u = \mathbb{E}^x\, N_t^u \overset{(8.3b)}{=} \mathbb{E}^x\, f(B_t),$$

which shows that the initial condition f uniquely determines the solution.

Existence: Set $u(t,x) := \mathbb{E}^x\, f(B_t) = P_t f(x)$ and assume that $f\in\mathfrak{D}(A)$. Lemma 7.10 together with an obvious modification of the proof of Proposition 7.3 g) show that u is in $\mathcal{C}^{1,2}((0,\infty)\times\mathbb{R}^d)\cap\mathcal{C}([0,\infty)\times\mathbb{R}^d)$ and satisfies (8.1). Therefore, (8.2) tells us that

$$M_s^u := u(t-s,B_s) - u(t,B_0) + \int_0^s \left(\frac{\partial}{\partial t} - \frac{1}{2}\Delta_x\right) u(t-r,B_r)\,dr$$

is a martingale. On the other hand, the Markov property reveals that

$$u(t-s,B_s) = \mathbb{E}^{B_s} f(B_{t-s}) = \mathbb{E}^x(f(B_t)\mid\mathscr{F}_s), \quad 0\leqslant s\leqslant t,$$

Ex. 8.2 is a martingale, and so is $\int_0^s \left(\frac{\partial}{\partial t} - \frac{1}{2}\Delta_x\right) u(t-r,B_r)\,dr$, $s\leqslant t$. Since it has continuous paths of bounded variation, it is identically zero on $[0,t)$, cf. Proposition A.22, and (8.3a) follows. □

The proof of Proposition 7.3 g) shows that $(t,x)\mapsto P_t f(x) = \mathbb{E}^x\, f(B_t)$, $t>0$, $x\in\mathbb{R}^d$, is smooth. This allows us to extend the probabilistic approach using a localization technique described in Remark 8.4 below.

8.3 Theorem. *Let $(B_t)_{t\geqslant 0}$ be a BM^d and $f\in\mathcal{B}(\mathbb{R}^d)$ such that $|f(x)|\leqslant C\,e^{C|x|}$. Then $u(t,x) := \mathbb{E}^x\, f(B_t)$ is the unique solution to the heat equation* (8.3) *satisfying the growth estimate $|u(t,x)|\leqslant Ce^{C|x|}$.*

The growth estimate in the statement of Theorem 8.3 is essential for the uniqueness, cf. John [109, Chapter 7.1].

8.4 Remark (Localization). Let $(B_t)_{t\geqslant 0}$ be a BM^d, $u\in\mathcal{C}^{1,2}((0,\infty)\times\mathbb{R}^d)\cap\mathcal{C}([0,\infty)\times\mathbb{R}^d)$, fix $t>0$, and consider the following sequence of stopping times

$$\tau_n := \inf\{s>0 : |B_s|\geqslant n\}\wedge(t-n^{-1}), \quad n\geqslant 1/t. \tag{8.4}$$

For every $n \geqslant 1/t$ we pick some cut-off function $\chi_n \in \mathcal{C}^{1,2}([0,\infty) \times \mathbb{R}^d)$ with compact support, $\chi_n|_{[0,1/2n]\times\mathbb{R}^d} \equiv 0$ and $\chi_n|_{[1/n,t]\times\mathbb{B}(0,n)} \equiv 1$. Clearly, for M_s^u as in (8.2),

$$M^u_{s\wedge\tau_n} = M^{\chi_n u}_{s\wedge\tau_n}, \quad \text{for all } s \in [0,t) \quad \text{and} \quad n \geqslant 1/t.$$

Since $\chi_n u \in \mathcal{C}_b^{1,2}$ satisfies (8.1), $(M_s^{\chi_n u}, \mathscr{F}_s)_{s\in[0,t)}$ is, for each $n \geqslant 1/t$, a bounded martingale. By optional stopping, Remark A.21,

$$\left(M^u_{s\wedge\tau_n}, \mathscr{F}_s\right)_{s\in[0,t)} \quad \text{is a bounded martingale for all } u \in \mathcal{C}^{1,2} \text{ and } n \geqslant 1/t. \tag{8.5}$$

Assume that $u \in \mathcal{C}^{1,2}((0,\infty) \times \mathbb{R}^d)$. Replacing in the probabilistic proof 8.2 B_s, M_s^u and N_s^u by the stopped versions $B_{s\wedge\tau_n}$, $M^u_{s\wedge\tau_n}$ and $N^u_{s\wedge\tau_n}$, we see that[2]

$$M^u_{s\wedge\tau_n} = u(t - s\wedge\tau_n, B_{s\wedge\tau_n}) - u(t, B_0) + \underbrace{\int_0^{s\wedge\tau_n} \left(\frac{\partial}{\partial t} - \frac{1}{2}\Delta_x\right) u(t-r, B_r)\,dr}_{=0 \text{ by (8.3a)}},$$

and $N^u_{s\wedge\tau_n} = u(t - s\wedge\tau_n, B_{s\wedge\tau_n})$, $s \in [0,t)$, are for every $n \geqslant 1/t$ bounded martingales. In particular, $\mathbb{E}^x N_0^u = \mathbb{E}^x N^u_{s\wedge\tau_n}$ for all $s \in [0,t)$, and by the martingale convergence theorem we get

$$u(t,x) = \mathbb{E}^x u(t, B_0) = \mathbb{E}^x N_0^u = \lim_{s\to t} \mathbb{E}^x N^u_{s\wedge\tau_n} = \mathbb{E}^x N^u_{t\wedge\tau_n}.$$

As $\lim_{n\to\infty} \tau_n = t$ a. s., we see

$$u(t,x) = \mathbb{E}^x N^u_{t\wedge\tau_n} = \lim_{k\to\infty} \mathbb{E}^x u(t - t\wedge\tau_k, B_{t\wedge\tau_k}) = \mathbb{E}^x u(0, B_t) \overset{(8.3b)}{=} \mathbb{E}^x f(B_t),$$

which shows that the solution is unique.

To see that $u(t,x) := P_t f(x)$ is a solution, we assume that $|f(x)| \leqslant Ce^{C|x|}$. We can adapt the proof of Proposition 7.3 g) to see that $u \in \mathcal{C}^{1,2}((0,\infty) \times \mathbb{R}^d)$. Therefore, (8.5) shows that $(M^u_{s\wedge\tau_n}, \mathscr{F}_s)_{s\in[0,t)}$ is, for every $n \geqslant 1/t$, a bounded martingale. By the Markov property

$$u(t-s, B_s) = \mathbb{E}^{B_s} f(B_{t-s}) = \mathbb{E}^x(f(B_t) \mid \mathscr{F}_s), \quad 0 \leqslant s \leqslant t,$$

and by optional stopping $(u(t - s\wedge\tau_n, B_{s\wedge\tau_n}), \mathscr{F}_s)_{s\in[0,t)}$ is for every $n \geqslant 1/t$ a martingale. This implies that $\int_0^{s\wedge\tau_n} \left(\frac{\partial}{\partial t} - \frac{1}{2}\Delta_x\right) u(t-r, B_r)\,dr$ is a martingale with continuous paths of bounded variation. Hence, it is identically zero on $[0, t\wedge\tau_n)$, cf. Proposition A.22. As $\lim_{n\to\infty} \tau_n = t$, (8.3a) follows. Ex. 8.2

2 Note that $\int_0^{s\wedge\tau_n} \cdots\, dr = \int_{[0,s\wedge\tau_n)} \cdots\, dr$.

8.2 The inhomogeneous initial value problem

If we replace the right-hand side of the heat equation (8.3a) with some function $g(t,x)$, we get the following *inhomogeneous initial value problem*.

$$\frac{\partial}{\partial t}v(t,x) - \frac{1}{2}\Delta_x v(t,x) = g(t,x), \tag{8.6a}$$

$$v(0,x) = f(x). \tag{8.6b}$$

Assume that u solves the corresponding homogeneous problem (8.3). If w solves the inhomogeneous problem (8.6) with zero initial condition $f = 0$, then $v = u + w$ is a solution of (8.6). Therefore, it is customary to consider

$$\frac{\partial}{\partial t}v(t,x) - \frac{1}{2}\Delta_x v(t,x) = g(t,x), \tag{8.7a}$$

$$v(0,x) = 0 \tag{8.7b}$$

instead of (8.6). Let us use the probabilistic approach of Section 8.1 to develop an educated guess about the solution to (8.7).

Let $v(t,x)$ be a solution to (8.7) which is bounded and sufficiently smooth ($\mathcal{C}^{1,2}$ and (8.1) will do). Then, for $s \in [0,t)$,

$$M_s^v := v(t-s,B_s) - v(t,B_0) + \int_0^s \underbrace{\left(\frac{\partial}{\partial t}v(t-r,B_r) - \frac{1}{2}\Delta_x v(t-r,B_r)\right)}_{=g(t-r,B_r) \text{ by (8.7a)}} dr$$

is a martingale, and so is $N_s^v := v(t-s,B_s) + \int_0^s g(t-r,B_r)\,dr$, $s \in [0,t)$; in particular, $\mathbb{E}^x N_0^v = \mathbb{E}^x N_s^v$ for all $s \in [0,t)$. If we can use the martingale convergence theorem, e. g. if the martingale is uniformly integrable, we can let $s \to t$ and get

$$v(t,x) = \mathbb{E}^x N_0^v = \lim_{s\to t} \mathbb{E}^x N_s^v = \mathbb{E}^x N_t^v \overset{(8.7b)}{=} \mathbb{E}^x\left(\int_0^t g(t-r,B_r)\,dr\right).$$

This consideration already yields uniqueness. For the existence we have to make sure that we may perform all manipulations in the above calculation. With some effort we can justify everything if $g(t,x)$ is bounded or if $g = g(x)$ is from the so-called Kato class. For the sake of clarity, we content ourselves with the following simple version.

Ex. 8.4 **8.5 Theorem.** *Let $(B_t)_{t\geqslant 0}$ be a BM^d and $g \in \mathfrak{D}(A)$. Then the unique solution of the inhomogeneous initial value problem (8.7) with $g = g(x)$ satisfying $|v(t,x)| \leqslant C\,t$ is given by $v(t,x) = \mathbb{E}^x \int_0^t g(B_s)\,ds$.*

Proof. Since $v(t,x) = \int_0^t \mathbb{E}^x g(B_s)\,ds = \int_0^t P_s g(x)\,ds$, Lemma 7.10 and a variation of the proof of Proposition 7.3 g) imply that $v \in \mathcal{C}^{1,2}((0,\infty)\times\mathbb{R}^d)$. This means that the heuristic derivation of $v(t,x)$ furnishes a uniqueness proof.

To see that $v(t,x)$ is indeed a solution, we use (8.2) and conclude that

$$\begin{aligned} M_s^v &= v(t-s,B_s) - v(t,B_0) + \int_0^s \left(\frac{\partial}{\partial t} - \frac{1}{2}\Delta_x\right) v(t-r,B_r)\,dr \\ &= v(t-s,B_s) + \int_0^s g(B_r)\,dr - v(t,B_0) \\ &\qquad + \int_0^s \left[\left(\frac{\partial}{\partial t} - \frac{1}{2}\Delta_x\right) v(t-r,B_r) - g(B_r)\right] dr \end{aligned}$$

is a martingale. On the other hand, the Markov property gives for all $s \leqslant t$,

$$\begin{aligned} \mathbb{E}^x\left(\int_0^t g(B_r)\,dr \,\middle|\, \mathscr{F}_s\right) &= \int_0^s g(B_r)\,dr + \mathbb{E}^x\left(\int_s^t g(B_r)\,dr \,\middle|\, \mathscr{F}_s\right) \\ &= \int_0^s g(B_r)\,dr + \mathbb{E}^x\left(\int_0^{t-s} g(B_{r+s})\,dr \,\middle|\, \mathscr{F}_s\right) \\ &= \int_0^s g(B_r)\,dr + \mathbb{E}^{B_s}\left(\int_0^{t-s} g(B_r)\,dr\right) \\ &= \int_0^s g(B_r)\,dr + v(t-s,B_s). \end{aligned}$$

This shows that the right-hand side is for $s \leqslant t$ a martingale.

Consequently,

$$\int_0^s \left(\frac{\partial}{\partial t} v(t-r,B_r) - \frac{1}{2}\Delta_x v(t-r,B_r) - g(B_r)\right) dr, \qquad s \in [0,t),$$

is a martingale with continuous paths of bounded variation; by Proposition A.22 it is zero on $[0,t)$, and (8.7a) follows. □ Ex. 8.2

If $g = g(x)$ does not depend on time, we can use the localization technique described in Remark 8.4 to show that Theorem 8.5 remains valid if $g \in \mathcal{B}_b(\mathbb{R}^d)$. This requires a modification of the proof of Proposition 7.3 g) to see that $v(t,x) := \mathbb{E}^x \int_0^t g(B_r)\,dr$ is in $\mathcal{C}^{1,2}((0,\infty)\times\mathbb{R}^d)$. Then the proof is similar to the one for the heat equation described in Remark 8.4. As soon as g depends on time, $g = g(t,x)$, things become messy, cf. [60, Chapter 8.2].

8.3 The Feynman–Kac formula

We have seen in Section 8.2 that the inhomogeneous initial value problem (8.7) is solved by the expectation of the random variable $G_t = \int_0^t g(B_r)\,dr$. It is an interesting question whether we can find out more about the probability distributions of G_t and (B_t, G_t). A natural approach would be to compute the joint characteristic function $\mathbb{E}^x[\exp(i\xi G_t + i\langle\eta, B_t\rangle)]$. Let us consider the more general problem to calculate

$$w(t,x) := \mathbb{E}^x\left(f(B_t)e^{C_t}\right) \quad \text{where} \quad C_t := \int_0^t c(B_r)\,dr$$

for some functions $f, c : \mathbb{R}^d \to \mathbb{C}$. Setting $f(x) = e^{i\langle\eta,x\rangle}$ and $c(x) = i\xi g(x)$ brings us back to the original question. We will show that, under suitable conditions on f and c, the function $w(t,x)$ is the unique solution to the following initial value problem

$$\frac{\partial}{\partial t} w(t,x) - \frac{1}{2}\Delta_x w(t,x) - c(x)w(t,x) = 0, \tag{8.8a}$$

$$w(t,x) \text{ is continuous and } w(0,x) = f(x). \tag{8.8b}$$

Considering real and imaginary parts separately, we may assume that f and c are real-valued.

Let us follow the strategy of Sections 8.1 and 8.2. Assume that $w \in \mathcal{C}^{1,2}$ solves (8.8) and satisfies (8.1). If $c(x)$ is bounded, we can apply (8.2) to $w(t-s,x)\exp(C_s)$ for $s \in [0,t)$. Since $\frac{d}{dt}\exp(C_t) = c(B_t)\exp(C_t)$, we see that for $s \in [0,t)$

$$M_s^w := w(t-s,B_s)e^{C_s} - w(t,B_0) + \int_0^s \underbrace{\left(\frac{\partial}{\partial t} - \frac{1}{2}\Delta_x - c(B_r)\right) w(t-r,B_r)}_{=0 \text{ by (8.8a)}} e^{C_r}\,dr$$

is a uniformly integrable martingale under the filtration $(\mathscr{F}_s)_{s\in[0,t)}$, and the same holds for $N^w := (w(t-s,B_s)\exp(C_s))_{s\in[0,t)}$. Since $\mathbb{E}^x N_0^w = \mathbb{E}^x N_s^w$ for all $s \in [0,t)$, the martingale convergence theorem, Theorem A.7, gives

$$w(t,x) = \mathbb{E}^x w(t,B_0) = \mathbb{E}^x N_0^w = \lim_{s\to t}\mathbb{E}^x N_s^w = \mathbb{E}^x N_t^w \overset{(8.8b)}{=} \mathbb{E}^x\left(f(B_t)e^{C_t}\right).$$

This reveals both the structure of the solution and its uniqueness. To show with martingale methods that $w(t,x) := \mathbb{E}^x\left(f(B_t)e^{C_t}\right)$ solves (8.8) requires *a priori* knowledge that w is smooth, e. g. $w \in \mathcal{C}^{1,2}$ – but this is quite hard to check, even if $f \in \mathfrak{D}(A)$ and $c \in \mathcal{C}_b(\mathbb{R}^d)$. Therefore, we use semigroup methods to show the next theorem.

8.6 Theorem (Kac 1949). *Let $(B_t)_{t\geqslant 0}$ be a* BM^d*, $f \in \mathfrak{D}(A)$ and $c \in \mathcal{C}_b(\mathbb{R}^d)$. The unique solution to the initial value problem* (8.8) *is given by*

$$w(t,x) = \mathbb{E}^x\left[f(B_t)\exp\left(\int_0^t c(B_r)\,dr\right)\right]. \tag{8.9}$$

This solution is bounded by $e^{\alpha t}$ where $\alpha = \sup_{x\in\mathbb{R}^d} c(x)$.

The representation formula (8.9) is called *Feynman–Kac* formula.

Proof of Theorem 8.6. Denote by P_t the Brownian semigroup. Set $C_t := \int_0^t c(B_r)\,dr$ and $T_t u(x) := \mathbb{E}^x(u(B_t)e^{C_t})$. If we can show that $(T_t)_{t\geqslant 0}$ is a Feller semigroup with generator $Lu = Au + cu$, we are done: Existence follows from Lemma 7.10, uniqueness from the fact that A has a resolvent which is uniquely determined by T_t, cf. Proposition 7.13.

Since

$$|T_t u(x)| \leqslant e^{t\sup_{x\in\mathbb{R}^d} c(x)}\, \mathbb{E}^x\,|u(B_t)| \leqslant e^{t\alpha}\,\|u\|_\infty,$$

the operator T_t is, in general, only a contraction if $\alpha \leqslant 0$. Let us assume this for the time being and show that $(T_t)_{t\geqslant 0}$ is a Feller semigroup.

a) *Positivity* is clear from the very definition.

b) *Feller property:* For all $u \in \mathcal{C}_\infty(\mathbb{R}^d)$ we have

$$\begin{aligned} T_t u(x) &= \mathbb{E}^x\left[u(B_t)\exp\left(\int_0^t c(B_r)\,dr\right)\right] \\ &= \mathbb{E}\left[u(B_t + x)\exp\left(\int_0^t c(B_r + x)\,dr\right)\right]. \end{aligned}$$

Using the dominated convergence theorem yields $\lim_{y\to x} T_t u(y) = T_t u(x)$ and $\lim_{|x|\to\infty} T_t u(x) = 0$. Thus, $T_t u \in \mathcal{C}_\infty(\mathbb{R}^d)$.

c) *Strong continuity:* Let $u \in \mathcal{C}_\infty(\mathbb{R}^d)$ and $t \geqslant 0$. Using the elementary inequality $1 - e^c \leqslant |c|$ for all $c \leqslant 0$, we get

$$\begin{aligned} |T_t u(x) - u(x)| &\leqslant \left|\mathbb{E}^x\left(u(B_t)(e^{C_t} - 1)\right)\right| + \left|\mathbb{E}^x\left(u(B_t) - u(x)\right)\right| \\ &\leqslant \|u\|_\infty\, \mathbb{E}^x\,|C_t| + \|P_t u - u\|_\infty \\ &\leqslant t\,\|u\|_\infty \|c\|_\infty + \|P_t u - u\|_\infty. \end{aligned}$$

Since the Brownian semigroup $(P_t)_{t\geqslant 0}$ is strongly continuous on $\mathcal{C}_\infty(\mathbb{R}^d)$, the strong continuity of $(T_t)_{t\geqslant 0}$ follows.

d) *Semigroup property:* Let $s, t \geqslant 0$ and $u \in \mathcal{C}_\infty(\mathbb{R}^d)$. Using the tower property of conditional expectation and the Markov property, we see

$$\begin{aligned} T_{t+s}u(x) &= \mathbb{E}^x\left[u(B_{t+s})e^{\int_0^{t+s} c(B_r)\,dr}\right] \\ &= \mathbb{E}^x\left[\mathbb{E}^x\left(u(B_{t+s})e^{\int_s^{t+s} c(B_r)\,dr}e^{\int_0^s c(B_r)\,dr} \;\Big|\; \mathscr{F}_s\right)\right] \\ &= \mathbb{E}^x\left[\mathbb{E}^x\left(u(B_{t+s})e^{\int_0^t c(B_{r+s})\,dr} \;\Big|\; \mathscr{F}_s\right)e^{\int_0^s c(B_r)\,dr}\right] \\ &= \mathbb{E}^x\left[\mathbb{E}^{B_s}\left(u(B_t)e^{\int_0^t c(B_r)\,dr}\right)e^{\int_0^s c(B_r)\,dr}\right] \\ &= \mathbb{E}^x\left[T_t u(B_s)e^{\int_0^s c(B_r)\,dr}\right] \\ &= T_s T_t u(x). \end{aligned}$$

Because of Theorem 7.22, we may calculate the generator L using pointwise convergence: For all $u \in \mathfrak{D}(A)$ and $x \in \mathbb{R}^d$

$$\lim_{t\to 0}\frac{T_t u(x)-u(x)}{t}=\lim_{t\to 0}\frac{T_t u(x)-P_t u(x)}{t}+\lim_{t\to 0}\frac{P_t u(x)-u(x)}{t}.$$

The second limit equals $Au(x)$; since $(e^{C_t}-1)/t \to c(B_0)$ boundedly – cf. the estimate in the proof of the strong continuity c) – we can use dominated convergence and find

$$\lim_{t\to 0}\mathbb{E}^x\left(u(B_t)\,\frac{e^{C_t}-1}{t}\right)=\mathbb{E}^x\left(u(B_0)c(B_0)\right)=u(x)c(x).$$

This completes the proof in the case $\alpha \leqslant 0$.

Ex. 7.11 If $\alpha > 0$ we can consider the following auxiliary problem: Replace in (8.8) the function $c(x)$ by $c(x)-\alpha$. From the first part we know that $e^{-\alpha t}T_t f = \mathbb{E}^x\left(f(B_t)e^{C_t-\alpha t}\right)$ is the unique solution of the auxiliary problem

$$\left(\frac{\partial}{\partial t}-\frac{1}{2}\Delta-c(\cdot)+\alpha\right)e^{-\alpha t}T_t f=0.$$

On the other hand,

$$\frac{d}{dt}e^{-\alpha t}T_t f=-\alpha e^{-\alpha t}T_t f+e^{-\alpha t}\frac{d}{dt}T_t f,$$

and this shows that $T_t f$ is the unique solution of the original problem. □

8.7 An Application: Lévy's arc-sine law (Lévy 1939. Kac 1949). Let $(B_t)_{t\geqslant 0}$ be a BM^1 and $G_t := \int_0^t \mathbb{1}_{(0,\infty)}(B_s)\,ds$ the total amount of time a Brownian motion spends in the positive half-axis up to time t. Lévy showed in [143] that G_t has an arc-sine distribution:

$$\mathbb{P}^0\left(\int_0^t \mathbb{1}_{(0,\infty)}(B_s)\,ds \leqslant x\right)=\frac{2}{\pi}\arcsin\sqrt{\frac{x}{t}},\qquad 0\leqslant x\leqslant t. \tag{8.10}$$

We calculate the Laplace transform $g_\alpha(t)=\mathbb{E}^0\,e^{-\alpha G_t}$, $\alpha \geqslant 0$, to identify the probability distribution of G_t. Define for $\lambda > 0$ and $x \in \mathbb{R}$

$$v_\lambda(x) := \int_0^\infty e^{-\lambda t}\,\mathbb{E}^x\,e^{-\alpha G_t}\,dt=\int_0^\infty e^{-\lambda t}\,\mathbb{E}^x\left(\mathbb{1}_{\mathbb{R}}(B_t)\cdot e^{-\alpha\int_0^t \mathbb{1}_{(0,\infty)}(B_s)\,ds}\right)dt.$$

Assume that we *could* apply Theorem 8.6 with $f(x)=\mathbb{1}_{\mathbb{R}}(x)$ and $c(x)=-\alpha\mathbb{1}_{(0,\infty)}(x)$. Then $\lambda \mapsto v_\lambda(x)$ is be the Laplace transform of $w(t,x)=\mathbb{E}^x\left(\mathbb{1}_{\mathbb{R}}(B_t)\cdot e^{-\alpha\int_0^t \mathbb{1}_{(0,\infty)}(B_s)\,ds}\right)$ which, by Theorem 8.6, is the solution of the initial value problem (8.8). A short calculation shows that the Laplace transform satisfies the following linear ODE:

$$\frac{1}{2}v_\lambda''(x)-\alpha\mathbb{1}_{(0,\infty)}(x)v_\lambda(x)=\lambda v_\lambda(x)-\mathbb{1}_{\mathbb{R}}(x)\quad\text{for all } x\in\mathbb{R}\setminus\{0\}. \tag{8.11}$$

This ODE is, with suitable initial conditions, uniquely solvable in $(-\infty,0)$ and $(0,\infty)$. It will become clear from the approximation below that v_λ is of class $\mathcal{C}^1 \cap \mathcal{C}_b$ on the whole real line. Therefore we can prescribe conditions in $x = 0$ such that v_λ is the unique bounded $\mathcal{C}^1$ solution of (8.11). With the usual *Ansatz* for linear second-order ODEs, we get

$$v_\lambda(x) = \begin{cases} \dfrac{1}{\lambda} + C_1 e^{\sqrt{2\lambda}x} + C_2 e^{-\sqrt{2\lambda}x}, & x < 0, \\[2ex] \dfrac{1}{\alpha+\lambda} + C_3 e^{\sqrt{2(\alpha+\lambda)}x} + C_4 e^{-\sqrt{2(\alpha+\lambda)}x}, & x > 0. \end{cases}$$

Let us determine the constants C_j, $j = 1, 2, 3, 4$. The boundedness of the function v_λ gives $C_2 = C_3 = 0$. Since v_λ and v_λ' are continuous at $x = 0$, we get

$$\begin{aligned} v_\lambda(0+) &= \frac{1}{\alpha+\lambda} + C_4 = \frac{1}{\lambda} + C_1 = v_\lambda(0-), \\ v_\lambda'(0+) &= -\sqrt{2(\alpha+\lambda)}\, C_4 = \sqrt{2\lambda}\, C_1 = v_\lambda'(0-). \end{aligned}$$

Solving for C_1 gives $C_1 = -\lambda^{-1} + (\lambda(\lambda+\alpha))^{-1/2}$, and so

$$v_\lambda(0) = \int_0^\infty e^{-\lambda t}\, \mathbb{E}^0(e^{-\alpha G_t})\, dt = \frac{1}{\sqrt{\lambda(\lambda+\alpha)}}.$$

On the other hand, by Fubini's theorem and the formula $\int_0^\infty \frac{1}{\sqrt{\pi t}} e^{-ct}\, dt = \frac{1}{\sqrt{c}}$:

$$\begin{aligned} \int_0^\infty e^{-\lambda t} \frac{2}{\pi} \int_0^{\pi/2} e^{-\alpha t \sin^2\theta}\, d\theta\, dt &= \int_0^\infty e^{-\lambda t} \int_0^t \frac{e^{-\alpha r}}{\pi\sqrt{r(t-r)}}\, dr\, dt \\ &= \int_0^\infty \frac{e^{-\alpha r}}{\sqrt{\pi r}} \int_r^\infty \frac{e^{-\lambda t}}{\sqrt{\pi(t-r)}}\, dt\, dr \\ &= \int_0^\infty \frac{e^{-(\alpha+\lambda)r}}{\sqrt{\pi r}}\, dr \int_0^\infty \frac{e^{-\lambda s}}{\sqrt{\pi s}}\, ds = \frac{1}{\sqrt{(\lambda+\alpha)\lambda}}. \end{aligned}$$

Because of the uniqueness of the Laplace transform, cf. [205, Proposition 1.2], we conclude

$$g_\alpha(t) = \mathbb{E}^0\, e^{-\alpha G_t} = \frac{2}{\pi} \int_0^{\pi/2} e^{-\alpha t \sin^2\theta}\, d\theta = \frac{2}{\pi} \int_0^t e^{-\alpha x}\, d_x \arcsin\sqrt{\frac{x}{t}},$$

and, again by the uniqueness of the Laplace transform, we get (8.10).

Let us now justify the formal manipulations that led to (8.11). In order to apply Theorem 8.6, we cut and smooth the functions $f(x) = \mathbb{1}_{\mathbb{R}}(x)$ and $g(x) = \mathbb{1}_{(0,\infty)}(x)$. Pick a sequence of cut-off functions $\chi_n \in \mathcal{C}_c(\mathbb{R})$ such that $\mathbb{1}_{[-n,n]} \leqslant \chi_n \leqslant 1$, and define $g_n(x) = 0 \vee (nx) \wedge 1$. Then $\chi_n \in \mathfrak{D}(A)$, $g_n \in \mathcal{C}_b(\mathbb{R})$, $\sup_{n\geqslant 1} \chi_n = \mathbb{1}_{\mathbb{R}}$ and $\sup_{n\geqslant 0} g_n = g$. By Theorem 8.6,

$$w_n(t,x) := \mathbb{E}^x\left(\chi_n(B_t) e^{-\alpha \int_0^t g_n(B_s)\, ds}\right), \quad t > 0,\ x \in \mathbb{R},$$

is, for each $n \geqslant 1$, the unique bounded solution of the initial value problem (8.8) with $f = f_n$ and $c = g_n$; note that the bound from Theorem 8.6 does not depend on $n \geqslant 1$. The Laplace transform $\lambda \mapsto v_{n,\lambda}(x)$ of $w_n(t,x)$ is the unique bounded solution of the ODE

$$\frac{1}{2}\,v''_{n,\lambda}(x) - \alpha\chi_n(x)v_{n,\lambda}(x) = \lambda v_{n,\lambda}(x) - g_n(x), \quad x \in \mathbb{R}. \tag{8.12}$$

Ex. 8.3 Integrating two times, we see that $\lim_{n\to\infty} v_{n,\lambda}(x) = v_\lambda(x)$, $\lim_{n\to\infty} v'_{n,\lambda}(x) = v'_\lambda(x)$ for all $x \in \mathbb{R}$ and $\lim_{n\to\infty} v''_{n,\lambda}(x) = v''_\lambda(x)$ for all $x \neq 0$; moreover, we have $v_\lambda \in \mathcal{C}^1(\mathbb{R})$, cf. Problem 8.3.

8.4 The Dirichlet problem

Up to now we have considered parabolic initial value problems. In this section we focus on the most basic elliptic boundary value problem, Dirichlet's problem for the Laplace operator with zero boundary data. Throughout this section $D \subset \mathbb{R}^d$ is a bounded open and connected domain, and we write $\partial D = \overline{D} \setminus D$ for its boundary; $(B_t)_{t\geqslant 0}$ is a BM^d starting from $B_0 = x$, and $\tau = \tau_{D^c} = \inf\{t > 0 : B_t \notin D\}$ is the first exit time from the set D. We assume that the $(B_t)_{t\geqslant 0}$ is adapted to a right-continuous filtration $(\mathscr{F}_t)_{t\geqslant 0}$.[3] We are interested in the following question: Find some function u
Ex. 8.5 on $\overline{D}$ with the following properties

$$\Delta u(x) = 0 \quad \text{for } x \in D \tag{8.13a}$$
$$u(x) = f(x) \quad \text{for } x \in \partial D \tag{8.13b}$$
$$u \text{ is continuous in } \overline{D}. \tag{8.13c}$$

To get a feeling for the problem, we use similar heuristic arguments as in Sections 8.1–8.3. Assume that u is a solution of the Dirichlet problem (8.13). If ∂D is smooth, we can extend u onto $\mathbb{R}^d$ in such a way that $u \in \mathcal{C}^2(\mathbb{R}^d)$ with $\operatorname{supp} u$ compact. Therefore, we can apply (8.2) with $u = u(x)$ to see that

$$M_t^u = u(B_t) - u(B_0) - \int_0^t \frac{1}{2}\Delta u(B_r)\,dr$$

is a bounded martingale.
By the optional stopping theorem, Theorem A.18 and Remark A.21, $(M^u_{t\wedge\tau}, \mathscr{F}_t)_{t\geqslant 0}$ is a martingale, and we see that[4]

$$M^u_{t\wedge\tau} = u(B_{t\wedge\tau}) - u(B_0) - \underbrace{\int_0^{t\wedge\tau} \frac{1}{2}\Delta u(B_r)\,dr}_{=0 \text{ by (8.13a)}} = u(B_{t\wedge\tau}) - u(B_0).$$

3 A condition for this can be found in Theorem 6.21.

4 Note that $\int_0^{t\wedge\tau} \cdots\, dr = \int_{[0,t\wedge\tau)} \cdots\, dr$.

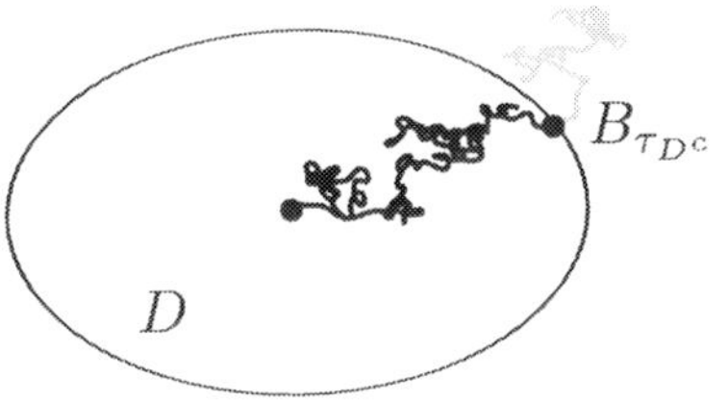

Fig. 8.1. Exit time and position of a Brownian motion from the set D.

Therefore, $N^u_{t\wedge\tau} := u(B_{t\wedge\tau})$ is a bounded martingale. In particular, $\mathbb{E}^x N^u_0 = \mathbb{E}^x N^u_{t\wedge\tau}$ for all $t \geqslant 0$. If $\tau = \tau_{D^c} < \infty$, the martingale convergence theorem yields

$$u(x) = \mathbb{E}^x N^u_0 = \lim_{t\to\infty} \mathbb{E}^x N^u_{t\wedge\tau} = \mathbb{E}^x N^u_\tau = \mathbb{E}^x u(B_\tau) \stackrel{(8.13b)}{=} \mathbb{E}^x f(B_\tau).$$

This shows that $\mathbb{E}^x f(B_\tau)$ is a good candidate for the solution of (8.13). But before we go on, let us quickly check that τ is indeed a. s. finite. The following result is similar to Lemma 7.33.

8.8 Lemma. *Let $(B_t)_{t\geqslant 0}$ be a BM^d and $D \subset \mathbb{R}^d$ be an open and bounded set. Then $\mathbb{E}^x \tau_{D^c} < \infty$ where τ_{D^c} is the first hitting time of D^c.* Ex. 8.6

Proof. Because of Lemma 5.8, $\tau = \tau_{D^c}$ is a stopping time. Since D is a bounded set, there exists some $r > 0$ such that $D \subset \mathbb{B}(0,r)$. Define $u(x) := -\exp(-|x|^2/4r^2)$. Then $u \in \mathcal{C}^2_\infty(\mathbb{R}^d) \subset \mathfrak{D}(A)$ and, for all $|x| < r$,

$$\frac{1}{2}\Delta u(x) = \frac{1}{2}\left(\frac{d}{2r^2} - \frac{|x|^2}{4r^4}\right)e^{-|x|^2/4r^2} \geqslant \frac{1}{2}\left(\frac{d}{2r^2} - \frac{1}{4r^2}\right)e^{-1/4} =: \kappa > 0.$$

Therefore, we can use Dynkin's formula (7.28) with $\sigma = \tau \wedge n$ to get

$$\kappa\, \mathbb{E}^x(\tau \wedge n) \leqslant \mathbb{E}^x\left(\int_0^{\tau\wedge n} \frac{1}{2}\Delta u(B_r)\,dr\right) = \mathbb{E}^x u(B_{\tau\wedge n}) - u(x) \leqslant 2\|u\|_\infty = 2.$$

By Fatou's lemma, $\mathbb{E}^x \tau \leqslant \underline{\lim}_{n\to\infty} \mathbb{E}^x(\tau \wedge n) \leqslant 2/\kappa < \infty$. □ Ex. 8.7

Let us now show that $u(x) = \mathbb{E}^x f(B_\tau)$ is a solution of (8.13). For the behaviour in the interior, i. e. (8.13a), the following classical notion turns out to be the key ingredient.

8.9 Definition. Let $D \subset \mathbb{R}^d$ be an open set. A function $u : D \to \mathbb{R}$ is *harmonic* (in the set D), if $u \in \mathcal{C}^2(D)$ and $\Delta u(x) = 0$.

In fact, harmonic functions are arbitrarily often differentiable. This is an interesting by-product of the proof of the next result.

8.10 Proposition. *Let $D \subset \mathbb{R}^d$ be an open and connected set and $u : D \to \mathbb{R}$ a locally bounded function. Then the following assertions are equivalent.*

a) *u is harmonic on D.*

b) *u has the* (spherical) mean value property*: For all* $x \in D$ *and* $r > 0$ *such that* $\overline{\mathbb{B}}(x,r) \subset D$ *one has*

$$u(x) = \int_{\partial\mathbb{B}(0,r)} u(x+z)\,\pi_r(dz) \tag{8.14}$$

where π_r *is the normalized uniform measure on the sphere* $\partial\mathbb{B}(0,r)$.

c) *u is continuous and* $(u(B_{t\wedge\tau_{G^c}}), \mathscr{F}_t)_{t\geqslant 0}$ *is a martingale w. r. t. the law* $\mathbb{P}^x$ *of a Brownian motion* $(B_t, \mathscr{F}_t)_{t\geqslant 0}$ *started at any* $x \in G$*; here G is a bounded open set such that* $\overline{G} \subset D$*, and* τ_{G^c} *is the first exit time from the set G.*

Proof. c)⇒b): Fix $x \in D$ and $r > 0$ such that $\overline{\mathbb{B}}(x,r) \subset D$, and set $\sigma := \tau_{\mathbb{B}^c(x,r)}$. Clearly, there is some bounded open set G such that $\overline{\mathbb{B}}(x,r) \subset \overline{G} \subset D$. By assumption, $(u(B_{t\wedge\tau_{G^c}}), \mathscr{F}_t)_{t\geqslant 0}$ is a $\mathbb{P}^x$-martingale.

Since $\sigma < \infty$, cf. Lemma 8.8, we find by optional stopping (Theorem A.18) that $(u(B_{t\wedge\sigma\wedge\tau_{G^c}}))_{t\geqslant 0}$ is a martingale. Using $\sigma \leqslant \tau_{G^c}$ we get

$$u(x) = \mathbb{E}^x\, u(B_0) = \mathbb{E}^x\, u(B_{t\wedge\sigma}).$$

Now $|u(B_{t\wedge\sigma})| \leqslant \sup_{y\in\mathbb{B}(x,r)} |u(y)| < \infty$, and by dominated convergence and the continuity of Brownian paths

$$\begin{aligned} u(x) = \lim_{t\to\infty} \mathbb{E}^x\, u(B_{t\wedge\sigma}) = \mathbb{E}^x\, u(B_\sigma) &= \int u(y)\,\mathbb{P}^x(B_\sigma \in dy) \\ &= \int u(z+x)\,\mathbb{P}^0(B_\tau \in dz), \end{aligned}$$

where $\tau = \tau_{\mathbb{B}^c(0,r)}$. By the rotational symmetry of Brownian motion, cf. Lemma 2.8,
Ex. 2.12 $\mathbb{P}^0(B_\tau \in dz)$ is a uniformly distributed probability measure on the sphere $\partial\mathbb{B}(0,r)$, and (8.14) follows.[5]

b)⇒a): Fix $x \in D$ and $\delta > 0$ such that $\overline{\mathbb{B}}(x,\delta) \subset D$. Pick $\phi_\delta \in \mathcal{C}^\infty(0,\infty)$ such that $\phi_\delta|_{[\delta^2,\infty)} \equiv 0$ and set $c_d := \int_0^\delta \phi_\delta(r^2)\,r^{d-1}\,dr$. By Fubini's theorem and changing to polar coordinates

$$\begin{aligned} u(x) = \frac{1}{c_d}\int_0^\delta \phi_\delta(r^2)\,r^{d-1}\,u(x)\,dr &\overset{(8.14)}{\underset{b)}{=}} \frac{1}{c_d}\int_0^\delta\int_{\partial\mathbb{B}(0,r)} \phi_\delta(r^2)\,r^{d-1}\,u(z+x)\,\pi_r(dz)\,dr \\ &= \frac{1}{c_d'}\int_{\mathbb{B}(0,\delta)} \phi_\delta(|y|^2)\,u(y+x)\,dy \\ &= \frac{1}{c_d'}\int_{\mathbb{B}(x,\delta)} \phi_\delta(|y-x|^2)\,u(y)\,dy. \end{aligned}$$

---

5 For a rotation $Q : \mathbb{R}^d \times \mathbb{R}^d$, $(QB_t)_{t\geqslant 0}$ is again a BM^d. Let $\tau = \tau_{\mathbb{B}^c(0,r)}$ and $C \in \mathscr{B}(\partial\mathbb{B}(0,r))$. Then $\mathbb{P}^0(B_\tau \in C) = \mathbb{P}^0(QB_\tau \in C) = \mathbb{P}^0(B_\tau \in Q^{-1}C)$ which shows that $\mathbb{P}^0(B_\tau \in dz)$ must be the (normalized) surface measure on the sphere $\partial\mathbb{B}(0,r)$.

As u is locally bounded, the differentiation lemma for parameter dependent integrals, e. g. [204, Theorem 11.5], shows that the last integral is arbitrarily often differentiable at the point x. Since $x \in D$ is arbitrary, we have $u \in \mathcal{C}^\infty(D)$.

Fix $x_0 \in D$ and assume that $\Delta u(x_0) > 0$. Because D is open and Δu continuous, there is some $\delta > 0$ such that $\overline{\mathbb{B}}(x_0, \delta) \subset D$ and $\Delta u|_{\mathbb{B}(x_0,\delta)} > 0$. Set $\sigma := \tau_{\mathbb{B}^c(x_0,\delta)}$ and take $\chi \in \mathcal{C}_c^\infty(D)$ such that $\mathbb{1}_{\mathbb{B}(x_0,\delta)} \leqslant \chi \leqslant \mathbb{1}_D$. Then we can use (8.2) for the smooth and bounded function $u_\chi := \chi u$. Using an optional stopping argument we conclude that

$$M_{t\wedge\sigma}^{u_\chi} = u_\chi(B_{t\wedge\sigma}) - u_\chi(B_0) - \int_0^{t\wedge\sigma} \frac{1}{2}\Delta u_\chi(B_r)\,dr$$
$$= u(B_{t\wedge\sigma}) - u(B_0) - \int_0^{t\wedge\sigma} \frac{1}{2}\Delta u(B_r)\,dr$$

(use $\chi|_{\mathbb{B}(x_0,\delta)} \equiv 1$) is a martingale. Therefore, $\mathbb{E}\, M_{t\wedge\sigma}^{u_\chi} = 0$, and letting $t \to \infty$ we get

$$\mathbb{E}^{x_0} u(B_\sigma) - u(x_0) = \frac{1}{2}\,\mathbb{E}^{x_0}\left(\int_0^\sigma \Delta u(B_r)\,dr\right) > 0.$$

The mean-value property shows that the left-hand side equals 0 and we have a contradiction. Similarly one shows that $\Delta u(x_0) < 0$ gives a contradiction, thus $\Delta u(x) = 0$ for all $x \in D$.

a)⇒c): Fix any open and bounded set G such that $\overline{G} \subset D$ and pick a cut-off function $\chi \in \mathcal{C}_c^\infty$ such that $\mathbb{1}_G \leqslant \chi \leqslant \mathbb{1}_D$. Set $\sigma := \tau_{G^c}$ and $u_\chi(x) := \chi(x)u(x)$. By (8.2),

$$M_t^{u_\chi} = u_\chi(B_t) - u_\chi(B_0) - \int_0^t \frac{1}{2}\Delta u_\chi(B_r)\,dr$$

is a uniformly integrable martingale. Therefore, we can use optional stopping and the fact that $\chi|_G \equiv 1$ to see that

$$M_{t\wedge\sigma}^{u_\chi} = u(B_{t\wedge\sigma}) - u(B_0) - \int_0^{t\wedge\sigma} \frac{1}{2}\Delta u(B_r)\,dr \overset{\text{a)}}{=} u(B_{t\wedge\sigma}) - u(B_0)$$

is a martingale. This proves c). □

Let us note another classical consequence of the mean-value property.

8.11 Corollary (Maximum principle). *Let $D \subset \mathbb{R}^d$ be a bounded, open and connected domain and $u \in \mathcal{C}(\overline{D}) \cap \mathcal{C}^2(D)$ be a harmonic function. Then*

$$\exists\, x_0 \in D \;:\; u(x_0) = \sup_{x\in D} u(x) \implies u \equiv u(x_0).$$

In other words: A non-constant harmonic function may attain its maximum only on the boundary of D.

Proof. Assume that u is a harmonic function in D and $u(x_0) = \sup_{x\in D} u(x)$ for some $x_0 \in D$. Since $u(x)$ enjoys the spherical mean-value property (8.14), we see that $u|_{\partial\mathbb{B}(x_0,r)} \equiv u(x_0)$ for all $r > 0$ such that $\overline{\mathbb{B}}(x_0,r) \subset D$. Thus, $u|_{\mathbb{B}(x_0,r)} \equiv u(x_0)$, and we see that the set $M := \{x \in D : u(x) = u(x_0)\}$ is open. Since u is continuous, M is also closed, and the connectedness of D shows that $D = M$ as M is not empty. □

When we solve the Dirichlet problem, cf. Theorem 8.17 below, we use Proposition 8.10 to get (8.13a), and Corollary 8.11 for the uniqueness of the solution; the condition (8.13b), i. e. $\lim_{D\ni x\to x_0} \mathbb{E}^x f(B_\tau) = f(x_0)$ for $x_0 \in \partial D$, requires some kind of regularity for the hitting time $\tau = \tau_{D^c}$.

8.12 Definition. A point $x_0 \in \partial D$ is called *regular (for D^c)* if $\mathbb{P}^{x_0}(\tau_{D^c} = 0) = 1$. Every non-regular point is called *singular*.

At a regular point x_0, Brownian motion leaves the domain D immediately; typical situations are shown in Example 8.15 below.

The point $x_0 \in \partial D$ is singular for D^c if, and only if, $\mathbb{P}^{x_0}(\tau_{D^c} > 0) = 1$. If $E \supset D$ such that $x_0 \in \partial E$, the point x_0 is also singular for E^c. This follows from $E^c \subset D^c$ and $\tau_{E^c} \geqslant \tau_{D^c}$.

8.13 Lemma. *$x_0 \in \partial D$ is singular for D^c if, and only if, $\mathbb{P}^{x_0}(\tau_{D^c} = 0) = 0$.*

Proof. By definition, any singular point x_0 satisfies $\mathbb{P}^{x_0}(\tau_{D^c} = 0) < 1$. On the other hand,

$$\{\tau_{D^c} = 0\} = \bigcap_{n=1}^{\infty} \{\tau_{D^c} \leqslant 1/n\} \in \bigcap_{n=1}^{\infty} \mathscr{F}_{1/n} = \mathscr{F}_{0+},$$

and by Blumenthal's 0-1–law, Corollary 6.22, $\mathbb{P}^{x_0}(\tau_{D^c} = 0) \in \{0, 1\}$. □

The following simple regularity criterion is good enough for most situations. Recall that a *truncated cone* in $\mathbb{R}^d$ with opening $r > 0$, vertex x_0 and direction x_1 is the set $V = \{x \in \mathbb{R}^d : x = x_0 + h\,\mathbb{B}(x_1, r),\ 0 < h < \epsilon\}$ for some $\epsilon > 0$.

8.14 Lemma (Outer cone condition. Poincaré 1890, Zaremba 1911). *Suppose $D \subset \mathbb{R}^d$, $d \geqslant 2$, is an open domain and $(B_t)_{t\geqslant 0}$ a BM^d. If we can touch $x_0 \in \partial D$ with the vertex of a truncated cone V such that $V \subset D^c$, then x_0 is a regular point for D^c.*

Proof. Assume that x_0 is singular. Denote by $\tau_V := \inf\{t > 0 : B_t \in V\}$ the first hitting time of the open cone V. Since $V \subset D^c$, we have $\tau_V \geqslant \tau_{D^c}$, i. e.

$$\mathbb{P}^{x_0}(\tau_V > 0) \geqslant \mathbb{P}^{x_0}(\tau_{D^c} > 0) = 1.$$

Pick $\epsilon > 0$ so small that $\mathbb{B}(x_0, \epsilon) \setminus \{x_0\}$ is covered by finitely many copies of the cone $V = V_0$ and its rotations $V_1, \ldots, V_n$ about the vertex x_0. Since a Brownian motion is

Fig. 8.2. The outer cone condition. A point $x \in D$ is regular, if we can touch it with a (truncated) cone which is in D^c.

invariant under rotations – this follows immediately from Lemma 2.8 – , we conclude that $\mathbb{P}^{x_0}(\tau_{V_j} > 0) = \mathbb{P}^{x_0}(\tau_{V_0} > 0) = 1$ for all $j = 1, \dots, n$.

Observe that $\bigcup_{j=0}^n V_j \supset \mathbb{B}(x_0, \epsilon) \setminus \{x_0\}$; thus, $\min_{0 \leqslant j \leqslant n} \tau_{V_j} \leqslant \tau_{\mathbb{B}(x_0,\epsilon)\setminus\{x_0\}}$, i. e.

$$\mathbb{P}^{x_0}\left(\tau_{\mathbb{B}(x_0,\epsilon)\setminus\{x_0\}} > 0\right) \geqslant \mathbb{P}^{x_0}\left(\min_{0\leqslant j\leqslant n} \tau_{V_j} > 0\right) = 1$$

which is impossible since $\mathbb{P}^{x_0}(B_t = x_0) = 0$ for every $t > 0$. □

8.15 Example. Unless indicated otherwise, we assume that $d \geqslant 2$. The following pictures show typical examples of singular points.

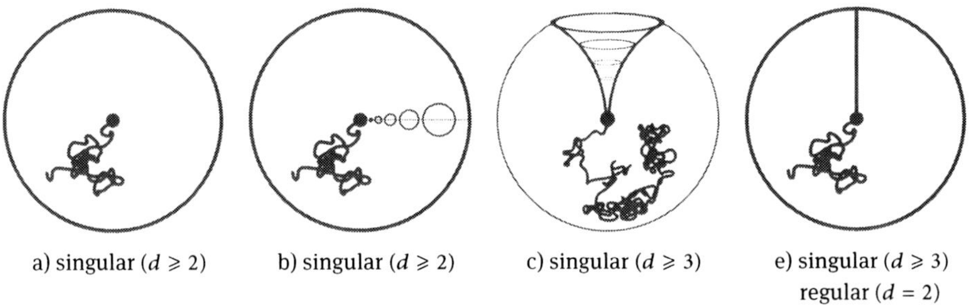

Fig. 8.3. Examples of singular points.

a) The boundary point 0 of $D = \mathbb{B}(0,1) \setminus \{0\}$ is singular. This was the first example of a singular point and it is due to Zaremba [242]. This is easy to see since $\tau_{\{0\}} = \infty$ almost surely: By the Markov property

$$\begin{aligned}\mathbb{P}^0(\tau_{\{0\}} < \infty) &= \mathbb{P}^0(\exists t > 0 : B_t = 0)\\ &= \lim_{n\to\infty} \mathbb{P}^0\left(\exists t > \tfrac{1}{n} : B_t = 0\right)\\ &= \lim_{n\to\infty} \mathbb{E}^0\Big(\underbrace{\mathbb{P}^{B_{1/n}}(\exists t > 0 : B_t = 0)}_{=0,\ \text{Corollary 6.16}}\Big) = 0.\end{aligned}$$

Therefore, $\mathbb{P}^0(\tau_{D^c} = 0) = \mathbb{P}^0(\tau_{\{0\}} \wedge \tau_{\mathbb{B}^c(0,1)} = 0) = \mathbb{P}^0(\tau_{\mathbb{B}^c(0,1)} = 0) = 0$.

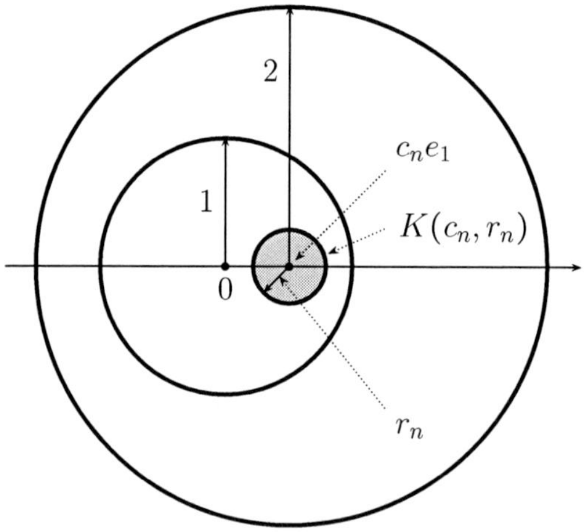

Fig. 8.4. Brownian motion has to exit $\mathbb{B}(0,1)$ before it leaves $\mathbb{B}(c_n e_1, 2)$.

b) Let $(r_n)_{n\geqslant 1}$ and $(c_n)_{n\geqslant 1}$ be sequences with $0 < r_n, c_n < 1$ which decrease to 0. We set $K(c_n, r_n) := \overline{\mathbb{B}}(c_n e_1, r_n) \subset \mathbb{R}^2$ and $e_1 = (1,0) \in \mathbb{R}^2$. Run a Brownian motion in the set $D := \mathbb{B}(0,1) \setminus \bigcup_{n\geqslant 1} K(c_n, r_n)$ obtained by removing the circles $K(c_n, r_n)$ from the unit ball. Clearly, the Brownian motion has to leave $\mathbb{B}(0,1)$ before it leaves $\mathbb{B}(c_n e_1, 2)$, see Figure 8.4. Thus,

$$\begin{aligned}\mathbb{P}^0\big(\tau_{K(c_n,r_n)} < \tau_{\mathbb{B}^c(0,1)}\big) &\leqslant \mathbb{P}^0\big(\tau_{K(c_n,r_n)} < \tau_{\mathbb{B}^c(c_n e_1,2)}\big)\\ &= \mathbb{P}^{-c_n e_1}\big(\tau_{K(0,r_n)} < \tau_{\mathbb{B}^c(0,2)}\big).\end{aligned}$$

Since $K(0, r_n) = \overline{\mathbb{B}}(0, r_n)$, this probability is known from Theorem 6.15:

$$\mathbb{P}^0\big(\tau_{K(c_n,r_n)} < \tau_{\mathbb{B}^c(0,1)}\big) \leqslant \frac{\log 2 - \log c_n}{\log 2 - \log r_n}.$$

Take $c_n = 2^{-n}$, $r_n = 2^{-n^3}$, $n \geqslant 2$, and sum the respective probabilities; then

$$\sum_{n=2}^{\infty} \mathbb{P}^0\big(\tau_{K(c_n,r_n)} < \tau_{\mathbb{B}^c(0,1)}\big) \leqslant \sum_{n=2}^{\infty} \frac{n+1}{n^3+1} < \infty.$$

The Borel–Cantelli lemma shows that Brownian motion visits at most finitely many of the balls $K(c_n, r_n)$ before it exits $\mathbb{B}(0,1)$. Since $0 \notin K(c_n, r_n)$, this means that $\mathbb{P}^0(\tau_{D^c} > 0) = 1$, i. e. 0 is a singular point.

c) (Lebesgue's spine) In 1913 Lebesgue gave the following example of a singular point. Consider in $d = 3$ the open unit ball and remove a(n exponentially) sharp spine, e. g. a closed cusp S which is obtained by rotating an exponentially decaying curve $y = h(x)$ about the positive x–axis: $E := \mathbb{B}(0,1) \setminus S$. Then 0 is a singular point for E^c.
For this, we use a similar construction as in b): We remove from $\mathbb{B}(0,1)$ the sets K_n, $D := \mathbb{B}(0,1) \setminus \bigcup_{n=1}^{\infty} K_n$, but the sets K_n are chosen in such a way that they cover the spine S, i. e. $D \subset E$.

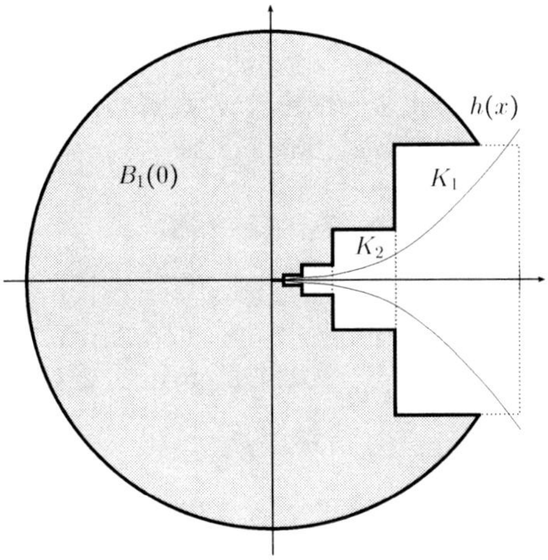

Fig. 8.5. A 'cubistic' version of Lebesgue's spine.

Let $K_n = [2^{-n}, 2^{-n+1}] \times \overline{\mathbb{B}}(0, r_n)$ be a cylinder in $\mathbb{R}^3$. The probability that a Brownian motion hits K_n can be made arbitrarily small, if we choose r_n small enough, see the technical wrap-up at the end of this paragraph. So we can arrange things in such a way that

$$\mathbb{P}^0(\tau_{K_n} < \tau_{\mathbb{B}^c(0,1)}) \leqslant p_n \quad \text{and} \quad \sum_{n=1}^{\infty} p_n < \infty.$$

As in b) we see that a Brownian motion visits at most finitely many of the cylinders K_n before exiting $\mathbb{B}(0,1)$, i. e. $\mathbb{P}^0(\tau_{D^c} > 0) = 1$. Therefore, 0 is singular for D^c and for $E^c \subset D^c$.

Technical wrap-up. Set $K_\epsilon^n = [2^{-n}, 2^{-n+1}] \times \overline{\mathbb{B}}(0, \epsilon)$, $\tau_{n,\epsilon} := \tau_{K_\epsilon^n}$, $B_t = (w_t, b_t, \beta_t)$ and $\sigma_n = \inf\{t > 0 : w_t \geqslant 2^{-n}\}$.

From Corollary 6.17 we deduce that $0 < \sigma_n, \tau_{\mathbb{B}^c(0,1)} < \infty$ a. s. Moreover, for fixed $n \geqslant 1$ and any $\epsilon' \leqslant \epsilon$ we have $\sigma_n \leqslant \tau_{n,\epsilon} \leqslant \tau_{n,\epsilon'}$, and so

$$\{\sigma_n \leqslant \tau_{n,\epsilon'} \leqslant \tau_{\mathbb{B}^c(0,1)}\} \subset \{\sigma_n \leqslant \tau_{n,\epsilon} \leqslant \tau_{\mathbb{B}^c(0,1)}\}.$$

Since the interval $[\sigma_n, \tau_{\mathbb{B}^c(0,1)}]$ is compact we get

$$\bigcap_{\epsilon>0} \{\sigma_n \leqslant \tau_{n,\epsilon} \leqslant \tau_{\mathbb{B}^c(0,1)}\} \subset \{\exists t \in [\sigma_n, \tau_{\mathbb{B}^c(0,1)}] : b_t = \beta_t = 0\}.$$

As
$$\mathbb{P}^0\left(\tau_{n,\epsilon} \leqslant \tau_{\mathbb{B}^c(0,1)}\right) = \mathbb{P}^0\left(\sigma_n \leqslant \tau_{n,\epsilon} \leqslant \tau_{\mathbb{B}^c(0,1)}\right)$$

and
$$\mathbb{P}^0\left(\exists t > 0 : b_t = \beta_t = 0\right) = 0,$$

see Corollary 6.16, we conclude that $\lim_{\epsilon\to 0} \mathbb{P}^0(\tau_{n,\epsilon} \leqslant \tau_{\mathbb{B}^c(0,1)}) = 0$ and this is all we need. □

Our construction shows that the spine will be exponentially sharp. Lebesgue used the function $h(x) = \exp(-1/x)$, $x > 0$, see [123, p. 285, p. 334].

d) Let $(b(t))_{t\geqslant 0}$ be a BM^1 and $\tau_{\{0\}} = \inf\{t > 0 : b(t) = 0\}$ the first return time to the origin. Then $\mathbb{P}^0(\tau_{\{0\}} = 0) = 1$. This means that in $d = 1$ a Brownian motion path
Ex. 8.8 oscillates infinitely often around its starting point. To see this, we begin with the remark that, because of the symmetry of a Brownian motion,

$$\mathbb{P}^0(b(t) \geqslant 0) = \mathbb{P}^0(b(t) \leqslant 0) = \frac{1}{2} \quad \text{for all } t \in (0, \epsilon],\ \epsilon > 0.$$

Thus, $\mathbb{P}^0(b(t) \geqslant 0\ \forall t \in [0,\epsilon]) \leqslant 1/2$ and $\mathbb{P}^0(\tau_{(-\infty,0]} \leqslant \epsilon) \geqslant 1/2$ for all $\epsilon > 0$. Letting $\epsilon \to 0$ we get $\mathbb{P}^0(\tau_{(-\infty,0]} = 0) \geqslant 1/2$ and, by Blumenthal's 0-1–law, Corollary 6.22,
Ex. 5.12 $\mathbb{P}^0(\tau_{(-\infty,0]} = 0) = 1$. The same argument applies to the interval $[0,\infty)$ and we get

$$\mathbb{P}^0(\tau_{\{0\}} = 0) = \mathbb{P}^0(\tau_{(-\infty,0]} \vee \tau_{[0,\infty)} = 0) = 1.$$

! e) (Zaremba's needle) Let $e_2 = (0,1,0,\dots,0) \in \mathbb{R}^d$, $d \geqslant 2$. The argument from c) shows that for the set $D = \mathbb{B}(0,1) \setminus [0,\infty)e_2$ the point 0 is singular if $d \geqslant 3$. In two
Ex. 8.9 dimensions the situation is different: 0 is regular!
Indeed: Let $B(t) = (b(t), \beta(t))$, $t \geqslant 0$, be a BM^2; we know that the coordinate processes $b = (b(t))_{t\geqslant 0}$ and $\beta = (\beta(t))_{t\geqslant 0}$ are independent one-dimensional Brownian motions. Set $\sigma_n = \inf\{t > 1/n : b(t) = 0\}$. Since $0 \in \mathbb{R}$ is regular for $\{0\} \subset \mathbb{R}$, see d), we get that $\lim_{n\to\infty} \sigma_n = \tau_{\{0\}} = 0$ almost surely with respect to $\mathbb{P}^0$. Since $\beta \perp\!\!\!\perp b$, the
Ex. 8.10 random variable $\beta(\sigma_n)$ is symmetric, and so

$$\mathbb{P}^0(\beta(\sigma_n) \geqslant 0) = \mathbb{P}^0(\beta(\sigma_n) \leqslant 0) \geqslant \frac{1}{2}.$$

Clearly, $B(\sigma_n) = (b(\sigma_n), \beta(\sigma_n)) = (0, \beta(\sigma_n))$ and $\{\beta(\sigma_n) \geqslant 0\} \subset \{\tau_{D^c} \leqslant \sigma_n\}$, so

$$\mathbb{P}^0(\tau_{D^c} = 0) = \lim_{n\to\infty} \mathbb{P}^0(\tau_{D^c} \leqslant \sigma_n) \geqslant \varlimsup_{n\to\infty} \mathbb{P}^0(\beta(\sigma_n) \geqslant 0) \geqslant \frac{1}{2}.$$

Now Blumenthal's 0–1-law, Corollary 6.22, applies and gives $\mathbb{P}^0(\tau_{D^c} = 0) = 1$.

8.16 Theorem. *Let $x_0 \in \partial D$ be a regular point and $\tau = \tau_{D^c}$. Then*

$$\lim_{D \ni x \to x_0} \mathbb{P}^x(\tau > h) = 0 \quad \textit{for all } h > 0.$$

Proof. Fix $h > 0$. Since D is open, we find for every $x \in \mathbb{R}^d$

$$\begin{aligned}
\mathbb{P}^x(\tau > h) &= \mathbb{P}^x(\forall s \in (0,h] : B(s) \in D) \\
&= \inf_n \mathbb{P}^x(\forall s \in (1/n, h] : B(s) \in D) \\
&= \inf_n \mathbb{E}^x\left(\mathbb{P}^{B(1/n)}(\forall s \in (0, h-1/n] : B(s) \in D)\right) \\
&= \inf_n \mathbb{E}^x\left(\mathbb{P}^{B(1/n)}(\tau > h - 1/n)\right).
\end{aligned} \tag{8.15}$$

Ex. 6.3 Set $\phi(y) := \mathbb{P}^y(\tau > h - 1/n)$. Since $\phi \in \mathcal{B}_b(\mathbb{R}^d)$, the strong Feller property, Proposition 7.3 g), shows that

$$x \mapsto \mathbb{E}^x\left(\mathbb{P}^{B(1/n)}(\tau > h - 1/n)\right) = \mathbb{E}^x\,\phi(B(1/n))$$

is a continuous function. Therefore,

$$\begin{aligned}
\overline{\lim_{D\ni x\to x_0}} \mathbb{P}^x(\tau > h) &= \overline{\lim_{D\ni x\to x_0}} \inf_n \mathbb{E}^x\left(\mathbb{P}^{B(1/n)}(\tau > h - 1/n)\right) \\
&\leqslant \inf_k \overline{\lim_{D\ni x\to x_0}} \mathbb{E}^x\left(\mathbb{P}^{B(1/k)}(\tau > h - 1/k)\right) \\
&\overset{\text{cts.}}{=} \inf_k \mathbb{E}^{x_0}\left(\mathbb{P}^{B(1/k)}(\tau > h - 1/k)\right) \\
&\overset{(8.15)}{=} \mathbb{P}^{x_0}(\tau > h) = 0,
\end{aligned}$$

since x_0 is regular. □

We are now ready for the main theorem of this section.

8.17 Theorem. *Let $(B_t)_{t\geqslant 0}$ be a d-dimensional Brownian motion, D a connected, open and bounded domain and $\tau = \tau_{D^c}$ the first hitting time of D^c. If every point in ∂D is regular for D^c and if $f : \partial D \to \mathbb{R}$ is continuous, then $u(x) = \mathbb{E}^x f(B_\tau)$ is the unique bounded solution of the Dirichlet problem* (8.13).

Proof. We know already that $u(x) = \mathbb{E}^x f(B_\tau)$ is well-defined for $x \in \overline{D}$.

1° *u is harmonic in D.* Let $x \in D$. Since D is open, there is some $\delta > 0$ such that $\mathbb{B}(x,\delta) \subset D$. Set $\sigma := \sigma_r := \inf\{t > 0 : B_t \notin \mathbb{B}(x,r)\}$ for $0 < r < \delta$. By the strong Markov property (6.9)

$$u(x) = \mathbb{E}^x f(B_\tau) = \mathbb{E}^x\left(\mathbb{E}^{B_\sigma} f(B_{\tau'})\right) = \mathbb{E}^x u(B_\sigma)$$

where $\tau' := \inf\{t > \sigma : B_t \notin D\}$ is the first exit time from D of an independent Brownian motion $(B_t)_{t\geqslant 0}$ started at B_σ. Therefore, u enjoys the spherical mean-value property (8.14) and we infer with Proposition 8.10 that u is harmonic in D, i. e. (8.13a) holds.

2° $\lim_{x\to x_0} u(x) = f(x_0)$ *for all $x_0 \in \partial D$:* Fix $\epsilon > 0$. Since f is continuous, there is some $\delta > 0$ such that

$$|f(x) - f(x_0)| \leqslant \epsilon \quad \text{for all } x \in \mathbb{B}(x_0,\delta) \cap \partial D.$$

Therefore, we find for every $h > 0$

$$\begin{aligned}
|u(x) - f(x_0)| &= \left|\mathbb{E}^x\left(f(B_\tau) - f(x_0)\right)\right| \\
&\leqslant \mathbb{E}^x\left(\left|f(B_\tau) - f(x_0)\right| \cdot \mathbb{1}_{\{\tau<h\}}\right) + 2\|f\|_\infty \mathbb{P}^x(\tau \geqslant h) \\
&\leqslant \mathbb{E}^x\left(\left|f(B_\tau) - f(x_0)\right| \cdot \mathbb{1}_{\{\tau<h\}} \mathbb{1}_{\{\sup_{t\leqslant h}|B_t - x_0| < \delta\}}\right) \\
&\quad + \mathbb{E}^x\left(\left|f(B_\tau) - f(x_0)\right| \cdot \mathbb{1}_{\{\tau<h\}} \mathbb{1}_{\{\sup_{t\leqslant h}|B_t - x_0| \geqslant \delta\}}\right) \\
&\quad + 2\|f\|_\infty \mathbb{P}^x(\tau \geqslant h).
\end{aligned}$$

Write $I_1(x,h)$, $I_2(x,h)$ and $I_3(x,h)$ for the terms appearing in the last line. Let $|x - x_0| \leqslant \delta/2$. By the triangle inequality we see that

$$\left\{\sup_{t\leqslant h}|B_t + x - x_0| \geqslant \delta\right\} \subset \left\{\sup_{t\leqslant h}|B_t| \geqslant \delta/2\right\}.$$

Thus, by Doob's maximal inequality (A.13),

$$\begin{aligned} |I_2(x,h)| &\leqslant 2\,\|f\|_\infty\,\mathbb{P}^0\Big(\sup_{t\leqslant h}|B_t+x-x_0|\geqslant\delta\Big)\\ &\overset{|x-x_0|\leqslant\delta/2}{\leqslant} 2\,\|f\|_\infty\,\mathbb{P}^0\Big(\sup_{t\leqslant h}|B_t|\geqslant\delta/2\Big)\\ &\overset{(A.13)}{\leqslant} \|f\|_\infty\,\frac{8\,\mathbb{E}^0\,|B_h|^2}{\delta^2}=\|f\|_\infty\,\frac{8hd}{\delta^2}. \end{aligned}$$

Because of the continuity of f we get

$$|I_1(x,h)|\leqslant\mathbb{E}^x\left(\epsilon\cdot\mathbb{1}_{\{\tau<h\}}\mathbb{1}_{\{\sup_{t\leqslant h}|B_t-x_0|<\delta\}}\right)\leqslant\epsilon.$$

Therefore,

$$|u(x)-f(x_0)|\leqslant 2\epsilon+2\,\|f\|_\infty\,\mathbb{P}^x(\tau\geqslant h)\quad\text{for}\quad|x-x_0|\leqslant\frac{\delta}{2},\quad h\leqslant\frac{\epsilon\,\delta^2}{8d\,\|f\|_\infty}.$$

Using Theorem 8.16, we see $\overline{\lim}_{x\to x_0}|u(x)-f(x_0)|\leqslant 2\epsilon$, and since $\epsilon>0$ is arbitrary, this proves (8.13b) as well as (8.13c).

3° *Uniqueness.* This follows from the maximum principle, Corollary 8.11. Assume that u and w are two solutions of (8.13) with the same boundary value f. Then the functions $\pm(u-w)$ solve (8.13) with $f\equiv 0$, i. e. $\pm(u-w)\leqslant 0$, hence $u\equiv w$. □

8.18 Further reading. The study of the Laplace operator and PDEs connected with the Laplacian is a central part of classical and probabilistic potential theory. A good starting point is the non-technical survey paper [168]. On the analytic side, one finds a very neat expositions in [95] on potential theory, and on general PDEs in [73]. Purely probabilistic treatments are [31], its sequel [33] and the monograph [60]. The booklet [66] has a very accessible exposition of the Dirichlet problem starting from random walks. The lecture notes [185] are a good introduction to probabilistic potential theory. Both sides, analysis and probability, are explained in the encyclopedic [52], but this is not an easy read. More general connections to analysis are explored in [9]

[9] Bass: *Probabilistic Techniques in Analysis.*
[31] Chung: *Lectures from Markov Processes to Brownian Motion.*
[33] Chung, Zhao: *From Brownian Motion to Schrödinger's Equation.*
[52] Doob: *Classical Potential Theory and Its Probabilistic Counterpart.*
[60] Durrett: *Brownian Motion and Martingales in Analysis.*
[66] Dynkin, Yushkevich: *Markov Processes. Theorems and Problems.*
[73] Evans: *Partial Differential Equations.*
[95] Helms: *Potential Theory.*
[168] Orey: Probabilistic methods in partial differential equations.
[185] Rao: *Brownian Motion and Classical Potential Theory.*

Problems

1. Assume that, in the setting of Lemma 8.1, the boundary function $f \in \mathcal{C}_\infty(\mathbb{R}^d)$ but not necessarily $f \in \mathfrak{D}(A)$. Consider (8.3) with $P_\epsilon f$ instead of f. Discuss the limit $\epsilon \to 0$.

2. Let $(B_t)_{t\geqslant 0}$ be a BM^d, $f : \mathbb{R}^d \to \mathbb{R}$ be a continuous function such that $\int_0^t f(B_s)\,ds = 0$ for all $t > 0$. Show that $f(B_s) = 0$ for all $s > 0$, and conclude that $f \equiv 0$.

3. Complete the approximation argument for Lévy's arc-sine law from Paragraph 8.7:
   (a) Show, by a direct calculation, that $v_{n,\lambda}(x)$ converges as $n \to \infty$. Conclude from (8.12) that $v''_{n,\lambda}$ converges.
   (b) Integrate the ODE (8.12) to see that $v_{n,\lambda}(x) - v'_{n,\lambda}(0)$ has a limit. Integrating once again shows that $v'_{n,\lambda}(0)$ converges, too.
   (c) Identify the limits of $v_{n,\lambda}$, $v'_{n,\lambda}$ and $v''_{n,\lambda}$.

4. Show Theorem 8.5 with semigroup methods.
   Hint: Observe that $A\int_0^t P_s g\,ds = P_t g - g = \frac{d}{dt}\int_0^t P_s g\,ds - g$.

5. Find the solution to the Dirichlet problem in dimension $d = 1$: $u''(x) = 0$ for all $x \in (0,1)$, $u(0) = a$, $u(1) = b$ and u is continuous in $[0,1]$. Compare your findings with Wald's identities, cf. Theorem 5.10 and Corollary 5.11.

6. Use Lemma 7.33 and give an alternative derivation of the result of Lemma 8.8.

7. Show that Lemma 8.8 remains true for any d-dimensional Feller process X_t with continuous paths and generator $L = \sum_{j,k=1}^d a_{jk}(x)\,\partial_j\partial_k + \sum_{j=1}^d b_j(x)\,\partial_j$ such that $a_{11}(x) \geqslant a_0 > 0$ and $|b_1(x)| \leqslant b_0 < \infty$.
   Hint: Set $u(x) = e^{-x_1^2/\gamma r^2}$ and multiply it with a smooth cut-off function $\chi \in \mathcal{C}_c^2(\mathbb{R}^d)$, $\chi|_{\mathbb{B}(0,r)} \equiv 1$, to make sure it is in $\mathfrak{D}(L)$. Then calculate $Lu(x) = L(u\chi)(x)$ for $x \in \mathbb{B}(0,r)$.

8. Use the LIL (Corollary 12.2) to give an alternative proof for the fact that a one-dimensional Brownian motion oscillates in every time interval $[0,\epsilon]$ infinitely often around its starting point.

9. (Flat cone condition; Chung 1982) Let $d \geqslant 2$. A *flat cone* in $\mathbb{R}^d$ is a cone in $\mathbb{R}^{d-1}$. Adapt the argument of Example 8.15 e) to show the following useful regularity criterion for a BM^d: *The boundary point x_0 is regular for D^c if there is a truncated flat cone with vertex at x_0 and lying entirely in D^c.*
   Hint: Example 8.15 e) for $d = 2$ and Chung [31, p. 165].

10. Let (b,β) be a BM^2 and $\sigma_n = \inf\{t > 1/n : b(t) = 0\}$. Show that the random variable $\beta(\sigma_n)$ has a probability density.

9 The variation of Brownian paths

In this chapter we start our study of the regularity of the Brownian paths. One possibility to measure the regularity of a function is to look at its oscillations.

On a small interval $[t_{j-1}, t_j]$ the oscillation of a continuous function f should roughly be $|f(t_j) - f(t_{j-1})|$ and summing up the oscillation over consecutive intervals should give a number quantifying the oscillatory behaviour of f. In this way we would measure the increase of a monotone function and, for a Lipschitz continuous functions, the oscillations would be bounded by the Lipschitz constant times the interval length. In order to capture Hölder continuity we will introduce a weight on the oscillations.

Recall the definition of p-variation: Let $f : [0, \infty) \to \mathbb{R}^d$ be a (non-random) function and let $\Pi = \{t_0 = 0 < t_1 < \cdots < t_n = t\}$ be a finite partition of the interval $[0, t]$. By $|\Pi| := \max_{1\leqslant j\leqslant n}(t_j - t_{j-1})$ we denote the *mesh* of the partition. For $p > 0$ we call

$$S_p^{\Pi}(f;t) := \sum_{t_{j-1},t_j\in\Pi} |f(t_j) - f(t_{j-1})|^p = \sum_{j=1}^{n} |f(t_j) - f(t_{j-1})|^p \tag{9.1}$$

the *p-variation sum* for Π. The supremum over all finite partitions is the *strong* or *total p-variation*,

$$\mathrm{VAR}_p(f;t) := \sup\left\{S_p^{\Pi}(f;t) \;:\; \Pi \text{ finite partition of } [0,t]\right\}. \tag{9.2}$$

Ex. 9.1 If $\mathrm{VAR}_1(f;t) < \infty$, the function f is said to be of *bounded* or *finite (total) variation*. Often the following notion of p-variation along a given sequence of finite partitions $(\Pi_n)_{n\geqslant 1}$ of $[0, t]$ with $\lim_{n\to\infty} |\Pi_n| = 0$ is used:

$$\mathrm{var}_p(f;t) := \lim_{|\Pi_n|\to 0} S_p^{\Pi_n}(f;t). \tag{9.3}$$

Note that $\mathrm{VAR}_p(f;t)$ is always defined; if $\mathrm{var}_p(f;t)$ exists, $\mathrm{VAR}_p(f;t) \geqslant \mathrm{var}_p(f;t)$. When a random function $X(s, \omega)$ is used in place of $f(s)$, we can speak of $\mathrm{var}_p(X(\cdot, \omega); t)$ as a
! limit in law, in probability, in kth mean, and a. s. Note, however, that the sequence of partitions $(\Pi_n)_{n\geqslant 1}$ must not depend on ω.

If f is continuous, it is no restriction to require that the endpoints $0, t$ of the interval $[0, t]$ are contained in every finite partition Π. Moreover, we may restrict ourselves to partitions with rational points or points from any other dense subset (plus the end-
Ex. 9.3 point t). This follows from the fact that we find for every $\epsilon > 0$ and every partition
Ex. 9.4 $\Pi = \{t_0 = 0 < t_1 < \cdots < t_n = t\}$ some $\Pi' = \{q_0 = 0 < q_1 < \cdots < q_n = t\}$ with points from the dense subset such that

$$\sum_{j=0}^{n} |f(t_j) - f(q_j)|^p \leqslant \epsilon.$$

Thus, $|S_p^{\Pi}(f;t) - S_p^{\Pi'}(f;t)| \leqslant c_{p,d}\,\epsilon$ with a constant depending only on p and the dimension d; since

$$\mathrm{VAR}_p(f;t) = \sup_n \sup \left\{ S_p^{\Pi}(f;t) \;:\; \Pi \text{ partition of } [0,t] \text{ with } \#\Pi = n \right\},$$

we see that it is enough to calculate $\mathrm{VAR}_p(f;t)$ along rational points.

9.1 The quadratic variation

Let us determine the quadratic variation $\mathrm{var}_2(B;t) = \mathrm{var}_2(B(\cdot,\omega);t)$ of a Brownian motion $B(t), t \geqslant 0$. In view of the results of Section 2.2 it is enough to consider a one-dimensional Brownian motion throughout this section. Ex. 9.2

9.1 Theorem. *Let $(B_t)_{t\geqslant 0}$ be a* BM[1] *and $(\Pi_n)_{n\geqslant 1}$ be any sequence of finite partitions of $[0,t]$ satisfying $\lim_{n\to\infty} |\Pi_n| = 0$. Then the following mean-square limit exists:*

$$\mathrm{var}_2(B;t) = L^2(\mathbb{P})\text{-}\lim_{n\to\infty} S_2^{\Pi_n}(B;t) = t.$$

Usually $\mathrm{var}_2(B;t)$ is called the *quadratic variation* of a Brownian motion.

Proof. Let $\Pi = \{t_0 = 0 < t_1 < \cdots < t_n \leqslant t\}$ be some partition of $[0,t]$. Then we have

$$\mathbb{E}\, S_2^{\Pi}(B;t) = \sum_{j=1}^{n} \mathbb{E}\left[(B(t_j) - B(t_{j-1}))^2\right] = \sum_{j=1}^{n}(t_j - t_{j-1}) = t.$$

Therefore,

$$\mathbb{E}\left[(S_2^{\Pi}(B;t) - t)^2\right] = \mathbb{V}\left[S_2^{\Pi}(B;t)\right] = \mathbb{V}\left[\sum_{j=1}^{n}(B(t_j) - B(t_{j-1}))^2\right].$$

By (B1) the random variables $(B(t_j) - B(t_{j-1}))^2$, $j = 1, \ldots, n$, are independent with mean $t_j - t_{j-1}$. With Bienaymé's identity we find

$$\begin{aligned}
\mathbb{E}\left[(S_2^{\Pi}(B;t) - t)^2\right] &\overset{\text{(B1)}}{=} \sum_{j=1}^{n} \mathbb{V}\left[(B(t_j) - B(t_{j-1}))^2\right] \\
&= \sum_{j=1}^{n} \mathbb{E}\left[\left\{(B(t_j) - B(t_{j-1}))^2 - (t_j - t_{j-1})\right\}^2\right] \\
&\overset{\text{(B2)}}{=} \sum_{j=1}^{n} \mathbb{E}\left[\left\{B(t_j - t_{j-1})^2 - (t_j - t_{j-1})\right\}^2\right] \\
&\overset{2.16}{=} \sum_{j=1}^{n}(t_j - t_{j-1})^2 \underbrace{\mathbb{E}\left[\left\{B(1)^2 - 1\right\}^2\right]}_{=2 \text{ cf. } 2.3} \\
&\leqslant 2\,|\Pi| \sum_{j=1}^{n}(t_j - t_{j-1}) = 2\,|\Pi|\,t \xrightarrow[|\Pi|\to 0]{} 0.
\end{aligned}$$

□

9.2 Corollary. *Almost all Brownian paths are of infinite total variation. In fact, we have* $\mathrm{VAR}_p(B;t) = \infty$ *a. s. for all* $p < 2$.

Proof. Let $p = 2 - \delta$ for some $\delta > 0$. Let Π_n be any sequence of partitions of $[0, t]$ with $|\Pi_n| \to 0$. Then

$$\begin{aligned}\sum_{t_{j-1},t_j\in\Pi_n} (B(t_j) - B(t_{j-1}))^2 &\leqslant \max_{t_{j-1},t_j\in\Pi_n} |B(t_j) - B(t_{j-1})|^\delta \sum_{t_{j-1},t_j\in\Pi_n} |B(t_j) - B(t_{j-1})|^{2-\delta} \\ &\leqslant \max_{t_{j-1},t_j\in\Pi_n} |B(t_j) - B(t_{j-1})|^\delta \, \mathrm{VAR}_{2-\delta}(B;t).\end{aligned}$$

The left-hand side converges, at least for a subsequence, almost surely to t. On the other hand, $\lim_{|\Pi_n|\to 0} \max_{t_{j-1},t_j\in\Pi_n} |B(t_j) - B(t_{j-1})|^\delta = 0$ since the Brownian paths are (uniformly) continuous on $[0, t]$. This shows that $\mathrm{VAR}_{2-\delta}(B;t) = \infty$ almost surely. □

It is not hard to adapt the above results for intervals of the form $[a, b]$. Either we argue directly, or we observe that if $(B_t)_{t\geqslant 0}$ is a Brownian motion, then so is the process $W(t) := B(t + a) - B(a)$, $t \geqslant 0$, see Paragraph 2.13.

Recall that a function $f : [0,\infty) \to \mathbb{R}$ is called *(locally) Hölder continuous* of order $\alpha > 0$, if for every compact interval $[a, b]$ there exists a constant $c = c(f, \alpha, [a, b])$ such that

$$|f(t) - f(s)| \leqslant c\,|t - s|^\alpha \quad \text{for all } s, t \in [a, b]. \tag{9.4}$$

Ex. 9.6 **9.3 Corollary.** *Almost all Brownian paths are nowhere (locally) Hölder continuous of order* $\alpha > \frac{1}{2}$.

Proof. Fix some interval $[a, b] \subset [0,\infty)$, $a < b$ are rational, $\alpha > \frac{1}{2}$ and assume that (9.4) holds with $f(t) = B(t,\omega)$ and some constant $c = c(\omega, \alpha, [a, b])$.

Then we find for all finite partitions Π of $[a, b]$

$$\begin{aligned}S_2^\Pi(B(\cdot,\omega); [a, b]) &= \sum_{t_{j-1},t_j\in\Pi} (B(t_j,\omega) - B(t_{j-1},\omega))^2 \\ &\leqslant c(\omega)^2 \sum_{t_{j-1},t_j\in\Pi} (t_j - t_{j-1})^{2\alpha} \leqslant c(\omega)^2 (b - a)|\Pi|^{2\alpha-1}.\end{aligned}$$

In view of Theorem 9.1 we know that the left-hand side converges almost surely to $b-a$ for some subsequence Π_n with $|\Pi_n| \to 0$. Since $2\alpha - 1 > 0$, the right-hand side tends to 0, and we have a contradiction.

This shows that (9.4) can only hold on a $\mathbb{P}$-null set $N_{a,b} \subset \Omega$. But the union of null sets $\bigcup_{0\leqslant a<b, a,b\in\mathbb{Q}} N_{a,b}$ is an exceptional set which is uniform for all intervals. □

9.2 Almost sure convergence of the variation sums

Getting almost sure convergence of the variation sums is more difficult. One of the first results in this direction is due to P. Lévy for refining partitions [144, Theorème 5] and the following simple argument for dyadic and other partitions with rapidly decaying mesh-size [145, Theorème 41.1]. In both cases, however, we need an additional condition on the sequence of partitions.

9.4 Theorem (Lévy 1948). *Let $(B_t)_{t\geqslant 0}$ be a one-dimensional Brownian motion. Along any sequence of finite partitions Π_n of $[0,t]$ such that $\sum_{n=1}^{\infty}|\Pi_n|<\infty$ we have*

$$\operatorname{var}_2(B;t)=\lim_{n\to\infty}S_2^{\Pi_n}(B;t)=t \quad \textit{almost surely.}$$

Proof. As in the proof of Theorem 9.1 we find

$$\mathbb{E}\left[\left(S_2^{\Pi_n}(B;t)-t\right)^2\right]=2\sum_{t_{j-1},t_j\in\Pi_n}(t_j-t_{j-1})^2\leqslant 2\,|\Pi_n|\,t.$$

Now we use the Chebyshev–Markov inequality to get for every $\epsilon>0$

$$\sum_{n=1}^{\infty}\mathbb{P}\left(\left|S_2^{\Pi_n}(B;t)-t\right|>\epsilon\right)\leqslant\frac{2t}{\epsilon^2}\sum_{n=1}^{\infty}|\Pi_n|<\infty.$$

By the (easy direction of the) Borel–Cantelli lemma we conclude that there is a set Ω_ϵ such that $\mathbb{P}(\Omega_\epsilon)=1$ and for all $\omega\in\Omega_\epsilon$ there is some $N(\omega)$ such that

$$\left|S_2^{\Pi_n}(B(\cdot,\omega);t)-t\right|\leqslant\epsilon \quad \text{for all } n\geqslant N(\omega).$$

This proves that $\operatorname{var}_2(B;t)=t$ on $\Omega_0:=\bigcap_{k\geqslant 1}\Omega_{1/k}$ along the sequence $(\Pi_n)_{n\geqslant 1}$. Since $\mathbb{P}(\Omega_0)=1$, we are done. □

The summability condition $\sum_{n=1}^{\infty}|\Pi_n|<\infty$ used in Theorem 9.4 is essentially an assertion on the decay of the sequence $|\Pi_n|$. If the mesh sizes decrease, we have

$$n|\Pi_n|\leqslant\sum_{j=1}^{n}|\Pi_j|\leqslant\sum_{j=1}^{\infty}|\Pi_j|,$$

hence $|\Pi_n|=O(1/n)$. The following result of Dudley, [55, p. 89], shows that the condition $|\Pi_n|=o(1/\log n)$ is sufficient, i. e. $\lim_{n\to\infty}|\Pi_n|\log n=0$.

9.5 Theorem (Dudley 1973). *Let $(B_t)_{t\geqslant 0}$ be a BM^1 and assume that Π_n is a sequence of partitions of $[0,t]$ such that $|\Pi_n|=o(1/\log n)$. Then*

$$\operatorname{var}_2(B;t)=\lim_{n\to\infty}S_2^{\Pi_n}(B;t)=t \quad \textit{almost surely.}$$

Proof. To simplify notation, we write in this proof

$$S_2^{\Pi_n}=S_2^{\Pi_n}(B;t),\quad s_j=t_j-t_{j-1}\quad\text{and}\quad W_j=\frac{B(t_j)-B(t_{j-1})}{\sqrt{t_j-t_{j-1}}}.$$

Then

$$\begin{aligned}S_2^{\Pi_n}-t&=\sum_{t_{j-1},t_j\in\Pi_n}\left[(B(t_j)-B(t_{j-1}))^2-(t_j-t_{j-1})\right]\\&=\sum_{t_{j-1},t_j\in\Pi_n}(t_j-t_{j-1})\left[\frac{(B(t_j)-B(t_{j-1}))^2}{t_j-t_{j-1}}-1\right]=\sum_{t_{j-1},t_j\in\Pi_n}s_j\left[W_j^2-1\right].\end{aligned}$$

Note that the W_j are iid $\mathsf{N}(0, 1)$ random variables. For every $\epsilon > 0$

$$\mathbb{P}\left(\left|S_2^{\Pi_n} - t\right| > \epsilon\right) \leqslant \mathbb{P}\left(S_2^{\Pi_n} - t > \epsilon\right) + \mathbb{P}\left(t - S_2^{\Pi_n} > \epsilon\right);$$

we estimate the right-hand side with the Markov inequality. For the first term we find

$$\begin{aligned}\mathbb{P}\left(S_2^{\Pi_n} - t > \epsilon\right) &\leqslant e^{-\epsilon\theta}\, \mathbb{E}\, e^{\theta(S_2^{\Pi_n} - t)} \\ &= e^{-\epsilon\theta}\, \mathbb{E}\, e^{\theta \sum_j s_j (W_j^2 - 1)} = e^{-\epsilon\theta} \prod_j \mathbb{E}\, e^{\theta s_j (W_j^2 - 1)}.\end{aligned}$$

(Below we will pick θ such that the right-hand side is finite.) If we combine the elementary inequality $|e^{\lambda x} - \lambda x - 1| \leqslant \lambda^2 x^2 e^{|\lambda x|}$, $x, \lambda \in \mathbb{R}$, with the fact that $\mathbb{E}(W^2 - 1) = 0$, we get

$$\begin{aligned}\mathbb{P}\left(S_2^{\Pi_n} - t > \epsilon\right) &\leqslant e^{-\epsilon\theta} \prod_j \left[1 + \mathbb{E}\left(e^{\theta s_j (W_j^2 - 1)} - \theta s_j (W_j^2 - 1) - 1\right)\right] \\ &\leqslant e^{-\epsilon\theta} \prod_j \left[1 + \mathbb{E}\left(\theta^2 s_j^2 (W_j^2 - 1)^2 e^{|\theta s_j| \cdot |W_j^2 - 1|}\right)\right].\end{aligned}$$

Ex. 9.7 A direct calculation reveals that for each $0 < \lambda_0 < 1/2$ there is a constant $C = C(\lambda_0)$ such that

$$\mathbb{E}\left[(W_j^2 - 1)^2 e^{|\lambda| \cdot |W_j^2 - 1|}\right] \leqslant C < \infty \quad \text{for all } 1 \leqslant j \leqslant n,\ |\lambda| \leqslant \lambda_0 < \frac{1}{2}.$$

Using the estimate $1 + y \leqslant e^y$ we arrive at

$$\mathbb{P}(S_2^{\Pi_n} - t > \epsilon) \leqslant e^{-\epsilon\theta} \prod_j \left[C\theta^2 s_j^2 + 1\right] \leqslant e^{-\epsilon\theta} \prod_j e^{C\theta^2 s_j^2} = e^{-\epsilon\theta + C\theta^2 s^2}$$

where $s^2 := \sum_j s_j^2$. If we choose

$$\theta = \theta_0 := \min\left\{\frac{\epsilon}{2Cs^2}, \frac{\lambda_0}{|\Pi_n|}\right\},$$

then $|s_j \theta_0| \leqslant \lambda_0$, and all expectations in the calculations above are finite. Moreover,

$$\mathbb{P}(S_2^{\Pi_n} - t > \epsilon) \leqslant e^{-\theta_0(\epsilon - C\theta_0 s^2)} \leqslant e^{-\frac{1}{2}\epsilon\theta_0} \leqslant e^{-\frac{\epsilon\lambda_0}{2|\Pi_n|}}.$$

The same estimate holds for $\mathbb{P}(t - S_2^{\Pi_n} > \epsilon)$, and we see

$$\mathbb{P}\left(\left|S_2^{\Pi_n} - t\right| > \epsilon\right) \leqslant 2 \exp\left(-\frac{\epsilon\lambda_0}{2\,|\Pi_n|}\right).$$

Since $|\Pi_n| = o(1/\log n)$, we have $|\Pi_n| = \epsilon_n / \log n$ where $\epsilon_n \to 0$ as $n \to \infty$. Thus,

$$\mathbb{P}\left(\left|S_2^{\Pi_n} - t\right| > \epsilon\right) \leqslant 2n^{-\frac{\epsilon\lambda_0}{2\epsilon_n}} \leqslant \frac{2}{n^2} \quad \text{for all } n \geqslant n_0.$$

The conclusion follows now, as in the proof of Theorem 9.4, from the Borel–Cantelli lemma. □

Yet another possibility to get almost sure convergence of the variation sums is to consider nested partitions. The proof relies on the a. s. martingale convergence theorem for backwards martingales, cf. Corollary A.8 in the appendix. We begin with an auxiliary result.

9.6 Lemma. *Let* $(B_t)_{t\geqslant 0}$ *be a* BM[1] *and* $\mathscr{G}_s := \sigma\big(B_r,\ r \leqslant s;\ (B_u - B_v)^2,\ u, v \geqslant s\big)$. *Then*

$$\mathbb{E}\left(B_t - B_s \mid \mathscr{G}_s\right) = 0 \quad \textit{for all } 0 \leqslant s < t.$$

Proof. From Lemma 2.14 we know that $\mathscr{F}_s^B = \sigma(B_r,\ r \leqslant s)$ and $\mathscr{F}_{[s,\infty)}^B := \sigma(B_t - B_s,\ t \geqslant s)$ are independent σ-algebras.

A typical set from a $\cap$-stable generator of $\mathscr{G}_s$ is of the form

$$\Gamma = \bigcap_{j=1}^{M} \{B_{r_j} \in C_j\} \cap \bigcap_{k=1}^{N} \{(B_{v_k} - B_{u_k})^2 \in D_k\}$$

where $M, N \geqslant 1$, $r_j \leqslant s \leqslant u_k \leqslant v_k$ and $C_j, D_k \in \mathscr{B}(\mathbb{R})$. Since $\mathscr{F}_s^B \perp\!\!\!\perp \mathscr{F}_{[s,\infty)}^B$ and $\tilde{B} := -B$ is again a Brownian motion, we get

$$\begin{aligned}
\int_\Gamma (B_t - B_s)\, d\mathbb{P} &= \int \underbrace{(B_t - B_s) \prod_{k=1}^{N} \mathbb{1}_{D_k}((B_{v_k} - B_{u_k})^2)}_{\mathscr{F}_{[s,\infty)}^B \text{ measurable}} \cdot \underbrace{\prod_{j=1}^{M} \mathbb{1}_{C_j}(B_{r_j})}_{\mathscr{F}_s^B \text{ measurable}}\, d\mathbb{P} \\
&= \int (B_t - B_s) \prod_{k=1}^{N} \mathbb{1}_{D_k}((B_{v_k} - B_{u_k})^2)\, d\mathbb{P} \cdot \int \prod_{j=1}^{M} \mathbb{1}_{C_j}(B_{r_j})\, d\mathbb{P} \\
&= \int (\tilde{B}_t - \tilde{B}_s) \prod_{k=1}^{N} \mathbb{1}_{D_k}((\tilde{B}_{v_k} - \tilde{B}_{u_k})^2)\, d\mathbb{P} \cdot \int \prod_{j=1}^{M} \mathbb{1}_{C_j}(B_{r_j})\, d\mathbb{P} \\
&= \int (B_s - B_t) \prod_{k=1}^{N} \mathbb{1}_{D_k}((B_{u_k} - B_{v_k})^2)\, d\mathbb{P} \cdot \int \prod_{j=1}^{M} \mathbb{1}_{C_j}(B_{r_j})\, d\mathbb{P} \\
&= \cdots = -\int_\Gamma (B_t - B_s)\, d\mathbb{P}.
\end{aligned}$$

This yields $\mathbb{E}(B_t - B_s \mid \mathscr{G}_s) = 0$. □

9.7 Theorem (Lévy 1940). *Let* $(B_t)_{t\geqslant 0}$ *be a one-dimensional Brownian motion and let* Π_n *be a sequence of nested partitions of* $[0, t]$, *i. e.* $\Pi_n \subset \Pi_{n+1} \subset \ldots$, *with* $\lim_{n\to\infty} |\Pi_n| = 0$. *Then*

$$\operatorname{var}_2(B; t) = \lim_{n\to\infty} S_2^{\Pi_n}(B; t) = t \quad \textit{almost surely.}$$

Proof. We can assume that $\#(\Pi_{n+1} \setminus \Pi_n) = 1$, otherwise we add the missing intermediate partitions.

We show that the variation sums $S_{-n} := S_2^{\Pi_n}(B, t)$ are a martingale for the filtration $\mathscr{H}_{-n} := \sigma(S_{-n}, S_{-n-1}, S_{-n-2}, \ldots)$. Fix $n \geqslant 1$; then $\Pi_{n+1} = \Pi_n \cup \{s\}$ and there are two consecutive points $t_j, t_{j+1} \in \Pi_n$ such that $t_j < s < t_{j+1}$. Then

$$\begin{aligned} S_{-n} - S_{-n-1} &= (B_{t_{j+1}} - B_{t_j})^2 - (B_s - B_{t_j})^2 - (B_{t_{j+1}} - B_s)^2 \\ &= 2(B_{t_{j+1}} - B_s)(B_s - B_{t_j}). \end{aligned}$$

Thus, with Lemma 9.6,

$$\mathbb{E}\left((B_{t_{j+1}} - B_s)(B_s - B_{t_j}) \mid \mathscr{G}_s\right) = (B_s - B_{t_j})\,\mathbb{E}\left((B_{t_{j+1}} - B_s) \mid \mathscr{G}_s\right) = 0$$

and, since $\mathscr{H}_{-n-1} \subset \mathscr{G}_s$, we can use the tower property to see

$$\mathbb{E}\left(S_{-n} - S_{-n-1} \mid \mathscr{H}_{-n-1}\right) = \mathbb{E}\left(\mathbb{E}\left(S_{-n} - S_{-n-1} \mid \mathscr{G}_s\right) \mid \mathscr{H}_{-n-1}\right) = 0.$$

Thus $(S_{-n}, \mathscr{H}_{-n})_{n\geqslant 1}$ is a (backwards) martingale and by the martingale convergence theorem, Corollary A.8, we see that the $\lim_{n\to\infty} S_{-n} = \lim_{n\to\infty} S_2^{\Pi_n}(B;t)$ exists almost surely. Since the L^2 limit of the variation sums is t, cf. Theorem 9.1, we get that the a. s. limit is again t. □

9.3 Almost sure divergence of the variation sums

So far we have only considered variation sums where the underlying partition Π does not depend on ω, i. e. on the particular Brownian path. This means, in particular, that we cannot make assertions on the *strong* quadratic variation $\mathrm{VAR}_2(B(\cdot,\omega);1)$ of a Brownian motion. We are going to show that, in fact, the strong quadratic variation of almost all Brownian paths is infinite. This is an immediate consequence of the following result.

9.8 Theorem (Lévy 1948). *Let $(B_t)_{t\geqslant 0}$ be a BM^1. For almost all $\omega \in \Omega$ there is a nested sequence of random partitions $(\Pi_n(\omega))_{n\geqslant 1}$ such that*

$$|\Pi_n(\omega)| \xrightarrow[n\to\infty]{} 0 \quad \text{and} \quad \mathrm{var}_2(B(\cdot,\omega);1) = \lim_{n\to\infty} S_2^{\Pi_n(\omega)}(B(\cdot,\omega);1) = +\infty.$$

The strong variation is, by definition, the supremum of the variation sums for *all* finite partitions. Thus,

9.9 Corollary (Lévy 1948). *Let $(B_t)_{t\geqslant 0}$ be a one-dimensional Brownian motion. Then $\mathrm{VAR}_2(B(\cdot,\omega);1) = \infty$ for almost all $\omega \in \Omega$.*

Proof of Theorem 9.8 (Freedman 1971). Let $w(t) := B(t,\omega)$, $t \in [0,1]$, be a fixed trajectory, write $\Pi_n := \{t_j : t_j = j/n,\ j = 0,1,\dots,n\}$ and set $\mathfrak{I}_n := \{[t_{j-1},t_j] : j = 1,\dots,n\}$. An interval $I = [a,b]$ is a *k-fast interval* for w, if

$$(w(b) - w(a))^2 \geqslant 2k(b-a).$$

1° If we can show that

$$\forall k \geqslant 1\ \exists n = n(k) \geqslant k\ \forall I \in \mathfrak{I}_n : \ I \text{ is a } k\text{-fast interval for } w \tag{9.5}$$

then

$$S_2^{\Pi_{n(k)}}(w;1) = \sum_{j=1}^{n(k)} |w(t_j) - w(t_{j-1})|^2 \geqslant 2k \sum_{j=1}^{n(k)} \frac{1}{n(k)} = 2k,$$

and so $\lim_{k\to\infty} S_2^{\Pi_{n(k)}}(w;1) = \infty$.

On the other hand, Theorem 9.1 shows that for a Brownian motion we *cannot expect* that *all* intervals are k-fast intervals. We will, however, show that – depending on ω – there are still sufficiently many k-fast intervals. For this we have to relax the condition (9.5).

2° The following condition

$$\begin{gathered}\exists\,(\alpha_k)_{k\geqslant 1} \subset (0,1)\ \ \forall k \geqslant 1,\ m \geqslant 1\ \ \exists n = n(k,m) \geqslant k\ \ \forall I \in \mathfrak{I}_m\ : \\ I \text{ contains at least } \lfloor \alpha_k n \rfloor + 1 \text{ many } k\text{-fast intervals from } \mathfrak{I}_{mn}\end{gathered} \tag{9.6}$$

allows us to construct a nested sequence of partitions $(\Pi_n)_{n\geqslant 1}$ such that $|\Pi_n| \to 0$ and $S_2^{\Pi_n}(w;1) \to \infty$.

Assume that (9.6) holds. Fix $k \geqslant 1$, $m = m_k \geqslant 1$ and pick the smallest integer $s_k \geqslant 1$ such that $(1-\alpha_k)^{s_k} \leqslant \frac{1}{2}$. For $I \in \mathfrak{I}_m$ we find some $n_1 = n_1(k,m)$ such that

$$\text{each } I \in \mathfrak{I}_m \text{ contains at least } \lfloor \alpha_k n_1 \rfloor + 1\ k\text{-fast intervals } I' \in \mathfrak{I}_{mn_1}.$$

Denote these intervals by $\mathfrak{I}^*_{(m,n_1)}$. By our construction, $\mathfrak{I}_{mn_1} \setminus \mathfrak{I}^*_{(m,n_1)}$ has no more than $(n_1 - \lfloor \alpha_k n_1 \rfloor - 1)m$ elements.

Applying (9.6) to the intervals from $\mathfrak{I}_{mn_1} \setminus \mathfrak{I}^*_{(m,n_1)}$ yields that there is some n_2 such that

$$\text{each } I \in \mathfrak{I}_{mn_1} \setminus \mathfrak{I}^*_{(m,n_1)} \text{ contains at least } \lfloor \alpha_k n_2 \rfloor + 1\ k\text{-fast intervals } I' \in \mathfrak{I}_{mn_1n_2}.$$

Denote these intervals by $\mathfrak{I}^*_{(m,n_1,n_2)}$. By construction, $\mathfrak{I}_{mn_1n_2} \setminus \mathfrak{I}^*_{(m,n_1,n_2)}$ has no more than $(n_2 - \lfloor \alpha_k n_2 \rfloor - 1)(n_1 - \lfloor \alpha_k n_1 \rfloor - 1)m$ elements.

If we repeat this procedure s_k times, there are no more than

$$m \prod_{j=1}^{s_k} (n_j - \lfloor \alpha_k n_j \rfloor - 1)$$

intervals left in $\mathfrak{I}_{mn_1\cdots n_{s_k}} \setminus \mathfrak{I}^*_{(m,n_1,\cdots,n_{s_k})}$; their total length is at most

$$m \prod_{j=1}^{s_k} (n_j - \lfloor \alpha_k n_j \rfloor - 1) \frac{1}{m \prod_{j=1}^{s_k} n_j} \leqslant (1-\alpha_k)^{s_k} \leqslant \frac{1}{2}.$$

The partition Π_k contains all points of the form $t = j/m$, $j = 0,\dots,m$, as well as the endpoints of the intervals from

$$\mathfrak{I}^*_{(m,n_1)} \cup \mathfrak{I}^*_{(m,n_1,n_2)} \cup \cdots \cup \mathfrak{I}^*_{(m,n_1,n_2,\dots,n_{s_k})}.$$

Since all intervals are k-fast and since their total length is at least $\frac{1}{2}$, we get

$$S_2^{\Pi_k}(w;1) \geqslant 2k\,\frac{1}{2} = k.$$

3° In order to construct Π_{k+1}, we repeat the construction from the second step with $m = m_k \cdot n_1 \cdot n_2 \cdot \ldots \cdot n_{s_k}$. This ensures that $\Pi_{k+1} \supset \Pi_k$.

4° All that remains is to show that almost all Brownian paths satisfy the condition (9.6) for a suitable sequence $(\alpha_k)_{k\geqslant 1} \subset (0,1)$. Since a Brownian motion has stationary and independent increments, the random variables

$$Z_l^{(n)} = \#\left\{1 \leqslant j \leqslant n : \left[B(\tfrac{l}{m} + \tfrac{j}{mn}, \cdot) - B(\tfrac{l}{m} + \tfrac{j-1}{mn}, \cdot)\right]^2 \geqslant 2k\,\tfrac{1}{mn}\right\}$$

$(l = 0, 1, \ldots, m-1)$ are iid binomial random variables with the parameters n and

$$p_k = \mathbb{P}\left(\left[B(\tfrac{l}{m} + \tfrac{j}{mn}, \cdot) - B(\tfrac{l}{m} + \tfrac{j-1}{mn}, \cdot)\right]^2 \geqslant 2k\,\tfrac{1}{mn}\right) = \mathbb{P}\left(|B(1)|^2 \geqslant 2k\right).$$

In the last equality we used the scaling property 2.16 of a Brownian motion. Observe that $\mathbb{E}\,Z_l^{(n)} = p_k n$; by the central limit theorem,

$$\mathbb{P}\left(\bigcap_{l=0}^{m-1}\{Z_l^{(n)} \geqslant \lfloor \tfrac{1}{2}\,p_k n\rfloor + 1\}\right) = \prod_{l=0}^{m-1}\mathbb{P}\left(Z_l^{(n)} \geqslant \lfloor \tfrac{1}{2}\,p_k n\rfloor + 1\right) \xrightarrow[n\to\infty]{} 1,$$

and so

$$\mathbb{P}\left(\bigcup_{n=1}^{\infty}\bigcap_{l=0}^{m-1}\{Z_l^{(n)} \geqslant \lfloor \tfrac{1}{2}\,p_k n\rfloor + 1\}\right) = 1.$$

This shows that

$$\mathbb{P}\left(\bigcap_{k=1}^{\infty}\bigcap_{m=1}^{\infty}\bigcup_{n=1}^{\infty}\bigcap_{l=0}^{m-1}\{Z_l^{(n)} \geqslant \lfloor \tfrac{1}{2}\,p_k n\rfloor + 1\}\right) = 1$$

which is just (9.6) for the sequence $\alpha_k = \frac{1}{2}\,\mathbb{P}(|B(1)|^2 \geqslant 2k)$. □

9.4 Lévy's characterization of Brownian motion

Theorem 9.1 remains valid for all continuous martingales $(X_t, \mathscr{F}_t)_{t\geqslant 0}$ with the property that $(X_t^2 - t, \mathscr{F}_t)_{t\geqslant 0}$ is a martingale. In fact, this is equivalent to saying that

$$\mathbb{E}\left[X_t - X_s \mid \mathscr{F}_s\right] = 0 \quad \text{for all } s \leqslant t, \tag{9.7}$$

and, since $\mathbb{E}[X_s X_t \mid \mathscr{F}_s] = X_s\,\mathbb{E}[X_t \mid \mathscr{F}_s] = X_s^2$,

$$\mathbb{E}\left[(X_t - X_s)^2 \mid \mathscr{F}_s\right] = t - s \quad \text{for all } s \leqslant t. \tag{9.8}$$

9.10 Lemma. *Let $(X_t, \mathscr{F}_t)_{t\geqslant 0}$ be a real-valued martingale with continuous sample paths such that $(X_t^2 - t, \mathscr{F}_t)_{t\geqslant 0}$ is a martingale. Then $\mathbb{E}\,X_t^4 < \infty$ and*

$$\mathbb{E}\left[(X_t - X_s)^4 \mid \mathscr{F}_s\right] \leqslant 4(t-s)^2 \quad \textit{for all } s < t. \tag{9.9}$$

Proof. Set $t_j := s + (t-s)\frac{j}{n}$, $j = 0, 1, \dots, n$, and $\delta_j := X_{t_j} - X_{t_{j-1}}$. Obviously,[1]

$$
\begin{aligned}
(X_t - X_s)^4 &= \Big(\sum_{j=1}^n \delta_j\Big)^2 \Big(\sum_{l=1}^n \delta_l\Big)^2 \\
&= \Big(\sum_{j=1}^n \delta_j^2 + 2\sum_{k=1}^n \sum_{j<k} \delta_j\delta_k\Big)\Big(\sum_{l=1}^n \delta_l^2 + 2\sum_{m=1}^n \sum_{l<m} \delta_l\delta_m\Big) \\
&= \sum_{j=1}^n \delta_j^4 + 2\sum_{l=1}^n \sum_{j<l} \delta_j^2\delta_l^2 + 4\sum_{j=1}^n \sum_{m=1}^n \sum_{l<m} \delta_j^2\delta_l\delta_m + 4\Big(\sum_{k=1}^n \sum_{j<k} \delta_j\delta_k\Big)^2 \\
&= \sum_{j=1}^n \delta_j^4 + 2\sum_{l=1}^n \sum_{j<l} \delta_j^2\delta_l^2 + 4\sum_{j=1}^n \delta_j \sum_{l=1}^n \delta_l^3 + 4\sum\sum\sum_{(j,l,m):\, l\neq j, m\neq j, l<m} \delta_j^2\delta_l\delta_m + 4\Big(\sum_{k=1}^n \sum_{j<k} \delta_j\delta_k\Big)^2.
\end{aligned}
$$

Using the Markov inequality we see

$$
\mathbb{P}\Big(\sum_{j=1}^n \delta_j^2 > N\Big) \leqslant \frac{1}{N}\,\mathbb{E}\Big(\sum_{j=1}^n \delta_j^2\Big) \overset{(9.8)}{=} \frac{1}{N}\sum_{j=1}^n (t_j - t_{j-1}) = \frac{t-s}{N} \xrightarrow[N\to\infty]{} 0.
$$

Thus, the sequence $(\sum_{j=1}^n \delta_j^2)_{n\geqslant 1}$ is bounded in probability; moreover

$$
\sum_{j=1}^n |\delta_j|^{2+\epsilon} \leqslant \max_{1\leqslant k\leqslant n} |\delta_k|^\epsilon \cdot \sum_{j=1}^n \delta_j^2 \xrightarrow[n\to\infty]{\mathbb{P}} 0
$$

where we used the the continuity of $t \mapsto X_t$. Selecting a subsequence, if necessary, we can assume that this convergence holds also a. s.

Since $\sum_{j=1}^n \delta_j = X_t - X_s$, we thus get

$$
\begin{aligned}
(X_t - X_s)^4 &\leqslant \varliminf_{n\to\infty} \Big[2\sum_{l=1}^n \sum_{j<l} \delta_j^2\delta_l^2 + 4\sum\sum\sum_{(j,l,m):\, l\neq j, m\neq j, l<m} \delta_j^2\delta_l\delta_m + 4\Big(\sum_{k=1}^n \sum_{j<k} \delta_j\delta_k\Big)^2\Big] \\
&\leqslant \varliminf_{n\to\infty} \Big[2\sum_{j=1}^n \delta_j^2 \Big(\sum_{k:\, k\neq j} \delta_k\Big)^2 + 4\Big(\sum_{k=1}^n \sum_{j<k} \delta_j\delta_k\Big)^2\Big].
\end{aligned}
$$

Since the expression in the square brackets is positive, the conditional version of Fatou's lemma yields

$$
\mathbb{E}\left[(X_t - X_s)^4 \mid \mathscr{F}_s\right] \leqslant \varliminf_{n\to\infty} \mathbb{E}\left[2\sum_{j=1}^n \delta_j^2 \Big(\sum_{k:k\neq j} \delta_k\Big)^2 + 4\Big(\sum_{k=1}^n \sum_{j<k} \delta_j\delta_k\Big)^2 \,\Big|\, \mathscr{F}_s\right].
$$

If we expand the squares on the right-hand side, we see that it is of the form

$$
c\sum_j \sum\sum_{(k,l):k\neq l, k\neq j, l\neq j} \delta_j^2\delta_k\delta_l + c' \sum\sum\sum\sum_{(j,k,l,m):j<k<l<m} \delta_j\delta_k\delta_l\delta_m + 8\sum\sum_{(j,k):j<k} \delta_j^2\delta_k^2.
$$

---

1 We learned this simplification of the original argument from Ms. Franziska Kühn.

Recall that $s \leqslant t_j < t_k \leqslant t$ for $j < k$ and $\delta_j = X_{t_j} - X_{t_{j-1}}$, $\delta_k = X_{t_k} - X_{t_{k-1}}$. Using the tower property of conditional expectations and (9.8), we find

$$\begin{aligned}
\mathbb{E}\left[\delta_k^2\delta_j^2 \mid \mathscr{F}_s\right] = \mathbb{E}\left[\mathbb{E}\left(\delta_k^2\delta_j^2 \mid \mathscr{F}_{t_j}\right) \mid \mathscr{F}_s\right] &= \mathbb{E}\left[\mathbb{E}\left(\delta_k^2 \mid \mathscr{F}_{t_j}\right)\delta_j^2 \mid \mathscr{F}_s\right] \\
&\overset{(9.8)}{=} \mathbb{E}\left[(t_k - t_{k-1})\,\delta_j^2 \mid \mathscr{F}_s\right] \\
&\overset{(9.8)}{=} (t_k - t_{k-1})(t_j - t_{j-1}).
\end{aligned}$$

If we also use (9.7), a very similar reasoning shows that

$$\mathbb{E}\left[\delta_j^2\delta_k\delta_l \mid \mathscr{F}_s\right] = \mathbb{E}\left[\delta_j\delta_k\delta_l\delta_m \mid \mathscr{F}_s\right] = 0$$

if the indices j, k, l and m are mutually different. Finally,

$$\begin{aligned}
\mathbb{E}\left[(X_t - X_s)^4 \mid \mathscr{F}_s\right] &\leqslant \lim_{n\to\infty} 8\sum_{k=1}^{n}\sum_{j<k}(t_j - t_{j-1})(t_k - t_{k-1}) \\
&= \lim_{n\to\infty} 8\,\frac{(t-s)^2}{n^2}\,\frac{n(n-1)}{2} = 4(t-s)^2,
\end{aligned}$$

and the claim follows. □

9.11 Corollary. *Let $(X_t, \mathscr{F}_t)_{t\geqslant 0}$ be a martingale with continuous sample paths such that $(X_t^2 - t, \mathscr{F}_t)_{t\geqslant 0}$ is a martingale. For any sequence $(\Pi_n)_{n\geqslant 1}$ of finite partitions of $[0, t]$ satisfying $\lim_{n\to\infty}|\Pi_n| = 0$, the mean-square limit exists:*

$$\operatorname{var}_2(X; t) = L^2(\mathbb{P})\text{-}\lim_{n\to\infty} S_2^{\Pi_n}(X; t) = t.$$

Proof. Let $\Pi := \{0 = t_0 < t_1 < \cdots < t_n = t\}$ be some partition of $[0, t]$, set $\delta_j := X_{t_j} - X_{t_{j-1}}$ and $\Delta t_j := t_j - t_{j-1}$, $j = 1, \dots, n$.

For any two increments δ_j and δ_k we find,

$$\begin{aligned}
\mathbb{E}\left[(\delta_j^2 - \Delta t_j)(\delta_k^2 - \Delta t_k)\right] &= \mathbb{E}\left[\delta_j^2\delta_k^2\right] - \Delta t_j\,\mathbb{E}\left[\delta_k^2\right] - \Delta t_k\,\mathbb{E}\left[\delta_j^2\right] + \Delta t_j\Delta t_k \\
&\overset{(9.8)}{=} \mathbb{E}\left[\delta_j^2\delta_k^2\right] - \Delta t_j\Delta t_k.
\end{aligned}$$

If $j \neq k$, the proof of Lemma 9.10 shows $\mathbb{E}[\delta_j^2\delta_k^2] = \Delta t_j\Delta t_k$, i. e. the mixed terms are zero. If $j = k$ we use (9.9) to get $\mathbb{E}[\delta_j^2\delta_k^2] = \mathbb{E}[\delta_j^4] \leqslant 4(\Delta t_j)^2$. Thus,

$$\begin{aligned}
\mathbb{E}\left[(S_2^{\Pi}(X; t) - t)^2\right] &= \sum_{j=1}^{n}\sum_{k=1}^{n}\mathbb{E}\left[(\delta_j^2 - \Delta t_j)(\delta_k^2 - \Delta t_k)\right] \\
&= \sum_{j=1}^{n}\mathbb{E}\left[(\delta_j^2 - \Delta t_j)^2\right] \leqslant 3\sum_{j=1}^{n}\Delta t_j^2 \leqslant 3|\Pi|\,t \xrightarrow[|\Pi|\to 0]{} 0,
\end{aligned}$$

where $|\Pi| := \max_{1\leqslant k\leqslant n}\Delta t_k$ is the mesh of the partition. □

If a process satisfies the assumptions of Lemma 9.10, it is already a Brownian motion. This is Lévy's martingale characterization of Brownian motion.

9.12 Theorem (Lévy 1948). *Let $(X_t, \mathscr{F}_t)_{t\geqslant 0}$, $X_0 = 0$, be a martingale with continuous sample paths such that $(X_t^2 - t, \mathscr{F}_t)_{t\geqslant 0}$ is a martingale. Then $(X_t)_{t\geqslant 0}$ is a Brownian motion.*

Proof (Gikhman, Skorokhod 1972). We have to show that

$$X_t - X_s \perp\!\!\!\perp \mathscr{F}_s \quad\text{and}\quad X_t - X_s \sim \mathsf{N}(0, t-s) \quad \text{for all } 0 \leqslant s \leqslant t.$$

This follows from

$$\mathbb{E}\left(e^{i\xi(X_t - X_s)} \;\middle|\; \mathscr{F}_s\right) = e^{-\frac{1}{2}(t-s)\xi^2} \quad \text{for all } s \leqslant t. \tag{9.10}$$

In fact, taking expectations in (9.10) yields

$$\mathbb{E}\left(e^{i\xi(X_t - X_s)}\right) = e^{-\frac{1}{2}(t-s)\xi^2},$$

i. e. $X_t - X_s \sim X_{t-s} \sim \mathsf{N}(0, t-s)$, and (9.10) becomes Ex. 9.8

$$\mathbb{E}\left(e^{i\xi(X_t - X_s)}\mathbb{1}_F\right) = \mathbb{E}\left(e^{i\xi(X_t - X_s)}\right)\mathbb{P}(F) \quad \text{for all } s \leqslant t,\ F \in \mathscr{F}_s;$$

this proves that $X_t - X_s \perp\!\!\!\perp F$ for all $F \in \mathscr{F}_s$.

For (9.10) we use a Taylor expansion up to order 3. Set $\Delta X = X_t - X_s$ and $h = t - s$. Then

$$e^{i\xi(X_t - X_s)} - e^{-\frac{1}{2}\xi^2(t-s)} = i\xi\Delta X + \frac{\xi^2}{2}h - \frac{\xi^2}{2}(\Delta X)^2 - \frac{\xi^4}{8}h^2 - \frac{i\xi^3}{6}(\eta\Delta X)^3 + \frac{\xi^6}{48}(\theta h)^3$$

where $\eta = \eta(\omega) \in (0,1)$ and $\theta \in (0,1)$. If we take conditional expectations with respect to $\mathscr{F}_s$ and observe that $\mathbb{E}(\Delta X \mid \mathscr{F}_s) = 0$ and $\mathbb{E}((\Delta X)^2 \mid \mathscr{F}_s) = h$ – this follows from (9.7) and (9.8) – , we get

$$\mathbb{E}\left(e^{i\xi(X_t - X_s)} - e^{-\frac{1}{2}\xi^2(t-s)} \;\middle|\; \mathscr{F}_s\right) = -\frac{\xi^4}{8}h^2 - \frac{i\xi^3}{6}\mathbb{E}\left((\eta\Delta X)^3 \mid \mathscr{F}_s\right) + \frac{\xi^6}{48}(\theta h)^3.$$

By (the conditional version of) Hölder's inequality with $p = 4/3$ and $q = 4$ we find

$$\left|\mathbb{E}\left((\eta\Delta X)^3 \mid \mathscr{F}_s\right)\right| \leqslant \mathbb{E}\left(|\Delta X|^3 \mid \mathscr{F}_s\right) \leqslant \left[\mathbb{E}\left((\Delta X)^4 \mid \mathscr{F}_s\right)\right]^{3/4} \overset{(9.9)}{\leqslant} 4^{3/4}h^{3/2}.$$

Therefore, we have for all $s < t$ with $h = t - s < 1$ and some constant c_ξ

$$\left|\mathbb{E}\left(e^{i\xi(X_t - X_s)} - e^{-\frac{1}{2}\xi^2(t-s)} \;\middle|\; \mathscr{F}_s\right)\right| \leqslant c_\xi (t-s)^{3/2}. \tag{9.11}$$

Fix $s < t$ and set $t_j := s + (t-s)j/n$, $j = 0, 1, \ldots, n$, $n \geqslant 1$. Observe that

$$e^{i\xi(X_t - X_s)} = \prod_{j=1}^{n} e^{i\xi(X_{t_j} - X_{t_{j-1}})} \quad\text{and}\quad e^{-\frac{1}{2}\xi^2(t-s)} = \prod_{j=1}^{n} e^{-\frac{1}{2}\xi^2(t_j - t_{j-1})}.$$

Using the elementary estimate

$$\left|\mathbb{E}(a\cdot A - b\cdot B \mid \mathscr{F}_s)\right| \leqslant \mathbb{E}(|a-b|\cdot|A| \mid \mathscr{F}_s) + \left|\mathbb{E}(b\cdot(A-B)\mid\mathscr{F}_s)\right|$$

with

$$a = \mathbb{E}\left(e^{i\xi(X_t - X_{t_{n-1}})} \,\Big|\, \mathscr{F}_{t_{n-1}}\right), \quad A = e^{i\xi(X_{t_{n-1}} - X_s)},$$
$$b = e^{-\frac{1}{2}\xi^2(t-t_{n-1})}, \qquad B = e^{-\frac{1}{2}\xi^2(t_{n-1}-s)},$$

yields, because of the tower property,

$$\begin{aligned}
&\left|\mathbb{E}\left(e^{i\xi(X_t-X_s)} - e^{-\frac{1}{2}\xi^2(t-s)} \,\Big|\, \mathscr{F}_s\right)\right| \\
&\quad\leqslant \mathbb{E}\left(\left|\mathbb{E}\left\{e^{i\xi(X_t-X_{t_{n-1}})} - e^{-\frac{1}{2}\xi^2(t-t_{n-1})} \,\Big|\, \mathscr{F}_{t_{n-1}}\right\}\right| \,\Big|\, \mathscr{F}_s\right) \\
&\qquad + \left|\mathbb{E}\left(e^{i\xi(X_{t_{n-1}}-X_s)} - e^{-\frac{1}{2}\xi^2(t_{n-1}-s)} \,\Big|\, \mathscr{F}_s\right)\right| \\
&\quad\leqslant \cdots \leqslant \sum_{j=1}^{n} \mathbb{E}\left(\left|\mathbb{E}\left\{e^{i\xi(X_{t_j}-X_{t_{j-1}})} - e^{-\frac{1}{2}\xi^2(t_j-t_{j-1})} \,\Big|\, \mathscr{F}_{t_{j-1}}\right\}\right| \,\Big|\, \mathscr{F}_s\right).
\end{aligned}$$

If $n \geqslant 1$ is large enough, we get $h := t_j - t_{j-1} < 1$. This means that we can use (9.11) with $(s,t) = (t_{j-1}, t_j)$ and find

$$\left|\mathbb{E}\left(e^{i\xi(X_t-X_s)} - e^{-\frac{1}{2}\xi^2(t-s)} \,\Big|\, \mathscr{F}_s\right)\right| \leqslant c_\xi \sum_{j=1}^{n}(t_j - t_{j-1})^{3/2} = \frac{c_\xi}{\sqrt{n}} \xrightarrow[n\to\infty]{} 0.$$

This proves (9.10). □

9.13 Further reading. Almost sure convergence of the p-variation of stochastic processes is quite difficult, and no sharp conditions are known. Some of the questions raised in the landmark paper [55] are still open. Up-to-date treatments of p-variation in analysis and probability are the Aarhus lecture notes [58] and the monograph [59].

[55] Dudley: Sample functions of the Gaussian process.
[58] Dudley, Norvaiša: *An Introduction to p-variation and Young integrals.*
[59] Dudley, Norvaiša: *Differentiability of Six Operators on Nonsmooth Functions and p-Variation.*

Problems

1. Let $(\Pi_n)_{n\geqslant 1}$ be a sequence of refining (i. e. $\Pi_n \subset \Pi_{n+1}$) partitions of $[0,1]$ such that $\lim_{n\to\infty}|\Pi_n| = 0$. Show that for every function $f \in \mathcal{C}[0,1]$ the limit

$$\lim_{n\to\infty} S_1^{\Pi_n}(f,1) = \mathrm{VAR}_1(f,1) := \sup\left\{S_1^{\Pi}(f,1) \;:\; \Pi \text{ finite partition of } [0,1]\right\}$$

exists in $[0,\infty]$ (and is, in particular, independent of the sequence $(\Pi_n)_{n\geqslant 1}$).

2. Let $f = (g, h) : [0, \infty) \to \mathbb{R}^2$ and $p > 0$. Show that $\mathrm{VAR}_p(f;t) < \infty$ if, and only if, $\mathrm{VAR}_p(g;t) + \mathrm{VAR}_p(h;t) < \infty$.

3. Let $f \in \mathcal{C}[0, 1]$. For every partition $\Pi = \{t_0 = 0 < t_1 < \cdots < t_n = 1\}$ and $\epsilon > 0$ there is some rational partition $\Pi' = \{q_0 = 0 < q_1 < \cdots < q_n = 1\}$ such that

$$\sum_{j=0}^{n} |f(t_j) - f(q_j)|^p \leqslant \epsilon.$$

Deduce from this that we may calculate $\mathrm{VAR}_p(f;1)$ along rational points.
Hint: Show $|S_p^{\Pi}(f;t) - S_p^{\Pi'}(f,t)| \leqslant c_{p,d}\epsilon$ using $(a+b)^p \leqslant 2^p(a^p + b^p)$, $a, b \geqslant 0$, $p \geqslant 0$.

4. Let f be continuous. Show that it does not affect the finiteness of $\mathrm{VAR}_p(f;t)$ and the numerical value of $\mathrm{var}_p(f;t)$ if we restrict ourselves to partitions not containing the endpoints 0 and t of the interval. (In this case we have to define the mesh as mesh of $\Pi \cup \{0, t\}$).

5. (Quadratic variation) Let $(B(t))_{t\geqslant 0}$ be a one-dimensional Brownian motion. Consider the random variables $Y_n := \sum_{k=1}^{n} \left(B(\frac{k}{n}) - B(\frac{k-1}{n})\right)^2$.
   (a) Find $\mathbb{E}\, Y_n$ and $\mathbb{V}\, Y_n$.
   (b) Determine the probability density of Y_n.
   (c) Find the characteristic function $\phi_n(\xi) = \mathbb{E}\, e^{i\xi Y_n}$, $\xi \in \mathbb{R}$, and determine the limit $\lim_{n\to\infty} \phi_n(\xi)$.
   (d) Show that $\lim \mathbb{E}\left[(Y_n - c)^2\right] = 0$ and determine the constant c.

6. Show that BM^1 is almost surely not $1/2$-Hölder continuous:
   (a) For all $Z \sim \mathsf{N}(0, 1)$ and $x > 0$ we have

$$\frac{1}{\sqrt{2\pi}} \frac{xe^{-x^2/2}}{x^2 + 1} < \mathbb{P}(Z > x) < \frac{1}{\sqrt{2\pi}} \frac{e^{-x^2/2}}{x}.$$

   (b) Set $A_{k,n} = \left\{|B((k+1)2^{-n}) - B(k2^{-n})| > c\sqrt{n\,2^{-n}}\right\}$ and show that for $c < \sqrt{2\log 2}$ one has $\mathbb{P}\left(\overline{\lim}_{n\to\infty} \bigcup_{k=0}^{2^n-1} A_{k,n}\right) = 1$. For this, use the lower bound of the first part and consider the complement of the set $\overline{\lim}_{n\to\infty} \bigcup_{k=0}^{2^n-1} A_{k,n}$.
   (c) Conclude from the previous part, that BM^1 is a. s. not $1/2$-Hölder continuous.

7. Let $X \sim \mathsf{N}(0, 1)$. Show that for every $\lambda_0 \in (0, \frac{1}{2})$ there is a constant $C = C(\lambda_0)$ such that $\sup_{\lambda\leqslant\lambda_0} \mathbb{E}\left((X^2 - 1)^2 e^{|\lambda|(X^2-1)}\right) \leqslant C < \infty$.

8. Let X be a real-valued random variable on $(\Omega, \mathscr{A}, \mathbb{P})$ and $\mathscr{F} \subset \mathscr{A}$ be σ-algebra. Show that

$$X \perp\!\!\!\perp \mathscr{F} \iff \forall \xi \in \mathbb{R},\ F \in \mathscr{F} : \mathbb{E}(e^{i\xi X} \mathbb{1}_F) = \mathbb{E}(e^{i\xi X})\mathbb{P}(F).$$

Hint: Show that $\mathbb{E}(e^{i\xi X} e^{i\eta \mathbb{1}_F}) = \mathbb{E}(e^{i\xi X})\,\mathbb{E}(e^{i\eta \mathbb{1}_F})$ for all $\xi, \eta \in \mathbb{R}$ and use the characterization of independence by characteristic functions.

10 Regularity of Brownian paths

We have seen in the previous chapter that Brownian paths are not Hölder continuous of any order $\alpha > 1/2$. In particular, they are not Lipschitz continuous or differentiable. In this chapter we continue the study of smoothness properties and show that the paths are actually a. s. nowhere (in t) differentiable.

10.1 Hölder continuity

We begin with a rather general criterion for the continuity of a stochastic process. Later on, in connection with stochastic differential equations, we need the following version for random fields, i. e. stochastic processes with a multi-dimensional index set.

Ex. 10.1 **10.1 Theorem** (Kolmogorov 1934; Slutsky 1937; Chentsov 1956). *Denote by $(\xi(x))_{x\in\mathbb{R}^n}$ a stochastic process on $(\Omega, \mathscr{A}, \mathbb{P})$ with values in $\mathbb{R}^d$ and index set $\mathbb{R}^n$. If*

$$\mathbb{E}\left(|\xi(x)-\xi(y)|^\alpha\right) \leqslant c\,|x-y|^{n+\beta} \quad \text{for all } x, y \in \mathbb{R}^n \tag{10.1}$$

holds for some constants $c > 0$ and $\alpha, \beta > 0$, then $(\xi(x))_{x\in\mathbb{R}^n}$ has a modification $(\tilde\xi(x))_{x\in\mathbb{R}^n}$ with exclusively continuous sample paths. Moreover,

$$\mathbb{E}\left[\left(\sup_{\substack{0<|x-y|<1\\ x,y\in[-T,T)^n}} \frac{|\tilde\xi(x)-\tilde\xi(y)|}{|x-y|^\gamma}\right)^\alpha\right] < \infty, \tag{10.2}$$

for all $T > 0$ and $0 \leqslant \gamma < \beta/\alpha$. In particular, $x \mapsto \tilde\xi(x)$ is a. s. locally Hölder continuous of order γ.

Proof. Let $\mathbb{N}_0 = \{0,1,2,\dots\}$ and $D_m := 2^{-m}\mathbb{N}_0^n \cap [0,1)^n$; then $D = \bigcup_{m\geqslant 0} D_m$ are the dyadic numbers in $[0,1)^n$. Set

$$\Delta_m := \{(x,y) \in D_m \times D_m : |x-y|_\infty \leqslant 2^{-m}\},$$

where $|x-y|_\infty = \max_{1\leqslant k\leqslant n} |x^k - y^k|$. Since each $x \in D_m$ has at most $3^n - 1$ nearest neighbours in D_m, the set Δ_m contains no more than $3^n \cdot 2^{mn}$ elements. Fix $\gamma \in (0, \beta/\alpha)$
Ex. 10.2 and define $\sigma_j = \sup_{(x,y)\in\Delta_j} |\xi(x)-\xi(y)|$. Then (10.1) shows

$$\mathbb{E}(\sigma_j^\alpha) \leqslant \sum_{(x,y)\in\Delta_j} \mathbb{E}\left(|\xi(x)-\xi(y)|^\alpha\right) \leqslant 3^n \cdot 2^{jn} \cdot c\,2^{-j(n+\beta)} = c\,3^n\,2^{-j\beta}.$$

Assume that $x, y \in D$ with $|x-y| < 2^{-m}$. For each $j \geqslant 0$ there are unique elements $x_j, y_j \in D_j$ such that x and y are in the 2^{-j}-cells $x_j + [0,2^{-j})^n$ and $y_j + [0,2^{-j})^n$, respectively. By construction, $\lim_{j\to\infty} x_j = x$ and $\lim_{j\to\infty} y_j = y$. For $j = m$ the condition $|x-y| < 2^{-m}$ ensures that the closed cells $x_m + [0,2^{-m}]^n$ and $y_m + [0,2^{-m}]^n$ have at least

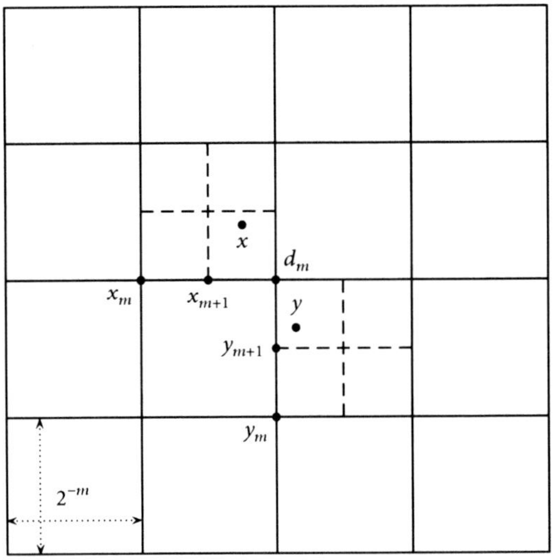

Fig. 10.1. Position of the points x and y and their dyadic approximations.

one element $d_m \in D_m$ in common. A possible situation is shown in Figure 10.1. Thus, $(x_m, d_m), (y_m, d_m) \in \Delta_m$ and

$$|\xi(x_m) - \xi(y_m)| \leqslant |\xi(x_m) - \xi(d_m)| + |\xi(y_m) - \xi(d_m)| \leqslant 2\sigma_m.$$

Moreover, our construction shows that $(x_j, x_{j+1}), (y_j, y_{j+1}) \in \Delta_{j+1}$, $j \geqslant m$. Therefore,

$$\xi(x) - \xi(y) = \sum_{j\geqslant m} [\xi(x_{j+1}) - \xi(x_j)] + \xi(x_m) - \xi(y_m) - \sum_{j\geqslant m} [\xi(y_{j+1}) - \xi(y_j)]$$

and

$$\begin{aligned}|\xi(x) - \xi(y)| &\leqslant 2\sigma_m + \sum_{j\geqslant m} [|\xi(x_{j+1}) - \xi(x_j)| + |\xi(y_{j+1}) - \xi(y_j)|] \\ &\leqslant 2\sigma_m + 2\sum_{j\geqslant m+1} \sigma_j = 2\sum_{j\geqslant m} \sigma_j.\end{aligned}$$

This entails that

$$\begin{aligned}\sup_{x,y\in D,\, x\neq y} \frac{|\xi(x) - \xi(y)|}{|x-y|^\gamma} &= \sup_{m\geqslant 0} \sup_{\substack{x,y\in D \\ 2^{-m-1}\leqslant|x-y|<2^{-m}}} \frac{|\xi(x) - \xi(y)|}{2^{-(m+1)\gamma}} \\ &\leqslant \sup_{m\geqslant 0}\Big(2\cdot 2^{(m+1)\gamma}\sum_{j\geqslant m}\sigma_j\Big) \\ &= 2^{1+\gamma}\sup_{m\geqslant 0}\sum_{j\geqslant m} 2^{m\gamma}\sigma_j \\ &\leqslant 2^{1+\gamma}\sum_{j=0}^{\infty} 2^{j\gamma}\sigma_j.\end{aligned}$$

For $\alpha \geqslant 1$ and $\gamma < \beta/\alpha$ we get

$$\begin{aligned}\left\| \sup_{x\neq y,\, x,y\in D} \frac{|\xi(x)-\xi(y)|}{|x-y|^\gamma} \right\|_{L^\alpha(\mathbb{P})} &\leqslant 2^{1+\gamma} \sum_{j=0}^{\infty} 2^{j\gamma} \|\sigma_j\|_{L^\alpha(\mathbb{P})} \\ &\leqslant c^{1/\alpha}\, 2^{1+\gamma} \sum_{j=0}^{\infty} 2^{j\gamma} 3^{n/\alpha}\, 2^{-j\beta/\alpha} \\ &= c^{1/\alpha}\, 2^{1+\gamma} 3^{n/\alpha} \sum_{j=0}^{\infty} 2^{j(\gamma-\beta/\alpha)} < \infty.\end{aligned}$$

Ex. 10.3 For $\alpha < 1$, the argument is similar, but we use $\mathbb{E}(|Z|^\alpha)$ instead of $\|Z\|_{L^\alpha(\mathbb{P})}$.

Since the expression inside the L^α-norm is almost surely finite, there exists a set $\Omega_1 \subset \Omega$ with $\mathbb{P}(\Omega_1) = 1$ such that for all $\omega \in \Omega_1$ the paths $D \ni x \mapsto \xi(x,\omega)$ are uniformly continuous:

$$|\xi(x,\omega) - \xi(y,\omega)| \leqslant c(\omega)\, |x-y|^\gamma \quad \text{for all } \ x, y \in D, \omega \in \Omega_1.$$

The same construction works if we replace $[0,1)^n$ by $[-T,T)^n$, and D by the dyadic numbers $D^T \subset [-T,T)^n$. The corresponding exceptional set is now T-dependent. We define $\Omega \setminus \Omega_T$. Set $\Omega_0 = \bigcap_{T=1}^{\infty} \Omega_T$ and $D = \bigcup_{T=1}^{\infty} D^T$. Then $\mathbb{P}(\Omega_0) = 1$ and the mapping $D \ni x \mapsto \xi(x,\omega)$ is continuous if $\omega \in \Omega_0$. Since D is dense in $\mathbb{R}^n$, we set for all $x \in \mathbb{R}^n$

$$\tilde{\xi}(x,\omega) := \begin{cases} \lim_{D\ni y\to x} \xi(y,\omega), & \omega \in \Omega_0, \\ 0, & \omega \notin \Omega_0. \end{cases} \tag{10.3}$$

Let $x_j \in D$ such that $x_j \to x$. By Fatou's lemma and the assumption (10.1)

$$\begin{aligned}\mathbb{E}\left(|\xi(x) - \tilde{\xi}(x)|^\alpha\right) &= \mathbb{E}\Big(\varliminf_{j\to\infty} |\xi(x) - \xi(x_j)|^\alpha \Big) \\ &\leqslant \varliminf_{j\to\infty} \mathbb{E}\left(|\xi(x) - \xi(x_j)|^\alpha\right) \leqslant c \varliminf_{j\to\infty} |x - x_j|^{n+\beta} = 0.\end{aligned}$$

Thus, $\mathbb{P}(\xi(x) = \tilde{\xi}(x)) = 1$ for all $x \in \mathbb{R}^n$. Finally,

$$\begin{aligned}\sup_{\substack{0<|x-y|<1\\ x,y\in[-T,T)^n}} \frac{|\tilde{\xi}(x)-\tilde{\xi}(y)|}{|x-y|^\gamma} &= \sup_{\substack{0<|x-y|<1\\ x,y\in[-T,T)^n}} \lim_{\substack{D^T\ni u\to x\\ D^T\ni v\to y}} \frac{|\xi(u)-\xi(v)|}{|u-v|^\gamma} \\ &\leqslant \sup_{\substack{0<|u-v|<1\\ u,v\in D^T}} \frac{|\xi(u)-\xi(v)|}{|u-v|^\gamma} \leqslant c_T(\omega)\end{aligned}$$

and (10.2) follows. Since the expression under the expectation in (10.2) is almost surely finite, we get

$$|\tilde{\xi}(x) - \tilde{\xi}(y)| \leqslant c(\omega)\, |x-y|^\gamma \quad \text{for all } \ |x-y| < 1,\ x, y \in [-T,T)^n. \qquad \square$$

10.2 Corollary (Wiener 1930; Paley–Wiener 1934). *A d-dimensional Brownian motion $(B_t)_{t\geqslant 0}$ is almost surely Hölder continuous up to order $\gamma < 1/2$.*

Proof. Since an $\mathbb{R}^d$-valued function is Hölder-continuous of order γ if its components are, it is enough to consider $d = 1$. Since $t \mapsto B(t)$ is continuous, there is no need to take a modification in (the proof of) Theorem 10.1. Because of (2.8) we have

$$\mathbb{E}\left(|B(t) - B(s)|^{2k}\right) = c_k\,|t-s|^k,$$

i. e. we can apply Theorem 10.1 with $d = n = 1$, $\alpha = 2k$ and $\beta = k - 1$. Then

$$\frac{\beta}{\alpha} = \frac{k-1}{2k} \xrightarrow[k\to\infty]{} \frac{1}{2}. \tag{10.4}$$

Theorem 10.1 gives for each $\gamma < (k-1)/2k$ an exceptional set $N_k = \Omega \setminus \Omega_k$ where $\mathbb{P}(N_k) = 0$. Since a γ-Hölder continuous function is also γ'-Hölder continuous for all $\gamma' < \gamma$, the assertion follows if we use the null set $N := \bigcup_{k\geqslant 1} N_k$. □

10.2 Non-differentiability

We show that almost all paths of a Brownian motion are nowhere differentiable. In view of the results from Section 2.2 it is enough to consider BM^1. The following theorem was discovered by Paley, Wiener and Zygmund. Our proof follows the classic exposition by Dvoretzky, Erdös and Kakutani.

10.3 Theorem (Paley, Wiener, Zygmund 1931). *Let $(B_t)_{t\geqslant 0}$ be a BM^1. Then the sample path $t \mapsto B_t(\omega)$ is for almost all $\omega \in \Omega$ nowhere differentiable.* Ex. 10.4

Proof (Dvoretzky, Erdös, Kakutani 1961).. Set for every $n \geqslant 1$

$$A_n := \{\omega \in \Omega \;:\; B(\cdot,\omega) \text{ nowhere differentiable in } [0,n)\}.$$

It is not clear if the set A_n is measurable. We will show that $\Omega \backslash A_n \subset N_n$ for a measurable null set N_n.

Assume that the function f is differentiable at $t_0 \in [0,n)$. Then

$$\exists\,\delta > 0\ \exists\,L > 0\ \forall\,t \in \mathbb{B}(t_0,\delta) \;:\; |f(t) - f(t_0)| \leqslant L\,|t - t_0|.$$

Consider for sufficiently large values of $k \geqslant 1$ the grid $\{\frac{j}{k} \;:\; j = 1,\dots,nk\}$. Then there exists a smallest index $j = j(k)$ such that

$$t_0 \leqslant \frac{j}{k} \quad\text{and}\quad \frac{j}{k},\dots,\frac{j+3}{k} \in \mathbb{B}(t_0,\delta).$$

For $i = j+1, j+2, j+3$ we get therefore

$$\begin{aligned}
\left|f(\tfrac{i}{k}) - f(\tfrac{i-1}{k})\right| &\leqslant \left|f(\tfrac{i}{k}) - f(t_0)\right| + \left|f(t_0) - f(\tfrac{i-1}{k})\right| \\
&\leqslant L\left(\left|\tfrac{i}{k} - t_0\right| + \left|\tfrac{i-1}{k} - t_0\right|\right) \\
&\leqslant L\left(\tfrac{4}{k} + \tfrac{3}{k}\right) = \tfrac{7L}{k}.
\end{aligned}$$

If f is a Brownian path, this implies that for the sets

$$C_m^L := \bigcap_{k=m}^{\infty} \bigcup_{j=1}^{kn} \bigcap_{i=j+1}^{j+3} \left\{ |B(\tfrac{i}{k}) - B(\tfrac{i-1}{k})| \leqslant \tfrac{7L}{k} \right\}$$

we have

$$\Omega \setminus A_n \subset \bigcup_{L=1}^{\infty} \bigcup_{m=1}^{\infty} C_m^L.$$

The assertion follows if we can show that $\mathbb{P}(C_m^L) = 0$ for all $m, L \geqslant 1$. If $k \geqslant m$,

$$\begin{aligned}
\mathbb{P}(C_m^L) &\leqslant \mathbb{P}\left(\bigcup_{j=1}^{kn} \bigcap_{i=j+1}^{j+3} \left\{ |B(\tfrac{i}{k}) - B(\tfrac{i-1}{k})| \leqslant \tfrac{7L}{k} \right\} \right) \\
&\leqslant \sum_{j=1}^{kn} \mathbb{P}\left(\bigcap_{i=j+1}^{j+3} \left\{ |B(\tfrac{i}{k}) - B(\tfrac{i-1}{k})| \leqslant \tfrac{7L}{k} \right\} \right) \\
&\overset{\text{(B1)}}{=} \sum_{j=1}^{kn} \prod_{i=j+1}^{j+3} \mathbb{P}\left(|B(\tfrac{i}{k}) - B(\tfrac{i-1}{k})| \leqslant \tfrac{7L}{k} \right) \\
&\overset{\text{(B2)}}{=} kn\, \mathbb{P}\left(|B(\tfrac{1}{k})| \leqslant \tfrac{7L}{k} \right)^3 \\
&\leqslant kn \left(\frac{c}{\sqrt{k}} \right)^3 \xrightarrow[k\to\infty]{} 0.
\end{aligned}$$

For the last estimate we use $B(\frac{1}{k}) \sim k^{-1/2} B(1)$, cf. 2.16, and therefore

$$\mathbb{P}\left(|B(\tfrac{1}{k})| \leqslant \tfrac{7L}{k} \right) = \mathbb{P}\left(|B(1)| \leqslant \tfrac{7L}{\sqrt{k}} \right) = \frac{1}{\sqrt{2\pi}} \int_{-7L/\sqrt{k}}^{7L/\sqrt{k}} \underbrace{e^{-x^2/2}}_{\leqslant 1} dx \leqslant \frac{c}{\sqrt{k}}. \qquad \square$$

10.3 Lévy's modulus of continuity

A strictly increasing, continuous function $\sigma : [0, \infty) \to [0, \infty)$ such that $\sigma(0) = 0$ is called a *modulus of continuity*. We can use σ in order to describe (local) smoothness properties of a (random or non-random) function $f : [0, \infty) \to \mathbb{R}^d$ in terms of the following estimate:

$$|f(s) - f(t)| \leqslant \sigma(|s - t|) \quad \text{for all } s, t \geqslant 0. \tag{10.5}$$

The most important special case, $\sigma(t) = t^\alpha$, leads to Hölder ($0 < \alpha < 1$) and Lipschitz ($\alpha = 1$) continuity. For Brownian motion we have discussed this in Theorem 10.1.

If $f = (f_1, \dots, f_d)$ is d-dimensional, $|f|^2 = \sum_{j=1}^d f_j^2$, and if each coordinate function f_j satisfies (10.5) with some modulus of continuity σ_j, then f satisfies (10.5) with $\sigma^2(t) := \sum_{j=1}^d \sigma_j^2(t)$. Therefore, it is enough to consider real-valued functions.

The following theorem shows that $\sigma(t) := \sqrt{2t|\log t|}$ is the exact modulus of continuity for one-dimensional Brownian motion. We begin with two auxiliary results.

10.4 Lemma. *The function $\sigma(t) = \sqrt{2t|\log t|}$ is monotone increasing on the interval $(0, e^{-1})$. For all $\kappa \in \mathbb{R}$ and $0 < h < \frac{1}{2}$ we have*

$$\sigma(2^\kappa h) \leqslant \sqrt{(|\kappa| + 1)2^\kappa}\, \sigma(h).$$

Proof. A short calculation shows that $\frac{1}{2}\frac{d}{dt}\sigma^2(t) = |\log t| - 1$ is positive on $(0, e^{-1})$. Therefore $\sigma(t)$ increases on this interval. The estimate follows from

$$\frac{\sigma^2(2^\kappa h)}{\sigma^2(h)} = \frac{2 \cdot 2^\kappa h |\log 2^\kappa h|}{2h|\log h|} \leqslant \frac{2^\kappa(|\log 2^\kappa| + |\log h|)}{|\log h|} \leqslant 2^\kappa(|\kappa| + 1)$$

since $\log 2 < |\log h|$ for $h < \frac{1}{2}$. □

10.5 Lemma. *Let $G \sim \mathsf{N}(0, 1)$ be a standard normal random variable. Then*

$$\frac{1}{\sqrt{2\pi}} \frac{x}{x^2 + 1} e^{-x^2/2} \leqslant \mathbb{P}(G > x) = \frac{1}{\sqrt{2\pi}} \int_x^\infty e^{-y^2/2}\, dy \leqslant \frac{1}{\sqrt{2\pi}} \frac{1}{x} e^{-x^2/2} \quad \textit{for all } x > 0.$$

Proof. The upper bound follows from

$$\int_x^\infty e^{-y^2/2}\, dy \leqslant \int_x^\infty \frac{y}{x} e^{-y^2/2}\, dy = \frac{1}{x} e^{-x^2/2}.$$

To see the lower bound we use integration by parts

$$\frac{1}{x^2} \int_x^\infty e^{-y^2/2}\, dy \geqslant \int_x^\infty \frac{1}{y^2} e^{-y^2/2}\, dy = \frac{1}{x} e^{-x^2/2} - \int_x^\infty e^{-y^2/2}\, dy$$

which gives

$$\int_x^\infty e^{-y^2/2}\, dy \geqslant \left(\frac{1}{x^2} + 1\right)^{-1} \frac{1}{x} e^{-x^2/2} = \frac{x}{x^2 + 1} e^{-x^2/2}.$$ □

10.6 Theorem (Lévy 1937). *Let $(B_t)_{t \geqslant 0}$ be a BM^1. Then* Ex. 10.5

$$\mathbb{P}\left(\varlimsup_{h \to 0} \frac{\sup_{0 \leqslant t \leqslant 1-h} |B(t + h) - B(t)|}{\sqrt{2h|\log h|}} = 1 \right) = 1.$$

Proof. We split the proof into two steps. We begin with the easier part and show '$\varlimsup \ldots \geqslant 1$'. Since the limes superior is the largest accumulation point it is enough to consider a particular sequence $h_n = 2^{-n} \to 0$. In fact,

$$\begin{aligned}
\left\{ \varlimsup_{h \to 0} \frac{\sup_{0 \leqslant t \leqslant 1-h} |B(t + h) - B(t)|}{\sqrt{2h|\log h|}} \geqslant 1 \right\} &\supset \left\{ \varlimsup_{n \to \infty} \frac{\max_{0 \leqslant j \leqslant 2^n - 1} |B(\frac{j+1}{2^n}) - B(\frac{j}{2^n})|}{\sqrt{2 \cdot 2^{-n} \log 2^n}} \geqslant 1 \right\} \\
&= \bigcap_{r \in \mathbb{Q} \cap (0,1)} \underbrace{\left\{ \varlimsup_{n \to \infty} \frac{\max_{0 \leqslant j \leqslant 2^n - 1} |B(\frac{j+1}{2^n}) - B(\frac{j}{2^n})|}{\sqrt{2 \cdot 2^{-n} \log 2^n}} \geqslant r \right\}}_{=: A_r}.
\end{aligned}$$

We are going to show that $\mathbb{P}(A_r) = 1$ or, equivalently, $\mathbb{P}(\Omega \setminus A_r) = 0$. Observe that for $\omega \notin A_r$ we have necessarily $\omega \in C_n$ for infinitely many n where

$$\begin{aligned} C_n &= \left\{ \max_{0 \leqslant j \leqslant 2^n - 1} \left|B(\tfrac{j+1}{2^n}) - B(\tfrac{j}{2^n})\right| \leqslant r\sqrt{2 \cdot 2^{-n} \log 2^n} \right\} \\ &= \bigcap_{j=0}^{2^n-1} \left\{ \left|B(\tfrac{j+1}{2^n}) - B(\tfrac{j}{2^n})\right| \leqslant r\sqrt{2 \cdot 2^{-n} \log 2^n} \right\}. \end{aligned}$$

By the independence and stationarity of increments (B1), (B2) we see

$$\mathbb{P}(C_n) = \Bigg[\underbrace{1 - \frac{1}{\sqrt{2\pi 2^{-n}}} \int_{r\sqrt{2\cdot 2^{-n}\log 2^n}}^{\infty} e^{-x^2/(2\cdot 2^{-n})}\,dx}_{=:I} \Bigg]^{2^n} = (1-I)^{2^n} \leqslant e^{-2^n I}.$$

The last estimate follows from the elementary convexity inequality $1 - e^{-x} \leqslant x$. From the lower estimate in Lemma 10.5 we get

$$\begin{aligned} 2^n I &= \frac{2^n}{\sqrt{2\pi 2^{-n}}} \int_{r\sqrt{2\cdot 2^{-n}\log 2^n}}^{\infty} e^{-x^2/(2\cdot 2^{-n})}\,dx \\ &= \frac{2^n}{\sqrt{2\pi}} \int_{r\sqrt{2\log 2^n}}^{\infty} e^{-y^2/2}\,dy \\ &\geqslant \frac{1}{\sqrt{2\pi}} \frac{rn\sqrt{2\log 2}}{1 + r^2 n 2\log 2} \frac{2^n}{\sqrt{n}} e^{-r^2\log 2^n} \\ &\geqslant c\,\frac{1}{\sqrt{n}}\,2^{(1-r^2)n} \end{aligned}$$

for some constant $c = c_r$ and all $n \geqslant n_0$. Consequently,

$$\sum_{n=n_0}^{\infty} \mathbb{P}(C_n) \leqslant \sum_{n=n_0}^{\infty} \exp\left(-2^n I\right) \leqslant \sum_{n=n_0}^{\infty} \exp\left(-\frac{c\,2^{(1-r^2)n}}{\sqrt{n}} \right) < \infty,$$

and the Borel–Cantelli lemma implies that

$$\mathbb{P}(\Omega \setminus A_r) \leqslant \mathbb{P}(C_n \text{ for infinitely many } n) = 0.$$

Let us now prove that '$\overline{\lim} \ldots \leqslant 1$'. We have to show that for almost all $\omega \in \Omega$ and every $r > 1$ there is some $h_0 = h_0(r, \omega)$ such that

$$\sup_{0 \leqslant t \leqslant 1-h} |B(t+h,\omega) - B(t,\omega)| \leqslant r\sqrt{2h|\log h|} \quad \text{for all } 0 < h \leqslant h_0. \tag{10.6}$$

Fix $r > 1$, write $r = \frac{1+\delta}{1-\delta}$ with $\delta \in (0,1)$ and denote by $\lfloor n\delta \rfloor$ the integer part of $n\delta$. Using the upper estimate from Lemma 10.5 with $x = r\sqrt{2\log(2^n/\ell)}$, we get for all $n \geqslant 1$

$$\begin{aligned}
&\mathbb{P}\left(\max_{1\leqslant \ell \leqslant 2^{\lfloor n\delta\rfloor}} \max_{0\leqslant j\leqslant 2^n-\ell} \frac{|B(\frac{j+\ell}{2^n}) - B(\frac{j}{2^n})|}{\sqrt{2\frac{\ell}{2^n}\log\frac{2^n}{\ell}}} > r \right) \\
&\leqslant \sum_{\ell=1}^{2^{\lfloor n\delta\rfloor}} \sum_{j=0}^{2^n-\ell} \mathbb{P}\left(\frac{|B(\frac{j+\ell}{2^n}) - B(\frac{j}{2^n})|}{\sqrt{\frac{\ell}{2^n}}} > r\sqrt{2\log\frac{2^n}{\ell}} \right) \\
&\leqslant \sum_{\ell=1}^{2^{\lfloor n\delta\rfloor}} 2^n \frac{2}{\sqrt{2\pi}} \frac{\exp\left[-r^2\log(\frac{2^n}{\ell})\right]}{\sqrt{2\log\frac{2^n}{\ell}}} \\
&\leqslant c \sum_{\ell=1}^{2^{\lfloor n\delta\rfloor}} \frac{2^n}{\left(2^{n-n\delta}\right)^{r^2}\sqrt{2\log 2^{n-n\delta}}} \\
&\leqslant c_\delta\, 2^{n+n\delta} 2^{-(n-n\delta)r^2} \\
&= c_\delta\, 2^{n(1+\delta)-nr^2(1-\delta)}.
\end{aligned} \tag{10.7}$$

By construction, $1 + \delta < r^2(1-\delta)$, so that (10.7) is the general term of a convergent series. By the Borel–Cantelli lemma there is some $\Omega_0 \subset \Omega$ with $\mathbb{P}(\Omega_0) = 1$, such that for all $\omega \in \Omega_0$ there is some $m = m(\delta,\omega) \geqslant 1$ with

$$\max_{1\leqslant \ell \leqslant 2^{\lfloor n\delta\rfloor}} \max_{0\leqslant j\leqslant 2^n-\ell} \frac{|B(\frac{j+\ell}{2^n},\omega) - B(\frac{j}{2^n},\omega)|}{\sigma(\frac{\ell}{2^n})} \leqslant r \quad \text{for all } n \geqslant m. \tag{10.8}$$

Without loss of generality we can assume that $m\delta > 4$. From now on we fix $\omega \in \Omega_0$. Set $h_0 := 2^{-m(1-\delta)}$, pick $h < h_0$ and choose $n \geqslant m \geqslant 1$ in such a way that

$$2^{-(n+1)(1-\delta)} \leqslant h < 2^{-n(1-\delta)}. \tag{10.9}$$

For t and $s = t + h$ there are sequences $(s_j)_{j\geqslant n}$ and $(t_j)_{j\geqslant n}$ of dyadic numbers such that $s_j, t_j \in \{k/2^j : k \geqslant 0\}$, $s_j \uparrow s$, $t_j \downarrow t$ and

$$|s_n - t_n| \leqslant h, \quad |s_j - s_{j+1}| \leqslant 2^{-j-1} \quad \text{and} \quad |t_j - t_{j+1}| \leqslant 2^{-j-1}.$$

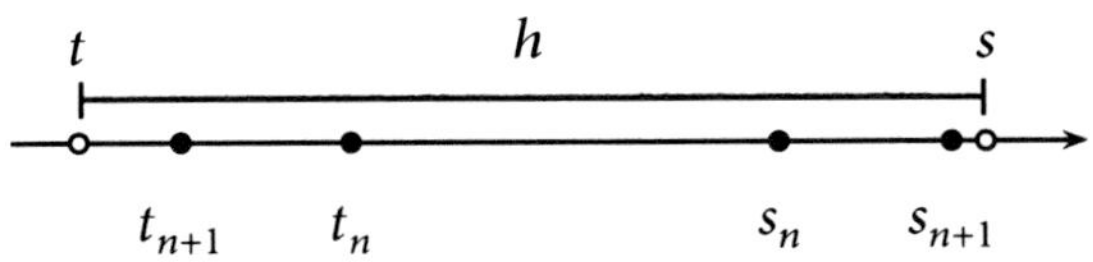

Fig. 10.2. Position of the points t_n, t_{n+1} and s_n, s_{n+1} relative to t and s.

Since $B(t)$ has continuous sample paths, we get

$$\begin{aligned}|B(t+h)-B(t)| &\leqslant |B(s)-B(s_n)|+|B(s_n)-B(t_n)|+|B(t_n)-B(t)|\\ &\leqslant \sum_{j=n}^{\infty}|B(s_{j+1})-B(s_j)|+|B(s_n)-B(t_n)|+\sum_{j=n}^{\infty}|B(t_{j+1})-B(t_j)|\\ &\overset{(10.8)}{\leqslant} r\sum_{j=n}^{\infty}\sigma(2^{-j-1})+r\,\sigma(h)+r\sum_{j=n}^{\infty}\sigma(2^{-j-1})\\ &= 2r\sum_{j=1}^{\infty}\sigma(2^{-j}2^{-n})+r\,\sigma(h)\\ &\overset{10.4}{\leqslant} 2r\sum_{j=1}^{\infty}\sqrt{j+1}\,2^{-j/2}\sigma(2^{-n})+r\,\sigma(h).\end{aligned}$$

By (10.9) and Lemma 10.4 with $\kappa=(n+1)(1-\delta)-n=1-\delta-n\delta$ we see that

$$\sigma(2^{-n})\leqslant\sigma(2^{-n}2^{(n+1)(1-\delta)}h)\leqslant\underbrace{2^{\kappa/2}\sqrt{|\kappa|+1}}_{\to 0,\ n\to\infty}\sigma(h)\leqslant(r-1)\sigma(h)$$

for sufficiently large values of $n\geqslant N(r)$, i. e. sufficiently small values of $h<h_0$. Thus, $2\sum_{j=1}^{\infty}\sqrt{j+1}\,2^{-j/2}\sigma(2^{-n})\leqslant c(r-1)\sigma(h)$, and we find for all $h<h_0$

$$\sup_{0\leqslant h\leqslant h_0}\ \sup_{0\leqslant t\leqslant 1-h}\frac{|B(t+h)-B(t)|}{\sigma(h)}\leqslant cr(r-1)+r\xrightarrow[r\downarrow 1]{}1.\qquad\square$$

10.7 Corollary. *Almost all sample paths of one-dimensional Brownian motion are not Hölder continuous of order $\alpha=1/2$.*

Proof. Note that

$$\frac{\sup_{0\leqslant t\leqslant 1-h}|B(t+h)-B(t)|}{\sqrt{h}}=\frac{\sup_{0\leqslant t\leqslant 1-h}|B(t+h)-B(t)|}{\sqrt{2h\,|\log h|}}\sqrt{2\,|\log h|}.$$

By Theorem 10.6 the lim sup of the right-hand side tends to $+\infty$ as $h\to 0$. This means that B_t is at some point $t\in[0,1)$ not Hölder continuous.

The same argument, applied to $W(t):=B(t+k)-B(k)$ – this is again a BM[1], see 2.13 – , completes the proof. □

10.8 Further reading. There is a whole zoo of function spaces which can be used to measure the regularity of (random) functions. A beautiful result is the estimate [86] which can be used for many embeddings. Regularity in Besov spaces is studied in [36]. The connection of the Kolmogorov–Chentsov theorem and Sobolev embeddings is the topic of the note [203]. A related investigation for fractional Brownian motion is [77]. Moduli of continuity and (non-)differentiability are presented in [40].

[36] Ciesielski, Kerkyacharian, Roynette: Quelques espaces fonctionnels associés à des processus gaussienes.

[40] Csörgő, Révész: *Strong Approximations in Probability and Statistics.*
[77] Feyel, de La Pradelle: On fractional Brownian processes.
[86] Garsia, Rodemich, Rumsey: A real variable lemma and the continuity of paths of some Gaussian processes.
[203] Schilling: Sobolev embedding for stochastic processes.

Problems

1. A *Poisson process* is a real-valued stochastic process $(N_t)_{t\geqslant 0}$ such that $N_0 = 0$, $N_t - N_s \sim N_{t-s}$ and for $t_0 = 0 < t_1 < \cdots < t_n$ the increments $N_{t_j} - N_{t_{j-1}}$ are independent and each N_t is a Poisson random variable with parameter $t \cdot \lambda$ for some $\lambda > 0$. (In particular, $(N_t)_{t\geqslant 0}$ satisfies (B0), (B1), (B2)).
   (a) Show that the process $(N_t)_{t\geqslant 0}$ does *not* satisfy the assumptions of the Kolmogorov–Slutsky–Chentsov theorem, Theorem 10.1.
   (b) Show that (10.1) holds true for $n = 1$, $\alpha > 0$ and $\beta = 0$. Discuss the role of β for Theorem 10.1.
   (c) Let $\lambda = 1$. Determine for the process $X_t = N_t - t$ the mean value $m(t)$ and the covariance $C(s,t) = \mathbb{E}(X_s X_t)$, $s,t \geqslant 0$.

2. Prove that in $\mathbb{R}^n$ all ℓ^p-norms ($1 \leqslant p \leqslant \infty$) are equivalent:

$$\max_{1\leqslant j\leqslant n} |x_j| \leqslant \Big(\sum_{j=1}^n |x_j|^p\Big)^{1/p} \leqslant n^{1/p} \max_{1\leqslant j\leqslant n} |x_j| \quad \text{for all } x = (x_1,\dots,x_n) \in \mathbb{R}^n.$$

3. Show that for $\alpha \in (0,1)$ the function $Z \mapsto \mathbb{E}(|Z|^\alpha)$ is subadditive and complete the argument in the proof of Theorem 10.1 for this case.

4. The proof of Theorem 10.3 actually shows, that almost all Brownian paths are nowhere Lipschitz continuous. Modify the argument of this proof to show that almost all Brownian paths are nowhere Hölder continuous for all $\alpha > 1/2$. Why does the argument break down for $\alpha = 1/2$?
   Hint: It is enough to consider rational α since α-Hölder continuity implies β-Hölder continuity for all $\beta < \alpha$.

5. Use Theorem 10.6 to show that the strong p-variation $\mathrm{VAR}_p(B;1)$ of BM^1 is for $p > 2$ finite.

11 Brownian motion as a random fractal

The Lévy–Ciesielski construction of Brownian motion, cf. Chapter 3, indicates that the trajectories $t \mapsto B_t(\omega)$ of a Brownian motion $(B_t)_{t\geqslant 0}$ are rather complicated functions. Although being Hölder continuous (up to order $\frac{1}{2}$), they are nowhere differentiable, see Chapter 10. This means that classical geometric approaches, which are often based on tangents, are likely to fail. Much better suited are the notion of (random) fractals and tools from fractal geometry in the sense of Mandelbrot [153].

Most assertions in this chapter hold for almost all paths of a Brownian motion, and the exceptional sets, i. e. null sets, play an important role. Throughout this chapter we assume that $(B_t)_{t\geqslant 0}$ is a BM^d which is defined on a complete probability space $(\Omega, \mathscr{A}, \mathbb{P})$ and equipped with a complete admissible filtration $(\mathscr{F}_t)_{t\geqslant 0}$.

11.1 Hausdorff measure and dimension

We give a very brief introduction to Hausdorff measure and Hausdorff dimension. These concepts were introduced by Hausdorff [92] in 1919 who wanted to extend Carathéodory's simple approach to Lebesgue measure as canonical volume measure on the d-dimensional Euclidean space to more general sets of *fractional* dimension, e. g. perfect sets of Cantor-type. Modern standard references include the classic monograph by Rogers [194] and the books by Mattila [157] or Falconer [75; 76]. For general results on measures we refer to [204]. Let

$$|E| := \sup_{x,y\in E} |x-y|$$

denote the *diameter* of the set $E \subset \mathbb{R}^d$. Any (at most) countable family of sets $E_j \subset E$ with $|E_j| \leqslant \delta$ and $E \subset \bigcup_{j\geqslant 1} E_j$ is called a *δ-cover* of E. For $\alpha > 0$ define

$$\mathscr{H}^\alpha_\delta(E) := \inf\left\{\sum_{j\geqslant 1} |E_j|^\alpha \;:\; (E_j)_{j\geqslant 1} \text{ is a } \delta\text{-cover of } E\right\}. \tag{11.1}$$

Since the diameter of a set, of its closure and of its closed convex hull coincide, we may as well consider in (11.1) only δ-covers comprising closed (convex) sets or (convex) Borel sets.

11.1 Definition (Hausdorff 1919). The set function

$$\mathscr{H}^\alpha(E) := \lim_{\delta\to 0} \mathscr{H}^\alpha_\delta(E) = \sup_{\delta>0} \mathscr{H}^\alpha_\delta(E), \quad E \subset \mathbb{R}^d,$$

is called *α-dimensional Hausdorff measure*.

Ex. 11.1 It is not hard to see that $\mathscr{H}^\alpha$ is an outer measure on the power set of $\mathbb{R}^d$, i. e. a monotone and σ-subadditive set function with $\mathscr{H}^\alpha(\emptyset) = 0$.

11.2 Lemma. *$\mathscr{H}^\alpha$ is a metric outer measure, i. e. $\mathscr{H}^\alpha(E \cup F) = \mathscr{H}^\alpha(E) + \mathscr{H}^\alpha(F)$ holds for all sets $E, F \subset \mathbb{R}^d$ with strictly positive distance $\inf_{x\in E,\, y\in F} |x-y| > 0$.*

Proof. Assume that $E, F \subset \mathbb{R}^d$ are any two sets such that $\delta_0 := \inf_{x\in E,\, y\in F} |x-y| > 0$. Because of the subadditivity of an outer measure it is clearly enough to show that $\mathscr{H}^\alpha(E\cup F) \geqslant \mathscr{H}^\alpha(E) + \mathscr{H}^\alpha(F)$; in particular, we may assume that $\mathscr{H}^\alpha(E \cup F) < \infty$.

Take any δ-cover $(A_j)_{j\geqslant 1}$ of $E \cup F$ with $\delta < \delta_0$. By the definition of δ_0 it is clear that each A_j can only intersect either E or F, so there is a natural split into δ-covers of E and F, respectively. We get

$$\sum_{j\geqslant 1} |A_j|^\alpha = \sum_{A_j\cap E\neq\emptyset} |A_j|^\alpha + \sum_{A_j\cap F\neq\emptyset} |A_j|^\alpha \geqslant \mathscr{H}^\alpha_\delta(E) + \mathscr{H}^\alpha_\delta(F).$$

Thus, $\mathscr{H}^\alpha_\delta(E\cup F) \geqslant \mathscr{H}^\alpha_\delta(E) + \mathscr{H}^\alpha_\delta(F)$, and the claim follows as $\delta \to 0$. □

This, combined with Carathéodory's extension theorem for measures, shows that $\mathscr{H}^\alpha$, restricted to the Borel sets $\mathscr{B}(\mathbb{R}^d)$, is a Borel measure. Since $\mathscr{H}^\alpha$ is invariant under translations and $0 < \mathscr{H}^d(\mathbb{B}(0,1)) < \infty$, it is clear that $\mathscr{H}^d$ is a constant multiple of d-dimensional Lebesgue measure λ^d. In order to work out the exact constant – we have to show that $\mathscr{H}^d(\mathbb{B}(0,1)) = 1$ – we need the isodiametric inequality, cf. [74, Chapter 2.2, Theorem 1, p. 69]. Ex. 11.2

11.3 Lemma. *$\mathscr{H}^\alpha$ is outer regular, i. e.*

$$\mathscr{H}^\alpha(E) = \inf\left\{\mathscr{H}^\alpha(B) \,:\, B \supset E,\ B \in \mathscr{B}(\mathbb{R}^d)\right\} \quad \textit{for all } E \subset \mathbb{R}^d.$$

Proof. We will construct a set $B \in \mathscr{B}(\mathbb{R}^d)$ such that $B \supset E$ and $\mathscr{H}^\alpha(B) = \mathscr{H}^\alpha(E)$. If $\mathscr{H}^\alpha(E) = \infty$, take $B = \mathbb{R}^d$. In all other cases there is for each $n \geqslant 1$ some δ-cover $(E_j^n)_{j\geqslant 1}$ of E satisfying

$$\sum_{j\geqslant 1} |E_j^n|^\alpha \leqslant \mathscr{H}^\alpha_\delta(E) + \frac{1}{n} \leqslant \mathscr{H}^\alpha(E) + \frac{1}{n}.$$

Since $|E_j^n| = |\bar{E}_j^n|$ we may assume that the sets E_j^n are closed, hence Borel sets. Then

$$B_n := \bigcup_{j\geqslant 1} E_j^n \quad \text{and} \quad B := \bigcap_{n\geqslant 1} B_n := \bigcap_{n\geqslant 1}\bigcup_{j\geqslant 1} E_j^n \supset E,$$

and we see

$$\mathscr{H}^\alpha_\delta(E) \leqslant \mathscr{H}^\alpha_\delta(B) \leqslant \mathscr{H}^\alpha_\delta(B_n) \leqslant \sum_{j\geqslant 1} |E_j^n|^\alpha \leqslant \mathscr{H}^\alpha(E) + \frac{1}{n}.$$

The claim follows letting first $n \to \infty$ and then $\delta \to 0$. □

The function $\alpha \mapsto \mathscr{H}^\alpha(E)$ is most of the time either 0 or $+\infty$. Indeed, if $(E_j)_{j\geqslant 1}$ is a δ-cover of E, then we get

$$\mathscr{H}^{\alpha+\epsilon}_\delta(E) \leqslant \sum_{j\geqslant 1} |E_j|^{\alpha+\epsilon} \leqslant \delta^\epsilon \sum_{j\geqslant 1} |E_j|^\alpha, \quad \text{hence,} \quad \mathscr{H}^{\alpha+\epsilon}_\delta(E) \leqslant \delta^\epsilon \cdot \mathscr{H}^\alpha_\delta(E).$$

Letting $\delta \to 0$ proves the next lemma.

11.4 Lemma. *If $\mathscr{H}^\alpha(E) < \infty$, then $\mathscr{H}^{\alpha+\epsilon}(E) = 0$ for all $\alpha, \epsilon > 0$.*

In particular, $\#\{\alpha : 0 < \mathscr{H}^\alpha(E) < \infty\} \leqslant 1$. The value where $\mathscr{H}^\alpha(E)$ jumps from $+\infty$ to 0 is the dimension of E.

Ex. 11.3 **11.5 Definition** (Hausdorff 1919). The *Hausdorff (or Hausdorff–Besicovitch) dimension* of a set $E \subset \mathbb{R}^d$ is the unique number

$$\dim E = \sup\left\{\alpha \geqslant 0 : \mathscr{H}^\alpha(E) = \infty\right\} = \inf\left\{\alpha \geqslant 0 : \mathscr{H}^\alpha(E) = 0\right\}$$

($\sup \emptyset := 0$). If $\alpha = \dim E$, the Hausdorff measure $\mathscr{H}^\alpha(E)$ can have any value in $[0, \infty]$.

Later on, we want to calculate the Hausdorff dimension of certain (random) sets. In the following remark we collect some useful techniques for this.

11.6 Remark (Determining the Hausdorff dimension). Let $E \subset \mathbb{R}^d$ and $f : \mathbb{R}^d \to \mathbb{R}^n$.

a) *Upper bound:* $\mathscr{H}^\alpha(E) = 0 \implies \dim E \leqslant \alpha$.
Showing that $\mathscr{H}^\alpha(E) = 0$ is often simple, since it is enough to construct a sequence of δ-covers (with $\delta = \delta_n \to 0$) such that $\sum_{j\geqslant 1} |E_j^n|^\alpha$ becomes small as $n \to \infty$.

b) *Lower bound:* $\mathscr{H}^\beta(E) > 0 \implies \dim E \geqslant \beta$.
It is usually hard to check that $\mathscr{H}^\beta(E) > 0$ since we have to show that $\sum_{j\geqslant 1} |E_j|^\beta$ is uniformly bounded from below for all possible δ-covers of E. This requires clever covering strategies or capacitary methods.

Ex. 11.6 c) *If f is Hölder continuous with exponent $\gamma \in (0, 1]$, i.e. $|f(x) - f(y)| \leqslant L|x - y|^\gamma$ for all $x, y \in \mathbb{R}^d$, then $\dim f(E) \leqslant \gamma^{-1} \dim E$.*

Proof. By the Hölder continuity $|f(E_j)| \leqslant L\,|E_j|^\gamma$, and for any δ-cover $(E_j)_{j\geqslant 1}$ of E, $(f(E_j))_{j\geqslant 1}$ is an $L\delta^\gamma$-cover of $f(E)$. If $\alpha\gamma > \dim E$, then $\mathscr{H}^{\alpha\gamma}(E) = 0$ and, for some suitable δ-cover $(E_j)_{j\geqslant 1}$, we see

$$L^{-\alpha} \sum_{j\geqslant 1} |f(E_j)|^\alpha \leqslant \sum_{j\geqslant 1} |E_j|^{\alpha\gamma} \leqslant \epsilon.$$

Using part a), we get $\dim f(E) \leqslant \alpha$. The claim follows if we let $\alpha \to \gamma^{-1} \dim E$. □

d) *Mass distribution principle.* Let $E \subset \mathbb{R}^d$, $\alpha > 0$ and assume that there is a non-trivial finite measure μ supported in E. If there exist constants $c > 0$ and $\epsilon > 0$ such that

$$\mu(F) \leqslant c\,|F|^\alpha \quad \text{for all closed sets } F \subset \mathbb{R}^d \text{ with } |F| \leqslant \epsilon,$$

then $\mathscr{H}^\alpha(E) > 0$ and $\dim E \geqslant \alpha$.

Proof. Let $\delta \leqslant \epsilon$ and consider any δ-covering of E by closed sets $(F_j)_{j\geqslant 1}$. By assumption

$$0 < \mu(E) \leqslant \sum_{j=1}^\infty \mu(F_j) \leqslant \sum_{j=1}^\infty c\,|F_j|^\alpha,$$

and this implies $\mathscr{H}_\delta^\alpha(E) \geqslant c^{-1}\mu(E)$ and $\mathscr{H}^\alpha(E) \geqslant c^{-1}\mu(E) > 0$. □

e) *Self-similarity.* A set $E \subset \mathbb{R}^d$ is said to be self-similar, if there are finitely many functions $f_j : \mathbb{R}^d \to \mathbb{R}^d$, $j = 1, \dots, n$, such that $|f_j(x) - f_j(y)| = c_j|x - y|$, $x, y \in \mathbb{R}^d$, and $E = f_1(E) \cup \dots \cup f_n(E)$. If the sets $f_j(E)$ do not overlap, then Ex. 11.7

$$\mathscr{H}^\alpha(E) = \sum_{j=1}^n c_j^\alpha \mathscr{H}^\alpha(E).$$

Assume that $\alpha = \dim E$ and $0 < \mathscr{H}^\alpha(E) < \infty$. Then we can calculate α from the equality $1 = \sum_{j=1}^n c_j^\alpha$.

f) *Energy criterion* (Frostman 1935; McKean 1955; Blumenthal & Getoor 1960). Let μ be a probability measure on E. Then

$$\iint_{E\times E} \frac{\mu(dx)\,\mu(dy)}{|x-y|^\alpha} < \infty \implies \mathscr{H}^\alpha(E) > 0 \implies \dim E \geqslant \alpha.$$

Proof. See Section A.8.1 in the appendix.

g) *Frostman's theorem* (Frostman 1935). Let K be a compact set. Then

$$\begin{aligned}\alpha < \dim K &\implies \mathscr{H}^\alpha(K) > 0\\ &\implies \exists \text{ finite measure } \sigma \quad \forall \beta < \alpha : \iint_{K\times K} \frac{\sigma(dx)\,\sigma(dy)}{|x-y|^\beta} < \infty.\end{aligned}$$

Proof. See Section A.8.1 in the appendix.

11.2 The Hausdorff dimension of Brownian paths

Recall that almost all Brownian motion paths satisfy a Hölder condition of any index $0 < \gamma < \frac{1}{2}$:

$$|B_t(\omega) - B_s(\omega)| \leqslant L_\gamma(\omega)|t-s|^\gamma, \quad 0 < \gamma < \tfrac{1}{2}, \quad \omega \in \Omega_0. \tag{11.2}$$

Here $\mathbb{P}(\Omega_0) = 1$ independently of γ and $\mathbb{P}(L_\gamma < \infty) = 1$. This follows from Corollary 10.2, but for the readers' convenience we provide a different proof which is closer to the original arguments by Wiener [229, §13] and Paley & Wiener [171, Theorem XLVII, pp. 160–162].

11.7 Lemma (Wiener 1930; Paley & Wiener 1934). BM^d *is almost surely Hölder continuous for any index* $\gamma \in (0, \frac{1}{2})$.

Proof. The coordinates of a d-dimensional Brownian motion are one-dimensional Brownian motions, cf. Corollary 2.9, and so we may assume that $(B_t)_{t\geqslant 0}$ is real-valued.

Since $(B_{t+1} - B_1)_{t\geqslant 0}$ is again a BM^1, it is enough to show Hölder continuity for every $t \in [0,1]$. Fix $\gamma \in (0, \frac{1}{2})$; using $B_{t+h} - B_t \sim B_h \sim \sqrt{h}\, B_1$, we get

$$\begin{aligned}\mathbb{P}\left(|B_{t+h} - B_t| > h^\gamma\right) = \mathbb{P}\left(|B_1| > h^{\gamma-\frac{1}{2}}\right) &= \sqrt{\frac{2}{\pi}} \int_{h^{\gamma-1/2}}^{\infty} e^{-\frac{1}{2}x^2}\, dx \\ &\leqslant \sqrt{\frac{2}{\pi}}\, h^{\frac{1}{2}-\gamma} \int_{h^{\gamma-1/2}}^{\infty} x e^{-\frac{1}{2}x^2}\, dx \\ &= \sqrt{\frac{2}{\pi}}\, h^{\frac{1}{2}-\gamma} e^{-\frac{1}{2}h^{2\gamma-1}} \leqslant ch^2\end{aligned}$$

for some constant $c > 0$. For $h = 2^{-n}$, $n \geqslant 1$, we find

$$\mathbb{P}\left(\bigcup_{n\geqslant k} \bigcup_{1\leqslant j\leqslant 2^n} \left\{\left|B_{j2^{-n}} - B_{(j-1)2^{-n}}\right| > 2^{-n\gamma}\right\}\right) \leqslant c \sum_{n=k}^{\infty} 2^n 2^{-2n} = c2^{-k+1} \xrightarrow[k\to\infty]{} 0.$$

By a standard Borel–Cantelli argument, there is a set Ω_γ with $\mathbb{P}(\Omega_\gamma) = 1$ such that

$$\left|B_{j2^{-n}}(\omega) - B_{(j-1)2^{-n}}(\omega)\right| \leqslant 2^{-n\gamma} \quad \text{for all } n \geqslant N(\omega),\ j = 1, \ldots, 2^n,\ \omega \in \Omega_\gamma.$$

If $[t, t+h] \subset [0,1]$, $h < 2^{-N(\omega)}$, we see that $(t, t+h) \subset \bigcup_{k\geqslant 1} I_k$ where the I_k are non-overlapping intervals of length $2^{-n(k)}$ and with dyadic end-points. Note that at most two intervals of any one length appear in the union. Indeed, $I_1 = [r, s]$ is the largest

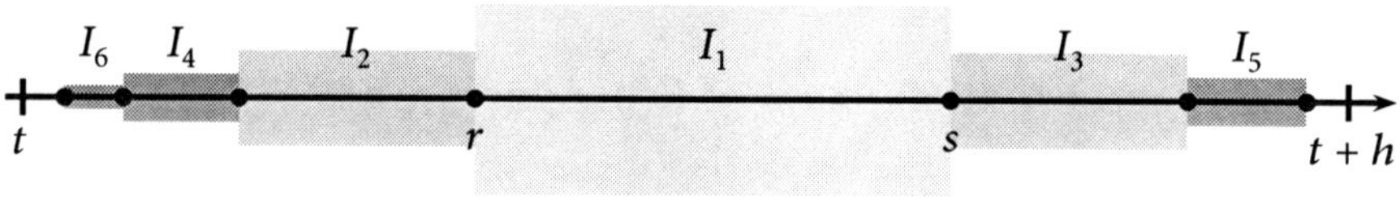

Fig. 11.1. Exhausting the interval $[t, t+h]$ by non-overlapping dyadic intervals. Dots "•" denote dyadic numbers. Observe that at most two dyadic intervals of any one length can occur.

possible dyadic interval fitting into $[t, t+h]$; its length is $s - r = 2^{-n(1)}$, where $n(1)$ is the smallest integer with $2^{-n(1)} < h$. Expanding $r - t$ and $t + h - s$ as dyadic series shows how to construct all further dyadic intervals to the left and right, respectively: They are directly adjacent to I_1 and their length is given by the first (second, third etc.) non-zero coefficient in the dyadic expansion, see Figure 11.1.

Applying the above estimate to the (endpoints of the) intervals I_k we see, by the continuity of Brownian motion, that for all $\omega \in \Omega_\gamma$

$$|B_{t+h}(\omega) - B_t(\omega)| \leqslant 2 \sum_{j=n(1)}^{\infty} 2^{-j\gamma} = \frac{2}{1-2^{-\gamma}}\, 2^{-n(1)\gamma} \leqslant \frac{2}{1-2^{-\gamma}}\, h^\gamma.$$

This proves Hölder continuity on $[0, 1]$ of order $\gamma < \frac{1}{2}$. Since $\Omega_0 := \bigcap_{k\geqslant 3} \Omega_{\frac{1}{2}-\frac{1}{k}}$ has full measure, $\mathbb{P}(\Omega_0) = 1$, and since Hölder continuity is inherited by smaller exponents, the above argument shows that all Brownian paths $t \mapsto B_t(\omega)$, $\omega \in \Omega_0$, are Hölder continuous of all orders $\gamma < \frac{1}{2}$. □

We are interested in the Hausdorff dimension of the *range* (or *image*) $B(E, \omega)$ and the graph $\operatorname{Gr} B(E, \omega)$ of a d-dimensional Brownian motion:

$$\begin{aligned} B(E, \omega) &:= \{B_t(\omega) \,:\, t \in E\}, \\ \operatorname{Gr} B(E, \omega) &:= \{(t, B_t(\omega)) \,:\, t \in E\}, \end{aligned} \qquad E \in \mathscr{B}[0, 1],$$

Dimension results for the range with $E = [0, 1]$ are due to Taylor [218], the general case is due to McKean [159].

11.8 Theorem (Taylor 1953; McKean 1955). *Let $(B_t)_{t\geqslant 0}$ be a BM^d and $F \subset [0, 1]$ a closed set. Then*

$$\dim B(F) = \min\{d,\ 2\dim F\} = \begin{cases} 2\dim F, & d \geqslant 2, \\ \min\{1, 2\dim F\}, & d = 1. \end{cases} \qquad a.s. \tag{11.3}$$

Proof. **1°** The *upper bound* in (11.3) follows at once from the Hölder continuity of Brownian motion (11.2) using Remark 11.6 c). For all $\gamma < \frac{1}{2}$ we have for any Borel set $E \subset [0, 1]$

$$\dim B(E) \leqslant \frac{1}{\gamma} \dim E \quad \text{a.s.}$$

Letting $\gamma \to \frac{1}{2}$ along a countable sequence, we get $\dim B(E) \leqslant 2\dim E$ a.s. Since $\dim B(E) \leqslant d$ is trivial, the upper bound on Hausdorff dimension follows. Note that, in this case, the exceptional null set is the same for all $E \subset [0, 1]$.

2° The proof of the *lower bound* in (11.3) requires methods from potential theory, cf. Remark 11.6 f) and g). Let $F \subset [0, 1]$ be a closed set and pick $\alpha < \dim F$. By definition, $\mathscr{H}^\alpha(F) > 0$ and by Frostman's theorem, Remark 11.6 g), there is a finite measure σ on F such that

$$\iint_{F\times F} \frac{\sigma(dt)\,\sigma(ds)}{|t - s|^\alpha} < \infty.$$

In order to determine the Hausdorff dimension of $B(F)$, we consider the *occupation measure*, i. e. the time (measured with σ) $(B_t)_{t\geqslant 0}$ spends in a Borel set $A \in \mathscr{B}(\mathbb{R}^d)$:

$$\mu_\sigma^B(A) = \int_F \mathbb{1}_A(B_t)\,\sigma(dt).$$

Assume, in addition, that $\lambda := 2\alpha < \min\{d, 2\dim F\}$. Then, by Tonelli's theorem

$$\mathbb{E}\left(\iint_{B(F)\times B(F)} \frac{\mu_\sigma^B(dx)\,\mu_\sigma^B(dy)}{|x - y|^\lambda}\right) = \mathbb{E}\left(\iint_{F\times F} \frac{\sigma(dt)\,\sigma(ds)}{|B_t - B_s|^\lambda}\right) = \iint_{F\times F} \mathbb{E}\left(\frac{1}{|B_t - B_s|^\lambda}\right)\sigma(dt)\,\sigma(ds).$$

Using stationary increments and scaling, $B_t - B_s \sim B_{t-s} \sim \sqrt{t-s}\,B_1$ we see

$$\mathbb{E}\left(\iint\limits_{B(F)\times B(F)} \frac{\mu_\sigma^B(dx)\,\mu_\sigma^B(dy)}{|x-y|^\lambda}\right) = \mathbb{E}\left(|B_1|^{-\lambda}\right) \iint\limits_{F\times F} \frac{\sigma(dt)\,\sigma(ds)}{|t-s|^{\lambda/2}}.$$

Ex. 11.8 It is not hard to check that $\mathbb{E}\left(|B_1|^{-\lambda}\right) < \infty$ for $\lambda < d$, while the double integral appearing on the right is finite since $\alpha = \frac{1}{2}\lambda < \dim F$. The energy criterion, see Remark 11.6 f), now proves that $\dim B(F) \geqslant \lambda$ a. s. Letting $\lambda \to \min\{d, 2\dim F\}$ along a countable sequence, gives the lower bound of (11.3). Note, however, that the null set may depend on F. □

The graph $\operatorname{Gr} B(E,\omega)$ of Brownian motion was studied by Taylor [219] for $E = [0,1]$, for general $E \in \mathscr{B}[0,1]$ we refer to Kahane [115, Chapter 18, §7].

11.9 Theorem (Taylor 1955). *Let $(B_t)_{t\geqslant 0}$ be a BM^d and $F \subset [0,1]$ a closed set. Then*

$$\dim \operatorname{Gr} B(F) = \min\left\{2\dim F, \dim F + \tfrac{1}{2}d\right\} = \begin{cases} \dim F + \frac{1}{2}, & d=1,\ \dim F \geqslant \frac{1}{2}, \\ 2\dim F, & d=1,\ \dim F \leqslant \frac{1}{2}, \\ 2\dim F, & d \geqslant 2, \end{cases} \quad a.s.$$

Proof. **1°** The *upper bound* uses two different coverings of the graph. Pick $\alpha > \dim F$. Then there is some δ-covering $(F_j)_{j\geqslant 1}$ of F by closed sets with diameters $|F_j| =: \delta_j \leqslant \delta$. Observe that $\operatorname{Gr} B(F_j) \subset F_j \times B(F_j)$; using the Hölder continuity (11.2) of Brownian motion, we get for all $\gamma \in \left(0, \frac{1}{2}\right)$

$$\left|\operatorname{Gr} B(F_j)\right| \leqslant |F_j| + |B(F_j)| \leqslant |F_j| + L_\gamma |F_j|^\gamma \leqslant (L_\gamma + 1)\,\delta^\gamma.$$

This shows that the closed sets $(F_j \times B(F_j))_{j\geqslant 1}$ are an $(L_\gamma+1)\delta^\gamma$-covering of $\operatorname{Gr} B(F)$ and

$$\sum_{j=1}^\infty \left|\operatorname{Gr} B(F_j)\right|^{\alpha/\gamma} \leqslant (L_\gamma + 1)^{\alpha/\gamma} \sum_{j=1}^\infty |F_j|^\alpha < \infty$$

i. e. $\dim \operatorname{Gr} B(F) \leqslant \gamma^{-1}\dim F$ a. s. Letting $\gamma \to \frac{1}{2}$ via a countable sequence, shows $\dim \operatorname{Gr} B(F) \leqslant 2\dim F$ a. s.

On the other hand, using the Hölder continuity of $t \mapsto B_t$, it is not hard to see that each $F_j \times B(F_j)$ is also covered by $c\delta_j^{(\gamma-1)d}$ many cubes of diameter δ_j. This gives a further δ-covering of $\operatorname{Gr} B(F)$ satisfying

$$\sum_{j=1}^\infty \left(c\delta_j^{(\gamma-1)d}\right)\delta_j^{\alpha+(1-\gamma)d} = c\sum_{j=1}^\infty \delta_j^\alpha < \infty,$$

thus $\mathscr{H}^{\alpha+(1-\gamma)d}(\operatorname{Gr} B(F)) < \infty$ and $\dim \operatorname{Gr} B(F) \leqslant \alpha + (1-\gamma)d$ a. s. Letting $\alpha \to \dim F$ and $\gamma \to \frac{1}{2}$ along countably many values proves $\dim \operatorname{Gr} B(F) \leqslant \dim F + \frac{1}{2}d$ a. s.

Combining both estimates gives the upper bound.

2° For the *lower bound* we use potential theoretic arguments.

Assume first that $\dim F \leqslant \frac{1}{2}d$. Since $B(F)$ is the projection of $\operatorname{Gr} B(F)$ onto $\mathbb{R}^d$, and since projections are Lipschitz continuous (i. e. Hölder continuous with $\gamma = 1$), we see with Theorem 11.8

$$\dim \operatorname{Gr} B(F) \geqslant \dim B(F) \geqslant 2\dim F \geqslant \min\left\{2\dim F,\ \dim F + \tfrac{1}{2}d\right\} \quad \text{a. s.}$$

The situation $\dim F > \frac{1}{2}d$ can only occur if $d = 1$; if so, $\dim \operatorname{Gr} B(F) \geqslant \dim F + \frac{1}{2}$ a. s. In order to see this, fix $\gamma \in (1, \dim F + \frac{1}{2})$. Since $\gamma - \frac{1}{2} < \dim F$, we conclude from Frostman's theorem, Remark 11.6 g), that there is some finite measure σ on F with

$$\iint_{F\times F} \frac{\sigma(dt)\,\sigma(ds)}{|t-s|^{\gamma-1/2}} < \infty.$$

Consider the occupation measure

$$\mu_\sigma^{\operatorname{Gr} B}(A) = \int_F \mathbb{1}_A(t, B_t)\,\sigma(dt), \quad A \in \mathscr{B}(\mathbb{R}^2).$$

As in the proof of Theorem 11.8 we see

$$\mathbb{E}\left(\iint \frac{\mu_\sigma^{\operatorname{Gr} B}(dx)\,\mu_\sigma^{\operatorname{Gr} B}(dy)}{|x-y|^\gamma}\right) = \mathbb{E}\left(\iint_{F\times F} \frac{\sigma(ds)\,\sigma(dt)}{\left(|t-s|^2 + |B_t - B_s|^2\right)^{\gamma/2}}\right) \leqslant \mathbb{E}\left(\iint_{F\times F} \frac{\sigma(ds)\,\sigma(dt)}{|B_t - B_s|^\gamma}\right).$$

Using Brownian scaling $|B_t - B_s| \sim \sqrt{|t-s|}|B_1|$ yields, as in the proof of Theorem 11.8,

$$\mathbb{E}\left(\iint \frac{\mu_\sigma^{\operatorname{Gr} B}(dx)\,\mu_\sigma^{\operatorname{Gr} B}(dy)}{|x-y|^\gamma}\right) \leqslant c_\gamma \iint_{F\times F} \frac{\sigma(ds)\,\sigma(dt)}{|t-s|^{\gamma/2}} \leqslant c_\gamma \iint_{F\times F} \frac{\sigma(ds)\,\sigma(dt)}{|t-s|^{\gamma-1/2}} < \infty.$$

The last inequality follows from the fact that $|t-s| \leqslant 1$ and $\frac{1}{2}\gamma \leqslant \gamma - \frac{1}{2}$.

The energy criterion, cf. Remark 11.6 f), now shows that $\dim \operatorname{Gr} B(F) \geqslant \gamma$ a. s., and the claim follows as $\gamma \to \dim F + \frac{1}{2}$ along countably many values. □

The exceptional set in Theorem 11.8 may depend on the set of times F. Kaufman [120] showed that for Brownian motion in dimension $d \geqslant 2$ this null set can be chosen independently of F. Kaufman's beautiful proof does not rely on capacitary arguments; at the same time it holds for arbitrary Borel sets $E \in \mathscr{B}[0,1]$. The key ingredient of the proof is the following technical lemma. Denote by $D(r) \subset \mathbb{R}^d$ an open ball of radius $r > 0$ and by $B^{-1}(D, \omega) = \{t : B_t(\omega) \in D\}$ the pre-image of D.

11.10 Lemma (Kaufman 1969). *Let $(B_t)_{t\geqslant 0}$ be a BM^d, $d \geqslant 2$, set $I_n^k = [(k-1)4^{-n}, k4^{-n}]$, $k = 1, \dots, 4^n$, and denote by $A_n \subset \Omega$ the event*

$$A_n = \left\{\omega \in \Omega \;\middle|\; \exists D = D(2^{-n}) \;:\; \#\left\{k \;:\; B^{-1}(D,\omega) \cap I_n^k \neq \emptyset\right\} \geqslant n^{2d}\right\}.$$

Then $N_0 := \limsup_{n\to\infty} A_n$ is a $\mathbb{P}$-null set.

Proof. For sufficiently large $n \geqslant 1$, e. g. $n \geqslant 4d^2$, we set

$$A_n^* = \left\{\omega \in \Omega \;\middle|\; \exists D = D(2\sqrt{n}\,2^{-n}) \;:\; \#\{k \;:\; B_{k4^{-n}}(\omega) \in D\} \geqslant n^{2d}\right\}.$$

Let us show that $A_n \subset A_n^*$. Write y_0 for the centre of the ball $D = D(2^{-n})$ from the definition of A_n. Using Lévy's modulus of continuity, Theorem 10.6, we find for every $t_0 \in B^{-1}(D, \omega) \cap I_n^k$ and any constant $c > 1$

$$|B_{k4^{-n}} - B_{t_0}| \leqslant c\sqrt{2 \cdot 4^{-n} \log 4^n} \leqslant \tfrac{3}{2}\sqrt{n}\,2^{-n}.$$

Thus, $|B_{k4^{-n}} - y_0| \leqslant |B_{k4^{-n}} - B_{t_0}| + |B_{t_0} - y_0| < 2\sqrt{n}\,2^{-n}$ and $B_{k4^{-n}} \in D(2\sqrt{n}\,2^{-n})$; this proves $A_n \subset A_n^*$. Set

$$A^*_{(k_1,\dots,k_n)} := \bigcap_{j=1}^{n-1} \left\{\left|B_{k_{j+1}4^{-n}} - B_{k_j 4^{-n}}\right| < 4\sqrt{n}\,2^{-n}\right\}, \tag{11.4}$$

where $1 \leqslant k_1 < k_2 < \ldots < k_n \leqslant 4^n$ are arbitrary integers. The probability of each set in the intersection on the right-hand side can be estimated by

$$\begin{aligned}\mathbb{P}\left(\left|B_{k_{j+1}4^{-n}} - B_{k_j 4^{-n}}\right| < 4\sqrt{n}\,2^{-n}\right) &= \mathbb{P}\left(|B_1| < 4\sqrt{n}/\sqrt{k_{j+1} - k_j}\right)\\ &\leqslant 4^d c_d n^{d/2}\left(k_{j+1} - k_j\right)^{-d/2}.\end{aligned}$$

Here we used $B_t - B_s \sim \sqrt{t-s}\,B_1$, and the estimate follows from

$$\mathbb{P}(|B_1| < r) = \frac{1}{(2\pi)^{d/2}} \int\limits_{|x|<r} e^{-\frac{1}{2}|x|^2}\,dx = \frac{r^d}{(2\pi)^{d/2}} \int\limits_{|y|<1} e^{-\frac{1}{2}r^2|y|^2}\,dy \leqslant r^d\,\frac{\mathrm{Leb}(\mathbb{B}(0,1))}{(2\pi)^{d/2}}.$$

By independence, we get for the probability of the set in (11.4)

$$\mathbb{P}(A^*_{(k_1,\dots,k_n)}) \leqslant (4^d c_d)^{n-1}\, n^{(n-1)d/2} \prod_{j=1}^{n-1}\left(k_{j+1} - k_j\right)^{-d/2}.$$

From the very definition of the set A_n^* we see[1] that

$$\mathbb{1}_{A_n^*}(\omega) \sum_{\substack{(k_1,\dots,k_n)\in\mathbb{N}^n\\ 1\leqslant k_1<\dots<k_n\leqslant 4^n}} \!\!\cdots\!\! \sum \mathbb{1}_{A^*_{(k_1,\dots,k_n)}}(\omega) = \sum_{\substack{(k_1,\dots,k_n)\in\mathbb{N}^n\\ 1\leqslant k_1<\dots<k_n\leqslant 4^n}} \!\!\cdots\!\! \sum \mathbb{1}_{A_n^*\cap A^*_{(k_1,\dots,k_n)}}(\omega) \geqslant \binom{n^{2d}}{n}\mathbb{1}_{A_n^*}(\omega)$$

for every $\omega \in \Omega$, and this gives

$$\begin{aligned}\mathbb{P}(A_n^*) &\leqslant \frac{1}{\binom{n^{2d}}{n}} \sum_{\substack{(k_1,\dots,k_n)\in\mathbb{N}^n\\ 1\leqslant k_1<\dots<k_n\leqslant 4^n}} \!\!\cdots\!\! \sum \mathbb{P}(A^*_{(k_1,\dots,k_n)})\\ &\leqslant (4^d c_d)^n \frac{n^{nd/2}}{\binom{n^{2d}}{n}} \sum_{\substack{(k_1,\dots,k_n)\in\mathbb{N}^n\\ 1\leqslant k_1<\dots<k_n\leqslant 4^n}} \!\!\cdots\!\! \sum \prod_{j=1}^{n-1}\left(k_{j+1} - k_j\right)^{-d/2}.\end{aligned}$$

1 We learned this elegant argument from Ms. Franziska Kühn

Using the integral comparison test for series, see e. g. [204, E.34, p. 359],

$$\sum_{k=k_{n-1}+1}^{4^n} (k-k_{n-1})^{-d/2} \leqslant \sum_{k=1}^{4^n} k^{-d/2} \leqslant \begin{cases} 1+\log 4^n = 1+n\log 4, & d=2, \\ \dfrac{d}{d-2}, & d \geqslant 3. \end{cases}$$

Moreover, by Stirling's formula

$$\binom{N}{n} \approx \frac{N^N}{n^n\, N^{N-n}} \frac{\sqrt{N}}{\sqrt{2\pi n N}} = \left(\frac{N}{n}\right)^n \frac{1}{\sqrt{2\pi n}} \overset{N=n^{2d}}{=} n^{(2d-1)n} \frac{1}{\sqrt{2\pi n}}.$$

This shows that, for all $d \geqslant 2$,

$$\mathbb{P}(A_n^*) \leqslant (C_d)^n\, n^n\, n^{\frac{1}{2}dn-(2d-1)n+\frac{1}{2}}$$

which proves that $\sum_{n=1}^\infty \mathbb{P}(A_n^*) < \infty$. The claim follows now with the easy direction of the Borel–Cantelli lemma. □

11.11 Theorem. *Let $(B_t)_{t\geqslant 0}$ be a BM^d, $d \geqslant 2$ and $E \subset [0,1]$ a Borel set. Then*

$$\mathbb{P}\big(\dim B(E) = 2\dim E \quad \forall E \in \mathscr{B}[0,1]\big) = 1. \tag{11.5}$$

Proof. The upper bound, $\dim B(E) \leqslant 2\dim E$ a. s. with the same exceptional set for all $E \in \mathscr{B}[0,1]$, follows from the first part of the proof of Theorem 11.8.

For the lower bound, let $E \in \mathscr{B}[0,1]$ and set $I_n^k := [(k-1)4^{-n}, k4^{-n}]$, $k = 1,2,\dots,4^n$. Let $(S_j)_{j\geqslant 1}$ be any δ-cover of the random set $B(E)$ with $2^{-n(j)+1} > |S_j| \geqslant 2^{-n(j)}$. Then S_j is contained in a ball D_j of radius $2^{-n(j)}$. Without loss of generality we can assume that $n(1) \leqslant n(2) \leqslant \dots$, and so $\delta \leqslant 2^{-n(1)+1}$.

Denote by N_0 the exceptional set from Lemma 11.10. For $\omega \in \Omega_0 := \Omega \setminus N_0$ and sufficiently large $n(1)$, i. e. small δ, we can cover D_j, hence S_j, by at most $n(j)^{2d}$ sets of the form $B(I_{n(j)}^k, \omega)$. The corresponding intervals in $[0,1]$,

$$I_{n(j)}^{k(1,j)}, \dots, I_{n(j)}^{k(i,j)}, \quad i = i(j) \leqslant n(j)^{2d}, \; j \geqslant 1,$$

are a δ^2-cover of the set E. If $\beta < \alpha < \dim E$, $\mathscr{H}^\alpha(E) = \infty$. Therefore, there is some $c > 0$ such that

$$0 < c \leqslant \sum_{j=1}^\infty \sum_{l=1}^i \left|I_{n(j)}^{k(l,j)}\right|^\alpha \leqslant \sum_{j=1}^\infty n(j)^{2d} 4^{-\alpha n(j)} \leqslant C \sum_{j=1}^\infty 4^{-\beta n(j)} \leqslant C \sum_{j=1}^\infty |S_j|^{2\beta} \quad \text{on } \Omega_0.$$

Since the δ-cover of $B(E,\omega)$ is arbitrary, we conclude that $\mathscr{H}^{2\beta}(B(E,\omega)) \geqslant c/C > 0$. Thus, $\dim B(E,\omega) \geqslant 2\beta$ for all $\omega \in \Omega_0$ and all Borel sets $E \subset [0,1]$. Letting $\beta \to \dim E$ finishes the proof. □

The fact that in Theorem 11.11 the exceptional set does not depend on E allows us to consider random time sets $E = E(\omega) \subset [0,1]$. Here is a typical application: Let $A \in \mathscr{B}(\mathbb{R}^d)$ and consider the pre-image

$$E(\omega) := B^{-1}(A,\omega) := \{t \in [0,1] \;:\; B_t(\omega) \in A\}.$$

Then $B(B^{-1}(A,\omega),\omega) \subset A$, and (11.5) shows that $2\dim B^{-1}(A) = \dim B(B^{-1}(A)) \leqslant \dim A$.

11.12 Corollary. *Let* $(B_t)_{t\geqslant 0}$ *be a* BM^d, $d \geqslant 2$. *Then*

$$\dim B^{-1}(A) \leqslant \min\left\{1, \tfrac{1}{2}\dim A\right\} \quad a.s.\ \textit{for all}\ A \in \mathscr{B}(\mathbb{R}^d).$$

In particular, the zero set $\mathrm{Z}(\omega) := B^{-1}(\{0\}, \omega)$ *satisfies* $\dim \mathrm{Z}(\omega) = 0$ *a.s.*

Corollary 11.12 shows why the lower bound in the uniform result of Theorem 11.11 cannot hold in dimension $d = 1$: The zero set of a one-dimensional Brownian motion $\mathrm{Z}(\omega)$ has Hausdorff dimension $\frac{1}{2}$, cf. Theorem 11.22, and Corollary 11.12 would lead to a contradiction.

Surprisingly enough, there are versions of Theorem 11.11 and Corollary 11.12 for one-dimensional Brownian motions. Both are due to Kaufman [122] and [121]. The proof of the first result relies on a technical lemma similar to Lemma 11.10 and potential theoretic methods, cf. Remark 11.6 f) and g); the second result uses the Hölder continuity of Brownian local time. We mention them here without proof.

11.13 Theorem (Kaufman 1989). *Let* $(B_t)_{t\geqslant 0}$ *be a* BM^1. *Then there exist subsets* $\Omega_0 \subset \Omega$ *and* $T \subset [0,1]$ *with* $\mathbb{P}(\Omega_0) = 1$ *and* $\mathrm{Leb}(T) = 1$, *such that*

$$\dim B(F + t, \omega) = \min\{1, 2\dim F\} \quad \textit{for all closed sets}\ F \subset [0,1],\ \omega \in \Omega_0,\ t \in T.$$

Ex. 11.9 **11.14 Theorem** (Kaufman 1985). *Let* $(B_t)_{t\geqslant 0}$ *be a* BM^1 *and* $A \in \mathscr{B}(\mathbb{R})$. *Then*

$$\dim B^{-1}(A) = \tfrac{1}{2} + \tfrac{1}{2}\dim A \quad a.s.$$

11.3 Local maxima of a Brownian motion

In this section $(B_t)_{t\geqslant 0}$ will denote a one-dimensional Brownian motion. It is impossible to imagine the position of the local extrema of a Brownian path $t \mapsto B_t(\omega)$: For almost all paths the set of times where a local maximum is attained is dense in $(0, \infty)$. Because of the symmetry of the Wiener measure, this also applies to the local minima.

Recall that a function $f : \mathbb{R} \to \mathbb{R}$ is *increasing* [resp. *decreasing*] in $(a, b) \subset \mathbb{R}$, if $f(s) \leqslant f(t)$ [resp. $f(s) \geqslant f(t)$] holds for all $a < s < t < b$. A *monotone function* is either in- or decreasing. The function f has a *local maximum* at $t \in (a, b)$ if there is some $\epsilon > 0$ such that $f(s) \leqslant f(t)$ for all $s \in (t-\epsilon, t+\epsilon)$; if $f(s) < f(t)$ for all $s \in (t-\epsilon, t+\epsilon)$, $s \neq t$, we have a *strict local maximum*.

11.15 Lemma. *Assume that* $f \in \mathcal{C}[a, b]$ *is nowhere monotone, i.e. there is no interval* $I \subset (a, b)$ *such that* $f|_I$ *is monotone. Then* f *has at least one (local) maximum in* (a, b).

Proof. Assume that $f(b) > f(a)$ and set $t = \max\{s \in [a, b] : f(s) = f(a)\}$. Since f is nowhere monotone, there exist $t < t_1 < t_2 < b$ such that $f(t_1) > f(t)$ and $f(t_2) < f(t_1)$ (draw a picture!). By the continuity of f, there is at least one (local) maximum in $[t, t_2]$.

A similar argument works if $f(b) < f(a)$. If $f(b) = f(a)$, then there is some $c \in (a, b)$ such that $f(c) \neq f(a)$, and we are in one of the previous cases. □

11.16 Lemma. *Let $(B_t)_{t\geqslant 0}$ be a BM^1. Then $t \mapsto B_t(\omega)$ is for almost all ω nowhere monotone, i. e. with probability 1 there is no open interval $(a,b) \subset [0,\infty)$ where B_t is monotone.*

In particular, the local maxima are dense in $[0,\infty)$.

Proof. Clearly, $\Omega \setminus \{\omega \in \Omega : t \mapsto B_t(\omega) \text{ is nowhere monotone}\}$

$$= \bigcup_{\substack{0<a<b<\infty\\ a,b\in\mathbb{Q}}} \{\omega \in \Omega : t \mapsto B_t(\omega) \text{ is monotone in } (a,b)\}.$$

For each $n = 1, 2, \dots$ and $0 < a < b < \infty$ we have the estimate

$$\begin{aligned}
\mathbb{P}(B|_{(a,b)} \text{ monotone}) &\leqslant \mathbb{P}(B|_{(a,b)} \text{ increasing}) + \mathbb{P}(B|_{(a,b)} \text{ decreasing})\\
&\leqslant \mathbb{P}\left(B_{a+(b-a)\frac{k-1}{n}} \leqslant B_{a+(b-a)\frac{k}{n}} \quad \forall k = 1,\dots,n\right)\\
&\quad + \mathbb{P}\left(B_{a+(b-a)\frac{k-1}{n}} \geqslant B_{a+(b-a)\frac{k}{n}} \quad \forall k = 1,\dots,n\right)\\
&= 2\mathbb{P}\left(B_{\frac{1}{n}} \geqslant 0\right)^n = 2^{1-n}.
\end{aligned}$$

The claim follows as $n \to \infty$. The density of the local maxima now follows from Lemma 11.15 applied to any interval $(a,b) \subset [0,\infty)$ with $a, b \in \mathbb{Q}$. □

11.17 Lemma. *Let $(B_t)_{t\geqslant 0}$ be a BM^1. For almost all ω, every local maximum of $t \mapsto B_t(\omega)$ is a strict local maximum.*

Proof. Denote by I a closed interval in $[0,\infty)$, set $M(I) := \sup_{t\in I} B_t$, and write Ω_0 for the set of all ω where every local maximum of $t \mapsto B_t(\omega)$ is a strict local maximum. Then

$$\Omega_1 := \bigcap_{\substack{I\cap I'=\emptyset,\ I,I'\subset[0,\infty)\\ I,I' \text{ rational intervals}}} \{M(I) \neq M(I')\} \subset \Omega_0 \tag{11.6}$$

(a *rational interval* is an interval with rational endpoints). Indeed, if a continuous function $f : [0,\infty) \to \mathbb{R}$ has a local maximum at $t > 0$ which is not a strict local maximum, then there is some $\epsilon > 0$ such that $f(s) \leqslant f(t)$ for all $s \in (t-\epsilon, t+\epsilon)$ and a sequence $(s_j)_{j\geqslant 1}$ such that $\lim_{j\to\infty} s_j = t$ and $f(s_j) = f(t)$ for all $j \geqslant 1$. Thus, there are rational intervals I, I' satisfying $I \cap I' = \emptyset$ and $M(I) = \sup_I f = \sup_{I'} f = M(I')$.

Let $I = [a,b]$, $I' = [c,d]$ and assume that $a < b < c < d$. By the independent increments property of Brownian motion,

$$M(I') - M(I) = \sup_{s\in I'} B_s - \sup_{s\in I} B_s = \sup_{s\in I'}(B_s - B_c) + (B_c - B_b) + \inf_{s\in I}(B_b - B_s)$$

is the sum of three independent random variables. Since $B_c - B_b$ has a probability density, this also holds for the sum and we get $\mathbb{P}(M(I') - M(I) \neq 0) = 1$. The intersection in (11.6) being countable, we conclude that $\mathbb{P}(\Omega_1) = 1$ and, by the completeness of the underlying probability space, $\mathbb{P}(\Omega_0) = 1$. □

11.18 Corollary. *Let $(a,b) \subset [0,\infty)$. For almost all ω the (global) maximum of $t \mapsto B_t(\omega)$, $t \in [a,b]$, is attained at exactly one point $\theta = \theta(\omega)$.*

Proof. We use the notation introduced in the proof of Lemma 11.17. Since

$$\mathbb{P}\left(B_a \geqslant B_t\ \forall t \in [a,b]\right) = \mathbb{P}\left(B_b \geqslant B_t\ \forall t \in [a,b]\right) = 0,$$

cf. Lemma 11.16, we may assume that the global maximum of $t \mapsto B_t$ is attained inside (a,b). By Lemma 11.17, the path $t \mapsto B_t(\omega)$ has only strict local maxima for all $\omega \in \Omega_0$ where $\mathbb{P}(\Omega_0) = 1$. Then

$$\begin{aligned}\Omega_0 \cap \left\{\omega \;\middle|\; \exists s,t \in [a,b],\ s \neq t\ :\ B_s(\omega) = B_t(\omega) = \sup_{u\in[a,b]} B_u(\omega)\right\} \\ \subset \Omega_0 \cap \bigcup_{\substack{I\cap I'=\emptyset,\ I,I'\subset[0,\infty)\\ I,I' \text{ rational intervals}}} \left\{M(I) = M(I')\right\}.\end{aligned}$$

The right-hand side is, however, contained in Ω_1^c, see the proof of Lemma 11.17, and $\mathbb{P}(\Omega_1^c) = 0$. □

Let us note that any $f \in \mathcal{C}[0,\infty)$ which admits only strict local maxima has at most countably many (strict local) maxima. This follows easily from the observation that the points t where a maximum is attained are contained in the countable set

$$\bigcup_{n\geqslant 1} A_n := \bigcup_{n\geqslant 1} \left\{s \in [0,\infty)\ :\ f(s) > f(t)\ \forall t \in \left(s-\tfrac{1}{n},\ s+\tfrac{1}{n}\right),\ t \neq s\right\}$$

(note that $s_1, s_2 \in A_n$ implies $|s_1 - s_2| \geqslant \frac{1}{n}$).

11.4 On the level sets of a Brownian motion

Let $(B_t)_{t\geqslant 0}$ be a BM^1. The *level sets* are the following random sets

$$T(x,\omega) = B^{-1}(\{x\},\omega) = \left\{t \in [0,\infty)\ :\ B_t(\omega) = x\right\}, \quad x \in \mathbb{R},\ \omega \in \Omega.$$

11.19 Lemma. *Let $(B_t)_{t\geqslant 0}$ be a BM^1. Then $\mathbb{P}(0$ is an accumulation point of $T(0,\cdot)) = 1$.*

Proof. Write $A_n := \left\{\omega \in \Omega\ :\ T(0,\omega) \cap \left[0,\frac{1}{n}\right) = \{0\}\right\}$, $n \geqslant 1$. Then

$$\Omega_0 := \left\{\omega \in \Omega\ :\ 0 \text{ is an accumulation point of } T(0,\omega)\right\} \quad \text{and} \quad \Omega \setminus \Omega_0 \subset \bigcup_{n\geqslant 1} A_n.$$

From Theorem 6.9 we know that the running maximum $M_t := \sup_{s\leqslant t} B_s$ and minimum $m_t := \inf_{s\leqslant t} B_s$ of a Brownian motion are distributed like $|B_t|$ and $-|B_t|$, respectively. Since $A_n \subset \{M_{1/n} = 0\} \cup \{m_{1/n} = 0\}$ we get

$$\mathbb{P}\left(A_n\right) \leqslant \mathbb{P}\left(M_{1/n} = 0\right) + \mathbb{P}\left(m_{1/n} = 0\right) = \mathbb{P}\left(|B_{1/n}| = 0\right) + \mathbb{P}\left(-|B_{1/n}| = 0\right) = 0$$

for all $n \geqslant 1$, and so $\mathbb{P}(\Omega_0) = 1$. □

The projectively reflected process $W_t := tB_{1/t}$, $(t > 0)$, and $W_0 = 0$ is again a BM^1, cf. Paragraph 2.17. Write $T^W(0,\omega)$ for the 0-level set of this process. Clearly, 0 is an accumulation point of $T^W(0,\omega)$ if, and only if, there exists a sequence $(t_j)_{j\geqslant 1}$ with $B_{t_j}(\omega) = 0$, $\lim_{j\to\infty} t_j = \infty$ and $\lim_{j\to\infty} B_{t_j}(\omega) = 0$. Therefore, Lemma 11.19 proves

11.20 Corollary. *The* 0*-level set* $T(0,\cdot)$ *of a* BM^1 *is a. s. unbounded.*

A set $F \subset \mathbb{R}$ is *perfect* if it is closed and dense in itself, i. e. each point of F is an accumulation point of F. A typical example of a (compact) perfect set is the Cantor middle thirds set. Every non-void perfect set is uncountable. We are going to show that Brownian level sets are perfect. Ex. 11.10

11.21 Theorem. *Let* $(B_t)_{t\geqslant 0}$ *be a* BM^1*. For every* $x \in \mathbb{R}$ *there is a null set* N_x *such that for every* $\omega \in \Omega \setminus N_x$

a) $T(x,\omega)$ *is closed;*
b) $T(x,\omega)$ *is unbounded;*
c) $T(x,\omega)$ *has Lebesgue measure* 0*;*
d) $T(x,\omega)$ *is dense in itself;*
e) $T(x,\omega)$ *has Hausdorff dimension* $\frac{1}{2}$.

Proof. a) This follows from $T(x,\omega) = B^{-1}(\{x\},\omega)$ and the a. s. continuity of $t \mapsto B_t(\omega)$.

b) Set $\Omega_0 = \{\omega \in \Omega : T(x,\omega) \text{ is unbounded}\}$ and write $\tau_x := \inf\{s > 0 : B_s = x\}$ for the first passage time of the level x. Clearly,

$$\Omega_0 = \left\{\tau_x < \infty,\ B_{t_n+\tau_x} - x = 0 \ \text{ for a sequence } (t_n)_{n\geqslant 1},\ t_n \to \infty\right\}.$$

By the strong Markov property $B_{t+\tau_x} - x = B_{t+\tau_x} - B_{\tau_x}$, $t \geqslant 0$, is again a Brownian motion which is independent of $\mathscr{F}_{\tau_x}$, cf. Theorem 6.5. Thus,

$$\begin{aligned}
\mathbb{P}(\Omega_0) &= \mathbb{P}(\underbrace{\{\tau_x < \infty\}}_{\in\mathscr{F}_{\tau_x}\ \text{cf. A.15}} \cap \{B_{t_n+\tau_x} - x = 0 \ \text{ for a sequence } (t_n)_{n\geqslant 1},\ t_n \to \infty\})\\
&= \mathbb{P}(\tau_x < \infty)\,\mathbb{P}(B_{t_n} = 0 \ \text{ for a sequence } (t_n)_{n\geqslant 1},\ t_n \to \infty)\\
&= \mathbb{P}(\tau_x < \infty)\,\mathbb{P}(T(0,\cdot) \text{ is unbounded}) = 1,
\end{aligned}$$

by Corollary 11.20 and (6.11) or Theorem 6.9.

c) By Lemma 6.24, the map $(t,\omega) \mapsto B_t(\omega)$, $(t,\omega) \in [0,\infty) \times \Omega$, is measurable. Using Fubini's theorem we see

$$\mathbb{E}\left[\int_0^\infty \mathbb{1}_{\{x\}}(B_t)\,dt\right] = \int_0^\infty \mathbb{E}\left[\mathbb{1}_{\{x\}}(B_t)\right]dt = \int_0^\infty \mathbb{P}(B_t = x)\,dt = 0.$$

On the other hand, $\mathbb{1}_{T(x,\omega)}(t) = \mathbb{1}_{\{x\}}(B_t(\omega))$, i. e.

$$\mathbb{E}\left[\int_0^\infty \mathbb{1}_{\{x\}}(B_t)\,dt\right] = \mathbb{E}\left[\mathrm{Leb}[T(x)]\right] = 0,$$

from which we conclude that $\operatorname{Leb}[T(x)] = 0$ a. s.

d) We will show that the set $T(x,\omega)$ has a. s. no isolated points. In fact, each $t \in T(x,\omega)$ can be approximated from the left (by a sequence of random times which are not stopping times) and from the right (by a sequence of stopping times). We are going to show the second assertion.

Let $q \geqslant 0$ be a rational number and denote by

$$\tau_x^q := \inf\{t > q : B_t = x\} \qquad (\inf \emptyset := \infty)$$

the first passage time of the level x after time q. From

$$\{\tau_x^q \leqslant t\} = \begin{cases} \emptyset, & \text{if } t < q, \\ \bigcap_{n=1}^{\infty} \bigcup_{s \in (q,t] \cap \mathbb{Q}} \left\{B_s > x - \frac{1}{n}\right\} \cap \left\{B_s < x + \frac{1}{n}\right\}, & \text{if } t \geqslant q, \end{cases}$$

we see that τ_x^q is a stopping time, and from part b) we know that $\mathbb{P}(\tau_x^q < \infty) = 1$.

Set $A_q := \{\omega \in \Omega : \tau_x^q(\omega) \text{ is an accumulation point of } T(x,\omega)\}$. Then

$$\bigcap_{q \in [0,\infty) \cap \mathbb{Q}} A_q \subset \{T(x,\cdot) \text{ has no isolated points}\}.$$

In order to see this inclusion, we assume that for fixed ω the point $t_0 \in T(x,\omega)$ is an isolated point. By its very definition, there exists some rational $q \geqslant 0$ such that $(q,t_0) \cap T(x,\omega) = \emptyset$, so $\tau_x^q(\omega) = t_0$; since t_0 is no accumulation point, $\omega \notin A_q$.

By the strong Markov property $B_{\tau_x^q + t} - x$ is again a Brownian motion and from Lemma 11.19 we see that $\mathbb{P}(A_q) = 1$ for all rational $q \geqslant 0$. Since the above intersection is countable, the claim follows.

e) This follows from Theorem 11.14 or Theorem 11.22 below. □

! The role of the exceptional sets appearing in Theorem 11.21 must not be neglected. For instance, consider the set

$$N_x := \{\omega \in \Omega : T(x,\omega) \text{ has isolated points}\}.$$

Then $\mathbb{P}\left(\bigcup_{x \in \mathbb{R}} N_x\right) = 1$, since every $t_0(\omega)$ where a local maximum $x_0 = x_0(\omega)$ is reached is necessarily an isolated point of $T(x_0,\omega)$: This is a consequence of Lemma 11.17 as Brownian local maxima are *strict* local maxima.

11.5 Roots and records

Let $(B_t)_{t \geqslant 0}$ be a BM^1. As in Chapter 6 we write $\mathbb{P}^x$ for the law of $B + x = (B_t + x)_{t \geqslant 0}$ (the Brownian motion started at x) and $\mathbb{P} = \mathbb{P}^0$. The 0-level set $T^{B+x}(0,\omega)$ of $B + x$ is the

$(-x)$-level set $T^B(-x,\omega)$ of B. This means that we can reduce the study of the level sets to studying the zero set

$$\mathsf{Z}(\omega) = B^{-1}(\{0\},\omega) = \{t \in [0,\infty) : B_t(\omega) = 0\}$$

under $\mathbb{P}^x$. Closely related to the zeros of a Brownian motion are the *records*, i. e. the times where a Brownian motion reaches its running maximum $M_t = \sup_{s\leqslant t} B_s$,

$$\mathsf{R}(\omega) = \{t \in [0,\infty) : B_t(\omega) = M_t(\omega)\}.$$

From the reflection principle, Theorem 6.9, we know that $M_t - B_t \sim |B_t|$. Both $M-B$ and $|B|$ are Markov processes, cf. Problem 6.1, and we find that the processes $(M_t - B_t)_{t\geqslant 0}$ Ex. 6.1
and $((|B_t|)_{t\geqslant 0})$ have the same distributions. This means, however, that the roots $\mathsf{Z}(\cdot)$ and the records $\mathsf{R}(\cdot)$ of a BM^1 have the same distributional properties. This was first observed by Lévy [143] who gave a penetrating study of the structure of $\mathsf{Z}(\omega)$ using $\mathsf{R}(\omega)$. Here is a typical example of this strategy.

11.22 Theorem. *Let $(B_t)_{t\geqslant 0}$ be a BM^1. Then $\mathbb{P}\left(\dim(\mathsf{Z}(\cdot)\cap[0,1]) = \frac{1}{2}\right) = 1$.*

Proof. **1°** (*Lower bound*). Since $\mathsf{Z}(\cdot)$ and $\mathsf{R}(\cdot)$ follow the same probability law, we show that $\dim(\mathsf{R}(\cdot)\cap[0,1]) \geqslant \frac{1}{2}$ a. s.

Define a measure μ_ω on $\mathsf{R}(\omega)$ by setting $\mu_\omega(s,t] = M_t(\omega) - M_s(\omega)$. Since $(B_t)_{t\geqslant 0}$ is a. s. nowhere monotone, μ_ω is supported in $\mathsf{R}(\omega)$. Brownian motion is a. s. Hölder continuous of order γ for any $\gamma \in (0,\frac{1}{2})$, cf. (11.2), and so

$$\mu_\omega(s,t] = M_t(\omega) - M_s(\omega) \leqslant \sup_{0\leqslant h\leqslant t-s} (B_{s+h}(\omega) - B_s(\omega)) \leqslant L_\gamma(\omega)(t-s)^\gamma$$

for all $s,t \in [0,1]$ and $\omega \in \Omega_\gamma$ where $\mathbb{P}(\Omega_\gamma) = 1$. Now we can use Example 11.6 d) to conclude that $\dim(\mathsf{R}(\omega)\cap[0,1]) \geqslant \gamma$ for all $\omega \in \Omega_\gamma$. Letting $\gamma \to \frac{1}{2}$ along countably many values yields the lower bound.

2° (*Upper bound*) Let $(\beta_t)_{t\geqslant 0}$ be a further BM^1 independent of $(B_t)_{t\geqslant 0}$. Applying Kaufman's theorem in the form of Corollary 11.12, to the two-dimensional Brownian motion $W_t = (B_t,\beta_t)$ we find

$$\dim(\mathsf{Z}(\omega)\cap[0,1]) \leqslant \dim B^{-1}(\{0\},\omega) = \dim W^{-1}(\{0\}\times\mathbb{R},\omega) \leqslant \tfrac{1}{2}\dim(\{0\}\times\mathbb{R}) = \tfrac{1}{2}. \qquad \square$$

For the rest of this section we will not follow Lévy's idea, but we use a direct attack on $\mathsf{Z}(\omega)$ without taking recourse to $\mathsf{R}(\omega)$. The trouble with $\mathsf{Z}(\omega)$ is that each Brownian zero is an accumulation point of other zeros, cf. Theorem 11.21, which means that we cannot move from a fixed zero to the 'next' zero, see Figure 11.2. Since $\mathsf{Z}(\omega)$ is a closed set, its complement is open and $\mathsf{Z}^c(\omega) = \bigcup_{n\geqslant 1} U_n(\omega)$ is a countable union of open intervals $U_n(\omega) \subset [0,\infty)$. We call the intervals $U_n(\omega)$ *excursion intervals*. In order to describe $U_n(\omega)$, we introduce the following random times

$$\xi_t(\omega) := \sup\{s \leqslant t : B_s(\omega) = 0\} \quad \text{and} \quad \eta_t(\omega) := \inf\{s \geqslant t : B_s(\omega) = 0\}$$

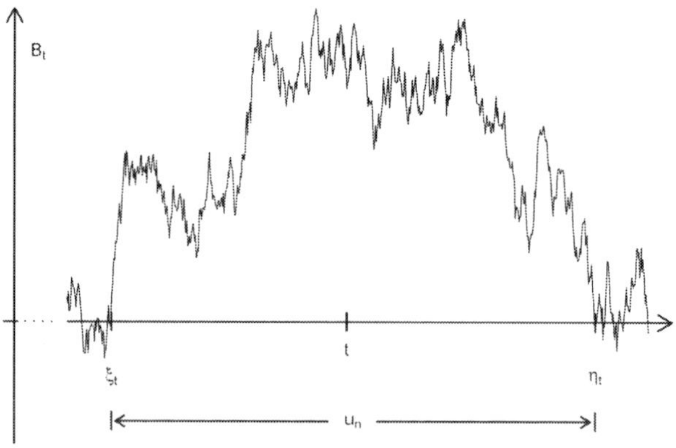

Fig. 11.2. A typical excursion interval: ξ_t and η_t are the last zero before $t > 0$ and the first zero after $t > 0$.

which are the last zero before t and the first zero after t; this is shown in Figure 11.2. Since $\mathbb{P}(B_t = 0) = 0$ we see that $\xi_t < t < \eta_t$ almost surely. While η_t is a stopping time, ξ_t is not since it depends on the knowledge of the whole path in $[0, t]$ (otherwise we could not decide if a given zero is the *last* zero before time t), see also the discussion in Section 6.6. Clearly, (ξ_t, η_t) is exactly the interval $U_n(\omega)$ which contains $t > 0$, hence $\mathsf{Z}^c(\omega) = \bigcup_{q \in \mathbb{Q}, q>0} (\xi_q(\omega), \eta_q(\omega))$.

One of the main ingredients in studying $\mathsf{Z}(\omega)$ is the law of the first passage distribution τ_x of the level $x \in \mathbb{R}$, cf. Theorem 6.9:

$$\mathbb{P}^0(\tau_x \in ds) = \frac{|x|}{\sqrt{2\pi s^3}}\, e^{-x^2/(2s)}\, ds = \mathbb{P}^x(\tau_0 \in ds)$$

(the second equality follows from the symmetry and translation invariance of Brownian motion).

Ex. 11.11 **11.23 Lemma** (Lévy 1948). *Let $(B_t)_{t\geqslant 0}$ be a one-dimensional Brownian motion. Then*

$$\mathbb{P}\left(B_s = 0 \text{ for some } s \in (t, t+h)\right) = \mathbb{P}\left(\mathsf{Z}(\cdot) \cap (t, t+h) \neq \emptyset\right) = \frac{2}{\pi} \arccos \sqrt{\frac{t}{t+h}}.$$

Proof. By the Markov property we get

$$\begin{aligned}
\mathbb{P}\left(B_s = 0 \text{ for some } s \in (t, t+h)\right) &= \mathbb{E}\left(\mathbb{P}^{B_t}\left(\tau_0 < h\right)\right) \\
&= \mathbb{E}\left(\int_0^h \frac{|B_t|}{\sqrt{2\pi s^3}}\, e^{-B_t^2/(2s)}\, ds\right) \\
&= \frac{1}{2\pi} \int_0^h \int_{-\infty}^{\infty} \frac{|y|}{\sqrt{t s^3}}\, e^{-\frac{1}{2}\left(\frac{1}{s}+\frac{1}{t}\right) y^2}\, dy\, ds \\
&= \frac{1}{\pi} \int_0^h \frac{\sqrt{t}}{(s+t)\sqrt{s}}\, ds.
\end{aligned}$$

Changing variables according to $s = tu^2$ yields

$$\mathbb{P}\left(B_s = 0 \text{ for some } s \in (t, t+h)\right) = \frac{2}{\pi} \int_0^{\sqrt{h/t}} \frac{du}{1+u^2} = \frac{2}{\pi} \arctan \sqrt{\frac{h}{t}}.$$

From $t = (t+h)\cos^2\left(\arctan\sqrt{\frac{h}{t}}\right)$ we conclude that $\arctan\sqrt{\frac{h}{t}} = \arccos\sqrt{\frac{t}{t+h}}$. □

11.24 Lemma. *Let $(B_t)_{t\geqslant 0}$ be a BM^1 and η_t the first zero after time $t > 0$. Then* Ex. 11.13

$$\mathbb{P}\left(B_t \in dy,\ \eta_t \in ds\right) = f_{(B_t,\eta_t)}(y,s)\,dy\,ds = \frac{|y|}{2\pi\sqrt{t(s-t)^3}}\, e^{-\frac{sy^2}{2t(s-t)}}\, \mathbb{1}_{(t,\infty)}(s)\,ds\,dy.$$

Proof. By definition, $\eta_t - t$ is just the first passage time τ_0 of the Brownian motion restarted at time $t > 0$. Using the Markov property we get for any $A \in \mathscr{B}(\mathbb{R})$ and $a > t$

$$\begin{aligned}
\mathbb{P}(B_t \in A,\ \tau_t \geqslant a) &= \mathbb{E}\left(\mathbb{1}_A(B_t)\,\mathbb{P}^{B_t}(\tau_0 \geqslant a - t)\right)\\
&= \mathbb{E}\left(\mathbb{1}_A(B_t) \int_{a-t}^{\infty} \frac{|B_t|}{\sqrt{2\pi s^3}}\, e^{-B_t^2/(2s)}\, ds\right)\\
&= \int_A \int_a^{\infty} \frac{|y|}{2\pi\sqrt{t(s-t)^3}}\, e^{-y^2/(2(s-t))}\, e^{-y^2/(2t)}\, ds\,dy.
\end{aligned}$$
□

If $(B_t)_{t\geqslant 0}$ is a Brownian motion, so is its projective reflection $W_t := tB_{1/t}$, $(t > 0)$, and $W_0 = 0$, cf. Paragraph 2.17. Since time is running backwards, the roles of the last zero before t and the first zero after t are reversed. This is the idea behind the following fundamental relation.

11.25 Theorem (Lévy 1939; Chung 1976). *Let $(B_t)_{t\geqslant 0}$ be a BM^1 and ξ_t the last zero before time $t > 0$. Then*

$$\mathbb{P}(|B_t| \in dy,\ \xi_t \in ds) = \frac{y}{\pi\sqrt{s(t-s)^3}}\, e^{-y^2/(2(t-s))}\, ds\,dy, \quad s < t,\ y > 0.$$

Proof. Let W_t be the projective reflection of B_t and denote by ξ_t^W the last zero of W before time $t > 0$. We will use the following abbreviations: $t^* := 1/t$, $s^* = 1/s$ and $u^* := 1/u$. For all $x > 0$ and $s < t$ we have

$$\begin{aligned}
\mathbb{P}\left(|W_t| \geqslant x,\ \xi_t^W \leqslant s\right) &= \mathbb{P}\left(|W_t| \geqslant x,\ W_u \neq 0 \quad \forall u \in (s,t]\right)\\
&= \mathbb{P}\left(|B_{t^*}| \geqslant xt^*,\ B_{u^*} \neq 0 \quad \forall u^* \in [t^*, s^*)\right)\\
&= \mathbb{P}\left(|B_{t^*}| \geqslant xt^*,\ \eta_{t^*} \geqslant s^*\right)\\
&= 2\int_{xt^*}^{\infty}\int_{s^*}^{\infty} f_{(B_{t^*},\eta_{t^*})}(y,r)\,dr\,dy
\end{aligned}$$

where $f_{(B_{t^*},\eta_{t^*})}(y,r)$ is the joint density from Lemma 11.24. Differentiating the right-hand side with respect to d/dx and d/ds gives

$$g_{(|W_t|,\xi_t^W)}(y,s) = 2f_{(B_{t^*},\eta_{t^*})}(yt^*,s^*)\,\frac{t^*}{s^2}.$$

Since $(W,\xi^W) \sim (B,\xi)$, the claim follows if we substitute $s^* = 1/s$ and $t^* = 1/t$. □

From Theorem 11.25 we can get various distributional identities for the zero set of a Brownian motion.

Ex. 11.12 **11.26 Corollary.** *Let ξ_t and η_t the last zero before and the first zero after time $t > 0$. Then*

$$\mathbb{P}\left(\xi_t \in ds,\ \eta_t \in du\right) = \frac{ds\,du}{2\pi\sqrt{s(u-s)^3}}, \qquad 0 < s < t < u.$$

Proof. Using a standard formula for conditional densities, we get

$$\begin{aligned}\mathbb{P}\left(\xi_t \in ds,\ \eta_t \in du,\ |B_t| \in dy\right) &= \mathbb{P}\left(\eta_t \in du \mid \xi_t = s,\ |B_t| = y\right)\mathbb{P}\left(\xi_t \in ds,\ |B_t| \in dy\right)\\ &= \mathbb{P}\left(\eta_t \in du \mid |B_t| = y\right)\mathbb{P}\left(\xi_t \in ds,\ |B_t| \in dy\right).\end{aligned}$$

The last equality follows from the fact that ξ_t is $\mathscr{F}_t$ measurable and the Markov property. Since for fixed t, $W_t := B_{t+s} - B_t$ is a Brownian motion, we can easily work out the conditional probability:

$$\mathbb{P}\left(\eta_t \in du \mid |B_t| = y\right) = \mathbb{P}^y\left(\tau_0 \in du - t\right) = \frac{y}{\sqrt{2\pi(u-t)^3}}\,e^{-y^2/(2(u-t))}\,du, \qquad u > t,$$

while Theorem 11.25 gives the density for $\mathbb{P}\left(\xi_t \in ds,\ |B_t| \in dy\right)$. Integrating out dy we thus get for $0 < s < t < u$

$$\begin{aligned}\frac{\mathbb{P}\left(\xi_t \in ds,\ \eta_t \in du\right)}{ds\,du} &= \int_0^\infty \frac{y}{\sqrt{2\pi(u-t)^3}}\,e^{-y^2/(2(u-t))}\,\frac{y}{\pi\sqrt{s(t-s)^3}}\,e^{-y^2/(2(t-s))}\,dy\\ &= \frac{1}{2\pi}\,\frac{1}{\sqrt{2\pi}}\,\frac{1}{\sqrt{s(t-s)^3}\sqrt{(u-t)^3}}\int_{-\infty}^\infty y^2 e^{-\frac12 y^2\big/\frac{(t-s)(u-t)}{u-s}}\,du\\ &= \frac{1}{2\pi(t-s)(u-t)}\,\frac{1}{\sqrt{s(u-s)}}\,\underbrace{\frac{1}{\sqrt{2\pi}\sqrt{\frac{(t-s)(u-t)}{u-s}}}\int_{-\infty}^\infty y^2 e^{-\frac12 y^2\big/\frac{(t-s)(u-t)}{u-s}}\,du}_{=\mathbb{E}\,B_c^2=c,\quad c=\frac{(t-s)(u-t)}{u-s}}\\ &= \frac{1}{2\pi}\,\frac{1}{\sqrt{s(u-s)^3}}\end{aligned}$$

as we have claimed. □

11.27 Further reading. The interplay between Hausdorff dimension and capacity is a classic topic of potential theory, going back to Frostman's PhD thesis [83]. Good modern introductions can be found in [26] and [1], see also [115, Chapter 10]. Much more is known on Hausdorff dimension and (the exact) Hausdorff measure of Brownian paths. Excellent surveys on these topics are [236] and [221]. Fractals and fractal geometry, in general, are nicely covered in [76] (sometimes a bit handwaving, though) and [157]. In the hands of Itô and Williams, excursion theory became a most powerful tool. Lévy's paper [143] and Chapter VI in [145] – which started all this – are more visionary than rigorous. A lucid introduction is given in [30], for a thorough study we recommend [189, Chapter XII].

[1] Aikawa, Essén: *Potential Theory – Selected Topics*.
[26] Carleson: *Selected Problems on Exceptional Sets*.
[30] Chung: Excursions in Brownian motion.
[76] Falconer: *Fractal Geometry*.
[83] Frostman: *Potentiel d'équilibre et capacité des ensembles avec quelques applications a la théorie des fonctions*.
[115] Kahane: *Some Random Series of Functions*.
[157] Mattila: *Geometry of Sets and Measures in Euclidean Spaces*.
[189] Revuz, Yor: *Continuous Martingales and Brownian Motion*.
[221] Taylor: The measure theory of random fractals.
[236] Xiao: Random fractals and Markov processes.

Problems

1. Verify that s-dimensional Hausdorff measure is an outer measure.
2. Show that the d-dimensional Hausdorff measure of a bounded open set $U \subset \mathbb{R}^d$ is finite and positive: $0 < \mathcal{H}^d(U) < \infty$. In particular, $\dim U = d$.
   Hint: Since Hausdorff measure is monotone, it is enough to show that for an axis-parallel cube $Q \subset \mathbb{R}^d$ we have $0 < \mathcal{H}^d(Q) < \infty$. Without loss, assume that $Q = [0,1]^d$. The upper bound follows by covering Q with small cubes of side-length $1/n$. The lower bound follows since every δ-cover $(E_j)_{j\geqslant 1}$ of Q gives rise to a cover by cubes whose side-length is at most 2δ.
3. Let $E \subset \mathbb{R}^d$. Show that Hausdorff dimension $\dim E$ coincides with the numbers
$$\sup\{\alpha \geqslant 0 : \mathcal{H}^\alpha(E) = \infty\}, \quad \sup\{\alpha \geqslant 0 : \mathcal{H}^\alpha(E) > 0\}$$
$$\inf\{\alpha \geqslant 0 : \mathcal{H}^\alpha(E) < \infty\}, \quad \inf\{\alpha \geqslant 0 : \mathcal{H}^\alpha(E) = 0\}.$$
4. Let $(E_j)_{j\geqslant 1}$ be subsets of $\mathbb{R}^d$ and $E := \bigcup_{j\geqslant 1} E_j$. Show that $\dim E = \sup_{j\geqslant 1} \dim E_j$.
5. Let $E \subset \mathbb{R}^d$. Show that $\dim(E \times \mathbb{R}^n) = \dim E + n$. Does a similar formula hold for $E \times F$ where $E \subset \mathbb{R}^d$ and $F \subset \mathbb{R}^n$?
6. Let $f : \mathbb{R}^d \to \mathbb{R}^n$ be a bi-Lipschitz map, i. e. both f and f^{-1} are Lipschitz continuous. Show that $\dim f(E) = \dim E$. Is this also true for a Hölder continuous map with index $\gamma \in (0,1)$?

7. Let C denote Cantor's discontinuum, which is obtained if we remove recursively the open middle third of any remaining interval:

$$[0,1] \rightsquigarrow [0,\tfrac{1}{3}] \cup [\tfrac{2}{3},1] \rightsquigarrow [0,\tfrac{1}{9}] \cup [\tfrac{2}{9},\tfrac{1}{3}] \cup [\tfrac{2}{3},\tfrac{7}{9}] \cup [\tfrac{8}{9},1] \rightsquigarrow \dots,$$

see e. g. [204, Problem 7.10]. Show that $C = f_1(C) \cup f_2(C)$ with $f_1(x) = \frac{1}{3}x$ and $f_2(x) = \frac{1}{3}x + \frac{2}{3}$ and $\dim C = \log 2/\log 3$.

8. Let $(B_t)_{t\geqslant 0}$ be a BM^d. Show that for $\lambda < d$ the expectation $\mathbb{E}\left(|B_1|^{-\lambda}\right)$ is finite.

9. Let $(B_t)_{t\geqslant 0}$ be a one-dimensional Brownian motion. Show that

$$\dim B^{-1}(A) \leqslant \tfrac{1}{2} + \tfrac{1}{2}\dim A \quad \text{a. s. for all } A \in \mathscr{B}(\mathbb{R}).$$

Hint: Apply Corollary 11.12 to $W_t = (B_t, \beta_t)$ where β is a further, independent BM^1; observe that $B^{-1}(A) = W^{-1}(A \times \mathbb{R})$.

10. Let $F \subset \mathbb{R}$ be a non-void perfect set, i. e. a closed set such that each point in F is an accumulation point of F. Show that a perfect set is uncountable.

11. Let $(B_t)_{t\geqslant 0}$ be a BM^1 and denote by ξ_t the last zero before time $t > 0$.
   (a) Use Theorem 11.25 to give a further proof of Lévy's arc-sine law (cf. Theorem 6.19): $\mathbb{P}(\xi_t < s) = \frac{2}{\pi}\arcsin\sqrt{\frac{s}{t}}$ for all $s \in [0,t]$.
   (b) Find the conditional density $\mathbb{P}(|B_t| \in dy \mid \xi_t = s)$.
   (c) Integrate the formula appearing in Theorem 11.25 to get

$$\sqrt{\frac{2}{\pi t}}\, e^{-y^2/(2t)} = \int_0^t \sqrt{\frac{2}{\pi s}}\, \frac{y}{\sqrt{2\pi(t-s)^3}}\, e^{-y^2/(2(t-s))}\, ds$$

   and give a probabilistic interpretation of this identity.

12. Show that Corollary 11.26 also follows from Lemma 11.23.
Hint: Calculate $\dfrac{\partial^2}{\partial u\,\partial s}\left(1 - \dfrac{2}{\pi}\arccos\sqrt{\dfrac{s}{u}}\right)$.

13. Let $(B_t)_{t\geqslant 0}$, η_t and ξ_t be as in Corollary 11.26. Define

$$L_t^- := t - \xi_t \quad \text{and} \quad L_t := \eta_t - \xi_t.$$

Find the laws of (L_t^-, L_t), L_t^-, L_t and L_t conditional on $L_t^- = r$.
Find the probability that $\mathbb{P}(L_t > r + s \mid L_t^- = r)$, i. e. the probability that an excursion continues for s units of time if we know that it is already r units of time under way.

12 The growth of Brownian paths

In this chapter we study the question how fast a Brownian path grows as $t \to \infty$. Since for a BM[1] $(B(t))_{t\geqslant 0}$, the process $(tB(t^{-1}))_{t>0}$, is again a Brownian motion, we get from the growth at infinity automatically local results for small t and vice versa. A first estimate follows from the strong law of large numbers. Write for $n \geqslant 1$

$$B(nt) = \sum_{j=1}^{n} \left[B(jt) - B((j-1)t)\right],$$

and observe that the increments of a Brownian motion are stationary and independent random variables. By the strong law of large numbers we get that $B(t,\omega) \leqslant \epsilon t$ for any $\epsilon > 0$ and all $t \geqslant t_0(\epsilon,\omega)$. Lévy's modulus of continuity, Theorem 10.6, gives a better bound. For all $\epsilon > 0$ and almost all $\omega \in \Omega$ there is some $h_0 = h_0(\epsilon,\omega)$ such that Ex. 12.1

$$|B(h,\omega)| \leqslant \sup_{0\leqslant t\leqslant 1-h} |B(t+h,\omega) - B(t,\omega)| \leqslant (1+\epsilon)\sqrt{2h|\log h|}$$

for all $h < h_0$. Since $tB(\frac{1}{t})$ is again a Brownian motion, cf. 2.17, we see that

$$|B(t,\omega)| \leqslant (1+\epsilon)\sqrt{2t\log t}$$

for all $t > t_0 = h_0^{-1}$. But also this bound is not optimal.

12.1 Khintchine's law of the iterated logarithm

We will show now that the function $\Lambda(t) := \sqrt{2t\log\log t}$, $t \geqslant 3$, neatly describes the behaviour of the sample paths as $t \to \infty$. Still better results can be obtained from Kolmogorov's integral test which we state (without proof) at the end of this section.

The key for the proof of the next theorem is a tail estimate for the normal distribution which we know from Lemma 10.5:

$$\frac{1}{\sqrt{2\pi}}\frac{x}{x^2+1}\,e^{-x^2/2} \leqslant \mathbb{P}(B(1) > x) \leqslant \frac{1}{\sqrt{2\pi}}\frac{1}{x}\,e^{-x^2/2}, \qquad x > 0, \tag{12.1}$$

and the following maximal estimate which is, for example, a consequence of the reflection principle, Theorem 6.9: Ex. 12.2

$$\mathbb{P}\Big(\sup_{s\leqslant t} B(s) > x\Big) \leqslant 2\mathbb{P}(B(t) > x), \qquad x > 0,\ t > 0. \tag{12.2}$$

Because of the form of the numerator the next result is often called the *law of the iterated logarithm* (LIL).

Ex. 12.3 **12.1 Theorem** (LIL. Khintchine 1933). *Let $(B(t))_{t\geqslant 0}$ be a* BM^1. *Then*

$$\mathbb{P}\left(\varlimsup_{t\to\infty} \frac{B(t)}{\sqrt{2t\log\log t}} = 1\right) = 1. \tag{12.3}$$

Since $(-B(t))_{t\geqslant 0}$ and $(tB(t^{-1}))_{t>0}$ are again Brownian motions, we can immediately get versions of the Khintchine LIL for the limit inferior and for $t \to 0$.

12.2 Corollary. *Let $(B_t)_{t\geqslant 0}$ be a* BM^1. *Then, almost surely,*

$$\varlimsup_{t\to\infty} \frac{B(t)}{\sqrt{2t\log\log t}} = 1, \qquad \varlimsup_{t\to 0} \frac{B(t)}{\sqrt{2t\log|\log t|}} = 1,$$
$$\varliminf_{t\to\infty} \frac{B(t)}{\sqrt{2t\log\log t}} = -1, \qquad \varliminf_{t\to 0} \frac{B(t)}{\sqrt{2t\log|\log t|}} = -1.$$

Proof of Theorem 12.1. $\mathbf{1^o}$ (Upper bound). Fix $\epsilon > 0$, $q > 1$ and set

$$A_n := \left\{ \sup_{0\leqslant s\leqslant q^n} B(s) \geqslant (1+\epsilon)\sqrt{2q^n \log\log q^n} \right\}.$$

From the maximal estimate (12.2), we see that

$$\begin{aligned} \mathbb{P}(A_n) &\leqslant 2\mathbb{P}\left(B(q^n) \geqslant (1+\epsilon)\sqrt{2q^n\log\log q^n}\right) \\ &= 2\mathbb{P}\left(\frac{B(q^n)}{\sqrt{q^n}} \geqslant (1+\epsilon)\sqrt{2\log\log q^n}\right). \end{aligned}$$

Since $B(q^n)/\sqrt{q^n} \sim B(1)$, we can use the upper bound from (12.1) and find

$$\mathbb{P}(A_n) \leqslant \frac{1}{(1+\epsilon)\sqrt{\pi}\,\sqrt{\log\log q^n}}\, e^{-(1+\epsilon)^2 \log\log q^n} \leqslant c\,(n\log q)^{-(1+\epsilon)^2}.$$

This shows that $\sum_{n=1}^{\infty} \mathbb{P}(A_n) < \infty$. Therefore, we can apply the Borel–Cantelli lemma and deduce that

$$\varlimsup_{n\to\infty} \frac{\sup_{0\leqslant s\leqslant q^n} B(s)}{\sqrt{2q^n\log\log q^n}} \leqslant (1+\epsilon) \quad \text{a. s.}$$

Every $t > 1$ is in some interval of the form $[q^{n-1}, q^n]$, and since $\Lambda(t) = \sqrt{2t\log\log t}$ is an increasing function for $t > 3$, we find for all $t \geqslant q^{n-1} > 3$

$$\frac{B(t)}{\sqrt{2t\log\log t}} \leqslant \frac{\sup_{s\leqslant q^n} B(s)}{\sqrt{2q^n\log\log q^n}} \frac{\sqrt{2q^n\log\log q^n}}{\sqrt{2q^{n-1}\log\log q^{n-1}}}.$$

Therefore

$$\varlimsup_{t\to\infty} \frac{B(t)}{\sqrt{2t\log\log t}} \leqslant (1+\epsilon)\sqrt{q} \quad \text{a. s.}$$

Letting $\epsilon \to 0$ and $q \to 1$ along countable sequences, we get the upper bound in (12.3).

2° If we use the estimate of step 1° for the Brownian motion $(-B(t))_{t\geqslant 0}$ and $\epsilon = 1$, we see that

$$-B(q^{n-1}) \leqslant 2\sqrt{2q^{n-1}\log\log q^{n-1}} \leqslant \frac{2}{\sqrt{q}}\sqrt{2q^n \log\log q^n} \quad \text{a. s.}$$

for all sufficiently large $n \geqslant n_0(\epsilon, \omega)$.

3° (Lower bound). So far we have not used the independence and stationarity of the increments of a Brownian motion. The independence of the increments shows that the sets

$$C_n := \left\{B(q^n) - B(q^{n-1}) \geqslant \sqrt{2(q^n - q^{n-1})\log\log q^n}\right\}, \quad n \geqslant 1,$$

are independent, and the stationarity of the increments implies

$$\mathbb{P}(C_n) = \mathbb{P}\left(\frac{B(q^n) - B(q^{n-1})}{\sqrt{q^n - q^{n-1}}} \geqslant \sqrt{2\log\log q^n}\right) = \mathbb{P}\left(\frac{B(q^n - q^{n-1})}{\sqrt{q^n - q^{n-1}}} \geqslant \sqrt{2\log\log q^n}\right).$$

Since $B(q^n - q^{n-1})/\sqrt{q^n - q^{n-1}} \sim B(1)$ and $2x/(1+x^2) \geqslant x^{-1}$ for all $x > 1$, we get from the lower tail estimate of (12.1) that

$$\mathbb{P}(C_n) \geqslant \frac{1}{\sqrt{8\pi}}\frac{1}{\sqrt{2\log\log q^n}}e^{-\log\log q^n} \geqslant \frac{c}{n\sqrt{\log n}}.$$

Therefore, $\sum_{n=1}^{\infty}\mathbb{P}(C_n) = \infty$, and since the sets C_n are independent, the Borel–Cantelli lemma applies and shows that, for infinitely many $n \geqslant 1$,

$$\begin{aligned} B(q^n) &\geqslant \sqrt{2(q^n - q^{n-1})\log\log q^n} + B(q^{n-1}) \\ &\overset{2^\circ}{\geqslant} \sqrt{2(q^n - q^{n-1})\log\log q^n} - \frac{2}{\sqrt{q}}\sqrt{2q^n\log\log q^n}. \end{aligned}$$

Now we divide both sides by $\sqrt{2q^n\log\log q^n}$ and get for all $q > 1$

$$\varlimsup_{t\to\infty}\frac{B(t)}{\sqrt{2t\log\log t}} \geqslant \varlimsup_{n\to\infty}\frac{B(q^n)}{\sqrt{2q^n\log\log q^n}} \geqslant \sqrt{1 - \frac{1}{q}} - \sqrt{\frac{4}{q}}$$

almost surely. Letting $q \to \infty$ along a countable sequence concludes the proof of the lower bound. □

12.3 Remark (Fast and slow points). For every $t > 0$ $W(h) := B(t+h) - B(t)$ is again a Brownian motion. Therefore, Corollary 12.2 shows that

$$\varlimsup_{h\to 0}\frac{|B(t+h) - B(t)|}{\sqrt{2h\log|\log h|}} = 1 \tag{12.4}$$

almost surely for every fixed $t \in [0,1]$. Using Fubini's theorem, we can choose the same exceptional set for (Lebesgue) almost all $t \in [0,1]$. It is instructive to compare (12.4) with Lévy's modulus of continuity, Theorem 10.6:

$$\varlimsup_{h\to 0}\frac{\sup_{0\leqslant t\leqslant 1-h}|B(t+h) - B(t)|}{\sqrt{2h|\log h|}} = 1. \tag{12.5}$$

By (12.4), the local modulus of continuity of a Brownian motion is $\sqrt{2h \log|\log h|}$; on the other hand, (12.5) means that globally (and at some necessarily random points) the modulus of continuity will be larger than $\sqrt{2h|\log h|}$. Therefore, a Brownian path must violate the LIL at some random points.

The random set of all times where the LIL fails,

$$E(\omega) := \left\{ t \geqslant 0 \;:\; \overline{\lim_{h \to 0}} \frac{|B(t+h,\omega) - B(t,\omega)|}{\sqrt{2h \log|\log h|}} \neq 1 \right\},$$

is almost surely uncountable and dense in $[0,\infty)$; this surprising result is due to S. J. Taylor [220], see also Orey and Taylor [169].

At the points $\tau \in E(\omega)$, Brownian motion is faster than usual. Therefore, these points are also called *fast points*. On the other hand, there are also *slow points*. A result by Kahane [113; 114] says that almost surely there exists a $t \in [0,1]$ such that

$$\overline{\lim_{h \to 0}} \frac{|B(t+h) - B(t)|}{\sqrt{h}} < \infty.$$

12.4 Remark. Blumenthal's 0-1–law, Corollary 6.22, implies that for any positive continuous function $\kappa(t)$

$$\mathbb{P}\big(B(t) < \kappa(t) \;\text{ for all sufficiently small }\; t\big) \in \{0,1\}.$$

If the probability is 1, we call κ an *upper function*. The following result due to Kolmogorov gives a sharp criterion whether κ is an upper function.

Ex. 12.4 **12.5 Theorem** (Kolmogorov's test). *Let $(B_t)_{t\geqslant 0}$ be a* BM^1 *and $\kappa : (0,1] \to (0,\infty)$ a con-*
Ex. 12.5 *tinuous function. If $\kappa(t)$ is increasing and $\kappa(t)/\sqrt{t}$ decreasing, then κ is an upper function if, and only if,*

$$\int_0^1 \frac{\kappa(t)}{t^{3/2}} e^{-\kappa^2(t)/2t}\,dt \quad \textit{converges.} \tag{12.6}$$

Kolmogorov's original proof was never published, and the first proof of this result appeared in Petrovsky [175]. Motoo's elegant approach from 1959 [163] can be found in Itô and McKean [105, 1.8, p. 33–36 and 4.12 pp. 161–164]. Here we show only the easier part of the assertion: The condition (12.6) is sufficient for κ to be an upper function.

Proof (Itô–McKean). Obviously, we have

$$\{B(t) < \kappa(t) \quad \text{for all small } t > 0\} \subset \bigcup_{\delta > 0} \{B(t) < \kappa(t) \quad \forall t \in (0,\delta]\}$$

and we are going to prove that Kolmogorov's condition (12.6) implies that

$$\lim_{\delta \to 0} \mathbb{P}\left(\exists t \in (0,\delta] \;:\; B(t) \geqslant \kappa(t)\right) = 0.$$

Pick any $n \geqslant 1$ and $0 < a = t_1 < t_2 < \ldots < t_n = b \leqslant 1$. The monotonicity of κ shows that the first passage times

$$\tau_{\kappa(t)} = \inf\left\{s > 0 \;:\; B(s) \geqslant \kappa(t)\right\}, \quad t > 0,$$

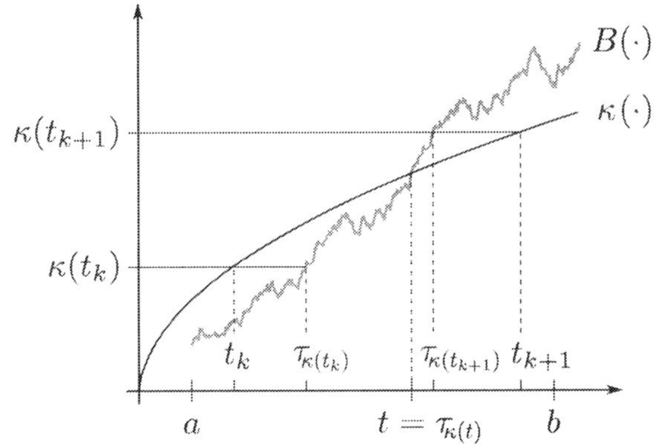

Fig. 12.1. The time $t = \tau_{\kappa(t)}$ is the first instance when the Brownian path $[a,b] \ni s \mapsto B(s)$ reaches the graph of $[a,b] \ni s \mapsto \kappa(s)$.

are increasing as t increases. Note that $t = \tau_{\kappa(t)}$ is the first instance when the Brownian path $s \mapsto B(s)$ reaches the boundary defined by $s \mapsto \kappa(s)$. This is shown in Figure 12.1.Assume that $B(t) = \kappa(t)$ for some $t = t(\omega) \in [a,b]$; by construction there exists some $k = k(\omega)$ such that $t \in (t_k, t_{k+1}]$ or $t = a$. In the latter case, we have $\tau_{\kappa(t_1)} = \tau_{\kappa(a)} \leqslant a$. Otherwise, we get from the definition of the first passage time that $\tau_{\kappa(t)} \leqslant t \leqslant t_{k+1}$. By the monotonicity of κ we see $\tau_{\kappa(t_k)} \leqslant \tau_{\kappa(t)} \leqslant t_{k+1}$. On the other hand, $t_k \geqslant \tau_{\kappa(t_k)}$ is not possible, since then the path would have crossed the $\kappa(\cdot)$-threshold before time t! Thus $t_k < \tau_{\kappa(t_k)} \leqslant t_{k+1}$. Therefore,

$$\{\exists t \in [a,b] \;:\; B(t) \geqslant \kappa(t)\} \subset \left\{\tau_{\kappa(t_1)} \leqslant a\right\} \cup \bigcup_{j=1}^{n-1} \{t_j < \tau_{\kappa(t_j)} \leqslant t_{j+1}\}$$

and, since we know from Theorem 6.9 that $|\kappa|(2\pi s^3)^{-1/2} e^{-\kappa^2/2s}\,ds$ is the density of τ_κ,

$$\begin{aligned}
\mathbb{P}\Big(\exists t \in [a,b] \;:\; B(t) \geqslant \kappa(t)\Big) &\leqslant \mathbb{P}\left(\tau_{\kappa(t_1)} \leqslant a\right) + \sum_{j=1}^{n-1} \mathbb{P}\left(t_j < \tau_{\kappa(t_j)} \leqslant t_{j+1}\right) \\
&= \int_0^a \frac{\kappa(a)}{\sqrt{2\pi s^3}} e^{-\kappa^2(a)/2s}\,ds + \sum_{j=1}^{n-1} \int_{t_j}^{t_{j+1}} \frac{\kappa(t_j)}{\sqrt{2\pi s^3}} e^{-\kappa^2(t_j)/2s}\,ds \\
&\leqslant \int_0^a \frac{\kappa(a)}{\sqrt{2\pi s^3}} e^{-\kappa^2(a)/2s}\,ds + \sum_{j=1}^{n-1} \int_{t_j}^{t_{j+1}} \frac{\kappa(s)}{\sqrt{2\pi s^3}} e^{-\kappa^2(t_j)/2s}\,ds.
\end{aligned}$$

If the mesh $\max_{1 \leqslant j \leqslant n-1}(t_{j+1} - t_j) \to 0$ as $n \to \infty$, we can use dominated convergence and the continuity of κ to get

$$\mathbb{P}\Big(\exists t \in [a,b] \;:\; B(t) \geqslant \kappa(t)\Big) \leqslant \int_0^a \frac{\kappa(a)}{\sqrt{2\pi s^3}} e^{-\kappa^2(a)/2s}\,ds + \int_a^b \frac{\kappa(s)}{\sqrt{2\pi s^3}} e^{-\kappa^2(s)/2s}\,ds.$$

Since $\kappa^2(s)/s$ decreasing and $\int_0^1 \kappa(s)s^{-3/2}\, e^{-\kappa^2(s)/2s}\, ds < \infty$, we see $\lim_{t\to 0} \frac{1}{t}\kappa^2(t) = \infty$, hence

$$\int_0^a \frac{\kappa(a)}{\sqrt{2\pi s^3}}\, e^{-\kappa^2(a)/2s}\, ds = \int_0^{a/\kappa^2(a)} \frac{1}{\sqrt{2\pi u^3}}\, e^{-1/2u}\, du \xrightarrow[a\to 0]{} 0,$$

and

$$\mathbb{P}\Big(\exists t \in (0,b] \,:\, B(t) \geqslant \kappa(t)\Big) \leqslant \int_0^b \frac{\kappa(s)}{\sqrt{2\pi s^3}}\, e^{-\kappa^2(s)/2s}\, ds.$$

The claim follows as $b \to 0$. □

12.2 Chung's 'other' law of the iterated logarithm

Let $(X_j)_{j\geqslant 1}$ be iid random variables such that $\mathbb{E}\, X_1 = 0$ and $\mathbb{E}\, X_1^2 < \infty$. K. L. Chung [29] proved for the maximum of the absolute value of the partial sums the following law of the iterated logarithm

$$\varliminf_{n\to\infty} \frac{\max_{1\leqslant j\leqslant n} |X_1 + \cdots + X_j|}{\sqrt{n/\log\log n}} = \frac{\pi}{\sqrt{8}}.$$

A similar result holds for the running maximum of the modulus of a Brownian path. The proof resembles Khintchine's LIL for Brownian motion, but we need a more subtle estimate for the distribution of the maximum.

Using Lévy's triple law (6.19) for $a = -x$, $b = x$ and $t = 1$, we see

$$\begin{aligned}
\mathbb{P}\Big(\sup_{s\leqslant 1} |B_s| < x\Big) &= \mathbb{P}\Big(\inf_{s\leqslant 1} B_s > -x,\ \sup_{s\leqslant 1} B_s < x,\ B_1 \in (-x,x)\Big)\\
&= \frac{1}{\sqrt{2\pi}} \int_{-x}^{x} \sum_{j=-\infty}^{\infty} (-1)^j e^{-(y+2jx)^2/2}\, dy\\
&= \frac{1}{\sqrt{2\pi}} \sum_{j=-\infty}^{\infty} (-1)^j \int_{(2j-1)x}^{(2j+1)x} e^{-y^2/2}\, dy\\
&= \frac{1}{\sqrt{2\pi}} \int_{\mathbb{R}} \underbrace{\sum_{j=-\infty}^{\infty} (-1)^j \mathbb{1}_{((2j-1)x,\,(2j+1)x)}(y)}_{=f(y)}\, e^{-y^2/2}\, dy.
\end{aligned}$$

The $4x$-periodic even function f is given by the following Fourier cosine series[1]

$$f(y) = \frac{4}{\pi} \sum_{k=0}^{\infty} \frac{(-1)^k}{2k+1} \cos\Big(\frac{2k+1}{2x}\pi y\Big).$$

---

1 Although we have used Lévy's formula (6.19) for the joint distribution of (m_t, M_t, B_t), it is not suitable to study the asymptotic behaviour of $\mathbb{P}(m_t < a, M_t > b, B_t \in [c,d])$ as $t \to \infty$. Our approach based on a Fourier expansion leads essentially to the following expression which is also due to Lévy [145,

Therefore, we find

$$\mathbb{P}\left(\sup_{s\leqslant 1}|B_s| < x\right) = \frac{4}{\pi}\sum_{k=0}^{\infty}\frac{1}{\sqrt{2\pi}}\frac{(-1)^k}{2k+1}\int_{\mathbb{R}}\cos\left(\frac{2k+1}{2x}\pi y\right)e^{-y^2/2}\,dy$$

$$\overset{(2.5)}{=} \frac{4}{\pi}\sum_{k=0}^{\infty}\frac{(-1)^k}{2k+1}e^{-\pi^2(2k+1)^2/(8x^2)}.$$

This proves the following lemma.

12.6 Lemma. *Let $(B(t))_{t\geqslant 0}$ be a* BM^1. *Then*

$$\lim_{x\to 0}\mathbb{P}\left(\sup_{s\leqslant 1}|B(s)| < x\right)\Big/\frac{4}{\pi}e^{-\pi^2/(8x^2)} = 1. \tag{12.7}$$

12.7 Theorem (Chung 1948). *Let $(B(t))_{t\geqslant 0}$ be a* BM^1. *Then*

$$\mathbb{P}\left(\varliminf_{t\to\infty}\frac{\sup_{s\leqslant t}|B(s)|}{\sqrt{t/\log\log t}} = \frac{\pi}{\sqrt{8}}\right) = 1. \tag{12.8}$$

Proof. **1°** (Upper bound) Set $t_n := n^n$ and

$$C_n := \left\{\sup_{t_{n-1}\leqslant s\leqslant t_n}|B(s)-B(t_{n-1})| < \frac{\pi}{\sqrt{8}}\sqrt{\frac{t_n}{\log\log t_n}}\right\}.$$

By the stationarity of the increments we see

$$\frac{\sup_{t_{n-1}\leqslant s\leqslant t_n}|B(s)-B(t_{n-1})|}{\sqrt{t_n}} \sim \frac{\sup_{s\leqslant t_n-t_{n-1}}|B(s)|}{\sqrt{t_n}} \leqslant \frac{\sup_{s\leqslant t_n-t_{n-1}}|B(s)|}{\sqrt{t_n-t_{n-1}}};$$

using Lemma 12.6 for $x = \pi/\sqrt{8\log\log t_n}$, we get for sufficiently large $n\geqslant 1$

$$\mathbb{P}(C_n) \geqslant \mathbb{P}\left(\frac{\sup_{s\leqslant t_n-t_{n-1}}|B(s)|}{\sqrt{t_n-t_{n-1}}} < \frac{\pi}{\sqrt{8}}\frac{1}{\sqrt{\log\log t_n}}\right) \geqslant c\,\frac{4}{\pi}\frac{1}{n\log n}.$$

III.19–20, pp. 78–84]:

$$\mathbb{P}(m_t > a, M_t < b, B_t \in I) = \sum_{j=1}^{\infty}e^{-\lambda_j t}\phi_j(0)\int_I \phi_j(y)\,dy$$

where λ_j and ϕ_j, $j\geqslant 1$, are the eigenvalues and normalized eigenfunctions of the boundary value problem

$$\frac{1}{2}f''(x) = \lambda f(x) \text{ for } a<x<b \qquad\text{and}\qquad f(x)=0 \text{ for } x=a,\ x=b.$$

It is not difficult to see that $\lambda_j = j^2\pi^2/2(b-a)^2$ and

$$\phi_j(x) = \sqrt{\tfrac{2}{b-a}}\cos\left(\tfrac{j\pi}{b-a}\left[x-\tfrac{a+b}{2}\right]\right) \quad\text{and}\quad \phi_j(x) = \sqrt{\tfrac{2}{b-a}}\sin\left(\tfrac{j\pi}{b-a}\left[x-\tfrac{a+b}{2}\right]\right).$$

for odd and even indices j, respectively.

Therefore, $\sum_n \mathbb{P}(C_n) = \infty$; since the sets C_n are independent, the Borel–Cantelli lemma applies and shows that

$$\varliminf_{n\to\infty} \frac{\sup_{t_{n-1}\leqslant s\leqslant t_n} |B(s) - B(t_{n-1})|}{\sqrt{t_n/\log\log t_n}} \leqslant \frac{\pi}{\sqrt{8}}.$$

Ex. 12.3 The argument in Step 1° of the proof of Khintchine's LIL, Theorem 12.1, yields

$$\varlimsup_{n\to\infty} \frac{\sup_{s\leqslant t_{n-1}} |B(s)|}{\sqrt{t_n/\log\log t_n}} = \varlimsup_{n\to\infty} \frac{\sup_{s\leqslant t_{n-1}} |B(s)|}{\sqrt{t_{n-1}\log\log t_{n-1}}} \frac{\sqrt{t_{n-1}\log\log t_{n-1}}}{\sqrt{t_n/\log\log t_n}} = 0.$$

Finally, using $\sup_{s\leqslant t_n} |B(s)| \leqslant 2\sup_{s\leqslant t_{n-1}} |B(s)| + \sup_{t_{n-1}\leqslant s\leqslant t_n} |B(s) - B(t_{n-1})|$ we get

$$\varliminf_{n\to\infty} \frac{\sup_{s\leqslant t_n} |B(s)|}{\sqrt{t_n/\log\log t_n}} \leqslant \varlimsup_{n\to\infty} \frac{2\sup_{s\leqslant t_{n-1}} |B(s)|}{\sqrt{t_n/\log\log t_n}} + \varliminf_{n\to\infty} \frac{\sup_{t_{n-1}\leqslant s\leqslant t_n} |B(s) - B(t_{n-1})|}{\sqrt{t_n/\log\log t_n}} \leqslant \frac{\pi}{\sqrt{8}}$$

which proves that $\varliminf_{t\to\infty} \sup_{s\leqslant t} |B(s)|/\sqrt{t/\log\log t} \leqslant \pi/\sqrt{8}$ a. s.

2° (Lower bound) Fix $\epsilon > 0$ and $q > 1$ and set

$$A_n := \left\{\sup_{s\leqslant q^n} |B(s)| < (1-\epsilon)\frac{\pi}{\sqrt{8}}\sqrt{\frac{q^n}{\log\log q^n}}\right\}.$$

Lemma 12.6 with $x = (1-\epsilon)\pi/\sqrt{8\log\log q^n}$ gives for large $n \geqslant 1$

$$\mathbb{P}(A_n) = \mathbb{P}\left(\frac{\sup_{s\leqslant q^n} |B(s)|}{\sqrt{q^n}} < (1-\epsilon)\frac{\pi}{\sqrt{8}}\frac{1}{\sqrt{\log\log q^n}}\right) \leqslant c\,\frac{4}{\pi}\left(\frac{1}{n\log q}\right)^{1/(1-\epsilon)^2}.$$

Thus, $\sum_n \mathbb{P}(A_n) < \infty$, and the Borel–Cantelli lemma yields that

$$\mathbb{P}\left(\sup_{s\leqslant q^n} |B(s)| < (1-\epsilon)\,\frac{\pi}{\sqrt{8}}\sqrt{\frac{q^n}{\log\log q^n}} \quad \text{for infinitely many } n\right) = 0,$$

i. e. we have almost surely that

$$\varliminf_{n\to\infty} \frac{\sup_{s\leqslant q^n} |B(s)|}{\sqrt{q^n/\log\log q^n}} \geqslant (1-\epsilon)\,\frac{\pi}{\sqrt{8}}.$$

For $t \in [q^{n-1}, q^n]$ and large $n \geqslant 1$ we get

$$\frac{\sup_{s\leqslant t} |B(s)|}{\sqrt{t/\log\log t}} \geqslant \frac{\sup_{s\leqslant q^{n-1}} |B(s)|}{\sqrt{q^n/\log\log q^n}} = \frac{\sup_{s\leqslant q^{n-1}} |B(s)|}{\sqrt{q^{n-1}/\log\log q^{n-1}}} \underbrace{\frac{\sqrt{q^{n-1}/\log\log q^{n-1}}}{\sqrt{q^n/\log\log q^n}}}_{\to 1/\sqrt{q} \text{ as } n\to\infty}.$$

This proves

$$\varliminf_{t\to\infty} \frac{\sup_{s\leqslant t} |B(s)|}{\sqrt{t/\log\log t}} \geqslant \frac{1-\epsilon}{\sqrt{q}}\,\frac{\pi}{\sqrt{8}}$$

almost surely. Letting $\epsilon \to 0$ and $q \to 1$ along countable sequences, we get the lower bound in (12.8). □

If we replace in Theorem 12.7 $\log\log t$ by $\log|\log t|$, and if we use in the proof n^{-n} and $q < 1$ instead of n^n and $q > 1$, respectively, we get the analogue of (12.8) for $t \to 0$. Combining this with shifting $B(t+h) - B(h)$, we obtain

12.8 Corollary. *Let $(B(t))_{t\geqslant 0}$ be a BM^1. Then*

$$\varlimsup_{t\to 0} \frac{\sup_{s\leqslant t}|B(s)|}{\sqrt{t/\log|\log t|}} = \frac{\pi}{\sqrt{8}} \quad a.s.,$$

$$\varliminf_{t\to 0} \frac{\sup_{s\leqslant t}|B(s+h) - B(h)|}{\sqrt{t/\log|\log t|}} = \frac{\pi}{\sqrt{8}} \quad a.s.\ for\ all\ h \geqslant 0.$$

Using similar methods Csörgo and Révész [40, pp. 44–47] show that

$$\lim_{h\to 0} \inf_{s\leqslant 1-h} \sup_{t\leqslant h} \frac{|B(s+t) - B(s)|}{\sqrt{h/|\log h|}} = \frac{\pi}{\sqrt{8}}. \tag{12.9}$$

This is the exact *modulus of non-differentiability* of a Brownian motion.

12.9 Further reading. Path properties of the LIL-type are studied in [40], [79] and [188]. Apart from the original papers which are mentioned in Remark 12.3, exceptional sets are beautifully treated in [162]. **i**

[40] Csörgő, Révész: *Strong Approximations in Probability and Statistics.*
[79] Freedman: *Brownian Motion and Diffusion.*
[162] Mörters, Peres: *Brownian Motion.*
[188] Révész: *Random Walk in Random and Non-Random Environments.*

Problems

1. Let $(B_t)_{t\geqslant 0}$ be a BM^1. Use the Borel-Cantelli lemma to show that the running maximum $M_n := \sup_{0\leqslant t\leqslant n} B_t$ cannot grow faster than $C\sqrt{n\log n}$ for any $C > 2$. Use this to show that
$$\varlimsup_{t\to\infty} \frac{M_t}{C\sqrt{t\log t}} \leqslant 1 \quad \text{a.s.}$$
Hint: Show that $M_n(\omega) \leqslant C\sqrt{n\log n}$ for sufficiently large $n \geqslant n_0(\omega)$.

2. Let $(B_t)_{t\geqslant 0}$ be a BM^1. Apply Doob's maximal inequality (A.13) to the exponential martingale $M_t^\xi := e^{\xi B_t - \frac{1}{2}\xi^2 t}$ to show that for all $x, \xi > 0$
$$\mathbb{P}\left(\sup_{s\leqslant t}\left(B_s - \tfrac{1}{2}\xi s\right) > x\right) \leqslant e^{-x\xi}.$$
(This inequality can be used for the upper bound in the proof of Theorem 12.1, avoiding the combination of (12.2) and the upper bound in (12.1).)

3. Show that the proof of Khinchine's LIL, Theorem 12.1, can be modified to give

$$\overline{\lim_{t\to\infty}} \frac{\sup_{s\leqslant t} |B(s)|}{\sqrt{2t \log \log t}} \leqslant 1.$$

Hint: Use in Step 1° of the proof $\mathbb{P}\left(\sup_{s\leqslant t} |B(s)| \geqslant x\right) \leqslant 4\mathbb{P}(|B(t)| \geqslant x)$.

4. Let $(B_t)_{t\geqslant 0}$ be a BM^1. Use Theorem 12.5 to show that $\kappa(t) = (1+\epsilon)\sqrt{2t \log|\log t|}$ is an upper function for $t \to 0$.

5. Let $(B_t)_{t\geqslant 0}$ be a BM^1. Deduce from Theorem 12.5 the following test for upper functions in large time. Assume that $\kappa \in \mathcal{C}[1,\infty)$ is a positive function such that $\kappa(t)/t$ is decreasing and $\kappa(t)/\sqrt{t}$ is increasing. Then

$$\mathbb{P}\left(B(t) < \kappa(t) \text{ as } t \to \infty\right) = \begin{vmatrix} 0 \\ 1 \end{vmatrix} \quad \text{according to} \quad \int_1^\infty \frac{\kappa(s)}{s^{3/2}} e^{-\kappa^2(s)/2s}\,ds \begin{vmatrix} = \infty \\ < \infty \end{vmatrix}.$$

6. (Wentzell [226, p. 176]) Let $(B_t)_{t\geqslant 0}$ be a one-dimensional Brownian motion, $a, b > 0$ and $\tau := \inf\left\{t \geqslant 0 \,:\, |B_t| = b\sqrt{a+t}\right\}$. Show that
   (a) $\mathbb{P}(\tau < \infty) = 1$;
   (b) $\mathbb{E}\,\tau = \infty \quad (b \geqslant 1)$;
   (c) $\mathbb{E}\,\tau < \infty \quad (b < 1)$.

Hint: Use in b) Wald's identities. For c) show that $\mathbb{E}(\tau \wedge n) \leqslant ab^2(1-b^2)^{-1}$ for $n \geqslant 1$.

13 Strassen's functional law of the iterated logarithm

From Khintchine's law of the iterated logarithm (LIL), Corollary 12.2, we know that for a one-dimensional Brownian motion

$$-1 = \varliminf_{s\to\infty} \frac{B(s)}{\sqrt{2s\log\log s}} < \varlimsup_{s\to\infty} \frac{B(s)}{\sqrt{2s\log\log s}} = 1.$$

Since Brownian paths are a. s. continuous, it is clear that *the set of limit points of* $\{B(s)/\sqrt{2s\log\log s} \,:\, s > e\}$ *as* $s \to \infty$ *is the whole interval* $[-1, 1]$.

This observation is the starting point for an extension of Khintchine's LIL, due to Strassen [211], which characterizes the set of limit points as $s \to \infty$ of the family

$$\{Z_s(\cdot) \,:\, s > e\} \subset \mathcal{C}_{(o)}[0,1] \quad \text{where} \quad Z_s(t) := \frac{B(st)}{\sqrt{2s\log\log s}}, \quad 0 \leqslant t \leqslant 1.$$

($\mathcal{C}_{(o)}[0,1]$ denotes the set of all continuous functions $w \,:\, [0,1] \to \mathbb{R}$ with $w(0) = 0$.) If $w_0 \in \mathcal{C}_{(o)}[0,1]$ is a limit point it is necessary that for every fixed $t \in [0,1]$ the number Ex. 13.1 $w_0(t)$ is a limit point of the family $\{Z_s(t) \,:\, s > e\} \subset \mathbb{R}$. Using Khintchine's LIL and the scaling property 2.16 of Brownian motion it is straightforward that the limit points of $\{Z_s(t) \,:\, s > e\}$ are the whole interval $[-\sqrt{t}, \sqrt{t}]$, i. e. $-\sqrt{t} \leqslant w_0(t) \leqslant \sqrt{t}$. In fact, we have

13.1 Theorem (Strassen 1964). *Let* $(B_t)_{t\geqslant 0}$ *be a one-dimensional Brownian motion and set* $Z_s(t,\omega) := B(st,\omega)/\sqrt{2s\log\log s}$. *For almost all* $\omega \in \Omega$,

$$\{Z_s(\cdot,\omega) \,:\, s > e\}$$

is a relatively compact subset of $(\mathcal{C}_{(o)}[0,1], \|\cdot\|_\infty)$, *and the set of all limit points (as* $s \to \infty$*) is given by*

$$\mathcal{K} = \left\{ w \in \mathcal{C}_{(o)}[0,1] \,:\, w \text{ is absolutely continuous and } \int_0^1 |w'(t)|^2\,dt \leqslant 1 \right\}.$$

Using the Cauchy-Schwarz inequality it is not hard to see that every $w \in \mathcal{K}$ satisfies Ex. 13.2 $|w(t)| \leqslant \sqrt{t}$ for all $0 \leqslant t \leqslant 1$. Theorem 13.1 contains two statements: With probability one,

a) the family $\{[0,1] \ni t \mapsto B(st,\omega)/\sqrt{2s\log\log s} \,:\, s > e\} \subset \mathcal{C}_{(o)}[0,1]$ contains a sequence which converges uniformly for all $t \in [0,1]$ to a function in the set $\mathcal{K}$;
b) every function from the (non-random) set $\mathcal{K}$ is the uniform limit of a suitable sequence $\left(B(s_n t,\omega)/\sqrt{2s_n\log\log s_n}\right)_{n\geqslant 1}$.

For a limit point w_0 it is necessary that the probability

$$\mathbb{P}\left(\|Z_s(\cdot) - w_0(\cdot)\|_\infty < \delta\right) \quad \text{as} \quad s \to \infty$$

is sufficiently large for any $\delta > 0$. From the scaling property of a Brownian motion we see that

$$Z_s(\cdot) - w_0(\cdot) \sim \frac{B(\cdot)}{\sqrt{2\log\log s}} - w_0(\cdot).$$

This means that we have to calculate probabilities involving a Brownian motion with a nonlinear drift, $B(t) - \sqrt{2\log\log s}\, w_0(t)$ and estimate it at an exponential scale, as we will see later. Both topics are interesting in their own right, and we will treat them separately in the following two sections. After that we return to our original problem, the proof of Strassen's Theorem 13.1.

13.1 The Cameron–Martin formula

In Chapter 4 we have introduced the Wiener space $(\mathcal{C}_{(0)}, \mathscr{B}(\mathcal{C}_{(0)}), \mu)$ as canonical probability space of a Brownian motion indexed by $[0,\infty)$. In the same way we can study a Brownian motion with index set $[0,1]$ and the Wiener space $(\mathcal{C}_{(0)}[0,1], \mathscr{B}(\mathcal{C}_{(0)}[0,1]),\mu)$ where $\mathcal{C}_{(0)} = \mathcal{C}_{(0)}[0,1]$ denotes the space of continuous functions $w : [0,1] \to \mathbb{R}$ such that $w(0) = 0$. The Cameron–Martin theorem is about the absolute continuity of the Wiener measure under shifts in the space $\mathcal{C}_{(0)}[0,1]$. For $w_0 \in \mathcal{C}_{(0)}[0,1]$ we set

$$Uw(t) := w(t) + w_0(t), \quad t \in [0,1],\ w \in \mathcal{C}_{(0)}[0,1].$$

Clearly, U maps $\mathcal{C}_{(0)}[0,1]$ into itself and induces an image measure of the Wiener measure

$$\mu_U(A) = \mu(U^{-1}A) \quad \text{for all } A \in \mathscr{B}(\mathcal{C}_{(0)}[0,1]);$$

this is equivalent to saying

$$\int_{\mathcal{C}_{(0)}} F(w + w_0)\,\mu(dw) = \int_{\mathcal{C}_{(0)}} F(Uw)\,\mu(dw) = \int_{\mathcal{C}_{(0)}} F(w)\,\mu_U(dw) \tag{13.1}$$

for all bounded continuous functionals $F : \mathcal{C}_{(0)}[0,1] \to \mathbb{R}$.

We will see that μ_U is absolutely continuous with respect to μ if the shifts w_0 are absolutely continuous (a. c.) functions such that the (Lebesgue almost everywhere existing) derivative w_0' is square integrable. This is the *Cameron–Martin space* which we denote by

$$\mathcal{H}^1 := \left\{ u \in \mathcal{C}_{(0)}[0,1] \;:\; u \text{ is a. c. and } \int_0^1 |u'(t)|^2\,dt < \infty \right\}. \tag{13.2}$$

13.2 Lemma. *For $u : [0,1] \to \mathbb{R}$ the following assertions are equivalent:*

a) $u \in \mathcal{H}^1$;

b) $u \in \mathcal{C}_{(0)}[0,1]$ *and* $M := \sup_\Pi \sum_{t_{j-1},t_j \in \Pi} |u(t_j) - u(t_{j-1})|^2/(t_j - t_{j-1}) < \infty$ *where the supremum is taken over all finite partitions* $\Pi = \{0 = t_0 < t_1 < \cdots < t_n = 1\}$ *of* $[0,1]$.

If a) or b) is true, then $\int_0^1 |u'(t)|^2\,dt = \lim_{|\Pi|\to 0} \sum_{t_{j-1},t_j\in\Pi} |u(t_j) - u(t_{j-1})|^2/(t_j - t_{j-1})$.

Proof. a)⇒ b): Assume that $u \in \mathcal{H}^1$ and let $0 = t_0 < t_1 < \cdots < t_n = 1$ be any partition of $[0, 1]$. Then

$$\sum_{j=1}^{n} \left[\frac{u(t_j) - u(t_{j-1})}{t_j - t_{j-1}} \right]^2 (t_j - t_{j-1}) = \int_0^1 f_n(t)\,dt$$

with

$$f_n(t) = \sum_{j=1}^{n} \left[\frac{u(t_j) - u(t_{j-1})}{t_j - t_{j-1}} \right]^2 \mathbb{1}_{[t_{j-1}, t_j)}(t).$$

By the Cauchy–Schwarz inequality

$$f_n(t) = \sum_{j=1}^{n} \left[\frac{1}{t_j - t_{j-1}} \int_{t_{j-1}}^{t_j} u'(s)\,ds \right]^2 \mathbb{1}_{[t_{j-1}, t_j)}(t) \leqslant \sum_{j=1}^{n} \frac{1}{t_j - t_{j-1}} \int_{t_{j-1}}^{t_j} |u'(s)|^2\,ds\, \mathbb{1}_{[t_{j-1}, t_j)}(t).$$

If we integrate this inequality with respect to t, we get

$$\int_0^1 f_n(t)\,dt \leqslant \sum_{j=1}^{n} \int_{t_{j-1}}^{t_j} |u'(s)|^2\,ds = \int_0^1 |u'(s)|^2\,ds,$$

hence $M \leqslant \int_0^1 |u'(s)|^2\,ds < \infty$, and the estimate in b) follows.

b)⇒ a): Conversely, assume that b) holds and denote by M the value of the supremum in b). Let $0 \leqslant r_1 < s_1 \leqslant r_2 < s_2 \leqslant \cdots \leqslant r_m < s_m \leqslant 1$ be the endpoints of the intervals (r_k, s_k), $k = 1, \ldots, m$. By the Cauchy-Schwarz inequality

$$\sum_{k=1}^{m} |u(s_k) - u(r_k)| \leqslant \left[\sum_{k=1}^{m} \frac{|u(s_k) - u(r_k)|^2}{s_k - r_k} \right]^{\frac{1}{2}} \left[\sum_{k=1}^{m} (s_k - r_k) \right]^{\frac{1}{2}} \leqslant \sqrt{M} \left[\sum_{k=1}^{m} (s_k - r_k) \right]^{\frac{1}{2}}.$$

This means that u is absolutely continuous and u' exists a. e. Let f_n be as in the first part of the proof and assume that the underlying partitions $\{0 = t_0 < t_1 < \cdots < t_n = 1\}$ have decreasing mesh. Then the sequence $(f_n)_{n\geqslant 1}$ converges to $(u')^2$ at all points where u' exists. By Fatou's lemma Ex. 13.3

$$\int_0^1 |u'(t)|^2\,dt \leqslant \varliminf_{n\to\infty} \int_0^1 f_n(t)\,dt = \varliminf_{n\to\infty} \sum_{j=1}^{n} \frac{|u(t_j) - u(t_{j-1})|^2}{t_j - t_{j-1}} \leqslant M.$$

Finally,

$$\varlimsup_{n\to\infty} \int_0^1 f_n(t)\,dt \leqslant M \leqslant \int_0^1 |u'(t)|^2\,dt \leqslant \varliminf_{n\to\infty} \int_0^1 f_n(t)\,dt.$$ □

In a first step we prove the absolute continuity of μ_U for shifts from a subset $\mathcal{H}^1_\circ$ of $\mathcal{H}^1$,

$$\mathcal{H}^1_\circ := \left\{ u \in \mathcal{C}_{(0)}[0,1] \;:\; u \in \mathcal{C}^1[0,1] \text{ and } u' \text{ has bounded variation} \right\}. \tag{13.3}$$

13.3 Lemma. *Let $w_0 : [0,1] \to \mathbb{R}$ be a function in $\mathcal{H}^1_\circ$, $n \geqslant 1$ and $t_j = j/n$. Then the linear functionals*

$$G_n(w) := \sum_{j=1}^{n} (w(t_j) - w(t_{j-1})) \frac{w_0(t_j) - w_0(t_{j-1})}{t_j - t_{j-1}}, \quad n \geqslant 1,\ w \in \mathcal{C}_{(o)}[0,1],$$

are uniformly bounded, i. e. $|G_n(w)| \leqslant c\|w\|_\infty$ for all $n \geqslant 1$ and $w \in \mathcal{C}_{(o)}[0,1]$; moreover, the following Riemann–Stieltjes integral exists:

$$G(w) := \lim_{n\to\infty} G_n(w) = \int_0^1 w_0'(t)\,dw(t).$$

Proof. The mean-value theorem shows that

$$G_n(w) = \sum_{j=1}^{n} (w(t_j) - w(t_{j-1})) \frac{w_0(t_j) - w_0(t_{j-1})}{t_j - t_{j-1}} = \sum_{j=1}^{n} (w(t_j) - w(t_{j-1}))\, w_0'(\theta_j)$$

with suitable intermediate points $\theta_j \in [t_{j-1}, t_j]$. It is not hard to see that

$$G_n(w) = w_0'(\theta_{n+1})w(t_n) - \sum_{j=1}^{n} w(t_j)(w_0'(\theta_{j+1}) - w_0'(\theta_j)) \qquad (\theta_{n+1} := 1)$$

From this we conclude that

$$|G_n(w)| \leqslant (\|w_0'\|_\infty + \mathrm{VAR}_1(w_0'; 1))\|w\|_\infty$$

as well as

$$\lim_{n\to\infty} G_n(w) = w_0'(1)w(1) - \int_0^1 w(t)\,dw_0'(t) = \int_0^1 w_0'(t)\,dw(t). \qquad \square$$

We can now state and prove the Cameron–Martin formula for shifts in $\mathcal{H}^1_\circ$.

13.4 Theorem (Cameron, Martin 1944). *Let $(\mathcal{C}_{(o)}[0,1], \mathscr{B}(\mathcal{C}_{(o)}[0,1]), \mu)$ be the Wiener space and $Uw := w + w_0$ the shift by $w_0 \in \mathcal{H}^1_\circ$. Then the image measure $\mu_U(A) := \mu(U^{-1}A)$, $A \in \mathscr{B}(\mathcal{C}_{(o)}[0,1])$, is absolutely continuous with respect to μ, and for every bounded continuous functional $F : \mathcal{C}_{(o)}[0,1] \to \mathbb{R}$ we have*

$$\begin{aligned}\int_{\mathcal{C}_{(o)}[0,1]} F(w)\,\mu_U(dw) &= \int_{\mathcal{C}_{(o)}[0,1]} F(Uw)\,\mu(dw) \\ &= \int_{\mathcal{C}_{(o)}[0,1]} F(w)\,\frac{d\mu_U(w)}{d\mu}\,\mu(dw)\end{aligned} \tag{13.4}$$

with the Radon-Nikodým density

$$\frac{d\mu_U(w)}{d\mu} = \exp\left(-\frac{1}{2}\int_0^1 |w_0'(t)|^2\,dt + \int_0^1 w_0'(t)\,dw(t)\right). \tag{13.5}$$

Proof. In order to deal with the infinite dimensional integrals appearing in (13.4) we use finite dimensional approximations. Set $t_j = j/n$, $j = 0, 1, \dots, n$, $n \geqslant 1$, and denote by

$$\Pi_n w(t) := \begin{cases} w(t_j), & \text{if } t = t_j,\ j = 0, 1, \dots, n \\ \text{linear}, & \text{for all other } t \end{cases}$$

the piecewise linear approximation of w on the grid $t_0, t_1, \dots, t_n$. Obviously, we have $\Pi_n w \in \mathcal{C}_{(0)}[0, 1]$ and $\lim_{n\to\infty} \Pi_n w(t) = w(t)$ uniformly for all t.

Let $F : \mathcal{C}_{(0)}[0, 1] \to \mathbb{R}$ be a bounded continuous functional. Since $\Pi_n w$ is uniquely determined by $x = (w(t_1), \dots, w(t_n)) \in \mathbb{R}^n$, there exists a bounded continuous function $F_n : \mathbb{R}^n \to \mathbb{R}$ such that

$$F_n(x) = F(\Pi_n w) \quad \text{where} \quad x_j = w(t_j).$$

Write $x^0 = (w_0(t_1), \dots, w_0(t_n))$ and $x_0 = x_0^0 := 0$. Then

$$\begin{aligned}
&\int\limits_{\mathcal{C}_{(0)}[0,1]} F(\Pi_n U w)\,\mu(dw) \\
&\quad = \mathbb{E}\left[F_n(B(t_1) + w_0(t_1), \dots, B(t_n) + w_0(t_n))\right] \\
&\quad \overset{(2.10a)}{=} \frac{1}{(2\pi/n)^{n/2}} \int\limits_{\mathbb{R}^n} F_n(x + x^0)\, e^{-\frac{1}{2} n \sum_{j=1}^n (x_j - x_{j-1})^2}\, dx \\
&\quad = \frac{1}{(2\pi/n)^{n/2}} \int\limits_{\mathbb{R}^n} F_n(y)\, e^{-\frac{1}{2} n \sum_{j=1}^n ((y_j - y_{j-1}) - (x_j^0 - x_{j-1}^0))^2}\, dy \\
&\quad = \frac{e^{-\frac{1}{2} n \sum_{j=1}^n (x_j^0 - x_{j-1}^0)^2}}{(2\pi/n)^{n/2}} \int\limits_{\mathbb{R}^n} F_n(y)\, e^{n \sum_{j=1}^n (y_j - y_{j-1})(x_j^0 - x_{j-1}^0)}\, e^{-\frac{1}{2} n \sum_{j=1}^n (y_j - y_{j-1})^2}\, dy.
\end{aligned}$$

If we set

$$\begin{aligned}
G_n(w) &:= \sum_{j=1}^n (w(t_j) - w(t_{j-1}))\, \frac{w_0(t_j) - w_0(t_{j-1})}{\frac{1}{n}}, \\
\alpha_n(w_0) &:= \sum_{j=1}^n \frac{(w_0(t_j) - w_0(t_{j-1}))^2}{\frac{1}{n}},
\end{aligned}$$

we find

$$\int\limits_{\mathcal{C}_{(0)}[0,1]} F(\Pi_n U w)\,\mu(dw) = e^{-\frac{1}{2}\alpha_n(w_0)} \int\limits_{\mathcal{C}_{(0)}[0,1]} F(\Pi_n w) e^{G_n(w)}\,\mu(dw). \tag{13.6}$$

Since F is bounded and continuous and $\lim_{n\to\infty} \Pi_n U w = U w$, the left-hand side converges to $\int_{\mathcal{C}_{(0)}[0,1]} F(Uw)\,\mu(dw)$. By construction, $t_j - t_{j-1} = \frac{1}{n}$; therefore, Lemmas 13.2 and 13.3 show that

$$\lim_{n\to\infty} \alpha_n(w_0) = \int_0^1 |w_0'(t)|^2\, dt \quad \text{and} \quad \lim_{n\to\infty} G_n(w) = \int_0^1 w_0'(t)\, dw(t).$$

Moreover, by Lemma 13.3,

$$e^{G_n(w)} \leqslant e^{c\|w\|_\infty} \leqslant e^{c \sup_{t\leqslant 1} w(t)} + e^{-c \inf_{t\leqslant 1} w(t)}.$$

By the reflection principle, Theorem 6.9, $\sup_{t\leqslant 1} B(t) \sim -\inf_{t\leqslant 1} B(t) \sim |B(1)|$. Thus

$$\int_{\mathcal{C}_{(o)}[0,1]} e^{G_n(w)} \mu(dw) \leqslant \mathbb{E}\left[e^{c \sup_{t\leqslant 1} B(t)}\right] + \mathbb{E}\left[e^{-c \inf_{t\leqslant 1} B(t)}\right] = 2\,\mathbb{E}\left[e^{c|B(1)|}\right].$$

Therefore, we can use the dominated convergence theorem to deduce that the right-hand side of (13.6) converges to

$$e^{-\frac{1}{2}\int_0^1 |w_0'(t)|^2\,dt} \int_{\mathcal{C}_{(o)}[0,1]} F(w)\, e^{\int_0^1 w_0'(t)\,dw(t)}\, \mu(dw). \qquad \square$$

We want to show that Theorem 13.4 remains valid for $w_0 \in \mathcal{H}^1$. This is indeed possible, if we use a stochastic approximation technique. The drawback, however, will be that we do not have any more an explicit formula for the limit $\lim_{n\to\infty} G_n(w)$ as a Riemann–Stieltjes integral. The trouble is that the integral

$$\int_0^1 w_0(t)\,dw(t) = w(1)w_0(1) - \int_0^1 w(t)\,dw_0(t)$$

is for *all* integrands $w \in \mathcal{C}_{(o)}[0,1]$ only defined in the Riemann–Stieltjes sense, if w_0 is of bounded variation, cf. Corollary A.41.

13.5 Paley–Wiener–Zygmund integrals (Paley, Wiener, Zygmund 1933). Denote by $\mathsf{BV}[0,1]$ the set of all functions of bounded variation on $[0,1]$. For $\phi \in \mathsf{BV}[0,1]$ the Riemann–Stieltjes integral

$$G^\phi(w) = \int_0^1 \phi(s)\,dw(s) = \phi(1)w(1) - \int_0^1 w(s)\,d\phi(s), \qquad w \in \mathcal{C}_{(o)}[0,1],$$

is well-defined. If we write G^ϕ as the limit of a Riemann–Stieltjes sum, it is not hard to
Ex. 13.4 see that, on the Wiener space $(\mathcal{C}_{(o)}[0,1], \mathscr{B}(\mathcal{C}_{(o)}[0,1]), \mu)$,

$$G^\phi \text{ is a normal random variable, mean } 0\text{, variance } \int_0^1 \phi^2(s)\,ds; \tag{13.7a}$$

$$\int_{\mathcal{C}_{(o)}[0,1]} G^\phi(w) G^\psi(w)\,\mu(dw) = \int_0^1 \phi(s)\psi(s)\,ds \text{ if } \phi,\psi \in \mathsf{BV}[0,1]. \tag{13.7b}$$

Therefore, the map

$$L^2[0,1] \supset \mathsf{BV}[0,1] \ni \phi \mapsto G^\phi \in L^2(\mathcal{C}_{(o)}[0,1], \mathscr{B}(\mathcal{C}_{(o)}[0,1]), \mu)$$

is a linear isometry. As $\mathsf{BV}[0,1]$ is a dense subset of $L^2[0,1]$, we find for every function $\phi \in L^2[0,1]$ a sequence $(\phi_n)_{n\geqslant 1} \subset \mathsf{BV}[0,1]$ such that $L^2\text{-}\lim_{n\to\infty} \phi_n = \phi$. In particular, $(\phi_n)_{n\geqslant 1}$ is a Cauchy sequence in $L^2[0,1]$ and, therefore, $(G^{\phi_n})_{n\geqslant 1}$ is a Cauchy sequence in $L^2(\mathcal{C}_{(0)}[0,1], \mathscr{B}(\mathcal{C}_{(0)}[0,1]), \mu)$. Thus,

$$G^{\phi}(w) := \int_0^1 \phi(s)\,dw(s) := L^2\text{-}\lim_{n\to\infty} G^{\phi_n}(w), \qquad w \in \mathcal{C}_{(0)}[0,1]$$

defines a linear functional on $\mathcal{C}_{(0)}[0,1]$. Note that the integral notation is *symbolical* and has *no meaning as a Riemann–Stieltjes integral* unless $\phi \in \mathsf{BV}[0,1]$. Obviously the properties (13.7) are preserved under L^2 limits. The functional G^{ϕ} is often called the *Paley–Wiener–Zygmund integral*. It is a special case of the Itô integral which we will encounter in Chapter 15. Ex. 13.5 Ex. 13.6 Ex. 15.13

13.6 Corollary (Cameron, Martin 1944). *Let $(\mathcal{C}_{(0)}[0,1], \mathscr{B}(\mathcal{C}_{(0)}[0,1]), \mu)$ be the Wiener space and $Uw := w + w_0$ the shift by $w_0 \in \mathcal{H}^1$. Then $\mu_U(A) := \mu(U^{-1}A)$, $A \in \mathscr{B}(\mathcal{C}_{(0)}[0,1])$ is absolutely continuous with respect to μ and for every bounded Borel measurable functional $F : \mathcal{C}_{(0)}[0,1] \to \mathbb{R}$ we have*

$$\begin{aligned}\int_{\mathcal{C}_{(0)}[0,1]} F(w)\,\mu_U(dw) &= \int_{\mathcal{C}_{(0)}[0,1]} F(Uw)\,\mu(dw)\\ &= \int_{\mathcal{C}_{(0)}[0,1]} F(w)\,\frac{d\mu_U(w)}{d\mu}\,\mu(dw)\end{aligned} \tag{13.8}$$

with the Radon-Nikodým density

$$\frac{d\mu_U(w)}{d\mu} = \exp\left(-\frac{1}{2}\int_0^1 |w_0'(t)|^2\,dt + \int_0^1 w_0'(t)\,dw(t)\right) \tag{13.9}$$

where the second integral is a Paley–Wiener–Zygmund integral.

Proof. Let $w_0 \in \mathcal{H}^1$ and $(\phi_n)_{n\geqslant 0}$ be a sequence in $\mathcal{H}^1_\circ$ with $L^2\text{-}\lim_{n\to\infty} \phi_n' = w_0'$. If F is a continuous functional, Theorem 13.4 shows for all $n \geqslant 1$

$$\int_{\mathcal{C}_{(0)}[0,1]} F(w+\phi_n)\,\mu(dw) = \int_{\mathcal{C}_{(0)}[0,1]} F(w)\,\exp\left(-\frac{1}{2}\int_0^1 |\phi_n'(t)|^2\,dt + \int_0^1 \phi_n'(t)\,dw(t)\right)\mu(dw).$$

It is, therefore, enough to show that the right-hand side converges as $n \to \infty$. Set

$$G(w) := \int_0^1 w_0'(s)\,dw(s) \quad\text{and}\quad G_n(w) := \int_0^1 \phi_n'(s)\,dw(s).$$

Since G_n and G are normal random variables on the Wiener space, we find for all $c > 0$

$$\begin{aligned}
&\int\limits_{\mathcal{C}_{(o)}[0,1]} \left|e^{G_n(w)} - e^{G(w)}\right| \mu(dw) \\
&\quad = \int\limits_{\mathcal{C}_{(o)}[0,1]} \left|e^{G_n(w)-G(w)} - 1\right| e^{G(w)}\, \mu(dw) \\
&\quad = \Bigg(\int\limits_{\{|G|>c\}} + \int\limits_{\{|G|\leqslant c\}} \Bigg) \left|e^{G_n(w)-G(w)} - 1\right| e^{G(w)}\, \mu(dw) \\
&\quad \leqslant \Bigg(\int\limits_{\mathcal{C}_{(o)}[0,1]} \left|e^{G_n(w)-G(w)} - 1\right|^2 \mu(dw) \Bigg)^{\frac{1}{2}} \Bigg(\int\limits_{\{|G|>c\}} e^{2G(w)}\, \mu(dw) \Bigg)^{\frac{1}{2}} \\
&\qquad + e^c \int\limits_{\mathcal{C}_{(o)}[0,1]} \left|e^{G_n(w)-G(w)} - 1\right| \mu(dw) \\
&\quad \leqslant C \Bigg(\int\limits_{|y|>c} e^{2y}\, e^{-y^2/(2a)}\, dy \Bigg)^{\frac{1}{2}} + \frac{e^c}{\sqrt{2\pi}} \int\limits_{\mathbb{R}} \left|e^{\sigma_n y} - 1\right| e^{-y^2/2}\, dy
\end{aligned}$$

where $\sigma_n^2 = \int_0^1 |\phi_n'(s) - w_0'(s)|^2\, ds$ and $a = \int_0^1 |w_0'(s)|^2\, ds$. Thus,

$$\lim_{n\to\infty} \int\limits_{\mathcal{C}_{(o)}[0,1]} \left|e^{G_n(w)} - e^{G(w)}\right| \mu(dw) = 0.$$

Using standard approximation arguments from measure theory, the formula (13.8) is easily extended from continuous F to all bounded measurable F. □

13.7 Remark. It is useful to rewrite Corollary 13.6 in the language of stochastic processes. Let $(B_t)_{t\geqslant 0}$ be a BM^1 and define a new stochastic process by

$$X_t(\omega) := U^{-1}B_t(\omega) = B_t(\omega) - w_0(t), \quad w_0 \in \mathcal{H}^1.$$

Then X_t is a Brownian motion under the new probability measure

$$\mathbb{Q}(d\omega) = \exp\Bigg(-\frac{1}{2} \int\limits_0^1 |w_0'(s)|^2\, ds + \int\limits_0^1 w_0'(s)\, dB_s(\omega) \Bigg) \mathbb{P}(d\omega).$$

This is a deterministic version of the Girsanov transformation, cf. Section 18.3.

In fact, the converse of Corollary 13.6 is also true.

13.8 Theorem. *Let* $(\mathcal{C}_{(o)}[0,1], \mathscr{B}(\mathcal{C}_{(o)}[0,1]), \mu)$ *be the Wiener space and* $Uw := w + h$ *the shift by* $h \in \mathcal{C}_{(o)}[0,1]$. *If the image measure* $\mu_U(A) := \mu(U^{-1}A)$, $A \in \mathscr{B}(\mathcal{C}_{(o)}[0,1])$, *is absolutely continuous with respect to* μ, *then* $h \in \mathcal{H}^1$.

Proof. Assume that $h \notin \mathcal{H}^1$. Then there exists a sequence of functions $(\phi_n)_{n\geqslant 1} \subset \mathcal{H}^1_\circ$ such that

$$\|\phi_n\|^2_{\mathcal{H}^1} = \int_0^1 |\phi_n'(s)|^2\,ds = 1 \quad\text{and}\quad \langle \phi_n, h\rangle_{\mathcal{H}^1} = \int_0^1 \phi_n'(s)\,dh(s) \geqslant 2n$$ Ex. 13.5

where $\langle \phi_n, h\rangle_{\mathcal{H}^1} = \int_0^1 \phi_n'(s)\,dh(s)$ is a Paley–Wiener–Zygmund integral $G^{\phi_n'}(h)$.

This follows easily from the well-known formula for the norm in a Hilbert space $\|h\|_{\mathcal{H}^1} = \sup_{\phi\in\mathcal{H}^1_\circ,\ \|\phi\|_{\mathcal{H}^1}=1} \langle \phi, h\rangle_{\mathcal{H}^1}$ since $\mathcal{H}^1_\circ$ is a dense subset of $\mathcal{H}^1$.

We will construct a set $A \in \mathscr{B}(\mathcal{C}_{(0)}[0,1])$ such that $\mu(A) = 1$ and $\mu_U(A) = 0$, i. e. μ and μ_U are singular. Set

$$A_n := \left\{ w \in \mathcal{C}_{(0)}[0,1] \;:\; \int_0^1 \phi_n'(s)\,dw_s \leqslant n \right\} \quad\text{and}\quad A := \bigcap_{n=1}^\infty \bigcup_{k=n}^\infty A_k.$$

By 13.5, the Paley–Wiener–Zygmund integral $G^{\phi_n'}(w) = \int_0^1 \phi_n'(s)\,dw_s$ is a normal random variable with mean zero and variance one on the Wiener space. Thus, $\mu(A_n) \to 1$ as $n \to \infty$. Therefore,

$$\mu(A) = \lim_{n\to\infty} \mu\left(\bigcup\nolimits_{k\geqslant n} A_k\right) \geqslant \lim_{n\to\infty} \mu(A_n) = 1.$$

On the other hand,

$$\begin{aligned}
\mu_U(A_n) &= \mu\left\{ w \;:\; \int_0^1 \phi_n'(s)\,d(w+h)_s \leqslant n \right\} \\
&= \mu\left\{ w \;:\; \int_0^1 \phi_n'(s)\,dw_s \leqslant \underbrace{n - \int_0^1 \phi_n'(s)\,dh_s}_{\leqslant n-2n=-n} \right\} \\
&\leqslant \mu\left\{ w \;:\; -\int_0^1 \phi_n'(s)\,dw_s \geqslant n \right\} \\
&\leqslant \frac{1}{n^2}.
\end{aligned}$$

In the last step we used again that $G^{\phi_n'}(w) = \int_0^1 \phi_n'(s)\,dw_s$ has a normal distribution under μ with mean zero and variance one, as well as the Markov inequality. Thus,

$$\mu_U(A) = \lim_{n\to\infty} \mu_U\left(\bigcup\nolimits_{k\geqslant n} A_k\right) \leqslant \lim_{n\to\infty} \sum_{k\geqslant n} \mu_U(A_k) \leqslant \lim_{n\to\infty} \sum_{k\geqslant n} \frac{1}{k^2} = 0.$$ □

13.2 Large deviations (Schilder's theorem)

The term *large deviations* generally refers to small probabilities on an exponential scale. We are interested in lower and upper (exponential) estimates of the probability that *the whole Brownian path* is in the neighbourhood of a deterministic function $u \neq 0$. Estimates of this type have been studied by Schilder and were further developed by Borovkov, Varadhan, Freidlin and Wentzell; our presentation follows the monograph [81] by Freidlin and Wentzell.

13.9 Definition. Let $u \in \mathcal{C}_{(o)}[0,1]$. The functional $I : \mathcal{C}_{(o)}[0,1] \to [0,\infty]$,

$$I(u) := \begin{cases} \frac{1}{2}\int_0^1 |u'(s)|^2\,ds, & \text{if } u \text{ is absolutely continuous} \\ +\infty, & \text{otherwise} \end{cases}$$

is the *action functional* of the Brownian motion $(B_t)_{t\in[0,1]}$.

The action functional I will be the exponent in the large deviation estimates for B_t. For this we need a few properties.

13.10 Lemma. *The action functional $I : \mathcal{C}_{(o)}[0,1] \to [0,\infty]$ is lower semicontinuous, i. e. for every sequence $(w_n)_{n\geqslant 1} \subset \mathcal{C}_{(o)}[0,1]$ which converges uniformly to $w \in \mathcal{C}_{(o)}[0,1]$ we have $\underline{\lim}_{n\to\infty} I(w_n) \geqslant I(w)$.*

Proof. Let $(w_n)_{n\geqslant 1}$ and w be as in the statement of the lemma. We may assume that $\underline{\lim}_{n\to\infty} I(w_n) < \infty$. For any finite partition $\Pi = \{0 = t_0 < t_1 < \cdots < t_m = 1\}$ of $[0,1]$ we have

$$\begin{aligned}\sup_{\Pi}\sum_{t_j\in\Pi}\frac{|w(t_j)-w(t_{j-1})|^2}{t_j-t_{j-1}} &= \sup_{\Pi}\lim_{n\to\infty}\sum_{t_j\in\Pi}\frac{|w_n(t_j)-w_n(t_{j-1})|^2}{t_j-t_{j-1}} \\ &\leqslant \underline{\lim}_{n\to\infty}\sup_{\Pi}\sum_{t_j\in\Pi}\frac{|w_n(t_j)-w_n(t_{j-1})|^2}{t_j-t_{j-1}}.\end{aligned}$$

Using Lemma 13.2, we see that $I(w) \leqslant \underline{\lim}_{n\to\infty} I(w_n)$. □

As usual, we write

$$d(w,A) := \inf\{\|w-u\|_\infty : u \in A\}, \quad w \in \mathcal{C}_{(o)}[0,1] \quad \text{and} \quad A \subset \mathcal{C}_{(o)}[0,1]$$

for the distance of the point w to the set A. Denote by

$$\Phi(r) := \{w \in \mathcal{C}_{(o)}[0,1] : I(w) \leqslant r\}, \quad r \geqslant 0,$$

the *sub-level sets* of the action functional I.

13.11 Lemma. *The sub-level sets $\Phi(r)$, $r \geqslant 0$, of the action functional I are compact subsets of $\mathcal{C}_{(o)}[0,1]$.*

Proof. Since I is lower semicontinuous, the sub-level sets are closed. For any $r > 0$, $0 \leqslant s < t \leqslant 1$ and $w \in \Phi(r)$ we find using the Cauchy-Schwarz inequality

$$\begin{aligned}|w(t)-w(s)| = \left|\int_s^t w'(x)\,dx\right| &\leqslant \left(\int_s^t |w'(x)|^2\,dx\right)^{\frac{1}{2}}\sqrt{t-s} \\ &\leqslant \sqrt{2I(w)}\,\sqrt{t-s} \leqslant \sqrt{2r}\,\sqrt{t-s}.\end{aligned}$$

This shows that the family $\Phi(r)$ is equi-bounded and equicontinuous. By Ascoli's theorem, $\Phi(r)$ is compact. □

We are now ready for the upper large deviation estimate.

13.12 Theorem (Schilder 1966). *Let $(B_t)_{t\in[0,1]}$ be a BM^1 and denote by $\Phi(r)$ the sub-level sets of the action functional I. For every $\delta > 0$, $\gamma > 0$ and $r_0 > 0$ there exists some $\epsilon_0 = \epsilon_0(\delta, \gamma, r_0)$ such that*

$$\mathbb{P}\big(d(\epsilon B, \Phi(r)) \geqslant \delta\big) \leqslant \exp\left[-\frac{r-\gamma}{\epsilon^2}\right] \tag{13.10}$$

for all $0 < \epsilon \leqslant \epsilon_0$ and $0 \leqslant r \leqslant r_0$.

Proof. Set $t_j = j/n$, $j = 0, 1, \dots, n$, $n \geqslant 1$, and denote by

$$\Pi_n[\epsilon B](t, \omega) := \begin{cases} \epsilon B(t_j, \omega), & \text{if } t = t_j,\ j = 0, 1, \dots, n, \\ \text{linear}, & \text{for all other } t, \end{cases}$$

the piecewise linear approximation of the continuous function $\epsilon B_t(\omega)$ on the grid $t_0, t_1, \dots, t_n$. Obviously, for all $t \in [t_{k-1}, t_k]$ and $k = 1, \dots, n$,

$$\Pi_n[\epsilon B](t) = \epsilon B(t_{k-1}) + (t - t_{k-1}) \frac{\epsilon B(t_k) - \epsilon B(t_{k-1})}{t_k - t_{k-1}}.$$

Since the approximations $\Pi_n[\epsilon B](t)$ are absolutely continuous, we can calculate the action functional

$$I(\Pi_n[\epsilon B]) = \frac{1}{2} \sum_{k=1}^{n} n\epsilon^2 (B(t_k) - B(t_{k-1}))^2 = \frac{\epsilon^2}{2} \sum_{k=1}^{n} \xi_k^2 \tag{13.11}$$

where $\xi_k = \sqrt{n}(B(t_k) - B(t_{k-1})) \sim \mathsf{N}(0,1)$ are iid standard normal random variables. Note that

$$\begin{aligned}
\{d(\epsilon B, \Phi(r)) \geqslant \delta\} &= \{d(\epsilon B, \Phi(r)) \geqslant \delta,\ d(\Pi_n[\epsilon B], \epsilon B) < \delta\} \\
&\quad \cup \{d(\epsilon B, \Phi(r)) \geqslant \delta,\ d(\Pi_n[\epsilon B], \epsilon B) \geqslant \delta\} \\
&\subset \underbrace{\{\Pi_n[\epsilon B] \notin \Phi(r)\}}_{=:A_n} \cup \underbrace{\{d(\Pi_n[\epsilon B], \epsilon B) \geqslant \delta\}}_{=:C_n}.
\end{aligned}$$

We will estimate $\mathbb{P}(A_n)$ and $\mathbb{P}(C_n)$ separately. Combining (13.11) with Chebychev's inequality gives for all $0 < \alpha < 1$

$$\begin{aligned}
\mathbb{P}(A_n) = \mathbb{P}\big(I(\Pi_n[\epsilon B]) > r\big) &= \mathbb{P}\left(\frac{1}{2}\sum_{k=1}^{n} \xi_k^2 > \frac{r}{\epsilon^2}\right) \\
&= \mathbb{P}\left(\exp\left[\frac{1-\alpha}{2}\sum_{k=1}^{n} \xi_k^2\right] > \exp\left[\frac{r}{\epsilon^2}(1-\alpha)\right]\right) \\
&\overset{\text{iid}}{\leqslant} \exp\left[-\frac{r}{\epsilon^2}(1-\alpha)\right] \underbrace{\left(\mathbb{E}\exp\left[\frac{1-\alpha}{2}\xi_1^2\right]\right)^n}_{=e^{n(1-\alpha)^2/8} \text{ by (2.6)}}.
\end{aligned}$$

Therefore, we find for all $n \geqslant 1$, $r_0 > 0$ and $\gamma > 0$ some $\epsilon_0 > 0$ such that

$$\mathbb{P}(A_n) \leqslant \frac{1}{2} \exp\left[-\frac{r}{\epsilon^2} + \frac{\gamma}{\epsilon^2}\right] \quad \text{for all } 0 < \epsilon \leqslant \epsilon_0,\ 0 \leqslant r \leqslant r_0.$$

In order to estimate $\mathbb{P}(C_n)$ we use the fact that Brownian motion has stationary increments.

$$\begin{aligned}
\mathbb{P}(C_n) &= \mathbb{P}\left(\sup_{0\leqslant t\leqslant 1} |\epsilon B(t) - \Pi_n[\epsilon B](t)| \geqslant \delta\right)\\
&\leqslant \sum_{k=1}^{n} \mathbb{P}\left(\sup_{t_{k-1}\leqslant t\leqslant t_k} \left|\epsilon B(t) - \epsilon B(t_{k-1}) - (t-t_{k-1})n\epsilon(B(t_k)-B(t_{k-1}))\right| \geqslant \delta\right)\\
&= n\mathbb{P}\left(\sup_{t\leqslant 1/n} \left|\epsilon B(t) - n\epsilon t B(1/n)\right| \geqslant \delta\right)\\
&\leqslant n\mathbb{P}\left(\sup_{t\leqslant 1/n} |B(t)| + |B(1/n)| \geqslant \frac{\delta}{\epsilon}\right)\\
&\leqslant n\mathbb{P}\left(\sup_{t\leqslant 1/n} |B(t)| \geqslant \frac{\delta}{2\epsilon}\right)\\
&\leqslant 2n\mathbb{P}\left(\sup_{t\leqslant 1/n} B(t) \geqslant \frac{\delta}{2\epsilon}\right).
\end{aligned}$$

We can now use the following maximal estimate which is, for example, a consequence of the reflection principle, Theorem 6.9, and the Gaussian tail estimate from Lemma 10.5

$$\mathbb{P}\left(\sup_{s\leqslant t} B(s) \geqslant x\right) \leqslant 2\mathbb{P}(B(t) \geqslant x) \leqslant \frac{2}{\sqrt{2\pi t}}\,\frac{1}{x}\, e^{-x^2/(2t)}, \qquad x > 0,\ t > 0,$$

to deduce

$$\mathbb{P}(C_n) \leqslant 4n\mathbb{P}\left(B(1/n) \geqslant \frac{\delta}{2\epsilon}\right) \leqslant 4n \frac{2\epsilon}{\sqrt{2\pi/n}\,\delta} \exp\left[-\frac{\delta^2 n}{8\epsilon^2}\right].$$

Finally, pick $n \geqslant 1$ such that $\delta^2 n/8 > r_0$ and ϵ_0 small enough such that $\mathbb{P}(C_n)$ is less than $\frac{1}{2}\exp\left[-\frac{r}{\epsilon^2} + \frac{\gamma}{\epsilon^2}\right]$ for all $0 < \epsilon \leqslant \epsilon_0$ and $0 \leqslant r \leqslant r_0$. □

13.13 Corollary. *Let $(B_t)_{t\in[0,1]}$ be a BM^1 and I the action functional. Then*

$$\overline{\lim_{\epsilon\to 0}}\, \epsilon^2 \log \mathbb{P}\left(\epsilon B \in F\right) \leqslant -\inf_{w\in F} I(w) \tag{13.12}$$

for every closed set $F \subset \mathcal{C}_{(o)}[0,1]$.

Proof. By the definition of $\Phi(r)$ we have $\Phi(r) \cap F = \emptyset$ for all $r < \inf_{w\in F} I(w)$. Since $\Phi(r)$ is compact,

$$d(\Phi(r), F) = \inf_{w\in\Phi(r)} d(w, F) =: \delta_r > 0.$$

From Theorem 13.12 we know for all $0 < \epsilon \leqslant \epsilon_0$ that

$$\mathbb{P}[\epsilon B \in F] \leqslant \mathbb{P}\big[d(\epsilon B, \Phi(r)) > \delta_r\big] \leqslant \exp\left[-\frac{r-\gamma}{\epsilon^2}\right],$$

and so

$$\overline{\lim_{\epsilon\to 0}}\, \epsilon^2 \log \mathbb{P}(\epsilon B \in F) \leqslant -r.$$

Since $r < \inf_{w\in F} I(w)$ is arbitrary, the claim follows. □

Next we prove the lower large deviation inequality.

13.14 Theorem (Schilder 1966). *Let $(B_t)_{t\in[0,1]}$ be a BM^1 and I the action functional. For every $\delta > 0$, $\gamma > 0$ and $c > 0$ there is some $\epsilon_0 = \epsilon_0(\delta,\gamma,c)$ such that*

$$\mathbb{P}(\|\epsilon B - w_0\|_\infty < \delta) > \exp\big(-(I(w_0)+\gamma)/\epsilon^2\big) \tag{13.13}$$

for all $0 < \epsilon \leqslant \epsilon_0$ and $w_0 \in \mathcal{C}_{(0)}[0,1]$ with $I(w_0) \leqslant c$.

Proof. By the Cameron–Martin formula (13.8), (13.9) we see that

$$\mathbb{P}\big(\|\epsilon B - w_0\|_\infty < \delta\big) = e^{-I(w_0)/\epsilon^2} \int_A e^{-\frac{1}{\epsilon}\int_0^1 w_0'(s)dw(s)}\,\mu(dw)$$

where $A := \{w \in \mathcal{C}_{(0)}[0,1] \,:\, \|w\|_\infty < \delta/\epsilon\}$.

Let $C := \Big\{w \in \mathcal{C}_{(0)}[0,1] \,:\, \int_0^1 w_0'(s)dw(s) \leqslant 2\sqrt{2I(w_0)}\Big\}$. Then

$$\int_A e^{-\frac{1}{\epsilon}\int_0^1 w_0'(s)dw(s)}\,\mu(dw) \geqslant \int_{A\cap C} e^{-\frac{1}{\epsilon}\int_0^1 w_0'(s)dw(s)}\,\mu(dw) \geqslant e^{-\frac{2}{\epsilon}\sqrt{2I(w_0)}}\mu(A\cap C).$$

Pick ϵ_0 so small that

$$\mathbb{P}(A) = \mathbb{P}\left(\sup_{t\leqslant 1}|B(t)| < \frac{\delta}{\epsilon}\right) \geqslant \frac{3}{4} \quad \text{for all } 0 < \epsilon \leqslant \epsilon_0.$$

Now we can use Chebyshev's inequality to find

$$\begin{aligned}
\mu(\Omega\setminus C) &= \mu\left\{w \in \mathcal{C}_{(0)}[0,1] \,:\, \int_0^1 w_0'(s)\,dw(s) > 2\sqrt{2I(w_0)}\right\}\\
&\leqslant \frac{1}{8I(w_0)} \int_{\mathcal{C}_{(0)}[0,1]} \left[\int_0^1 w_0'(s)\,dw(s)\right]^2 \mu(dw)\\
&\overset{(13.7a)}{=} \frac{1}{8I(w_0)} \int_0^1 |w_0'(s)|^2\,ds = \frac{1}{4}.
\end{aligned}$$

Thus, $\mu(C) \geqslant \frac{3}{4}$ and, therefore, $\mu(A\cap C) = \mu(A) + \mu(C) - \mu(A\cup C) \geqslant 1/2$. If we make ϵ_0 even smaller, we can make sure that

$$\frac{1}{2}\exp\left[-\frac{2}{\epsilon}\sqrt{2I(w_0)}\right] \geqslant \exp\left[-\frac{\gamma}{\epsilon^2}\right]$$

for all $0 < \epsilon \leqslant \epsilon_0$ and w_0 with $I(w_0) \leqslant c$. □

13.15 Corollary. *Let $(B_t)_{t\in[0,1]}$ be a* BM^1 *and I the action functional. Then*

$$\varliminf_{\epsilon\to 0} \epsilon^2 \log \mathbb{P}\,(\epsilon B \in U) \geqslant -\inf_{w\in U} I(w) \tag{13.14}$$

for every open set $U \subset \mathcal{C}_{(0)}[0,1]$.

Proof. For every $w_0 \in U$ there is some $\delta_0 > 0$ such that

$$\{w \in \mathcal{C}_{(0)}[0,1] \,:\, \|w - w_0\|_\infty < \delta_0\} \subset U.$$

Theorem 13.14 shows that

$$\mathbb{P}(\epsilon B \in U) \geqslant \mathbb{P}(\|\epsilon B - w_0\| < \delta_0) \geqslant \exp\left(-\frac{1}{\epsilon^2}\,(I(w_0) + \gamma)\right),$$

and so

$$\varliminf_{\epsilon\to 0} \epsilon^2 \log \mathbb{P}(\epsilon B \in U) \geqslant -I(w_0);$$

since $w_0 \in U$ is arbitrary, the claim follows. □

13.3 The proof of Strassen's theorem

Before we start with the rigorous argument we want to add

13.16 Some heuristic considerations. Let $(B_t)_{t\geqslant 0}$ be a BM^1 and

$$Z_s(t,\omega) = \frac{B(st,\omega)}{\sqrt{2s\log\log s}} \sim \frac{B(t,\omega)}{\sqrt{2\log\log s}}, \quad t \in [0,1],\ s > e.$$

In order to decide whether $w_0 \in \mathcal{C}_{(0)}[0,1]$ is a limit point of $\{Z_s \,:\, s > e\}$, we have to estimate the probability

$$\mathbb{P}(\|Z_s(\cdot) - w_0(\cdot)\|_\infty < \delta) \quad \text{as} \quad s \to \infty$$

for all $\delta > 0$. Using the canonical version of the Brownian motion on the Wiener space $(\mathcal{C}_{(0)}[0,1], \mathscr{B}(\mathcal{C}_{(0)}[0,1]), \mu)$ we see for $\epsilon := 1/\sqrt{2\log\log s}$.

$$\begin{aligned}\mathbb{P}(\|Z_s(\cdot) - w_0(\cdot)\|_\infty < \delta) &= \mathbb{P}(\|\epsilon B(\cdot) - w_0(\cdot)\|_\infty < \delta)\\ &= \mu\left\{w \in \mathcal{C}_{(0)}[0,1] \,:\, \|w - \epsilon^{-1} w_0\|_\infty < \delta/\epsilon\right\}.\end{aligned}$$

Now assume that $w_0 \in \mathcal{H}^1$ and set $A := \{w \in \mathcal{C}_{(0)}[0,1] \,:\, \|w\|_\infty < \delta/\epsilon\}$. The Cameron–Martin formula (13.8), (13.9) shows

$$\begin{aligned}&\mathbb{P}(\|Z_s(\cdot) - w_0(\cdot)\|_\infty < \delta)\\ &= \exp\left(-\frac{1}{2\epsilon^2}\int_0^1 |w_0'(t)|^2\,dt\right)\int_A \exp\left(-\frac{1}{\epsilon}\int_0^1 w_0'(s)\,dw(s)\right)\mu(dw).\end{aligned}$$

Using the large deviation estimates from Corollaries 13.13 and 13.15 we see that the first term on the right-hand side is the dominating factor as $\epsilon \to 0$ (i. e. $s \to \infty$) and that

$$\mathbb{P}\left(\|Z_s(\cdot) - w_0(\cdot)\|_\infty < \delta\right) \approx (\log s)^{-\int_0^1 |w_0'(t)|^2\,dt} \quad \text{as} \quad s \to \infty. \tag{13.15}$$

Thus, the probability on the left is only larger than zero if $\int_0^1 |w_0'(t)|^2\,dt < \infty$, i. e. if $w_0 \in \mathcal{H}^1$. Consider now, as in the proof of Khintchine's LIL, sequences of the form $s = s_n = q^n$, $n \geqslant 1$, where $q > 1$. Then (13.15) shows that

$$\sum_n \mathbb{P}\left(\|Z_{s_n}(\cdot) - w_0(\cdot)\|_\infty < \delta\right) = +\infty \quad \text{or} \quad < +\infty$$

according to $\int_0^1 |w_0'(t)|^2\,dt \leqslant 1$ or $\int_0^1 |w_0'(t)|^2\,dt > 1$, respectively. By the Borel–Cantelli lemma we infer that the family

$$\mathcal{K} = \left\{ w \in \mathcal{H}^1 \ : \ \int_0^1 |w'(t)|^2\,dt \leqslant 1 \right\}$$

is indeed a good candidate for the set of limit points. □

We will now give a rigorous proof of Theorem 13.1. As in Section 13.2 we write $I(u)$ for the action functional and $\Phi(r) = \{w \in \mathcal{C}_{(o)}[0,1] \ : \ I(w) \leqslant r\}$ for its sub-level sets; with this notation $\mathcal{K} = \Phi\left(\frac{1}{2}\right)$. Finally, set

$$\mathcal{L}(\omega) := \left\{w \in \mathcal{C}_{(o)}[0,1] \ : \ w \text{ is a limit point of } \{Z_s(\cdot,\omega) \ : \ s > e\} \text{ as } s \to \infty\right\}.$$

We have to show that $\mathcal{L}(\omega) = \Phi\left(\frac{1}{2}\right)$. We split the argument in a series of lemmas.

13.17 Lemma. *$\mathcal{L}(\omega) \subset \Phi\left(\frac{1}{2}\right)$ for almost all $\omega \in \Omega$.*

Proof. Since $\Phi\left(\frac{1}{2}\right) = \bigcap_{\eta>0} \Phi\left(\frac{1}{2} + \eta\right)$ it is clearly enough to show that

$$\mathcal{L}(\omega) \subset \Phi\left(\tfrac{1}{2} + \eta\right) \quad \text{for all } \eta > 0.$$

1° Consider the sequence $(Z_{q^n}(\cdot,\omega))_{n\geqslant 1}$ for some $q > 1$. From Theorem 13.12 we find for any $\delta > 0$ and $\gamma < \eta$

$$\begin{aligned}\mathbb{P}\left[d\left(Z_{q^n}, \Phi\left(\tfrac{1}{2} + \eta\right)\right) > \delta\right] &= \mathbb{P}\left[d\left((2\log\log q^n)^{-1/2} B, \Phi\left(\tfrac{1}{2} + \eta\right)\right) > \delta\right] \\ &\leqslant \exp\left[-2(\log\log q^n)\left(\tfrac{1}{2} + \eta - \gamma\right)\right].\end{aligned}$$

Consequently,

$$\sum_n \mathbb{P}\left[d\left(Z_{q^n}, \Phi\left(\tfrac{1}{2} + \eta\right)\right) > \delta\right] < \infty,$$

and the Borel–Cantelli lemma implies that, for almost all $\omega \in \Omega$,

$$d\left(Z_{q^n}(\cdot,\omega), \Phi\left(\tfrac{1}{2} + \eta\right)\right) \leqslant \delta \quad \text{for all } n \geqslant n_0(\omega).$$

2° Let us now fill the gaps in the sequence $(q^n)_{n\geqslant 1}$. Set $L(t) = \sqrt{2t\log\log t}$. Then

$$\begin{aligned}\sup_{q^{n-1}\leqslant t\leqslant q^n} \|Z_t - Z_{q^n}\|_\infty &= \sup_{q^{n-1}\leqslant t\leqslant q^n}\ \sup_{0\leqslant r\leqslant 1} \left|\frac{B(rt)}{L(t)} - \frac{B(rq^n)}{L(q^n)}\right| \\ &\leqslant \underbrace{\sup_{q^{n-1}\leqslant t\leqslant q^n}\ \sup_{0\leqslant r\leqslant 1} \frac{|B(rt)-B(rq^n)|}{L(q^n)}}_{=:X_n} + \underbrace{\sup_{q^{n-1}\leqslant t\leqslant q^n}\ \sup_{0\leqslant r\leqslant 1} \frac{|B(rt)|}{L(q^n)}\left(\frac{L(q^n)}{L(t)} - 1\right)}_{=:Y_n}.\end{aligned}$$

Since $L(t)$, $t > 3$, is increasing, we see that for sufficiently large $n \geqslant 1$

$$Y_n \leqslant \sup_{s\leqslant q^n} \frac{|B(s)|}{L(q^n)}\left(\frac{L(q^n)}{L(q^{n-1})} - 1\right).$$

Khintchine's LIL, Theorem 12.1, shows that the first factor is almost surely bounded for all $n \geqslant n_0(\omega)$. Moreover, $\lim_{n\to\infty} L(q^n)/L(q^{n-1}) = 1$, which means that we have

$$Y_n \leqslant \frac{\delta}{2} \quad \text{for all } n \geqslant n_0(\omega,\delta,q)$$

with $\delta > 0$ as in step 1° . For the estimate of X_n we use the scaling property of a Brownian motion to get

$$\begin{aligned}\mathbb{P}\left(X_n > \frac{\delta}{2}\right) &= \mathbb{P}\left(\sup_{q^{n-1}\leqslant t\leqslant q^n}\ \sup_{0\leqslant r\leqslant 1} \left|B(tq^{-n}r) - B(r)\right| > \frac{\delta}{2}\sqrt{2\log\log q^n}\right) \\ &= \mathbb{P}\left(\sup_{q^{-1}\leqslant c\leqslant 1}\ \sup_{0\leqslant r\leqslant 1} \left|B(cr) - B(r)\right| > \frac{\delta}{2}\sqrt{2\log\log q^n}\right) \\ &\leqslant \mathbb{P}(\gamma B \in F)\end{aligned}$$

where $\gamma := 2\delta^{-1}(2\log\log q^n)^{-1/2}$ and

$$F := \left\{ w \in \mathcal{C}_{(o)}[0,1] \ :\ \sup_{q^{-1}\leqslant c\leqslant 1}\ \sup_{0\leqslant r\leqslant 1} |w(cr) - w(r)| \geqslant 1\right\}.$$

Ex. 13.7 Since $F \subset \mathcal{C}_{(o)}[0,1]$ is a closed subset, we can use Corollary 13.13 and find

$$\mathbb{P}\left(X_n > \frac{\delta}{2}\right) \leqslant \exp\left(-\frac{1}{\gamma^2}\left(\inf_{w\in F} I(w) - h\right)\right)$$

for any $h > 0$. As $\inf_{w\in F} I(w) = q/(2q-2)$, cf. Lemma 13.18 below, we get

$$\mathbb{P}\left(X_n > \frac{\delta}{2}\right) \leqslant \exp\left(-\frac{\delta^2}{4}\log\log q^n \cdot \left(\frac{q}{q-1} - 2h\right)\right)$$

and, therefore, $\sum_n \mathbb{P}\left[X_n > \delta/2\right] < \infty$ if q is close to 1. Now the Borel–Cantelli lemma implies that $X_n \leqslant \delta/2$ almost surely if $n \geqslant n_0(\omega)$ and so

$$\sup_{q^{n-1}\leqslant s\leqslant q^n} \|Z_s(\cdot,\omega) - Z_{q^n}(\cdot,\omega)\|_\infty \leqslant \delta \quad \text{for all } n \geqslant n_0(\omega,\delta,q).$$

3° Steps 1° and 2° show that for every $\delta > 0$ there is some $q > 1$ such that there is for almost all $\omega \in \Omega$ some $s_0 = s_0(\omega) = q^{n_0(\omega)+1}$ with

$$d\left(Z_s(\cdot,\omega), \Phi\left(\tfrac{1}{2}+\eta\right)\right) \leqslant \|Z_s(\cdot,\omega) - Z_{q^n}(\cdot,\omega)\|_\infty + d\left(Z_{q^n}(\cdot,\omega), \Phi\left(\tfrac{1}{2}+\eta\right)\right) \leqslant 2\delta$$

for all $s \geqslant s_0$. Since $\delta > 0$ is arbitrary and $\Phi\left(\frac{1}{2}+\eta\right)$ is closed, we get

$$\mathcal{L}(\omega) \subset \overline{\Phi\left(\tfrac{1}{2}+\eta\right)} = \Phi\left(\tfrac{1}{2}+\eta\right). \qquad \square$$

13.18 Lemma. *Let* $F = \left\{w \in \mathcal{C}_{(0)}[0,1] : \sup_{q^{-1}\leqslant c\leqslant 1} \sup_{0\leqslant r\leqslant 1} |w(cr) - w(r)| \geqslant 1\right\}$, $q > 1$, *and* $I(\cdot)$ *be the action functional of a Brownian motion. Then*

$$\inf_{w\in F} I(w) = \frac{1}{2}\frac{q}{q-1}.$$

Proof. The function $w_0(t) := \frac{qt-1}{q-1}\mathbb{1}_{[1/q,1]}(t)$ is in F, and

$$2I(w_0) = \int_{1/q}^{1} \left(\frac{q}{q-1}\right)^2 dt = \frac{q}{q-1}.$$

Therefore, $2\inf_{w\in F} I(w) \leqslant 2I(w_0) = q/(q-1)$. On the other hand, for every $w \in F$ there are $x \in [0,1]$ and $c \geqslant q^{-1}$ such that

$$1 \leqslant |w(x) - w(cx)|^2 = \left|\int_{cx}^{x} w'(s)\,ds\right|^2 \leqslant (x - cx)\int_0^1 |w'(s)|^2\,ds \leqslant 2(1-q^{-1})I(w). \qquad \square$$

13.19 Corollary. *The set* $\{Z_s(\cdot,\omega) : s > e\}$ *is for almost all* $\omega \in \Omega$ *a relatively compact subset of* $\mathcal{C}_{(0)}[0,1]$.

Proof. Let $(s_j)_{j\geqslant 1}$ be a sequence in $(0,\infty)$. Without loss of generality we can assume that $(s_j)_{j\geqslant 1}$ is unbounded. Pick a further sequence $(\delta_n)_{n\geqslant 1} \subset (0,\infty)$ such that $\lim_{n\to\infty}\delta_n = 0$. Because of step 3° in the proof of Lemma 13.17 there exists for every $n \geqslant 1$ some $w_n \in \Phi\left(\frac{1}{2}+\eta\right)$ and an index $j(n)$ such that $\|Z_{s_{j(n)}}(\cdot,\omega) - w_n\|_\infty \leqslant \delta_n$. Since $\Phi\left(\frac{1}{2}+\eta\right)$ is a compact set, there is a subsequence $(s'_{j(n)})_{n\geqslant 1} \subset (s_{j(n)})_{n\geqslant 1}$ such that $(Z_{s'_{j(n)}}(\cdot,\omega))_{n\geqslant 1}$ converges in $\mathcal{C}_{(0)}[0,1]$. $\square$

13.20 Lemma. $\mathcal{L}(\omega) \supset \Phi\left(\frac{1}{2}\right)$ *for almost all* $\omega \in \Omega$.

Proof. **1°** The sub-level sets $\Phi(r)$ are compact, cf. Lemma 13.11, and $\Phi(r) \subset \Phi(r')$ if $r \leqslant r'$; it follows that $\overline{\bigcup_{r<1/2}\Phi(r)} \subset \Phi\left(\frac{1}{2}\right)$.

Since for every $w \in \Phi\left(\frac{1}{2}\right)$ we have $\sqrt{2r}w \in \Phi(r)$, $r < \frac{1}{2}$, the converse inclusion '$\supset$' is easy to see. By the very definition, the set of limit points $\mathcal{L}(\omega)$ is a closed set. Therefore, it is enough to show that $\Phi(r) \subset \mathcal{L}(\omega)$ for all $r < 1/2$.

Being subsets of $\mathcal{C}_{(0)}[0,1]$, the sub-level sets $\Phi(r)$ contain a countable dense subset. This means that it is enough to show that for every $w \in \Phi(r)$, every rational $r < 1/2$

and every $\epsilon > 0$ there is a sequence $(s_n)_{n\geqslant 1}$, $s_n \to \infty$, such that

$$\mathbb{P}\left[\|Z_{s_n}(\cdot) - w(\cdot)\|_\infty < \epsilon \quad \text{for all } n \geqslant 1\right] = 1.$$

2° As in the proof of Khintchine's LIL we set $s_n = q^n$, $n \geqslant 1$, for some $q > 1$, and $L(s) = \sqrt{2s \log\log s}$. Then

$$\begin{aligned}
\|Z_{s_n}(\cdot) - w(\cdot)\|_\infty &= \sup_{0\leqslant t\leqslant 1} \left|\frac{B(ts_n)}{L(s_n)} - w(t)\right| \\
&\leqslant \sup_{q^{-1}\leqslant t\leqslant 1} \left|\frac{B(ts_n) - B(s_{n-1})}{L(s_n)} - w(t)\right| \\
&\quad + \frac{|B(s_{n-1})|}{L(s_n)} + \sup_{t\leqslant q^{-1}} |w(t)| + \sup_{t\leqslant q^{-1}} \frac{|B(ts_n)|}{L(s_n)}.
\end{aligned}$$

We will estimate the terms separately.

Since Brownian motion has independent increments, we see that the sets

$$A_n = \left\{\sup_{q^{-1}\leqslant t\leqslant 1} \left|\frac{B(ts_n) - B(s_{n-1})}{L(s_n)} - w(t)\right| < \frac{\epsilon}{4}\right\}, \quad n \geqslant 1,$$

are independent. Combining the stationarity of the increments, the scaling property 2.16 of a Brownian motion and Theorem 13.14 yields for all $n \geqslant n_0$

$$\begin{aligned}
\mathbb{P}(A_n) &= \mathbb{P}\left(\sup_{q^{-1}\leqslant t\leqslant 1} \left|\frac{B(t - q^{-1})}{\sqrt{2\log\log s_n}} - w(t)\right| < \frac{\epsilon}{4}\right) \\
&\geqslant \mathbb{P}\left(\sup_{0\leqslant t\leqslant 1-q^{-1}} \left|\frac{B(t)}{\sqrt{2\log\log s_n}} - v(t)\right| < \frac{\epsilon}{8}\right) \\
&\geqslant \mathbb{P}\left(\sup_{0\leqslant t\leqslant 1} \left|\frac{B(t)}{\sqrt{2\log\log s_n}} - v(t)\right| < \frac{\epsilon}{8}\right) \\
&\geqslant \exp\left(-2\log\log s_n \cdot (I(v) + \eta)\right) \\
&= \left(\frac{1}{n\log q}\right)^{2(I(v)+\eta)};
\end{aligned}$$

here we choose $q > 1$ so large that $|w(q^{-1})| \leqslant \epsilon/8$, and use

$$v(t) = \begin{cases} w(t + \frac{1}{q}) - w(\frac{1}{q}), & \text{if } 0 \leqslant t \leqslant 1 - \frac{1}{q}, \\ w(1) - w(\frac{1}{q}), & \text{if } 1 - \frac{1}{q} \leqslant t \leqslant 1. \end{cases}$$

Since $\eta > 0$ is arbitrary and

$$I(v) = \frac{1}{2}\int_0^{1-1/q} |v'(t)|^2\,dt = \frac{1}{2}\int_{1/q}^1 |w'(s)|^2\,ds \leqslant \frac{1}{2}\int_0^1 |w'(s)|^2\,ds \leqslant r < \frac{1}{2},$$

we find $\sum_n \mathbb{P}(A_n) = \infty$. The Borel–Cantelli lemma implies that for large $q > 1$

$$\varliminf_{n\to\infty} \sup_{q^{-1}\leqslant t\leqslant 1} \left| \frac{B(ts_n) - B(s_{n-1})}{\sqrt{2s_n \log\log s_n}} - w(t) \right| \leqslant \frac{\epsilon}{4}.$$

Selecing a subsequence of $(s_n)_{n\geqslant 1}$, if necessary, we can assume that the lim inf is actually a limit.

3° Finally, Ex. 13.8

$$\mathbb{P}\left(\frac{|B(s_{n-1})|}{L(s_n)} > \frac{\epsilon}{4} \right) = \mathbb{P}\left(|B(1)| > \frac{\epsilon}{4}\sqrt{2q\log\log q^n} \right),$$

$$\sup_{t\leqslant q^{-1}} |w(t)| \leqslant \int_0^{1/q} |w'(s)|\,ds \leqslant \frac{\sqrt{2r}}{\sqrt{q}} < \frac{1}{\sqrt{q}},$$

$$\mathbb{P}\left(\sup_{0\leqslant t\leqslant q^{-1}} \frac{|B(ts_n)|}{L(s_n)} > \frac{\epsilon}{4} \right) \leqslant 2\mathbb{P}\left(|B(1)| > \frac{\epsilon}{4}\sqrt{2q\log\log q^n} \right),$$

and the Borel–Cantelli lemma shows that for sufficiently large $q > 1$

$$\varlimsup_{n\to\infty} \left(\frac{|B(s_{n-1})|}{L(s_n)} + \sup_{t\leqslant q^{-1}} |w(t)| + \sup_{t\leqslant q^{-1}} \frac{|B(ts_n)|}{L(s_n)} \right) < \frac{3}{4}\epsilon.$$

This finishes the proof. □

13.21 Further reading. The Cameron–Martin formula is the basis for many further developments in the analysis in Wiener and abstract Wiener spaces, see the *Further reading* section in Chapter 4; worth reading is the early account in [230]. There is an extensive body of literature on the topic of large deviations. A good starting point are the lecture notes [224] to get an idea of the topics and methods behind 'large deviations'. Since the topic is quite technical, it is very difficult to find easy introductions. Our favourites are [45] and [81] (which we use here), a standard reference is the classic monograph [46].

[45] Dembo, Zeitouni: *Large Deviations Techniques and Applications.*

[46] Deuschel, Stroock: *Large Deviations.*

[81] Freidlin, Wentzell: *Random Perturbations of Dynamical Systems.*

[224] Varadhan: *Large Deviations and Applications.*

[230] Wiener et al.: *Differential Space, Quantum Systems, and Prediction.*

Problems

1. Let $w \in \mathcal{C}_{(0)}[0,1]$ and assume that for every fixed $t \in [0,1]$ the number $w(t)$ is a limit point of the family $\{Z_s(t) : s > e\} \subset \mathbb{R}$. Show that this is not sufficient for w to be a limit point of $\mathcal{C}_{(0)}[0,1]$.

2. Let $\mathcal{K}$ be the set from Theorem 13.1. Show that for $w \in \mathcal{K}$ the estimate $|w(t)| \leqslant \sqrt{t}$, $t \in [0,1]$ holds.

3. Let $u \in \mathcal{H}^1$ and denote by Π_n, $n \geqslant 1$, a sequence of partitions of $[0,1]$ such that $\lim_{n\to\infty} |\Pi_n| = 0$. Show that the functions
$$f_n(t) = \sum_{t_j, t_{j-1} \in \Pi_n} \left[\frac{1}{t_j - t_{j-1}} \int_{t_{j-1}}^{t_j} u'(s)\,ds\right]^2 \mathbb{1}_{[t_{j-1}, t_j)}(t), \quad n \geqslant 1,$$
converge for $n \to \infty$ to $\left[u'(t)\right]^2$ for Lebesgue almost all t.

4. Let $\phi \in \mathsf{BV}[0,1]$ and consider the following Riemann-Stieltjes integral
$$\mathrm{G}^\phi(w) = \phi(1)w(1) - \int_0^1 w(s)\,d\phi(s), \quad w \in \mathcal{C}_{(0)}[0,1].$$
Show that G^ϕ is a linear functional on $(\mathcal{C}_{(0)}[0,1], \mathscr{B}(\mathcal{C}_{(0)}[0,1]), \mu)$ satisfying
(a) G^ϕ is a normal random variable, mean 0, variance $\int_0^1 \phi^2(s)\,ds$;
(b) $\int_{\mathcal{C}_{(0)}[0,1]} \mathrm{G}^\phi(w)\mathrm{G}^\psi(w)\,\mu(dw) = \int_0^1 \phi(s)\psi(s)\,ds$ for all $\phi, \psi \in \mathsf{BV}[0,1]$;
(c) If $(\phi_n)_{n\geqslant 1} \subset \mathsf{BV}[0,1]$ converges in L^2 to ϕ, then $\mathrm{G}^\phi := \lim_{n\to\infty} \mathrm{G}^{\phi_n}$ exists as the L^2-limit and defines a linear functional which satisfies a) and b).

5. Denote by $\mathcal{H}^1$ the Cameron–Martin space (13.2) and by $\mathcal{H}^1_\circ$ the set defined in (13.3).
(a) Show that $\mathcal{H}^1$ is a Hilbert space with the canonical norm $\|h\|^2_{\mathcal{H}^1} := \int_0^1 |h'(s)|^2\,ds$ and scalar product $\langle \phi, h\rangle_{\mathcal{H}^1} = \int_0^1 \phi'(s)\,dh(s)$ (this is a Paley–Wiener–Zygmund integral, cf. 13.5).
(b) Show that $\mathcal{H}^1_\circ$ is a dense subset of $\mathcal{H}^1$.
(c) Show that $\|h\|_{\mathcal{H}^1} = \sup_{\phi\in\mathcal{H}^1,\,\|\phi\|_{\mathcal{H}^1}=1} \langle \phi, h\rangle_{\mathcal{H}^1} = \sup_{\phi\in\mathcal{H}^1_\circ,\,\|\phi\|_{\mathcal{H}^1}=1} \langle \phi, h\rangle_{\mathcal{H}^1}$.
(d) Let $h \in \mathcal{C}_{(0)}[0,1]$. Show that $h \notin \mathcal{H}^1$ if, and only if, there exists a sequence $(\phi_n)_{n\geqslant 1} \subset \mathcal{H}^1_\circ$ such that $\|\phi_n\|_{\mathcal{H}^1} = 1$ and $\langle \phi_n, h\rangle_{\mathcal{H}^1} \geqslant 2n$.

6. Let $w \in \mathcal{C}_{(0)}[0,1]$. Find the densities of the following bi-variate random variables:
(a) $\left(\int_{1/2}^t s^2\,dw(s), w(1/2)\right)$ for $1/2 \leqslant t \leqslant 1$;
(b) $\left(\int_{1/2}^t s^2\,dw(s), w(u+1/2)\right)$ for $0 \leqslant u \leqslant t/2$, $1/2 \leqslant t \leqslant 1$;
(c) $\left(\int_{1/2}^t s^2\,dw(s), \int_{1/2}^t s\,dw(s)\right)$ for $1/2 \leqslant t \leqslant 1$;
(d) $\left(\int_{1/2}^1 e^s\,dw(s), w(1) - w(1/2)\right)$.

7. Show that $F := \left\{w \in \mathcal{C}_{(0)}[0,1] \,:\, \sup_{q^{-1}\leqslant c\leqslant 1} \sup_{0\leqslant r\leqslant 1} |w(cr) - w(r)| \geqslant 1\right\}$ is for every $q > 1$ a closed subset of $\mathcal{C}_{(0)}[0,1]$.

8. Check the (in)equalities from step 3° in the proof of Lemma 13.20.

14 Skorokhod representation

Let $(B_t)_{t\geqslant 0}$ be a one-dimensional Brownian motion. Denote by $\tau = \tau^\circ_{(-a,b)^c}$, $a, b > 0$, the first entry time into the interval $(-a,b)^c$. We know from Corollary 5.11 that

$$\begin{aligned} B_\tau &\sim \frac{b}{a+b}\,\delta_{-a} + \frac{a}{a+b}\,\delta_b \quad \text{and} \\ F_{-a,b}(x) &= \mathbb{P}(B_\tau \leqslant x) = \frac{b}{a+b}\,\mathbb{1}_{[-a,b)}(x) + \mathbb{1}_{[b,\infty)}(x). \end{aligned} \tag{14.1}$$

We will now study the problem which probability distributions can be obtained in this way. More precisely, let X be a random variable with probability distribution function $F(x) = \mathbb{P}(X \leqslant x)$. We want to know if there is a Brownian motion $(B_t)_{t\geqslant 0}$ and a stopping time τ such that $X \sim B_\tau$. If this is the case, we say that X (or F) can be *embedded into a Brownian motion*. The question and its solution (for continuous F) go back to Skorokhod [207, Chapter 7].

To simplify things, we do not require that τ is a stopping time for the natural filtration $(\mathscr{F}_t^B)_{t\geqslant 0}$, $\mathscr{F}_t^B = \sigma(B_s : s \leqslant t)$, but for a larger admissible filtration $(\mathscr{F}_t)_{t\geqslant 0}$, cf. Definition 5.1.

If X can be embedded into $(B_t)_{t\geqslant 0}$, it is clear that

$$\mathbb{E}(X^k) = \mathbb{E}(B_\tau^k)$$

whenever the moments exist. Moreover, if $\mathbb{E}\,\tau < \infty$, Wald's identities, cf. Theorem 5.10, entail

$$\mathbb{E}\,B_\tau = 0 \quad \text{and} \quad \mathbb{E}(B_\tau^2) = \mathbb{E}\,\tau.$$

Thus, X is necessarily centred: $\mathbb{E}\,X = 0$. On the other hand, every centred distribution function F can be obtained as a mixture of two-point distribution functions $F_{-a,b}$. This allows us to reduce the general embedding problem to the case considered in (14.1). The elegant proof of the following lemma is due to Breiman [19].

14.1 Lemma. *Let F be a distribution function with $\int_{\mathbb{R}} |x|\,dF(x) < \infty$ and $\int_{\mathbb{R}} x\,dF(x) = 0$. Then*

$$F(x) = \iint_{\mathbb{R}^2} F_{u,w}(x)\,\nu(du, dw)$$

with the mixing probability distribution

$$\nu(du, dw) = \frac{1}{\alpha}\,(w-u)\,\mathbb{1}_{\{u\leqslant 0<w\}}(u,w)\,dF(u)\,dF(w)$$

and $\alpha := \frac{1}{2}\int_{\mathbb{R}} |x|\,dF(x)$.

Proof. Since F is centred, we see that

$$\alpha = \frac{1}{2}\int_{\mathbb{R}} |x|\,dF(x) = \int_{(0,\infty)} x\,dF(x) = -\int_{(-\infty,0]} x\,dF(x).$$

Using the formula (14.1) for $F_{u,w}$ we find

$$\begin{aligned}
\alpha \iint F_{u,w}(x)\, \nu(du, dw)
&= \iint\limits_{u\leqslant 0<w} \Big(\frac{w}{w-u}(w-u)\mathbb{1}_{[u,w)}(x) + (w-u)\mathbb{1}_{[w,\infty)}(x)\Big) dF(u)\, dF(w)\\
&= \iint\limits_{u\leqslant 0<w} \Big(w\mathbb{1}_{[u,w)}(x) + (w-u)\mathbb{1}_{[w,\infty)}(x)\Big)\, dF(u)\, dF(w)\\
&= \iint\limits_{u\leqslant 0<w} \Big(w\mathbb{1}_{[u,\infty)}(x) - u\mathbb{1}_{[w,\infty)}(x)\Big)\, dF(u)\, dF(w)\\
&= \iint\limits_{u\leqslant 0<w} \Big(w\mathbb{1}_{(-\infty,x]}(u) - u\mathbb{1}_{(-\infty,x]}(w)\Big)\, dF(u)\, dF(w)\\
&= \underbrace{\int\limits_{(0,\infty)} w\, dF(w)}_{=\alpha} \int\limits_{(-\infty,0]} \mathbb{1}_{(-\infty,x]}(u)\, dF(u) - \underbrace{\int\limits_{(-\infty,0]} u\, dF(u)}_{=\alpha} \int\limits_{(0,\infty)} \mathbb{1}_{(-\infty,x]}(w)\, dF(w)\\
&= \alpha \int\limits_{\mathbb{R}} \mathbb{1}_{(-\infty,x]}(z)\, dF(z) = \alpha F(x).
\end{aligned}$$

Finally, the monotone convergence theorem shows that ν is a probability measure:

$$1 = \lim_{x\to\infty} F(x) = \iint\limits_{\mathbb{R}^2} \lim_{x\to\infty} F_{u,w}(x)\, \nu(du, dw) = \nu(\mathbb{R}^2). \qquad \square$$

We can now prove Skorokhod's embedding theorem.

14.2 Theorem (Skorokhod 1965). *Let X be a real random variable with distribution function F. If $\mathbb{E}|X| < \infty$ and $\mathbb{E}X = 0$, then there is a* BM[1] $(B_t)_{t\geqslant 0}$, *an admissible filtration $(\mathscr{F}_t)_{t\geqslant 0}$ and a stopping time τ such that $B_\tau \sim X$. Moreover, $\mathbb{E}X^2 = \mathbb{E}\tau \in [0, \infty]$.*

Proof. Let $(B_t)_{t\geqslant 0}$ be a BM[1] and (U, W) be a random vector with values in $\mathbb{R}^2$ and distribution ν as in Lemma 14.1. We can assume that $(B_t)_{t\geqslant 0}$ and (U, W) are independent.
Ex. 14.1 Therefore, the filtration $\mathscr{F}_t := \sigma(B_s : s \leqslant t;\ U;\ W)$ is admissible in the sense of Definition 5.1.

Consider the first exit time of B from the random interval (U, W),

$$\tau^\circ_{(U,W)^c} = \inf\{t \geqslant 0 : B_t \notin (U, W)\},$$

and write $F_{u,w}$, $u \leqslant 0 < w$, for the distribution function of $B_{\tau^\circ_{(u,w)^c}}$, see (14.1). Since we have

$$\left\{\tau^\circ_{(U,W)^c} > t\right\} = \left\{\min_{0\leqslant s\leqslant t} B_s - U > 0\right\} \cap \left\{\max_{0\leqslant s\leqslant t} B_s - W < 0\right\},$$

$\tau^\circ_{(U,W)^c}$ is an $\mathscr{F}_t$ stopping time. Because of the independence of (U, W) and $(B_t)_{t\geqslant 0}$, we can use Lemma 14.1 and find

$$\begin{aligned}
\mathbb{P}(B_{\tau^\circ_{(U,W)^c}} \leqslant x) &\overset{\text{tower}}{=} \mathbb{E}\left[\mathbb{E}\left(\mathbb{1}_{(-\infty,x]}(B_{\tau^\circ_{(U,W)^c}}) \,\Big|\, U, W\right)\right] \\
&\overset{\substack{\text{indep.}\\ A.3}}{=} \mathbb{E}\left[\mathbb{E}\left(\mathbb{1}_{(-\infty,x]}(B_{\tau^\circ_{(u,w)^c}})\right)\Big|_{u=U,w=W}\right] \\
&= \iint \mathbb{E}\left(\mathbb{1}_{(-\infty,x]}(B_{\tau^\circ_{(u,w)^c}})\right) \nu(du, dw) \\
&\overset{(14.1)}{=} \iint F_{u,w}(x)\, \nu(du, dw) \\
&\overset{14.1}{=} F(x).
\end{aligned}$$

Since $B_{\tau^\circ_{(U,W)^c}} \sim F$, we get by Tonelli's theorem, Wald's identities and (5.12)

$$\begin{aligned}
\int x^2\, dF(x) = \mathbb{E}\left[B^2_{\tau^\circ_{(U,W)^c}}\right] &= \iint \mathbb{E}\left[B^2_{\tau^\circ_{(u,w)^c}}\right] \nu(du, dw) \\
&\overset{5.10}{=} \iint \mathbb{E}\left[\tau^\circ_{(u,w)^c}\right] \nu(du, dw) \\
&\overset{(5.12)}{=} \iint uw\, \nu(du, dw). \qquad \square
\end{aligned}$$

Our simple proof of Skorokhod's embedding theorem comes at the cost that we have to use an *exogenous randomization* procedure which requires the enlargement of the natural Brownian filtration. There are several proofs which avoid this, but they are more complicated, see e. g. Sawyer [200].

Because of the strong Markov property of a Brownian motion we can use Theorem 14.2 to embed any random walk $(S_n)_{n\geqslant 1}$, $S_n = X_1 + \cdots + X_n$, where the X_j are independent centred random variables, in a Brownian motion. For simplicity we will consider only the case where the X_j are iid.

14.3 Corollary. *Let $(X_j)_{j\geqslant 1}$ be a sequence of iid real random variables with common distribution F such that $\mathbb{E}\,|X_1| < \infty$ and $\mathbb{E}\,X_1 = 0$. Then there is a one-dimensional Brownian motion $(B_t)_{t\geqslant 0}$, an admissible filtration $(\mathscr{F}_t)_{t\geqslant 0}$ and a sequence $(\tau_n)_{n\geqslant 1}$ of $\mathscr{F}_t$ stopping times such that the following statements hold:*

a) *the increments $(\tau_n - \tau_{n-1})_{n\geqslant 1}$ are iid positive random variables ($\tau_0 := 0$);*
b) $\mathbb{E}(\tau_n - \tau_{n-1}) = \mathbb{E}(X_1^2)$;
c) *the sequences $(B_{\tau_n})_{n\geqslant 1}$ and $(S_n)_{n\geqslant 1}$, $S_n := X_1 + \cdots + X_n$, have the same distribution.*

Proof. Let $(B_t)_{t\geqslant 0}$ be a one-dimensional Brownian motion and (U_n, W_n), $n \geqslant 0$, iid random vectors such that $(U_0, W_0) \sim \nu$ with ν as in Lemma 14.1. We can assume that $(B_t)_{t\geqslant 0}$ and $(U_n, W_n)_{n\geqslant 0}$ are independent. Then

$$\begin{aligned}
\mathscr{F}^{(n)}_t &:= \sigma\big(B_s \,:\, s \leqslant t;\ (U_0, W_0), \ldots, (U_n, W_n)\big), \quad t \geqslant 0, \\
\mathscr{F}_t &:= \sigma\big(B_s \,:\, s \leqslant t;\ (U_n, W_n),\ n \geqslant 0\big), \quad t \geqslant 0,
\end{aligned}$$

are, for every $n \geqslant 1$, admissible filtrations for $(B_t)_{t\geqslant 0}$.

Set $\tau_0 := 0$ and, recursively,

$$\tau_{n+1} := \inf\left\{t \geqslant \tau_n \;:\; B_t - B_{\tau_n} \notin (U_n, W_n)\right\}, \quad n = 0, 1, 2, \ldots.$$

As in Theorem 14.2 one can see that τ_n is an $\mathscr{F}_t^{(n)}$ stopping time and that B_{τ_n} is $\mathscr{F}_{\tau_n}^{(n)}$ measurable, cf. Lemma A.17 or Corollary 6.25.

Observe that

$$\tau_{n+1} - \tau_n = \inf\left\{s \geqslant 0 \;:\; B_{s+\tau_n} - B_{\tau_n} \notin (U_n, W_n)\right\}.$$

By the strong Markov property, Theorem 6.5, $(B_{s+\tau_n} - B_{\tau_n})_{s\geqslant 0}$ is independent of $\mathscr{F}_{\tau_n}^{(n)}$; in particular $B_{\tau_{n+1}} - B_{\tau_n} \perp\!\!\!\perp \mathscr{F}_{\tau_n}^{(n)}$. Since $(B_{s+\tau_n} - B_{\tau_n})_{s\geqslant 0}$ and (U_n, W_n) have the same distributions as $(B_t)_{t\geqslant 0}$ and (U_1, W_1), respectively, we find that $\tau_{n+1} - \tau_n \sim \tau_1$ and

$$B_{\tau_{n+1}} - B_{\tau_n} = B_{(\tau_{n+1}-\tau_n)+\tau_n} - B_{\tau_n} \sim B_{\tau_1}.$$

By Wald's identity we get, in particular, $\mathbb{E}(\tau_{n+1} - \tau_n) = \mathbb{E}\,\tau_1 = \mathbb{E}(X_1^2)$.

Finally, note that $\tau_1, \ldots, \tau_n$ and $B_{\tau_1}, \ldots, B_{\tau_n}$ are $\mathscr{F}_{\tau_n}^{(n)}$ measurable, and that τ_1 and B_{τ_1} satisfy the basic embedding Theorem 14.2. □

14.4 Remark. We can use the embedding of random walks in Brownian motion to derive asymptotic results for random walks from corresponding results for Brownian motion, cf. [40]; these are often easier to prove. We use the law of the iterated logarithm to illustrate this point.

Let $(X_j)_{j\geqslant 1}$ be a sequence of iid random variables with mean zero and variance σ^2. By the Skorokhod embedding theorem there are stopping times τ_n such that

$$S_n := X_1 + \cdots + X_n \sim B_{\tau_n}.$$

Corollary 14.3 and Wald's identities show that $(\tau_n)_{n\geqslant 1}$ is a random walk with mean value $\mathbb{E}\,\tau_n = \mathbb{E}(B_{\tau_n}^2) = \mathbb{E}(S_n^2) < \infty$. By the strong law of large numbers,

$$\lim_{n\to\infty} \frac{\tau_n}{n} = \mathbb{E}\,\tau_1 = \sigma^2 \quad \text{a.s.}$$

Set $\phi(s) := \sqrt{2s\log\log s}$. Then $\lim_{n\to\infty} \phi(\tau_n)/\phi(n) = \sigma$. We will show that

$$\varlimsup_{n\to\infty} \frac{S_n}{\sqrt{2n\log\log n}} = \varlimsup_{n\to\infty} \frac{S_n}{\phi(n)} = \sigma.$$

For this we claim that

$$\sup_{0\leqslant r\leqslant \tau_n-\tau_{n-1}} |B_{r+\tau_{n-1}} - B_{\tau_n}| = O(\sqrt{n}) \quad \text{a.s. as } n \to \infty. \tag{14.2}$$

Accepting this for the time being, we find for all $s \in [\tau_{n-1}, \tau_n]$

$$\begin{aligned}\left|\frac{B_s}{\phi(s)} - \frac{B_{\tau_n}}{\phi(\tau_n)}\right| &\leqslant \left|\frac{B_s}{\phi(s)} - \frac{B_s}{\phi(\tau_n)}\right| + \left|\frac{B_s - B_{\tau_n}}{\phi(\tau_n)}\right| \\ &\leqslant \left|\frac{B_s}{\phi(s)}\right|\left|1 - \frac{\phi(s)}{\phi(\tau_n)}\right| + \frac{1}{\phi(\tau_n)} \sup_{0\leqslant r\leqslant \tau_n-\tau_{n-1}} |B_{r+\tau_{n-1}} - B_{\tau_n}| \\ &\leqslant \left|\frac{B_s}{\phi(s)}\right|\left|1 - \frac{\phi(s)}{\phi(\tau_n)}\right| + \frac{C\sqrt{n}}{\phi(\tau_n)}.\end{aligned}$$

By the law of the iterated logarithm, Theorem 12.1, $B_s/\phi(s)$ stays bounded while the other terms tend to zero as $n \to \infty$. Thus,

$$1 = \overline{\lim_{s\to\infty}} \frac{B_s}{\phi(s)} = \overline{\lim_{n\to\infty}} \frac{B_{\tau_n}}{\phi(\tau_n)} = \overline{\lim_{n\to\infty}} \frac{S_n}{\phi(n)} \underbrace{\frac{\phi(n)}{\phi(\tau_n)}}_{\to 1/\sigma,\ n\to\infty} = \frac{1}{\sigma} \overline{\lim_{n\to\infty}} \frac{S_n}{\phi(n)}.$$

All that remains to be done is to check (14.2). For this we use a Borel–Cantelli argument. Observe that, by the strong Markov property,

$$\sup_{0\leqslant s\leqslant \tau_n-\tau_{n-1}} (B_{s+\tau_{n-1}} - B_{\tau_{n-1}}) \overset{\text{Thm. 6.5}}{\sim} \sup_{0\leqslant s\leqslant \tau_1} B_s.$$

Using the fact that $\sum_{n=1}^{\infty} \mu(f > n) \leqslant \int f\,d\mu$ and Doob's maximal inequality (A.14) for the stopped Brownian motion $(B_{t\wedge\tau_1})_{t\geqslant 0}$, we see

$$\begin{aligned}\sum_{n=1}^{\infty} \mathbb{P}\Big(\sup_{0\leqslant s\leqslant \tau_n-\tau_{n-1}} |B_{s+\tau_{n-1}} - B_{\tau_{n-1}}| > \sqrt{n}\Big) &= \sum_{n=1}^{\infty} \mathbb{P}\Big(\sup_{0\leqslant s\leqslant \tau_1} B_s^2 > n\Big) \\ &\leqslant \mathbb{E}\Big(\sup_{0\leqslant s\leqslant \tau_1} B_s^2\Big) = \mathbb{E}\Big(\sup_{s\geqslant 0} B_{s\wedge\tau_1}^2\Big) \overset{\text{Doob}}{\underset{\text{(A.14)}}{\leqslant}} 4\,\mathbb{E}\big(B_{\tau_1}^2\big) = 4\,\mathbb{E}\big(X_1^2\big) < \infty.\end{aligned}$$

Now we can use the (simple direction of the) Borel–Cantelli lemma to deduce, for sufficiently large values of n,

$$\sup_{0\leqslant s\leqslant \tau_n-\tau_{n-1}} |B_{s+\tau_{n-1}} - B_{\tau_{n-1}}| \leqslant C\sqrt{n}.$$

This inequality holds in particular for $s = \tau_n - \tau_{n-1}$, and the triangle inequality gives (14.2).

We have seen in Section 3.5 that Brownian motion can be constructed as the limit of random walks. For every *finite* family $(B_{t_1}, \dots, B_{t_k})$, $t_1 < \cdots < t_k$, this follows from the classical central limit theorem: Let $(\epsilon_j)_{j\geqslant 1}$ be iid Bernoulli random variables such that $\mathbb{P}(\epsilon_1 = \pm 1) = \frac{1}{2}$ and $S_n := \epsilon_1 + \cdots + \epsilon_n$, then[1]

$$\frac{S_{\lfloor nt\rfloor}}{\sqrt{n}} \xrightarrow[n\to\infty]{\text{weakly}} B_t \quad \text{for all } t \in (0, 1].$$

1 In fact, it is enough to assume that the ϵ_j are iid random variables with mean zero and finite variance. This universality is the reason why Theorem 14.5 is usually called *invariance principle*.

We are interested in approximations of the whole path $[0,1] \ni t \mapsto B_t(\omega)$ of a Brownian motion, and the above convergence will give pointwise but not uniform results. To get uniform results, it is useful to interpolate the partial sums $S_{\lfloor nt \rfloor}/\sqrt{n}$

$$S^n(t) := \frac{1}{\sqrt{n}}\left(S_{\lfloor nt \rfloor} - (nt - \lfloor nt \rfloor)\epsilon_{\lfloor nt \rfloor + 1}\right), \quad t \in [0,1].$$

This is a piecewise linear continuous function which is equal to $S_j/\sqrt{n}$ if $t = j/n$. We are interested in the convergence of

$$\Phi(S^n(\cdot)) \xrightarrow[n\to\infty]{} \Phi(B(\cdot))$$

where $\Phi : (\mathcal{C}[0,1], \mathbb{R}) \to \mathbb{R}$ is a (not necessarily linear) bounded functional which is uniformly continuous with respect to uniform convergence.

Although the following theorem implies Theorem 3.10, we cannot use it to construct Brownian motion since the proof already uses the existence of a Brownian motion. Nevertheless the result is important as it helps to understand Brownian motion intuitively and permits simple proofs of various limit theorems.

14.5 Theorem (Invariance principle. Donsker 1951). *Let $(B_t)_{t\in[0,1]}$ be a one-dimensional Brownian motion, $(S^n(t))_{t\in[0,1]}$, $n \geqslant 1$, be the sequence of processes from above, and $\Phi : \mathcal{C}([0,1], \mathbb{R}) \to \mathbb{R}$ a uniformly continuous bounded functional. Then*

$$\lim_{n\to\infty} \mathbb{E}\,\Phi(S^n(\cdot)) = \mathbb{E}\,\Phi(B(\cdot)),$$

i. e. $S^n(\cdot)$ converges weakly to $B(\cdot)$.

Proof. Denote by $(W_t)_{t\geqslant 0}$ the Brownian motion into which we can embed the random walk $(S_n)_{n\geqslant 1}$, i. e. $W_{\tau_n} \sim S_n$ for an increasing sequence of stopping times $\tau_1 \leqslant \tau_2 \leqslant \cdots$, cf. Corollary 14.3. Since our claim is a statement on the probability distributions, we can assume that $S_n = W_{\tau_n}$.

By assumption, we find for all $\epsilon > 0$ some $\eta > 0$ such that $|\Phi(f) - \Phi(g)| < \epsilon$ for all $\|f - g\|_\infty < \eta$.

The assertion follows if we can prove, for all $\epsilon > 0$ and for sufficiently large values of $n \geqslant N_\epsilon$, that the Brownian motions defined by $B^n(t) := n^{-1/2} W_{nt} \sim B_t$, $t \in [0,1]$, satisfy

$$\mathbb{P}\left(\left|\Phi(S^n(\cdot)) - \Phi(B^n(\cdot))\right| > \epsilon\right) \leqslant \mathbb{P}\left(\sup_{0\leqslant t\leqslant 1} \left|S^n(t) - B^n(t)\right| > \eta\right) \leqslant \epsilon.$$

This means that $\Phi(S^n(\cdot)) - \Phi(B^n(\cdot))$ converges in probability to 0, therefore

$$|\,\mathbb{E}\,\Phi(S^n(\cdot)) - \mathbb{E}\,\Phi(B(\cdot))| \overset{B^n\sim B}{=} |\,\mathbb{E}\,\Phi(S^n(\cdot)) - \mathbb{E}\,\Phi(B^n(\cdot))| \xrightarrow[n\to\infty]{} 0.$$

In order to see the estimate we need three simple observations. Pick $\epsilon, \eta > 0$ as above.

1° Since a Brownian motion has continuous sample paths, there is some $\delta > 0$ such that

$$\mathbb{P}\Big(\exists\, s, t \in [0,1],\ |s-t| \leqslant \delta\ :\ |W_t - W_s| > \eta\Big) \leqslant \frac{\epsilon}{2}.$$

2° From Corollary 14.3 and Wald's identities it follows that $(\tau_n)_{n\geqslant 1}$ is a random walk with mean $\mathbb{E}\,\tau_n = \mathbb{E}(B^2_{\tau_n}) = \mathbb{E}(S_n^2) < \infty$. Therefore, the strong law of large numbers tells us that

$$\lim_{n\to\infty} \frac{\tau_n}{n} = \mathbb{E}\,\tau_1 = 1 \quad \text{a. s.}$$

Thus, there is some $M(\eta)$ such that for all $n > m \geqslant M(\eta)$

$$\begin{aligned}\frac{1}{n}\max_{1\leqslant k\leqslant n}|\tau_k - k| &\leqslant \frac{1}{n}\max_{1\leqslant k\leqslant m}|\tau_k - k| + \max_{m\leqslant k\leqslant n}\left|\frac{\tau_k}{k} - 1\right| \\ &\leqslant \frac{1}{n}\max_{1\leqslant k\leqslant m}|\tau_k - k| + \eta \xrightarrow[n\to\infty]{} \eta \xrightarrow[\eta\to 0]{} 0.\end{aligned}$$

Consequently, for some $N(\epsilon,\eta)$ and all $n \geqslant N(\epsilon,\eta)$

$$\mathbb{P}\left(\frac{1}{n}\max_{1\leqslant k\leqslant n}|\tau_k - k| > \frac{\delta}{3}\right) \leqslant \frac{\epsilon}{2}.$$

3° For every $t \in [k/n, (k+1)/n]$ there is a $u \in [\tau_k/n, \tau_{k+1}/n]$, $k = 0, 1, \dots, n-1$, such that

$$S^n(t) = n^{-1/2} W_{nu} = B^n(u).$$

Indeed: $S^n(t)$ interpolates $B^n(\tau_k/n)$ and $B^n(\tau_{k+1}/n)$ linearly, i. e. $S^n(t)$ is between the minimum and the maximum of $B^n(s)$, $\tau_k \leqslant s\cdot n \leqslant \tau_{k+1}$. Since $s \mapsto B^n(s)$ is continuous, there is some intermediate value u such that $S^n(t) = B^n(u)$. This situation is shown in Figure 14.1.

Clearly, we have

$$\begin{aligned}&\mathbb{P}\left(\sup_{0\leqslant t\leqslant 1}\left|S^n(t) - B^n(t)\right| > \eta\right) \\ &\leqslant \mathbb{P}\left(\sup_{0\leqslant t\leqslant 1}\left|S^n(t) - B^n(t)\right| > \eta,\ \max_{1\leqslant l\leqslant n}\left|\frac{\tau_l}{n} - \frac{l}{n}\right| \leqslant \frac{\delta}{3}\right) + \mathbb{P}\left(\frac{1}{n}\max_{1\leqslant l\leqslant n}|\tau_l - l| > \frac{\delta}{3}\right).\end{aligned}$$

If $\left|S^n(t) - B^n(t)\right| > \eta$ for some $t \in [k/n, (k+1)/n]$, then we find by Step 3° some value $u \in [\tau_k/n, \tau_{k+1}/n]$ with $\left|B^n(u) - B^n(t)\right| > \eta$. Moreover,

$$|u - t| \leqslant \underbrace{\left|u - \frac{\tau_k}{n}\right|}_{\to 0,\ n\to\infty \text{ by } 2^\circ} + \left|\frac{\tau_k}{n} - \frac{k}{n}\right| + \left|\frac{k}{n} - t\right| \leqslant \frac{\delta}{3} + \frac{\delta}{3} + \frac{1}{n} \leqslant \delta$$

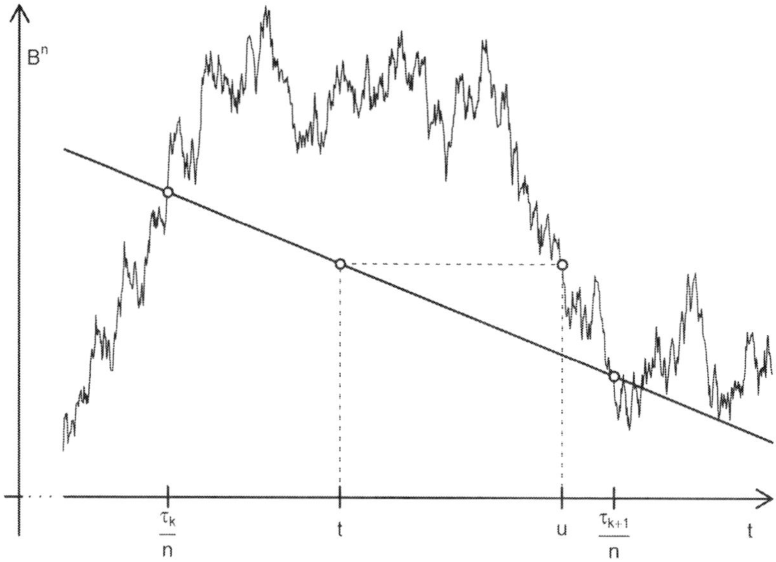

Fig. 14.1. For each value $S^n(t)$ of the line segment connecting $B^n(\tau_k/n)$ and $B^n(\tau_{k+1}/n)$ there is some u such that $S^n(t) = B^n(u)$.

if $n \geqslant N(\epsilon, \eta) \vee (3/\delta)$ and $\max_{1 \leqslant l \leqslant n} \left|\frac{\tau_l}{n} - \frac{l}{n}\right| \leqslant \delta/3$. This, 2° and 1° give for all sufficiently large n

$$\mathbb{P}\Big(\sup_{0 \leqslant t \leqslant 1} \left|S^n(t) - B^n(t)\right| > \eta\Big)$$
$$\leqslant \mathbb{P}\left(\exists\, u, t \in [0,1],\ |u-t| \leqslant \delta\ :\ \left|B^n(u) - B^n(t)\right| > \eta\right) + \frac{\epsilon}{2} \leqslant \epsilon. \qquad \square$$

Typical examples of uniformly continuous functionals are

$$\Phi_0(f) = f(1), \qquad \Phi_1(f) = \sup_{0 \leqslant t \leqslant 1} |f(t) - at|,$$
$$\Phi_2(f) = \int_0^1 |f(t)|^p\, dt, \qquad \Phi_3(f) = \int_0^1 |f(t) - tf(1)|^4\, dt.$$

With a bit of extra work one can prove Theorem 14.5 for all $\Phi : \mathcal{C}([0,1], \mathbb{R}) \to \mathbb{R}$ which are continuous with respect to uniform convergence *at almost every Brownian path.* This version includes the following functionals

$$\Phi_4(f) = \mathbb{1}_{\{\sup_{0 \leqslant t \leqslant 1} |f(t)| \leqslant 1\}}, \qquad \Phi_5(f) = \mathrm{Leb}\{f > 0\} - \mathrm{Leb}\{f < 0\},$$
$$\Phi_6(f) = \mathbb{1}_{\{\forall t\, :\, a(t) < f(t) < b(t)\}} \quad (a, b : [0,1] \to \mathbb{R} \text{ smooth functions}).$$

Using Φ_0 we could deduce the central limit theorem (without using characteristic function methods), Φ_2 gives the limit theorem

$$\lim_{n\to\infty} \mathbb{P}\left(\sum_{k=1}^{n} |S_k|^p \leqslant \lambda n^p\right) = \mathbb{P}\left(\int_0^1 |B_t|^p \, dt \leqslant \lambda\right)$$

while Φ_4 corresponds to the gambler's ruin problem, and Φ_6 is Khintchine's Problem, see Sawyer [200] for a more detailed discussion.

14.6 Further reading. More applications and variants of Skorokhod's embedding theorem can be found in the monograph [79] and in the survey papers [200] and [166].

[79] Freedman: *Brownian Motion and Diffusion.*
[166] Obłój: The Skorokhod embedding problem and its offspring.
[200] Sawyer: The Skorokhod Representation.

Problems

1. In the proof of Theorem 14.2 we assumed that $(B_t)_{t\geqslant 0}$ and (U, W) are independent. Show that $\mathscr{F}_t := \sigma(B_s,\ s \leqslant t; U, W)$ is an admissible filtration for a Brownian motion, cf. Definition 5.1.

15 Stochastic integrals: L^2-Theory

A function $f : [0, T] \to \mathbb{R}$ is *Riemann integrable* if the *Riemann sums*

$$S^{\Pi}(f) := \sum_{j=1}^{n} f(\xi_j^{\Pi})(s_j^{\Pi} - s_{j-1}^{\Pi}), \quad \xi_j^{\Pi} \in [s_{j-1}^{\Pi}, s_j^{\Pi}], \tag{15.1}$$

converge to a finite limit I. The limit is taken along all finite partitions of the interval $[0, T]$, $\Pi = \{0 = s_0^{\Pi} < s_1^{\Pi} < \cdots < s_n^{\Pi} = T\}$, with $|\Pi| := \max_j(s_j^{\Pi} - s_{j-1}^{\Pi}) \to 0$ and it is independent of the particular sequence and the choice of the intermediate points ξ_j^{Π}. The number I is called the *definite Riemann integral* of f in $[0, T]$ and one writes $I = \int_0^T f(s)\,ds$.

Riemann's definition has been modified in many ways. T. J. Stieltjes used another function g, replacing $s_j^{\Pi} - s_{j-1}^{\Pi}$ in (15.1) by $g(s_j^{\Pi}) - g(s_{j-1}^{\Pi})$. The resulting integral $\int_0^T f(s)\,dg(s)$ is the *Riemann–Stieltjes integral.* The requirement that the Riemann–Stieltjes sums converge imposes some restrictions on f and g. For example, if we want that $\int_0^T f(s)\,dg(s)$ exists for all $f \in \mathcal{C}[0, T]$, we can show that g is necessarily of bounded variation: $\mathrm{VAR}_1(g; t) < \infty$, see Corollary A.41. Since we know from Chapter 9 that a Brownian motion is of unbounded variation on any compact interval, it is clear that we cannot define $\int_0^T f(s)\,dB_s(\omega)$ as a Riemann–Stieltjes integral.

Using the notion of L^2 martingales K. Itô succeeded 1942 to give a consistent and powerful meaning to $\int_0^T f(s, \omega)\,dB_s(\omega)$ where the class of integrand processes $f(s, \omega)$ contains all bounded processes with continuous sample paths.

15.1 Discrete stochastic integrals

Let $(\mathscr{F}_n)_{n\geqslant 0}$ be a filtration. An L^2 martingale $X = (X_n, \mathscr{F}_n)_{n\geqslant 0}$ is a martingale such that $X_n \in L^2(\mathbb{P})$ for all $n \geqslant 0$. We write $\mathcal{M}^2$ for the family of L^2 martingales. If $X \in \mathcal{M}^2$, then $X^2 = (X_n^2, \mathscr{F}_n)_{n\geqslant 0}$ is a submartingale and, by the Doob decomposition,

$$X_n^2 = X_0^2 + M_n + A_n \tag{15.2}$$

where $(M_n, \mathscr{F}_n)_{n\geqslant 0}$ is a martingale and $(A_n)_{n\geqslant 0}$ is a *previsible* (i. e. A_n is $\mathscr{F}_{n-1}$ measurable, $n \geqslant 1$) and *increasing* (i. e.: $A_n \leqslant A_{n+1}$, $n \geqslant 0$) process. This decomposition is unique. Therefore, the following definition makes sense.

15.1 Definition. Let $(X_n, \mathscr{F}_n)_{n\geqslant 0}$ be an L^2 martingale. The *quadratic variation* of X, $\langle X\rangle = (\langle X\rangle_n)_{n\geqslant 0}$, is the unique increasing and previsible process with $\langle X\rangle_0 = 0$ appearing in the Doob decomposition (15.2).

The next lemma explains why $\langle X\rangle$ is called the quadratic variation.

15.2 Lemma. *Let $X = (X_n, \mathscr{F}_n)_{n\geqslant 0}$ be an L^2 martingale and $\langle X\rangle$ the quadratic variation. Then, for all $n > m \geqslant 0$,*

$$\mathbb{E}\left[(X_n - X_m)^2 \mid \mathscr{F}_m\right] = \mathbb{E}\left[X_n^2 - X_m^2 \mid \mathscr{F}_m\right] = \mathbb{E}\left[\langle X\rangle_n - \langle X\rangle_m \mid \mathscr{F}_m\right]. \tag{15.3}$$

In particular,

$$\langle X\rangle_n = \sum_{j=1}^{n} \mathbb{E}\left[(X_j - X_{j-1})^2 \mid \mathscr{F}_{j-1}\right] = \sum_{j=1}^{n} \mathbb{E}\left[X_j^2 - X_{j-1}^2 \mid \mathscr{F}_{j-1}\right]. \tag{15.4}$$

Proof. Note that for all $n > m \geqslant 0$

$$\begin{aligned}
&\mathbb{E}\left[(X_n - X_m)^2 \mid \mathscr{F}_m\right] \\
&\quad = \mathbb{E}\left[X_n^2 - 2X_nX_m + X_m^2 \mid \mathscr{F}_m\right] \\
&\quad = \mathbb{E}\left[X_n^2 \mid \mathscr{F}_m\right] - 2X_m \underbrace{\mathbb{E}\left[X_n \mid \mathscr{F}_m\right]}_{=X_m,\ \text{martingale}} + X_m^2 \\
&\quad = \mathbb{E}\left[X_n^2 \mid \mathscr{F}_m\right] - X_m^2 \\
&\quad = \mathbb{E}\left[X_n^2 - X_m^2 \mid \mathscr{F}_m\right] \\
&\quad = \mathbb{E}\Big[\underbrace{X_n^2 - \langle X\rangle_n}_{\text{martingale}} - (X_m^2 - \langle X\rangle_m) \mid \mathscr{F}_m\Big] + \mathbb{E}\left[\langle X\rangle_n - \langle X\rangle_m \mid \mathscr{F}_m\right] \\
&\quad = \mathbb{E}\left[\langle X\rangle_n - \langle X\rangle_m \mid \mathscr{F}_m\right].
\end{aligned}$$

Since $\langle X\rangle_j$ is $\mathscr{F}_{j-1}$ measurable, we get (15.4) from (15.3) if we set $n = j$, $m = j-1$, and sum over $j = 1, \dots, n$. □

15.3 Definition. Let $(X_n, \mathscr{F}_n)_{n\geqslant 0}$ be a martingale and $(C_n, \mathscr{F}_n)_{n\geqslant 0}$ an adapted process. Then

$$C \bullet X_n := \sum_{j=1}^{n} C_{j-1}(X_j - X_{j-1}), \quad C \bullet X_0 := 0, \quad n \geqslant 1, \tag{15.5}$$

is called a *martingale transform*.

Let us review a few properties of the martingale transform.

15.4 Theorem. *Let $(M_n, \mathscr{F}_n)_{n\geqslant 0}$ be an L^2 martingale and $(C_n, \mathscr{F}_n)_{n\geqslant 0}$ be a bounded adapted process. Then*

a) *$C \bullet M$ is an L^2 martingale, i. e. $C\bullet : \mathcal{M}^2 \to \mathcal{M}^2$;*
b) *$C \mapsto C \bullet M$ is linear;*
c) *$\langle C \bullet M\rangle_n = C^2 \bullet \langle M\rangle_n := \sum_{j=1}^n C_{j-1}^2(\langle M\rangle_j - \langle M\rangle_{j-1})$ for all $n \geqslant 1$;*
d) *$\mathbb{E}\left[(C \bullet M_n - C \bullet M_m)^2 \mid \mathscr{F}_m\right] = \mathbb{E}\left[C^2 \bullet \langle M\rangle_n - C^2 \bullet \langle M\rangle_m \mid \mathscr{F}_m\right]$ for $n > m \geqslant 0$. In particular,*

$$\mathbb{E}\left[(C \bullet M_n)^2\right] = \mathbb{E}\left[C^2 \bullet \langle M\rangle_n\right] \tag{15.6}$$

$$= \mathbb{E}\left[\sum_{j=1}^{n} C_{j-1}^2 (M_j - M_{j-1})^2\right]. \tag{15.7}$$

Proof. a) By assumption, there is some constant $K > 0$ such that $|C_j| \leqslant K$ a. s. for all $n \geqslant 0$. Thus, $|C_{j-1}(M_j - M_{j-1})| \leqslant K(|M_{j-1}| + |M_j|) \in L^2$, which means that $C \bullet M_n \in L^2$. Let us check that $C \bullet M$ is a martingale. For all $n \geqslant 1$

$$\begin{aligned}\mathbb{E}\left(C \bullet M_n \mid \mathscr{F}_{n-1}\right) &= \mathbb{E}\left(C \bullet M_{n-1} + C_{n-1}(M_n - M_{n-1}) \mid \mathscr{F}_{n-1}\right)\\ &= C \bullet M_{n-1} + C_{n-1} \underbrace{\mathbb{E}\left((M_n - M_{n-1}) \mid \mathscr{F}_{n-1}\right)}_{=0,\text{ martingale}}\\ &= C \bullet M_{n-1}.\end{aligned}$$

b) The linearity of $C \mapsto C \bullet M$ is obvious.

c) If we apply Lemma 15.2 to the L^2 martingale $X = C \bullet M$, we get

$$\begin{aligned}\langle C \bullet M\rangle_n &\overset{(15.4)}{=} \sum_{j=1}^n \mathbb{E}\left[(C \bullet M_j - C \bullet M_{j-1})^2 \mid \mathscr{F}_{j-1}\right]\\ &= \sum_{j=1}^n \mathbb{E}\left[C_{j-1}^2 (M_j - M_{j-1})^2 \mid \mathscr{F}_{j-1}\right]\\ &= \sum_{j=1}^n C_{j-1}^2\, \mathbb{E}\left[(M_j - M_{j-1})^2 \mid \mathscr{F}_{j-1}\right]\\ &\overset{(15.3)}{=} \sum_{j=1}^n C_{j-1}^2\, \mathbb{E}\left[\langle M\rangle_j - \langle M\rangle_{j-1} \mid \mathscr{F}_{j-1}\right]\\ &= \sum_{j=1}^n C_{j-1}^2 (\langle M\rangle_j - \langle M\rangle_{j-1})\\ &= C^2 \bullet \langle M\rangle_n.\end{aligned}$$

d) Using c) and Lemma 15.2 for the L^2 martingale $C \bullet M$ gives

$$\begin{aligned}\mathbb{E}\left((C \bullet M_n - C \bullet M_m)^2 \mid \mathscr{F}_m\right) &= \mathbb{E}\left((C \bullet M_n)^2 - (C \bullet M_m)^2 \mid \mathscr{F}_m\right)\\ &= \mathbb{E}\left(C^2 \bullet \langle M\rangle_n - C^2 \bullet \langle M\rangle_m \mid \mathscr{F}_m\right).\end{aligned}$$

If we take $m = 0$, we get (15.6). If we use Lemma 15.2 for the L^2 martingale M, we see $\mathbb{E}(\langle M\rangle_j - \langle M\rangle_{j-1} \mid \mathscr{F}_{j-1}) = \mathbb{E}((M_j - M_{j-1})^2 \mid \mathscr{F}_{j-1})$. Therefore, (15.7) follows from (15.6) by the tower property. □

Note that $M \mapsto \langle M\rangle$ is a quadratic form. Therefore, we can use polarization to get the *quadratic covariation* which is a bilinear form:

$$\langle M, N\rangle := \frac{1}{4}(\langle M + N\rangle - \langle M - N\rangle) \quad \text{for all } M, N \in \mathcal{M}^2. \tag{15.8}$$

Ex. 15.1 By construction, $\langle M, M\rangle = \langle M\rangle$. Using polarization, we see that $(M_n N_n - \langle M, N\rangle_n)_{n\geqslant 0}$ is a martingale. Moreover

15.5 Lemma. *Let $(M_n, \mathscr{F}_n)_{n\geqslant 0}$, $(N_n, \mathscr{F}_n)_{n\geqslant 0}$ be L^2 martingales and $(C_n, \mathscr{F}_n)_{n\geqslant 0}$ a bounded adapted process. Then*

$$\langle C \bullet M, N\rangle_n = C \bullet \langle M, N\rangle_n = \sum_{j=1}^{n} C_{j-1}(\langle M, N\rangle_j - \langle M, N\rangle_{j-1}). \tag{15.9}$$

Moreover, if for some $I \in \mathcal{M}^2$ with $I_0 = 0$

$$\langle I, N\rangle_n = C \bullet \langle M, N\rangle_n \quad \textit{for all } N \in \mathcal{M}^2,\ n \geqslant 1, \tag{15.10}$$

then $I = C \bullet M$.

Lemma 15.5 shows that $C \bullet M$ is the only L^2 martingale with $C \bullet M_0 = 0$ satisfying (15.10).

Proof. Since $MN - \langle M, N\rangle$, M and N are martingales, we find for all $j = 1, \dots, n$

$$\begin{aligned}\langle M, N\rangle_j - \langle M, N\rangle_{j-1} &= \mathbb{E}\left(M_jN_j - M_{j-1}N_{j-1} \mid \mathscr{F}_{j-1}\right)\\ &= \mathbb{E}\left(M_jN_j - M_{j-1}N_j - M_jN_{j-1} + M_{j-1}N_{j-1} \mid \mathscr{F}_{j-1}\right)\\ &= \mathbb{E}\left((M_j - M_{j-1})(N_j - N_{j-1}) \mid \mathscr{F}_{j-1}\right)\end{aligned}$$

(compare this calculation with the proof of Lemma 15.2). Replacing M by $C \bullet M$ yields

$$\begin{aligned}\langle C \bullet M, N\rangle_j - \langle C \bullet M, N\rangle_{j-1} &= \mathbb{E}\left((C \bullet M_j - C \bullet M_{j-1})(N_j - N_{j-1}) \mid \mathscr{F}_{j-1}\right)\\ &= \mathbb{E}\left(C_{j-1}(M_j - M_{j-1})(N_j - N_{j-1}) \mid \mathscr{F}_{j-1}\right)\\ &\overset{\substack{\text{pull}\\ \text{out}}}{=} C_{j-1}\, \mathbb{E}\left((M_j - M_{j-1})(N_j - N_{j-1}) \mid \mathscr{F}_{j-1}\right)\\ &= C_{j-1}(\langle M, N\rangle_j - \langle M, N\rangle_{j-1}).\end{aligned}$$

If we sum over $j = 1, \dots, n$ we get (15.9). [1]

In particular, (15.10) holds for $I = C \bullet M$.

Now assume that (15.10) is true. Then

$$\langle I, N\rangle \overset{(15.10)}{=} C \bullet \langle M, N\rangle \overset{(15.9)}{=} \langle C \bullet M, N\rangle$$

or $\langle I - C \bullet M, N\rangle = 0$ for all $N \in \mathcal{M}^2$. Since $I - C \bullet M \in \mathcal{M}^2$, we can set $N = I - C \bullet M$ and find

$$\langle I - C \bullet M, I - C \bullet M\rangle = \langle I - C \bullet M\rangle = 0.$$

Thus, $\mathbb{E}\left[(I_n - C \bullet M_n)^2\right] = \mathbb{E}\left[\langle I - C \bullet M\rangle_n\right] = 0$, which shows that $I_n = C \bullet M_n$ a. s. for all $n \geqslant 0$. □

1 We learned this short proof of (15.9) from Ms. Franziska Kühn.

15.2 Simple integrands

We can now consider stochastic integrals with respect to a Brownian motion. Throughout the rest of this chapter $(B_t)_{t\geqslant 0}$ is a BM^1, $(\mathscr{F}_t)_{t\geqslant 0}$ is an admissible filtration such that each $\mathscr{F}_t$ contains all $\mathbb{P}$-null sets, e. g. the completed natural filtration $\mathscr{F}_t = \overline{\mathscr{F}}_t^B$ (cf. Theorem 6.21), and $[0,T]$, $T<\infty$, a finite interval. Although we do not explicitly state it, most results remain valid for $T=\infty$ and $[0,\infty)$. In line with the notation introduced in Section 15.1 we use

- X^τ and $X_s^\tau := X_{\tau\wedge s}$ for the stopped process $(X_t)_{t\geqslant 0}$;
- $\mathcal{M}_T^2$ for the family of $\mathscr{F}_t$ martingales $(M_t)_{0\leqslant t\leqslant T}\subset L^2(\mathbb{P})$;
- $\mathcal{M}_T^{2,c}$ for the L^2 martingales with (almost surely) continuous sample paths;
- $L^2(\lambda_T\otimes\mathbb{P}) = L^2([0,T]\times\Omega,\lambda_T\otimes\mathbb{P})$ for the family of (equivalence classes of) all $\mathscr{B}[0,T]\otimes\mathscr{A}$ measurable random functions $f:[0,T]\times\Omega\to\mathbb{R}$ equipped with the norm $\|f\|^2_{L^2(\lambda_T\otimes\mathbb{P})} = \int_0^T[\mathbb{E}(|f(s)|^2)]\,ds<\infty$.

15.6 Definition. A real-valued stochastic process $(f(t,\cdot))_{t\in[0,T]}$ of the form

$$f(t,\omega) = \sum_{j=1}^n \phi_{j-1}(\omega)\mathbb{1}_{[s_{j-1},s_j)}(t) \tag{15.11}$$

where $n\geqslant 1$, $0=s_0\leqslant s_1\leqslant\cdots\leqslant s_n\leqslant T$ and $\phi_j\in L^\infty(\mathscr{F}_{s_j})$ are bounded $\mathscr{F}_{s_j}$ measurable random variables, $j=0,\dots,n-1$, is called a *(right-continuous) simple process*. We write $\mathcal{S}_T$ for the family of all simple processes on $[0,T]$.

Note that we can, a bit tautologically, rewrite (15.11) as

$$f(t,\omega) = \sum_{j=1}^n f(s_{j-1},\omega)\mathbb{1}_{[s_{j-1},s_j)}(t). \tag{15.12}$$

15.7 Definition. Let $M\in\mathcal{M}_T^{2,c}$ be a continuous L^2 martingale and $f\in\mathcal{S}_T$. Then

$$f\bullet M_T = \sum_{j=1}^n f(s_{j-1})(M(s_j)-M(s_{j-1})) \tag{15.13}$$

is called the *stochastic integral* of $f\in\mathcal{S}_T$. Instead of $f\bullet M_T$ we also write $\int_0^T f(s)\,dM_s$.

Ex. 15.2 Since (15.13) is linear, it is obvious that (15.13) does not depend on the particular representation of the simple process $f\in\mathcal{S}_T$. It is not hard to see that $(f\bullet M_{s_j})_{j=0,\dots,n}$ is a martingale transform in the sense of Definition 15.3: Just take $C_j = f(s_j) = \phi_j$ and $M_j = M(s_j)$.

In order to use the results of Section 15.1 we note that for a Brownian motion

$$\begin{aligned}\langle B\rangle_{s_j} - \langle B\rangle_{s_{j-1}} &\overset{(15.3)}{=} \mathbb{E}\left[(B(s_j)-B(s_{j-1}))^2 \mid \mathscr{F}_{s_{j-1}}\right] \\ &\underset{(5.1)}{\overset{(B1)}{=}} \mathbb{E}\left[(B(s_j)-B(s_{j-1}))^2\right] = s_j - s_{j-1}.\end{aligned} \tag{15.14}$$

Since $\langle B\rangle_0 = 0$, and since $0 = s_0 < s_1 < \cdots < s_n \leqslant T$ is an arbitrary partition, we see that $\langle B\rangle_t = t$. We can now use Theorem 15.4 to describe the properties of the stochastic integral for simple processes.

15.8 Theorem. *Let $(B_t)_{t\geqslant 0}$ be a* BM^1, $(M_t)_{t\leqslant T} \in \mathcal{M}_T^{2,c}$, $(\mathscr{F}_t)_{t\geqslant 0}$ *be an admissible filtration and $f \in \mathcal{S}_T$. Then*

a) $f \mapsto f \bullet M_T \in L^2(\mathbb{P})$ *is linear;*

b) $\langle f \bullet B\rangle_T = f^2 \bullet \langle B\rangle_T = \int_0^T |f(s)|^2\,ds$;

c) *For all $f \in \mathcal{S}_T$*

$$\|f \bullet B_T\|^2_{L^2(\mathbb{P})} = \mathbb{E}\left[(f \bullet B)_T^2\right] = \mathbb{E}\left[\int_0^T |f(s)|^2\,ds\right] = \|f\|^2_{L^2(\lambda_T\otimes\mathbb{P})}. \tag{15.15}$$

Proof. Let f be a simple process given by (15.11). Without loss of generality we can assume that $s_n = T$. All assertions follow immediately from the corresponding statements for the martingale transform, cf. Theorem 15.4. For b) we observe that

$$\sum_{j=1}^n |f(s_{j-1})|^2(\langle B\rangle_{s_j} - \langle B\rangle_{s_{j-1}}) \overset{(15.14)}{=} \sum_{j=1}^n |f(s_{j-1})|^2(s_j - s_{j-1}) \overset{(15.12)}{=} \int_0^{s_n} |f(s)|^2\,ds,$$

and c) follows if we take expectations and use that $\left((f \bullet B_{s_j})^2 - \langle f \bullet B\rangle_{s_j}\right)_{j=0,\dots,n}$ is a martingale, i. e. for $s_n = T$ we have $\mathbb{E}(f \bullet B_T^2) = \mathbb{E}\langle f \bullet B\rangle_T$. □

Note that for $T < \infty$ we always have $\mathcal{S}_T \subset L^2(\lambda_T \otimes \mathbb{P})$, i. e. (15.15) is always finite and $f \bullet B_T \in L^2(\mathbb{P})$ always holds true. For $T = \infty$ this remains true if we *assume* that $f \in L^2(\lambda_T \otimes \mathbb{P})$ and $s_n < T = \infty$. Therefore one should understand the equality in Theorem 15.8 c) as an *isometry* between $(\mathcal{S}_T, \|\cdot\|_{L^2(\lambda_T\otimes\mathbb{P})})$ and a subspace of $L^2(\mathbb{P})$.

It is natural to study $T \mapsto f \bullet M_T = \int_0^T f(s)\,dM_s$ as a function of time. Since for every $\mathscr{F}_t$ stopping time τ the stopped martingale $M^\tau = (M_{s\wedge\tau})_{s\leqslant T}$ is again in $\mathcal{M}_T^{2,c}$, see Theorem A.18 and Corollary A.20, we can define

$$f \bullet M_\tau := f \bullet (M^\tau)_T = \sum_{j=1}^n f(s_{j-1})(M_{\tau\wedge s_j} - M_{\tau\wedge s_{j-1}}). \tag{15.16}$$

As usual, we write

$$f \bullet M_\tau \overset{\text{def}}{=} \int_0^T f(s)\,dM_s^\tau = \int_0^\tau f(s)\,dM_s.$$

This holds, in particular, for the constant stopping time $\tau \equiv t$, $0 \leqslant t \leqslant T$.

15.9 Theorem. *Let $(B_t)_{t\geqslant 0}$ be a* BM^1, $(M_t)_{t\leqslant T} \in \mathcal{M}_T^{2,c}$, $(\mathscr{F}_t)_{t\geqslant 0}$ *be an admissible filtration and $f \in \mathcal{S}_T$. Then*

a) $(f \bullet M_t)_{t\leqslant T}$ *is a continuous L^2 martingale; in particular*

$$\mathbb{E}\left[(f \bullet B_t - f \bullet B_s)^2 \;\middle|\; \mathscr{F}_s\right] = \mathbb{E}\left[\int_s^t |f(r)|^2\,dr \;\middle|\; \mathscr{F}_s\right], \qquad s < t \leqslant T.$$

b) *Localization in time. For every $\mathscr{F}_t$ stopping time τ with finitely many values*

$$(f \bullet M)^\tau_t = f \bullet (M^\tau)_t = (f \mathbb{1}_{[0,\tau)}) \bullet M_t.$$

c) *Localization in space. If $f, g \in \mathcal{S}_T$ are simple processes with $f(s,\omega) = g(s,\omega)$ for all $(s,\omega) \in [0,t] \times F$ for some $t \leqslant T$ and $F \in \mathscr{F}_\infty$, then*

$$(f \bullet M_s)\mathbb{1}_F = (g \bullet M_s)\mathbb{1}_F, \quad 0 \leqslant s \leqslant t.$$

Proof. a) Let $0 \leqslant s < t \leqslant T$ and $f \in \mathcal{S}_T$. We may assume that $s, t \in \{s_0, \dots, s_n\}$ where the s_j are as in (15.11); otherwise we add s and t to the partition. From Theorem 15.4 we know that $(f \bullet M_{s_j})_j$ is a martingale, thus

$$\mathbb{E}\,(f \bullet M_t \mid \mathscr{F}_s) = f \bullet M_s.$$

Since s, t are arbitrary, we conclude that $(f \bullet M_t)_{t \leqslant T} \in \mathcal{M}^2_T$.

The very definition of $t \mapsto f \bullet M_t$, see (15.16), shows that $(f \bullet M_t)_{t \leqslant T}$ is a continuous process. Using Lemma 15.2 or Theorem 15.4 d) we get

$$\begin{aligned}\mathbb{E}\left[(f \bullet B_t - f \bullet B_s)^2 \mid \mathscr{F}_s\right] &= \mathbb{E}\left[\langle f \bullet B\rangle_t - \langle f \bullet B\rangle_s \mid \mathscr{F}_s\right] \\ &\overset{\substack{15.8\\ \text{b)}}}{=} \mathbb{E}\left[\int_0^t |f(r)|^2\,dr - \int_0^s |f(r)|^2\,dr \;\Big|\; \mathscr{F}_s\right].\end{aligned}$$

b) Write $f \in \mathcal{S}_T$ in the form (15.12). Then we have for *all* stopping times τ

$$\begin{aligned}(f \bullet M)^\tau_t = (f \bullet M)_{t\wedge\tau} &= \sum_{j=1}^n f(s_{j-1})(M_{s_j\wedge\tau\wedge t} - M_{s_{j-1}\wedge\tau\wedge t}) \\ &= \sum_{j=1}^n f(s_{j-1})(M^\tau_{s_j\wedge t} - M^\tau_{s_{j-1}\wedge t}) = f \bullet (M^\tau)_t.\end{aligned}$$

If τ takes only finitely many values, we may assume that $\tau(\omega) \in \{s_0, s_1, \dots, s_n\}$ with s_j from the representation (15.11) of f. Thus, $\tau(\omega) > s_{j-1}$ implies $\tau(\omega) \geqslant s_j$, and

$$\begin{aligned}f(s)\mathbb{1}_{[0,\tau)}(s) &= \sum_{j=1}^n f(s_{j-1})\mathbb{1}_{[0,\tau)}(s)\mathbb{1}_{[s_{j-1},s_j)}(s) \\ &= \sum_{j=1}^n \underbrace{\left(f(s_{j-1})\mathbb{1}_{\{\tau > s_{j-1}\}}\right)}_{\mathscr{F}_{s_{j-1}}\text{ mble}} \underbrace{\mathbb{1}_{[s_{j-1},\tau\wedge s_j)}}_{\tau\wedge s_j = s_j}(s).\end{aligned}$$

This shows that $f\mathbb{1}_{[0,\tau)} \in \mathcal{S}_T$ as well as

$$\begin{aligned}(f\mathbb{1}_{[0,\tau)}) \bullet M_t &= \sum_{j=1}^n f(s_{j-1})\mathbb{1}_{\{\tau > s_{j-1}\}}(M_{s_j\wedge t} - M_{s_{j-1}\wedge t}) \\ &= \sum_{j=1}^n f(s_{j-1})\underbrace{(M_{s_j\wedge\tau\wedge t} - M_{s_{j-1}\wedge\tau\wedge t})}_{=0,\text{ if } \tau \leqslant s_{j-1}} = f \bullet (M^\tau)_t.\end{aligned}$$

c) Because of b) we can assume that $t = T$, otherwise we replace f and g by $f\mathbb{1}_{[0,t)}$ and $g\mathbb{1}_{[0,t)}$, respectively. Refining the partitions, if necessary, $f, g \in \mathcal{S}_T$ can have the same support points $0 = s_0 < s_1 < \cdots < s_n \leqslant T$ in their representation (15.11). Since we know that $f(s_j)\mathbb{1}_F = g(s_j)\mathbb{1}_F$, we get

$$
\begin{aligned}
(f \bullet M_T)\mathbb{1}_F &= \sum_{j=1}^{n} f(s_{j-1})\mathbb{1}_F \cdot (M_{s_j} - M_{s_{j-1}}) \\
&= \sum_{j=1}^{n} g(s_{j-1})\mathbb{1}_F \cdot (M_{s_j} - M_{s_{j-1}}) = (g \bullet M_T)\mathbb{1}_F.
\end{aligned}
$$

□

15.3 Extension of the stochastic integral to $\mathcal{L}_T^2$

We have seen in the previous section that the map $f \mapsto f \bullet B$ is a linear isometry from $\mathcal{S}_T$ to $\mathcal{M}_T^{2,c}$. Therefore, we can use standard arguments to extend this mapping to the closure of $\mathcal{S}_T$ with respect to the norm in $L^2(\lambda_T \otimes \mathbb{P})$. As before, most results remain valid for $T = \infty$ and $[0, \infty)$ but, for simplicity, we restrict ourselves to the case $T < \infty$.

15.10 Lemma. *Let $M \in \mathcal{M}_T^{2,c}$. Then $\|M\|_{\mathcal{M}_T^2} := \left(\mathbb{E}\left[\sup_{s\leqslant T}|M_s|^2\right]\right)^{\frac{1}{2}}$ is a norm on $\mathcal{M}_T^{2,c}$ and $\mathcal{M}_T^{2,c}$ is closed under this norm. Moreover,*

$$
\mathbb{E}\left[\sup_{s\leqslant T}|M_s|^2\right] \asymp \mathbb{E}\left[|M_T|^2\right] = \sup_{s\leqslant T}\mathbb{E}\left[|M_s|^2\right]. \tag{15.17}
$$

Proof. By Doob's maximal inequality A.10 we get that

$$
\mathbb{E}\left[|M_T|^2\right] \leqslant \mathbb{E}\left[\sup_{s\leqslant T}|M_s|^2\right] \overset{(A.14)}{\leqslant} 4\sup_{s\leqslant T}\mathbb{E}\left[|M_s|^2\right] \overset{(A.14)}{=} 4\,\mathbb{E}\left[|M_T|^2\right].
$$

This proves (15.17) since $(|M_t|^2)_{t\leqslant T}$ is a submartingale, i. e. $t \mapsto \mathbb{E}(|M_t|^2)$ is increasing. Since $\|\cdot\|_{\mathcal{M}_T^2}$ is the $L^2(\Omega, \mathbb{P}; \mathcal{C}[0,T])$-norm, i. e. the $L^2(\mathbb{P})$ norm (of the supremum norm) of $\mathcal{C}[0,T]$-valued random variables, it is clear that it is again a norm. Ex. 15.3

Let $M_n \in \mathcal{M}_T^{2,c}$ with $\lim_{n\to\infty}\|M_n - M\|_{\mathcal{M}_T^2} = 0$. In particular we see that for a subsequence $n(j)$

$$
\lim_{j\to\infty}\sup_{s\leqslant T}|M_{n(j)}(s,\omega) - M(s,\omega)| = 0 \quad \text{for almost all } \omega.
$$

This shows that $s \mapsto M(s,\omega)$ is continuous. Since $L^2(\mathbb{P})\text{-}\lim_{n\to\infty} M_n(s) = M(s)$ for every $s \leqslant T$, we find !

$$
\int_F M(s)\,d\mathbb{P} = \lim_{n\to\infty}\int_F M_n(s)\,d\mathbb{P} \overset{\text{martin-gale}}{=} \lim_{n\to\infty}\int_F M_n(T)\,d\mathbb{P} = \int_F M(T)\,d\mathbb{P}
$$

for all $s \leqslant T$ and $F \in \mathscr{F}_s$. Thus, M is a martingale, and $M \in \mathcal{M}_T^{2,c}$. □

Itô's isometry (15.15) allows us to extend the stochastic integral from the simple integrands $\mathcal{S}_T \subset L^2(\lambda_T \otimes \mathbb{P})$ to the closure $\overline{\mathcal{S}}_T$ in $L^2(\lambda_T \otimes \mathbb{P})$.

15.11 Definition. By $\mathcal{L}_T^2$ we denote the closure $\overline{\mathcal{S}}_T$ of the simple processes $\mathcal{S}_T$ with respect to the norm in $L^2(\lambda_T \otimes \mathbb{P})$.

Using (15.15) we find for any sequence $(f_n)_{n\geqslant 1} \subset \mathcal{S}_T$ with limit $f \in \mathcal{L}_T^2$ that

$$\begin{aligned}\|f_n \bullet B_T - f_m \bullet B_T\|^2_{L^2(\mathbb{P})} &= \mathbb{E}\left[\int_0^T |f_n(s,\cdot) - f_m(s,\cdot)|^2\, ds\right] \\ &= \|f_n - f_m\|^2_{L^2(\lambda_T\otimes\mathbb{P})} \xrightarrow[m,n\to\infty]{} 0.\end{aligned}$$

Therefore, for each $0 \leqslant t \leqslant T$, the limit

$$L^2(\mathbb{P})\text{-}\lim_{n\to\infty} f_n \bullet B_t \tag{15.18}$$

Ex. 15.4 exists and does not depend on the approximating sequence. By Lemma 15.10, (15.18) defines an element in $\mathcal{M}_T^{2,c}$. We will see later on (cf. Section 15.5) that $\mathcal{L}_T^2$ is the set of all (equivalence classes of) processes from $L^2(\lambda_T \otimes \mathbb{P})$ which have an progressively measurable representative.

15.12 Definition. Let $(B_t)_{t\geqslant 0}$ be a BM^1 and let $f \in \mathcal{L}_T^2$. Then the *stochastic integral* (or *Itô integral*) is defined as

$$f \bullet B_t = \int_0^t f(s)\, dB_s := L^2(\mathbb{P})\text{-}\lim_{n\to\infty} f_n \bullet B_t, \quad 0 \leqslant t \leqslant T, \tag{15.19}$$

where $(f_n)_{n\geqslant 1} \subset \mathcal{S}_T$ is any sequence approximating $f \in \mathcal{L}_T^2$ in $L^2(\lambda_T \otimes \mathbb{P})$.

The stochastic integral is constructed in such a way that all properties from Theorems 15.8 and 15.9 carry over to integrands from $f \in \mathcal{L}_T^2$.

Ex. 15.6 **15.13 Theorem.** *Let $(B_t)_{t\geqslant 0}$ be a BM^1, $(\mathscr{F}_t)_{t\geqslant 0}$ be an admissible filtration and $f \in \mathcal{L}_T^2$.*
Ex. 15.7 *Then we have for all $t \leqslant T$*
Ex. 15.9
Ex. 15.10
a) *$f \mapsto f \bullet B$ is a linear map from $\mathcal{L}_T^2$ into $\mathcal{M}_T^{2,c}$, i.e. $(f \bullet B_t)_{t\leqslant T}$ is a continuous L^2 martingale;*

b) *$(f \bullet B)^2 - \langle f \bullet B\rangle$ is a martingale where $\langle f \bullet B\rangle_t := f^2 \bullet \langle B\rangle_t = \int_0^t |f(s)|^2\, ds$;[2] in particular,*

$$\mathbb{E}\left[(f \bullet B_t - f \bullet B_s)^2 \,\middle|\, \mathscr{F}_s\right] = \mathbb{E}\left[\int_s^t |f(r)|^2\, dr \,\middle|\, \mathscr{F}_s\right], \quad s < t \leqslant T. \tag{15.20}$$

c) *Itô's isometry.*

$$\|f \bullet B_T\|^2_{L^2(\mathbb{P})} = \mathbb{E}\left[(f \bullet B)_T^2\right] = \mathbb{E}\left[\int_0^T |f(s)|^2\, ds\right] = \|f\|^2_{L^2(\lambda_T\otimes\mathbb{P})}. \tag{15.21}$$

2 Observe that we have defined the angle bracket $\langle\cdot\rangle$ *only* for discrete martingales. Therefore we have to *define* the quadratic variation here. For each fixed t this new definition is in line with 15.8 b).

d) *Maximal inequalities.*

$$\mathbb{E}\left[\int_0^T |f(s)|^2\,ds\right] \leqslant \mathbb{E}\left[\sup_{t\leqslant T}\left(\int_0^t f(s)\,dB_s\right)^2\right] \leqslant 4\,\mathbb{E}\left[\int_0^T |f(s)|^2\,ds\right]. \tag{15.22}$$

e) *Localization in time. Let τ be an $\mathscr{F}_t$ stopping time. Then*

$$(f \bullet B)_t^\tau = f \bullet (B^\tau)_t = (f\mathbb{1}_{[0,\tau)}) \bullet B_t \quad \textit{for all } t \leqslant T.$$

f) *Localization in space. If $f, g \in \mathcal{L}_T^2$ are integrands such that $f(s,\omega) = g(s,\omega)$ for all $(s,\omega) \in [0,t] \times F$ where $t \leqslant T$ and $F \in \mathscr{F}_\infty$, then*

$$(f \bullet B_s)\mathbb{1}_F = (g \bullet B_s)\mathbb{1}_F, \quad 0 \leqslant s \leqslant t.$$

Proof. Throughout the proof we fix $f \in \mathcal{L}_T^2$ and some sequence $f_n \in \mathcal{S}_T$, such that $L^2(\lambda_T \otimes \mathbb{P})\text{-}\lim_{n\to\infty} f_n = f$. Clearly, all statements are true for $f_n \bullet B$ and we only have to see that the L^2 limits preserve them.

a) follows from Theorem 15.9, Lemma 15.10 and the linearity of L^2 limits.

b) Note that

$$(f_n \bullet B_t)^2 - \langle f_n \bullet B\rangle_t = (f_n \bullet B_t)^2 - \int_0^t |f_n(s)|^2\,ds$$

is a sequence of martingales, cf. Theorem 15.8 b). Note that

$$\begin{aligned} &\left|\langle f_n \bullet B\rangle_t - \langle f \bullet B\rangle_t\right| \\ &\quad\leqslant \int_0^t \left||f_n(s)|^2 - |f(s)|^2\right| ds \\ &\quad= \int_0^t |f_n(s) + f(s)| \cdot |f_n(s) - f(s)|\,ds \\ &\quad\leqslant \left(\int_0^t |f_n(s) + f(s)|^2\,ds\right)^{\frac12} \left(\int_0^t |f_n(s) - f(s)|^2\,ds\right)^{\frac12}. \end{aligned}$$

Taking expectations on both sides reveals that the second expression tends to zero while the first bracket stays uniformly bounded. This follows from our assumption as $L^2(\lambda_T \otimes \mathbb{P})\text{-}\lim_{n\to\infty} f_n = f$.

With essentially the same calculation we get that $L^2(\mathbb{P})\text{-}\lim_{n\to\infty} f_n \bullet B_t = f \bullet B_t$ Ex. 15.11 implies $L^1(\mathbb{P})\text{-}\lim_{n\to\infty}(f_n \bullet B_t)^2 = (f \bullet B_t)^2$. As L^1-limits preserve martingales, we get that $(f \bullet B)^2 - f^2 \bullet \langle B\rangle$ is again a martingale.

Finally, noting that $f \bullet B$ is a martingale, the technique used in the proof of Lemma 15.2 applies and yields (15.20).

c) is a consequence of the same equalities in Theorem 15.8 and the completion.

d) Apply Lemma 15.10 to the L^2 martingale $(f \bullet B_t)_{t\leqslant T}$ and use c):

$$\mathbb{E}\left[\sup_{t\leqslant T}\left(\int_0^t f(s)\,dB_s\right)^2\right] \overset{(15.17)}{\asymp} \mathbb{E}\left[\left(\int_0^T f(s)\,dB_s\right)^2\right] \overset{c)}{=} \mathbb{E}\left[\int_0^T |f(s)|^2\,ds\right].$$

e) From the proof of Theorem 15.9 b) we know that $f_n \bullet (B^\tau)_t = (f_n \bullet B)_t^\tau$ holds for *all* stopping times τ and $f_n \in \mathcal{S}_T$. Using the maximal inequality d) we find

$$\begin{aligned}\mathbb{E}\left[\sup_{s\leqslant T}\left|f_n \bullet (B^\tau)_s - f_m \bullet (B^\tau)_s\right|^2\right] &= \mathbb{E}\left[\sup_{s\leqslant T}\left|f_n \bullet B_{s\wedge\tau} - f_m \bullet B_{s\wedge\tau}\right|^2\right]\\ &\leqslant \mathbb{E}\left[\sup_{s\leqslant T}\left|f_n \bullet B_s - f_m \bullet B_s\right|^2\right]\\ &\overset{d)}{\leqslant} 4\,\mathbb{E}\left[\int_0^T \left|f_n(s) - f_m(s)\right|^2 ds\right]\\ &\xrightarrow[m,n\to\infty]{} 0.\end{aligned}$$

Since $\mathcal{M}_T^{2,c}$ is complete, we conclude that

$$(f_n \bullet B)^\tau = f_n \bullet (B^\tau) \xrightarrow[n\to\infty]{\mathcal{M}_T^{2,c}} f \bullet (B^\tau).$$

On the other hand, $\lim_{n\to\infty} f_n \bullet B = f \bullet B$, hence $\lim_{n\to\infty}(f_n \bullet B)^\tau = (f \bullet B)^\tau$. This gives the first equality.

For the second equality we assume that τ is a *discrete* stopping time. Then we know from Theorem 15.9 b)

$$(f_n \mathbb{1}_{[0,\tau)}) \bullet B = (f_n \bullet B)^\tau \xrightarrow[n\to\infty]{\mathcal{M}_T^{2,c}} (f \bullet B)^\tau.$$

By the maximal inequality d) we get

$$\begin{aligned}&\mathbb{E}\left[\sup_{s\leqslant T}\left|(f_n \mathbb{1}_{[0,\tau)}) \bullet B_s - (f \mathbb{1}_{[0,\tau)}) \bullet B_s\right|^2\right]\\ &\qquad\leqslant 4\,\mathbb{E}\left[\int_0^T \left|f_n(s) - f(s)\right|^2 \underbrace{\mathbb{1}_{[0,\tau)}(s)}_{\leqslant 1}\,ds\right] \xrightarrow[n\to\infty]{} 0.\end{aligned}$$

This shows that $f \mathbb{1}_{[0,\tau)} \in \mathcal{L}_T^2$ and $\mathcal{M}_T^{2,c}$-$\lim_{n\to\infty}(f_n \mathbb{1}_{[0,\tau)}) \bullet B = (f \mathbb{1}_{[0,\tau)}) \bullet B$.

Finally, let $\tau_j \downarrow \tau$ be discrete stopping times, e. g. $\tau_j = (\lfloor 2^j\tau\rfloor + 1)/2^j$ as in Lemma A.16. With calculations similar to those above we get

$$\begin{aligned}\mathbb{E}\left[\sup_{s\leqslant T}\left|(f\mathbb{1}_{[0,\tau)})\bullet B_s - (f\mathbb{1}_{[0,\tau_j)})\bullet B_s\right|^2\right] &\leqslant 4\,\mathbb{E}\left[\int_0^T \left|f(s)\mathbb{1}_{[0,\tau)}(s) - f(s)\mathbb{1}_{[0,\tau_j)}(s)\right|^2 ds\right]\\ &\leqslant 4\,\mathbb{E}\left[\int_0^T \left|f(s)\mathbb{1}_{[\tau,\tau_j)}(s)\right|^2 ds\right]\end{aligned}$$

and the last expression tends to zero as $j \to \infty$ by dominated convergence. Since $f\bullet B_t^{\tau_j} \to f\bullet B^\tau$ is obvious (by continuity of the paths), the assertion follows.

f) Let $f_n, g_n \in \mathcal{S}_T$ be approximations of $f, g \in \mathcal{L}^2_T$. We will see in the proof of Theorem 15.20 below, that we can choose f_n, g_n in such a way that

$$f_n(s,\omega) = g_n(s,\omega) \quad \text{for all } n \geqslant 1 \quad \text{and} \quad (s,\omega) \in [0,t]\times F.$$

Since

$$\begin{aligned}\mathbb{E}\left[\sup_{s\leqslant t}\left|(f_n\bullet B_s)\mathbb{1}_F - (f\bullet B_s)\mathbb{1}_F\right|^2\right] &\leqslant \mathbb{E}\left[\sup_{s\leqslant t}\left|f_n\bullet B_s - f\bullet B_s\right|^2\right]\\ &\overset{\text{d)}}{\leqslant} 4\,\mathbb{E}\left[\int_0^t \left|f_n(s)-f(s)\right|^2 ds\right],\end{aligned}$$

we see that the corresponding assertion for step functions, Theorem 15.9 c), remains valid for the L^2 limit. □

15.4 Evaluating Itô integrals

Using integrands from $\mathcal{S}_T$ allows us to calculate only very few Itô integrals. For example, up to now we have the natural formula $\int_0^T \xi\cdot\mathbb{1}_{[a,b)}(s)\,dB_s = \xi\cdot(B_b - B_a)$ only for *bounded* $\mathscr{F}_a$ measurable random variables ξ.

15.14 Lemma. *Let* $(B_t)_{t\geqslant 0}$ *be a* BM^1, $0 \leqslant a < b \leqslant T$ *and* $\xi \in L^2(\mathbb{P})$ *be* $\mathscr{F}_a$ *measurable. Then* $\xi\cdot\mathbb{1}_{[a,b)} \in \mathcal{L}^2_T$ *and*

$$\int_0^T \xi\cdot\mathbb{1}_{[a,b)}(s)\,dB_s = \xi\cdot(B_b - B_a).$$

Proof. Clearly, $[(-k)\vee\xi\wedge k]\mathbb{1}_{[a,b)} \in \mathcal{S}_T$ for every $k \geqslant 1$. Dominated convergence yields

$$\mathbb{E}\left[\int_0^T \left|(-k)\vee\xi\wedge k - \xi\right|^2\mathbb{1}_{[a,b)}(s)\,ds\right] = (b-a)\,\mathbb{E}\left[\left|(-k)\vee\xi\wedge k - \xi\right|^2\right] \xrightarrow[k\to\infty]{} 0,$$

i. e. $\xi \cdot \mathbb{1}_{[a,b)} \in \mathcal{L}^2_T$. By the definition of the Itô integral we find

$$\begin{aligned}\int_0^T \xi \cdot \mathbb{1}_{[a,b)}(s)\,dB_s &= L^2(\mathbb{P})\text{-}\lim_{k\to\infty} \int_0^T ((-k) \vee \xi \wedge k)\mathbb{1}_{[a,b)}(s)\,dB_s \\ &= L^2(\mathbb{P})\text{-}\lim_{k\to\infty} ((-k) \vee \xi \wedge k)(B_b - B_a) \\ &= \xi \cdot (B_b - B_a).\end{aligned}$$

□

Let us now discuss an example that shows how to calculate a stochastic integral explicitly.

Ex. 15.12 **15.15 Example.** Let $(B_t)_{t\geqslant 0}$ be a BM^1 and $T > 0$. Then

$$\int_0^T B_t\,dB_t = \frac{1}{2}\,(B_T^2 - T). \tag{15.23}$$

As a by-product of our proof we will see that $(B_t)_{t\leqslant T} \in \mathcal{L}^2_T$. Note that (15.23) differs from what we expect from calculus. If $u(t)$ is differentiable and $u(0) = 0$, we get

$$\int_0^T u(t)\,du(t) = \int_0^T u(t)u'(t)\,dt = \frac{1}{2}\,u^2(t)\Big|_{t=0}^T = \frac{1}{2}\,u^2(T),$$

i. e. we would expect $\frac{1}{2}\,B_T^2$ rather than $\frac{1}{2}\,(B_T^2 - T)$. On the other hand, the stochastic integral must be a martingale, see Theorem 15.13 a); and the correction '$-T$' turns the submartingale B_T^2 into a bone fide martingale.

Let us now verify (15.23). Pick any partition $\Pi = \{0 = s_0 < s_1 < \cdots < s_n = T\}$ with mesh $|\Pi| = \max_{1\leqslant j\leqslant n}(s_j - s_{j-1})$ and set

$$f^\Pi(t,\omega) := \sum_{j=1}^n B(s_{j-1},\omega)\mathbb{1}_{[s_{j-1},s_j)}(t).$$

Applying Lemma 15.14 n times, we get

$$f^\Pi \in \mathcal{L}^2_T \quad\text{and}\quad \int_0^T f^\Pi(t)\,dB_t = \sum_{j=1}^n B(s_{j-1})(B(s_j) - B(s_{j-1})).$$

Observe that

$$\begin{aligned}\mathbb{E}\left[\int_0^T \left|f^\Pi(t) - B(t)\right|^2 dt\right] &= \sum_{j=1}^n \mathbb{E}\left[\int_{s_{j-1}}^{s_j} \left|B(s_{j-1}) - B(t)\right|^2 dt\right] \\ &= \sum_{j=1}^n \int_{s_{j-1}}^{s_j} \mathbb{E}\left[\left|B(s_{j-1}) - B(t)\right|^2\right] dt \\ &= \sum_{j=1}^n \int_{s_{j-1}}^{s_j} (t - s_{j-1})\,dt = \frac{1}{2}\sum_{j=1}^n (s_j - s_{j-1})^2 \xrightarrow[|\Pi|\to 0]{} 0.\end{aligned}$$

Since $\mathcal{L}_T^2$ is by definition closed (w. r. t. the norm in $L^2(\lambda_T \otimes \mathbb{P})$), we conclude that $(B_t)_{t\leqslant T} \in \mathcal{L}_T^2$. By Itô's isometry (15.21)

$$\mathbb{E}\left[\left(\int_0^T \left(f^\Pi(t) - B_t\right) dB_t\right)^2\right] = \mathbb{E}\left[\int_0^T \left|f^\Pi(t) - B(t)\right|^2 dt\right]$$

$$= \frac{1}{2}\sum_{j=1}^n (s_j - s_{j-1})^2 \xrightarrow[|\Pi|\to 0]{} 0.$$

Therefore,

$$\int_0^T B_t\,dB_t = L^2(\mathbb{P})\text{-}\lim_{|\Pi|\to 0}\sum_{j=1}^n B(s_{j-1})(B(s_j) - B(s_{j-1})),$$

i. e. we can approximate the stochastic integral in a concrete way by Riemann–Stieltjes sums. Finally,

$$B_T^2 = \sum_{j=1}^n (B_{s_j} - B_{s_{j-1}})^2 + 2\sum_{j=1}^n\sum_{k=1}^{j-1}(B_{s_j} - B_{s_{j-1}})(B_{s_k} - B_{s_{k-1}})$$

$$= \underbrace{\sum_{j=1}^n (B_{s_j} - B_{s_{j-1}})^2}_{\to T,\ |\Pi|\to 0,\ \text{Thm. 9.1}} + 2\underbrace{\sum_{j=1}^n (B_{s_j} - B_{s_{j-1}})B_{s_{j-1}}}_{\to \int_0^T B_t\,dB_t,\ |\Pi|\to 0}$$

in $L^2(\mathbb{P})$ and, therefore, $B_T^2 - T = 2\int_0^T B_t\,dB_t$.

In the calculation of Example 15.15 we used (implicitly) the following result which is interesting on its own.

15.16 Proposition. *Let $f \in \mathcal{L}_T^2$ be a process which is, as a function of t, mean-square continuous, i. e.* Ex. 15.14

$$\lim_{[0,T]\ni s\to t} \mathbb{E}\left[|f(s,\cdot) - f(t,\cdot)|^2\right] = 0, \quad t \in [0,T].$$

Then

$$\int_0^T f(t)\,dB_t = L^2(\mathbb{P})\text{-}\lim_{|\Pi|\to 0}\sum_{j=1}^n f(s_{j-1})(B_{s_j} - B_{s_{j-1}})$$

holds for any sequence of partitions $\Pi = \{s_0 = 0 < s_1 < \cdots < s_n = T\}$ with mesh $|\Pi| \to 0$.

Proof. As in Example 15.15 we consider the following discretization of the integrand f

$$f^\Pi(t,\omega) := \sum_{j=1}^n f(s_{j-1},\omega)\mathbb{1}_{[s_{j-1},s_j)}(t).$$

Since $f(t) \in L^2(\mathbb{P})$, we can use Lemma 15.14 to see $f^\Pi \in \mathcal{L}_T^2$. By Itô's isometry,

$$\begin{aligned}\mathbb{E}\left[\left|\int_0^T \left[f(t) - f^\Pi(t)\right] dB_t\right|^2\right] &= \int_0^T \mathbb{E}\left[\left|f(t) - f^\Pi(t)\right|^2\right] dt \\ &= \sum_{j=1}^n \int_{s_{j-1}}^{s_j} \mathbb{E}\left[\left|f(t) - f(s_{j-1})\right|^2\right] dt \\ &\leqslant \sum_{j=1}^n \int_{s_{j-1}}^{s_j} \underbrace{\sup_{u,v\in[s_{j-1},s_j]} \mathbb{E}\left[\left|f(u) - f(v)\right|^2\right]}_{\to 0,\ |\Pi|\to 0} dt \\ &\xrightarrow[|\Pi|\to 0]{} 0.\end{aligned}$$

Combining this with Lemma 15.14 finally gives

$$\sum_{j=1}^n f(s_{j-1}) \cdot (B_{s_j} - B_{s_{j-1}}) = \int_0^T f^\Pi(t)\, dB_t \xrightarrow[|\Pi|\to 0]{L^2(\mathbb{P})} \int_0^T f(t)\, dB_t. \qquad \square$$

15.5 What is the closure of $\mathcal{S}_T$?

Let us now see how big $\mathcal{L}_T^2 = \overline{\mathcal{S}}_T$ really is. For this we need some preparations

Ex. 15.16 **15.17 Definition.** Let $(B_t)_{t\geqslant 0}$ be a BM[1] and $(\mathscr{F}_t)_{t\geqslant 0}$ an admissible filtration. The *progres-*
Ex. 15.17 *sive σ-algebra* $\mathscr{P}$ is the family of all sets $\Gamma \subset [0, T] \times \Omega$ such that

$$\Gamma \cap ([0, t] \times \Omega) \in \mathscr{B}[0, t] \otimes \mathscr{F}_t \quad \text{for all } t \leqslant T. \tag{15.24}$$

Ex. 15.18 By $L^2_{\mathscr{P}}(\lambda_T \otimes \mathbb{P})$ we denote those equivalence classes in $L^2([0, T] \times \Omega, \lambda_T \otimes \mathbb{P})$ which have a $\mathscr{P}$ measurable representative.

In Lemma 6.24 we have already seen that Brownian motion itself is $\mathscr{P}$ measurable.

15.18 Lemma. *Let $T > 0$. Then $(L^2_{\mathscr{P}}(\lambda_T \otimes \mathbb{P}), \|\cdot\|_{L^2(\lambda_T\otimes\mathbb{P})})$ is a Hilbert space.*

Proof. By definition, $L^2_{\mathscr{P}}(\lambda_T \otimes \mathbb{P}) \subset L^2(\lambda_T \otimes \mathbb{P})$ is a subspace and $L^2(\lambda_T \otimes \mathbb{P})$ is a Hilbert space for the norm $\|\cdot\|_{L^2(\lambda_T\otimes\mathbb{P})}$. It is, therefore, enough to show that for any sequence $(f_n)_{n\geqslant 0} \subset L^2_{\mathscr{P}}(\lambda_T \otimes \mathbb{P})$ such that $L^2(\lambda_T \otimes \mathbb{P})\text{-}\lim_{n\to\infty} f_n = f$, the limit f has a $\mathscr{P}$ measurable representative. We may assume that the processes f_n are $\mathscr{P}$ measurable. Since f_n converges in L^2, there exists a subsequence $(f_{n(j)})_{j\geqslant 1}$ such that $\lim_{j\to\infty} f_{n(j)} = f$ $\lambda \otimes \mathbb{P}$ almost everywhere. Therefore, the upper limit $\phi(t, \omega) := \overline{\lim}_{j\to\infty} f_{n(j)}(t, \omega)$ is a $\mathscr{P}$ measurable representative of f. $\square$

Our aim is to show that $\mathcal{L}_T^2 = L^2_{\mathscr{P}}(\lambda_T \otimes \mathbb{P})$. For this we need a simple deterministic lemma. As before we set $L^2(\lambda_T) = L^2([0, T], \mathscr{B}[0, T], \lambda)$.

15.19 Lemma. *Let $\phi \in L^2(\lambda_T)$ and $T > 0$. Then*

$$\Theta_n[\phi](t) := \sum_{j=1}^{n-1} \xi_{j-1} \mathbb{1}_{[t_{j-1},t_j)}(t) \tag{15.25}$$

where

$$t_j = \frac{jT}{n}, \quad j = 0, \ldots, n, \qquad \xi_0 = 0 \quad \text{and} \quad \xi_{j-1} = \frac{n}{T} \int_{t_{j-2}}^{t_{j-1}} \phi(s)\, ds, \; j \geqslant 2,$$

is a sequence of step functions with $L^2(\lambda_T)$-$\lim_{n\to\infty} \Theta_n[\phi] = \phi$.

Proof. If ϕ is a step function, it has only finitely many jumps, say $\sigma_1 < \sigma_2 < \cdots < \sigma_m$, and set $\sigma_0 := 0$. Choose n so large that in each interval $[t_{j-1}, t_j)$ there is at most one jump σ_k. From the definition of $\Theta_n[\phi]$ it is easy to see that ϕ and $\Theta_n[\phi]$ differ exactly in the first, the last and in those intervals $[t_{j-1}, t_j)$ which contain some σ_k or whose predecessor $[t_{j-2}, t_{j-1})$ contains some σ_k. Formally,

$$\phi(t) = \Theta_n[\phi](t) \quad \text{for all } t \notin \bigcup_{j=0}^{m} [\sigma_j - \tfrac{T}{n}, \sigma_j + \tfrac{2T}{n}) \cup [0, \tfrac{T}{n}) \cup [\tfrac{n-1}{n}T, T].$$

This is shown in Figure 15.1. Therefore, $\lim_{n\to\infty} \Theta_n[\phi] = \phi$ Lebesgue almost everywhere and in $L^2(\lambda_T)$.

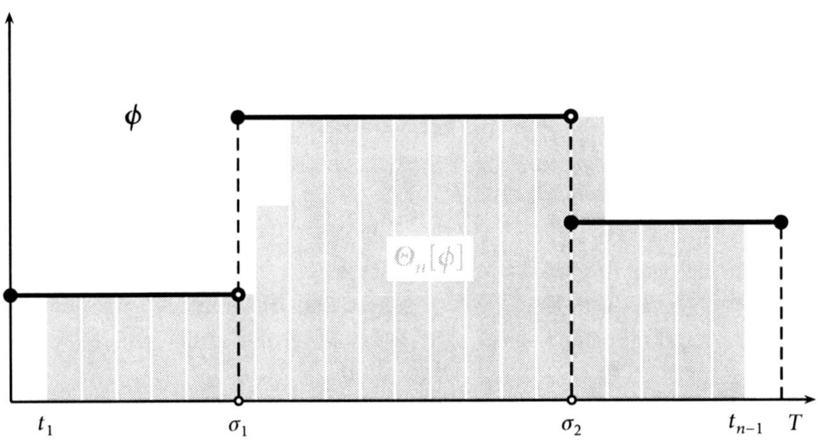

Fig. 15.1. Approximation of a simple function by a left-continuous simple function.

Since the step functions are dense in $L^2(\lambda_T)$, we get for every $\phi \in L^2(\lambda_T)$ a sequence of step functions $L^2(\lambda_T)$-$\lim_{k\to\infty} \phi^k = \phi$. Now let $\Theta_n[\phi]$ and $\Theta_n[\phi^k]$ be the step functions defined in (15.25). Then $\Theta_n[\cdot]$ is a contraction,

$$\begin{aligned}
\int_0^T |\Theta_n[\phi](s)|^2\,ds &= \sum_{j=1}^{n-1} \xi_{j-1}^2(t_j - t_{j-1}) \\
&= \sum_{j=2}^{n-1} (t_j - t_{j-1}) \left[\frac{1}{t_{j-1} - t_{j-2}} \int_{t_{j-2}}^{t_{j-1}} \phi(s)\,ds\right]^2 \\
&\overset{\text{Jensen}}{\leqslant} \sum_{j=2}^{n-1} \underbrace{\frac{t_j - t_{j-1}}{t_{j-1} - t_{j-2}}}_{=1} \int_{t_{j-2}}^{t_{j-1}} |\phi(s)|^2\,ds \\
&\leqslant \int_0^T |\phi(s)|^2\,ds,
\end{aligned}$$

and, by the triangle inequality and the contractivity of $\Theta_n[\cdot]$,

$$\begin{aligned}
\|\phi - \Theta_n[\phi]\|_{L^2} &\leqslant \|\phi - \phi^k\|_{L^2} + \|\phi^k - \Theta_n[\phi^k]\|_{L^2} + \|\Theta_n[\phi^k] - \Theta_n[\phi]\|_{L^2} \\
&= \|\phi - \phi^k\|_{L^2} + \|\phi^k - \Theta_n[\phi^k]\|_{L^2} + \|\Theta_n[\phi^k - \phi]\|_{L^2} \\
&\leqslant \|\phi - \phi^k\|_{L^2} + \|\phi^k - \Theta_n[\phi^k]\|_{L^2} + \|\phi^k - \phi\|_{L^2}.
\end{aligned}$$

Letting $n \to \infty$ (with k fixed) we get $\overline{\lim}_{n\to\infty} \|\phi - \Theta_n[\phi]\|_{L^2} \leqslant 2\|\phi^k - \phi\|_{L^2}$; now $k \to \infty$ shows that the whole expression tends to zero. □

15.20 Theorem. *The $L^2(\lambda_T \otimes \mathbb{P})$ closure of $\mathcal{S}_T$ is $L^2_{\mathscr{P}}(\lambda_T \otimes \mathbb{P})$.*

Proof. Write $\mathcal{L}_T^2 = \overline{\mathcal{S}}_T$. Obviously, $\mathcal{S}_T \subset L^2_{\mathscr{P}}(\lambda_T \otimes \mathbb{P})$. By Lemma 15.18, the set $L^2_{\mathscr{P}}(\lambda_T \otimes \mathbb{P})$ is closed under the norm $\|\cdot\|_{\mathcal{L}_T^2}$, thus $\overline{\mathcal{S}}_T = \mathcal{L}_T^2 \subset L^2_{\mathscr{P}}(\lambda_T \otimes \mathbb{P})$. All that remains to be shown is $L^2_{\mathscr{P}}(\lambda_T \otimes \mathbb{P}) \subset \mathcal{L}_T^2$.

Let $f(t,\omega)$ be (the progressive representative of) a process in $L^2_{\mathscr{P}}(\lambda_T \otimes \mathbb{P})$. By dominated convergence we see that the truncated process $f_C := -C \vee f \wedge C$ converges in $L^2(\lambda_T \otimes \mathbb{P})$ to f as $C \to \infty$. Therefore, we can assume that f is bounded.

Define for each ω a simple process $\Theta_n[f](t,\omega)$, where Θ_n is as in Lemma 15.19, i. e.

$$\xi_{j-1}(\omega) := \begin{cases} \frac{n}{T} \int_{t_{j-2}}^{t_{j-1}} f(s,\omega)\,ds, & \text{if the integral exists} \\ 0, & \text{else.} \end{cases}$$

Note that the set of ω for which the integral does not exist is an $\mathscr{F}_{t_{j-1}}$ measurable null set, since $f(\cdot,\omega)$ is $\mathscr{P}$ measurable and

$$\mathbb{E}\left[\int_{t_{j-2}}^{t_{j-1}} |f(s,\cdot)|^2\,ds\right] \leqslant \mathbb{E}\left[\int_0^T |f(s,\cdot)|^2\,ds\right] < \infty, \qquad j \geqslant 2.$$

Thus, $\Theta_n[f] \in \mathcal{S}_T$. By the proof of Lemma 15.19, we have for $\mathbb{P}$ almost all ω

$$\int_0^T |\Theta_n[f](t,\omega)|^2 \, dt \leqslant \int_0^T |f(t,\omega)|^2 \, dt, \tag{15.26}$$

$$\int_0^T |\Theta_n[f](t,\omega) - f(t,\omega)|^2 \, dt \xrightarrow[n\to\infty]{} 0. \tag{15.27}$$

If we take expectations in (15.27), we get by dominated convergence – the majorant is guaranteed by (15.26) – that $L^2(\lambda_T \otimes \mathbb{P})\text{-}\lim_{n\to\infty} \Theta_n[f] = f$, thus $f \in \mathcal{L}^2_T$. □

15.6 The stochastic integral for martingales

Up to now we have defined the stochastic integral with respect to a Brownian motion and for integrands in the closure of the simple processes $\mathcal{L}^2_T = \overline{\mathcal{S}}_T$. For $f \in \mathcal{S}_T$, we can define the stochastic integral for any continuous L^2 martingale $M \in \mathcal{M}^{2,c}_T$, cf. Definition 15.6; only in the completion procedure, see Section 15.3, we used that $(B_t^2 - t, \mathscr{F}_t)_{t\leqslant T}$ is a martingale, that is $\langle B \rangle_t = t$. If we *assume*[3] that $(M_t, \mathscr{F}_t)_{t\leqslant T}$ is a continuous square-integrable martingale such that

$$\begin{gathered}(M_t^2 - A_t, \mathscr{F}_t)_{t\leqslant T} \quad \text{is a martingale for some} \\ \text{adapted, continuous and increasing process } (A_t)_{t\leqslant T}\end{gathered} \tag{$\star$}$$

then we can define $f \bullet M$ for f from a suitable completion of $\mathcal{S}_T$.

As before, $(\mathscr{F}_t)_{t\leqslant T}$ is a complete filtration. Unless otherwise stated, we assume that all processes are adapted to this filtration and that all martingales are martingales w. r. t. this filtration. Let $M \in \mathcal{M}^{2,c}_T$ be such that $(\star)$ holds and $0 = t_0 < t_1 < \cdots < t_n = T$ be any partition of $[0,T]$. Then $(M_{t_j})_j$ is a martingale and $(\star)$ shows, because of the uniqueness of the Doob decomposition, that $A_{t_j} = \langle M \rangle_{t_j}$. This allows us to use the proof of Theorem 15.8 c) and to see that the following Itô isometry holds for all simple processes $f \in \mathcal{S}_T$:

$$\|f \bullet M_T\|^2_{L^2(\mathbb{P})} = \mathbb{E}\left[(f \bullet M)_T^2\right] = \mathbb{E}\left[\int_0^T |f(s)|^2 \, dA_s\right] =: \|f\|^2_{L^2(\mu_T \otimes \mathbb{P})}. \tag{15.28}$$

For fixed ω we denote by $\mu_T(\omega, ds) = dA_s(\omega)$ the measure on $[0,T]$ induced by the increasing function $s \mapsto A_s(\omega)$, i. e. $\mu_T(\omega, [s,t)) = A_t(\omega) - A_s(\omega)$ for all $s \leqslant t \leqslant T$.

This observation allows us to follow the procedure set out in Section 15.3 to define a stochastic integral for continuous square-integrable martingales.

---

3 In fact, *every* martingale $M \in \mathcal{M}^{2,c}_T$ satisfies $(\star)$, see Section A.6 in the appendix.

15.21 Definition. Let $M \in \mathcal{M}_T^{2,c}$ satisfying ($\star$) and set $\mathcal{L}_T^2(M) := \overline{\mathcal{S}}_T$ where the closure is taken in the space $L^2(\mu_T \otimes \mathbb{P})$. Then the *stochastic integral* (or *Itô integral*) is defined as

$$f \bullet M_t = \int_0^t f(s)\, dM_s := L^2(\mathbb{P})\text{-}\lim_{n\to\infty} f_n \bullet M_t, \quad 0 \leqslant t \leqslant T, \tag{15.29}$$

where $(f_n)_{n\geqslant 1} \subset \mathcal{S}_T$ is any sequence which approximates $f \in \mathcal{L}_T^2(M)$ in the $L^2(\mu_T \otimes \mathbb{P})$-norm.

Exactly the same arguments as in Section 15.3 can be used to show the analogue of Theorem 15.13.

Ex. 15.19 **15.22 Theorem.** *Let $(M_t, \mathscr{F}_t)_{t\leqslant T}$ be in $\mathcal{M}_T^{2,c}$ satisfying ($\star$), write μ_T for the measure induced by the increasing process A_s, and let $f \in \mathcal{L}_T^2(M)$. Then we have for all $t \leqslant T$*

a) *$f \mapsto f \bullet M$ is a linear map from $\mathcal{L}_T^2(M)$ into $\mathcal{M}_T^{2,c}$, i. e. $(f \bullet M_t)_{t\leqslant T}$ is a continuous L^2 martingale;*

b) *$(f \bullet M)^2 - f^2 \bullet A$ is a martingale where $f^2 \bullet A_t := \int_0^t |f(s)|^2\, dA_s$; in particular*

$$\mathbb{E}\left[(f \bullet M_t - f \bullet M_s)^2 \mid \mathscr{F}_s\right] = \mathbb{E}\left[\int_s^t |f(r)|^2\, dA_r \,\middle|\, \mathscr{F}_s\right], \quad s < t \leqslant T. \tag{15.30}$$

c) *Itô's isometry.*

$$\|f \bullet M_T\|^2_{L^2(\mathbb{P})} = \mathbb{E}\left[(f \bullet M)_T^2\right] = \mathbb{E}\left[\int_0^T |f(s)|^2\, dA_s\right] = \|f\|^2_{L^2(\mu_T\otimes\mathbb{P})}. \tag{15.31}$$

d) *Maximal inequalities.*

$$\mathbb{E}\left[\int_0^T |f(s)|^2\, dA_s\right] \leqslant \mathbb{E}\left[\sup_{t\leqslant T}\left(\int_0^t f(s)\, dM_s\right)^2\right] \leqslant 4\,\mathbb{E}\left[\int_0^T |f(s)|^2\, dA_s\right]. \tag{15.32}$$

e) *Localization in time. Let τ be an $\mathscr{F}_t$ stopping time. Then*

$$(f \bullet M)_t^\tau = f \bullet (M^\tau)_t = (f \mathbb{1}_{[0,\tau)}) \bullet M_t \quad \text{for all } t \leqslant T.$$

f) *Localization in space. Assume that $f, g \in \mathcal{L}_T^2(M)$ satisfy $f(s,\omega) = g(s,\omega)$ for all $(s,\omega) \in [0,t] \times F$ for some $t \leqslant T$ and $F \in \mathscr{F}_\infty$. Then*

$$(f \bullet M_s)\mathbb{1}_F = (g \bullet M_s)\mathbb{1}_F, \quad 0 \leqslant s \leqslant t.$$

The characterization of $\overline{\mathcal{S}}_T$, see Section 15.5, relies on a deterministic approximation result, Lemma 15.19, and this remains valid in the general martingale setting.

15.23 Theorem. *The $L^2(\mu_T \otimes \mathbb{P})$ closure of $\mathcal{S}_T$ is $L^2_{\mathcal{P}}(\mu_T \otimes \mathbb{P})$, i. e. the family of all equivalence classes in $L^2(\mu_T \otimes \mathbb{P})$ which have a progressively measurable representative.*

Sketch of the proof. Replace in Lemma 15.19 the measure λ_T by μ_T and the partition points t_j by the random times $\tau_j(\omega) := \inf\{s \geqslant 0 : A_s(\omega) > jT/n\} \wedge T$, $j \geqslant 1$, and $\tau_0(\omega) = 0$.

The continuity of $s \mapsto A_s$ ensures that $A_{\tau_j} - A_{\tau_{j-1}} = T/n$ for all $1 \leqslant j \leqslant \lfloor nA_T/T \rfloor$. Therefore, the proof of Lemma 15.19 remains valid.

Moreover, $\{\tau_j \leqslant t\} = \{A_t \geqslant jT/n\} \in \mathscr{F}_t$, i. e. the τ_j are stopping times. Since $L^2_{\mathcal{P}}(\mu_T \otimes \mathbb{P})$ is closed (Lemma 15.18), the inclusion $\overline{\mathcal{S}}_T \subset L^2_{\mathcal{P}}(\mu_T \otimes \mathbb{P})$ is clear. For the other direction we can argue as in Theorem 15.20 and see that every $f \in L^2_{\mathcal{P}}(\mu_T \otimes \mathbb{P})$ can be approximated by a sequence of random step functions of the form

$$\sum_{j=1}^{\infty} \xi_{j-1}(\omega)\mathbb{1}_{[\tau_{j-1}(\omega),\tau_j(\omega))}(t), \qquad \xi_{j-1} \text{ is } \mathscr{F}_{\tau_{j-1}} \text{ mble}, \quad |\xi_{j-1}| \leqslant C.$$

We are done if we can show that any process $\xi(\omega)\mathbb{1}_{[\sigma(\omega),\tau(\omega))}(t)$, where $\sigma \leqslant \tau \leqslant T$ are stopping times and ξ is a bounded and $\mathscr{F}_\sigma$ measurable random variable, can be approximated in $L^2(\mu_T \otimes \mathbb{P})$ by a sequence from $\mathcal{S}_T$.

Set $\sigma^m := \sum_{k=1}^{\lfloor mT \rfloor} \frac{k}{m} \mathbb{1}_{[(k-1)/m,\,k/m)}(\sigma) + \frac{\lfloor mT \rfloor}{m} \mathbb{1}_{[\lfloor mT \rfloor/m,\,T]}(\sigma)$ and define τ^m in the same way. Then

$$\xi(\omega)\mathbb{1}_{[\sigma^m(\omega),\tau^m(\omega))}(t) = \sum_{k=1}^{\lfloor mT \rfloor} \xi(\omega)\mathbb{1}_{\{\sigma<\frac{k-1}{m}\leqslant\tau\}}(\omega)\mathbb{1}_{[\frac{k-1}{m},\frac{k}{m})}(t) + \xi(\omega)\mathbb{1}_{\{\sigma<\frac{\lfloor mT \rfloor}{m}\leqslant\tau\}}(\omega)\mathbb{1}_{[\frac{\lfloor mT \rfloor}{m},T]}(t).$$

Because for all Borel sets $B \subset \mathbb{R}$ and $k = 1, 2, \ldots, \lfloor mT \rfloor + 1$

$$\underbrace{\overbrace{\{\xi \in B\}}^{\in \mathscr{F}_\sigma} \cap \{\sigma < (k-1)/m\}}_{\in \mathscr{F}_{(k-1)/m}} \cap \{(k-1)/m \leqslant \tau\} \in \mathscr{F}_{(k-1)/m},$$

we see that $\xi(\omega)\mathbb{1}_{[\sigma^m(\omega),\tau^m(\omega))}(t)$ is in $\mathcal{S}_T$. Finally,

$$\mathbb{E}\left[\int_0^T \left|\xi\mathbb{1}_{[\sigma^m,\tau^m)}(t) - \xi\mathbb{1}_{[\sigma,\tau)}(t)\right|^2 dA_t\right] \leqslant 2C^2\left[\mathbb{E}(A_{\tau^m} - A_\tau) + \mathbb{E}(A_{\sigma^m} - A_\sigma)\right],$$

and this tends to zero as $m \to \infty$. □

15.24 Further reading. Most books consider stochastic integrals immediately for continuous martingales. A particularly nice and quite elementary presentation is [134]. The approach of [189] is short and mathematically elegant. For discontinuous martingales standard references are [100] who use a point-process perspective, and [184] who starts out with semi-martingales as 'good' stochastic integrators. [132] and [133] construct the stochastic integral using random orthogonal measures. This leads to a unified treatment of Brownian motion and stationary processes.

[100] Ikeda, Watanabe: *Stochastic Differential Equations and Diffusion Processes.*
[132] Krylov: *Introduction to the Theory of Diffusion Processes.*
[133] Krylov: *Introduction to the Theory of Random Processes.*
[134] Kunita: *Stochastic Flows and Stochastic Differential Equations.*
[184] Protter: *Stochastic Integration and Differential Equations.*
[189] Revuz, Yor: *Continuous Martingales and Brownian Motion.*

Problems

1. Let $(M_n, \mathscr{F}_n)_{n\geqslant 0}$ and $(N_n, \mathscr{F}_n)_{n\geqslant 0}$ be L^2 martingales; then $(M_nN_n - \langle M, N\rangle_n, \mathscr{F}_n)_{n\geqslant 0}$ is a martingale.

2. Let $f \in \mathcal{S}_T$ be a simple process and $M \in \mathcal{M}_T^{2,c}$. Show that the definition of the stochastic integral $\int_0^T f(s)\,dM_s$ (cf. Definition 15.7) does not depend on the particular representation of the simple process (15.11).

3. Show that $\|M\|_{\mathcal{M}_T^2} := \left(\mathbb{E}\left[\sup_{s\leqslant T}|M_s|^2\right]\right)^{\frac{1}{2}}$ is a norm in the family $\mathcal{M}_T^2$ of equivalence classes of L^2 martingales (M and $\widetilde{M}$ are said to be equivalent if, and only if, $\sup_{0\leqslant t\leqslant T}|M_t - \widetilde{M}_t| = 0$ a. s.).

4. Show that the limit (15.18) does not depend on the approximating sequence.

5. Let $(B_t, \mathscr{F}_t)_{t\geqslant 0}$ be a BM^1 and τ a stopping time. Find $\langle B^\tau\rangle_t$.

6. Let $(B_t)_{t\geqslant 0}$ be a BM^1 and $f, g \in \mathcal{L}_T^2$. Show that
   (a) $\mathbb{E}(f \bullet B_t \cdot g \bullet B_t \mid \mathscr{F}_s) = \mathbb{E}\left[\int_s^t f(u,\cdot)g(u,\cdot)\,du \mid \mathscr{F}_s\right]$ if f, g vanish on $[0, s]$.
   (b) $\mathbb{E}\left[f \bullet B_t \mid \mathscr{F}_s\right] = 0$ if f vanishes on $[0, s]$.
   (c) $f(t, \omega) = 0$ for all $\omega \in A \in \mathscr{F}_T$ and all $t \leqslant T$ implies $f \bullet B(\omega) = 0$ for all $\omega \in A$.

7. Let $(B_t)_{t\geqslant 0}$ be a BM^1 and $(f_n)_{n\geqslant 1}$ a sequence in $\mathcal{L}_T^2$ such that $f_n \to f$ in $L^2(\lambda_T \otimes \mathbb{P})$. Then $f_n \bullet B \to f \bullet B$ in $\mathcal{M}_T^{2,c}$.

8. Let $(B_t)_{t\geqslant 0}$ be a BM^1 and $f \in L^2(\lambda_T \otimes \mathbb{P})$ for some $T > 0$. Assume that the limit $\lim_{\epsilon\to 0} f(\epsilon) = f(0)$ exists in $L^2(\mathbb{P})$. Show that
$$\frac{1}{B_\epsilon}\int_0^\epsilon f(s)\,dB_s \xrightarrow[\epsilon\to 0]{\mathbb{P}} f(0).$$

9. Use the fact that any continuous, square-integrable martingale with bounded variation paths is constant (cf. Proposition A.22) to show the following: $\langle f \bullet B\rangle_t := \int_0^t |f(s)|^2\,ds$ is the unique continuous and increasing process such that $(f \bullet B)^2 - f^2 \bullet \langle B\rangle$ is a martingale.

10. The quadratic covariation of two *continuous* L^2 martingales $M, N \in \mathcal{M}_T^{2,c}$ is defined as in the discrete case by the polarization formula (15.8).
Let $(B_t)_{t\geqslant 0}$ be a BM^1 and $f, g \in \mathcal{L}_T^2$. Show that $\langle f \bullet B, g \bullet B\rangle_t = \int_0^t f(s)g(s)\,ds$ for all $t \in [0,T]$.

11. Let $(X_n)_{n\geqslant 0}$ be a sequence of r. v. in $L^2(\mathbb{P})$. Show that $L^2\text{-}\lim_{n\to\infty} X_n = X$ implies $L^1\text{-}\lim_{n\to\infty} X_n^2 = X^2$.

12. Let $(B_t)_{t\geqslant 0}$ be a BM^1. Show that $\int_0^T B_s^2\,dB_s = \frac{1}{3}B_T^3 - \int_0^T B_s\,ds$, $T > 0$.

13. Let $(B_t)_{t\geqslant 0}$ be a BM^1 and $f \in \mathrm{BV}[0,T]$, $T < \infty$, a non-random function. Prove that $\int_0^T f(s)\,dB_s = f(T)B_T - \int_0^T B_s\,df(s)$ (the latter integral is understood as the mean-square limit of Riemann–Stieltjes sums). Conclude from this that the Itô integral extends the Paley–Wiener–Zygmund integral, cf. Paragraph 13.5.

14. Adapt the proof of Proposition 15.16 and prove that we also have

$$\lim_{|\Pi|\to 0} \mathbb{E}\left[\sup_{t\leqslant T}\left|\int_0^t f(s)\,dB_s - \sum_{j=1}^n f(s_{j-1})(B_{s_j\wedge t} - B_{s_{j-1}\wedge t})\right|^2\right] = 0.$$

15. Let $(B_t, \mathscr{F}_t)_{t\geqslant 0}$ be a one-dimensional Brownian motion, $T < \infty$ and Π_n, $n \geqslant 1$, a sequence of partitions of $[0,T]$: $0 = t_{n,0} < t_{n,1} < \cdots < t_{n,k(n)} = T$. Assume that $\lim_{n\to\infty} |\Pi_n| = 0$ and define intermediate points $\theta_{n,k}^\alpha := t_{n,k} + \alpha(t_{n,k+1} - t_{n,k})$ for some fixed $0 \leqslant \alpha \leqslant 1$. Show that the Riemann–Stieltjes sums

$$\sum_{k=0}^{k(n)-1} B(\theta_{n,k}^\alpha)(B(t_{n,k+1}) - B(t_{n,k}))$$

converge in $L^2(\mathbb{P})$ and find the limit $L_T(\alpha)$. For which α is $L_t(\alpha)$, $0 \leqslant t \leqslant T$ a martingale?

16. Let $(B_t, \mathscr{F}_t)_{t\geqslant 0}$ be a one-dimensional Brownian motioin and τ a stopping time. Show that $f(s,\omega) := \mathbb{1}_{[0,T\wedge\tau(\omega))}(s)$, $0 \leqslant s \leqslant T < \infty$ is progressively measurable. Use the very definition of the Itô integral to calculate $\int_0^T f(s)\,dB_s$ and compare the result with the localization principle of Theorem 15.13.

17. Show that

$$\mathscr{P} = \{\Gamma : \Gamma \subset [0,T]\times\Omega,\ \Gamma \cap ([0,t]\times\Omega) \in \mathscr{B}[0,t]\otimes\mathscr{F}_t \quad \text{for all } t \leqslant T\}$$

is a σ-algebra.

18. Show that every right- or left-continuous adapted process is $\mathscr{P}$ measurable.

19. Show that the process $f^2 \bullet A_t := \int_0^t |f(s)|^2\,dA_s$ appearing in Theorem 15.22 b) is adapted.

16 Stochastic integrals: beyond $\mathcal{L}_T^2$

Let $(B_t)_{t\geqslant 0}$ be a one-dimensional Brownian motion and $\mathscr{F}_t$ an admissible, right-continuous complete filtration, e. g. $\mathscr{F}_t = \overline{\mathscr{F}}_t^B$, cf. Theorem 6.21. We have seen in Chapter 15 that the Itô integral $\int_0^T f(s)\,dB_s$ exists for all integrands $f \in \mathcal{L}_T^2 = L^2_{\mathscr{P}}(\lambda_T \otimes \mathbb{P})$, i. e. for all stochastic processes f which have a representative which is $\mathscr{P}$ measurable, cf. Definition 15.17, and satisfy

$$\mathbb{E}\left[\int_0^T |f(s,\cdot)|^2\,ds\right] < \infty. \tag{16.1}$$

It is possible to weaken (16.1) and to extend the family of possible integrands.

16.1 Definition. We say that $f \in \mathcal{L}_{T,\mathrm{loc}}^2$, if there exists a sequence $(\sigma_n)_{n\geqslant 0}$ of $\mathscr{F}_t$ stopping times such that $\sigma_n \uparrow \infty$ a. s. and $f\mathbb{1}_{[0,\sigma_n)} \in \mathcal{L}_T^2$ for all $n \geqslant 0$. The sequence $(\sigma_n)_{n\geqslant 0}$ is called a *localizing sequence.*

The family $\mathcal{L}_T^0$ consists of all $\mathscr{P}$ measurable processes f such that

$$\int_0^T |f(s,\cdot)|^2\,ds < \infty \quad \text{a. s.} \tag{16.2}$$

16.2 Lemma. *Let $f \in \mathcal{L}_T^0$. Then $\int_0^t |f(s,\cdot)|^2\,ds$ is $\mathscr{F}_t$ measurable for all $t \leqslant T$.*

Proof. For every $f \in \mathcal{L}_T^0$ we see that $|f|^2$ is $\mathscr{P}$ measurable.

Therefore, $(s,\omega) \mapsto |f(s,\omega)|^2\mathbb{1}_{[0,t]}(s)$ is $\mathscr{B}[0,t] \otimes \mathscr{F}_t$ measurable for all $t \in [0,T]$, and Tonelli's theorem shows that $\int_0^t |f(s,\cdot)|^2\,ds$ is $\mathscr{F}_t$ measurable. □

Ex. 16.1 **16.3 Lemma.** $\mathcal{L}_T^2 \subset \mathcal{L}_T^0 \subset \mathcal{L}_{T,\mathrm{loc}}^2$.

Proof. Since a square-integrable random variable is a. s. finite, the first inclusion is clear.

To see the other inclusion, we assume that $f \in \mathcal{L}_T^0$ and define

$$\sigma_n = \inf\left\{t \leqslant T : \int_0^t |f(s,\cdot)|^2\,ds \geqslant n\right\}, \quad n \geqslant 1, \tag{16.3}$$

($\inf \emptyset = \infty$). Clearly, $\sigma_n \uparrow \infty$. Moreover,

$$\sigma_n(\omega) > t \iff \int_0^t |f(s,\omega)|^2\,ds < n.$$

By Lemma 16.2, the integral is $\mathscr{F}_t$ measurable; thus, $\{\sigma_n \leqslant t\} = \{\sigma_n > t\}^c \in \mathscr{F}_t$. This means that σ_n is a stopping time.

Note that $f\mathbb{1}_{[0,\sigma_n)}$ is $\mathscr{P}$ measurable since it is the product of two $\mathscr{P}$ measurable random variables. From Ex. 15.16

$$\mathbb{E}\left[\int_0^T |f(s)\mathbb{1}_{[0,\sigma_n)}(s)|^2\,ds\right] = \mathbb{E}\left[\int_0^{\sigma_n} |f(s)|^2\,ds\right] \leqslant \mathbb{E}\,n = n,$$

we conclude that $f\mathbb{1}_{[0,\sigma_n)} \in \mathcal{L}_T^2$. □

16.4 Lemma. *Let $f \in \mathcal{L}_{T,\text{loc}}^2$ and set $f_n := f\mathbb{1}_{[0,\sigma_n)}$ for a localizing sequence $(\sigma_n)_{n\geqslant 0}$. Then*

$$(f_n \bullet B)_t^{\sigma_m} = f_m \bullet B_t \quad \text{for all } m \leqslant n \quad \text{and} \quad t \leqslant \sigma_m.$$

Proof. Using Theorem 15.13 e) for $f_n, f_m \in \mathcal{L}_T^2$ we get for all $t \leqslant \sigma_m$

$$(f_n \bullet B)_t^{\sigma_m} = (f_n\mathbb{1}_{[0,\sigma_m)}) \bullet B_t = (f\mathbb{1}_{[0,\sigma_n)}\mathbb{1}_{[0,\sigma_m)}) \bullet B_t = (f\mathbb{1}_{[0,\sigma_m)}) \bullet B_t = f_m \bullet B_t. \qquad \square$$

Lemma 16.4 allows us to define the stochastic integral for integrands in $\mathcal{L}_{T,\text{loc}}^2$.

16.5 Definition. Let $f \in \mathcal{L}_{T,\text{loc}}^2$ with localizing sequence $(\sigma_n)_{n\geqslant 0}$. Then

$$f \bullet B_t(\omega) := (f\mathbb{1}_{[0,\sigma_n)}) \bullet B_t(\omega) \quad \text{for all } t \leqslant \sigma_n(\omega) \tag{16.4}$$

is called the *(extended) stochastic integral* or *(extended) Itô integral*. We also write $\int_0^t f(s)\,dB_s := f \bullet B_t$.

Lemma 16.4 guarantees that this definition is consistent and independent of the localizing sequence: Indeed for any two localizing sequences $(\sigma_n)_{n\geqslant 0}$ and $(\tau_n)_{n\geqslant 0}$ we see easily that $\rho_n := \sigma_n \wedge \tau_n$ is again such a sequence and then the telescoping argument of Lemma 16.4 applies.

We can also localize the notion of a martingale.

16.6 Definition. A *local martingale* is an adapted right-continuous process $(M_t, \mathscr{F}_t)_{t\leqslant T}$ for which there exists a sequence of stopping times $(\sigma_n)_{n\geqslant 0}$ such that $\sigma_n \uparrow \infty$ a. s. and for every $n \geqslant 0$, $(M_t^{\sigma_n}\mathbb{1}_{\{\sigma_n>0\}}, \mathscr{F}_t)_{t\leqslant T}$ is a martingale. The sequence $(\sigma_n)_{n\geqslant 0}$ is called a *localizing sequence*.

We denote by $\mathcal{M}_{T,\text{loc}}^c$ the continuous local martingales and by $\mathcal{M}_{T,\text{loc}}^{2,c}$ those local martingales where $M^{\sigma_n}\mathbb{1}_{\{\sigma_n>0\}} \in \mathcal{M}_T^{2,c}$.

Let us briefly review the properties of the extended stochastic integral. The following theorem should be compared with its $\mathcal{L}_T^2$ analogue, Theorem 15.13

16.7 Theorem. *Let $(B_t)_{t\geqslant 0}$ be a* BM[1]*, $(\mathscr{F}_t)_{t\geqslant 0}$ an admissible filtration and $f \in \mathcal{L}_{T,\text{loc}}^2$. Then we have for all $t \leqslant T$*

a) *$f \mapsto f \bullet B$ is a linear map from $\mathcal{L}_{T,\text{loc}}^2$ to $\mathcal{M}_{T,\text{loc}}^{2,c}$, i. e. $(f \bullet B_t)_{t\leqslant T}$ is a continuous local L^2 martingale;*

b) *$(f \bullet B)^2 - \langle f \bullet B\rangle$ is a local martingale where*

$$\langle f \bullet B\rangle_t := f^2 \bullet \langle B\rangle_t = \int_0^t |f(s)|^2\,ds;$$

c) *⟨no equivalent of Itô's isometry⟩*

d) $\mathbb{P}\left(\sup_{t\leqslant T}\left|\int_0^t f(s)\,dB_s\right| > \epsilon\right) \leqslant \frac{4C}{\epsilon^2} + \mathbb{P}\left(\int_0^T |f(s)|^2\,ds \geqslant C\right), \qquad \epsilon > 0,\ C > 0.$

e) *Localization in time. Let τ be a $\mathscr{F}_t$ stopping time. Then*

$$(f \bullet B)_t^\tau = f \bullet (B^\tau)_t = (f\mathbb{1}_{[0,\tau)}) \bullet B_t \quad \textit{for all } t \leqslant T.$$

f) *Localization in space. If the processes $f, g \in \mathcal{L}^2_{T,\text{loc}}$ satisfy $f(s,\omega) = g(s,\omega)$ for all $(s,\omega) \in [0,t] \times F$ for some $t \leqslant T$ and $F \in \mathscr{F}_\infty$, then*

$$(f \bullet B_s)\mathbb{1}_F = (g \bullet B_s)\mathbb{1}_F, \quad 0 \leqslant s \leqslant t.$$

Proof. Throughout the proof we fix some localizing sequence $(\sigma_n)_{n\geqslant 1}$.

We begin with the assertion e). By the definition of the extended Itô integral and Theorem 15.13 e) we have for all $n \geqslant 1$

$$((f\mathbb{1}_{[0,\sigma_n)}) \bullet B)^\tau = \underbrace{(f\mathbb{1}_{[0,\sigma_n)}\mathbb{1}_{[0,\tau)}) \bullet B}_{=((f\mathbb{1}_{[0,\tau)})\mathbb{1}_{[0,\sigma_n)})\bullet B} = (f\mathbb{1}_{[0,\sigma_n)}) \bullet (B^\tau).$$

Now a) follows directly from its counterpart in Theorem 15.13. To see b), we note that by Theorem 15.13 b), applied to $f\mathbb{1}_{[0,\sigma_n)}$, $(f \bullet B^{\sigma_n})^2 - \int_0^{\sigma_n\wedge\bullet} |f(s)|^2\,ds$, is a martingale for each $n \geqslant 1$.

For d) we define $\tau := \inf\left\{t \leqslant T : \int_0^t |f(s,\cdot)|^2\,ds \geqslant C\right\}$. As in Lemma 16.3 we see that τ is a stopping time. Therefore we get by e) and the maximal inequality (15.22)

$$\begin{aligned}
\mathbb{P}\left(\sup_{t\leqslant T}|f \bullet B_t| > \epsilon\right) &= \mathbb{P}\left(\sup_{t\leqslant T}|f \bullet B_t| > \epsilon,\ T < \tau\right) + \mathbb{P}\left(\sup_{t\leqslant T}|f \bullet B_t| > \epsilon,\ T \geqslant \tau\right)\\
&\overset{\text{e)}}{\leqslant} \mathbb{P}\left(\sup_{t\leqslant T}|(f\mathbb{1}_{[0,\tau)}) \bullet B_t| > \epsilon\right) + \mathbb{P}(T \geqslant \tau)\\
&\leqslant \frac{4}{\epsilon^2}\,\mathbb{E}\left[|(f\mathbb{1}_{[0,\tau)}) \bullet B_T|^2\right] + \mathbb{P}\left(\int_0^T |f(s)|^2\,ds \geqslant C\right)\\
&= \frac{4}{\epsilon^2}\,\mathbb{E}\left[\int_0^{T\wedge\tau} |f(s)|^2\,ds\right] + \mathbb{P}\left(\int_0^T |f(s)|^2\,ds \geqslant C\right)\\
&\leqslant \frac{4C}{\epsilon^2} + \mathbb{P}\left(\int_0^T |f(s)|^2\,ds \geqslant C\right),
\end{aligned}$$

and d) follows.

Let $f, g \in \mathcal{L}^2_{T,\text{loc}}$ with localizing sequences $(\sigma_n)_{n\geqslant 0}$ and $(\tau_n)_{n\geqslant 0}$, respectively. Then $\sigma_n \wedge \tau_n$ localizes both f and g, and by Theorem 15.13 f),

$$(f\mathbb{1}_{[0,\sigma_n\wedge\tau_n)}) \bullet B\mathbb{1}_F = (g\mathbb{1}_{[0,\sigma_n\wedge\tau_n)}) \bullet B\mathbb{1}_F \quad \text{for all } n \geqslant 0.$$

Because of the definition of the extended Itô integral this is just f). □

For the extended Itô integral we also have a Riemann–Stieltjes-type approximation, cf. Proposition 15.16.

16.8 Theorem. *Let f be an $\mathscr{F}_t$ adapted right-continuous process with left limits. Then $f \in \mathcal{L}^2_{T,\mathrm{loc}}$.*

For every sequence $(\Pi_k)_{k\geqslant 1}$ of partitions of $[0,T]$ such that $\lim_{k\to\infty}|\Pi_k| = 0$, the Riemann–Stieltjes sums

$$Y_t^{\Pi_k} := \sum_{s_j,s_{j+1}\in\Pi_k} f(s_j)(B_{s_{j+1}\wedge t} - B_{s_j\wedge t})$$

converge uniformly (on compact t-sets) in probability to $f \bullet B_t$, $t \leqslant T$:

$$\lim_{k\to\infty} \mathbb{P}\left(\sup_{0\leqslant t\leqslant T}\left|Y_t^{\Pi_k} - \int_0^t f(s)\,dB_s\right| > \epsilon\right) = 0 \quad \textit{for all } \epsilon > 0. \tag{16.5}$$

Proof. Since f is right-continuous with left limits, it is bounded on compact t-sets. Therefore

$$\tau_n := \inf\{s \geqslant 0 \,:\, |f(s)|^2 \geqslant n\} \wedge n$$

is a sequence of finite stopping times with $\tau_n \uparrow \infty$ a. s.

Let $f^{\Pi_k}(t) := \sum_{s_{j-1},s_j\in\Pi_k} f(s_{j-1})\mathbb{1}_{[s_{j-1},s_j)}(t)$. Using the stopping times $(\tau_n)_{n\geqslant 1}$ we infer from Lemma 15.14 that $f^{\Pi_k} \in \mathcal{L}^2_{T,\mathrm{loc}}$ and $Y_t^{\Pi_k} = f^{\Pi_k} \bullet B_t$.

Moreover,

$$\mathbb{E}\left[\int_0^{\tau_n\wedge T} |f(s)|^2\,ds\right] \leqslant \mathbb{E}\left[\int_0^{\tau_n\wedge T} n\,ds\right] \leqslant \mathbb{E}\left[\int_0^{T} n\,ds\right] \leqslant Tn,$$

i. e. τ_n is also a localizing sequence for the stochastic integral. Therefore, by Theorem 16.7 a) and e) and Theorem 15.13 c) and d),

$$\begin{aligned}
\mathbb{E}\left[\sup_{t\leqslant T}\left|Y_{t\wedge\tau_n}^{\Pi_k} - \int_0^{t\wedge\tau_n} f(s)\,dB_s\right|^2\right] &= \mathbb{E}\left[\sup_{t\leqslant T}\left|(f^{\Pi_k} - f)\bullet B_{t\wedge\tau_n}\right|^2\right] \\
&\overset{(15.22)}{\leqslant} 4\,\mathbb{E}\left[\left|(f^{\Pi_k} - f)\bullet B_{T\wedge\tau_n}\right|^2\right] \\
&\overset{(15.21)}{=} 4\,\mathbb{E}\left[\int_0^{\tau_n\wedge T} |f^{\Pi_k}(s) - f(s)|^2\,ds\right].
\end{aligned}$$

For $0 \leqslant s \leqslant \tau_n \wedge T$ the integrand is bounded by $4n$ and converges, as $k \to \infty$, to 0. By dominated convergence we see that the integral tends to 0 as $k \to \infty$. Finally, f is right-continuous with finite left-hand limits and, therefore, the set of discontinuity points $\{s \in [0,T] \,:\, f(s,\omega) \neq f(s-,\omega)\}$ is a Lebesgue null set.

Since $\lim_{n\to\infty}\tau_n = \infty$ a. s., we see that

$$\forall\,\delta > 0 \quad \exists\,N_\delta \quad \forall\,n \geqslant N_\delta \,:\, \mathbb{P}(\tau_n < T) < \delta.$$

Thus, by Chebyshev's inequality, Theorem 16.7 e) and the above calculation

$$\begin{aligned}
&\mathbb{P}\left(\sup_{t\leqslant T}\left|Y_t^{\Pi_k} - \int_0^t f(s)\,dB_s\right| > \epsilon\right)\\
&\leqslant \mathbb{P}\left(\sup_{t\leqslant T}\left|Y_t^{\Pi_k} - \int_0^t f(s)\,dB_s\right| > \epsilon,\ \tau_n \geqslant T\right) + \mathbb{P}(\tau_n < T)\\
&\leqslant \mathbb{P}\left(\sup_{t\leqslant T}\left|Y_{t\wedge\tau_n}^{\Pi_k} - \int_0^{t\wedge\tau_n} f(s)\,dB_s\right| > \epsilon\right) + \mathbb{P}(\tau_n < T)\\
&\leqslant \frac{4}{\epsilon^2}\,\mathbb{E}\left[\int_0^{T\wedge\tau_n} |f^{\Pi_k}(s) - f(s)|^2\,ds\right] + \delta \xrightarrow[k\to\infty]{} \delta \xrightarrow[\delta\to 0]{} 0. \qquad \square
\end{aligned}$$

Let $(g_n)_{n\geqslant1} \subset \mathcal{L}^2_{T,\mathrm{loc}}$ be a sequence of adapted processes and *assume* that there is a common localizing sequence $(\tau_k)_{k\geqslant1}$ for all g_n. A sufficient condition for this is, e. g. a majorization condition $|g_n| \leqslant G$ where $G \in \mathcal{L}^2_{T,\mathrm{loc}}$ is independent of $n \geqslant 1$.

The sequence $(g_n)_{n\geqslant1}$ converges locally in $\mathcal{L}^2_{T,\mathrm{loc}}$, if for each $k \geqslant 1$ the sequence $(g_n \mathbb{1}_{[0,\tau_k)})_{n\geqslant1}$ converges in $\mathcal{L}_T^2$. This is equivalent to saying that

$$\lim_{m,n\to\infty} \mathbb{E}\left[\int_0^{T\wedge\tau_k} |g_n(s) - g_m(s)|^2\,ds\right] = 0 \quad \text{for all } k \geqslant 1.$$

This shows that the limit g is in $\mathcal{L}^2_{T,\mathrm{loc}}$ and that $(\tau_k)_{k\geqslant1}$ is a localizing sequence for g. We will see that convergence in $\mathcal{L}^2_{T,\mathrm{loc}}$ turns into convergence in probability of the stochastic integrals.

16.9 Corollary. *Let $(B_t)_{t\geqslant0}$ be a BM^1, $(\mathscr{F}_t)_{t\geqslant0}$ an admissible filtration and $(g_n)_{n\geqslant1} \subset \mathcal{L}^2_{T,\mathrm{loc}}$ with a common localizing sequence $(\tau_k)_{k\geqslant1}$. If the sequence $(g_n)_{n\geqslant1}$ converges locally in $\mathcal{L}_T^2$ to $g \in \mathcal{L}^2_{T,\mathrm{loc}}$, then the sequence of stochastic integrals $(g_n \bullet B)_{n\geqslant1}$ converges uniformly in probability to $g \bullet B$, i. e.*

$$\lim_{n\to\infty} \mathbb{P}\left(\sup_{t\leqslant T}\left|\int_0^t (g_n(s) - g(s))\,dB_s\right| > \epsilon\right) = 0 \quad \textit{for all } \epsilon > 0. \tag{16.6}$$

Proof. Denote by $(\tau_k)_{k\geqslant1}$ the common localizing sequence for $(g_n)_{n\geqslant1}$ and g. Using Theorem 16.7 d) we get for all $\epsilon > 0$, $C > 0$ and $k, n \geqslant 1$

$$\begin{aligned}
&\mathbb{P}\left(\sup_{t\leqslant T}\left|\int_0^t (g_n(s) - g(s))\,dB_s\right| > \epsilon\right)\\
&\leqslant \frac{4C}{\epsilon^2} + \mathbb{P}\left(\int_0^T |g_n(s) - g(s)|^2\,ds \geqslant C\right)
\end{aligned}$$

$$
\begin{aligned}
&\leqslant \frac{4C}{\epsilon^2} + \mathbb{P}\Bigg(\int_0^T |g_n(s) - g(s)|^2 \, ds \geqslant C,\ \tau_k > T \Bigg) + \mathbb{P}(\tau_k \leqslant T) \\
&\leqslant \frac{4C}{\epsilon^2} + \mathbb{P}\Bigg(\int_0^{T\wedge\tau_k} |g_n(s) - g(s)|^2 \, ds \geqslant C \Bigg) + \mathbb{P}(\tau_k \leqslant T) \\
&\leqslant \frac{4C}{\epsilon^2} + \frac{1}{C}\mathbb{E}\Bigg(\int_0^{T\wedge\tau_k} |g_n(s) - g(s)|^2 \, ds \Bigg) + \mathbb{P}(\tau_k \leqslant T)
\end{aligned}
$$

keeping ϵ, C and k fixed, we can let $n \to \infty$ and get

$$
\xrightarrow[n\to\infty]{} \frac{4C}{\epsilon^2} + \mathbb{P}(\tau_k \leqslant T) \xrightarrow[k\to\infty]{} \frac{4C}{\epsilon^2} \xrightarrow[C\to 0]{} 0,
$$

where we used in the last line the fact that τ_k converges to infinity as $k \to \infty$. □

16.10 Further reading. Localized stochastic integrals are covered by most of the books mentioned in the *Further reading* section of Chapter 15. Theorem 16.8 has variants where the driving noise is an approximation of Brownian motion. Such *Wong–Zakai* results are important for practitioners. A good source are the original papers [235] and [99].

[99] Ikeda, Nakao, Yamato: A class of approximations of Brownian motion.

[235] Wong, Zakai: Riemann–Stieltjes approximations of stochastic integrals.

Problems

1. Let $T < \infty$. Do we have $\mathcal{L}_T^0 = \mathcal{L}_{T,\mathrm{loc}}^2$?

2. Let τ be a stopping time. Show that $\mathbb{1}_{[0,\tau)}$ is $\mathscr{P}$ measurable.

3. If σ_n is localizing for a local martingale, so is $\tau_n := \sigma_n \wedge n$.

4. Let $(I_t, \mathscr{F}_t)_{t\geqslant 0}$ be a continuous adapted process with values in $[0, \infty)$ and a. s. increasing sample paths. Set for $u \geqslant 0$

   $$\sigma_u(\omega) := \inf\{t \geqslant 0 : I_t(\omega) > u\} \quad \text{and} \quad \tau_u(\omega) := \inf\{t \geqslant 0 : I_t(\omega) \geqslant u\}.$$

   Show that

   (a) $\sigma_u \geqslant t \iff I_t \leqslant u$ and $\tau_u > t \iff I_t < u$.

   (b) τ_u is an $\mathscr{F}_t$ stopping time and σ_u is an $\mathscr{F}_{t+}$ stopping time.

   (c) $u \mapsto \sigma_u$ is right-continuous and $u \mapsto \tau_u$ is left-continuous.

   (d) $\tau_u = \sigma_{u-} = \lim_{\epsilon\downarrow 0} \sigma_{u-\epsilon}$.

   (e) $u \mapsto \tau_u$, $u \mapsto \sigma_u$ are continuous, if $t \mapsto I_t$ is strictly increasing.

17 Itô's formula

An important consequence of the fundamental theorem of integral and differential calculus is the fact every differentiation rule has an integration counterpart. Consider, for example, the chain rule $(f \circ g)'(t) = f'(g(t)) \cdot g'(t)$. Then

$$f(g(t)) - f(g(0)) = \int_0^t f'(g(s)) \cdot g'(s)\, ds = \int_0^t f'(g(s))\, dg(s)$$

which is just the substitution or change-of-variable rule for integrals. Moreover, it provides a useful calculus formula if we want to evaluate integrals.

Let $(B_t)_{t \geqslant 0}$ be a BM^1 and $(\mathscr{F}_t)_{t \geqslant 0}$ an admissible, right-continuous complete filtration, e. g. $\mathscr{F}_t = \overline{\mathscr{F}}_t^B$, cf. Theorem 6.21. Itô's formula is the stochastic counterpart of the chain rule; it is also known as the *change-of-variable-formula* or *Itô's Lemma*. The unbounded variation of Brownian sample paths leaves, again, its traces: The chain rule has a correction term involving the quadratic variation of $(B_t)_{t \geqslant 0}$ and the second derivative of the function f. Here is the statement of the basic result.

17.1 Theorem (Itô 1942). *Let $(B_t)_{t \geqslant 0}$ be a BM^1 and let $f : \mathbb{R} \to \mathbb{R}$ be a $\mathcal{C}^2$-function. Then we have for all $t \geqslant 0$ almost surely*

$$f(B_t) - f(B_0) = \int_0^t f'(B_s)\, dB_s + \frac{1}{2} \int_0^t f''(B_s)\, ds. \tag{17.1}$$

Before we prove Theorem 17.1, we need to discuss Itô processes and Itô differentials.

17.1 Itô processes and stochastic differentials

Let $T \in [0, \infty]$. Stochastic processes of the form

$$X_t = X_0 + \int_0^t \sigma(s)\, dB_s + \int_0^t b(s)\, ds, \quad t < T, \tag{17.2}$$

are called *Itô processes*. If the integrands satisfy $\sigma \in \mathcal{L}^2_{T,\text{loc}}$ and, for almost all ω, $b(\cdot, \omega) \in L^1([0, T], ds)$ (or $b(\cdot, \omega) \in L^1_{\text{loc}}([0, \infty), ds)$ if $T = \infty$) then all integrals in (17.2) make sense. Roughly speaking, an Itô process is *locally* a Brownian motion with standard deviation $\sigma(s, \omega)$ plus a drift with mean value $b(s, \omega)$. Note that these parameters change infinitesimally over time.

In dimensions other than $d = 1$, the stochastic integral is *defined* for each coordinate and we can read (17.2) as matrix and vector equality, i. e.

$$X_t^j = X_0^j + \sum_{k=1}^d \int_0^t \sigma_{jk}(s)\, dB_s^k + \int_0^t b_j(s)\, ds, \quad t < T, \ j = 1, \ldots, m, \tag{17.3}$$

where $X_t = (X_t^1, \dots, X_t^m)^\top$, $B_t = (B_t^1, \dots, B_t^d)^\top$ is a BM^d, $\sigma(t,\omega) = (\sigma_{jk}(t,\omega))_{jk} \in \mathbb{R}^{m\times d}$ is a matrix-valued stochastic process and $b(t,\omega) = (b_1(t,\omega), \dots, b_m(t,\omega))^\top \in \mathbb{R}^m$ is a vector-valued stochastic process; the coordinate processes σ_{jk} and b_j are chosen in such a way that all (stochastic) integrals in (17.3) are defined.

17.2 Definition. Let $(B_t)_{t\geqslant 0}$ be a BM^d, $(\sigma_t)_{t\geqslant 0}$ and $(b_t)_{t\geqslant 0}$ be two progressively measurable $\mathbb{R}^{m\times d}$ and $\mathbb{R}^m$-valued locally bounded processes. Then

$$X_t = X_0 + \int_0^t \sigma_s \, dB_s + \int_0^t b_s \, ds, \quad t \geqslant 0,$$

is an *m-dimensional Itô process*.

17.3 Remark. a) If we know that σ and b are locally bounded, i. e. there is some $\Omega_0 \subset \Omega$ with $\mathbb{P}(\Omega_0) = 1$ and

$$\max_{j,k} \sup_{t\leqslant T} |\sigma_{jk}(t,\omega)| + \max_{j} \sup_{t\leqslant T} |b_j(t,\omega)| < \infty \quad \text{for all } T > 0,\ \omega \in \Omega_0,$$

then $b_j(\cdot,\omega) \in L^1([0,T], dt)$, $T > 0$, almost surely; moreover, the stopping times

$$\tau_n(\omega) := \min_{j,k} \inf \left\{ t \geqslant 0 : \int_0^t |\sigma_{jk}(s,\omega)|^2 \, ds \geqslant n \right\}$$

satisfy $\tau_n \uparrow \infty$ almost surely, i. e.

$$\mathbb{E}\left[\int_0^{\tau_n \wedge T} |\sigma_{jk}(s,\cdot)|^2 \, ds \right] \leqslant n$$

which proves that $\sigma_{jk} \in \mathcal{L}^2_{T,\mathrm{loc}}$ for all $T > 0$.

b) From Chapter 16 we know that $X_t = M_t + A_t$ where M_t is a (vector-valued) continuous local martingale and the coordinates of the continuous process A_t satisfy

$$|A_t^j(\omega) - A_s^j(\omega)| \leqslant \int_s^t |b_j(s,\omega)| \, ds \leqslant C_T(\omega)(t-s) \quad \text{for all } s,t < T,$$

i. e. A_t is locally of bounded variation.

It is convenient to rewrite (17.2) in *differential form*, i. e. as

$$dX_t(\omega) = \sigma(t,\omega)\, dB_t(\omega) + b(t,\omega)\, dt \tag{17.4}$$

where we use again, if appropriate, vector and matrix notation.

In this new notation, Example 15.15 reads

$$(B_t)^2 = 2\int_0^t B_s \, dB_s + t, \quad \text{hence,} \quad d(B_t)^2 = 2B_t \, dB_t + dt$$

and we see that this is a special case of Itô's formula (17.1) with $f(x) = x^2$:

$$df(B_t) = f'(B_t)\,dB_t + \frac{1}{2} f''(B_t)\,dt.$$

17.2 The heuristics behind Itô's formula

Let us give an intuitive argument why stochastic calculus differs from the usual calculus. Note that

$$\Delta B_t = \sqrt{\Delta t}\,\frac{B_{t+\Delta t} - B_t}{\sqrt{\Delta t}} \sim \sqrt{\Delta t}\,G \quad \text{and} \quad (\Delta B_t)^2 = \Delta t\,G^2$$

where $G \sim \mathsf{N}(0,1)$ is a standard normal random variable. Since $\mathbb{E}\,|G| = \sqrt{2/\pi}$ and $\mathbb{E}\,G^2 = 1$, we get from $\Delta t \to 0$

$$|dB_t| \approx \sqrt{\frac{2}{\pi}\,dt} \quad \text{and} \quad (dB_t)^2 \approx dt.$$

This means that Itô's formula is essentially a second-order Taylor formula while in ordinary calculus we only go up to order one. Terms of higher order, for example $(dB_t)^3 \approx (dt)^{3/2}$ and $dt\,dB_t \approx (dt)^{3/2}$, do not give any contributions since integrals against $(dt)^{3/2}$ correspond to an integral sum with $(\Delta t)^{-1}$-many terms which are all
Ex. 17.2 of the order $(\Delta t)^{3/2}$ – and such a sum tends to zero as $\Delta t \to 0$. The same argument also explains why we do not have terms of second order, i. e. $(dt)^2$, in ordinary calculus.

In Section 17.5 we consider Itô's formula for a d-dimensional Brownian motion. Then we will encounter differentials of the form $dt\,dB_t^k$ and $dB_t^j\,dB_t^k$.

Clearly, $dt\,dB_t^k \approx (dt)^{3/2}$ is negligible, while for $j \neq k$

$$(B_{t+\Delta t}^j - B_t^j)(B_{t+\Delta t}^k - B_t^k) = \Delta t\,\frac{(B_{t+\Delta t}^j - B_t^j)}{\sqrt{\Delta t}}\,\frac{(B_{t+\Delta t}^k - B_t^k)}{\sqrt{\Delta t}} \sim \Delta t\,GG'$$

where G and G' are independent $\mathsf{N}(0,1)$ distributed random variables. Note that

$$\Delta t\,GG' = \frac{1}{2}\Delta t\left[\left(\frac{G+G'}{\sqrt{2}}\right)^2 - \left(\frac{G-G'}{\sqrt{2}}\right)^2\right].$$

Ex. 1.7 Since $\frac{1}{\sqrt{2}}(G \pm G')$ are independent $\mathsf{N}(0,1)$ random variables (cf. Problem 1.7), both terms yield, as $\Delta t \to 0$, the same contribution towards the quadratic variation, and we get

$$dB_t^j\,dB_t^k \approx \delta_{jk}\,dt = \begin{cases} 0, & j \neq k, \\ dt, & j = k, \end{cases}$$

(cf. the argument at end of the proof of Theorem 17.8). This can be nicely summed up in a multiplication table, Table 17.1.

Table 17.1. Multiplication table for stochastic differentials.

$j \neq k$	dt	dB_t^j	dB_t^k
dt	0	0	0
dB_t^j	0	dt	0
dB_t^k	0	0	dt

17.3 Proof of Itô's formula (Theorem 17.1)

We begin with an auxiliary result on Riemann–Stieltjes-type approximations of (stochastic) integrals. As a by-product we get a rigorous proof for $(dB_t)^2 \approx dt$, compare Section 17.2.

17.4 Lemma. *Let $(B_t)_{t\geqslant 0}$ be a BM^1, $\Pi = \{0 = t_0 < t_1 < \ldots < t_N = T\}$ a generic partition of the interval $[0, T]$ with mesh $|\Pi| = \max_l |t_l - t_{l-1}|$ and $g \in \mathcal{C}_b(\mathbb{R})$. Then*

$$\sum_{l=1}^{N} g(B_{t_{l-1}})(B_{t_l} - B_{t_{l-1}}) \xrightarrow[|\Pi|\to 0]{\mathbb{P}} \int_0^T g(B_s)\,dB_s \tag{17.5}$$

and

$$\sum_{l=1}^{N} g(\xi_l)(B_{t_l} - B_{t_{l-1}})^2 \xrightarrow[|\Pi|\to 0]{\mathbb{P}} \int_0^T g(B_s)\,ds \tag{17.6}$$

where $\xi_l(\omega) = B_{t_{l-1}}(\omega) + \theta_l(\omega)(B_{t_l}(\omega) - B_{t_{l-1}}(\omega))$, $0 \leqslant \theta_l \leqslant 1$, $l = 1, \ldots, N$, are intermediate values such that $g(\xi_l)$ is measurable.

Proof. By Proposition 15.16 (or Theorem 16.8) we may use a Riemann-sum approximation of the stochastic integral provided that $s \mapsto g(B_s)$ is continuous in $L^2(\mathbb{P})$-sense. This, however, follows from $\|g\|_\infty < \infty$ and the continuity of $x \mapsto g(x)$ and $s \mapsto B_s$; for all sequences $(s_n)_{n\geqslant 1}$ with $s_n \to s$, we have by dominated convergence

$$\lim_{n\to\infty} \mathbb{E}\left\{\left(g(B_{s_n}) - g(B_s)\right)^2\right\} = 0;$$

this proves (17.5).

The integral appearing on the right-hand side of (17.6) is, for each ω, a Riemann integral. Since $s \mapsto g(B_s)$ is bounded and continuous, we may approximate the integral by Riemann sums

$$\lim_{|\Pi|\to 0} \sum_{l=1}^{N} g(B_{t_{l-1}})(t_l - t_{l-1}) = \int_0^T g(B_s)\,ds$$

almost surely, hence in probability. Therefore, the assertion follows if we can show that the following expression converges to 0 in probability:

$$\begin{aligned}
\sum_{l=1}^{N} g(\xi_l)(B_{t_l} - B_{t_{l-1}})^2 - \sum_{l=1}^{N} g(B_{t_{l-1}})(t_l - t_{l-1}) \\
= \sum_{l=1}^{N} g(B_{t_{l-1}})\{(B_{t_l} - B_{t_{l-1}})^2 - (t_l - t_{l-1})\} \\
+ \sum_{l=1}^{N} (g(\xi_l) - g(B_{t_{l-1}}))(B_{t_l} - B_{t_{l-1}})^2 =: \mathrm{J}_1 + \mathrm{J}_2.
\end{aligned}$$

For the second term we have

$$|\mathrm{J}_2| \leqslant \sum_{l=1}^{N} |g(\xi_l) - g(B_{t_{l-1}})|(B_{t_l} - B_{t_{l-1}})^2 \leqslant \max_{1\leqslant l\leqslant N} |g(\xi_l) - g(B_{t_{l-1}})| \underbrace{\sum_{l=1}^{N} (B_{t_l} - B_{t_{l-1}})^2}_{=S_2^{\Pi}(B;T) \text{ cf. } (9.1)}.$$

Taking expectations and using the Cauchy-Schwarz inequality we get

$$\mathbb{E}\,|\mathrm{J}_2| \leqslant \sqrt{\mathbb{E}\left(\max_{1\leqslant l\leqslant N} |g(\xi_l) - g(B_{t_{l-1}})|^2\right)}\,\sqrt{\mathbb{E}\,[(S_2^{\Pi}(B;T))^2]}. \tag{17.7}$$

By Theorem 9.1, $L^2(\mathbb{P})\text{-}\lim_{|\Pi|\to 0} S_2^{\Pi}(B;T) = T$; since $s \mapsto g(B_s)$ is uniformly continuous on $[0,T]$ and bounded by $\|g\|_\infty$, we can use dominated convergence to get $\lim_{|\Pi|\to 0} \mathbb{E}\,|\mathrm{J}_2| = 0$; therefore $\mathrm{J}_2 \to 0$ in $L^1(\mathbb{P})$ and in probability.

For the first term we get

$$\begin{aligned}
\mathbb{E}(\mathrm{J}_1^2) &= \mathbb{E}\left[\left(\sum_{l=1}^{N} g(B_{t_{l-1}})[(B_{t_l} - B_{t_{l-1}})^2 - (t_l - t_{l-1})]\right)^2\right] \\
&= \mathbb{E}\left[\sum_{l=1}^{N} |g(B_{t_{l-1}})|^2[(B_{t_l} - B_{t_{l-1}})^2 - (t_l - t_{l-1})]^2\right].
\end{aligned}$$

In the last equality only the pure squares survive when we multiply out the outer square. This is due to the fact that $(B_t^2 - t)_{t\geqslant 0}$ is a martingale: Indeed, if $l < m$, then $t_{l-1} < t_l \leqslant t_{m-1} < t_m$. For brevity we write

$$g_{l-1} := g(B_{t_{l-1}}), \quad \Delta_l B := B_{t_l} - B_{t_{l-1}}, \quad \Delta_l t := t_l - t_{l-1}.$$

By the tower property,

$$\begin{aligned}
&\mathbb{E}\left(g_{l-1}\{(\Delta_l B)^2 - \Delta_l t\}g_{m-1}\{(\Delta_m B)^2 - \Delta_m t\}\right) \\
&= \mathbb{E}\left(\mathbb{E}\left[g_{l-1}\{(\Delta_l B)^2 - \Delta_l t\}g_{m-1}\{(\Delta_m B)^2 - \Delta_m t\} \,\middle|\, \mathscr{F}_{t_{m-1}}\right]\right) \\
&= \mathbb{E}\left(\underbrace{g_{l-1}\{(\Delta_l B)^2 - \Delta_l t\}g_{m-1}}_{\mathscr{F}_{t_{m-1}} \text{ measurable}} \cdot \underbrace{\mathbb{E}\left[\{(\Delta_m B)^2 - \Delta_m t\} \,\middle|\, \mathscr{F}_{t_{m-1}}\right]}_{=0 \text{ martingale, cf. 5.2 c)}}\right) = 0,
\end{aligned}$$

i. e. the mixed terms break away. From (the proof of) Theorem 9.1 we get

$$\begin{aligned}\mathbb{E}(\mathrm{J}_1^2) &\leqslant \|g\|_\infty^2 \ \mathbb{E}\left[\sum_{l=1}^N \left[(B_{t_l} - B_{t_{l-1}})^2 - (t_l - t_{l-1})\right]^2\right] \\ &\leqslant 2\|g\|_\infty^2 \,|\Pi| \sum_{l=1}^N (t_l - t_{l-1}) = 2\|g\|_\infty^2 \,|\Pi|\, T \xrightarrow[|\Pi|\to 0]{} 0.\end{aligned} \tag{17.8}$$

This proves $L^2(\mathbb{P})$-convergence and, consequently, convergence in probability. □

Proof of Theorem 17.1. **1°** Let us first assume that $\operatorname{supp} f \subset [-K, K]$ is compact. Then

$$C_f := \|f\|_\infty + \|f'\|_\infty + \|f''\|_\infty < \infty.$$

For any $\Pi := \{t_0 = 0 < t_1 < \ldots < t_N = t\}$ with mesh $|\Pi| = \max_l (t_l - t_{l-1})$ we get by Taylor's theorem

$$\begin{aligned}f(B_t) - f(B_0) &= \sum_{l=1}^N \left(f(B_{t_l}) - f(B_{t_{l-1}})\right) \\ &= \sum_{l=1}^N f'(B_{t_{l-1}})(B_{t_l} - B_{t_{l-1}}) + \frac{1}{2}\sum_{l=1}^N f''(\xi_l)(B_{t_l} - B_{t_{l-1}})^2 \\ &=: \mathrm{J}_1 + \mathrm{J}_2.\end{aligned}$$

Here $\xi_l = \xi_l(\omega)$ is an intermediate point between $B_{t_{l-1}}$ and B_{t_l}. From the integral representation of the remainder term in the Taylor formula,

$$\frac{1}{2} f''(\xi_l) = \int_0^1 f''(B_{t_{l-1}} + r(B_{t_l} - B_{t_{l-1}}))(1-r)\,dr,$$

we see that $f''(\xi_l)$ is measurable.

2° By (17.5) we know that $\mathrm{J}_1 \to \int_0^t f'(B_s)\,dB_s$ in probability as $|\Pi| \to 0$.

3° By (17.6) we know that $\mathrm{J}_2 \to \frac{1}{2}\int_0^t f''(B_s)\,ds$ in probability as $|\Pi| \to 0$.

4° So far we have shown (17.1) for functions $f \in \mathcal{C}^2$ such that $C_f < \infty$. For a general $f \in \mathcal{C}^2$ we fix $\ell \geqslant 0$, pick a cut-off function $\chi_\ell \in \mathcal{C}_c^2$ satisfying $\mathbb{1}_{\mathbb{B}(0,\ell)} \leqslant \chi_\ell \leqslant \mathbb{1}_{\mathbb{B}(0,\ell+1)}$ and set $f_\ell(x) := f(x)\chi_\ell(x)$. Clearly, $C_{f_\ell} < \infty$, and therefore

$$f_\ell(B_t) - f_\ell(B_0) = \int_0^t f_\ell'(B_s)\,dB_s + \frac{1}{2}\int_0^t f_\ell''(B_s)\,ds.$$

Consider the stopping times

$$\tau(\ell) := \inf\{s > 0 : |B_s| \geqslant \ell\}, \quad \ell \geqslant 1,$$

and note that $\frac{d^j}{dx^j} f_\ell(B_{s\wedge\tau(\ell)}) = f^{(j)}(B_{s\wedge\tau(\ell)})$, $j = 0, 1, 2$. Thus, Theorem 16.7 shows for all $t \geqslant 0$

$$\begin{aligned}
f(B_{t\wedge\tau(\ell)}) - f(B_0) &= \int_0^{t\wedge\tau(\ell)} f_\ell'(B_s)\,dB_s + \frac{1}{2}\int_0^{t\wedge\tau(\ell)} f_\ell''(B_s)\,ds \\
&\overset{16.7\,e)}{=} \int_0^t f_\ell'(B_s)\mathbb{1}_{[0,\tau(\ell))}(s)\,dB_s + \frac{1}{2}\int_0^t f_\ell''(B_s)\mathbb{1}_{[0,\tau(\ell))}(s)\,ds \\
&= \int_0^t f'(B_s)\mathbb{1}_{[0,\tau(\ell))}(s)\,dB_s + \frac{1}{2}\int_0^t f''(B_s)\mathbb{1}_{[0,\tau(\ell))}(s)\,ds \\
&\overset{16.7\,e)}{=} \int_0^{t\wedge\tau(\ell)} f'(B_s)\,dB_s + \frac{1}{2}\int_0^{t\wedge\tau(\ell)} f''(B_s)\,ds.
\end{aligned}$$

Since $\lim_{\ell\to\infty}\tau(\ell) = \infty$ almost surely – Brownian motion does not explode in finite time – and since the (stochastic) integrals are continuous as functions of their upper boundary, the proof of Theorem 17.1 is complete. □

17.4 Itô's formula for stochastic differentials

We will now derive Itô's formula for one-dimensional Itô processes, i. e. processes of the form

$$dX_t = \sigma(t)\,dB_t + b(t)\,dt$$

where σ, b are locally bounded (in t) and progressively measurable. We have seen in Remark 17.3 that under this assumption $b \in L^1([0,T], dt)$ a. s. and $\sigma \in \mathcal{L}^2_{T,\text{loc}}$ for all $T > 0$. We can also show that $b \in \mathcal{L}^2_{T,\text{loc}}$. Fix some common localization sequence τ_n. Since $\overline{\mathcal{S}}_T = \mathcal{L}^2_T$ we find, for every n, sequences of simple processes $(b_\Pi)_\Pi, (\sigma_\Pi)_\Pi \subset \mathcal{S}_T$ such that

$$\sigma_\Pi \mathbb{1}_{[0,\tau_n)} \xrightarrow[|\Pi|\to 0]{L^2(\lambda_T\otimes\mathbb{P})} \sigma\mathbb{1}_{[0,\tau_n)} \quad\text{and}\quad b_\Pi \mathbb{1}_{[0,\tau_n)} \xrightarrow[|\Pi|\to 0]{L^2(\lambda_T\otimes\mathbb{P})} b\mathbb{1}_{[0,\tau_n)}.$$

Using a diagonal procedure we can achieve that the sequences $(b_\Pi)_\Pi, (\sigma_\Pi)_\Pi$ are independent of n.

17.5 Lemma. *Let $X_t^\Pi = X_0 + \int_0^t \sigma_\Pi(s)\,dB_s + \int_0^t b_\Pi(s)\,ds$. Then $X_t^\Pi \to X_t$ uniformly in probability as $|\Pi| \to 0$, i. e. we have*

$$\lim_{|\Pi|\to 0} \mathbb{P}\left(\sup_{t\leqslant T}|X_t^\Pi - X_t| > \epsilon\right) = 0 \quad \textit{for all } \epsilon > 0,\ T > 0. \tag{17.9}$$

Proof. The argument is similar to the one used to prove Corollary 16.9. Let τ_n be some localizing sequence and fix $n \geqslant 0$, $T > 0$ and $\epsilon > 0$. Then

$$
\begin{aligned}
&\mathbb{P}\left(\sup_{t\leqslant T}\left|\int_0^t (\sigma_{\Pi}(s) - \sigma(s))\,dB_s\right| > \epsilon\right)\\
&\leqslant \mathbb{P}\left(\sup_{t\leqslant T}\left|\int_0^t (\sigma_{\Pi}(s) - \sigma(s))\,dB_s\right| > \epsilon,\ \tau_n > T\right) + \mathbb{P}(\tau_n \leqslant T)\\
&\leqslant \mathbb{P}\left(\sup_{t\leqslant T}\left|\int_0^{t\wedge\tau_n} (\sigma_{\Pi}(s) - \sigma(s))\,dB_s\right| > \epsilon\right) + \mathbb{P}(\tau_n \leqslant T)\\
&\underset{\text{Doob}}{\overset{\text{Chebyshev}}{\leqslant}} \frac{4}{\epsilon^2}\,\mathbb{E}\left[\left|\int_0^{T\wedge\tau_n} (\sigma_{\Pi}(s) - \sigma(s))\,dB_s\right|^2\right] + \mathbb{P}(\tau_n \leqslant T)\\
&\overset{(15.21)}{\leqslant} \frac{4}{\epsilon^2}\,\mathbb{E}\left[\int_0^{T\wedge\tau_n} |\sigma_{\Pi}(s) - \sigma(s)|^2\,ds\right] + \mathbb{P}(\tau_n \leqslant T)\\
&\xrightarrow[|\Pi|\to 0]{n\text{ fixed}} \mathbb{P}(\tau_n \leqslant T) \xrightarrow[n\to\infty]{} 0.
\end{aligned}
$$

A similar, but simpler, calculation yields

$$
\mathbb{P}\left(\sup_{t\leqslant T}\left|\int_0^t (b_{\Pi}(s) - b(s))\,ds\right| > \epsilon\right) \xrightarrow[|\Pi|\to 0]{} 0.
$$

Finally, for any two random variables $X, Y : \Omega \to [0, \infty)$,

$$
\mathbb{P}(X + Y > \epsilon) \leqslant \mathbb{P}(X > \epsilon/2) + \mathbb{P}(Y > \epsilon/2),
$$

and the claim follows since the supremum is subadditive. □

17.6 Lemma. *Let $X_t = X_0 + \int_0^t \sigma(s)\,dB_s + \int_0^t b(s)\,ds$ and $Y_t = Y_0 + \int_0^t \tau(s)\,dB_s + \int_0^t c(s)\,ds$, $t \in [0, T]$, be two Itô processes with $\sigma, b \in \mathcal{L}_T^2$ and $\tau, c \in \mathcal{S}_T$, respectively. Denote by $\Pi = \{0 = t_0 < t_1 < \ldots < t_N = T\}$ a partition of the interval $[0, T]$ and $g \in \mathcal{C}_b(\mathbb{R})$. Then*

$$
\sum_{l=1}^N g(X_{t_{l-1}})(Y_{t_l} - Y_{t_{l-1}}) \xrightarrow[|\Pi|\to 0]{\mathbb{P}} \int_0^T g(X_s)\tau(s)\,dB_s + \int_0^T g(X_s)c(s)\,ds \tag{17.10}
$$

$$
\sum_{l=1}^N g(\xi_l)(Y_{t_l} - Y_{t_{l-1}})^2 \xrightarrow[|\Pi|\to 0]{\mathbb{P}} \int_0^T g(X_s)\tau^2(s)\,ds \tag{17.11}
$$

where $\xi_l(\omega) = X_{t_{l-1}}(\omega) + \theta_l(\omega)(X_{t_l}(\omega) - X_{t_{l-1}}(\omega))$, $0 \leqslant \theta_l \leqslant 1$, $l = 1, \ldots, N$, are intermediate values such that $g(\xi_l)$ is measurable.

We *define* the stochastic integral for an Itô process $dX_t = \sigma(t)\,dB_t + b(t)\,dt$ as

$$\int_0^t f(s)\,dX_s := \int_0^t f(s)\sigma(s)\,dB_s + \int_0^t f(s)b(s)\,ds \tag{17.12}$$

whenever the right-hand side makes sense. It is possible to define the Itô integral for the integrator dX_t directly in the sense of Chapters 15 and 16 and to *derive* the equality (17.12) rather than to use it as a definition.

Proof of Lemma 17.6. Denote by Δ the common refinement of the partitions of the simple processes τ and c. Let $\Pi = \{0 = t_0 < t_1 < \dots < t_N = T\}$ be partitions of $[0, T]$. Without loss of generality we can assume that $\Delta \subset \Pi$. Since the processes $s \mapsto g(X_s)\tau(s)$ and $s \mapsto g(X_s)c(s)$ are bounded, right-continuous and have only finitely many discontinuities, we see from Theorem 16.8 (or Proposition 15.16) and an approximation by Riemann sums that

$$\begin{aligned}\sum_{l=1}^N g(X_{t_{l-1}})(Y_{t_l} - Y_{t_{l-1}}) &= \sum_{l=1}^N g(X_{t_{l-1}})\tau(t_{l-1})(B_{t_l} - B_{t_{l-1}}) + \sum_{l=1}^N g(X_{t_{l-1}})c(t_{l-1})(t_l - t_{l-1})\\ &\xrightarrow[|\Pi|\to 0]{\mathbb{P}} \int_0^T g(X_s)\tau(s)\,dB_s + \int_0^T g(X_s)c(s)\,ds;\end{aligned}$$

this proves (17.10).

In order to see (17.11) we observe that

$$\begin{aligned}\sum_{l=1}^N g(\xi_l)(Y_{t_l} - Y_{t_{l-1}})^2 &= \sum_{l=1}^N g(\xi_l)\big(\tau(t_{l-1})(B_{t_l} - B_{t_{l-1}}) + c(t_{l-1})(t_l - t_{l-1})\big)^2\\ &= \sum_{l=1}^N g(\xi_l)\tau^2(t_{l-1})(B_{t_l} - B_{t_{l-1}})^2\\ &\quad + 2\sum_{l=1}^N g(\xi_l)\tau(t_{l-1})c(t_{l-1})(t_l - t_{l-1})(B_{t_l} - B_{t_{l-1}})\\ &\quad + \sum_{l=1}^N g(\xi_l)c^2(t_{l-1})(t_l - t_{l-1})^2\\ &=: \mathrm{J}_1 + \mathrm{J}_2 + \mathrm{J}_3.\end{aligned}$$

The proof of (17.6) still works, almost literally, if we have $g(\xi_\ell)\tau(t_{l-1})$ instead of $g(\xi_\ell)$ where $\tau(t_{l-1})$ is bounded and $\mathscr{F}_{\tau_{l-1}}$ measurable. Since $s \mapsto g(X_s)\tau^2(s)$ is piecewise continuous with finitely many discontinuities, we see $\lim_{|\Pi|\to 0} \mathrm{J}_1 = \int_0^T g(X_s)\tau^2(s)\,ds$ in probability. Moreover,

$$\begin{aligned}|\mathrm{J}_2| &\leqslant 2\sum_{l=1}^N \|g\|_\infty\|\tau\|_\infty\|c\|_\infty(t_l - t_{l-1})\max_k |B_{t_k} - B_{t_{k-1}}|\\ &= 2\,\|g\|_\infty\|\tau\|_\infty\|c\|_\infty\, T\, \max_k |B_{t_k} - B_{t_{k-1}}|,\end{aligned}$$

and, as Brownian motion is a. s. uniformly continuous on $[0, T]$, we get $J_2 \to 0$ a. s. and in probability; a similar calculation gives $J_3 \to 0$ a. s. and in probability. □

17.7 Theorem (Itô 1942). *Let X_t be a one-dimensional Itô process in the sense of Definition 17.2 such that $dX_t = \sigma(t)\,dB_t + b(t)\,dt$ and let $f : \mathbb{R} \to \mathbb{R}$ be a $\mathcal{C}^2$-function. Then we have for all $t \geqslant 0$ almost surely*

$$
\begin{aligned}
f(X_t) - f(X_0) &= \int_0^t f'(X_s)\sigma(s)\,dB_s + \int_0^t f'(X_s)b(s)\,ds + \frac{1}{2}\int_0^t f''(X_s)\sigma^2(s)\,ds \\
&=: \int_0^t f'(X_s)\,dX_s + \frac{1}{2}\int_0^t f''(X_s)\sigma^2(s)\,ds.
\end{aligned}
\tag{17.13}
$$

Proof. **1°** As in the proof of Theorem 17.1 it is enough to show (17.13) locally for $t \leqslant \tau(\ell)$ where

$$\tau(\ell) = \inf\{s > 0 \,:\, |X_s| \geqslant \ell\}.$$

This means that we can assume, without loss of generality, that

$$\|f\|_\infty + \|f'\|_\infty + \|f''\|_\infty + \|\sigma\|_\infty + \|b\|_\infty \leqslant C < \infty.$$

2° Lemma 17.5 shows that we can approximate σ and b by elementary processes $\sigma_{_\Pi}$ and $b_{_\Pi}$. Denote the corresponding Itô process by X^Π. Since $f \in \mathcal{C}_b^2$, the following limits are uniform in probability:

$$X^\Pi \xrightarrow[|\Pi|\to 0]{} X, \quad f(X^\Pi) \xrightarrow[|\Pi|\to 0]{} f(X).$$

The right-hand side of (17.13) with $\sigma_{_\Pi}$, $b_{_\Pi}$ and X^Π also converges uniformly in probability by (an obvious modification of) Lemma 17.5. Ex. 17.15

This means that we have to show (17.13) only for Itô processes where σ and b are from $\mathcal{S}_T$ for some $T > 0$.

3° Assume that $\sigma, b \in \mathcal{S}_T$ and that $\Pi = \{t_0 = 0 < t_1 < \ldots < t_N = T\}$ is the joint refinement of the underlying partitions. Fix $t \leqslant T$ and assume, without loss of generality, that $t = t_k \in \Pi$. As in step 1° of the proof of Theorem 17.1, we get by Taylor's theorem

$$
\begin{aligned}
f(X_t) - f(X_0) &= \sum_{l=1}^{k} f'(X_{t_{l-1}})(X_{t_l} - X_{t_{l-1}}) + \frac{1}{2}\sum_{l=1}^{k} f''(\xi_l)(X_{t_l} - X_{t_{l-1}})^2 \\
&=: J_1 + J_2
\end{aligned}
$$

where ξ_l is an intermediate point between $X_{t_{l-1}}$ and X_{t_l} such that $f''(\xi_l)$ is measurable.

Since σ and b are simple processes, we can use Lemma 17.6 to conclude that

$$J_1 \xrightarrow[|\Pi|\to 0]{\mathbb{P}} \int_0^t f'(X_s)\,dX_s \quad\text{and}\quad J_2 \xrightarrow[|\Pi|\to 0]{\mathbb{P}} \frac{1}{2}\int_0^t f''(X_s)\sigma^2(s)\,ds.$$ □

17.5 Itô's formula for Brownian motion in $\mathbb{R}^d$

Assume that $B = (B^1, \dots, B^d)$ is a BM^d. Recall that the stochastic integral with respect to dB is defined for each coordinate, cf. Section 17.1. With this convention, the multi-dimensional version of Itô's formula becomes

17.8 Theorem (Itô 1942). *Let $(B_t)_{t\geqslant 0}$, $B_t = (B_t^1, \dots, B_t^d)$, be a d-dimensional Brownian motion and $X_t = X_0 + \int_0^t \sigma(s)\,dB_s + \int_0^t b(s)\,ds$ an m-dimensional Itô process. For every $\mathcal{C}^2$-function $f : \mathbb{R}^m \to \mathbb{R}$ be a $\mathcal{C}^2$-function*

$$f(X_t) - f(X_0) = \sum_{k=1}^d \int_0^t \left[\sum_{j=1}^m \partial_j f(X_s)\sigma_{jk}(s)\right] dB_s^k + \sum_{j=1}^m \int_0^t \partial_j f(X_s) b_j(s)\,ds$$
$$+ \frac{1}{2}\sum_{i,j=1}^m \int_0^t \partial_i\partial_j f(X_s) \sum_{k=1}^d \sigma_{ik}(s)\sigma_{jk}(s)\,ds. \tag{17.14}$$

Sketch of the proof. The argument is similar to the proof of the one-dimensional version in Theorem 17.7.

Since Lemma 17.5 remains valid for m-dimensional Itô processes, the first two steps, 1° and 2° of the proof of Theorem 17.7 remain the same. In step 3° we use now Taylor's formula for functions in $\mathbb{R}^m$.

Let $\sigma_{jk}, b_j \in \mathcal{S}_T$, $j = 1, \dots, m$, $k = 1, \dots, d$ and $\Pi = \{t_0 = 0 < t_1 < \dots < t_N = T\}$ be the joint refinement of all underlying partitions. Fix $t \leqslant T$ and assume, without loss of generality, that $t = t_n \in \Pi$. As in step 1° of the proof of Theorem 17.1, we have

$$f(X_t) - f(X_0) = \sum_{l=1}^n \left(f(X_{t_l}) - f(X_{t_{l-1}})\right)$$

and if we apply Taylor's theorem to each increment, we see for $l = 1, \dots, n$

$$f(X_{t_l}) - f(X_{t_{l-1}})$$
$$= \sum_{j=1}^m \partial_j f(X_{t_{l-1}})(X_{t_l}^j - X_{t_{l-1}}^j) + \frac{1}{2}\sum_{i,j=1}^m \partial_i\partial_j f(\xi_l)(X_{t_l}^i - X_{t_{l-1}}^i)(X_{t_l}^j - X_{t_{l-1}}^j)$$

where $\xi_l = \xi_l(\omega)$ is an intermediate point between $X_{t_{l-1}}$ and X_{t_l}. From the integral representation of the remainder term in the Taylor formula,

$$\frac{1}{2}\partial_i\partial_j f(\xi_l) = \int_0^1 \partial_i\partial_j f(X_{t_{l-1}} + r(X_{t_l} - X_{t_{l-1}}))(1-r)\,dr$$

we see that $\partial_i\partial_j f(\xi_l)$ is measurable.

Recall that

$$X_{t_l}^j - X_{t_{l-1}}^j = \sum_{k=1}^d \sigma_{jk}(t_{l-1})(B_{t_l}^k - B_{t_{l-1}}^k) + b_j(t_{l-1})(t_l - t_{l-1}).$$

Using the shorthand $\Delta_l B^k = B^k_{t_l} - B^k_{t_{l-1}}$ and $\Delta_l t = t_l - t_{l-1}$, we see

$$\begin{aligned}(X^i_{t_l} - X^i_{t_{l-1}})(X^j_{t_l} - X^j_{t_{l-1}}) = &\sum_{k,k'=1}^{d} \sigma_{jk}(t_{l-1})\sigma_{ik'}(t_{l-1})\,\Delta_l B^k \Delta_l B^{k'} \\ &+ \sum_{k=1}^{d} \sigma_{jk}(t_{l-1}) b_i(t_{l-1})\,\Delta_l B^k \Delta_l t \\ &+ \sum_{k'=1}^{d} \sigma_{ik'}(t_{l-1}) b_j(t_{l-1})\,\Delta_l B^{k'} \Delta_l t \\ &+ b_i(t_{l-1}) b_j(t_{l-1})(\Delta_l t)^2.\end{aligned}$$

Inserting this into Taylor's formula, we get six different groups of terms,

$$f(X_t) - f(X_0) = \mathsf{J}_0 + \frac{1}{2}(\mathsf{J}_1 + \cdots + \mathsf{J}_5),$$

which we consider separately. By Lemma 17.6, (17.10),

$$\mathsf{J}_0 = \sum_{j=1}^{m}\sum_{l=1}^{n} \partial_j f(X_{t_{l-1}})(X^j_{t_l} - X^j_{t_{l-1}}) \xrightarrow[|\Pi|\to 0]{\mathbb{P}} \sum_{j=1}^{m} \int_0^t \partial_j f(X_s)\,dX^j_s.$$

The last three groups,

$$\mathsf{J}_5 = \sum_{i,j=1}^{m}\sum_{l=1}^{n} \partial_i\partial_j f(\xi_l) b_i(t_{l-1}) b_j(t_{l-1})(\Delta_l t)^2$$

$$\mathsf{J}_3 + \mathsf{J}_4 = 2\sum_{i,j=1}^{m}\sum_{l=1}^{n} \partial_i\partial_j f(\xi_l) b_i(t_{l-1}) \sum_{k=1}^{d} \sigma_{jk}(t_{l-1})\,\Delta_l B^k \Delta_l t$$

tend to zero as $|\Pi| \to 0$ because of the argument used in the proof of (17.11) from Lemma 17.6. Again by (17.11),

$$\mathsf{J}_1 = \sum_{i,j=1}^{m}\sum_{l=1}^{n} \partial_i\partial_j f(\xi_l) \sum_{k} \sigma_{jk}(t_{l-1})\sigma_{ik}(t_{l-1})(\Delta_l B^k)^2 \xrightarrow[|\Pi|\to 0]{\mathbb{P}} \Sigma$$

where

$$\Sigma := \sum_{i,j=1}^{m} \int_0^t \partial_i\partial_j f(X_s) \sum_{k} \sigma_{jk}(s)\sigma_{ik}(s)\,ds.$$

Finally,

$$\mathsf{J}_2 = \sum_{i,j=1}^{m}\sum_{l=1}^{n} \partial_i\partial_j f(\xi_l) \sum_{k\neq k'} \sigma_{jk}(t_{l-1})\sigma_{ik'}(t_{l-1})\,\Delta_l B^k \Delta_l B^{k'} \xrightarrow[|\Pi|\to 0]{\mathbb{P}} \Sigma - \Sigma = 0.$$

This follows from the convergence of J_1 and the independence of B^k and $B^{k'}$: we have

$$2\left(\Delta_l B^k\right)\left(\Delta_l B^{k'}\right) = \left[\Delta_l\left(\frac{B^k + B^{k'}}{\sqrt{2}}\right)\right]^2 - \left[\Delta_l\left(\frac{B^k - B^{k'}}{\sqrt{2}}\right)\right]^2$$

where $\frac{1}{\sqrt{2}}(B^k \pm B^{k'})$ are independent one-dimensional Brownian motions. □

17.9 Remark. a) The calculations in the proof of Theorem 17.8 show that we can write Itô's formula using stochastic differentials which is much easier to memorize than (17.14)

$$df(X_t) = \sum_{j=1}^{m} \partial_j f(X_t)\, dX_t^j + \frac{1}{2}\sum_{i,j=1}^{m} \partial_i\partial_j f(X_t)\, dX_t^i\, dX_t^j \tag{17.15}$$

and the 'product' $dX_t^i\, dX_t^j$ can be formally expanded using Itô's multiplication table from Section 17.2. (Note that the proof of Theorem 17.8 justifies the rules of this table.)

b) In the literature often the following matrix and vector version of (17.14) is used:

$$\begin{aligned} f(X_t) - f(X_0) = \int_0^t \nabla f(X_s)^\top \sigma(s)\, dB_s + \int_0^t \nabla f(X_s)^\top b(s)\, ds \\ + \frac{1}{2}\int_0^t \operatorname{trace}(\sigma(s)^\top D^2 f(X_s)\sigma(s))\, ds \end{aligned} \tag{17.16}$$

where $\nabla = (\partial_1, \dots, \partial_m)^\top$ is the gradient, $D^2 f = (\partial_j \partial_k f)_{j,k=1}^m$ is the Hessian, and dB_s is understood as a column vector from $\mathbb{R}^d$. In differential notation this becomes

$$\begin{aligned} df(X_t) = \nabla f(X_t)^\top \sigma(t)\, dB_t + \nabla f(X_t)^\top b(t)\, dt \\ + \frac{1}{2}\operatorname{trace}(\sigma(t)^\top D^2 f(X_t)\sigma(t))\, dt. \end{aligned} \tag{17.17}$$

17.6 The time-dependent Itô formula

We will finally come to our final extension of Itô's formula where we allow the function f to depend also on time: $f = f(t, X_t)$. With a simple trick we can see that we can treat this in the framework of Section 17.5. If X is an m-dimensional Itô process,

$$X_t = X_0 + \int_0^t \sigma(s)\, dB_s + \int_0^t b(s)\, ds \text{ where } \sigma \in \mathbb{R}^{m\times d},\ b \in \mathbb{R}^m,$$

then we see that

$$\begin{pmatrix} t \\ X_t \end{pmatrix} = \int_0^t \hat\sigma(s)\, dB_s + \int_0^t \hat b(s)\, ds \text{ where } \hat\sigma = \begin{pmatrix} 0 \\ \sigma \end{pmatrix} \in \mathbb{R}^{(m+1)\times d},\ \hat b = \begin{pmatrix} 1 \\ b \end{pmatrix} \in \mathbb{R}^{m+1}$$

is an $(m+1)$-dimensional Itô process. Since $dt\,dB_t^k = 0$, cf. Remark 17.9 a), we get immediately the following result.

17.10 Corollary (Itô 1942). *Let $(B_t)_{t\geqslant 0}$, $B_t = (B_t^1, \dots, B_t^d)$, be a d-dimensional Brownian motion and $X_t = X_0 + \int_0^t \sigma(s)\,dB_s + \int_0^t b(s)\,ds$ an m-dimensional Itô process. For every $\mathcal{C}^2$-function $f : [0,\infty) \times \mathbb{R}^m \to \mathbb{R}$*

$$df(t, X_t) = \partial_t f(t, X_t)\,dt + \sum_{j=1}^{m} \partial_j f(t, X_t) b_j(t)\,dt + \sum_{k=1}^{d}\sum_{j=1}^{m} \partial_j f(t, X_t)\sigma_{jk}(t)\,dB_t^k$$

$$+ \frac{1}{2}\sum_{i,j=1}^{m} \partial_i\partial_j f(t, X_t) \sum_{k=1}^{d} \sigma_{ik}(t)\sigma_{jk}(t)\,dt. \tag{17.18}$$

In formula (17.18) the second derivative $\partial_t^2 f(t,x)$ does not appear, and it is natural to assume that (17.18) remains valid for functions $f \in \mathcal{C}^{1,2}([0,\infty) \times \mathbb{R}^m)$.

17.11 Theorem (Itô 1942). *The time-dependent Itô-formula* (17.18) *holds for all $f \in \mathcal{C}^{1,2}$, i. e. $f : [0,\infty) \times \mathbb{R}^m \to \mathbb{R}$, $(t,x) \mapsto f(t,x)$, which are continuously differentiable in t and twice continuously differentiable in x.*

Sketch of the proof.. We indicate two possibilities how we can obtain this improvement of Corollary 17.10.

Method 1: For every $f \in \mathcal{C}^{1,2}$ there is a sequence of functions $f_n : [0,\infty) \times \mathbb{R}^m \to \mathbb{R}$ such that

$$f_n \in \mathcal{C}^2 \quad\text{and}\quad f_n,\ \partial_t f_n,\ \partial_j f_n,\ \partial_i\partial_j f_n \xrightarrow[n\to\infty]{\text{locally uniformly}} f,\ \partial_t f,\ \partial_j f,\ \partial_i\partial_j f.$$

Set $g_n(s) := \partial_t f_n(s, X_s)$ (similarly: $\partial_j f_n(s, X_s), \partial_i\partial_j f_n(s, X_s)$) and observe that $g_n \in \mathcal{L}^2_{T,\text{loc}}$ (cf. Theorem 16.8) and $g_n(s) \to g(s) = \partial_t f(s, X_s)$ in $\mathcal{L}^2_{T,\text{loc}}$.

Since (17.18) holds for $f_n \in \mathcal{C}^2$, we can use Corollary 16.9 or the argument from Lemma 17.5 to see that the stochastic integrals appearing on the right-hand side of (17.18) converge in probability as $n \to \infty$. The left-hand side $f_n(t, X_t) - f_n(0, X_0)$ converges to $f(t, X_t) - f(0, X_0)$ a. s., and the proof is finished.

Method 2: We can avoid the approximation argument of the first proof if we inspect the proof of Theorem 17.8. The Taylor expansion for the increment $f(X_{t_l}) - f(X_{t_{l-1}})$ is replaced by the following calculation: for $l = 1, \dots, n$

$$\begin{aligned} f(t_l, X_{t_l}) - f(t_{l-1}, X_{t_{l-1}}) &= \big(f(t_l, X_{t_l}) - f(t_{l-1}, X_{t_l})\big) + \big(f(t_{l-1}, X_{t_l}) - f(t_{l-1}, X_{t_{l-1}})\big) \\ &= \partial_t f(\theta_l, X_{t_l})(t_l - t_{l-1}) + \sum_{j=1}^{m} \partial_j f(t_{l-1}, X_{t_{l-1}})(X_{t_l}^j - X_{t_{l-1}}^j) \\ &\quad + \frac{1}{2}\sum_{i,j=1}^{m} \partial_i\partial_j f(t_{l-1}, \xi_l)(X_{t_l}^i - X_{t_{l-1}}^i)(X_{t_l}^j - X_{t_{l-1}}^j) \end{aligned}$$

where $\theta_l = \theta_l(\omega)$ and $\xi_l = \xi_l(\omega)$ are convex combinations of t_{l-1}, t_l and $X_{t_{l-1}}, X_{t_l}$, respectively, such that $\partial_t f(\theta_l, X_{t_l})$ and $\partial_i\partial_j f(t_{l-1}, \xi_l)$ are measurable. The rest of the proof is now similar to the proof of Theorem 17.8. □

17.7 Tanaka's formula and local time

Itô's formula has been extended in various directions. Here we discuss the prototype of generalizations which relax the smoothness assumptions for the function f. The key point is that we retain some control on the (generalized) second derivative of f, e. g. if f is convex or if $f(x) = \int_0^x \phi(y)\,dy$ where $\phi \in \mathcal{B}_b(\mathbb{R})$ as in the Bouleau–Yor formula [184, Theorem IV.77]. In this section we consider one-dimensional Brownian motion $(B_t)_{t\geqslant 0}$ and the function $f(x) = |x|$.

Differentiating f (in the sense of generalized functions) yields $f'(x) = \operatorname{sgn}(x)$ and $f''(x) = \delta_0(x)$. A formal application of Itô's formula (17.1) gives

$$|B_t| = \int_0^t \operatorname{sgn}(B_s)\,dB_s + \frac{1}{2}\int_0^t \delta_0(B_s)\,ds. \tag{17.19}$$

In order to justify (17.19), we have to smooth out the function $f(x) = |x|$. For $\epsilon > 0$

$$f_\epsilon(x) := \begin{cases} |x| - \frac{1}{2}\epsilon, & |x| > \epsilon, \\ \frac{1}{2\epsilon}x^2, & |x| \leqslant \epsilon, \end{cases} \qquad f_\epsilon'(x) = \begin{cases} \operatorname{sgn}(x), & |x| > \epsilon, \\ \frac{1}{\epsilon}x, & |x| \leqslant \epsilon, \end{cases} \qquad f_\epsilon''(x) = \begin{cases} 0, & |x| > \epsilon, \\ \frac{1}{\epsilon}, & |x| < \epsilon. \end{cases}$$

The approximations f_ϵ are not yet $\mathcal{C}^2$ functions. Therefore, we use a Friedrichs mollifier: Pick any $\chi \in \mathcal{C}_c^\infty(-1,1)$ with $0 \leqslant \chi \leqslant 1$, $\chi(0) = 1$ and $\int \chi(x)\,dx = 1$, and set for $n \geqslant 1$

$$\chi_n(x) := n\chi(nx) \quad \text{and} \quad f_{\epsilon,n}(x) := f_\epsilon \star \chi_n(x) = \int \chi_n(x-y) f_\epsilon(y)\,dy.$$

Ex. 17.12 It is a not hard to see that $\lim_{n\to\infty} f_{\epsilon,n} = f_\epsilon$ and $\lim_{n\to\infty} f'_{\epsilon,n} = f'_\epsilon$ uniformly while $\lim_{n\to\infty} f''_{\epsilon,n}(x) = f''_\epsilon(x)$ for all $x \neq \pm\epsilon$.

If we apply Itô's formula (17.1) we get

$$f_{\epsilon,n}(B_t) = \int_0^t f'_{\epsilon,n}(B_s)\,dB_s + \frac{1}{2}\int_0^t f''_{\epsilon,n}(B_s)\,ds \quad \text{a. s.}$$

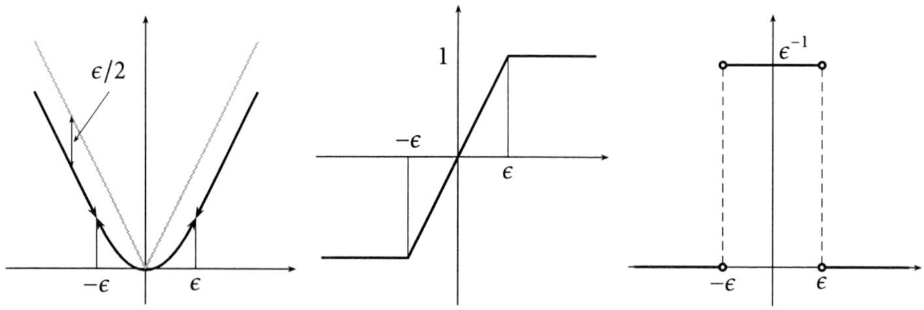

Fig. 17.1. Smooth approximation of the function $x \mapsto |x|$ and its derivatives.

On the left-hand side we see $\lim_{\epsilon\to 0}\lim_{n\to\infty} f_{\epsilon,n}(B_t) = |B_t|$ a. s. and in probability.

For the expression on the right we have

$$\mathbb{E}\left[\left|\int_0^t (f'_{\epsilon,n}(B_s) - \operatorname{sgn}(B_s))\,dB_s\right|^2\right] = \mathbb{E}\left[\int_0^t \left|f'_{\epsilon,n}(B_s) - \operatorname{sgn}(B_s)\right|^2 ds\right].$$

Letting first $n \to \infty$ and then $\epsilon \to 0$, the right-hand side converges to

$$\mathbb{E}\left[\int_0^t \left|f'_{\epsilon}(B_s) - \operatorname{sgn}(B_s)\right|^2 ds\right] \xrightarrow[\epsilon\to 0]{\text{dom. conv.}} 0;$$

in particular, $\mathbb{P}\text{-}\lim_{\epsilon\to 0}\int_0^t f'_\epsilon(B_s)\,dB_s = \int_0^t \operatorname{sgn}(B_s)\,dB_s$.

Finally, set

$$\Omega_s := \left\{\omega \,:\, \lim_{n\to\infty} f''_{\epsilon,n}(B_s(\omega)) = f''_\epsilon(B_s(\omega))\right\}.$$

As $\mathbb{P}(B_s = \pm\epsilon) = 0$, we know that $\mathbb{P}(\Omega_s) = 1$ for each $s \in [0, t]$. By Fubini's theorem

$$\mathbb{P}\otimes \operatorname{Leb}\{(\omega, s) \,:\, s\in[0,t],\ \omega\notin\Omega_s\} = \int_0^t \mathbb{P}(\Omega\setminus\Omega_s)\,ds = 0,$$

and so $\operatorname{Leb}\{s\in[0,t] \,:\, \omega\in\Omega_s\} = t$ for $\mathbb{P}$ almost all ω. Since $\|f''_{\epsilon,n}\|_\infty \leqslant \epsilon^{-1}$, the dominated convergence theorem yields

$$L^2\text{-}\lim_{n\to\infty}\int_0^t f''_{\epsilon,n}(B_s)\,ds = \int_0^t f''_\epsilon(B_s)\,ds = \frac{1}{\epsilon}\int_0^t \mathbb{1}_{(-\epsilon,\epsilon)}(B_s)\,ds.$$

Therefore, we have shown that

$$\mathbb{P}\text{-}\lim_{\epsilon\to 0}\frac{1}{2\epsilon}\int_0^t \mathbb{1}_{(-\epsilon,\epsilon)}(B_s)\,ds = |B_t| - \int_0^t \operatorname{sgn}(B_s)\,dB_s$$

exists and defines a stochastic process. The same argument applies for the shifted function $f(x) = |x-a|$, $a\in\mathbb{R}$. Therefore, the following definition makes sense.

17.12 Definition (Brownian local time. Lévy 1939). Let $(B_t)_{t\geqslant 0}$ be a BM^1 and $a\in\mathbb{R}$. The *local time at the level a up to time t* is the random process

$$L_t^a := \mathbb{P}\text{-}\lim_{\epsilon\to 0}\frac{1}{2\epsilon}\int_0^t \mathbb{1}_{(-\epsilon,\epsilon)}(B_s - a)\,ds. \tag{17.20}$$

The local time L_t^a represents the total amount of time Brownian motion spends at the level a up to time t:

$$\frac{1}{2\epsilon}\int_0^t \mathbb{1}_{(-\epsilon,\epsilon)}(B_s - a)\,ds = \frac{1}{2\epsilon}\operatorname{Leb}\{s\in[0,t] \,:\, B_s\in(a-\epsilon, a+\epsilon)\}.$$

We are now ready for the next result.

17.13 Theorem (Tanaka's formula. Tanaka 1963). *Let $(B_t)_{t\geqslant 0}$ denote a one-dimensional Brownian motion. Then we have for every $a \in \mathbb{R}$*

$$|B_t - a| = |a| + \int_0^t \operatorname{sgn}(B_s - a)\,dB_s + L_t^a \tag{17.21}$$

where L_t^a is Brownian local time at the level a. In particular, $(t,a) \mapsto L_t^a$ (has a version which) is continuous.

Proof. The formula (17.21) follows from the preceding discussion for the shifted function $f(x) = |x-a|$. The continuity of $(t,a) \mapsto L_t^a$ is now obvious since both Brownian motion and the stochastic integral have continuous versions. □

Tanaka's formula is the starting point for many further investigations. For example,
Ex. 17.13 it is possible to show that $\beta_t = \int_0^t \operatorname{sgn}(B_s)\,dB_s$ is again a one-dimensional Brownian motion and $L_t^0 = \sup_{s\leqslant t}(-\beta_s)$. Moreover, $\mathbb{P}^0(L_\infty^a = \infty) = 1$ for any $a \in \mathbb{R}$.

17.14 Further reading. All books mentioned in the *Further reading* section of Chapter 15 contain proofs of Itô's formula. Local times are treated in much greater detail in [119] and [189]. An interesting alter-
Ex. 17.4 native proof of Itô's formula is given in [132], see also Problem 17.4.

[119] Karatzas, Shreve: *Brownian Motion and Stochastic Calculus.*
[132] Krylov: *Introduction to the Theory of Diffusion Processes.*
[189] Revuz, Yor: *Continuous Martingales and Brownian Motion.*

Problems

1. Let $(B_t)_{t\geqslant 0}$ be a BM^1. Use Itô's formula to obtain representations of

$$X_t = \int_0^t \exp(B_s)\,dB_s \quad\text{and}\quad Y_t = \int_0^t B_s \exp(B_s^2)\,dB_s$$

which do not contain Itô integrals.

2. Let $\Pi = \{0 = t_0 < t_1 < \ldots < t_N = T\}$ be a generic partition of the interval $[0,T]$ and denote by $|\Pi| = \max_j |t_j - t_{j-1}|$ the mesh of the partition. Show that the limit

$$\lim_{|\Pi|\to 0} \sum_{j=1}^N (t_j - t_{j-1})^\gamma$$

is ∞, T or 0 according to $\gamma < 1$, $\gamma = 1$ or $\gamma > 1$.

3. Show that the limits (17.5) and (17.6) actually hold in $L^2(\mathbb{P})$ and uniformly for T from compact sets.

Hint: To show that the limits hold uniformly, use Doob's maximal inequality. For the L^2-version of (17.6) show first that $\mathbb{E}\left[\left(\sum_j (B_{t_j} - B_{t_{j-1}})^2\right)^4\right]$ converges to $c\,T^4$ for some constant c.

4. The following exercise contains an alternative proof of Itô's formula (17.1) for a one-dimensional Brownian motion $(B_t)_{t\geqslant 0}$.
(a) Let $f \in \mathcal{C}^1(\mathbb{R})$. Show that for $\Pi = \{t_0 = 0 < t_1 < \ldots < t_n = t\}$ and suitable intermediate values ξ_l

$$B_t f(B_t) = \sum_l B_{t_l}(f(B_{t_l}) - f(B_{t_{l-1}})) + \sum_l f(B_{t_{l-1}})(B_{t_l} - B_{t_{l-1}}) + \sum_l f'(\xi_l)(B_{t_l} - B_{t_{l-1}})^2.$$

Conclude from this that the differential expression $B_t\, df(B_t)$ is well-defined.
(b) Show that $d(B_t f(B_t)) = B_t\, df(B_t) + f(B_t)\, dB_t + f'(B_t)\, dt$.
(c) Use (b) to find a formula for dB_t^n, $n \geqslant 2$, and show that for every polynomial $p(x)$ Itô's formula holds:

$$dp(B_t) = p'(B_t)\, dB_t + \frac{1}{2} p''(B_t)\, dt.$$

(d) Show the following stability result for Itô integrals:

$$L^2(\mathbb{P})\text{-}\lim_{n\to\infty} \int_0^T g_n(s)\, dB_s = \int_0^T g(s)\, dB_s$$

for any sequence $(g_n)_{n\geqslant 1}$ which converges in $\mathcal{L}_T^2$.
(e) Use a suitable version of Weierstraß' approximation theorem and the stopping technique from the proof of Theorem 17.1, step 4° , to deduce from c) Itô's formula for $f \in \mathcal{C}^2$.

5. (a) Use the (two-dimensional, deterministic) chain rule $d(F \circ G) = F' \circ G\, dG$ to deduce the formula for integration by parts for Stieltjes integrals:

$$\int_0^t f(s)\, dg(s) = f(t)g(t) - f(0)g(0) - \int_0^t g(s)\, df(s) \quad \text{for all } f, g \in \mathcal{C}^1([0,\infty), \mathbb{R}).$$

(b) Use the Itô formula for a Brownian motion $(b_t, \beta_t)_{t\geqslant 0}$ in $\mathbb{R}^2$ to show that

$$\int_0^t b_s\, d\beta_s = b_t\beta_t - \int_0^t \beta_s\, db_s, \quad t \geqslant 0.$$

What happens if b and β are not independent?

6. Prove the following time-dependent version of Itô's formula: Let $(B_t)_{t\geqslant 0}$ be a BM^1 and $f : [0,\infty) \times \mathbb{R} \to \mathbb{R}$ be a function of class $\mathcal{C}^{1,2}$. Then

$$f(t, B_t) - f(0,0) = \int_0^t \frac{\partial f}{\partial x}(s, B_s)\, dB_s + \int_0^t \left(\frac{\partial f}{\partial t}(s, B_s) + \frac{1}{2}\frac{\partial^2 f}{\partial x^2}(s, B_s)\right) ds.$$

Prove and state the d-dimensional counterpart.

Hint: Use the Itô formula for the $(d+1)$-dimensional Itô process $(t, B_t^1, \dots, B_t^d)$.

7. Prove Theorem 5.6 using Itô's formula.

8. Let $(B_t, \mathscr{F}_t)_{t\geqslant 0}$ be a BM^1. Use Itô's formula to verify that the following processes are martingales:
$$X_t = e^{t/2}\cos B_t \quad\text{and}\quad Y_t = (B_t + t)\exp\left(-B_t - t/2\right).$$

9. Let $B_t = (b_t, \beta_t)$, $t \geqslant 0$ be a BM^2 and set $r_t := |B_t| = \sqrt{b_t^2 + \beta_t^2}$.
   (a) Show that the stochastic integrals $\int_0^t b_s/r_s\,db_s$ and $\int_0^t \beta_s/r_s\,d\beta_s$ exist.
   (b) Show that $W_t := \int_0^t b_s/r_s\,db_s + \int_0^t \beta_s/r_s\,d\beta_s$ is a BM^1.

10. Let $B_t = (b_t, \beta_t)$, $t \geqslant 0$ be a BM^2 and $f(x+iy) = u(x,y) + iv(x,y)$, $x, y \in \mathbb{R}$, an analytic function. If $u_x^2 + u_y^2 = 1$, then $\left(u(b_t,\beta_t), v(b_t,\beta_t)\right)$, $t \geqslant 0$, is a BM^2.

11. Show that the d-dimensional Itô formula remains valid if we replace the real-valued function $f : \mathbb{R}^d \to \mathbb{R}$ by a complex function $f = u + iv : \mathbb{R}^d \to \mathbb{C}$.

12. (Friedrichs mollifier) Let $\chi \in \mathcal{C}_c^\infty(-1,1)$ with $0 \leqslant \chi \leqslant 1$, $\chi(0) = 1$ and $\int \chi(x)\,dx = 1$. Set $\chi_n(x) := n\chi(nx)$.
   (a) Show that $\operatorname{supp}\chi_n \subset [-1/n, 1/n]$ and $\int \chi_n(x)\,dx = 1$.
   (b) Let $f \in \mathcal{C}(\mathbb{R})$ and $f_n := f \star \chi_n$.
   Show that $|\partial^k f_n(x)| \leqslant n^k \sup_{y\in\mathbb{B}(x,1/n)} |f(y)| \|\partial^k \chi\|_{L^1}$.
   (c) Let $f \in \mathcal{C}(\mathbb{R})$ uniformly continuous. Show that $\lim_{n\to\infty} \|f \star \chi_n - f\|_\infty = 0$. What can be said if f is continuous but not uniformly continuous?
   (d) Let f be piecewise continuous. Show that $\lim_{n\to\infty} f \star \chi_n(x) = f(x)$ at all x where f is continuous.

13. Show that $\beta_t = \int_0^t \operatorname{sgn}(B_s)\,dB_s$ is a BM^1.

Hint: Use Lévy's characterization of a BM^1, Theorem 9.12 or 18.5.

14. Let $(B_t)_{t\geqslant 0}$ be a BM^1, $\Phi \in \mathcal{C}_b^2(\mathbb{R})$, $\Phi(0) = 0$ and $f, g \in \mathcal{L}_T^2$ for some $T > 0$. Show that for all $t \in [0,T]$
$$\begin{aligned}
&\mathbb{E}\left[\int_0^t f(r)\,dB_r \cdot \Phi\left(\int_0^t g(r)\,dB_r\right)\right] \\
&\quad = \mathbb{E}\left[\int_0^t f(s)g(s)\Phi'\left(\int_0^s g(r)\,dB_r\right)ds\right] \\
&\qquad + \frac{1}{2}\,\mathbb{E}\left[\int_0^t\int_0^s f(r)\,dB_r\,\Phi''\left(\int_0^s g(r)\,dB_r\right)g^2(s)\,ds\right].
\end{aligned}$$

15. Let $\sigma_{_\Pi}, b_{_\Pi} \in \mathcal{S}_T$ which approximate $\sigma, b \in \mathcal{L}_T^2$ as $|\Pi| \to 0$, and consider the Itô processes $X_t^\Pi = X_0 + \int_0^t \sigma_{_\Pi}(s)\,dB_s + \int_0^t b_{_\Pi}(s)\,ds$ and $X_t = X_0 + \int_0^t \sigma(s)\,dB_s + \int_0^t b(s)\,ds$. Adapt the proof of Lemma 17.5 to show that

$$\int_0^T g(X_s^\Pi)\,dX_s^\Pi \xrightarrow[|\Pi|\to 0]{\mathbb{P}} \int_0^T g(X_s)\,dX_s \quad \text{for all } g \in \mathcal{C}_b(\mathbb{R}).$$

18 Applications of Itô's formula

Itô's formula has many applications and we restrict ourselves to a few of them. We will use it to obtain a characterization of a Brownian motion as a martingale (Lévy's theorem 18.5) and to describe the structure of 'Brownian' martingales (Theorems 18.11, 18.13 and 18.16). In some sense, this will show that a Brownian motion is *both* a very particular martingale *and* the typical martingale. Girsanov's theorem 18.8 allows us to change the underlying probability measure which will become important if we want to solve stochastic differential equations. Finally, the Burkholder–Davis–Gundy inequalities, Theorem 18.17, provide moment estimates for stochastic integrals.

Throughout this section $(B_t, \mathscr{F}_t)_{t\geqslant 0}$ is a BM with an admissible complete filtration, e. g. $\overline{\mathscr{F}}_t^B$, see Theorem 6.21. Recall that $\mathcal{L}_T^2 = L^2_{\mathscr{P}}(\lambda_T \otimes \mathbb{P})$.

18.1 Doléans–Dade exponentials

Let $(B_t, \mathscr{F}_t)_{t\geqslant 0}$ be a Brownian motion on $\mathbb{R}$. We have seen in Example 5.2 e) that $(M_t^\xi, \mathscr{F}_t)_{t\geqslant 0}$ is a martingale where $M_t^\xi := e^{\xi B_t - \frac{t}{2}\xi^2}$, $\xi \in \mathbb{R}$. Using Itô's formula we find that M_t^ξ satisfies the following integral equation

$$M_t^\xi = 1 + \int_0^t \xi\, M_s^\xi\, dB_s \quad \text{or} \quad dM_t^\xi = \xi\, M_t^\xi\, dB_t.$$

In the usual calculus the differential equation $dy(t) = y(t)\,dx(t)$, $y(0) = y_0$ has the unique solution $y(t) = y_0 e^{x(t)-x(0)}$. If we compare this with the Brownian setting, we see that there appears the additional factor $-\frac{1}{2}\xi^2 t = -\frac{1}{2}\langle \xi B\rangle_t$ which is due to the second order term in Itô's formula. On the other hand, it is this factor which makes M^ξ into a martingale. Because of this analogy it is customary to call

$$\mathcal{E}(M)_t := \exp\left(M_t - \frac{1}{2}\langle M\rangle_t\right), \quad t \geqslant 0, \tag{18.1}$$

for any continuous L^2 martingale $(M_t)_{t\geqslant 0}$ the *stochastic* or *Doléans–Dade exponential*. If we use Itô's formula for the stochastic integral for martingales from Section 15.6, it is not hard to see that $\mathcal{E}(M)_t$ is itself a local martingale. Here we consider only some special cases which can be directly written in terms of a Brownian motion.

Ex. 18.1 **18.1 Lemma.** *Let $(B_t, \mathscr{F}_t)_{t\geqslant 0}$ be a BM^1, $f \in L^2_{\mathscr{P}}(\lambda_T \otimes \mathbb{P})$ for all $T > 0$, and assume that $|f(s,\omega)| \leqslant C$ for some $C > 0$ and all $s \geqslant 0$ and $\omega \in \Omega$. Then*

$$\exp\left(\int_0^t f(s)\,dB_s - \frac{1}{2}\int_0^t f^2(s)\,ds\right), \quad t \geqslant 0, \tag{18.2}$$

is a martingale for the filtration $(\mathscr{F}_t)_{t\geqslant 0}$.

Proof. Set $X_t = \int_0^t f(s)\,dB_s - \frac{1}{2}\int_0^t f^2(s)\,ds$. Itô's formula, Theorem 17.7, yields[1]

$$\begin{aligned} e^{X_t} - 1 &= \int_0^t e^{X_s} f(s)\,dB_s - \frac{1}{2}\int_0^t e^{X_s} f^2(s)\,ds + \frac{1}{2}\int_0^t e^{X_s} f^2(s)\,ds \\ &= \int_0^t \exp\left(\int_0^s f(r)\,dB_r - \frac{1}{2}\int_0^s f^2(r)\,dr\right) f(s)\,dB_s. \end{aligned} \tag{18.3}$$

If we can show that the integrand is in $L^2_{\mathscr{P}}(\lambda_T \otimes \mathbb{P})$ for every $T > 0$, then Theorem 15.13 applies and shows that the stochastic integral, hence e^{X_t}, is a martingale.

We begin with an estimate for simple processes. Fix $T > 0$ and assume that $g \in \mathcal{S}_T$. Then $g(s,\omega) = \sum_{j=1}^n g(s_{j-1},\omega)\mathbb{1}_{[s_{j-1},s_j)}(s)$ for some partition $0 = s_0 < \cdots < s_n = T$ of $[0,T]$ and $|g(s,\omega)| \leqslant C$ for some constant $C < \infty$. Since $g(s_{j-1})$ is $\mathscr{F}_{s_{j-1}}$ measurable and $B_{s_j} - B_{s_{j-1}} \perp\!\!\!\perp \mathscr{F}_{s_{j-1}}$, we find

$$\begin{aligned} \mathbb{E}\left[e^{2\int_0^T g(r)\,dB_r}\right] &= \mathbb{E}\left[\prod_{j=1}^n e^{2g(s_{j-1})(B_{s_j}-B_{s_{j-1}})}\right] \\ &\overset{\text{tower}}{=} \mathbb{E}\left[\prod_{j=1}^{n-1} e^{2g(s_{j-1})(B_{s_j}-B_{s_{j-1}})}\,\underbrace{\mathbb{E}\left(e^{2g(s_{n-1})(B_{s_n}-B_{s_{n-1}})} \mid \mathscr{F}_{s_{n-1}}\right)}_{\overset{(A.4)}{\underset{2.2}{=}} \exp[2g^2(s_{n-1})(s_n-s_{n-1})]}\right] \\ &\leqslant e^{2C^2(s_n-s_{n-1})}\,\mathbb{E}\left[\prod_{j=1}^{n-1} e^{2g(s_{j-1})(B_{s_j}-B_{s_{j-1}})}\right]. \end{aligned}$$

Further applications of the tower property yield

$$\mathbb{E}\left[e^{2\int_0^T g(r)\,dB_r}\right] \leqslant \prod_{j=1}^n e^{2C^2(s_j-s_{j-1})} = e^{2C^2T}.$$

If $f \in L^2_{\mathscr{P}}(\lambda_T \otimes \mathbb{P})$, there is a sequence of simple processes $(f_n)_{n\geqslant 1} \subset \mathcal{S}_T$ such that $\lim_{n\to\infty} f_n = f$ in $L^2_{\mathscr{P}}(\lambda_T \otimes \mathbb{P})$ and $\lim_{n\to\infty} \int_0^t f_n(s)\,dB_s = \int_0^t f(s)\,dB_s$ for any $t \in [0,T]$ in $L^2(\mathbb{P})$ and, for a subsequence, almost surely. We can assume that $|f_n(s,\omega)| \leqslant C$, otherwise we would consider $-C \vee f_n \wedge C$. Then

$$\mathbb{E}\left[\left|e^{\int_0^t f(r)\,dB_r - \frac{1}{2}\int_0^t f^2(r)\,dr} f(t)\right|^2\right] \leqslant C^2\,\mathbb{E}\left[e^{2\int_0^t f(r)\,dB_r}\right]$$

By Fatou's lemma and the estimate for simple processes from above we get for $t \in [0,T]$

$$\mathbb{E}\left[\lim_{n\to\infty} e^{2\int_0^t f_n(r)\,dB_r}\right] \leqslant \varliminf_{n\to\infty} \mathbb{E}\left[e^{2\int_0^t f_n(r)\,dB_r}\right] \leqslant e^{2C^2t}.$$

1 Compare the following equality (18.3) with (18.1).

If we integrate the resulting inequality

$$\mathbb{E}\left[\left|e^{\int_0^t f(r)\,dB_r - \frac{1}{2}\int_0^t f^2(r)\,dr} f(t)\right|^2\right] \leqslant C^2 e^{2C^2 t}$$

with respect to $\int_0^T \dots dt$, the claim follows. □

It is clear that Lemma 18.1 also holds for BM^d and d-dimensional integrands. More interesting is the observation that we can replace the condition that the integrand is bounded by an integrability condition.

18.2 Theorem. *Let $(B_t, \mathscr{F}_t)_{t\geqslant 0}$, $B_t = (B_t^1, \dots, B_t^d)$, be a* BM^d *and $f = (f_1, \dots, f_d)$ be a d-dimensional $\mathscr{P}$ measurable process such that $f_j \in L^2_{\mathscr{P}}(\lambda_T \otimes \mathbb{P})$ for all $T > 0$ and $j = 1, \dots, d$. Then*

$$M_t = \exp\left(\sum_{j=1}^d \int_0^t f_j(s)\,dB_s^j - \frac{1}{2}\int_0^t |f(s)|^2\,ds\right) \tag{18.4}$$

is a martingale for the filtration $(\mathscr{F}_t)_{t\geqslant 0}$ if, and only if, $\mathbb{E}\,M_t = 1$.

Proof. The necessity is obvious since $M_0 = 1$. Since $f \in L^2_{\mathscr{P}}(\lambda_T \otimes \mathbb{P})$ for all $T > 0$, we see with the d-dimensional version of Itô's formula (17.14) applied to the process $X_t = \sum_{j=1}^d \int_0^t f_j(s)\,dB_s^j - \frac{1}{2}\int_0^t |f(s)|^2\,ds$ that

$$\begin{aligned} e^{X_t} &= 1 + \sum_{j=1}^d \int_0^t f_j(s)\,e^{X_s}\,dB_s^j \\ &= 1 + \sum_{j=1}^d \int_0^t f_j(s)\,\exp\left(\sum_{j=1}^d \int_0^s f_j(r)\,dB_r^j - \frac{1}{2}\int_0^s |f(r)|^2\,dr\right) dB_s^j. \end{aligned}$$

Let $\tau_n = \inf\{s \geqslant 0 : |\exp(X_s)| \geqslant n\} \wedge n$; since $s \mapsto \exp(X_s)$ is continuous, τ_n is a stopping time and $\exp(X_s^{\tau_n}) \leqslant n$. Therefore, $f^{\tau_n} \cdot \exp(X^{\tau_n})$ is in $L^2_{\mathscr{P}}(\lambda_T \otimes \mathbb{P})$ for all $T > 0$, and Theorem 15.13 e) shows that $M^{\tau_n} = \exp(X^{\tau_n})$ is a martingale. Clearly, $\lim_{n\to\infty} \tau_n = \infty$ which means that M_t is a local martingale, cf. Definition 16.6.

Proposition 18.3 below shows that any positive local martingale with $\mathbb{E}\,M_t = 1$ is already a martingale, and the claim follows. □

The last step in the proof of Theorem 18.2 is interesting on its own:

18.3 Proposition. a) *Every positive local martingale $(M_t, \mathscr{F}_t)_{t\geqslant 0}$ is a (positive) supermartingale.*

b) *Let $(M_t, \mathscr{F}_t)_{t\geqslant 0}$ be a supermartingale with $\mathbb{E}\,M_t = \mathbb{E}\,M_0$ for all $t \geqslant 0$, then $(M_t, \mathscr{F}_t)_{t\geqslant 0}$ is a martingale.*

Proof. a) Let τ_n bei a localizing sequence for $(M_t)_{t\geqslant 0}$. By Fatou's lemma we see that

$$\mathbb{E}\,M_t = \mathbb{E}\left(\varliminf_{n\to\infty} M_{t\wedge\tau_n}\right) \leqslant \varliminf_{n\to\infty} \mathbb{E}\,M_{t\wedge\tau_n} = \mathbb{E}\,M_0 < \infty,$$

since M^{τ_n} is a martingale. The conditional version of Fatou's lemma shows for all $s \leqslant t$

$$\mathbb{E}(M_t \mid \mathscr{F}_s) \leqslant \varliminf_{n\to\infty} \mathbb{E}\left(M_{t\wedge\tau_n} \mid \mathscr{F}_s\right) = \varliminf_{n\to\infty} M_{s\wedge\tau_n} = M_s,$$

i. e. $(M_t)_{t\geqslant 0}$ is a supermartingale.

b) Let $s \leqslant t$ and $F \in \mathscr{F}_s$. By assumption, $\int_F M_s\,d\mathbb{P} \geqslant \int_F M_t\,d\mathbb{P}$. Subtracting from both sides $\mathbb{E}\,M_s = \mathbb{E}\,M_t$ gives

$$\int_F M_s\,d\mathbb{P} - \mathbb{E}\,M_s \geqslant \int_F M_t\,d\mathbb{P} - \mathbb{E}\,M_t.$$

Therefore,

$$\int_{F^c} M_s\,d\mathbb{P} \leqslant \int_{F^c} M_t\,d\mathbb{P} \quad \text{for all } s \leqslant t,\ F^c \in \mathscr{F}_s.$$

This means that $(M_t)_{t\geqslant 0}$ is a submartingale, hence a martingale. □

We close this section with the Novikov condition which gives a sufficient criterion for $\mathbb{E}\,M_t = 1$. The following proof is essentially the proof which we used for the exponential Wald identity, cf. Theorem 5.14. We prove only the one-dimensional version, the extension to higher dimensions is obvious.

18.4 Theorem (Novikov 1972). *Let $(B_t)_{t\geqslant 0}$ be a* BM[1] *and $f \in L^2_{\mathscr{P}}(\lambda_T \otimes \mathbb{P})$ for all $T > 0$. If*

$$\mathbb{E}\left[\exp\left(\frac{1}{2}\int_0^\infty |f(s)|^2\,ds\right)\right] < \infty, \tag{18.5}$$

then $\mathbb{E}\,M_t = 1$ where M_t is the the stochastic exponential

$$M_t = \exp\left(\int_0^t f(s)\,dB_s - \frac{1}{2}\int_0^t |f(s)|^2\,ds\right).$$

Proof. We have seen in the proof of Theorem 18.2 that M_t is a positive local martingale, and by Proposition 18.3 a) it is a supermartingale. In particular, $\mathbb{E}\,M_t \leqslant \mathbb{E}\,M_0 = 1$.

For the converse inequality write $X_t = \int_0^t f(s)\,dB_s$ and $\langle X\rangle_t = \int_0^t |f(s)|^2\,ds$. Then $M_t = e^{X_t - \frac{1}{2}\langle X\rangle_t} = \mathcal{E}(X)_t$ and the condition (18.5) reads $\mathbb{E}[e^{\frac{1}{2}\langle X\rangle_\infty}] < \infty$.

Let $(\tau_n)_{n\geqslant 1}$ be a localizing sequence for the local martingale M, fix some $c \in (0,1)$ and pick $p = p(c) > 1$ such that $p < \frac{1}{c(2-c)}$ $\left(\leqslant \frac{1}{c}\right)$.

Then $M_t^{\tau_n} = \exp[X_t^{\tau_n} - \frac{1}{2}\langle X\rangle_t^{\tau_n}]$ is a martingale such that $\mathbb{E}\,M_{t\wedge\tau_n} = \mathbb{E}\,M_0 = 1$. So,

$$\mathbb{E}\left[\mathcal{E}(cX^{\tau_n})_t^p\right] = \mathbb{E}\left[e^{pcX_t^{\tau_n} - \frac{1}{2}pc^2\langle X\rangle_t^{\tau_n}}\right] = \mathbb{E}\left[e^{pc(X_t^{\tau_n} - \frac{1}{2}\langle X\rangle_t^{\tau_n})} e^{\frac{1}{2}pc(1-c)\langle X\rangle_t^{\tau_n}}\right],$$

and from the Hölder inequality for the exponents $1/pc$ and $1/(1-pc)$ we find

$$\leqslant \underbrace{\left[\mathbb{E}\,e^{X_t^{\tau_n} - \frac{1}{2}\langle X\rangle_t^{\tau_n}}\right]^{pc}}_{=[\mathbb{E}\,M_{t\wedge\tau_n}]^{pc}=1} \left[\mathbb{E}\,e^{\frac{1}{2}\frac{pc(1-c)}{1-pc}\langle X\rangle_t}\right]^{1-pc} \leqslant \left[\mathbb{E}\,e^{\frac{1}{2}\langle X\rangle_\infty}\right]^{1-pc}.$$

In the last step we used that $pc(1-c)/(1-pc) \leqslant 1$. This shows that the pth moment of the family $(M_t^{\tau_n})_{n\geqslant 1}$ is uniformly bounded, hence it is uniformly integrable.

Therefore, we can let $n \to \infty$ and find, using uniform integrability and Hölder's inequality for the exponents $1/c$ and $1/(1-c)$

$$1 = \mathbb{E}\left[e^{cX_t^{\tau_n} - \frac{1}{2}c^2\langle X\rangle_t^{\tau_n}}\right] = \mathbb{E}\left[e^{cX_t - \frac{1}{2}c\langle X\rangle_t} e^{\frac{1}{2}c(1-c)\langle X\rangle_t}\right]$$

$$\leqslant \left[\mathbb{E}\, e^{X_t - \frac{1}{2}\langle X\rangle_t}\right]^c \left[\mathbb{E}\, e^{\frac{1}{2}c\langle X\rangle_t}\right]^{1-c}.$$

Since the last factor is bounded by $(\mathbb{E}\, e^{\langle X\rangle_\infty})^{1-c}$, we find as $c \to 1$,

$$1 \leqslant \mathbb{E}\, e^{X_t - \frac{1}{2}\langle X\rangle_t} = \mathbb{E}\, M_t.$$

□

18.2 Lévy's characterization of Brownian motion

In the hands of Kunita and Watanabe [136] Itô's formula became a most powerful tool. We will follow their ideas and give an elegant proof of the following theorem due to P. Lévy.

Ex. 18.2 **18.5 Theorem** (Lévy 1948). *Let $(X_t, \mathscr{F}_t)_{t\geqslant 0}$, $X_0 = 0$, be a real-valued, adapted process*
Ex. 18.3 *with continuous sample paths. If both $(X_t, \mathscr{F}_t)_{t\geqslant 0}$ and $(X_t^2 - t, \mathscr{F}_t)_{t\geqslant 0}$ are martingales, then $(X_t)_{t\geqslant 0}$ is a Brownian motion with filtration $(\mathscr{F}_t)_{t\geqslant 0}$.*

Theorem 18.5 means, in particular, that BM^1 *is the only continuous martingale with quadratic variation $\langle B\rangle_t \equiv t$.*

! The way we have *stated* Itô's formula, we cannot use it for $(X_t)_{t\geqslant 0}$ as in Theorem 18.5. Nevertheless, a close inspection of the construction of the stochastic integral in Chapters 15 and 16 shows that all results remain valid if the integrator dB_t is replaced by dX_t where $(X_t)_{t\geqslant 0}$ is a continuous martingale with quadratic variation $\langle X\rangle_t = t$. For steps 2° and 3° of the proof of Theorem 17.1, cf. (17.7) and (17.8) in the the proof of Lemma 17.4, we have to replace Theorem 9.1 by Corollary 9.11. In fact, Theorem 18.5 is an ex-post vindication for this since such martingales *already are Brownian motions.* In order not to end up with a circular conclusion, we have to go through the proofs of Chapters 15–17 again, though.

After this word of caution we can use (17.1) in the setting of Theorem 18.5 and start with the

Proof of Theorem 18.5 (Kunita, Watanabe 1967). We have to show that

$$X_t - X_s \perp\!\!\!\perp \mathscr{F}_s \quad \text{and} \quad X_t - X_s \sim \mathsf{N}(0, t-s) \quad \text{for all } 0 \leqslant s \leqslant t.$$

This is equivalent to

$$\mathbb{E}\left(e^{i\xi(X_t - X_s)}\,\mathbb{1}_F\right) = e^{-\frac{1}{2}(t-s)\xi^2}\,\mathbb{P}(F) \quad \text{for all } s \leqslant t,\ F \in \mathscr{F}_s. \tag{18.6}$$

In fact, setting $F = \Omega$ yields

$$\mathbb{E}\left(e^{i\xi(X_t-X_s)}\right) = e^{-\frac{1}{2}(t-s)\xi^2},$$

i. e. $X_t - X_s \sim X_{t-s} \sim \mathsf{N}(0, t-s)$, and (18.6) becomes Ex. 9.8

$$\mathbb{E}\left(e^{i\xi(X_t-X_s)}\,\mathbb{1}_F\right) = \mathbb{E}\left(e^{i\xi(X_t-X_s)}\right)\mathbb{P}(F) \quad \text{for all } s \leqslant t,\ F \in \mathscr{F}_s;$$

this proves that $X_t - X_s \perp\!\!\!\perp F$ for all $F \in \mathscr{F}_s$.

Let us now prove (18.6). Note that $f(x) := e^{i\xi x}$ is a $\mathcal{C}^2$-function with

$$f(x) = e^{ix\xi}, \quad f'(x) = i\xi e^{ix\xi}, \quad f''(x) = -\xi^2 e^{ix\xi}.$$

Applying (17.1) to $f(X_t)$ and $f(X_s)$ and subtracting the respective results yields

$$e^{i\xi X_t} - e^{i\xi X_s} = i\xi \int_s^t e^{i\xi X_r}\,dX_r - \frac{\xi^2}{2}\int_s^t e^{i\xi X_r}\,dr. \tag{18.7}$$

Since $|e^{i\eta}| = 1$, the real and imaginary parts of the integrands are $L^2_{\mathscr{P}}(\lambda_T \otimes \mathbb{P})$-functions for any $t \leqslant T$ and, using Theorem 15.13 a) with X instead of B, we see that $\int_0^t e^{i\xi X_r}\,dX_r$ is a martingale. Hence,

$$\mathbb{E}\left(\int_s^t e^{i\xi X_r}\,dX_r \,\middle|\, \mathscr{F}_s\right) = 0 \quad \text{a. s.}$$

Multiplying both sides of (18.7) by $e^{-i\xi X_s}\,\mathbb{1}_F$, where $F \in \mathscr{F}_s$, and then taking expectations gives

$$\mathbb{E}\left(e^{i\xi(X_t-X_s)}\,\mathbb{1}_F\right) - \mathbb{P}(F) = -\frac{\xi^2}{2}\int_s^t \mathbb{E}\left(e^{i\xi(X_r-X_s)}\,\mathbb{1}_F\right)dr.$$

This means that $\phi(t) := \mathbb{E}\left(e^{i\xi(X_t-X_s)}\,\mathbb{1}_F\right)$, $s \leqslant t$, is a solution of the integral equation

$$\Phi(t) = \mathbb{P}(F) - \frac{\xi^2}{2}\int_s^t \Phi(r)\,dr. \tag{18.8}$$

On the other hand, it is easy to see that (18.8) is also solved by $\psi(t) := \mathbb{P}(F)\,e^{-\frac{1}{2}(t-s)\xi^2}$. Since

$$|\phi(t) - \psi(t)| = \frac{\xi^2}{2}\left|\int_s^t (\phi(r) - \psi(r))\,dr\right| \leqslant \frac{\xi^2}{2}\int_s^t |\phi(r) - \psi(r)|\,dr,$$

we can use Gronwall's lemma, Theorem A.47, with $u = |\phi - \psi|$, $a \equiv 0$ and $b \equiv \xi^2/2$, to get

$$|\phi(t) - \psi(t)| \leqslant 0 \quad \text{for all } t \geqslant 0.$$

This shows that $\phi(t) = \psi(t)$, i. e. (18.8) admits *only one solution*, and (18.6) follows. □

18.3 Girsanov's theorem

Let $(B_t)_{t\geqslant 0}$ be a real Brownian motion. The process $W_t := B_t - \ell t$, $\ell \in \mathbb{R}$, is called *Brownian motion with drift*. It describes a uniform motion with speed $-\ell$ which is perturbed by a Brownian motion B_t. Girsanov's theorem provides a very useful transformation of the underlying probability measure $\mathbb{P}$ such that, under the new measure, W_t is itself a Brownian motion. This is very useful if we want to calculate, e. g. the distribution of the random variable $\sup_{s\leqslant t} W_s$ and other functionals of W.

18.6 Example. Let $G \sim \mathsf{N}(m, \sigma^2)$ be on $(\Omega, \mathscr{A}, \mathbb{P})$ a real-valued normal random variable with mean m and variance σ^2. We want to transform G into a mean zero normal random variable. Obviously, $G \rightsquigarrow G - m$ would do the trick on $(\Omega, \mathscr{A}, \mathbb{P})$ but this would also change G. Another possibility is to change the probability measure $\mathbb{P}$.

Observe that $\mathbb{E}\, e^{i\xi G} = e^{-\frac{1}{2}\sigma^2\xi^2 + im\xi}$. Since G has exponential moments, cf. Corollary 2.2, we can insert $\xi = im/\sigma^2$ and find

$$\mathbb{E} \exp\left(-\tfrac{m}{\sigma^2}\, G\right) = \exp\left(-\tfrac{1}{2}\, \tfrac{m^2}{\sigma^2}\right).$$

Therefore, $\mathbb{Q}(d\omega) := \exp(-\frac{m}{\sigma^2}\, G(\omega) + \frac{1}{2}\, \frac{m^2}{\sigma^2})\, \mathbb{P}(d\omega)$ is a probability measure, and we find for the corresponding mathematical expectation $\mathbb{E}_{\mathbb{Q}} = \int_\Omega \cdots\, d\mathbb{Q}$

$$\begin{aligned}
\mathbb{E}_{\mathbb{Q}}\, e^{i\xi G} &= \int e^{i\xi G} \exp\left(-\tfrac{m}{\sigma^2}\, G + \tfrac{1}{2}\, \tfrac{m^2}{\sigma^2}\right)\, d\mathbb{P}\\
&= \exp\left(\tfrac{1}{2}\, \tfrac{m^2}{\sigma^2}\right) \int \exp\left(i\left(\xi + i\tfrac{m}{\sigma^2}\right) G\right)\, d\mathbb{P}\\
&= \exp\left(\tfrac{1}{2}\, \tfrac{m^2}{\sigma^2}\right) \exp\left(-\tfrac{1}{2}\, \left(\xi + i\tfrac{m}{\sigma^2}\right)^2 \sigma^2 + im(\xi + i\tfrac{m}{\sigma^2})\right)\\
&= \exp\left(-\tfrac{1}{2}\, \sigma^2\xi^2\right).
\end{aligned}$$

This means that, under $\mathbb{Q}$, we have $G \sim \mathsf{N}(0, \sigma^2)$.

The case of a Brownian motion with drift is similar. First we need a formula which shows how conditional expectations behave under changes of measures.

Let β be a positive random variable with $\mathbb{E}\,\beta = 1$. Then $\mathbb{Q}(d\omega) := \beta(\omega)\mathbb{P}(d\omega)$ is a new probability measure. If $\mathscr{F}$ is a σ-algebra in $\mathscr{A}$, then

$$\int_F \mathbb{E}(\beta \mid \mathscr{F})\, d\mathbb{P} = \int_F \beta\, d\mathbb{P} = \mathbb{Q}(F) \quad \text{i.e.} \quad \mathbb{Q}|_{\mathscr{F}} = \mathbb{E}(\beta \mid \mathscr{F})\mathbb{P}|_{\mathscr{F}}.$$

If we write $\mathbb{E}_{\mathbb{Q}}$ for the expectation w. r. t. the measure $\mathbb{Q}$, we see for all $F \in \mathscr{F}$

$$\begin{aligned}
\mathbb{E}_{\mathbb{Q}}(\mathbb{1}_F X) = \mathbb{E}(\mathbb{1}_F X\beta) = \mathbb{E}\left(\mathbb{1}_F\, \mathbb{E}(X\beta \mid \mathscr{F})\right) &= \mathbb{E}\Bigg(\overbrace{\mathbb{1}_F \frac{\mathbb{E}(X\beta \mid \mathscr{F})}{\mathbb{E}(\beta \mid \mathscr{F})}}^{\mathscr{F}\text{ mble}}\, \mathbb{E}(\beta \mid \mathscr{F})\Bigg)\\
&= \mathbb{E}_{\mathbb{Q}}\left(\mathbb{1}_F \frac{\mathbb{E}(X\beta \mid \mathscr{F})}{\mathbb{E}(\beta \mid \mathscr{F})}\right).
\end{aligned}$$

This means that $\mathbb{E}_{\mathbb{Q}}(X \mid \mathscr{F}) = \mathbb{E}(X\beta \mid \mathscr{F}) / \mathbb{E}(\beta \mid \mathscr{F})$.

18.7 Example. Let $(B_t, \mathscr{F}_t)_{t\geqslant 0}$ be a BM[1] and set $\beta = \beta_T$ where $\beta_t = e^{\ell B_t - \ell^2 t/2}$. From Example 5.2 e) we know that $(\beta_s)_{s\leqslant T}$ is a martingale for the filtration $(\mathscr{F}_t)_{t\geqslant 0}$. Therefore, $\mathbb{E}\,\beta_T = 1$ and $\mathbb{Q} := \beta_T \mathbb{P}$ is a probability measure. We want to calculate the conditional characteristic function of $W_t := B_t - \ell t$ under $\mathbb{E}_{\mathbb{Q}}$. Let $\xi \in \mathbb{R}$ and $s \leqslant t \leqslant T$. By the tower property we get

$$
\begin{aligned}
\mathbb{E}_{\mathbb{Q}}\left[e^{i\xi(W_t - W_s)} \mid \mathscr{F}_s\right] &= \mathbb{E}\left[e^{i\xi(W_t - W_s)}\beta_T \mid \mathscr{F}_s\right] \Big/ \mathbb{E}[\beta_T \mid \mathscr{F}_s] \\
&\overset{\text{tower}}{=} \mathbb{E}\left[e^{i\xi(W_t - W_s)}\,\mathbb{E}\left(\beta_T \mid \mathscr{F}_t\right) \mid \mathscr{F}_s\right] \Big/ \beta_s \\
&= \mathbb{E}\left[e^{i\xi(W_t - W_s)}\beta_t \mid \mathscr{F}_s\right] \Big/ \beta_s \\
&= e^{-i\ell\xi(t-s) - \frac{1}{2}\ell^2(t-s)}\,\mathbb{E}\left[e^{i\xi(B_t - B_s)}e^{\ell(B_t - B_s)} \mid \mathscr{F}_s\right] \\
&\overset{\text{(B1),(5.1)}}{=} e^{-i\ell\xi(t-s) - \frac{1}{2}\ell^2(t-s)}\,\mathbb{E}\left[e^{i\xi(B_t - B_s)}e^{\ell(B_t - B_s)}\right] \\
&\overset{(2.6)}{=} e^{-i\ell\xi(t-s) - \frac{1}{2}\ell^2(t-s)}\,e^{\frac{1}{2}(i\xi + \ell)^2(t-s)} \\
&= e^{-\frac{1}{2}(t-s)\xi^2}.
\end{aligned}
$$

Lemma 5.4 shows that $(W_t, \mathscr{F}_t)_{t\leqslant T}$ is a Brownian motion under $\mathbb{Q}$.

We will now consider the general case.

18.8 Theorem (Girsanov 1960). *Let $(B_t, \mathscr{F}_t)_{t\geqslant 0}$, $B_t = (B_t^1, \dots, B_t^d)$, be a* BM^d *and assume that $f = (f_1, \dots, f_d)$ is a $\mathscr{P}$ measurable process such that $f_j \in L^2_{\mathscr{P}}(\lambda_T \otimes \mathbb{P})$ for all $0 < T < \infty$ and $j = 1, \dots, d$. If* Ex. 18.4

$$
q_t = \exp\left(\sum_{j=1}^{d}\int_0^t f_j(s)\,dB_s^j - \frac{1}{2}\int_0^t |f(s)|^2\,ds\right) \tag{18.9}
$$

is integrable and $\mathbb{E}\,q_t \equiv 1$, then for every $T > 0$ the process

$$
W_t := B_t - \int_0^t f(s)\,ds, \quad t \leqslant T, \tag{18.10}
$$

is a BM^d *under the new probability measure $\mathbb{Q} = \mathbb{Q}_T$*

$$
\mathbb{Q}(d\omega) := q_T(\omega)\,\mathbb{P}(d\omega). \tag{18.11}
$$

Notice that the new probability measure $\mathbb{Q}$ in Theorem 18.8 depends on the right end-point T. !

Proof. Since $f_j \in L^2_{\mathscr{P}}(\lambda_T \otimes \mathbb{P})$ for all $T > 0$, the processes

$$
X_t^j := \int_0^t f_j(s)\,dB_s^j - \frac{1}{2}\int_0^t f_j^2(s)\,ds, \quad j = 1, \dots, d,
$$

are well-defined, and Theorem 18.2 shows that under the condition $\mathbb{E}\, q_t = 1$ the process $(q_t, \mathscr{F}_t)_{t \leqslant T}$ is a martingale. Set

$$\beta_t := \exp\left[i\langle \xi, W(t)\rangle + \frac{1}{2}\, t\, |\xi|^2\right]$$

where $W_t = B_t - \int_0^t f(s)\, ds$ is as in (18.10). If we *knew* that $(\beta_t)_{t \leqslant T}$ is a martingale on the probability space $(\Omega, \mathscr{F}_T, \mathbb{Q} = \mathbb{Q}_T)$, then

$$\begin{aligned}
\mathbb{E}_{\mathbb{Q}}\left(\exp[i\langle \xi, W(t) - W(s)\rangle] \mid \mathscr{F}_s\right) &= \mathbb{E}_{\mathbb{Q}}\left(\beta_t \exp\left[-i\langle \xi, W(s)\rangle\right] \exp\left[-\tfrac{1}{2}\, t\, |\xi|^2\right] \mid \mathscr{F}_s\right) \\
&= \mathbb{E}_{\mathbb{Q}}\left(\beta_t \mid \mathscr{F}_s\right) \exp\left[-i\langle \xi, W(s)\rangle\right] \exp\left[-\tfrac{1}{2}\, t\, |\xi|^2\right] \\
&= \beta_s \exp\left[-i\langle \xi, W(s)\rangle\right] \exp\left[-\tfrac{1}{2}\, t\, |\xi|^2\right] \\
&= \exp\left[-\tfrac{1}{2}\,(t-s)\, |\xi|^2\right].
\end{aligned}$$

Lemma 5.4 shows that $W(t)$ is a Brownian motion under the probability measure $\mathbb{Q}$.

Let us check that $(\beta_t)_{t \leqslant T}$ is a martingale. We have

$$\begin{aligned}
&\underbrace{\sum_{j=1}^{d} \int_0^t f_j(s)\, dB_s^j - \frac{1}{2} \int_0^t |f(s)|^2\, ds}_{\text{exponent of } q_t} + \underbrace{i \sum_{j=1}^{d} \xi_j B_t^j - i \sum_{j=1}^{d} \int_0^t \xi_j f_j(s)\, ds + \frac{1}{2}\, |\xi|^2\, t}_{\text{exponent of } \beta_t} \\
&\qquad = \sum_{j=1}^{d} \int_0^t (f_j(s) + i\xi_j)\, dB_s^j - \frac{1}{2} \int_0^t \langle f(s) + i\xi,\ f(s) + i\xi\rangle\, ds.
\end{aligned}$$

This shows that

$$q_t \beta_t = \exp\left[\sum_{j=1}^{d} \int_0^t (f_j(s) + i\xi_j)\, dB_s^j - \frac{1}{2} \int_0^t \langle f(s) + i\xi,\ f(s) + i\xi\rangle\, ds\right],$$

and, just as for q_t, we find for the product $q_t\beta_t$ by Itô's formula that

$$q_t \beta_t = 1 + \sum_{j=1}^{d} \int_0^t q_s \beta_s\, (f_j(s) + i\xi_j)\, dB_s^j.$$

Therefore, $(q_t\beta_t)_{t \geqslant 0}$ is a local martingale. If $(\tau_k)_{k \geqslant 1}$ is a localizing sequence, we get for $k \geqslant 1$, $F \in \mathscr{F}_s$ and $s \leqslant t \leqslant T$

$$\begin{aligned}
\mathbb{E}_{\mathbb{Q}}\left(\beta_{t \wedge \tau_k} \mathbb{1}_F\right) &\overset{\text{def}}{=} \mathbb{E}\left(q_T \beta_{t \wedge \tau_k} \mathbb{1}_F\right) \\
&\overset{\text{tower}}{=} \mathbb{E}\left[\mathbb{E}\left(q_T \beta_{t \wedge \tau_k} \mathbb{1}_F \mid \mathscr{F}_t\right)\right] \\
&= \mathbb{E}\Big[\underbrace{\mathbb{E}\left(q_T \mid \mathscr{F}_t\right)}_{= q_t,\ \text{martingale}} \beta_{t \wedge \tau_k} \mathbb{1}_F\Big] \\
&= \mathbb{E}\left[(q_t - q_{t \wedge \tau_k}) \beta_{t \wedge \tau_k} \mathbb{1}_F\right] + \mathbb{E}\left[q_{t \wedge \tau_k} \beta_{t \wedge \tau_k} \mathbb{1}_F\right] \\
&= \mathbb{E}\left[(q_t - q_{t \wedge \tau_k}) \beta_{t \wedge \tau_k} \mathbb{1}_F\right] + \mathbb{E}\left[q_{s \wedge \tau_k} \beta_{s \wedge \tau_k} \mathbb{1}_F\right].
\end{aligned}$$

In the last step we used the fact that $(q_t\beta_t, \mathscr{F}_t)_{t\geqslant 0}$ is a local martingale. Since $t \mapsto q_t$ is continuous and $q_{t\wedge\tau_k} = \mathbb{E}(q_t \mid \mathscr{F}_{t\wedge\tau_k})$, we know that $(q_{t\wedge\tau_k}, \mathscr{F}_{t\wedge\tau_k})_{k\geqslant 1}$ is a uniformly integrable martingale, thus, $\lim_{k\to\infty} q_{t\wedge\tau_k} = q_t$ a. s. and in $L^1(\mathbb{P})$. Moreover, we have that $|\beta_t| = e^{\frac{1}{2}|\xi|^2 t} \leqslant e^{\frac{1}{2}|\xi|^2 T}$. By dominated convergence,

$$\begin{aligned}\mathbb{E}_{\mathbb{Q}}\left(\beta_t \mathbb{1}_F\right) &= \lim_{k\to\infty} \mathbb{E}_{\mathbb{Q}}\left(\beta_{t\wedge\tau_k}\mathbb{1}_F\right)\\ &= \lim_{k\to\infty} \mathbb{E}\left[(q_t - q_{t\wedge\tau_k})\beta_{t\wedge\tau_k}\mathbb{1}_F\right] + \lim_{k\to\infty}\mathbb{E}\left[q_{s\wedge\tau_k}\beta_{s\wedge\tau_k}\mathbb{1}_F\right]\\ &= \mathbb{E}\left[q_s\beta_s\mathbb{1}_F\right].\end{aligned}$$

If we use the last equality with $t = s$, we get

$$\mathbb{E}_{\mathbb{Q}}\left(\beta_t\mathbb{1}_F\right) = \mathbb{E}\left[q_s\beta_s\mathbb{1}_F\right] \overset{t=s}{=} \mathbb{E}_{\mathbb{Q}}\left(\beta_s\mathbb{1}_F\right) \quad \text{for all } F \in \mathscr{F}_s,\ 0 \leqslant s \leqslant T,$$

i. e. $(\beta_t, \mathscr{F}_t)_{0\leqslant t\leqslant T}$ is a martingale under $\mathbb{Q}$ for every $T > 0$. □

18.4 Martingale representation – 1

Let $(B_t)_{t\geqslant 0}$ be a BM[1] and $(\mathscr{F}_t)_{t\geqslant 0}$ an admissible complete filtration. From Theorem 15.13 we know that for every $T \in [0, \infty]$ and $X \in L^2_{\mathscr{P}}(\lambda_T \otimes \mathbb{P})$ the stochastic integral

$$M_t = x + \int_0^t X_s\,dB_s, \quad t \in [0, T],\ x \in \mathbb{R}, \tag{18.12}$$

is a continuous L^2 martingale for the filtration $(\mathscr{F}_t)_{t\leqslant T}$ of the underlying Brownian motion. We will now show the converse: If $(M_t, \mathscr{F}_t)_{t\leqslant T}$ is an L^2 martingale for the complete Brownian filtration $(\mathscr{F}_t)_{t\geqslant 0}$, then M_t must be of the form (18.12) for some $X \in L^2_{\mathscr{P}}(\lambda_T \otimes \mathbb{P})$.

For this we define the set

$$\mathcal{H}^2_T := \left\{ M_T \,:\, M_T = x + \int_0^T X_s\,dB_s,\ x \in \mathbb{R},\ X \in L^2_{\mathscr{P}}(\lambda_T \otimes \mathbb{P}) \right\}.$$

18.9 Lemma. *The space $\mathcal{H}^2_T$ is a closed linear subspace of $L^2(\Omega, \mathscr{F}_T, \mathbb{P})$.*

Proof. $\mathcal{H}^2_T$ is obviously a linear space and, by Theorem 15.13, it is contained in $L^2(\Omega, \mathscr{F}_T, \mathbb{P})$. We show that it is a closed subspace. Let $(M^n_T)_{n\geqslant 1} \subset \mathcal{H}^2_T$,

$$M^n_T = x^n + \int_0^T X^n_s\,dB_s,$$

be an L^2 Cauchy sequence; since $L^2(\Omega, \mathscr{F}_T, \mathbb{P})$ is complete, the sequence converges to some limit $M \in L^2(\Omega, \mathscr{F}_T, \mathbb{P})$. We have to show that $M \in \mathcal{H}^2_T$. Using the Itô isometry

(15.21) we see for all $m, n \geqslant 1$

$$\mathbb{E}\left[((M_T^n - x^n) - (M_T^m - x^m))^2\right] = \mathbb{E}\left[\left(\int_0^T (X_s^n - X_s^m)\, dB_s\right)^2\right] = \mathbb{E}\left[\int_0^T |X_s^n - X_s^m|^2\, ds\right].$$

This shows that $(X_s^n)_{s\leqslant T}$ is a Cauchy sequence in $L^2(\lambda_T \otimes \mathbb{P})$. Because of the completeness of $L^2(\lambda_T \otimes \mathbb{P})$ the sequence converges to some process $(X_s)_{s\leqslant T}$ and, for some subsequence, we get

$$\lim_{k\to\infty} X_s^{n(k)}(\omega) = X_s(\omega) \quad \text{for } \lambda_T \otimes \mathbb{P} \text{ almost all } (s,\omega) \text{ and in } L^2(\lambda_T \otimes \mathbb{P}).$$

Since pointwise limits preserve measurability, we have $(X_s)_{s\leqslant T} \in L^2_{\mathscr{P}}(\lambda_T \otimes \mathbb{P})$. Using again Itô's isometry, we find

$$\int_0^T X_s^n\, dB_s \xrightarrow[n\to\infty]{L^2(\Omega,\mathscr{F}_T,\mathbb{P})} \int_0^T X_s\, dB_s.$$

As L^2-convergence implies L^1-convergence, we get

$$\lim_{n\to\infty} x^n \overset{15.13\,a)}{=} \lim_{n\to\infty} \mathbb{E}\left[x^n + \int_0^T X_s^n\, dB_s\right] = \mathbb{E}\, M,$$

and this shows that $M = \mathbb{E}\, M + \int_0^T X_s\, dB_s \in \mathcal{H}_T^2$. □

Ex. 18.7 **18.10 Lemma.** *Let $f \in L^2([0,T], dt)$ be a (non-random) function. Then $e^{i\int_0^T f(t)\, dB_t}$ is in $\mathcal{H}_T^2 \oplus i\mathcal{H}_T^2$.*

Proof. Clearly, $f \in \mathcal{L}_T^2$ so that all (stochastic) integrals are well-defined.

Set $M_t := i\int_0^t f(s)\, dB_s + \frac{1}{2}\int_0^t f^2(s)\, ds$ and apply Itô's formula to e^{M_t}:

$$\begin{aligned} e^{M_T} &= 1 + \int_0^T e^{M_s}\left(if(s)\, dB_s + \frac{1}{2}f^2(s)\, ds\right) - \frac{1}{2}\int_0^T e^{M_s} f^2(s)\, ds \\ &= 1 + \int_0^T ie^{M_s} f(s)\, dB_s. \end{aligned}$$

Thus,

$$e^{if\bullet B_T} = e^{-\frac{1}{2}\int_0^T f^2(s)\, ds} + \int_0^T if(t)e^{if\bullet B_t - \frac{1}{2}\int_t^T f^2(s)\, ds}\, dB_t$$

and the integrand is obviously in $L^2_{\mathscr{P}}(\lambda_T \otimes \mathbb{P}) \oplus iL^2_{\mathscr{P}}(\lambda_T \otimes \mathbb{P})$. □

We are going to use Lemma 18.10 for step functions of the form

$$g(s) = \sum_{j=1}^n (\xi_j + \ldots + \xi_n)\mathbb{1}_{[t_{j-1},t_j)}(s) \tag{18.13}$$

where $n \geqslant 1$, $\xi_1, \dots, \xi_n \in \mathbb{R}$ and $0 = t_0 < t_1 < \dots < t_n = T$. It is not hard to see that

$$g \bullet B_T = \int_0^T g(s)\,dB_s = \sum_{j=1}^n \xi_j B_{t_j}. \tag{18.14}$$

18.11 Theorem. *Let $(B_t)_{t\geqslant 0}$ be a* BM[1] *and $\mathscr{F}_t := \overline{\mathscr{F}}_t^B$ the completed natural filtration. Every $Y \in L^2(\Omega, \mathscr{F}_T, \mathbb{P})$ is of the form $Y = y + \int_0^T X_s\,dB_s$ for some $X \in L^2_{\mathscr{P}}(\lambda_T \otimes \mathbb{P})$ and $y \in \mathbb{R}$.*

Proof. We have to show that $\mathcal{H}_T^2 = L^2(\Omega, \mathscr{F}_T, \mathbb{P})$. Since $\mathcal{H}_T^2$ is a closed subspace of $L^2(\Omega, \mathscr{F}_T, \mathbb{P})$, it is enough to prove that $\mathcal{H}_T^2$ is dense in $L^2(\Omega, \mathscr{F}_T, \mathbb{P})$.

To do so, we take any $Y \in L^2(\Omega, \mathscr{F}_T, \mathbb{P})$ and show that

$$Y \perp \mathcal{H}_T^2 \implies Y = 0.$$

Recall that $Y \perp \mathcal{H}_T^2$ means that $\mathbb{E}(YH) = 0$ for all $H \in \mathcal{H}_T^2$. Take $H = e^{ig\bullet B_T}$ where g is of the form (18.13). Write $\xi = (\xi_1, \dots, \xi_n)$ and $\Gamma = (B_{t_1}, \dots, B_{t_n})$. As $H \in \mathcal{H}_T^2 \oplus i\mathcal{H}_T^2$, we get with (18.14)

$$0 = \mathbb{E}\left[e^{i\sum_{j=1}^n \xi_j B_{t_j}} Y\right] = \mathbb{E}\left[e^{i\langle\xi,\Gamma\rangle}\,\mathbb{E}(Y \mid \Gamma)\right] = \mathbb{E}\left[e^{i\langle\xi,\Gamma\rangle} u(\Gamma)\right].$$

Here we used the tower property and the fact that $\mathbb{E}(Y \mid \Gamma)$ can be expressed as a function u of $\Gamma = (B_{t_1}, \dots, B_{t_n})$. This shows

$$\int e^{i\langle\xi,x\rangle} u(x)\,\mathbb{P}(\Gamma \in dx) = \int_{\mathbb{R}^n} e^{i\langle\xi,x\rangle} u(x) p_\Gamma(x)\,dx = 0,$$

where $p_\Gamma(x)$ is the (strictly positive) density of the random vector $\Gamma = (B_{t_1}, \dots, B_{t_n})$, cf. (2.10b). Therefore, the Fourier transform of the function $u(x)p_\Gamma(x)$ is zero. This means that $u(x)p_\Gamma(x) = 0$, hence $u(x) = 0$ for Lebesgue almost all x.

We conclude that $\mathbb{E}(Y \mid B_{t_1}, \dots, B_{t_n}) = 0$ a.s. for all $0 \leqslant t_1 < \dots < t_n = T$, Ex. 18.8
and so – for example using Lévy's martingale convergence theorem and the fact that $\bigcup_{t_1<\dots<t_n=T,\ n\geqslant 1} \sigma(B_{t_1}, \dots, B_{t_n})$ generates $\mathscr{F}_T^B$ –, we get $Y = \mathbb{E}(Y \mid \mathscr{F}_T) = 0$. □

18.12 Corollary. *Let $(M_t)_{t\leqslant T}$ be any L^2 martingale with respect to the completed natural filtration $(\bar{\mathscr{F}}_t^B)_{t\leqslant T}$ of a* BM[1] *$(B_t)_{t\geqslant 0}$. Then there exist some $x \in \mathbb{R}$ and $X \in L^2_{\mathscr{P}}(\lambda_T \otimes \mathbb{P})$ such that*

$$M_t = x + \int_0^t X_s\,dB_s \quad \textit{for all } t \leqslant T.$$

In particular, any such martingale must be continuous.

Proof. Write $\mathscr{F}_t := \bar{\mathscr{F}}_t^B$. By Theorem 18.11 we know that $M_T = x + \int_0^T X_s\,dB_s$ for some $x \in \mathbb{R}$ and $X \in L^2_{\mathscr{P}}(\lambda_T \otimes \mathbb{P})$. Since $M_t = \mathbb{E}(M_T \mid \mathscr{F}_t)$, $t \leqslant T$, we see with Theorem 15.13 a) that

$$M_t = \mathbb{E}\left(M_T \mid \mathscr{F}_t\right) = \mathbb{E}\left[x + \int_0^T X_s\,dB_s \,\middle|\, \mathscr{F}_t\right] = x + \int_0^t X_s\,dB_s.$$ □

18.5 Martingale representation – 2

In the previous section we have seen that, given a Brownian motion $(B_t, \mathscr{F}_t)_{t\geqslant 0}$ with a complete filtration, every L^2 martingale $(M_t, \mathscr{F}_t)_{t\leqslant T}$ can be represented as a stochastic integral with respect to $(B_t)_{t\geqslant 0}$. In particular, every L^2 martingale with respect to a complete Brownian filtration is continuous.

We will now show that for many continuous L^2 martingales $(M_t, \mathscr{F}_t)_{t\leqslant T}$ there is a Brownian motion $(W_t)_{t\geqslant 0}$ such that M_t can be written as a stochastic integral with respect to $(W_t)_{t\geqslant 0}$. This requires a general martingale stochastic integral as in Section 15.6. Recall that the quadratic variation of an L^2 martingale is the unique, increasing process $\langle M\rangle = (\langle M\rangle_t)_{t\leqslant T}$ with $\langle M\rangle_0 = 0$ such that $(M_t^2 - \langle M\rangle_t, \mathscr{F}_t)_{t\leqslant T}$ is a martingale. In particular,

$$\mathbb{E}\left[(M_t - M_s)^2 \mid \mathscr{F}_s\right] = \mathbb{E}\left[\langle M\rangle_t - \langle M\rangle_s \mid \mathscr{F}_s\right] \quad \text{for all } s \leqslant t \leqslant T.$$

18.13 Theorem (Doob 1953). *Let $(M_t, \mathscr{F}_t)_{t\leqslant T}$ be a continuous L^2 martingale such that $\langle M\rangle_t = \int_0^t m^2(s,\cdot)\,ds$ for some $\lambda_T \otimes \mathbb{P}$-almost everywhere strictly positive process $m(s,\omega)$. Then there exists a Brownian motion $(W_t)_{t\leqslant T}$ such that $(\mathscr{F}_t)_{t\leqslant T}$ is admissible and*

$$M_t - M_0 = \int_0^t m(s,\cdot)\,dW_s.$$

Proof. Since the set $\{(s,\omega) \;:\; m(s,\omega) = 0\}$ is a $\lambda_T \otimes \mathbb{P}$ null set, we can *define* in all calculations below that $1/m(s,\omega) := 0$ if $m(s,\omega) = 0$. Note that

$$\mathbb{E}\int_0^t \frac{1}{m^2(s,\cdot)}\,d\langle M\rangle_s = \mathbb{E}\int_0^t \frac{1}{m^2(s,\cdot)}\,m^2(s,\cdot)\,ds = t.$$

Therefore, Theorem 15.22 shows that the stochastic integral

$$W_t := \int_0^t \frac{1}{m(s,\cdot)}\,dM_s, \quad t \leqslant T,$$

exists and defines a continuous L^2 martingale. Again by Theorem 15.22 the quadratic variation is

$$\langle W\rangle_t = \left\langle \int_0^\bullet \frac{1}{m(s,\cdot)}\,dM_s \right\rangle_t = \int_0^t \frac{1}{m^2(s,\cdot)}\,d\langle M\rangle_s = \int_0^t \frac{m^2(s,\cdot)}{m^2(s,\cdot)}\,ds = t.$$

Lévy's martingale characterization of a Brownian motion, Theorem 18.5, shows that $(W_t, \mathscr{F}_t)_{t\leqslant T}$ is indeed a Brownian motion. Finally,

$$\int_0^t m(s,\cdot)\,dW_s = \int_0^t m(s,\cdot)\,\frac{1}{m(s,\cdot)}\,dM_s = \int_0^t dM_s = M_t - M_0. \qquad \square$$

If $\{(s,\omega) : m(s,\omega) = 0\}$ is not a null set, we have to modify our argument. In this case the probability space is too small and we have to extend it by adding an independent Brownian motion $(B_t, \mathscr{G}_t)_{t\geqslant 0}$.[2] If U and V are mean zero $\mathscr{F}_t$ resp. $\mathscr{G}_t$ measurable random variables such that $\mathbb{E}(U \mid \mathscr{F}_s) = \mathbb{E}(V \mid \mathscr{G}_s) = 0$ for $s \leqslant t$, we find because of independence, that $\mathbb{E}(UV \mid \sigma(\mathscr{F}_s, \mathscr{G}_s)) = 0$. This follows from

$$\int_{F\cap G} UV\,d\mathbb{P} = \mathbb{E}(U\mathbb{1}_F V\mathbb{1}_G) = \int_F U\,d\mathbb{P}\int_G V\,d\mathbb{P} = 0$$

for all $F \in \mathscr{F}_s$, $G \in \mathscr{G}_s$, and the fact that $\{F\cap G : F \in \mathscr{F}_s,\ G \in \mathscr{G}_s\}$ is a $\cap$-stable generator of $\sigma(\mathscr{F}_s, \mathscr{G}_s)$.

18.14 Corollary (Doob 1953). *Let $(M_t, \mathscr{F}_t)_{t\leqslant T}$ be a continuous L^2 martingale such that $\langle M\rangle_t = \int_0^t m^2(s,\cdot)\,ds$. Then there is an enlargement $(\tilde\Omega, \tilde{\mathscr{A}}, \tilde{\mathbb{P}})$ of the underlying probability space $(\Omega, \mathscr{A}, \mathbb{P})$ and a Brownian motion $(W_t)_{t\leqslant T}$ on $(\tilde\Omega, \tilde{\mathscr{A}}, \tilde{\mathbb{P}})$ such that*

$$M_t - M_0 = \int_0^t m(s,\cdot)\,dW_s.$$

Proof. If necessary, we enlarge $(\Omega, \mathscr{A}, \mathbb{P})$ so that there is a further Brownian motion $(B_t, \mathscr{G}_t)_{t\leqslant T}$ which is independent of $(M_t, \mathscr{F}_t)_{t\leqslant T}$. Set $\mathscr{H}_t := \sigma(\mathscr{F}_t, \mathscr{G}_t)$. Because of the independence, both $(M_t, \mathscr{H}_t)_{t\leqslant T}$ and $(B_t, \mathscr{H}_t)_{t\leqslant T}$ are again L^2 martingales. Therefore, Ex. 18.9
Theorem 15.22 shows that the stochastic integrals

$$W_t := \int_0^t \frac{1}{m(s,\cdot)}\mathbb{1}_{\{m\neq 0\}}(s,\cdot)\,dM_s + \int_0^t \mathbb{1}_{\{m=0\}}(s,\cdot)\,dB_s, \qquad t \leqslant T,$$

exist and define a continuous L^2 martingale for the filtration $(\mathscr{H}_t)_{t\leqslant T}$. From (15.30) and the calculation just before the statement of Corollary 18.14 we see

$$\begin{aligned}
&\mathbb{E}\left[(W_t - W_s)^2 \mid \mathscr{H}_s\right]\\
&= \mathbb{E}\left[\left(\int_s^t \frac{1}{m(r,\cdot)}\mathbb{1}_{\{m\neq 0\}}(r,\cdot)\,dM_r\right)^2 \,\middle|\, \mathscr{H}_s\right] + \mathbb{E}\left[\left(\int_s^t \mathbb{1}_{\{m=0\}}(r,\cdot)\,dB_r\right)^2 \,\middle|\, \mathscr{H}_s\right]\\
&\qquad + 2\,\mathbb{E}\left[\int_s^t \frac{1}{m(r,\cdot)}\mathbb{1}_{\{m\neq 0\}}(r,\cdot)\,dM_r \int_s^t \mathbb{1}_{\{m=0\}}(r,\cdot)\,dB_r \,\middle|\, \mathscr{H}_s\right]\\
&= \mathbb{E}\left[\int_s^t \frac{1}{m^2(r,\cdot)}\mathbb{1}_{\{m\neq 0\}}(r,\cdot)\,d\langle M\rangle_r \,\middle|\, \mathscr{H}_s\right] + \mathbb{E}\left[\int_s^t \mathbb{1}_{\{m=0\}}(r,\cdot)\,dr \,\middle|\, \mathscr{H}_s\right]
\end{aligned}$$

2 This can be done by the usual product construction.

$$= \mathbb{E}\left[\int_s^t \frac{m^2(r,\cdot)}{m^2(r,\cdot)} \mathbb{1}_{\{m\neq 0\}}(r,\cdot)\,dr \;\Big|\; \mathscr{H}_s\right] + \mathbb{E}\left[\int_s^t \mathbb{1}_{\{m=0\}}(r,\cdot)\,dr \;\Big|\; \mathscr{H}_s\right]$$

$$= \int_s^t \left(\mathbb{1}_{\{m\neq 0\}}(r,\cdot) + \mathbb{1}_{\{m=0\}}(r,\cdot)\right) dr = t - s.$$

By Lévy's theorem 18.5, $(W_t, \mathscr{H}_t)_{t\leqslant T}$ is a Brownian motion, and the rest of the proof follows along the lines of Theorem 18.13, observing that $M_t - M_0 = \int_0^t \mathbb{1}_{\{m\neq 0\}}(s,\cdot)\,dM_s$ as $\int_0^t \mathbb{1}_{\{m=0\}}(s,\cdot)\,dM_s = 0$ by Theorem 15.22 c). □

18.6 Martingales as time-changed Brownian motion

Throughout this section $(M_t, \mathscr{F}_t)_{t\geqslant 0}$ is a real valued, continuous martingale such that $M_t \in L^2(\mathbb{P})$. We show that there is a Brownian motion $(B_t)_{t\geqslant 0}$ such that $M_t = B_{\sigma(t)}$ for some random time $\sigma(t)$.

We begin with a few preparations from analysis. Let $a : [0,\infty) \to [0,\infty]$ be a right-continuous, increasing function. Then

$$\tau(s) := \inf\{t \geqslant 0 : a(t) > s\}, \qquad \inf\emptyset = \infty, \tag{18.15}$$

is the *generalized inverse* of a.

18.15 Lemma. *Let τ be the generalized inverse of a right-continuous increasing function*
Ex. 18.10 $a : [0,\infty) \to [0,\infty]$.

a) $\tau : [0,\infty) \to [0,\infty]$ *is increasing and right-continuous;*
b) $a(\tau(s)) \geqslant s$;
c) $a(t) \geqslant s \iff \tau(s-) \leqslant t$;
d) $a(t) = \inf\{s \geqslant 0 : \tau(s) > t\}$;
e) $a(\tau(s)) = \sup\{u \geqslant s : \tau(u) = \tau(s)\}$;
f) $\tau(a(t)) = \sup\{w \geqslant t : a(w) = a(t)\}$;
g) *if a is continuous, $a(\tau(s)) = s$ for all $s \in a([0,\infty))$.*

Proof. Before we begin with the formal proof, it is helpful to draw a picture of the situation, cf. Figure 18.1. All assertions are clear if t is a point of strict increase for the function a. Moreover, if a is flat, the inverse τ makes a jump, and vice versa.

a) It is obvious from the definition that τ is increasing. Note that

$$\{t : a(t) > s\} = \bigcup_{\epsilon>0}\{t : a(t) > s+\epsilon\}.$$

Therefore, $\inf\{t \geqslant 0 : a(t) > s\} = \inf_{\epsilon>0} \inf\{t \geqslant 0 : a(t) > s+\epsilon\}$ proving right-continuity.

b) Since $a(u) \geqslant s$ for all $u > \tau(s)$, we see that $a(\tau(s)) \geqslant s$.

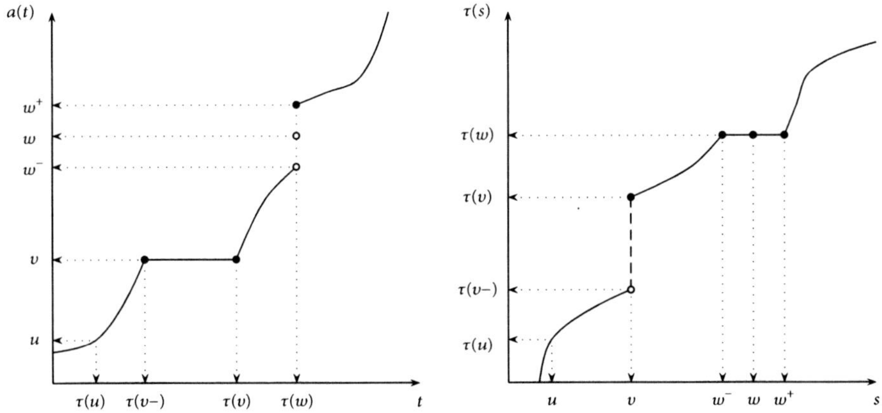

Fig. 18.1. An increasing right-continuous function and its generalized right-continuous inverse.

c) By the very definition of the generalized inverse τ we have

$$\begin{aligned} a(t) \geqslant s &\iff a(t) > s - \epsilon \ \ \forall\, \epsilon > 0 \\ &\iff \tau(s-\epsilon) \leqslant t \ \ \forall\, \epsilon > 0 \iff \tau(s-) \leqslant t. \end{aligned}$$

d) From c) we see that

$$\begin{aligned} a(t) = \sup\{s \geqslant 0 \,:\, a(t) \geqslant s\} &\overset{c)}{=} \sup\{s \geqslant 0 \,:\, \tau(s-) \leqslant t\} \\ &= \inf\{s \geqslant 0 \,:\, \tau(s) > t\}. \end{aligned}$$

e) Assume that $u \geqslant s$ and $\tau(u) = \tau(s)$. Then $a(\tau(s)) = a(\tau(u)) \geqslant u \geqslant s$, and therefore $a(\tau(s)) \geqslant \sup\{u \geqslant s \,:\, \tau(u) = \tau(s)\}$. For the converse, we use d) and find

$$a(\tau(s)) = \inf\{r \geqslant 0 \,:\, \tau(r) > \tau(s)\} \leqslant \sup\{u \geqslant s \,:\, \tau(u) = \tau(s)\}.$$

f) Follows as e) since a and τ play symmetric roles.

g) Because of b) it is enough to show that $a(\tau(s)) \leqslant s$. If $\tau(s) = 0$, this is trivial. Otherwise we find $\epsilon > 0$ such that $0 < \epsilon < \tau(s)$ and $a(\tau(s) - \epsilon) \leqslant s$ by the definition of $\tau(s)$. Letting $\epsilon \to 0$, we obtain $a(\tau(s)) \leqslant s$ by the continuity of a. □

The following theorem was discovered by Doeblin in 1940. Doeblin never published this result but deposited it in a *pli cacheté* with the Academy of Sciences in Paris. The *pli* was finally opened in the year 2000, see the historical notes by Bru and Yor [22]. Around 25 years after Doeblin, the result was in 1965 rediscovered by Dambis and by Dubins and Schwarz in a more general context.

18.16 Theorem (Doeblin 1940; Dambis 1965; Dubins–Schwarz 1965). *Let $(M_t, \mathscr{F}_t)_{t\geqslant 0}$, $M_t \in L^2(\mathbb{P})$, be a continuous martingale with a right-continuous filtration, i. e. $\mathscr{F}_t = \mathscr{F}_{t+}$,*

$t \geqslant 0$, and $\langle M\rangle_\infty = \lim_{t\to\infty}\langle M\rangle_t = \infty$. Denote by $\tau_s = \inf\{t \geqslant 0 : \langle M\rangle_t > s\}$ the generalized inverse of $\langle M\rangle$. Then $(B_t, \mathscr{G}_t)_{t\geqslant 0} := (M_{\tau_s}, \mathscr{F}_{\tau_s})_{s\geqslant 0}$ is a Brownian motion and

$$M_t = B_{\langle M\rangle_t} \quad \textit{for all } t \geqslant 0.$$

Proof. Define $B_s := M_{\tau_s}$ and $\mathscr{G}_t := \mathscr{F}_{\tau_t}$. Since τ_s is the first entry time into (s,∞) for the process $\langle M\rangle$, it is an $\mathscr{F}_{t+}$ stopping time, cf. Lemma 5.7, which is almost surely finite
Ex. 18.11 because of $\langle M\rangle_\infty = \infty$.

Let $\Omega^* = \{\omega \in \Omega : M_\bullet(\omega)$ and $\langle M\rangle_\bullet(\omega)$ are constant on the same intervals$\}$. From the definition of the quadratic variation, Theorem A.31, one can show that $\mathbb{P}(\Omega^*) = 1$, see Corollary A.32 in the appendix.

In particular, $s \mapsto M_{\tau_s}(\omega)$ is almost surely continuous: If s is a continuity point of τ_s, this is obvious. Otherwise $\tau_{s-} < \tau_s$, but for $\omega \in \Omega^*$ and all $\tau_{s-}(\omega) \leqslant r \leqslant \tau_s(\omega)$ we have $M_{\tau_{s-}}(\omega) = M_r(\omega) = M_{\tau_s}(\omega)$.

For all $k \geqslant 1$ and $s \geqslant 0$ we find

$$\mathbb{E}\left(M^2_{\tau_s\wedge k}\right) \leqslant \mathbb{E}\left(\sup\nolimits_{u\leqslant k} M_u^2\right) \underset{\text{(A.14)}}{\overset{\text{Doob}}{\leqslant}} 4\,\mathbb{E}\,M_k^2 < \infty.$$

The optional stopping Theorem A.18 and Corollary A.20 show that $(M_{\tau_s\wedge k})_{s\geqslant 0}$ is a martingale for the filtration $(\mathscr{G}_s)_{s\geqslant 0}$ with $\mathscr{G}_s := \mathscr{F}_{\tau_s}$. Thus,

$$\mathbb{E}\left(M^2_{\tau_s\wedge k}\right) = \mathbb{E}\left(\langle M\rangle_{\tau_s\wedge k}\right) \leqslant \mathbb{E}\left(\langle M\rangle_{\tau_s}\right) = s,$$

showing that for every $s \geqslant 0$ the family $(M_{\tau_s\wedge k})_{k\geqslant 1}$ is uniformly integrable. By the continuity of M we see

$$M_{\tau_s\wedge k} \xrightarrow[k\to\infty]{L^1 \text{ and a. e.}} M_{\tau_s} =: B_s.$$

In particular, $(B_s, \mathscr{G}_s)_{s\geqslant 0}$ is a martingale. Fatou's lemma shows that

$$\mathbb{E}\,B_s^2 \leqslant \varliminf_{k\to\infty} \mathbb{E}\,M^2_{\tau_s\wedge k} \leqslant s,$$

i. e. B_s is also in $L^2(\mathbb{P})$.

If we apply optional stopping to the martingale $(M_t^2 - \langle M\rangle_t)_{t\geqslant 0}$, we find

$$\mathbb{E}\left[(B_t - B_s)^2 \mid \mathscr{G}_s\right] = \mathbb{E}\left[(M_{\tau_t} - M_{\tau_s})^2 \mid \mathscr{F}_{\tau_s}\right] = \mathbb{E}\left[\langle M\rangle_{\tau_t} - \langle M\rangle_{\tau_s} \mid \mathscr{F}_{\tau_s}\right] = t - s.$$

Thus $(B_s, \mathscr{G}_s)_{s\geqslant 0}$ is a continuous martingale whose quadratic variation process is s. Lévy's characterization of a Brownian motion, Theorem 18.5, applies and we see that $(B_s, \mathscr{G}_s)_{s\geqslant 0}$ is a BM^1. Finally, $B_{\langle M\rangle_t} = M_{\tau_{\langle M\rangle_t}} = M_t$ for all $t \geqslant 0$. □

18.7 Burkholder–Davis–Gundy inequalities

The Burkholder–Davis–Gundy inequalities are estimates for the pth moment of a continuous martingale X in terms of the pth norm of $\sqrt{\langle X\rangle}$:

$$c_p\,\mathbb{E}\left[\langle X\rangle_T^{p/2}\right] \leqslant \mathbb{E}\left[\sup_{t\leqslant T} |X_t|^p\right] \leqslant C_p\,\mathbb{E}\left[\langle X\rangle_T^{p/2}\right], \quad 0 < p < \infty. \tag{18.16}$$

The constants can be made explicit and they depend only on p. In fact, (18.16) holds for continuous local martingales; we restrict ourselves to Brownian L^2 martingales, i. e. to martingales of the form

$$X_t = \int_0^t f(s)\,dB_s \quad\text{and}\quad \langle X\rangle_t = \int_0^t |f(s)|^2\,ds$$

where $(B_t)_{t\geqslant 0}$ is a BM^1 and $f \in L^2_{\mathscr{P}}(\lambda_T \otimes \mathbb{P})$ for all $T > 0$, see also Sections 18.4 and 18.5.

We will first consider the case where $p \geqslant 2$.

18.17 Theorem (Burkholder 1966). *Let $(B_t)_{t\geqslant 0}$ be a BM^1 and $f \in L^2_{\mathscr{P}}(\lambda_T \otimes \mathbb{P})$ for all $T > 0$. Then, we have for all $p \geqslant 2$ and $q = p/(p-1)$*

$$\mathbb{E}\left[\sup_{t\leqslant T}\left|\int_0^t f(s)\,dB_s\right|^p\right] \leqslant C_p\ \mathbb{E}\left[\left(\int_0^T |f(s)|^2\,ds\right)^{p/2}\right], \quad p \geqslant 2, \tag{18.17}$$

where $C_p = [q^p p(p-1)/2]^{p/2}$, and

$$\mathbb{E}\left[\sup_{t\leqslant T}\left|\int_0^t f(s)\,dB_s\right|^p\right] \geqslant c_p\ \mathbb{E}\left[\left(\int_0^T |f(s)|^2\,ds\right)^{p/2}\right], \quad p \geqslant 2, \tag{18.18}$$

where $c_p = 1/(2p)^{p/2}$.

Proof. Set $X_t := \int_0^t f(s)\,dB_s$ and $\langle X\rangle_t = \int_0^t |f(s)|^2\,ds$, cf. Theorem 15.13.

If we combine Doob's maximal inequality (A.14) and Itô's isometry (15.21) we get

$$\mathbb{E}\left[\langle X\rangle_T\right] = \mathbb{E}\left[|X_T|^2\right] \leqslant \mathbb{E}\left[\sup_{s\leqslant T}|X_s|^2\right] \leqslant 4\sup_{s\leqslant T}\mathbb{E}\left[|X_s|^2\right] = 4\ \mathbb{E}\left[\langle X\rangle_T\right].$$

This shows that (18.17) and (18.18) hold for $p = 2$ with $C_p = 4$ and $c_p = 1/4$, respectively.

Assume that $p > 2$. By stopping with the stopping time

$$\tau_n := \inf\left\{t \geqslant 0 : |X_t|^2 + \langle X\rangle_t \geqslant n\right\},$$

(note that $X^2 + \langle X\rangle$ is a continuous process) we can, without loss of generality, assume that both X and $\langle X\rangle$ are bounded. Since $x \mapsto |x|^p$, $p > 2$, is a $\mathcal{C}^2$-function, Itô's formula, Theorem 17.8, yields

$$|X_t|^p = p\int_0^t |X_s|^{p-1}\,dX_s + \frac{1}{2}\,p(p-1)\int_0^t |X_s|^{p-2}\,d\langle X\rangle_s$$

where we use that $d\langle X\rangle_s = |f(s)|^2\,ds$. Because of the boundedness of $|X_s|^{p-1}$, the first integral on the right-hand side is a martingale and we see with Doob's maximal inequality and the Hölder inequality for $p/(p-2)$ and $p/2$

$$\begin{aligned}\mathbb{E}\left[\sup_{s\leqslant T}|X_s|^p\right] &\overset{(A.14)}{\leqslant} q^p\,\frac{p(p-1)}{2}\,\mathbb{E}\left[\int_0^T |X_s|^{p-2}\,d\langle X\rangle_s\right]\\ &\leqslant q^p\,\frac{p(p-1)}{2}\,\mathbb{E}\left[\sup_{s\leqslant T}|X_s|^{p-2}\,\langle X\rangle_T\right]\\ &\leqslant q^p\,\frac{p(p-1)}{2}\left(\mathbb{E}\left[\sup_{s\leqslant T}|X_s|^p\right]\right)^{1-2/p}\left(\mathbb{E}\left[\langle X\rangle_T^{p/2}\right]\right)^{2/p}.\end{aligned}$$

If we divide by $(\mathbb{E}\left[\sup_{s\leqslant T}|X_s|^p\right])^{1-2/p}$ we find

$$\left(\mathbb{E}\left[\sup_{s\leqslant T}|X_s|^p\right]\right)^{2/p}\leqslant q^p\,\frac{p(p-1)}{2}\left(\mathbb{E}\left[\langle X\rangle_T^{p/2}\right]\right)^{2/p}$$

and (18.17) follows with $C_p = [q^p\,p(p-1)/2]^{p/2}$.

For the lower estimate (18.18) we set

$$Y_t := \int_0^t \langle X\rangle_s^{(p-2)/4}\,dX_s.$$

Ex. 18.12 From Theorem 15.22 (or with Theorem 15.13 observing that $dX_s = f(s)\,dB_s$) we see that

$$\langle Y\rangle_T = \int_0^T \langle X\rangle_s^{(p-2)/2}\,d\langle X\rangle_s = \frac{2}{p}\,\langle X\rangle_T^{p/2}$$

and

$$\mathbb{E}\left[\langle X\rangle_T^{p/2}\right] = \frac{p}{2}\,\mathbb{E}\left[\langle Y\rangle_T\right] = \frac{p}{2}\,\mathbb{E}\left[|Y_T|^2\right].$$

Using Itô's formula (17.14) for the function $f(x,y) = xy$ and the process $(X_t, \langle X\rangle_t^{(p-2)/4})$ we obtain

$$X_T\langle X\rangle_T^{(p-2)/4} = \int_0^T \langle X\rangle_s^{(p-2)/4}\,dX_s + \int_0^T X_s\,d\langle X\rangle_s^{(p-2)/4} = Y_T + \int_0^T X_s\,d\langle X\rangle_s^{(p-2)/4}.$$

If we rearrange this equality we get $|Y_T| \leqslant 2\sup_{s\leqslant T}|X_s|\,\langle X\rangle_T^{(p-2)/4}$. Finally, using Hölder's inequality with $p/2$ and $p/(p-2)$,

$$\begin{aligned}\frac{2}{p}\,\mathbb{E}\left[\langle X\rangle_T^{p/2}\right] = \mathbb{E}\left[|Y_T|^2\right] &\leqslant 4\,\mathbb{E}\left[\sup_{s\leqslant T}|X_s|^2\,\langle X\rangle_T^{(p-2)/2}\right]\\ &\leqslant 4\Big(\mathbb{E}\left[\sup_{s\leqslant T}|X_s|^p\right]\Big)^{2/p}\left(\mathbb{E}\left[\langle X\rangle_T^{p/2}\right]\right)^{1-2/p}.\end{aligned}$$

Dividing on both sides by $(\mathbb{E}[\langle X\rangle_T^{p/2}])^{1-2/p}$ gives (18.18) with $c_p = (2p)^{-p/2}$. □

The inequalities for $0 < p < 2$ can be proved with the following useful lemma.

18.18 Lemma. *Let X and A be positive random variables and $\rho \in (0,1)$. If*

$$\mathbb{P}(X > x, A \leqslant x) \leqslant \frac{1}{x}\,\mathbb{E}(A \wedge x) \quad \textit{for all } x > 0, \tag{18.19}$$

then

$$\mathbb{E}(X^\rho) \leqslant \frac{2-\rho}{1-\rho}\,\mathbb{E}(A^\rho).$$

Proof. Using (18.19) and Tonelli's theorem we find

$$\begin{aligned}
\mathbb{E}(X^\rho) &= \rho \int_0^\infty \mathbb{P}(X > x)\, x^{\rho-1}\,dx \\
&\leqslant \rho \int_0^\infty \big[\mathbb{P}(X > x, A \leqslant x) + \mathbb{P}(A > x)\big]\, x^{\rho-1}\,dx \\
&\leqslant \rho \int_0^\infty \left[\frac{1}{x}\,\mathbb{E}(A\wedge x) + \mathbb{P}(A > x)\right] x^{\rho-1}\,dx \\
&= \mathbb{E}\left[\int_0^\infty (A \wedge x)\rho x^{\rho-2}\,dx\right] + \mathbb{E}\left[A^\rho\right] \\
&= \mathbb{E}\left[\int_0^A \rho x^{\rho-1}\,dx + \int_A^\infty A\rho x^{\rho-2}\,dx\right] + \mathbb{E}\left[A^\rho\right] \\
&= \mathbb{E}\left[A^\rho - \frac{\rho A^\rho}{\rho - 1}\right] + \mathbb{E}\left[A^\rho\right] = \frac{2-\rho}{1-\rho}\,\mathbb{E}(A^\rho).
\end{aligned}$$

□

We are going to apply Lemma 18.18 with $X = \sup_{s\leqslant T} |X_s|^2 = \sup_{s\leqslant T} \left|\int_0^s f(r)\,dB_r\right|^2$ and $A = \langle X\rangle_T = \int_0^T |f(s)|^2\,ds$. Define $\tau := \inf\{t \geqslant 0 : \langle X\rangle_t \geqslant x\}$. Then

$$\begin{aligned}
P := \mathbb{P}\Big(\sup_{s\leqslant T} |X_s|^2 > x,\ \langle X\rangle_T \leqslant x\Big) &= \mathbb{P}\Big(\sup_{s\leqslant T} |X_s|^2 > x,\ \tau \geqslant T\Big) \\
&\leqslant \mathbb{P}\Big(\sup_{s\leqslant T\wedge\tau} |X_s|^2 > x\Big)
\end{aligned}$$

and with Doob's maximal inequality (A.13) we find

$$P \leqslant \frac{1}{x}\,\mathbb{E}\left(|X_{T\wedge\tau}|^2\right) \stackrel{(*)}{=} \frac{1}{x}\,\mathbb{E}\left(\langle X\rangle_{T\wedge\tau}\right) = \frac{1}{x}\,\mathbb{E}\left(\langle X\rangle_T \wedge x\right).$$

For the equality $(*)$ we used optional stopping and the fact that $(X_t^2 - \langle X\rangle_t)_{t\geqslant 0}$ is a martingale. Now we can apply Lemma 18.18 and get

$$\mathbb{E}\Big[\sup_{s\leqslant T} |X_s|^{2\rho}\Big] \leqslant \frac{2-\rho}{1-\rho}\,\mathbb{E}\left[\langle X\rangle_T^\rho\right]$$

which is the analogue of (18.17) for $0 < p < 2$.

In a similar way we can apply Lemma 18.18 with $X = \langle X\rangle_T = \int_0^T |f(s)|^2\,ds$ and $A = \sup_{s\leqslant T}|X_s|^2 = \sup_{s\leqslant T}\left|\int_0^s f(r)\,dB_r\right|^2$. Define $\sigma := \inf\{t>0 \,:\, |X_t|^2 \geqslant x\}$. Then

$$P' := \mathbb{P}\Big(\langle X\rangle_T > x,\ \sup_{s\leqslant T}|X_s|^2 \leqslant x\Big) \leqslant \mathbb{P}(\langle X\rangle_T > x,\ \sigma \geqslant T) \leqslant \mathbb{P}(\langle X\rangle_{T\wedge\sigma} > x).$$

Using the Markov inequality we get

$$P' \leqslant \frac{1}{x}\,\mathbb{E}\left[\langle X\rangle_{T\wedge\sigma}\right] \leqslant \frac{1}{x}\,\mathbb{E}\left[\sup_{s\leqslant T\wedge\sigma}|X_s|^2\right] \leqslant \frac{1}{x}\,\mathbb{E}\left[\sup_{s\leqslant T}|X_s|^2\wedge x\right],$$

and Lemma 18.18 gives

$$\mathbb{E}\left[\sup_{s\leqslant T}|X_s|^{2\rho}\right] \geqslant \frac{1-\rho}{2-\rho}\,\mathbb{E}\left[\langle X\rangle_T^{\rho}\right]$$

which is the analogue of (18.18) for $0 < p < 2$.

Combining this with Theorem 18.17 we have shown

Ex. 18.13 **18.19 Theorem** (Burkholder, Davis, Gundy 1972). *Let $(B_t)_{t\geqslant 0}$ denote a one-dimensional Brownian motion and $f \in L^2_{\mathscr{P}}(\lambda_T \otimes \mathbb{P})$ for all $T > 0$. Then, we have for $0 < p < \infty$*

$$\mathbb{E}\left[\left(\int_0^T |f(s)|^2\,ds\right)^{p/2}\right] \asymp \mathbb{E}\left[\sup_{t\leqslant T}\left|\int_0^t f(s)\,dB_s\right|^p\right] \tag{18.20}$$

with finite comparison constants which depend only on $p \in (0,\infty)$.

---

i **18.20 Further reading.** Further applications, e. g. to reflecting Brownian motion $|B_t|$, excursions, local times etc. are given in [100]. The monograph [132] proves Burkholder's inequalities in $\mathbb{R}^d$ with dimension-free constants. A concise introduction to Malliavin's calculus and stochastic analysis is [206].

[100] Ikeda, Watanabe: *Stochastic Differential Equations and Diffusion Processes.*
[132] Krylov: *Introduction to the Theory of Diffusion Processes.*
[206] Shigekawa: *Stochastic Analysis.*

Problems

1. State and prove the d-dimensional version of Lemma 18.1.
2. Let $(N_t)_{t\geqslant 0}$ be a Poisson process with intensity $\lambda = 1$ (see Problem 10.1 for the definition). Show that for $\mathscr{F}_t^N := \sigma(N_r \,:\, r \leqslant t)$ both $X_t := N_t - t$ and $X_t^2 - t$ are martingales. Explain why this does not contradict Theorem 18.5.

3. Let $B_t = (b_t, \beta_t), t \geqslant 0$, be a two-dimensional Brownian motion, i. e. $(b_t)_{t\geqslant 0}, (\beta_t)_{t\geqslant 0}$ are independent one-dimensional Brownian motions. Show that for all $\sigma_1, \sigma_2 > 0$

$$W_t := \frac{\sigma_1}{\sqrt{\sigma_1^2 + \sigma_2^2}}\, b_t + \frac{\sigma_2}{\sqrt{\sigma_1^2 + \sigma_2^2}}\, \beta_t, \quad t \geqslant 0,$$

is a one-dimensional Brownian motion.

4. Let $(B(t))_{t\geqslant 0}$ be a BM^1 and $T < \infty$. Define $\beta_T = \exp\left(\xi B(T) - \frac{1}{2}\xi^2 T\right)$, $\mathbb{Q} := \beta_T \mathbb{P}$ and $W(t) := B(t) - \xi t, t \in [0, T]$. Verify by a direct calculation that for all $0 < t_1 < \cdots < t_n$ and $A_1, \dots, A_n \in \mathscr{B}(\mathbb{R})$

$$\mathbb{Q}\big(W(t_1) \in A_1, \dots, W(t_n) \in A_n\big) = \mathbb{P}\big(B(t_1) \in A_1, \dots, B(t_n) \in A_n\big).$$

5. Let $(B_t)_{t\geqslant 0}$ be a BM^1, $\Phi(y) := \mathbb{P}(B_1 \leqslant y)$, and set $X_t := B_t + \alpha t$ for some $\alpha \in \mathbb{R}$. Use Girsanov's theorem to show for all $0 \leqslant x \leqslant y, t > 0$

$$\mathbb{P}\Big(X_t \leqslant x, \sup_{s\leqslant t} X_s \leqslant y\Big) = \Phi\left(\frac{x - \alpha t}{\sqrt{t}}\right) - e^{2\alpha y}\Phi\left(\frac{x - 2y - \alpha t}{\sqrt{t}}\right).$$

6. Let X_t be as in Problem 18.5 and set $\widehat{\tau}_b := \inf\{t \geqslant 0 : X_t = b\}$.
(a) Use Problem 18.5 to find the probability density of $\widehat{\tau}_b$ if $\alpha, b > 0$.
(b) Find $\mathbb{P}(\widehat{\tau}_b < \infty)$ for $\alpha < 0$ and $\alpha \geqslant 0$, respectively.

7. Let $(B_t)_{t\geqslant 0}$ be a BM^1, denote by $\mathcal{H}_T^2$ the space used in Lemma 18.9 and pick $\xi_1, \dots, \xi_n \in \mathbb{R}$ and $0 = t_0 < t_1 < \dots < t_n = T$. Show that $\exp\left[i \sum_{j=1}^n \xi_j B_{t_j}\right]$ is contained in $\mathcal{H}_T^2 \oplus i\mathcal{H}_T^2$.

8. Let $(B_t)_{t\geqslant 0}$ be a BM^1, $n \geqslant 1$, $0 \leqslant t_1 < \dots < t_n = T$ and let Y an $\mathscr{F}_T$ measurable, integrable random variable. Assume that $\mathbb{E}\left[Y \mid \sigma(B_{t_1}, \dots, B_{t_n})\right] = 0$ for any choice of $n \geqslant 1$ and $0 \leqslant t_1 < \dots t_n = T$. Show that $Y = 0$ a. s.

9. Assume that $(B_t, \mathscr{G}_t)_{t\geqslant 0}$ and $(M_t, \mathscr{F}_t)_{t\geqslant 0}$ are independent martingales. Show that $(M_t)_{t\leqslant T}$ and $(B_t)_{t\leqslant T}$ are martingales for the enlarged filtration $\mathscr{H}_t := \sigma(\mathscr{F}_t, \mathscr{G}_t)$.

10. Show the following addition to Lemma 18.15:

$$\tau(t-) = \inf\{s \geqslant 0 : a(s) \geqslant t\} \quad \text{and} \quad a(t-) = \inf\{s \geqslant 0 : \tau(s) \geqslant t\}$$

and $\tau(s) \geqslant t \iff a(t-) \leqslant s$.

11. Show that, in Theorem 18.16, the quadratic variation $\langle M \rangle_t$ is a $\mathscr{G}_{\tau_s}$ stopping time.
Hint: Direct calculation, use Lemma 18.15 c) and A.15

12. Let $f : [0, \infty) \to \mathbb{R}$ be absolutely continuous. Show that $\int f^{a-1}\, df = f^a / a$ for all $a > 0$.

13. State and prove a d-dimensional version of the Burkholder–Davis–Gundy inequalities (18.20) if $p \in [2, \infty)$.
Hint: Use the fact that all norms in $\mathbb{R}^d$ are equivalent

19 Stochastic differential equations

The ordinary differential equation $\frac{d}{ds}x_s = b(s, x_s)$, $x(0) = x_0$, describes the position x_t of a particle which moves with speed $b(s, x)$ depending on time and on the current position. One possibility to take into account random effects, e. g. caused by measurement errors or hidden parameters, is to add a random perturbation which may depend on the current position. This leads to an equation of the form

$$X(t + \Delta t) - X(t) = b(t, X(t))\Delta t + \sigma(t, X(t))(B(t + \Delta t) - B(t)).$$

Letting $\Delta t \to 0$ we get the following equation for *stochastic differentials*, cf. Section 17.1,

$$dX_t = b(t, X_t)\, dt + \sigma(t, X_t)\, dB_t. \tag{19.1}$$

Since (19.1) is a formal way of writing things, we need a precise definition of the solution of the stochastic differential equation (19.1).

19.1 Definition. Let $(B_t, \mathscr{F}_t)_{t\geqslant 0}$ be a d-dimensional Brownian motion with admissible filtration $(\mathscr{F}_t)_{t\geqslant 0}$, and let $b : [0, \infty) \times \mathbb{R}^n \to \mathbb{R}^n$ and $\sigma : [0, \infty) \times \mathbb{R}^n \to \mathbb{R}^{n\times d}$ be measurable functions. A *solution* of the *stochastic differential equation (SDE)* (19.1) with initial condition $X_0 = \xi$ is a progressively measurable stochastic process $(X_t)_{t\geqslant 0}$, $X_t = (X_t^1, \dots, X_t^n) \in \mathbb{R}^n$, such that the following integral equation holds

$$X_t = \xi + \int_0^t b(s, X_s)\, ds + \int_0^t \sigma(s, X_s)\, dB_s, \quad t \geqslant 0, \tag{19.2}$$

i. e. for all $j = 1, \dots, n$,

$$X_t^j = \xi^j + \int_0^t b_j(s, X_s)\, ds + \sum_{k=1}^d \int_0^t \sigma_{jk}(s, X_s)\, dB_s^k, \quad t \geqslant 0. \tag{19.3}$$

(It is implicit in the definition that all stochastic integrals make sense.)

Throughout this chapter we use the Euclidean norms $|b|^2 = \sum_{j=1}^n b_j^2$, $|x|^2 = \sum_{k=1}^d x_k^2$ and $|\sigma|^2 = \sum_{j=1}^n \sum_{k=1}^d \sigma_{jk}^2$ for vectors $b \in \mathbb{R}^n$, $x \in \mathbb{R}^d$ and matrices $\sigma \in \mathbb{R}^{n\times d}$, respectively. The Cauchy-Schwarz inequality shows that for these norms $|\sigma x| \leqslant |\sigma| \cdot |x|$ holds.

! The expectation $\mathbb{E}$ and the probability measure $\mathbb{P}$ are given by the driving Brownian motion; in particular $\mathbb{P}(B_0 = 0) = 1$.

19.1 The heuristics of SDEs

We have introduced the SDE (19.1) as a random perturbation of the ODE $\dot{x}_t = b(t, x_t)$ by a *multiplicative* noise: $\sigma(t, X_t)\, dB_t$. The following heuristic reasoning explains why multiplicative noise is actually a natural choice for many applications. We learned this argument from Michael Röckner.

Denote by x_t the position of a particle at time t. After Δt units of time, the particle is at $x_{t+\Delta t}$. We assume that the motion can be described by an equation of the form

$$\Delta x_t = x_{t+\Delta t} - x_t = f(t, x_t; \Delta t)$$

where f is a function which depends on the initial position x_t and the time interval $[t, t+\Delta t]$. Unknown influences on the movement, e. g. caused by hidden parameters or measurement errors, can be modelled by adding some random noise $\Gamma_{t,\Delta t}$. This leads to

$$X_{t+\Delta t} - X_t = F(t, X_t; \Delta t, \Gamma_{t,\Delta t})$$

where X_t is now a random process. Note that $F(t, X_t; 0, \Gamma_{t,0}) = 0$. In many applications it makes sense to assume that the random variable $\Gamma_{t,\Delta t}$ is independent of X_t, that it has zero mean, and that it depends only on the increment Δt. If we have no further information on the nature of the noise, we should strive for a random variable which is maximally uncertain. One possibility to measure uncertainty is entropy. We know from information theory that, among all probability densities $p(x)$ with a fixed variance σ^2, the Gaussian density maximizes the entropy $H(p) = -\int_{-\infty}^{\infty} p(x) \log p(x)\, dx$, cf. Ash [4, Theorem 8.3.3]. Therefore, we assume that $\Gamma_{t,\Delta t} \sim \mathsf{N}(0, \sigma^2(\Delta t))$. By the same argument, the random variables $\Gamma_{t,\Delta t}$ and $\Gamma_{t+\Delta t,\Delta s}$ are independent and $\Gamma_{t,\Delta t+\Delta s} \sim \Gamma_{t,\Delta t} + \Gamma_{t+\Delta t,\Delta s}$. Thus, $\sigma^2(\Delta t + \Delta s) = \sigma^2(\Delta t) + \sigma^2(\Delta s)$, and if $\sigma^2(\cdot)$ is continuous, we see that $\sigma^2(\cdot)$ is a linear function. Without loss of generality we can assume that $\sigma^2(t) = t$; therefore

$$\Gamma_{t,\Delta t} \sim \Delta B_t := B_{t+\Delta t} - B_t$$

where $(B_t)_{t\geqslant 0}$ is a one-dimensional Brownian motion and we can write

$$X_{t+\Delta t} - X_t = F(t, X_t; \Delta t, \Delta B_t).$$

If F is sufficiently smooth, a Taylor expansion of F in the last two variables (∂_j denotes the partial derivative w. r. t. the jth variable) yields

$$\begin{aligned} X_{t+\Delta t} - X_t &= \partial_4 F(t, X_t; 0, 0)\, \Delta B_t + \partial_3 F(t, X_t; 0, 0)\, \Delta t \\ &\quad + \frac{1}{2}\, \partial_4^2 F(t, X_t; 0, 0)(\Delta B_t)^2 + \frac{1}{2}\, \partial_3^2 F(t, X_t; 0, 0)(\Delta t)^2 \\ &\quad + \partial_3 \partial_4 F(t, X_t; 0, 0) \Delta t\, \Delta B_t + R(\Delta t, \Delta B_t). \end{aligned}$$

The remainder term R is given by

$$R(\Delta t, \Delta B_t) = \sum_{\substack{0\leqslant j,k\leqslant 3\\ j+k=3}} \frac{3}{j!\,k!} \int_0^1 (1-\theta)^2 \partial_3^j \partial_4^k F(t, X_t; \theta\Delta t, \theta\Delta B_t)\, d\theta \cdot (\Delta t)^j (\Delta B_t)^k.$$

Since $\Delta B_t \sim \sqrt{\Delta t} B_1$, we get $R = o(\Delta t)$ and $\Delta t\, \Delta B_t = o((\Delta t)^{3/2})$. Neglecting all terms of order $o(\Delta t)$ and using $(\Delta B_t)^2 \approx \Delta t$, we get for small increments Δt

$$X_{t+\Delta t} - X_t = \underbrace{\left(\partial_3 F(t, X_t; 0, 0) + \frac{1}{2}\partial_4^2 F(t, X_t; 0, 0)\right)}_{=:\, b(t, X_t)} \Delta t + \underbrace{\partial_4 F(t, X_t; 0, 0)}_{=:\, \sigma(t, X_t)}\ \Delta B_t.$$

Letting $\Delta t \to 0$, we get (19.1).

19.2 Some examples

Before we study general existence and uniqueness results for the SDE (19.1), we will review a few examples where we can find the solution explicitly. Of course, we must make sure that all integrals in (19.2) exist, i. e.

$$\int_0^T |\sigma(s, X_s)|^2\, ds \quad \text{and} \quad \int_0^T |b(s, X_s)|\, ds$$

should have an expected value or be at least almost surely finite, cf. Chapter 16. We will assume this throughout this section. In order to keep things simple we discuss only the one-dimensional situation $d = n = 1$.

Ex. 19.1 **19.2 Example.** Let $(B_t)_{t\geqslant 0}$ be a BM^1. We consider the SDE

$$X_t - x = \int_0^t b(s)\, ds + \int_0^t \sigma(s)\, dB_s \tag{19.4}$$

with deterministic coefficients $b, \sigma : [0, \infty) \to \mathbb{R}$. It is clear that X_t is a normal random variable and that $(X_t)_{t\geqslant 0}$ inherits the independent increments property from the driving Brownian motion $(B_t)_{t\geqslant 0}$. Therefore, it is enough to calculate $\mathbb{E}\, X_t$ and $\mathbb{V}\, X_t$ in order to characterize the solution:

$$\mathbb{E}(X_t - X_0) = \int_0^t b(s)\, ds + \mathbb{E}\left[\int_0^t \sigma(s)\, dB_s\right] \overset{15.13\text{ a)}}{=} \int_0^t b(s)\, ds$$

and

$$\mathbb{V}(X_t - X_0) = \mathbb{E}\left[\left(X_t - X_0 - \int_0^t b(s)\, ds\right)^2\right] = \mathbb{E}\left[\left(\int_0^t \sigma(s)\, dB_s\right)^2\right] \overset{(15.21)}{=} \int_0^t \sigma^2(s)\, ds.$$

We will now use the time-dependent Itô formula to solve SDEs.

19.3 An Ansatz. Recall, cf. (17.18) and Theorem 17.11, that for $f \in \mathcal{C}^{1,2}$ (i. e. $\partial_t f, \partial_x f, \partial_x^2 f$ exist and are continuous) we have

$$df(t, B_t) = \partial_t f(t, B_t)\, dt + \partial_x f(t, B_t)\, dB_t + \frac{1}{2}\partial_x^2 f(t, B_t)\, dt.$$

In order to solve the one-dimensional SDE (19.1)

$$dX_t = b(t, X_t)\,dt + \sigma(t, X_t)\,dB_t, \qquad X_0 = x,$$

we make the following *Ansatz*:

$$X_t = f(t, B_t),$$

apply Itô's formula, and compare coefficients. This leads to the following system of (partial) differential equations

$$\begin{aligned} b(t, f(t,x)) &= \partial_t f(t,x) + \frac{1}{2}\partial_x^2 f(t,x), \\ \sigma(t, f(t,x)) &= \partial_x f(t,x). \end{aligned} \tag{19.5}$$

This will often yield necessary conditions for the solution (hence, uniqueness) as well as a hint to the general form of the solution; *do not forget to check that the result $f(t, B_t)$ is indeed a solution by using Itô's formula.* !

19.4 Example (Geometric Brownian motion). A *geometric Brownian motion* is the solution of the following SDE Ex. 19.3

$$X_t = X_0 + b\int_0^t X_s\,ds + \sigma\int_0^t X_s\,dB_s, \quad X_0 = x_0 \in \mathbb{R}$$

where $b \in \mathbb{R}$ and $\sigma > 0$ are constants. It is given by

$$X_t = x_0 \exp\left[\left(b - \tfrac{1}{2}\sigma^2\right)t + \sigma B_t\right], \quad t \geqslant 0. \tag{19.6}$$

The easiest way to verify that (19.6) is indeed a solution to the SDE, is to use Itô's formula and to check directly that (19.6) satisfies the SDE. This, however, does not guarantee uniqueness and this method only works if we already know the solution.

Using the *Ansatz* 19.3 we can show that the solution is indeed unique and of the form (19.5): From (19.5) we get

$$bf(t,x) \stackrel{!}{=} \partial_t f(t,x) + \frac{1}{2}\partial_x^2 f(t,x); \tag{19.7}$$

$$\sigma f(t,x) \stackrel{!}{=} \partial_x f(t,x). \tag{19.8}$$

Now (19.8) yields

$$\partial_x^2 f(t,x) \stackrel{(19.8)}{=} \sigma\partial_x f(t,x) \stackrel{(19.8)}{=} \sigma^2 f(t,x)$$

and inserting this into (19.7) we get

$$\left(b - \tfrac{1}{2}\sigma^2\right)f(t,x) = \partial_t f(t,x) \quad \text{and} \quad \sigma f(t,x) = \partial_x f(t,x).$$

The usual separation method $f(t,x) = g(t)h(x)$ reveals

$$g(t) = g(0)e^{(b-\frac{1}{2}\sigma^2)t} \quad \text{and} \quad h(x) = h(0)e^{\sigma x}$$

and we get (19.6). The calculation clearly shows that the conditions are necessary and one still has to check that the result is indeed a solution. This is left to the reader.

! The following example shows that the solution of an SDE need not be of the form $f(t, B_t)$, i. e. the *Ansatz* 19.3 is not always suitable.

Ex. 19.5 **19.5 Example** (Ornstein–Uhlenbeck process). An *Ornstein–Uhlenbeck process* is the solution of *Langevin's SDE*[1]

$$X_t = X_0 + b\int_0^t X_s\,ds + \sigma\int_0^t dB_s, \quad X_0 = x_0 \in \mathbb{R}$$

where $b, \sigma \in \mathbb{R}$ are constants. The solution is given by

$$X_t = e^{bt}x_0 + \sigma e^{bt}\int_0^t e^{-bs}\,dB_s, \quad t \geqslant 0. \tag{19.9}$$

Example 19.2 shows that the Ornstein–Uhlenbeck process is a Gaussian process.

! Using the Ansatz 19.3 *naively*, yields

$$bf(t,x) \stackrel{!}{=} \partial_t f(t,x) + \frac{1}{2}\partial_x^2 f(t,x), \tag{19.10}$$

$$\sigma \stackrel{!}{=} \partial_x f(t,x). \tag{19.11}$$

The second equation implies $\partial_x^2 f = \partial_x\partial_t f = 0$, and so,

$$bf = \partial_t f \quad \text{and} \quad \sigma = \partial_x f \implies b\partial_x f = \partial_x\partial_t f = 0.$$

This gives

$$b\sigma \stackrel{(19.11)}{=} b\partial_x f \underset{(19.10)}{\stackrel{(19.11)}{=}} \partial_x\partial_t f = 0$$

or $b\sigma = 0$. Thus, $b = 0$ and $f(t,x) = \sigma x + \text{const.}$, and we get just a trivial solution.

In order to avoid this mistake, we consider first the *deterministic* equation

$$x_t = x_0 + b\int_0^t x_s\,ds$$

which is solved by $x_t = x_0 e^{bt}$ or $e^{-bt}x_t = x_0 = \text{const}$. This indicates that we should use the *Ansatz*

$$Y_t := e^{-bt}X_t = f(t, X_t) \quad \text{with} \quad f(t,x) = e^{-bt}x.$$

1 This shows, incidentally, that an Ornstein–Uhlenbeck process is the continuous-time analogue of an autoregressive time series. This follows if we formally set $dX_t = X_{t+dt} - X_t$, $dB_t = B_{t+dt} - B_t$ and $dt = 1$. Then

$$X_{t+1} - X_t = bX_t + \sigma(B_{t+1} - B_t) \iff X_{t+1} = (b+1)X_t + \underbrace{Z_t}_{\sim \mathsf{N}(0,\sigma^2),\ \text{iid}}.$$

It is easy to calculate the partial derivatives

$$\partial_t f(t,x) = -bf(t,x) = -bxe^{-bt}, \quad \partial_x f(t,x) = e^{-bt}, \quad \partial_x^2 f(t,x) = 0,$$

and we find, using the time-dependent Itô formula (17.18),

$$\begin{aligned} Y_t - Y_0 &= \int_0^t \Big(\underbrace{\partial_t f(s,X_s) + bX_s\partial_x f(s,X_s)}_{=0} + \underbrace{\tfrac{1}{2}\sigma^2\partial_x^2 f(s,X_s)}_{=0}\Big)\,ds + \int_0^t \sigma\partial_x f(s,X_s)\,dB_s \\ &= \int_0^t \sigma e^{-bs}\,dB_s, \end{aligned}$$

and this gives (19.9). Problem 19.5 extends this example to random variables $X_0 = \xi$. Ex. 19.5

19.3 The general linear SDE

We will now study linear SDEs in a more systematic way.

19.6 Example (Homogeneous linear SDEs). A *homogeneous linear SDE* is an SDE of the form

$$dX_t = \beta(t)X_t\,dt + \delta(t)X_t\,dB_t$$

with non-random coefficients $\beta, \delta : [0,\infty) \to \mathbb{R}$. Let us assume that the initial condition is positive: $X_0 > 0$. Since the solution has continuous sample paths, it is clear that $X_t > 0$ for sufficiently small values of $t > 0$. Therefore it is plausible to set $Z_t := \log X_t$. By Itô's formula (17.13) we get, at least formally,

$$\begin{aligned} dZ_t &= \frac{1}{X_t}\,\beta(t)X_t\,dt + \frac{1}{X_t}\,\delta(t)X_t\,dB_t - \frac{1}{2}\frac{1}{X_t^2}\,\delta^2(t)X_t^2\,dt \\ &= \left(\beta(t) - \tfrac{1}{2}\delta^2(t)\right)dt + \delta(t)\,dB_t. \end{aligned}$$

Thus,

$$X_t = X_0 \exp\left[\int_0^t \left(\beta(s) - \tfrac{1}{2}\delta^2(s)\right)ds\right] \exp\left[\int_0^t \delta(s)\,dB_s\right].$$

Now we can verify by a direct calculation that X_t is indeed a solution of the SDE – even if $X_0 < 0$.

If $\beta \equiv 0$, we get the formula for the stochastic exponential, see Section 18.1.

19.7 Example (Linear SDEs). A *linear SDE* is an SDE of the form Ex. 19.4

$$dX_t = (\alpha(t) + \beta(t)X_t)\,dt + (\gamma(t) + \delta(t)X_t)\,dB_t,$$

with non-random coefficients $\alpha, \beta, \gamma, \delta : [0,\infty) \to \mathbb{R}$. Let X_t° be a random process such that $1/X_t^\circ$ is the solution of the homogeneous equation (i. e. where $\alpha = \gamma = 0$) for the initial value $X_0^\circ = 1$. From the explicit formula for $1/X_t^\circ$ of Example 19.6 it is easily seen that

$$dX_t^\circ = (-\beta(t) + \delta^2(t))X_t^\circ\, dt - \delta(t)X_t^\circ\, dB_t$$

or

$$X_t^\circ = \exp\left[-\int_0^t \left(\beta(s) - \tfrac{1}{2}\delta^2(s)\right) ds - \int_0^t \delta(s)\, dB_s\right]$$

Set $Z_t := X_t^\circ X_t$; using Itô's formula (17.14) for the two-dimensional Itô process (X_t, X_t°), we see

$$dZ_t = X_t\, dX_t^\circ + X_t^\circ\, dX_t - (\gamma(t) + \delta(t)X_t)\delta(t)X_t^\circ\, dt$$

and this becomes, if we insert the formulae for dX_t and dX_t°,

$$dZ_t = (\alpha(t) - \gamma(t)\delta(t))X_t^\circ\, dt + \gamma(t)X_t^\circ\, dB_t.$$

Since we know X_t° from Example 19.6, we can thus get a formula for Z_t and then for $X_t = Z_t/X_t^\circ$:

$$\begin{aligned} X_t = \exp&\left[\int_0^t \left(\beta(s) - \tfrac{1}{2}\delta^2(s)\right) ds + \int_0^t \delta(s)\, dB_s\right] \times \\ &\times \left\{X_0 + \int_0^t \exp\left[-\int_0^s \left(\beta(r) - \tfrac{1}{2}\delta^2(r)\right) dr - \int_0^s \delta(r)\, dB_r\right] \gamma(s)\, dB_s + \right. \\ &\left. + \int_0^t \exp\left[-\int_0^s \left(\beta(r) - \tfrac{1}{2}\delta^2(r)\right) dr - \int_0^s \delta(r)\, dB_r\right] (\alpha(s) - \gamma(s)\delta(s))\, ds\right\} \end{aligned}$$

19.4 Transforming an SDE into a linear SDE

We will now study some transformations of an SDE which help us to reduce a general SDE (19.1) to a linear SDE or an SDE with non-random coefficients.

Assume that $(X_t)_{t\geqslant 0}$ is a solution of the SDE (19.1) and denote by $f : [0,\infty)\times\mathbb{R} \to \mathbb{R}$ a function such that

- $x \mapsto f(t,x)$ is monotone;
- the derivatives $f_t(t,x)$, $f_x(t,x)$ and $f_{xx}(t,x)$ exist and are continuous;
- $g(t,\cdot) := f^{-1}(t,\cdot)$ exists for each t, i. e. $f(t, g(t,x)) = g(t, f(t,x)) = x$.

We set $Z_t := f(t, X_t)$ and use Itô's formula (17.13) to find the stochastic differential equation corresponding to the process $(Z_t)_{t\geqslant 0}$: Ex. 17.6

$$\begin{aligned} dZ_t &= f_t(t, X_t)\, dt + f_x(t, X_t)\, dX_t + \frac{1}{2} f_{xx}(t, X_t)\sigma^2(t, X_t)\, dt \\ &= \underbrace{\left[f_t(t, X_t) + f_x(t, X_t) b(t, X_t) + \frac{1}{2} f_{xx}(t, X_t)\sigma^2(t, X_t) \right]}_{=: \overline{b}(t, Z_t)} dt + \underbrace{f_x(t, X_t)\sigma(t, X_t)}_{=: \overline{\sigma}(t, Z_t)}\, dB_t \end{aligned}$$

where the new coefficients $\overline{b}(t, Z_t)$ and $\overline{\sigma}(t, Z_t)$ are given by

$$\begin{aligned} \overline{b}(t, Z_t) &= f_t(t, X_t) + f_x(t, X_t) b(t, X_t) + \frac{1}{2} f_{xx}(t, X_t)\sigma^2(t, X_t), \\ \overline{\sigma}(t, Z_t) &= f_x(t, X_t)\sigma(t, X_t) \end{aligned}$$

with $X_t = g(t, Z_t)$. We will now try to choose f and g in such a way that the transformed SDE

$$dZ_t = \overline{b}(t, Z_t)\, dt + \overline{\sigma}(t, Z_t)\, dB_t \tag{19.12}$$

has an explicit solution from which we obtain the solution $X_t = g(t, Z_t)$ of (19.1). There are two obvious choices:

Variance transform. Assume that $\overline{\sigma}(t, z) \equiv 1$. This means that

$$f_x(t, x)\sigma(t, x) \equiv 1 \iff f_x(t, x) = \frac{1}{\sigma(t, x)} \iff f(t, x) = \int_0^x \frac{dy}{\sigma(t, y)} + c.$$

This can be done if $\sigma(t, x) > 0$ and if the derivatives σ_t and σ_x exist and are continuous.

Drift transform. Assume that $\overline{b}(t, z) \equiv 0$. In this case $f(t, x)$ satisfies the partial differential equation

$$f_t(t, x) + b(t, x) f_x(t, x) + \frac{1}{2}\sigma^2(t, x) f_{xx}(t, x) = 0.$$

If $b(t, x) = b(x)$ and $\sigma(t, x) = \sigma(x)$, we have $b_t = \sigma_t = f_t = 0$, and the partial differential equation becomes an ordinary differential equation

$$b(x) f'(x) + \frac{1}{2}\sigma^2(x) f''(x) = 0$$

which has the following explicit solution

$$f(x) = c_1 + c_2 \int_0^x \exp\left(- \int_0^y \frac{2b(v)}{\sigma^2(v)}\, dv \right) dy.$$

19.8 Lemma. *Let $(B_t)_{t\geqslant 0}$ be a* BM^1. *The SDE $dX_t = b(t, X_t)\,dt + \sigma(t, X_t)\,dB_t$ can be transformed into the form $dZ_t = \overline{b}(t)\,dt + \overline{\sigma}(t)\,dB_t$ if, and only if, the coefficients satisfy the condition*

$$0 = \frac{\partial}{\partial x}\left\{\sigma(t,x)\left(\frac{\sigma_t(t,x)}{\sigma^2(t,x)} - \frac{\partial}{\partial x}\frac{b(t,x)}{\sigma(t,x)} + \frac{1}{2}\sigma_{xx}(t,x)\right)\right\}. \tag{19.13}$$

Proof. If we use the transformation $Z_t = f(t, X_t)$ from above, it is necessary that

$$\overline{b}(t) = f_t(t,x) + b(t,x) f_x(t,x) + \frac{1}{2} f_{xx}(t,x)\sigma^2(t,x) \tag{19.14a}$$

$$\overline{\sigma}(t) = f_x(t,x)\sigma(t,x) \tag{19.14b}$$

with $x = g(t,z) = f^{-1}(t,z)$. Now we get from (19.14b)

$$f_x(t,x) = \frac{\overline{\sigma}(t)}{\sigma(t,x)}$$

and therefore

$$f_{xt}(t,x) = \frac{\overline{\sigma}_t(t)\sigma(t,x) - \overline{\sigma}(t)\sigma_t(t,x)}{\sigma^2(t,x)} \quad \text{and} \quad f_{xx}(t,x) = \frac{-\overline{\sigma}(t)\sigma_x(t,x)}{\sigma^2(t,x)}.$$

Differentiate (19.14a) in x,

$$0 = f_{tx}(t,x) + \frac{\partial}{\partial x}\left(b(t,x)f_x(t,x) + \frac{1}{2}\sigma^2(t,x) f_{xx}(t,x)\right),$$

and plug in the formulae for f_x, f_{tx} and f_{xx} to obtain

$$\frac{\overline{\sigma}_t(t)}{\overline{\sigma}(t)} = \sigma(t,x)\left(\frac{\sigma_t(t,x)}{\sigma^2(t,x)} - \frac{\partial}{\partial x}\frac{b(t,x)}{\sigma(t,x)} + \frac{1}{2}\sigma_{xx}(t,x)\right).$$

If we differentiate this expression in x we finally get (19.13).

Since we can, with suitable integration constants, perform all calculations in reversed order, it turns out that (19.13) is a necessary and sufficient condition to transform the SDE. □

Ex. 19.6 **19.9 Example.** Assume that $b(t,x) = b(x)$ and $\sigma(t,x) = \sigma(x)$ and that condition (19.13) holds, i. e.

$$0 = \frac{d}{dx}\left\{\sigma(x)\left(-\frac{d}{dx}\frac{b(x)}{\sigma(x)} + \frac{1}{2}\sigma''(x)\right)\right\}.$$

Then the transformation

$$Z_t = f(t, X_t) = e^{ct}\int_0^{X_t}\frac{dy}{\sigma(y)}$$

with the constant $c = \sigma(x)\left[\frac{1}{2}\sigma''(x) - \frac{d}{dx}\frac{b(x)}{\sigma(x)}\right]$ converts $dX_t = b(X_t)\,dt + \sigma(X_t)\,dB_t$ into $dZ_t = e^{ct}\,dB_t$. This equation can be solved as in Example 19.2.

19.10 Lemma. *Let $(B_t)_{t\geqslant 0}$ be a BM^1. The SDE $dX_t = b(X_t)\,dt + \sigma(X_t)\,dB_t$ with coefficients $b, \sigma : \mathbb{R} \to \mathbb{R}$ can be transformed into a linear SDE* Ex. 19.10

$$dZ_t = (\alpha + \beta Z_t)\,dt + (\gamma + \delta Z_t)\,dB_t, \quad \alpha, \beta, \gamma, \delta \in \mathbb{R},$$

if, and only if,

$$\frac{d}{dx}\left(\frac{\frac{d}{dx}(\kappa'(x)\sigma(x))}{\kappa'(x)}\right) = 0 \quad \text{where} \quad \kappa(x) = \frac{b(x)}{\sigma(x)} - \frac{1}{2}\sigma'(x). \tag{19.15}$$

Set $d(x) = \int_0^x dy/\sigma(y)$ and $\delta = -\left[\frac{d}{dx}(\kappa'(x)\sigma(x))\right]/\kappa'(x)$. The transformation $Z_t = f(X_t)$ is given by

$$f(x) = \begin{cases} e^{\delta d(x)}, & \text{if } \delta \neq 0, \\ \gamma d(x), & \text{if } \delta = 0. \end{cases}$$

Proof. Since the coefficients do not depend on t, it is clear that the transformation $Z_t = f(X_t)$ should not depend on t. Using the general transformation scheme from above we see that the following relations are necessary:

$$b(x)f'(x) + \frac{1}{2}\sigma^2(x)f''(x) = \alpha + \beta f(x) \tag{19.16a}$$

$$\sigma(x)f'(x) = \gamma + \delta f(x) \tag{19.16b}$$

where $x = g(z)$, $z = f(x)$, i. e. $g(z) = f^{-1}(z)$.

Case 1. If $\delta \neq 0$, we set $d(x) = \int_0^x dy/\sigma(y)$; then

$$f(x) = ce^{\delta d(x)} - \frac{\gamma}{\delta}$$

is a solution to the ordinary differential equation (19.16b). Inserting this into (19.16a) yields

$$\left[\frac{\delta b(x)}{\sigma(x)} + \frac{1}{2}\sigma^2(x)\left(\frac{\delta^2}{\sigma^2(x)} - \frac{\delta\sigma'(x)}{\sigma^2(x)}\right)\right] ce^{\delta d(x)} = \beta ce^{\delta d(x)} - \frac{\beta\gamma}{\delta} + \alpha$$

which becomes

$$\left[\frac{\delta b(x)}{\sigma(x)} + \frac{1}{2}\delta^2 - \beta - \frac{\delta}{2}\sigma'(x)\right] e^{\delta d(x)} = \frac{\alpha\delta - \gamma\beta}{c\delta}.$$

Set $\kappa(x) := b(x)/\sigma(x) - \frac{1}{2}\sigma'(x)$. Then this relation reads

$$\left[\delta\kappa(x) + \frac{1}{2}\delta^2 - \beta\right] e^{\delta d(x)} = \frac{\alpha\delta - \gamma\beta}{c\delta}.$$

Now we differentiate this equality in order to get rid of the constant c:

$$\delta\kappa'(x)e^{\delta d(x)} + \left[\delta\kappa(x) + \frac{1}{2}\delta^2 - \beta\right] e^{\delta d(x)} \frac{\delta}{\sigma(x)} = 0$$

which is the same as

$$\delta\kappa(x) + \frac{1}{2}\delta^2 - \beta = -\kappa'(x)\sigma(x).$$

Differentiating this equality again yields (19.15); the formulae for δ and $f(x)$ now follow by integration.

Case 2. If $\delta = 0$ then (19.16b) becomes

$$\sigma(x)f'(x) = \gamma \quad \text{and so} \quad f(x) = \gamma \int_0^x \frac{dy}{\sigma(y)} + c.$$

Inserting this into (19.16a) shows

$$\kappa(x)\gamma = \beta\gamma d(x) + c'$$

and if we differentiate this we get $\sigma(x)\kappa'(x) \equiv \beta$, i. e. $\frac{d}{dx}\left(\sigma(x)\kappa'(x)\right) = 0$, which implies (19.15).

The sufficiency of (19.15) is now easily verified by a (lengthy) direct calculation with the given transformation. □

19.5 Existence and uniqueness of solutions

The main question, of course, is whether an SDE has a solution and, if so, whether the solution is unique. If $\sigma \equiv 0$, the SDE becomes an ordinary differential equation, and it is well known that existence and uniqueness results for ordinary differential equations usually require a Lipschitz condition for the coefficients. Therefore it seems reasonable to require

$$\sum_{j=1}^{n} |b_j(t,x) - b_j(t,y)|^2 + \sum_{j=1}^{n}\sum_{k=1}^{d} |\sigma_{jk}(t,x) - \sigma_{jk}(t,y)|^2 \leqslant L^2\,|x-y|^2 \tag{19.17}$$

for all $x, y \in \mathbb{R}^n$, $t \in [0,T]$ and with a (global) Lipschitz constant $L = L_T$. If we pick $y = 0$, the *global* Lipschitz condition implies the following linear growth condition

$$\sum_{j=1}^{n} |b_j(t,x)|^2 + \sum_{j=1}^{n}\sum_{k=1}^{d} |\sigma_{jk}(t,x)|^2 \leqslant M^2\,(1+|x|)^2 \tag{19.18}$$

Ex. 19.11 for all $x \in \mathbb{R}^n$ and with some constant $M = M_T$ which can be estimated from below by $M^2 \geqslant 2L^2 + 2\sum_{j=1}^n \sup_{t\leqslant T} |b_j(t,0)|^2 + 2\sum_{j=1}^n \sum_{k=1}^d \sup_{t\leqslant T} |\sigma_{jk}(t,0)|^2$.

We begin with a stability and uniqueness result.

19.11 Theorem (Stability). *Let $(B_t, \mathscr{F}_t)_{t\geqslant 0}$ be a d-dimensional Brownian motion and assume that the coefficients $b : [0,\infty) \times \mathbb{R}^n \to \mathbb{R}^n$ and $\sigma : [0,\infty) \times \mathbb{R}^n \to \mathbb{R}^{n\times d}$ of the SDE* (19.1) *satisfy the Lipschitz condition* (19.17) *for all $x, y \in \mathbb{R}^n$ with the global Lipschitz constant $L = L_T$ for $T > 0$. If $(X_t)_{t\geqslant 0}$ and $(Y_t)_{t\geqslant 0}$ are any two solutions of the SDE* (19.1) *with $\mathscr{F}_0$ measurable initial conditions $X_0 = \xi \in L^2(\mathbb{P})$ and $Y_0 = \eta \in L^2(\mathbb{P})$, respectively; then*

$$\mathbb{E}\left[\sup_{t\leqslant T}|X_t - Y_t|^2\right] \leqslant 3\,e^{3L^2(T+4)T}\,\mathbb{E}\left[|\xi-\eta|^2\right]. \tag{19.19}$$

The expectation $\mathbb{E}$ is given by the Brownian motion started at $B_0 = 0$.

!

Proof. Note that

$$X_t - Y_t = (\xi - \eta) + \int_0^t (b(s,X_s) - b(s,Y_s))\,ds + \int_0^t (\sigma(s,X_s) - \sigma(s,Y_s))\,dB_s.$$

The elementary estimate $(a+b+c)^2 \leqslant 3a^2 + 3b^2 + 3c^2$ yields

$$\begin{aligned}\mathbb{E}\left[\sup_{t\leqslant T}|X_t - Y_t|^2\right] &\leqslant 3\,\mathbb{E}\left[|\xi-\eta|^2\right]\\ &\quad + 3\,\mathbb{E}\left[\sup_{t\leqslant T}\left|\int_0^t (b(s,X_s) - b(s,Y_s))\,ds\right|^2\right]\\ &\quad + 3\,\mathbb{E}\left[\sup_{t\leqslant T}\left|\int_0^t (\sigma(s,X_s) - \sigma(s,Y_s))\,dB_s\right|^2\right].\end{aligned}$$

Using the Cauchy-Schwarz inequality and (19.17), we find for the middle term

$$\begin{aligned}\mathbb{E}&\left[\sup_{t\leqslant T}\left|\int_0^t (b(s,X_s) - b(s,Y_s))\,ds\right|^2\right]\\ &\leqslant \mathbb{E}\left[\left(\int_0^T |b(s,X_s) - b(s,Y_s)|\,ds\right)^2\right]\\ &\leqslant T\,\mathbb{E}\left[\int_0^T |b(s,X_s) - b(s,Y_s)|^2\,ds\right]\\ &\leqslant L^2 T\int_0^T \mathbb{E}\left[|X_s - Y_s|^2\right]\,ds.\end{aligned} \tag{19.20}$$

For the third term we use the maximal inequality for stochastic integrals, (15.22) in Theorem 15.13 d), and (19.17) and get

$$\begin{aligned}\mathbb{E}\left[\sup_{t\leqslant T}\left|\int_0^t (\sigma(s,X_s) - \sigma(s,Y_s))\,dB_s\right|^2\right] &\leqslant 4\int_0^T \mathbb{E}\left[|\sigma(s,X_s) - \sigma(s,Y_s)|^2\right]\,ds\\ &\leqslant 4L^2\int_0^T \mathbb{E}\left[|X_s - Y_s|^2\right]\,ds.\end{aligned} \tag{19.21}$$

This proves

$$\mathbb{E}\left[\sup_{t\leqslant T}|X_t - Y_t|^2\right] \leqslant 3\,\mathbb{E}\left[|\xi-\eta|^2\right] + 3L^2(T+4)\int_0^T \mathbb{E}\left[|X_s - Y_s|^2\right] ds$$

$$\leqslant 3\,\mathbb{E}\left[|\xi-\eta|^2\right] + 3L^2(T+4)\int_0^T \mathbb{E}\left[\sup_{r\leqslant s}|X_r - Y_r|^2\right] ds.$$

Now Gronwall's inequality, Theorem A.47, applied to $u(T) = \mathbb{E}\left[\sup_{t\leqslant T}|X_t - Y_t|^2\right]$, and $a(s) = 3\,\mathbb{E}[|\xi-\eta|^2]$ and $b(s) = 3L^2(T+4)$, yields

$$\mathbb{E}\left[\sup_{t\leqslant T}|X_t - Y_t|^2\right] \leqslant 3\,e^{3L^2(T+4)T}\,\mathbb{E}\left[|\xi-\eta|^2\right]. \qquad \square$$

19.12 Corollary (Uniqueness). *Let $(B_t, \mathscr{F}_t)_{t\geqslant 0}$ be a* BM^d *and assume that the coefficients $b : [0,\infty)\times\mathbb{R}^n \to \mathbb{R}^n$ and $\sigma : [0,\infty)\times\mathbb{R}^n \to \mathbb{R}^{n\times d}$ of the SDE* (19.1) *satisfy the Lipschitz condition* (19.17) *for all $x, y \in \mathbb{R}^n$ with the global Lipschitz constant $L = L_T$ for $T > 0$. Any two solutions $(X_t)_{t\geqslant 0}$ and $(Y_t)_{t\geqslant 0}$ of the SDE* (19.1) *with the same $\mathscr{F}_0$ measurable initial condition $X_0 = Y_0 = \xi \in L^2(\mathbb{P})$ are indistinguishable.*

Proof. The estimate (19.19) shows that $\mathbb{P}(\forall\, t \in [0,n] : X_t = Y_t) = 1$ for any $n \geqslant 1$. Therefore, $\mathbb{P}(\forall\, t \geqslant 0 : X_t = Y_t) = 1$. $\square$

In order to show the existence of the solution of the SDE we can use the Picard iteration scheme from the theory of ordinary differential equations.

19.13 Theorem (Existence). *Let $(B_t, \mathscr{F}_t)_{t\geqslant 0}$ be a* BM^d *and assume that the coefficients $b : [0,\infty)\times\mathbb{R}^n \to \mathbb{R}^n$ and $\sigma : [0,\infty)\times\mathbb{R}^n \to \mathbb{R}^{n\times d}$ of the SDE* (19.1) *satisfy the Lipschitz condition* (19.17) *for all $x, y \in \mathbb{R}^n$ with the global Lipschitz constant $L = L_T$ for $T > 0$ and the linear growth condition* (19.18) *with the constant $M = M_T$ for $T > 0$. For every $\mathscr{F}_0$ measurable initial condition $X_0 = \xi \in L^2(\mathbb{P})$ there exists a unique solution $(X_t)_{t\geqslant 0}$; this solution satisfies*

$$\mathbb{E}\left[\sup_{s\leqslant T}|X_s|^2\right] \leqslant \kappa_T\,\mathbb{E}\left[(1+|\xi|)^2\right] \quad \textit{for all } T > 0. \tag{19.22}$$

Proof. The uniqueness of the solution follows from Corollary 19.12.

To prove the existence we use Picard's iteration scheme. Set

$$X_0(t) := \xi,$$

$$X_{n+1}(t) := \xi + \int_0^t \sigma(s, X_n(s))\,dB_s + \int_0^t b(s, X_n(s))\,ds.$$

1° Since $(a+b)^2 \leqslant 2a^2 + 2b^2$, we see that

$$|X_{n+1}(t) - \xi|^2 \leqslant 2\left|\int_0^t b(s, X_n(s))\,ds\right|^2 + 2\left|\int_0^t \sigma(s, X_n(s))\,dB_s\right|^2.$$

With the calculations (19.20) and (19.21) – cf. the proof of Theorem 19.11 – where we set $X_s = X_n(s)$ and omit the Y-terms, and with the linear growth condition (19.18) instead of the Lipschitz condition (19.17), we get

$$\begin{aligned}\mathbb{E}\left[\sup_{t\leqslant T}|X_{n+1}(t)-\xi|^2\right] &\leqslant 2M^2(T+4)\int_0^T \mathbb{E}\left[(1+|X_n(s)|)^2\right]ds\\ &\leqslant 2M^2(T+4)T\,\mathbb{E}\left[\sup_{s\leqslant T}(1+|X_n(s)|)^2\right].\end{aligned}$$

Starting with $n = 0$, we see recursively that the processes X_{n+1}, $n = 0, 1, \dots$, are well-defined.

2° As in step 1° we see

$$\begin{aligned}|X_{n+1}(t)-X_n(t)|^2 &\leqslant 2\left|\int_0^t [b(s,X_n(s))-b(s,X_{n-1}(s))]\,ds\right|^2\\ &\quad+2\left|\int_0^t [\sigma(s,X_n(s))-\sigma(s,X_{n-1}(s))]\,dB_s\right|^2.\end{aligned}$$

Using again (19.20) and (19.21) with $X = X_n$ and $Y = X_{n-1}$ we get

$$\mathbb{E}\left[\sup_{t\leqslant T}|X_{n+1}(t)-X_n(t)|^2\right] \leqslant 2L^2(T+4)\int_0^T \mathbb{E}\left[|X_n(s)-X_{n-1}(s)|^2\right]ds.$$

Set $\phi_n(T) := \mathbb{E}\left[\sup_{t\leqslant T}|X_{n+1}(t)-X_n(t)|^2\right]$ and $c_T = 2L^2(T+4)$. Several iterations of this inequality yield

$$\begin{aligned}\phi_n(T) &\leqslant c_T^n \int\!\cdots\!\int_{t_1\leqslant\cdots\leqslant t_{n-1}\leqslant T} \phi_0(t_0)\,dt_0\dots dt_{n-1}\\ &\leqslant c_T^n \int\!\cdots\!\int_{t_1\leqslant\cdots\leqslant t_{n-1}\leqslant T} dt_0\dots dt_{n-1}\,\phi_0(T) = c_T^n\frac{T^n}{n!}\phi_0(T).\end{aligned}$$

The estimate from 1° for $n = 0$ shows

$$\phi_0(T) = \mathbb{E}\left[\sup_{s\leqslant T}|X_1(s)-\xi|^2\right] \leqslant 2M^2(T+4)T\,\mathbb{E}\left[(1+|\xi|)^2\right]$$

and this means that for some constant $C_T = C(L, M, T)$

$$\sum_{n=0}^\infty\left\{\mathbb{E}\left[\sup_{t\leqslant T}|X_{n+1}(t)-X_n(t)|^2\right]\right\}^{\frac12} \leqslant \sum_{n=0}^\infty\left(c_T^n\frac{T^n}{n!}\phi_0(T)\right)^{\frac12} \leqslant C_T\sqrt{\mathbb{E}\left[(1+|\xi|)^2\right]}.$$

3° For $n \geqslant m$ we find

$$\left\{\mathbb{E}\left[\sup_{s\leqslant T}|X_n(s)-X_m(s)|^2\right]\right\}^{\frac{1}{2}} \leqslant \sum_{j=m+1}^{\infty}\left\{\mathbb{E}\left[\sup_{s\leqslant T}|X_j(s)-X_{j-1}(s)|^2\right]\right\}^{\frac{1}{2}}.$$

Therefore, $(X_n)_{n\geqslant 0}$ is a Cauchy sequence in L^2; denote its limit by X. The above inequality shows that there is a subsequence $m(k)$ such that

$$\lim_{k\to\infty}\sup_{s\leqslant T}|X(s)-X_{m(k)}(s)|^2 = 0 \quad \text{a.s.}$$

Since $X_{m(k)}(\cdot)$ is continuous and adapted, we see that $X(\cdot)$ is also continuous and adapted. Moreover,

$$\left\{\mathbb{E}\left[\sup_{s\leqslant T}|X(s)-\xi|^2\right]\right\}^{\frac{1}{2}} \leqslant \sum_{n=0}^{\infty}\left\{\mathbb{E}\left[\sup_{t\leqslant T}|X_{n+1}(t)-X_n(t)|^2\right]\right\}^{\frac{1}{2}} \leqslant C_T\sqrt{\mathbb{E}\left[(1+|\xi|)^2\right]},$$

and (19.22) follows with $\kappa_T = C_T^2 + 1$.

4° Recall that $X_{n+1}(t) = \xi + \int_0^t \sigma(s, X_n(s))\,dB_s + \int_0^t b(s, X_n(s))\,ds$. Let $n = m(k)$; because of the estimate (19.22) we can use the dominated convergence theorem to get

$$X(t) = \xi + \int_0^t \sigma(s, X(s))\,dB_s + \int_0^t b(s, X(s))\,ds$$

which shows that X is indeed a solution of the SDE. □

19.14 Remark. The proof of the existence of a solution of the SDE (19.1) shows, in particular, that the solution depends *measurably* on the initial condition. More precisely, if X_t^x denotes the unique solution of the SDE with initial condition $x \in \mathbb{R}^n$, then $(s, x, \omega) \mapsto X_s^x(\omega)$, $(s, x) \in [0, t] \times \mathbb{R}^n$ is for every $t \geqslant 0$ measurable with respect to $\mathscr{B}[0, t] \otimes \mathscr{B}(\mathbb{R}^n) \otimes \mathscr{F}_t$. It is enough to show this for the Picard approximations $X_n^x(t, \omega)$ since the measurability is preserved under pointwise limits. For $n = 0$ we have $X_n^x(t, \omega) = x$ and there is nothing to show. Assume we know already that $X_n^x(t, \omega)$ is $\mathscr{B}[0, t] \otimes \mathscr{B}(\mathbb{R}^n) \otimes \mathscr{F}_t$ measurable. Since

$$X_{n+1}^x(t) = x + \int_0^t b(s, X_n^x(s))\,ds + \int_0^t \sigma(s, X_n^x(s))\,dB_s$$

we are done, if we can show that both integral terms have the required measurability. For the Lebesgue integral this is obvious from the measurability of the integrand and the continuous dependence on the upper integral limit t. To see the measurability of the stochastic integral, it is enough to assume $d = n = 1$ and that σ is a step function. Then

$$\int_0^t \sigma(s, X_n^x(s))\,dB_s = \sum_{j=1}^{m}\sigma(s_{j-1}, X_n^x(s_{j-1}))(B(s_j) - B(s_{j-1}))$$

where $0 = s_0 < s_1 < \cdots < s_m = t$ and this shows the required measurability.

19.6 Further examples and counterexamples

Throughout this section we consider one-dimensional time-homogeneous SDEs of the form

$$dX_t = b(X_t)\,dt + \sigma(X_t)\,dB_t, \quad X_0 = x \in \mathbb{R} \tag{19.23}$$

where $(B_t, \mathscr{F}_t)_{t\geqslant 0}$ is a BM[1] with an admissible complete filtration. If the coefficients b and σ satisfy some (local) Hölder and global growth condition, then the equation (19.23) has a unique solution. We will present a few examples that illustrate what may happen if these conditions do not hold.

The first example is the classical example for an ODE without a solution; setting $\sigma \equiv 0$, we see that ODEs are particular SDEs.

19.15 Example. The one-dimensional ODE

$$dx_t = -\operatorname{sgn} x_t\,dt, \quad x_0 = 0, \quad \operatorname{sgn} x = \mathbb{1}_{(0,\infty)}(x) - \mathbb{1}_{(-\infty,0]}(x), \tag{19.24}$$

does not have a solution.

Proof. Assume that x_t is a solution of (19.24) and that there are $\xi \neq 0$ and $t_0 > 0$ such that $x_{t_0} = \xi$. Define

$$\tau := \inf\{t \geqslant 0 : x_t = \xi\} \quad \text{and} \quad \sigma := \sup\{0 \leqslant t \leqslant \tau : x_t = 0\}$$

Inserting these times into the ODE yields

$$\xi = x_\tau - x_\sigma = -\int_\sigma^\tau \operatorname{sgn} x_s\,ds = -(\tau - \sigma)\operatorname{sgn}\xi.$$

This leads to a contradiction. Thus, only $x_t \equiv 0$ qualifies as a solution, but this does not solve (19.24). □

If we replace dt by dB_t in (19.24) we get a similar result:

19.16 Example (Tanaka 1963). The one-dimensional SDE

$$dX_t = -\operatorname{sgn} X_t\,dB_t, \quad X_0 = 0, \quad \operatorname{sgn} x = \mathbb{1}_{(0,\infty)}(x) - \mathbb{1}_{(-\infty,0]}(x), \tag{19.25}$$

does not have a solution.

Proof. We may assume that $\mathscr{F}_t$ is the completion of the natural filtration $\mathscr{F}_t^B$, otherwise we could reduce the filtration. Let $(X_t, \mathscr{F}_t^X)_{t\geqslant 0}$ be a solution to (19.25); we may assume that $(\mathscr{F}_t^X)_{t\geqslant 0}$ is complete. By definition, $\mathscr{F}_t^X \subset \mathscr{F}_t$ and X is a continuous martingale. Its quadratic variation is given by

$$\langle X\rangle_t = \left\langle -\int_0^\bullet \operatorname{sgn} X_s\,dB_s \right\rangle_t = \int_0^t \operatorname{sgn}^2 X_s\,ds = t.$$

By Lévy's characterization of Brownian motion, Theorem 9.12 or Theorem 18.5, we see that $(X_t, \mathscr{F}_t^X)_{t\geqslant 0}$ is a BM^1. Applying Tanaka's formula (17.21) to X shows

$$|X_t| - L_t^0(X) = \int_0^t \operatorname{sgn} X_s \, dX_s = -\int_0^t (\operatorname{sgn} X_s)^2 \, dB_s = -B_t,$$

where $L_t^0(X)$ is the local time (at 0) of the Brownian motion X. Since

$$L_t^0(X) = \mathbb{P}\text{-}\lim_{\epsilon\to 0} \frac{1}{2\epsilon} \int_0^t \mathbb{1}_{[0,\epsilon)}(|X_s|) \, ds,$$

cf. (17.20), we see that $\mathscr{F}_t \subset \mathscr{F}_t^{|X|}$ where $\mathscr{F}_t^{|X|}$ is the completed canonical filtration of the process $(|X_t|)_{t\geqslant 0}$. Since $(X_t, \mathscr{F}_t^X)_{t\geqslant 0}$ is a solution of (19.25), we also have $\mathscr{F}_t^X \subset \mathscr{F}_t$, which means that $\mathscr{F}_t^X \subset \mathscr{F}_t^{|X|}$. This is impossible, since X is a Brownian motion. □

In fact, one can even show that there are SDEs of the form $dX_t = \sigma(X_t)\, dB_t$ without a solution, where σ is continuous, bounded and strictly positive, see Barlow [11].

The one-sided version of the SDE (19.25) admits, however, a solution.

19.17 Example. The one-dimensional SDE

$$dX_t = \mathbb{1}_{(0,\infty)}(X_t)\, dB_t, \quad X_0 = x \in \mathbb{R}, \tag{19.26}$$

has a solution.

Proof. Consider the stopping time $\tau = \inf\{t \geqslant 0 : x + B_t \leqslant 0\}$, and set $X_t := x + B_{t\wedge\tau}$. Obviously, $(X_t)_{t\geqslant 0}$ is continuous, $\mathscr{F}_t$-adapted, and we have

$$X_t - X_0 = \int_0^t \mathbb{1}_{\{\tau>s\}}\, dB_s = \int_0^t \mathbb{1}_{(0,\infty)}(x + B_{s\wedge\tau})\, dB_s = \int_0^t \mathbb{1}_{(0,\infty)}(X_s)\, dB_s.$$

This argument still holds if the initial condition X_0 is an independent (of $(B_t)_{t\geqslant 0}$) random variable. □

If we compare Example 19.15 with Example 19.16, we see that in the latter example the non-existence is a measurability issue: Use $W_t = -\int_0^t \operatorname{sgn} X_s\, dX_s$, where $(X_t, \mathscr{F}_t^X)_{t\geqslant 0}$ is an *arbitrary Brownian motion*, instead of the given Brownian motion $(B_t)_{t\geqslant 0}$. By Lévy's theorem, $(W_t, \mathscr{F}_t^X)_{t\geqslant 0}$ is also a Brownian motion, and we have $dW_t = -\operatorname{sgn} X_t\, dX_t$ or $dX_t = -\operatorname{sgn} X_t\, dW_t$. If we relax the Definition 19.1 of the solution to an SDE, we can take such situations into account.

19.18 Definition (Skorokhod 1965). A *weak solution* to the SDE (19.23) consists of some Brownian motion $(W_t, \mathscr{F}_t^W)_{t\geqslant 0}$ and a progressively measurable process $(X_t, \mathscr{F}_t^W)_{t\geqslant 0}$ such that $(\mathscr{F}_t^W)_{t\geqslant 0}$ is a complete admissible filtration and

$$X_t = x + \int_0^t b(X_s)\, ds + \int_0^t \sigma(X_s)\, dW_s.$$

(As in Definition 19.1, it is implicit that all (stochastic) integrals make sense).

Every (strong) solution in the sense of Definition 19.1 is also a weak solution. We do not want to follow this line of discussion, but refer to Karatzas & Shreve [119, Chapter 5.3], Ikeda & Watanabe [100, Chapter IV.1–3] and Cherny & Engelbert [28] for properties of weak solutions. In the following examples *solution* always means a strong solution in the sense of Definition 19.1.

19.19 Example (Zvonkin 1974). The one-dimensional SDE

$$dX_t = -\frac{1}{2X_t}\mathbb{1}_{\mathbb{R}\setminus\{0\}}(X_t)\,dt + dB_t, \quad X_0 = 0, \tag{19.27}$$

has no solution (and also no weak solution).

Proof. Assume that X_t is a (weak) solution of (19.27). Let $\phi \in \mathcal{C}_c^\infty(\mathbb{R})$ such that $0 \leqslant \phi \leqslant 1$ and $\phi|_{(-1,1)} = 1$. Set $\phi_\epsilon(x) := \phi(x/\epsilon)$; then

$$\begin{aligned}|\phi_\epsilon(x)| + |x\phi_\epsilon'(x)| + |x^2\phi_\epsilon''(x)| &= \left|\phi\left(\tfrac{x}{\epsilon}\right)\right| + \left|\tfrac{x}{\epsilon}\phi'\left(\tfrac{x}{\epsilon}\right)\right| + \left|\tfrac{x^2}{\epsilon^2}\phi''\left(\tfrac{x}{\epsilon}\right)\right| \\ &\leqslant \sup_{y\in\mathbb{R}}|\phi(y)| + \sup_{y\in\mathbb{R}}|y\phi'(y)| + \sup_{y\in\mathbb{R}}|y^2\phi''(y)| < \infty.\end{aligned}$$

By Itô's formula – which also holds for weak solutions with the Brownian motion specified in the definition of the weak solution – we get

$$\begin{aligned}&X_t^2\phi_\epsilon(X_t)\\ &= \int_0^t \left(2X_s\phi_\epsilon(X_s) + X_s^2\phi_\epsilon'(X_s)\right)dX_s + \frac{1}{2}\int_0^t \left(2\phi_\epsilon(X_s) + 4X_s\phi_\epsilon'(X_s) + X_s^2\phi_\epsilon''(X_s)\right)ds\\ &= \int_0^t \left(2X_s\phi_\epsilon(X_s) + X_s^2\phi_\epsilon'(X_s)\right)dB_s + \int_0^t\Big(\underbrace{\phi_\epsilon(X_s)\mathbb{1}_{\{0\}}(X_s)}_{=\mathbb{1}_{\{0\}}(X_s)} + \frac{3}{2}X_s\phi_\epsilon'(X_s) + \frac{1}{2}X_s^2\phi_\epsilon''(X_s)\Big)ds.\end{aligned}$$

Since all integrands are bounded, the stochastic integral is a martingale and we can take expectations:

$$\mathbb{E}\left[X_t^2\phi_\epsilon(X_t)\right] = \mathbb{E}\left[\int_0^t \mathbb{1}_{\{0\}}(X_s)\,ds\right] + \mathbb{E}\left[\int_0^t\left(\frac{3}{2}X_s\phi_\epsilon'(X_s) + \frac{1}{2}X_s^2\phi_\epsilon''(X_s)\right)ds\right].$$

Using dominated convergence, we can let $\epsilon \to 0$ to conclude that for all $t > 0$

$$0 = \mathbb{E}\left[\int_0^t \mathbb{1}_{\{0\}}(X_s)\,ds\right], \quad \text{and so,} \quad \int_0^t \mathbb{1}_{\{0\}}(X_s)\,ds = 0 \text{ a. s.}$$

Applying Itô's formula again yields

$$X_t^2 = \int_0^t \mathbb{1}_{\{0\}}(X_s)\,ds + \int_0^t 2X_s\,dB_s = \int_0^t 2X_s\,dB_s.$$

Thus, $(X_t^2)_{t\geqslant 0}$ is a positive local martingale. As such, it is a supermartingale, cf. Proposition 18.3 a), and we find for all $t > 0$

$$\mathbb{E}\left(X_t^2\right) \leqslant \mathbb{E}\left(X_0^2\right) = 0, \quad \text{hence,} \quad X_t = 0 \text{ a.s.}$$

Since $t \mapsto X_t$ is continuous, this contradicts the fact that $\int_0^t \mathbb{1}_{\{0\}}(X_s)\,ds = 0$. □

19.20 Example (Zvonkin 1974). The one-dimensional SDE

$$dX_t = -\operatorname{sgn} X_t\,dt + dB_t, \quad X_0 = x \in \mathbb{R} \tag{19.28}$$

has a unique solution. This is a non-trivial result. It is a special case of the fact that the SDE

$$dX_t = b(X_t)\,dt + dB_t, \quad X_0 = x \in \mathbb{R}, \quad b \in \mathcal{B}_b(\mathbb{R}), \tag{19.29}$$

has a unique (strong) solution, cf. Zvonkin [243].

19.21 Example (Maruyama's drift transformation; Maruyama 1954). Using Girsanov's theorem it is straightforward to show that the SDE (19.29) from Example 19.20 admits a weak solution in the sense of Definition 19.18. Let $(B_t, \mathscr{F}_t)_{t\geqslant 0}$ be any BM[1] with an admissible, complete filtration. We set

$$\begin{aligned} X_t &:= x + B_t, \\ M_t &:= \exp\left(\int_0^t b(X_s)\,dB_s - \frac{1}{2}\int_0^t b^2(X_s)\,ds\right), \\ \mathbb{Q}|_{\mathscr{F}_t} &:= M_t \cdot \mathbb{P}|_{\mathscr{F}_t}. \end{aligned}$$

By Girsanov's theorem, Theorem 18.8, $(W_t, \mathscr{F}_t)_{t\leqslant T}$ is a Brownian motion where $T > 0$ and $W_t := B_t - \int_0^t b(X_s)\,ds$ on the probability space $(\Omega, \mathscr{A}, \mathbb{Q} = \mathbb{Q}|_{\mathscr{F}_T})$. In particular, $((W_t, \mathscr{F}_t)_{t\geqslant 0}, (X_t, \mathscr{F}_t)_{t\geqslant 0})$ is a weak solution since

$$dX_t = dB_t = dW_t + d\int_0^t b(X_s)\,ds = dW_t + b(X_t)\,dt \quad \text{and} \quad X_0 = x.$$

If we replace the function $b(x)$ by a functional $\beta(t, \omega)$ which depends on the whole path $[0, t] \ni s \mapsto X_s(\omega)$ up to time t, then there need not be a solution, cf. Tsirelson [222].

19.22 Example. The one-dimensional SDE

$$dX_t = \mathbb{1}_{\mathbb{R}\setminus\{0\}}(X_t)\,dB_t, \quad X_0 = 0, \tag{19.30}$$

has more than one solution.

Proof. Note that both $X_t \equiv 0$ and $X_t = B_t$ solve (19.30). □

19.7 Solutions as Markov processes

Recall, cf. Remark 6.3 a), that a *Markov process* is an n-dimensional, adapted stochastic process $(X_t, \mathscr{F}_t)_{t\geqslant 0}$ with (right-)continuous paths such that

$$\mathbb{E}\left[u(X_t) \mid \mathscr{F}_s\right] = \mathbb{E}\left[u(X_t) \mid X_s\right] \quad \text{for all } t \geqslant s \geqslant 0,\ u \in \mathcal{B}_b(\mathbb{R}^n). \tag{19.31}$$

19.23 Theorem. *Let $(B_t, \mathscr{F}_t)_{t\geqslant 0}$ be a d-dimensional Brownian motion and assume that the coefficients $b(s,x) \in \mathbb{R}^n$ and $\sigma(s,x) \in \mathbb{R}^{n\times d}$ of the SDE* (19.1) *satisfy the global Lipschitz condition* (19.17). *Then the unique solution of the SDE is a Markov process.*

Proof. Consider for fixed $s \geqslant 0$ and $x \in \mathbb{R}^n$ the following SDE

$$X_t = x + \int_s^t b(r, X_r)\,dr + \int_s^t \sigma(r, X_r)\,dB_r, \quad t \geqslant s.$$

It is not hard to see that the existence and uniqueness results of Section 19.5 also hold for this SDE and we denote by $(X_t^{s,x})_{t\geqslant s}$ its unique solution. On the other hand, we find for every $\mathscr{F}_0$ measurable initial condition $\xi \in L^2(\mathbb{P})$ that

$$X_t^{0,\xi} = \xi + \int_0^t b(r, X_r^{0,\xi})\,dr + \int_0^t \sigma(r, X_r^{0,\xi})\,dB_r, \quad t \geqslant 0,$$

and so

$$X_t^{0,\xi} = X_s^{0,\xi} + \int_s^t b(r, X_r^{0,\xi})\,dr + \int_s^t \sigma(r, X_r^{0,\xi})\,dB_r, \quad t \geqslant s.$$

Because of the uniqueness of the solution, we get

$$X_t^{0,\xi} = X_t^{s,X_s^{0,\xi}} \quad \text{a. s. for all } t \geqslant s.$$

Moreover, $X_t^{0,\xi}(\omega)$ is of the form $\Phi(X_s^{0,\xi}(\omega), s, t, \omega)$ and the functional Φ is measurable with respect to $\mathscr{G}_s := \sigma(B_r - B_s : r \geqslant s)$. Since $(B_t)_{t\geqslant 0}$ has independent increments, we know that $\mathscr{F}_s \perp\!\!\!\perp \mathscr{G}_s$, cf. Lemma 2.14 or Theorem 6.1. Therefore, we get for all $s \leqslant t$ and $u \in \mathcal{B}_b(\mathbb{R}^n)$

$$\begin{aligned}
\mathbb{E}\left[u(X_t^{0,\xi}) \mid \mathscr{F}_s\right] &= \mathbb{E}\left[u(\Phi(X_s^{0,\xi}, s, t, \cdot)) \mid \mathscr{F}_s\right] \\
&\underset{\Phi \perp\!\!\!\perp \mathscr{F}_s}{\overset{\text{(A.5)}}{=}} \mathbb{E}\left[u(\Phi(y, s, t, \cdot))\right]\Big|_{y=X_s^{0,\xi}} \\
&= \mathbb{E}\left[u(X_t^{s,y})\right]\Big|_{y=X_s^{0,\xi}}.
\end{aligned}$$

Using the tower property for conditional expectations we see now

$$\mathbb{E}\left[u(X_t^{0,\xi}) \mid \mathscr{F}_s\right] = \mathbb{E}\left[u(X_t^{0,\xi}) \mid X_s^{0,\xi}\right]. \qquad \square$$

For $u = \mathbb{1}_B$, $B \in \mathscr{B}(\mathbb{R}^n)$ the proof of Theorem 19.23 actually shows

$$p(s, X_s^{0,\xi}; t, B) := \mathbb{P}(X_t^{0,\xi} \in B \mid \mathscr{F}_s) = \mathbb{P}(X_t^{s,y} \in B)\big|_{y=X_s^{0,\xi}}.$$

The family of measures $p(s, x; t, \cdot)$ is the *transition function* of the Markov process X. If $p(s, x; t, \cdot)$ depends only on the difference $t - s$, we say that the Markov process is *homogeneous*. In this case we write $p(t - s, x; \cdot)$.

19.24 Corollary. *Let $(B_t, \mathscr{F}_t)_{t\geqslant 0}$ be a BM^d and assume that the coefficients $b(x) \in \mathbb{R}^n$ and $\sigma(x) \in \mathbb{R}^{n\times d}$ of the SDE* (19.1) *are autonomous (i. e. they do not depend on time) and satisfy the global Lipschitz condition* (19.17). *Then the unique solution of the SDE is a (homogeneous) Markov process with the transition function*

$$p(t, x; B) = \mathbb{P}(X_t^x \in B), \quad t \geqslant 0,\ x \in \mathbb{R}^n,\ B \in \mathscr{B}(\mathbb{R}^n).$$

Proof. As in the proof of Theorem 19.23 we see for all $x \in \mathbb{R}^n$ and $s, t \geqslant 0$

$$\begin{aligned} X_{s+t}^{s,x} &= x + \int_s^{s+t} b(X_r^{s,x})\,dr + \int_s^{s+t} \sigma(X_r^{s,x})\,dB_r \\ &= x + \int_0^t b(X_{s+v}^{s,x})\,dv + \int_0^t \sigma(X_{s+v}^{s,x})\,dW_v \end{aligned}$$

where $W_v := B_{s+v} - B_s$, $v \geqslant 0$, is again a Brownian motion. Thus, $(X_{s+t}^{s,x})_{t\geqslant 0}$ is the unique solution of the SDE

$$Y_t = x + \int_0^t b(Y_v)\,dv + \int_0^t \sigma(Y_v)\,dW_v.$$

Since W and B have the same probability distribution, we conclude that

$$\mathbb{P}\left(X_{s+t}^{s,x} \in B\right) = \mathbb{P}\left(X_t^{0,x} \in B\right) \quad \text{for all } s \geqslant 0,\ B \in \mathscr{B}(\mathbb{R}^n).$$

The claim follows if we set $p(t, x; B) := \mathbb{P}(X_t^x \in B) := \mathbb{P}(X_t^{0,x} \in B)$. □

19.8 Localization procedures

We will show that the existence and uniqueness theorem for the solution of an SDE still holds under a *local* Lipschitz condition (19.17) and a *global* linear growth condition (19.18). We begin with a useful localization result for stochastic differential equations.

19.25 Lemma (Localization). *Let $(B_t, \mathscr{F}_t)_{t\geqslant 0}$ be a BM^d. Consider the SDEs*

$$dX_t^j = b_j(t, X_t^j)\,dt + \sigma_j(t, X_t^j)\,dB_t, \qquad j = 1, 2,$$

with initial condition $X_0^1 = X_0^2 = \xi \in L^2$. Assume that ξ is $\mathscr{F}_0$ measurable and that the coefficients $b_j : [0, \infty) \times \mathbb{R}^n \to \mathbb{R}^n$ and $\sigma_j : [0, \infty) \times \mathbb{R}^n \to \mathbb{R}^{n\times d}$ satisfy the Lipschitz *condition* (19.17) *for all* $x, y \in \mathbb{R}^n$ *with a global Lipschitz constant* L. *If for some* $T > 0$ *and* $R > 0$

$$b_1|_{[0,T]\times\mathbb{B}(0,R)} = b_2|_{[0,T]\times\mathbb{B}(0,R)} \quad \text{and} \quad \sigma_1|_{[0,T]\times\mathbb{B}(0,R)} = \sigma_2|_{[0,T]\times\mathbb{B}(0,R)},$$

then we have for the stopping times $\tau_j = \inf\{t \geqslant 0 : |X_t^j| \geqslant R\} \wedge T$

$$\mathbb{P}(\tau_1 = \tau_2) = 1 \quad \text{and} \quad \mathbb{P}\Big(\sup_{s\leqslant\tau_1} |X_s^1 - X_s^2| = 0\Big) = 1. \tag{19.32}$$

Proof. Since the solutions of the SDEs are continuous processes, τ_1 and τ_2 are indeed stopping times, cf. Lemma 5.8. Observe that

$$\begin{aligned} X^1_{t\wedge\tau_1} - X^2_{t\wedge\tau_1} &= \int_0^{t\wedge\tau_1} [b_1(s, X_s^1) - b_2(s, X_s^2)]\, ds + \int_0^{t\wedge\tau_1} [\sigma_1(s, X_s^1) - \sigma_2(s, X_s^2)]\, dB_s \\ &= \int_0^{t\wedge\tau_1} [\underbrace{b_1(s, X_s^1) - b_2(s, X_s^1)}_{=0} + b_2(s, X_s^1) - b_2(s, X_s^2)]\, ds \\ &\quad + \int_0^{t\wedge\tau_1} [\underbrace{\sigma_1(s, X_s^1) - \sigma_2(s, X_s^1)}_{=0} + \sigma_2(s, X_s^1) - \sigma_2(s, X_s^2)]\, dB_s. \end{aligned}$$

Now we can use the elementary inequality $(a+b)^2 \leqslant 2a^2 + 2b^2$ and (19.20), (19.21) from the proof of the stability Theorem 19.11, with $X = X^1$ and $Y = X^2$ to deduce

$$\mathbb{E}\left[\sup_{t\leqslant T} |X^1_{t\wedge\tau_1} - X^2_{t\wedge\tau_1}|^2\right] \leqslant 2L^2(T+4) \int_0^T \mathbb{E}\left[\sup_{r\leqslant s} |X^1_{r\wedge\tau_1} - X^2_{r\wedge\tau_1}|^2\right] ds.$$

Gronwall's inequality, Theorem A.47 with $a(s) = 0$ and $b(s) = 2L^2(T+4)$ yields

$$\mathbb{E}\left[\sup_{r\leqslant s} |X^1_{r\wedge\tau_1} - X^2_{r\wedge\tau_1}|^2\right] = 0 \quad \text{for all } s \leqslant T.$$

Therefore, $X^1_{\bullet\wedge\tau_1} = X^2_{\bullet\wedge\tau_1}$ almost surely and, in particular, $\tau_1 \leqslant \tau_2$. Swapping the roles of X^1 and X^2 we find $\tau_2 \leqslant \tau_1$ a. s., and the proof is finished. □

We will use Lemma 19.25 to prove local uniqueness and existence results.

19.26 Theorem. *Let* $(B_t, \mathscr{F}_t)_{t\geqslant 0}$ *be a d-dimensional Brownian motion and assume that the coefficients* $b : [0,\infty)\times\mathbb{R}^n \to \mathbb{R}^n$ *and* $\sigma : [0,\infty)\times\mathbb{R}^n \to \mathbb{R}^{n\times d}$ *of the SDE* (19.1) *satisfy the linear growth condition* (19.18) *for all* $x \in \mathbb{R}^n$ *and the Lipschitz condition* (19.17) *for all* $x, y \in K$ *for every compact set* $K \subset \mathbb{R}^n$ *with the local Lipschitz constants* $L_{T,K}$. *For every* $\mathscr{F}_0$ *measurable initial condition* $X_0 = \xi \in L^2(\mathbb{P})$ *there exists a unique solution* $(X_t)_{t\geqslant 0}$, *and this solution satisfies*

$$\mathbb{E}\left[\sup_{s\leqslant T} |X_s|^2\right] \leqslant \kappa_T\, \mathbb{E}\left[(1+|\xi|)^2\right] \quad \textit{for all } T > 0. \tag{19.33}$$

Proof. Fix $T > 0$. For $R > 0$ we find a cut-off function $\chi_R \in \mathcal{C}_c^\infty(\mathbb{R}^n)$ with $\mathbb{1}_{\mathbb{B}(0,R)} \leqslant \chi_R \leqslant 1$. Set

$$b_R(t,x) := b(t,x)\chi_R(x) \quad \text{and} \quad \sigma_R(t,x) := \sigma(t,x)\chi_R(x).$$

By Theorem 19.13 the SDE

$$dY_t = b_R(t, Y_t)\,dt + \sigma_R(t, Y_t)\,dB_t, \quad Y_0 = \chi_R(\xi)\xi$$

has a unique solution X_t^R. Set

$$\tau_R := \inf\left\{s \geqslant 0 : |X_s^R| \geqslant R\right\}.$$

For $S > R$ and all $t \geqslant 0$ we have

$$b_R|_{[0,t]\times\mathbb{B}(0,R)} = b_S|_{[0,t]\times\mathbb{B}(0,R)} \quad \text{and} \quad \sigma_R|_{[0,t]\times\mathbb{B}(0,R)} = \sigma_S|_{[0,t]\times\mathbb{B}(0,R)}.$$

With Lemma 19.25 we conclude that

$$\mathbb{P}\Big(\sup_{t\leqslant T}|X_t^R - X_t^S| > 0\Big) \leqslant \mathbb{P}(\tau_R < T) = \mathbb{P}\Big(\sup_{t\leqslant T}|X_t^R| \geqslant R\Big).$$

Therefore it is enough to show that

$$\lim_{R\to\infty} \mathbb{P}\left(\tau_R < T\right) = 0. \tag{19.34}$$

Once (19.34) is established, we see that $X_t := \overline{\lim}_{R\to\infty} X_t^R$ exists uniformly for $t \in [0,T]$ and $X_{t\wedge\tau_R} = X_{t\wedge\tau_R}^R$. Therefore, $(X_t)_{t\geqslant 0}$ is a continuous stochastic process. Moreover,

$$\begin{aligned} X_{t\wedge\tau_R}^R &= \chi_R(\xi)\xi + \int_0^{t\wedge\tau_R} b_R(s, X_s^R)\,ds + \int_0^{t\wedge\tau_R} \sigma_R(s, X_s^R)\,dB_s \\ &= \chi_R(\xi)\xi + \int_0^{t\wedge\tau_R} b(s, X_s)\,ds + \int_0^{t\wedge\tau_R} \sigma(s, X_s)\,dB_s. \end{aligned}$$

Letting $R \to \infty$ shows that X_t solves the SDE (19.1).

Let us now show (19.34). Since $\xi \in L^2(\mathbb{P})$, we can use Theorem 19.13 to get

$$\mathbb{E}\Big[\sup_{t\leqslant T}|X_t^R|^2\Big] \leqslant \kappa_T\,\mathbb{E}\left[(1 + |\chi_R(\xi)\xi|)^2\right] \leqslant \kappa_T\,\mathbb{E}\left[(1+|\xi|)^2\right] < \infty,$$

and with Markov's inequality,

$$\mathbb{P}(\tau_R < T) = \mathbb{P}\Big(\sup_{s\leqslant T}|X_s^R| \geqslant R\Big) \leqslant \frac{\kappa_T}{R^2}\,\mathbb{E}\left[(1+|\xi|)^2\right] \xrightarrow[R\to\infty]{} 0.$$

To see uniqueness, assume that X and Y are any two solutions. Since X^R is unique, we conclude that

$$X_{t\wedge\tau_R\wedge T_R} = Y_{t\wedge\tau_R\wedge T_R} = X_{t\wedge\tau_R\wedge T_R}^R$$

if $\tau_R := \inf\{s \geqslant 0 : |X_s| \geqslant R\}$ and $T_R := \inf\{s \geqslant 0 : |Y_s| \geqslant R\}$. Thus, $X_t = Y_t$ for all $t \leqslant \tau_R \wedge T_R$. Since $\tau_R, T_R \to \infty$ as $R \to \infty$, we see that $X = Y$. □

19.9 Dependence on the initial values

Often it is important to know how the solution $(X_t^x)_{t\geqslant 0}$ of the SDE (19.1) depends on the starting point $X_0 = x$. Typically, one is interested whether $x \mapsto X_t^x$ is continuous or differentiable. A first weak continuity result follows already from the stability Theorem 19.11.

19.27 Corollary (Feller continuity). *Let $(B_t, \mathscr{F}_t)_{t\geqslant 0}$ be a BM^d. Assume that the coefficients $b : [0,\infty) \times \mathbb{R}^n \to \mathbb{R}^n$ and $\sigma : [0,\infty) \times \mathbb{R}^n \to \mathbb{R}^{n\times d}$ of the SDE (19.1) satisfy the Lipschitz condition (19.17) for all $x, y \in \mathbb{R}^n$ with the global Lipschitz constant L. Denote by $(X_t^x)_{t\geqslant 0}$ the solution of the SDE (19.1) with initial condition $X_0 \equiv x \in \mathbb{R}^n$. Then*

$$x \mapsto \mathbb{E}\left[f(X_t^x)\right] \quad \textit{is continuous for all } f \in \mathcal{C}_b(\mathbb{R}^n).$$

Proof. Assume that $f \in \mathcal{C}_c(\mathbb{R}^n)$ is a continuous function with compact support. Since f is uniformly continuous, we find for any $\epsilon > 0$ some $\delta > 0$ such that $|\xi - \eta| < \delta$ implies $|f(\xi) - f(\eta)| \leqslant \epsilon$. Now consider for $x, y \in \mathbb{R}^n$

$$\begin{aligned}
&\mathbb{E}\left[|f(X_t^x) - f(X_t^y)|\right] \\
&\quad= \mathbb{E}\left[|f(X_t^x) - f(X_t^y)|\,\mathbb{1}_{\{|X_t^x - X_t^y|\geqslant\delta\}}\right] + \mathbb{E}\left[|f(X_t^x) - f(X_t^y)|\,\mathbb{1}_{\{|X_t^x - X_t^y|<\delta\}}\right] \\
&\quad\leqslant 2\|f\|_\infty\,\mathbb{E}\left[\mathbb{1}_{\{|X_t^x - X_t^y|\geqslant\delta\}}\right] + \epsilon \\
&\quad\leqslant \frac{2\|f\|_\infty}{\delta^2}\,\mathbb{E}\left[|X_t^x - X_t^y|^2\right] + \epsilon \xrightarrow[|x-y|\to 0]{(19.19)} \epsilon \xrightarrow[\epsilon\to 0]{} 0.
\end{aligned}$$

This shows that $x \mapsto \mathbb{E}\left[f(X_t^x)\right]$ is continuous for all $f \in \mathcal{C}_c(\mathbb{R}^n)$. If $f \in \mathcal{C}_b(\mathbb{R}^n)$, we choose a sequence of cut-off functions $\chi_k \in \mathcal{C}_c(\mathbb{R}^n)$ such that $\mathbb{1}_{\mathbb{B}(0,k)} \leqslant \chi_k \leqslant 1$. Then $\chi_k f \in \mathcal{C}_c(\mathbb{R}^n)$ and we see for all $x, y \in \mathbb{R}^n$

$$\begin{aligned}
&\mathbb{E}\left[|f(X_t^x) - f(X_t^y)|\right] \\
&\quad\leqslant \mathbb{E}\left[|\chi_k f(X_t^x) - \chi_k f(X_t^y)|\right] + \|f\|_\infty(\mathbb{P}(|X_t^x| \geqslant k) + \mathbb{P}(|X_t^y| \geqslant k)) \\
&\quad\leqslant \mathbb{E}\left[|\chi_k f(X_t^x) - \chi_k f(X_t^y)|\right] + \frac{\|f\|_\infty}{k^2}\left(\mathbb{E}\left[|X_t^x|^2\right] + \mathbb{E}\left[|X_t^y|^2\right]\right) \\
&\quad\leqslant \mathbb{E}\left[|\chi_k f(X_t^x) - \chi_k f(X_t^y)|\right] + \kappa_t\,\frac{\|f\|_\infty}{k^2}\left((1 + |x|)^2 + (1 + |y|)^2\right) \\
&\quad\xrightarrow[y\to x]{} 2\kappa_t\,\frac{\|f\|_\infty}{k^2}(1 + |x|)^2 \xrightarrow[k\to\infty]{} 0.
\end{aligned}$$

□

If we want to show that $x \mapsto X_t^x$ is almost surely continuous, we need the Kolmogorov–Chentsov Theorem 10.1 and Burkholder's inequalities, Theorem 18.17.

19.28 Theorem. *Let $(B_t, \mathscr{F}_t)_{t\geqslant 0}$ be a d-dimensional Brownian motion. Assume that the coefficients $b : [0,\infty) \times \mathbb{R}^n \to \mathbb{R}^n$ and $\sigma : [0,\infty) \times \mathbb{R}^n \to \mathbb{R}^{n\times d}$ of the SDE (19.1) satisfy the*

Lipschitz condition (19.17) for all $x, y \in \mathbb{R}^n$ with the global Lipschitz constant L. Denote by $(X_t^x)_{t\geqslant 0}$ the solution of the SDE (19.1) with initial condition $X_0 \equiv x \in \mathbb{R}^n$. Then we have for all $p \geqslant 2$

$$\mathbb{E}\left[\sup_{s\leqslant T}(1+|X_s^x|)^p\right] \leqslant c_{p,T}\,(1+|x|^p) \quad \textit{for all } T>0, \tag{19.35}$$

$$\mathbb{E}\left[\sup_{s\leqslant T}|X_s^x - X_s^y|^p\right] \leqslant c'_{p,T}\,|x-y|^p \quad \textit{for all } T>0. \tag{19.36}$$

In particular, $x \mapsto X_t^x$ (has a modification which) is continuous.

Proof. The proof of (19.35) is very similar to the proof of (19.36). Note that the global Lipschitz condition (19.17) implies the linear growth condition (19.18) which will be needed for (19.35). Therefore, we will only prove (19.36). Note that

$$X_t^x - X_t^y = (x-y) + \int_0^t (b(s,X_s^x) - b(s,X_s^y))\,ds + \int_0^t (\sigma(s,X_s^x) - \sigma(s,X_s^y))\,dB_s.$$

From Hölder's inequality we see that $(a+b+c)^p \leqslant 3^{p-1}(a^p+b^p+c^p)$. Thus,

$$\begin{aligned}
\mathbb{E}\left[\sup_{t\leqslant T}|X_t^x - Y_t^y|^p\right] &\leqslant 3^{p-1}|x-y|^p \\
&\quad + 3^{p-1}\,\mathbb{E}\left[\sup_{t\leqslant T}\left|\int_0^t (b(s,X_s^x)-b(s,X_s^y))\,ds\right|^p\right] \\
&\quad + 3^{p-1}\,\mathbb{E}\left[\sup_{t\leqslant T}\left|\int_0^t (\sigma(s,X_s^x)-\sigma(s,X_s^y))\,dB_s\right|^p\right].
\end{aligned}$$

Using the Hölder inequality and (19.17), we find for the middle term

$$\begin{aligned}
\mathbb{E}\left[\sup_{t\leqslant T}\left|\int_0^t (b(s,X_s^x)-b(s,X_s^y))\,ds\right|^p\right] &\leqslant \mathbb{E}\left[\left(\int_0^T |b(s,X_s^x)-b(s,X_s^y)|\,ds\right)^p\right] \\
&\leqslant T^{p-1}\,\mathbb{E}\left[\int_0^T |b(s,X_s^x)-b(s,X_s^y)|^p\,ds\right] \\
&\overset{(19.17)}{\leqslant} L^pT^{p-1}\int_0^T \mathbb{E}\left[|X_s^x - X_s^y|^p\right]ds \\
&\leqslant L^pT^{p-1}\int_0^T \mathbb{E}\left[\sup_{r\leqslant s}|X_r^x - X_r^y|^p\right]ds.
\end{aligned}$$

For the third term we combine Burkholder's inequality (applied to the coordinate processes), the estimate (19.17) and Hölder's inequality to get

$$\begin{aligned}\mathbb{E}\left[\sup_{t\leqslant T}\left|\int_0^t (\sigma(s,X_s^x)-\sigma(s,X_s^y))\,dB_s\right|^p\right] &\overset{(18.17)}{\leqslant} C_p\,\mathbb{E}\left[\left(\int_0^T |\sigma(s,X_s^x)-\sigma(s,X_s^y)|^2\,ds\right)^{p/2}\right]\\ &\overset{(19.17)}{\leqslant} C_pL^p\,\mathbb{E}\left[\left(\int_0^T |X_s^x-X_s^y|^2\,ds\right)^{p/2}\right]\\ &\leqslant C_pL^p\,\mathbb{E}\left[\left(\int_0^T \sup_{r\leqslant s}|X_r^x-X_r^y|^2\,ds\right)^{p/2}\right]\\ &\leqslant C_pL^p\,T^{p/2-1}\int_0^T \mathbb{E}\left[\sup_{r\leqslant s}|X_r^x-X_r^y|^p\right]ds.\end{aligned}$$

This shows that

$$\mathbb{E}\left[\sup_{t\leqslant T}|X_t^x-Y_t^y|^p\right]\leqslant c|x-y|^p+c\int_0^T \mathbb{E}\left[\sup_{r\leqslant s}|X_r^x-Y_r^y|^p\right]ds$$

with a constant $c = c(T,p,L)$. Gronwall's inequality, Theorem A.47, yields when applied to $u(T)=\mathbb{E}[\sup_{t\leqslant T}|X_t^x-X_t^y|^p]$, $a(s)=c\,|x-y|^p$ and $b(s)=c$,

$$\mathbb{E}\left[\sup_{t\leqslant T}|X_t^x-X_t^y|^p\right]\leqslant c\,e^{cT}\,|x-y|^p.$$

The assertion follows from the Kolmogorov–Chentsov theorem, Theorem 10.1, where we use $\alpha=\beta+n=p$. □

19.29 Remark. a) A weaker form of the boundedness estimate (19.35) from Theorem 19.28 holds for *all* $p\in\mathbb{R}$:

$$\mathbb{E}\left[(1+|X_t^x|)^p\right]\leqslant c_{p,T}\,(1+|x|^2)^{p/2}\quad\text{for all } x\in\mathbb{R}^n,\ t\leqslant T,\ p\in\mathbb{R}. \tag{19.37}$$

Let us sketch the argument for $d=n=1$.

Outline of the proof. Since $(1+|x|)^p\leqslant(1+2^{p/2})(1+|x|^2)^{p/2}$ for *all* $p\in\mathbb{R}$, it is enough to prove (19.37) for $\mathbb{E}\left[(1+|X_t^x|^2)^{p/2}\right]$. Set $f(x)=(1+x^2)^{p/2}$. Then $f\in\mathcal{C}^2(\mathbb{R})$ and we find that

$$f'(x)=px(1+x^2)^{p/2-1}\quad\text{and}\quad f''(x)=p(1-x^2+px^2)(1+x^2)^{p/2-2},$$

i. e. $|(1+x^2)^{1/2}f'(x)|\leqslant C_pf(x)$ and $|(1+x^2)f''(x)|\leqslant C_pf(x)$ for some absolute constant C_p. Itô's formula (17.1) yields

$$f(X_t^x)-f(x)=\int_0^t f'(X_r^x)b(r,X_r^x)\,dr+\int_0^t f'(X_r^x)\sigma(r,X_r^x)\,dB_r+\frac{1}{2}\int_0^t f''(X_r^x)\sigma^2(r,X_r^x)\,dr.$$

The second term on the right is a local martingale, cf. Theorem 16.7 and Remark 17.3. Using a localizing sequence $(\sigma_k)_{k\geqslant 0}$ and the fact that $b(r,x)$ and $\sigma(r,x)$ satisfy the linear growth condition (19.18), we get for all $k \geqslant 0$ and $t \leqslant T$

$$\mathbb{E}\, f(X^x_{t\wedge\sigma_k}) \leqslant f(x) + \mathbb{E}\left[\int_0^{t\wedge\sigma_k} \left(M(1+|X^x_r|)|f'(X^x_r)| + M^2(1+|X^x_r|)^2|f''(X^x_r)|\right) dr\right]$$

$$\leqslant f(x) + C(p,M)\int_0^t \mathbb{E}\, f(X^x_r)\, dr.$$

By Fatou's lemma $\mathbb{E}\, f(X^x_t) \leqslant \underline{\lim}_{k\to\infty} \mathbb{E}\, f(X^x_{t\wedge\sigma_k})$, and we can apply Gronwall's lemma, Theorem A.47. This shows for all x and $t \leqslant T$

$$\mathbb{E}\, f(X^x_t) \leqslant e^{C(p,M)T} f(x). \qquad \square$$

b) For the differentiability of the solution with respect to the initial conditions, we can still use the idea of the proof of Theorem 19.26. We consider the (directional) difference quotients

$$\frac{X^x_t - X^{x+he_j}_t}{h}, \quad h > 0, \quad e_j = (\overbrace{0,\dots,0,1}^{j},0,\dots) \in \mathbb{R}^n,$$

and show that, as $h \to 0$, the partial derivatives $\partial_j X^x_t$, $j = 1,\dots,n$, exist and satisfy the formally differentiated SDE

$$d\partial_j X^x_t = \overbrace{D_z b(t,z)}^{\in\mathbb{R}^{n\times n}}\Big|_{z=X^x_t} \underbrace{\partial_j X^x_t}_{\in\mathbb{R}^n}\, dt + \sum_{\ell=1}^d \overbrace{D_z \underbrace{\sigma_{\bullet,\ell}(t,z)}_{\ell\text{th column}, \in\mathbb{R}^n}}^{\in\mathbb{R}^{n\times n}\ \forall\ell}\Big|_{z=X^x_t} \overbrace{dB^\ell_t}^{\ell\text{th coordinate}, \in\mathbb{R}} \underbrace{\partial_j X^x_t}_{\in\mathbb{R}^n}.$$

As one would expect, we need now (global) Lipschitz and growth conditions also for the *derivatives* of the coefficients. We will not go into further details but refer to Kunita [134, Chapter 4.6] and [135, Chapter 3].

19.30 Corollary. *Let $(B_t,\mathscr{F}_t)_{t\geqslant 0}$ be a d-dimensional Brownian motion. Assume that the coefficients $b : [0,\infty)\times\mathbb{R}^n \to \mathbb{R}^n$ and $\sigma : [0,\infty)\times\mathbb{R}^n \to \mathbb{R}^{n\times d}$ of the SDE (19.1) satisfy the Lipschitz condition (19.17) for all $x, y \in \mathbb{R}^n$ with the global Lipschitz constant L. Denote by $(X^x_t)_{t\geqslant 0}$ the solution of the SDE (19.1) with initial condition $X_0 \equiv x \in \mathbb{R}^n$. Then we have for all $p \geqslant 2$, $s,t \in [0,T]$ and $x, y \in \mathbb{R}^n$*

$$\mathbb{E}\left[|X^x_s - X^y_t|^p\right] \leqslant C_{p,T}\left(|x-y|^p + (1+|x|+|y|)^p\, |s-t|^{p/2}\right). \tag{19.38}$$

In particular, $(t,x) \mapsto X^x_t$ (has a modification which) is continuous.

Proof. Assume that $s \leqslant t$. Then the inequality $(a+b)^p \leqslant 2^{p-1}(a^p+b^p)$ gives

$$\mathbb{E}\left[|X^x_s - X^y_t|^p\right] \leqslant 2^{p-1}\,\mathbb{E}\left[|X^x_s - X^y_s|^p\right] + 2^{p-1}\,\mathbb{E}\left[|X^y_s - X^y_t|^p\right].$$

The first term can be estimated by (19.36) with $c'_{p,T}|x-y|^p$. For the second term we note that

$$|X_s^y - X_t^y|^p = \left|\int_s^t b(r, X_r^y)\,dr + \int_s^t \sigma(r, X_r^y)\,dB_r\right|^p.$$

Now we argue as in Theorem 19.28: Because of $(a+b)^p \leqslant 2^{p-1}(a^p + b^p)$, it is enough to estimate the integrals separately. The linear growth condition (19.18) (which follows from the global Lipschitz condition) and similar arguments as in the proof of Theorem 19.28 yield

$$\begin{aligned}
\mathbb{E}\left[\left|\int_s^t b(r, X_r^y)\,dr\right|^p\right] &\leqslant |t-s|^{p-1}\,\mathbb{E}\left[\int_s^t |b(r, X_r^y)|^p\,dr\right]\\
&\overset{(19.18)}{\leqslant} |t-s|^{p-1} M^p\,\mathbb{E}\left[\int_s^t (1 + |X_r^y|)^p\,dr\right]\\
&\leqslant |t-s|^p M^p\,\mathbb{E}\left[\sup_{r\leqslant T}(1 + |X_r^y|)^p\right]\\
&\overset{(19.35)}{\leqslant} |t-s|^p M^p c_{p,T}(1 + |y|)^p.
\end{aligned} \tag{19.39}$$

For the second integral we find with Burkholder's inequality (applied to the coordinate processes)

$$\begin{aligned}
\mathbb{E}\left[\left|\int_s^t \sigma(r, X_r^y)\,dB_r\right|^p\right] &\overset{(18.17)}{\leqslant} C_p\,\mathbb{E}\left[\left(\int_s^t |\sigma(r, X_r^y)|^2\,dr\right)^{p/2}\right]\\
&\leqslant |t-s|^{p/2} C_p M^p c_{p,T}(1+|y|)^p
\end{aligned}$$

where the last estimate follows as (19.39) with $|b| \rightsquigarrow |\sigma|^2$ and $p \rightsquigarrow p/2$. Thus,

$$\mathbb{E}\left[|X_s^x - X_t^y|^p\right] \leqslant c''_{p,T}\left(|x-y|^p + |t-s|^p(1+|y|)^p + |t-s|^{p/2}(1+|y|)^p\right).$$

Since $t - s \leqslant T$, we have $|t-s|^p \leqslant T^{p/2}|t-s|^{p/2}$.

If $t \leqslant s$, the starting points x and y change their roles in the above calculation, and we see that (19.38) is just a symmetric (in x and y) formulation of the last estimate.

The existence of a jointly continuous modification $(t, x) \mapsto X_t^x$ follows from the Kolmogorov–Chentsov theorem, Theorem 10.1, for the index set $\mathbb{R}^n \times [0,\infty) \subset \mathbb{R}^{n+1}$ and with $p = \alpha = 2(\beta + n + 1)$. □

19.31 Corollary. *Let $(B_t, \mathscr{F}_t)_{t\geqslant 0}$ be a d-dimensional Brownian motion and assume that the coefficients $b : [0,\infty) \times \mathbb{R}^n \to \mathbb{R}^n$ and $\sigma : [0,\infty) \times \mathbb{R}^n \to \mathbb{R}^{n\times d}$ of the SDE (19.1) satisfy the Lipschitz condition (19.17) for all $x, y \in \mathbb{R}^n$ with the global Lipschitz constant L. Denote by $(X_t^x)_{t\geqslant 0}$ the solution of the SDE (19.1) with initial condition $X_0 \equiv x \in \mathbb{R}^n$. Then* Ex. 19.12

$$\lim_{|x|\to\infty} |X_t^x(\omega)| = \infty \quad \textit{almost surely for every } t \geqslant 0. \tag{19.40}$$

Proof. For $x \in \mathbb{R}^n \setminus \{0\}$ we set $\hat{x} := x/|x|^2 \in \mathbb{R}^n$, and define for every $t \geqslant 0$

$$Y_t^{\hat{x}} = \begin{cases} (1 + |X_t^x|)^{-1} & \text{if } \hat{x} \neq 0, \\ 0 & \text{if } \hat{x} = 0. \end{cases}$$

Observe that $|\hat{x}| \to 0$ if, and only if, $|x| \to \infty$. If $\hat{x}, \hat{y} \neq 0$ we find for all $p \geqslant 2$

$$|Y_t^{\hat{x}} - Y_t^{\hat{y}}|^p = |Y_t^{\hat{x}}|^p \cdot |Y_t^{\hat{y}}|^p \cdot \big||X_t^x| - |X_t^y|\big|^p \leqslant |Y_t^{\hat{x}}|^p \cdot |Y_t^{\hat{y}}|^p \cdot |X_t^x - X_t^y|^p,$$

and applying Hölder's inequality two times we get

$$\mathbb{E}\left[|Y_t^{\hat{x}} - Y_t^{\hat{y}}|^p\right] \leqslant \left(\mathbb{E}\left[|Y_t^{\hat{x}}|^{4p}\right]\right)^{1/4} \left(\mathbb{E}\left[|Y_t^{\hat{y}}|^{4p}\right]\right)^{1/4} \left(\mathbb{E}\left[|X_t^x - X_t^y|^{2p}\right]\right)^{1/2}.$$

From Theorem 19.28 and Remark 19.29 a) we infer

$$\mathbb{E}\left[|Y_t^{\hat{x}} - Y_t^{\hat{y}}|^p\right] \leqslant c_{p,T}(1 + |x|)^{-p}(1 + |y|)^{-p}|x - y|^p \leqslant c_{p,T}|\hat{x} - \hat{y}|^p.$$

Ex. 19.13 In the last estimate we used the inequality

$$\frac{|x - y|}{(1 + |x|)(1 + |y|)} \leqslant \left|\frac{x}{|x|^2} - \frac{y}{|y|^2}\right|, \quad x, y \in \mathbb{R}^n.$$

The case where $\hat{x} = 0$ is similar but easier as we use only (19.37) from Remark 19.29 a).

The Kolmogorov–Chentsov theorem, Theorem 10.1, shows that $\hat{x} \mapsto Y_t^{\hat{x}}$ is continuous at $\hat{x} = 0$, and the claim follows. □

19.32 Further reading. The classic book by [87] is still a very good treatise on the L^2 theory of strong solutions. Weak solutions and extensions of existence and uniqueness results are studied in [119] or [100]. For degenerate SDEs you should consult [71]. The dependence on the initial value and more advanced flow properties can be found in [134] for processes with continuous paths and, for jump processes, in [135]. The viewpoint from random dynamical systems is explained in [3].

[3] Arnold: *Random Dynamical Systems.*
[71] Engelbert, Cherny: *Singular Stochastic Differential Equations.*
[87] Gikhman, Skorokhod: *Stochastic Differential Equations.*
[100] Ikeda, Watanabe: *Stochastic Differential Equations and Diffusion Processes.*
[119] Karatzas, Shreve: *Brownian Motion and Stochastic Calculus.*
[134] Kunita: *Stochastic Flows and Stochastic Differential Equations.*
[135] Kunita: Stochastic differential equations based on Lévy processes and stochastic flows of diffeomorphisms.

Problems

1. Let $X = (X_t)_{t\geqslant 0}$ be the process from Example 19.2. Show that X is a Gaussian process with independent increments and find $C(s,t) = \mathbb{E}\, X_s X_t$, $s, t \geqslant 0$.

   Hint: Let $0 = t_0 < t_1 < \cdots < t_n = t$. Find the characteristic function of the random variables $(X_{t_1} - X_{t_0}, \ldots, X_{t_n} - X_{t_{n-1}})$ and $(X_{t_1}, \ldots, X_{t_n})$.

2. Let $(B(t))_{t\in[0,1]}$ be a BM^1 and set $t_k = k/2^n$ for $k = 0, 1, \dots, 2^n$ and $\Delta t = 2^{-n}$. Consider the difference equation

$$\Delta X_n(t_k) = -\frac{1}{2} X_n(t_k)\Delta t + \Delta B(t_k), \qquad X_n(0) = A,$$

where $\Delta X_n(t_k) = X_n(t_k+\Delta t) - X_n(t_k)$ and $\Delta B(t_k) = B(t_k+\Delta t) - B(t_k)$ for $0 \leqslant k \leqslant 2^n - 1$ and $A \sim \mathsf{N}(0,1)$ is independent of $(B_t)_{t\in[0,1]}$.

(a) Find an explicit formula for $X_n(t_k)$, determine the distribution of $X_n(1/2)$ and the asymptotic distribution as $n \to \infty$.

(b) Letting $n \to \infty$ turns the difference equation into the SDE

$$dX(t) = -\frac{1}{2} X(t)\,dt + dB(t), \qquad X(0) = A.$$

Solve this SDE and determine the distribution of $X(1/2)$ as well as the correlation function $C(s,t) = \mathbb{E}\, X_s X_t$, $s, t \in [0,1]$.

3. Let $B_t = (b_t, \beta_t)$ be a BM^2. Solve the SDE

$$X_t = x + b\int_0^t X_s\,ds + \sigma_1 \int_0^t X_s\,db_s + \sigma_2 \int_0^t X_s\,d\beta_s$$

with $x, b \in \mathbb{R}$ and $\sigma_1, \sigma_2 > 0$.

Hint: Rewrite the SDE using the result of Problem 18.3 and then apply Example 19.4.

4. Show that in Example 19.7

$$X_t^\circ = \exp\left(-\int_0^t (\beta(s) - \delta^2(s)/2)\,ds\right)\exp\left(-\int_0^t \delta(s)\,dB_s\right),$$

and verify the expression given for dX_t°. Moreover, show that

$$X_t = \frac{1}{X_t^\circ}\left(X_0 + \int_0^t (\alpha(s) - \gamma(s)\delta(s))X_s^\circ\,ds + \int_0^t \gamma(s)X_s^\circ\,dB_s\right).$$

5. Let $(B_t)_{t\geqslant 0}$ be a BM^1. The solution of the SDE

$$dX_t = -\beta X_t\,dt + \sigma dB_t, \qquad X_0 = \xi$$

with $\beta > 0$ and $\sigma \in \mathbb{R}$ is an Ornstein–Uhlenbeck process.

(a) Find X_t explicitly.

(b) Determine the distribution of X_t. Is $(X_t)_{t\geqslant 0}$ a Gaussian process? Find the correlation function $C(s,t) = \mathbb{E}\, X_s X_t$, $s, t \geqslant 0$.

(c) Determine the asymptotic distribution μ of X_t as $t \to \infty$.

(d) Assume that the initial condition ξ is a random variable which is independent of $(B_t)_{t\geqslant 0}$ and assume that $\xi \sim \mu$ with μ from above. Find the characteristic function of $(X_{t_1},\dots,X_{t_n})$ for all $0 \leqslant t_1 < t_2 < \dots < t_n < \infty$.

(e) Show that the process $(U_t)_{t\geqslant 0}$, where $U_t := \frac{\sigma}{\sqrt{2\beta}} e^{-\beta t} B(e^{2\beta t} - 1)$, has the same finite dimensional distributions as $(X_t)_{t\geqslant 0}$ with initial condition $X_0 = 0$.

6. Verify the claim made in Example 19.9 using Itô's formula.
Derive from the proof of Lemma 19.8 explicitly the form of the transformation and the coefficients in Example 19.9.
Hint: Integrate the condition (19.13) and retrace the steps of the proof of Lemma 19.8 from the end to the beginning.

7. Let $(B_t)_{t\geqslant 0}$ be a BM^1. Find an SDE which has $X_t = tB_t$, $t \geqslant 0$, as unique solution.

8. Let $(B_t)_{t\geqslant 0}$ be a BM^1. Find for each of the following processes an SDE which is solved by it:

$$\text{(a)}\quad U_t = \frac{B_t}{1+t}; \qquad \text{(b)}\quad V_t = \sin B_t; \qquad \text{(c)}\quad \begin{pmatrix} X_t \\ Y_t \end{pmatrix} = \begin{pmatrix} a\cos B_t \\ b \sin B_t \end{pmatrix}, \qquad a, b \neq 0.$$

9. Let $(B_t)_{t\geqslant 0}$ be a BM^1. Solve the following SDEs
(a) $dX_t = b\,dt + \sigma X_t\,dB_t$, $X_0 = x_0$ and $b, \sigma \in \mathbb{R}$;

(b) $dX_t = (m - X_t)\,dt + \sigma\,dB_t$, $X_0 = x_0$ and $m, \sigma > 0$.

10. Let $(B_t)_{t\geqslant 0}$ be a BM^1. Use Lemma 19.10 to find the solution of the following SDE:

$$dX_t = \left(\sqrt{1+X_t^2} + \tfrac{1}{2}X_t\right)dt + \sqrt{1+X_t^2}\,dB_t.$$

11. Show that the constant M in (19.18) can be chosen in the following way:

$$M^2 \geqslant 2L^2 + 2\sum_{j=1}^{n} \sup_{t\leqslant T} |b_j(t,0)|^2 + 2\sum_{j=1}^{n}\sum_{k=1}^{d} \sup_{t\leqslant T} |\sigma_{jk}(t,0)|^2$$

12. The linear growth of the coefficients is essential for Corollary 19.31.
(a) Consider the case where $d = n = 1$, $b(x) = -e^x$ and $\sigma(x) = 0$. Find the solution of this deterministic ODE and compare your findings for $x \to +\infty$ with Corollary 19.31

(b) Assume that $|b(x)|+|\sigma(x)| \leqslant M$ for all x. Find a simpler proof of Corollary 19.31.
Hint: Find an estimate for $\mathbb{P}\left(\left|\int_0^t \sigma(X_s^x)\,dB_s\right| > r\right)$.

(c) Assume that $|b(x)| + |\sigma(x)|$ grow, as $|x| \to \infty$, like $|x|^{1-\epsilon}$ for some $\epsilon > 0$. Show that one can still find a (relatively) simple proof of Corollary 19.31.

13. Show for all $x, y \in \mathbb{R}^n$ the inequality

$$|x-y|(1+|x|)^{-1}(1+|y|)^{-1} \leqslant \left||x|^{-2}x - |y|^{-2}y\right|.$$

Hint: Use a direct calculation after squaring both sides.

14. Let $(B_t)_{t\geqslant 0}$ be a BM^1 and $b(x), \sigma(x)$ autonomous and globally Lipschitz continuous coefficients. We have seen in Corollary 19.24 that the solution of the stochastic differential equation $dX_t = b(X_t)\,dt + \sigma(X_t)\,dB_t$ is a Markov process. Recall that $\mathcal{C}_b(\mathbb{R})$ and $\mathcal{C}_\infty(\mathbb{R})$ are the bounded continuous functions and the continuous functions vanishing at infinity.
    (a) Find the transition semigroup T_t of this Markov process and show that T_t maps $\mathcal{C}_b(\mathbb{R})$ into itself.
    (b) Use Corollary 19.31 to show that T_t maps $\mathcal{C}_\infty(\mathbb{R})$ into itself.
    (c) Find the infinitesimal generator A of T_t on $\mathcal{C}_c^2(\mathbb{R})$. Show that A has an extension onto $\mathcal{C}_b^2(\mathbb{R})$.
    Hint: Use Itô's formula.
    (d) Prove the following generalization of Dynkin's formula: if b, σ are bounded and if τ is a stopping time with $\mathbb{E}^x \tau < \infty$ and $u \in \mathcal{C}_b^{1,2}$, $u = u(t,x)$, then

$$\mathbb{E}^x\left[u(\tau, X_\tau)\right] - u(0,x) = \mathbb{E}^x\left[\int_0^\tau \left(\partial_s u(s, X_s) + A_x u(s, X_s)\right) ds\right].$$

20 Stratonovich's stochastic calculus

Itô's integral does not obey the usual rules from classical deterministic differential calculus. This is sometimes disconcerting and a small variation in the definition of the stochastic integral can change this. We will, however, loose the property that the stochastic integral is a martingale.

Ex. 15.15 A good starting point is Problem 15.15 (see also Example 15.15) where you have shown that

$$\sum_{t_{l-1},t_l\in\Pi} B_{\theta_l}(B_{t_l} - B_{t_{l-1}}) \xrightarrow[|\Pi|\to 0]{\mathbb{P}} \frac{1}{2}(B_T^2 + (2\alpha - 1)T)$$

with a partition $\Pi = \{0 = t_0 < t_1 < \ldots < t_N = T\}$ and $\theta_l = t_{l-1} + \alpha(t_l - t_{l-1})$ for some $\alpha \in [0,1]$. This means that the character of the integral depends on the supporting point $\theta_l \in [t_{l-1}, t_l]$ where we evaluate the integrand. For $\alpha = 0$ we get the left-end point and Itô's integral, for $\alpha = \frac{1}{2}$ we get the mid-point of the interval and an integral that will reproduce the usual calculus rules, but we loose the martingale property. This observation is due to Stratonovich, and we will now discuss the so-called Stratonovich integral and its relation to Itô's integral.

20.1 The Stratonovich integral

We begin with the following stronger version of Lemma 17.6.

20.1 Lemma. *Let $(B_t)_{t\geqslant 0}$ be a one-dimensional Brownian motion, $b, c, \sigma, \tau \in \mathcal{L}_T^2$ and $X_t := X_0 + \int_0^t \sigma(s)\,dB_s + \int_0^t b(s)\,ds$, $Y_t := Y_0 + \int_0^t \tau(s)\,dB_s + \int_0^t c(s)\,ds$ be the corresponding Itô processes. Denote by $\Pi = \{0 = t_0 < t_1 < \ldots < t_N = T\}$ a generic partition of the interval $[0,T]$. Then*

$$\sum_{l=1}^{N} g(X_{t_{l-1}})(B_{t_l} - B_{t_{l-1}}) \xrightarrow[|\Pi|\to 0]{\mathbb{P}} \int_0^T g(X_s)\,dB_s \tag{20.1}$$

holds for all $g \in \mathcal{C}_b(\mathbb{R})$. Moreover, for all $g \in \mathcal{C}_b(\mathbb{R})$

$$\sum_{l=1}^{N} g(\xi_l)(Y_{t_l} - Y_{t_{l-1}})^2 \xrightarrow[|\Pi|\to 0]{\mathbb{P}} \int_0^T g(X_s)\tau^2(s)\,ds \tag{20.2}$$

$$\sum_{l=1}^{N} g(\xi_l)(Y_{t_l} - Y_{t_{l-1}})(B_{t_l} - B_{t_{l-1}}) \xrightarrow[|\Pi|\to 0]{\mathbb{P}} \int_0^T g(X_s)\tau(s)\,ds \tag{20.3}$$

where $\xi_l(\omega) = X_{t_{l-1}}(\omega) + \theta_l(\omega)(X_{t_l}(\omega) - X_{t_{l-1}}(\omega))$, $0 \leqslant \theta_l \leqslant 1$, $l = 1, \ldots, N$ are intermediate values such that $g(\xi_l)$ is measurable.

Proof. Note that $Y \pm B$ are again Itô processes with the coefficients $\tau \pm \mathbb{1}_{[0,T]}$ and c. Thus, (20.3) follows from (20.2) by polarization:

$$4(Y_t - Y_s)(B_t - B_s) = ((Y+B)_t - (Y+B)_s)^2 - ((Y-B)_t - (Y-B)_s)^2$$
$$\text{and} \quad 4\tau(s) = (\tau(s) + \mathbb{1}_{[0,T]}(s))^2 - (\tau(s) - \mathbb{1}_{[0,T]}(s))^2, \quad s \in [0,T].$$

For simple coefficients $\tau, c \in \mathcal{S}_T$, (20.2) is exactly (17.11) of Lemma 17.6. For general $\tau, c \in \mathcal{L}_T^2$ we pick some approximating sequences $(\tau_\Delta)_\Delta, (c_\Delta)_\Delta \subset \mathcal{S}_T$ indexed by the partitions Δ of the simple processes. We set $Y_t^\Delta := Y_0 + \int_0^t \tau_\Delta(s)\,dB_s + \int_0^t c_\Delta(s)\,ds$. For any fixed Δ we may assume that $\Delta \subset \Pi$. Then

$$\begin{aligned}
&\sum_{l=1}^N g(\xi_l)(Y_{t_l} - Y_{t_{l-1}})^2 - \int_0^T g(X_s)\,\tau^2(s)\,ds \\
&\quad= \sum_{l=1}^N g(\xi_l)(Y_{t_l} - Y_{t_{l-1}} - Y_{t_l}^\Delta + Y_{t_{l-1}}^\Delta)^2 \\
&\qquad+ \sum_{l=1}^N g(\xi_l)(Y_{t_l}^\Delta - Y_{t_{l-1}}^\Delta)^2 - \int_0^T g(X_s)\tau_\Delta^2(s)\,ds \\
&\qquad+ 2\sum_{l=1}^N g(\xi_l)(Y_{t_l} - Y_{t_{l-1}} - Y_{t_l}^\Delta + Y_{t_{l-1}}^\Delta)(Y_{t_l}^\Delta - Y_{t_{l-1}}^\Delta) \\
&\qquad+ \int_0^T g(X_s)\tau_\Delta^2(s)\,ds - \int_0^T g(X_s)\tau^2(s)\,ds \\
&\quad=: \mathsf{J}_1(g) + \Big[\mathsf{J}_2(g) - \int_0^T g(X_s)\tau_\Delta^2(s)\,ds\Big] + \mathsf{J}_3(g) + \mathsf{J}_4(g).
\end{aligned} \tag{20.4}$$

Note that $|\mathsf{J}_k(g)| \leqslant \|g\|_\infty \mathsf{J}_k(1)$ for $k = 1, 2$. By definition, and with the elementary inequality $(a+b)^2 \leqslant 2(a^2+b^2)$,

$$\begin{aligned}
\mathsf{J}_1(1) &= \sum_{l=1}^N \Bigg(\int_{t_{l-1}}^{t_l} (\tau(s) - \tau_\Delta(s))\,dB_s + \int_{t_{l-1}}^{t_l} (c(s) - c_\Delta(s))\,ds \Bigg)^2 \\
&\leqslant 2\sum_{l=1}^N \Bigg[\Bigg(\int_{t_{l-1}}^{t_l} (\tau(s) - \tau_\Delta(s))\,dB_s \Bigg)^2 + \Bigg(\int_{t_{l-1}}^{t_l} (c(s) - c_\Delta(s))\,ds \Bigg)^2 \Bigg].
\end{aligned}$$

Taking expectations, we get using Itô's isometry, the Cauchy-Schwarz inequality and with $|\Pi| := \max_{1 \leqslant l \leqslant N}(t_l - t_{l-1})$

$$\mathbb{E}\,|\mathsf{J}_1(1)| \leqslant 2\,\mathbb{E}\Bigg(\int_0^T (\tau(s) - \tau_\Delta(s))^2\,ds \Bigg) + 2|\Pi|\,\mathbb{E}\Bigg(\int_0^T (c(s) - c_\Delta(s))^2\,ds \Bigg).$$

Since τ_Δ and c_Δ approximate τ and c, respectively, we see that

$$\overline{\lim_{|\Pi|\to 0}} \mathbb{E}\,|J_1(g)| = \epsilon(\Delta) \quad \text{where} \quad \lim_{|\Delta|\to 0} \epsilon(\Delta) = 0.$$

In particular, $\lim_{|\Delta|\to 0} \overline{\lim}_{|\Pi|\to 0} J_1(g) = 0$ in probability.

A similar, but simpler, calculation shows that

$$\mathbb{E}\,|J_2(1)| \leqslant 2\,\mathbb{E}\left(\int_0^T \tau_\Delta^2(s)\,ds\right) + 2|\Pi|\,\mathbb{E}\left(\int_0^T c_\Delta^2(s)\,ds\right) \leqslant M < \infty.$$

Since $|\Pi| \leqslant T$ and $(\tau_\Delta)_\Delta, (c_\Delta)_\Delta$ approximate τ and c, respectively, the bound M can be chosen independently of Π and Δ. Moreover, for fixed Δ, we see from (17.11) that $\lim_{|\Pi|\to 0} J_2(g) = \int_0^T g(X_s)\tau_\Delta^2(s)\,ds$ in probability.

By the Cauchy-Schwarz inequality we get uniformly in Π

$$\mathbb{E}\,|J_3(g)| \leqslant 2\,\|g\|_\infty \sqrt{\mathbb{E}\,|J_1(1)|}\sqrt{\mathbb{E}\,|J_2(1)|} \leqslant 2\,\|g\|_\infty \sqrt{\epsilon(\Delta)}\sqrt{M} \xrightarrow[|\Delta|\to 0]{} 0.$$

Finally, $\lim_{|\Delta|\to 0} J_4(g) = 0$ in L^1 and in probability by the construction of the sequence $(\tau_\Delta)_\Delta$.

If we combine the results for $J_1(g), \dots, J_4(g)$ and let in (20.4) first $|\Pi| \to 0$ and then $|\Delta| \to 0$, we get (20.2). □

20.2 Theorem. *Let $X_t := X_0 + \int_0^t \sigma(s)\,dB_s + \int_0^t b(s)\,ds$ be a one-dimensional Itô process and $b, \sigma \in \mathcal{L}_T^2$. Denote by $\Pi = \{0 = t_0 < t_1 < \dots < t_N = T\}$ a generic partition of the interval $[0, T]$. Then the limit*

$$\sum_{l=1}^N g\big(\tfrac{1}{2}X_{t_{l-1}} + \tfrac{1}{2}X_{t_l}\big)(B_{t_l} - B_{t_{l-1}}) \xrightarrow[|\Pi|\to 0]{\mathbb{P}} \int_0^T g(X_s)\,dB_s + \frac{1}{2}\int_0^T g'(X_s)\sigma(s)\,ds \tag{20.5}$$

exists for all $g \in \mathcal{C}_b^1(\mathbb{R})$.

Proof. For $s < t$ and some intermediate value ξ between X_s and $\frac{1}{2}(X_s + X_t)$, we find by Taylor's formula

$$g\big(\tfrac{1}{2}X_s + \tfrac{1}{2}X_t\big)(B_t - B_s) = g(X_s)(B_t - B_s) + \tfrac{1}{2}g'(\xi)(X_t - X_s)(B_t - B_s).$$

The integral form of the remainder term in Taylor's formula reveals that $\omega \mapsto g'(\xi(\omega))$ is measurable, compare step $\mathbf{1^o}$ in the proof of Theorem 17.1. Thus, the assertion follows from Lemma 20.1. □

20.3 Definition (Stratonovich 1964, 1966). Let $X_t := X_0 + \int_0^t \sigma(s)\,dB_s + \int_0^t b(s)\,ds$ be a one-dimensional Itô process with $b, \sigma \in \mathcal{L}_T^2$. Let $\Pi = \{0 = t_0 < t_1 < \dots < t_N = T\}$ be a generic partition of the interval $[0, T]$ and $g : [0, T] \times \mathbb{R} \to \mathbb{R}$ a continuous function. If the limit

$$\mathbb{P}\text{-}\lim_{|\Pi|\to 0} \sum_{l=1}^N g\big(t_{l-1}, \tfrac{1}{2}X_{t_{l-1}} + \tfrac{1}{2}X_{t_l}\big)(B_{t_l\wedge t} - B_{t_{l-1}\wedge t}), \quad t \in [0, T], \tag{20.6}$$

exists in probability, it is called the *Stratonovich* or *symmetrized stochastic integral*.

Notation: $\int_0^t g(s, X_s) \circ dB_s$; '$\circ$' is often called *Stratonovich's circle*.

With the usual stopping techniques, see Chapter 16, it is not hard to see that the results of this section still hold for $b, \sigma \in \mathcal{L}^2_{T,\mathrm{loc}}$.

20.4 Relation of Itô and Stratonovich integrals. Let $X_t = X_0 + \int_0^t \sigma(s)\, dB_s + \int_0^t b(s)\, ds$ be an Itô process with $\sigma, b \in \mathcal{L}^2_T$.

a) Theorem 20.2 shows that for every $g \in \mathcal{C}^1_b(\mathbb{R})$ which does not depend on t, the Stratonovich integral exists and

$$\int_0^t g(X_s) \circ dB_s = \int_0^t g(X_s)\, dB_s + \frac{1}{2} \int_0^t g'(X_s)\sigma(s)\, ds.$$

With a bit more effort, we can show that the Stratonovich integral exists for time-dependent $g : [0, T] \times \mathbb{R} \to \mathbb{R}$ such that $g(\cdot, x)$, $g(t, \cdot)$ and $\partial_x g(t, \cdot)$ are bounded continuous functions; in this case

$$\int_0^t g(s, X_s) \circ dB_s = \int_0^t g(s, X_s)\, dB_s + \frac{1}{2} \int_0^t \partial_x g(s, X_s)\sigma(s)\, ds. \tag{20.7}$$

For a proof we refer to Stratonovich [212] and [213, Chapter 2.1]. Using the stopping technique of step 4° in the proof of Theorem 17.1, we can remove the boundedness assumption on g and its derivative. Thus, (20.7) remains true for $g \in \mathcal{C}^{0,1}([0, T] \times \mathbb{R})$.

b) The correction term which describes the relation between Itô's and Stratonovich's integral in (20.7) can be formally written as

$$\partial_x g(t, X_t)\sigma(t)\, dt = dB_t\, dg(t, X_t)$$

$$\text{or} \quad \int_0^{\bullet} \partial_x g(s, X_s)\sigma(s)\, ds = \left\langle \int_0^{\bullet} \partial_x g(s, X_s)\, dB_s,\ X \right\rangle$$

where $\langle I, X \rangle$ denotes the quadratic co-variation, cf. (15.8).

20.5 Remark. a) The Stratonovich integral satisfies the classical chain rule. Indeed, if $f \in \mathcal{C}^2(\mathbb{R})$ and $dX_t = \sigma(t)\, dB_t + b(t)\, dt$ is an Itô process, we get Ex. 20.1

$$\begin{aligned} df(X_t) &\overset{(17.13)}{=} f'(X_t)\, dX_t + \tfrac{1}{2} f''(X_t)\, (dX_t)^2 \\ &= f'(X_t)\sigma(t)\, dB_t + f'(X_t)b(t)\, dt + \tfrac{1}{2} f''(X_t)\sigma^2(t)\, dt \\ &\overset{(20.7)}{=} f'(X_t)\sigma(t) \circ dB_t + f'(X_t)b(t)\, dt. \end{aligned}$$

On the other hand, if

$$\dot{x}(t) = \sigma(t)\dot{\beta}(t) + b(t) \quad \text{or} \quad dx(t) = \sigma(t)\, d\beta(t) + b(t)\, dt$$

for some continuously differentiable function $\beta(t)$ with $\beta(0) = 0$, then

$$df(x(t)) = f'(x(t))\, dx(t) = f'(x(t))\sigma(t)\, d\beta(t) + f'(x(t))b(t)\, dt.$$

Substituting $x(t) \rightsquigarrow X_t$ and $d\beta(t) \rightsquigarrow \circ dB_t$ shows that both expressions coincide formally.

! b) For an applied scientist or an engineer it is natural to sample a Brownian motion at the points $\Pi = \{0 = t_0 < t_1 < \ldots < t_n = T\}$ and to think of Brownian motion as a piecewise linear interpolation $(B_t^\Pi)_{t\in[0,T]}$ of the points $(t_l, B_{t_l})_{t_l\in\Pi}$. This is indeed a good approximation of a Brownian path which converges uniformly almost surely and in $L^2(\mathbb{P})$ to $(B_t)_{t\in[0,T]}$, see e. g. the Lévy–Ciesielski construction of Brownian motion in
Ex. 3.3 Chapter 3.2 or Problem 3.3, but with stochastic integrals some caution is needed. Since $t \mapsto B_t^\Pi$ is for any partition piecewise differentiable, we get

$$\int_0^T f(B_t^\Pi(\omega))\, dB_t^\Pi(\omega) = \int_0^T f(B_t^\Pi(\omega))\, \frac{dB_t^\Pi(\omega)}{dt}\, dt = F(B_t^\Pi(\omega))\Big|_0^T = F(B_T^\Pi(\omega))$$

where F is a primitive of f such that $F(0) = 0$. This shows that the limit

$$\lim_{|\Pi|\to 0} \int_0^T f(B_t^\Pi)\, dB_t^\Pi = \lim_{|\Pi|\to 0} F(B_T^\Pi) = F(B_T)$$

exists a. s. On the other hand, if $F \in \mathcal{C}^2(\mathbb{R})$, Itô's formula (17.1) yields

$$F(B_T) = \int_0^T f(B_t)\, dB_t + \frac{1}{2}\int_0^T f'(B_t)\, dt \overset{(20.7)}{=} \int_0^T f(B_t) \circ dB_t$$

which means that, in the limit, the integrals $\int_0^T f(B_t^\Pi(\omega))\, dB_t^\Pi(\omega)$ lead to a Stratonovich integral and not to an Itô integral.

20.2 Solving SDEs with Stratonovich's calculus

The observation from Remark 20.5 a) that the chain rule for Stratonovich integrals coincides (formally) with the classical chain rule can be used to find solutions to certain stochastic differential equations.

20.6 Some heuristic considerations. Let $(B_t)_{t\geqslant 0}$ be a BM^1 and consider the following SDE

$$dX_t = \sigma(t, X_t)\, dB_t + b(t, X_t)\, dt, \qquad X_0 = x_0. \tag{20.8}$$

We assume that it has a unique solution, cf. Chapter 19.5 for sufficient conditions.

1° Find the associated Stratonovich SDE: Using (20.7) we get

$$dX_t = \sigma(t, X_t) \circ dB_t - \frac{1}{2}\sigma(t, X_t)\partial_x\sigma(t, X_t)\,dt + b(t, X_t)\,dt, \qquad X_0 = x_0. \tag{20.9}$$

2° Find the associated ODE: Replace in (20.9) X_t by $x(t)$ and $\circ dB_t$ by $\dot\beta(t)\,dt$ where β is assumed to be continuously differentiable and $\beta(0) = 0$. This gives

$$\dot x(t) = \sigma(t, x(t))\dot\beta(t) - \frac{1}{2}\sigma(t, x(t))\partial_x\sigma(t, x(t)) + b(t, x(t)), \qquad x(0) = x_0. \tag{20.10}$$

3° Find the solution to the ODE: $x(t) = u(\beta(t), \circledast)$ and re-substitute $\beta(t) \rightsquigarrow B_t$ and $x(t) \rightsquigarrow X_t$.

Obviously, step 3° is problematic since we need to make sure that the ODE has a unique solution and we need to know the quantities, symbolically denoted by '$\circledast$', which determine this solution.

Before we give a general result in this direction, let us study a few examples.

20.7 Example. Let $(B_t)_{t\geqslant 0}$ be a BM[1] and assume that $\sigma \in \mathcal{C}_b^2(\mathbb{R})$ is either strictly positive or strictly negative. Then the SDE

$$dX_t = \sigma(X_t)\,dB_t + \frac{1}{2}\sigma(X_t)\sigma'(X_t)\,dt, \quad X_0 = x_0,$$

has a unique solution given by

$$X_t = u(B_t) \quad \text{where} \quad u(y) = g^{-1}(y) \quad \text{and} \quad g = \int \frac{dx}{\sigma(x)} \quad \text{such that} \quad g(x_0) = 0.$$

Proof. Without loss of generality we assume that $\sigma > 0$. Since σ, σ' and σ'' are bounded and continuous, it is obvious that the coefficients of the SDE satisfy the usual Lipschitz and linear growth conditions from Chapter 19.5. Thus there exists a unique solution. Rather than verifying that the solution provided in the example is indeed a solution, let us show how we can derive it using the Stratonovich chain rule as in the previous paragraph § 20.6. Since we use exclusively the chain rule, this already proves that the thus found expression is indeed the solution.

The associated Stratonovich SDE with initial condition $X_0 = x_0$ is

$$dX_t = \sigma(X_t) \circ dB_t - \frac{1}{2}\sigma(X_t)\sigma'(X_t)\,dt + \frac{1}{2}\sigma(X_t)\sigma'(X_t)\,dt = \sigma(X_t) \circ dB_t,$$

and the corresponding ODE is

$$\dot x(t) = \sigma(x(t))\dot\beta(t), \quad x(0) = x_0, \ \beta(0) = 0.$$

Since σ is continuously differentiable, we can use separation of variables to get the unique solution of the ODE

$$\beta(t) = \int_0^t \dot\beta(s)\,ds = \int_{x_0}^{x(t)} \frac{dy}{\sigma(y)} = g(x(t)) - g(x_0).$$

Without loss we may assume that the primitive satisfies $g(x_0) = 0$. Since $\sigma > 0$, we can invert g and get $x(t) = g^{-1}(\beta(t))$ or $X_t = u(B_t)$ with $u = g^{-1}$. □

Example 20.7 works because the drift term of the SDE has a very special structure. The next example shows that we can introduce small perturbations.

20.8 Example. Let $(B_t)_{t\geqslant 0}$ be a BM^1, $a : [0,\infty) \to \mathbb{R}$ be integrable, and assume that $\sigma \in \mathcal{C}_b^2(\mathbb{R})$ is either strictly positive or strictly negative. Then the SDE

$$dX_t = \sigma(X_t)\,dB_t + \Big(a(t)\sigma(X_t) + \frac{1}{2}\,\sigma(X_t)\sigma'(X_t)\Big)\,dt, \qquad X_0 = x_0,$$

has unique solution of the form $X_t = u(B_t + A(t))$ where

$$u(y) = g^{-1}(y), \qquad g = \int \frac{dx}{\sigma(x)}, \qquad A = \int a(s)\,ds \quad \text{and} \quad g(x_0) = A(0) = 0.$$

Proof. As in Example 20.7 we see that there is a unique solution and that the associated Stratonovich SDE and ODE are given by

$$\begin{aligned} dX_t &= a(t)\sigma(X_t)\,dt + \sigma(X_t)\circ dB_t,\\ \dot{x}(t) &= a(t)\sigma(x(t)) + \sigma(x(t))\dot{\beta}(t). \end{aligned}$$

Again by separation of variables we get

$$\beta(t) = \int_0^t \dot{\beta}(s)\,ds = \int_{x_0}^{x(t)} \frac{dy}{\sigma(y)} - \int_0^t a(s)\,ds = g(x(t)) - g(x_0) - A(t) + A(0).$$

Without loss we may assume that the primitives satisfy $A(0) = 0 = g(x_0)$ and $\sigma > 0$. Therefore, we can invert g and get $x(t) = g^{-1}(\beta(t) + A(t))$ or $X_t = u(B_t + A(t))$ with $u = g^{-1}$. □

The simple arguments in Examples 20.7 and 20.8 are due to the fact that the associated ODE can be explicitly solved by separation of variables and that the solution of the SDE is, essentially, a function of the driving Brownian motion $X_t = u(B_t)$ sampled at a single instance. The following theorem extends these examples; the price we have to pay is that now u depends on B_t and another 'control' process Y_t and that u is not any longer explicit. On the other hand, it is good to know that the solution X_t is a (still rather simple) function of the driving noise.

20.9 Theorem (Doss 1978, Sussmann 1977). *Let $(B_t)_{t\geqslant 0}$ be a BM^1, $\sigma \in \mathcal{C}(\mathbb{R})$ such that $\sigma', \sigma'' \in \mathcal{C}_b(\mathbb{R})$ and $b : \mathbb{R} \to \mathbb{R}$ be globally Lipschitz continuous. Then the SDE*

$$dX_t = \sigma(X_t)\,dB_t + \Big(b(X_t) + \frac{1}{2}\,\sigma(X_t)\sigma'(X_t)\Big)\,dt, \qquad X_0 = x_0, \tag{20.11}$$

has a unique solution given by

$$X_t(\omega) = u(B_t(\omega), Y_t(\omega)), \quad t \in [0,\infty),\ \omega \in \Omega,$$

where $u : \mathbb{R}^2 \to \mathbb{R}$ is a continuous function, and the control process $(Y_t(\omega))_{t\geqslant 0}$ is, for each $\omega \in \Omega$, the solution of an ODE.

Proof. By assumption, the coefficients b and σ satisfy the usual (global) Lipschitz conditions needed for the existence and uniqueness of the solution to (20.11), see Chapter 19.5.

Using (20.7) we rewrite the SDE (20.11) as an Stratonovich SDE:

$$dX_t = \sigma(X_t) \circ dB_t + b(X_t)\,dt, \qquad X_0 = x_0.$$

The corresponding ODE is

$$\dot{x}(t) = \sigma(x(t))\dot{\beta}(t) + b(x(t)), \qquad x(0) = x_0.$$

Since the coefficients are Lipschitz continuous, it has a unique solution. In order to find the solution, we make the *Ansatz* $x(t) = u(\beta(t), y(t))$ where $y(t)$ and $\beta(t)$ are differentiable with $\beta(0) = 0$ and $u : \mathbb{R}^2 \to \mathbb{R}$ is sufficiently regular. By the chain rule

$$\dot{x}(t) = \partial_x u(\beta(t), y(t))\,\dot{\beta}(t) + \partial_y u(\beta(t), y(t))\,\dot{y}(t),$$

and, comparing this with the previous formula for $\dot{x}(t)$, we get with $u = u(\beta(t), y(t))$

$$\partial_x u = \sigma(u) \quad \text{and} \quad \dot{y}(t) = \frac{b(u)}{\partial_y u}. \tag{20.12}$$

From $\partial_y \partial_x u = \partial_y \sigma(u) = \sigma'(u)\partial_y u$ we get

$$\partial_y u(x, y) = \partial_y u(0, y) \exp\left(\int_0^x \sigma'(u(z, y))\,dz\right) \tag{20.13}$$

which means that $\partial_y u(0, y) = 1$ or $u(0, y) = y$ is a very convenient choice of the initial condition in (20.12) which allows us to solve the ODE $\partial_x u = \sigma(u)$ uniquely (σ is globally Lipschitz continuous).

In order to solve the second ODE in (20.12) we need that

$$f(x, y) := \frac{b(u(x, y))}{\partial_y u(x, y)}$$

is, for fixed x, Lipschitz continuous and grows at most linearly. This we will check in Lemma 20.10 below. The correct initial condition is fixed by

$$x_0 = u(\beta(0), y(0)) = u(0, y(0)) = y(0).$$

Thus, there is a unique solution $y(t)$ of

$$\dot{y}(t) = f(\beta(t), y(t)), \qquad y(0) = x_0,$$

and this proves that $x(t) = u(\beta(t), y(t))$ solves the associated ODE.

Finally, $X_t(\omega) = u(B_t(\omega), Y_t(\omega))$ solves the Stratonovich SDE, hence (20.11); for fixed ω the control process is given by $\dot{Y}_t(\omega) = f(B_t(\omega), Y_t(\omega))$ and $Y_0(\omega) = x_0$. □

20.10 Lemma. *The function $y \mapsto f(x,y) := b(u(x,y))/\partial_y u(x,y)$ appearing in the proof of Theorem 20.9 is, for fixed x, locally Lipschitz continuous and grows at most linearly.*

Proof. Set $c := \|\sigma'\|_\infty + \|\sigma''\|_\infty$. By assumption, $c < \infty$ and $\partial_y u(0,y) = 1$, so

$$e^{-c|x|} \leqslant \frac{1}{\partial_y u(x,y)} = \exp\left(- \int_0^x \sigma'(u(z,y))\,dz \right) \leqslant e^{c|x|}, \qquad x \in \mathbb{R}.$$

If L is the (global) Lipschitz constant of b, we get for $x, y, y' \in \mathbb{R}$

$$|b(u(x,y)) - b(u(x,y'))| \leqslant L|u(x,y) - u(x,y')| \leqslant Le^{c|x|}|y - y'|.$$

Moreover, by the mean value theorem

$$\left|\frac{1}{\partial_y u(x,y)} - \frac{1}{\partial_y u(x,y')}\right| \leqslant \left\|\partial_y \left(\frac{1}{\partial_y u(x,\cdot)}\right)\right\|_\infty |y - y'|.$$

From the explicit formula (20.13) for $\partial_y u(x,y)$ we see

$$\left|\partial_y \left(\frac{1}{\partial_y u(x,y)}\right)\right| = \left|\frac{\int_0^x \sigma''(u(z,y))\,\partial_y u(z,y)\,dz}{\partial_y u(x,y)}\right| \leqslant e^{c|x|} c \int_0^{|x|} e^{cz}\,dz.$$

Finally, since $|\partial_y u(x,y)| \geqslant e^{-c|x|} > 0$ for every x, we see that $1/\partial_y u(x,y)$ is bounded as a function of y.

Ex. 20.3 Thus, $f(x,y)$ is the product of two globally Lipschitz continuous functions, one of which is bounded. As such it is locally Lipschitz continuous and grows at most linearly. □

20.11 Further reading. A good discussion of the Stratonovich integral and its applications can be found in [2]. The most comprehensive source for approximations of stochastic integrals and SDEs, including Stratonovich integrals and SDEs is [100, Chapter VI.7, pp. 480–517]. The original papers by Wong and Zakai [233; 234; 235] are very intuitively written and still make a formidable reading.

[2] Arnold: *Stochastic Differential Equations: Theory and Applications.*
[100] Ikeda, Watanabe: *Stochastic Differential Equations and Diffusion Processes.*
[213] Stratonovich: *Conditional Markov Processes and Their Application to the Theory of Optimal Control.*
[233] Wong, Zakai: On the convergence of ordinary integrals to stochastic integrals.
[234] Wong, Zakai: On the relation between ordinary and stochastic differential equations.
[235] Wong, Zakai: Riemann–Stieltjes approximations of stochastic integrals.

Problems

1. Show that Remark 20.5 a) remains valid if we replace $f(x)$ by $f(s,x)$.

2. Show that Example 20.7 remains valid if we assume that $\sigma(0) = 0$ and $\sigma(x) > 0$ for $x \neq 0$.

3. Let $f, g : \mathbb{R} \to \mathbb{R}$ be two globally Lipschitz continuous functions. If f is bounded, then $h(x) := f(x)g(x)$ is locally Lipschitz continuous and at most linearly growing.

21 On diffusions

Diffusion is a physical phenomenon which describes the tendency of two (or more) substances, e. g. gases or liquids, to reach an equilibrium: Particles of one type, say B, move or 'diffuse' in(to) another substance, say S, leading to a more uniform distribution of the type B particles within S. Since the particles are physical objects, it is reasonable to assume that their trajectories are continuous (in fact, differentiable) and that the movement is influenced by various temporal and spatial (in-)homogeneities which depend on the nature of the substance S. Diffusion phenomena are governed by Fick's law. Denote by $p = p(t,x)$ the concentration of the B particles at a space-time point (t,x), and by $J = J(t,x)$ the flow of particles. Then

$$J = -D\frac{\partial p}{\partial x} \quad \text{and} \quad \frac{\partial p}{\partial t} = -\frac{\partial J}{\partial x}$$

where D is called the diffusion constant which takes into account the geometry and the properties of the substances B and S. Einstein considered in 1905 the particle movement observed by Brown and – under the assumption that the movement is temporally and spatially homogeneous – he showed that it can be described as a diffusion phenomenon:

$$\frac{\partial p(t,x)}{\partial t} = D\frac{\partial^2 p(t,x)}{\partial x^2} \quad \text{hence} \quad p(t,x) = \frac{1}{\sqrt{4\pi D}}\, e^{-\frac{x^2}{4Dt}}.$$

The diffusion coefficient $D = \frac{RT}{N}\frac{1}{6\pi kP}$ depends on the absolute temperature T, the universal gas constant R, Avogadro's number N, the friction coefficient k and the radius of the particles (i. e. atoms) P.

If the diffusion coefficient depends on time and space, $D = D(t,x)$, Fick's law leads to a differential equation of the type

$$\frac{\partial p(t,x)}{\partial t} = D(t,x)\frac{\partial^2 p(t,x)}{\partial x^2} + \frac{\partial D(t,x)}{\partial x}\frac{\partial p(t,x)}{\partial x}.$$

In a mathematical model of a diffusion we could either use a macroscopic approach and model the particle density $p(t,x)$, or a microscopic point of view modelling the movement of the particles themselves and determine the particle density. Using probability theory we can describe the (random) position and trajectories of the particles by a stochastic process. In view of the preceding discussion it is reasonable to require that the stochastic process

- has continuous trajectories $t \mapsto X_t(\omega)$;
- has a generator which is a second-order differential operator.

Nevertheless there is no standard definition of a diffusion process. Depending on the model, one will prefer one assumption over the other or add further requirements.

Throughout this chapter we use the following definition.

21.1 Definition. A *diffusion process* is a Feller process $(X_t)_{t\geqslant 0}$ with values in $\mathbb{R}^d$, continuous trajectories and an infinitesimal generator $(A, \mathfrak{D}(A))$ such that $\mathcal{C}_c^\infty(\mathbb{R}^d) \subset \mathfrak{D}(A)$ and for $u \in \mathcal{C}_c^\infty(\mathbb{R}^d)$ Ex. 21.1

$$Au(x) = L(x, D)u(x) = \frac{1}{2}\sum_{i,j=1}^{d} a_{ij}(x)\frac{\partial^2 u(x)}{\partial x_i \partial x_j} + \sum_{j=1}^{d} b_j(x)\frac{\partial u(x)}{\partial x_j}. \tag{21.1}$$

The symmetric, positive semidefinite matrix $a(x) = (a_{ij}(x))_{i,j=1}^d \in \mathbb{R}^{d\times d}$ is called the *diffusion matrix*, and $b(x) = (b_1(x), \dots, b_d(x)) \in \mathbb{R}^d$ is called the *drift vector*.

Since $(X_t)_{t\geqslant 0}$ is a Feller process, A maps $\mathcal{C}_c^\infty(\mathbb{R}^d)$ into $\mathcal{C}_\infty(\mathbb{R}^d)$, therefore $a(\cdot)$ and $b(\cdot)$ have to be continuous functions.

21.2 Remark. Continuity of paths and local operators. Let $(X_t)_{t\geqslant 0}$ be a Feller process with generator $(A, \mathfrak{D}(A))$ such that $\mathcal{C}_c^\infty(\mathbb{R}^d) \subset \mathfrak{D}(A)$.

By Theorems 7.38 and 7.37, the continuity of the trajectories implies that A is a local operator, hence a second-order differential operator.

The converse is true if $(X_t)_{t\geqslant 0}$ takes values in a compact space. For this, we need the Dynkin–Kinney criterion for the continuity of the sample paths. A proof can be found in Dynkin [64, Chapter 6 § 5] or Wentzell [226, Chapter 9]. By $\mathbb{R}^d \cup \{\partial\}$ we denote the one-point compactification of $\mathbb{R}^d$.

21.3 Theorem (Dynkin 1952, Kinney 1953). *Let $(X_t)_{t\geqslant 0}$ be a strong Markov process with state space $E \subset \mathbb{R}^d \cup \{\partial\}$. If*

$$\forall\, \delta > 0 \quad \lim_{h\to 0} \frac{1}{h} \sup_{t\leqslant h} \sup_{x\in E} \mathbb{P}^x(|X_t - x| > \delta) = 0, \tag{21.2}$$

then there is a modification of $(X_t)_{t\geqslant 0}$ which has continuous trajectories in E.

The continuity of the trajectories implies that the generator is a local operator, cf. Theorem 7.38. If the state space E is compact, (an obvious modification of) Theorem 7.39 shows that this is the same as to say that

$$\forall\, \delta > 0 \quad \lim_{t\to 0} \frac{1}{t} \sup_{x\in E} \mathbb{P}^x(|X_t - x| > \delta) = 0. \tag{21.3}$$

Since $h^{-1} P(h) \leqslant h^{-1} \sup_{t\leqslant h} P(t) \leqslant \sup_{t\leqslant h} t^{-1} P(t)$ we see that (21.2) and (21.3) are equivalent conditions.

We might apply this to $(X_t)_{t\geqslant 0}$ with paths in $E = \mathbb{R}^d \cup \{\partial\}$, but the trouble is that continuity of $t \mapsto X_t$ in $\mathbb{R}^d$ and $\mathbb{R}^d \cup \{\partial\}$ are quite different notions: The latter allows *explosion in finite time* as well as *sample paths coming back from* ∂, while continuity in $\mathbb{R}^d$ means infinite life-time: $\mathbb{P}^x(\forall t \geqslant 0 : X_t \in \mathbb{R}^d) = 1$.

21.1 Kolmogorov's theory

Kolmogorov's seminal paper [128] *Über die analytischen Methoden der Wahrscheinlichkeitsrechnung* marks the beginning of the mathematically rigorous theory of stochastic processes in continuous time. In §§13, 14 of this paper Kolmogorov develops the foundations of the theory of diffusion processes and, we will now follow Kolmogorov's ideas.

21.4 Theorem (Kolmogorov 1933). *Let $(X_t)_{t\geqslant 0}$, $X_t = (X_t^1,\dots,X_t^d)$, be a Feller process with values in $\mathbb{R}^d$ such that for all $\delta > 0$ and $i,j = 1,\dots,d$*

$$\lim_{t\to 0}\frac{1}{t}\sup_{x\in\mathbb{R}^d}\mathbb{P}^x(|X_t - x| > \delta) = 0, \tag{21.4a}$$

$$\lim_{t\to 0}\frac{1}{t}\,\mathbb{E}^x\left((X_t^j - x_j)\,\mathbb{1}_{\{|X_t - x|\leqslant\delta\}}\right) = b_j(x), \tag{21.4b}$$

$$\lim_{t\to 0}\frac{1}{t}\,\mathbb{E}^x\left((X_t^i - x_i)(X_t^j - x_j)\,\mathbb{1}_{\{|X_t - x|\leqslant\delta\}}\right) = a_{ij}(x). \tag{21.4c}$$

Then $(X_t)_{t\geqslant 0}$ is a diffusion process in the sense of Definition 21.1.

Proof. The Dynkin–Kinney criterion, see the discussion in Remark 21.2, shows that (21.4a) guarantees the continuity of the trajectories.

We are going to show that $\mathcal{C}_c^\infty(\mathbb{R}^d)\subset\mathfrak{D}(A)$ and that $A|_{\mathcal{C}_c^\infty(\mathbb{R}^d)}$ is a differential operator of the form (21.1). The proof is similar to Example 7.9.

Let $u\in\mathcal{C}_c^\infty(\mathbb{R}^d)$. Since $\partial_j\partial_k u$ is uniformly continuous,

$$\forall\epsilon > 0\quad\exists\delta > 0 :\ \max_{1\leqslant i,j\leqslant d}\ \sup_{|x-y|\leqslant\delta}|\partial_i\partial_j u(x) - \partial_i\partial_j u(y)|\leqslant\epsilon.$$

Fix $\epsilon > 0$, choose $\delta > 0$ as above and set $T_t u(x) := \mathbb{E}^x u(X_t)$, $\Omega_\delta := \{|X_t - x|\leqslant\delta\}$. Then

$$\frac{T_t u(x) - u(x)}{t} = \underbrace{\frac{1}{t}\,\mathbb{E}^x\left([u(X_t) - u(x)]\,\mathbb{1}_{\Omega_\delta}\right)}_{=\mathrm{J}} + \underbrace{\frac{1}{t}\,\mathbb{E}^x\left([u(X_t) - u(x)]\,\mathbb{1}_{\Omega_\delta^c}\right)}_{=\mathrm{J}'}.$$

It is enough to show that $\lim_{t\to 0}(T_t u(x) - u(x))/t = Au(x)$ holds for every $x\in\mathbb{R}^d$, all $u\in\mathcal{C}_c^\infty(\mathbb{R}^d)$ and the operator A from (21.1), see Theorem 7.22.

The second term can be estimated by

$$|\mathrm{J}'|\leqslant 2\,\|u\|_\infty\,\frac{1}{t}\mathbb{P}^x(|X_t - x| > \delta)\xrightarrow[t\to 0]{(21.4a)} 0.$$

Now we consider J. By Taylor's formula there is some $\theta\in(0,1)$ such that for the intermediate point $\Theta = (1-\theta)x + \theta X_t$

$$\begin{aligned} u(X_t) - u(x) &= \sum_{j=1}^d (X_t^j - x_j)\,\partial_j u(x) + \frac{1}{2}\sum_{j,k=1}^d (X_t^j - x_j)(X_t^k - x_k)\,\partial_j\partial_k u(x)\\ &\quad + \frac{1}{2}\sum_{j,k=1}^d (X_t^j - x_j)(X_t^k - x_k)\left(\partial_j\partial_k u(\Theta) - \partial_j\partial_k u(x)\right). \end{aligned}$$

Denote the terms on the right-hand side by J_1, J_2 and J_3, respectively. Since $\partial_j\partial_k u$ is uniformly continuous, the elementary inequality $2|ab| \leqslant a^2 + b^2$ yields

$$\begin{aligned}\frac{1}{t}\,\mathbb{E}^x\left(|J_3|\,\mathbb{1}_{\Omega_\delta}\right) &\leqslant \frac{\epsilon}{2}\sum_{i,j=1}^{d}\frac{1}{t}\,\mathbb{E}^x\left[|(X_t^i - x_i)(X_t^j - x_j)|\,\mathbb{1}_{\Omega_\delta}\right]\\ &\leqslant \frac{\epsilon d}{2}\sum_{j=1}^{d}\frac{1}{t}\,\mathbb{E}^x\left[(X_t^j - x_j)^2\mathbb{1}_{\Omega_\delta}\right] \xrightarrow[t\to 0]{(21.4c)} \frac{\epsilon d}{2}\sum_{j=1}^{d} a_{jj}(x).\end{aligned}$$

Moreover,

$$\begin{aligned}\frac{1}{t}\,\mathbb{E}^x\left(J_2\mathbb{1}_{\Omega_\delta}\right) &= \frac{1}{2}\sum_{j,k=1}^{d}\frac{1}{t}\,\mathbb{E}^x\left[(X_t^j - x_j)(X_t^k - x_k)\mathbb{1}_{\Omega_\delta}\right]\partial_j\partial_k u(x)\\ &\xrightarrow[t\to 0]{(21.4c)} \frac{1}{2}\sum_{i,j=1}^{d} a_{ij}(x)\,\partial_i\partial_j u(x),\end{aligned}$$

and

$$\frac{1}{t}\,\mathbb{E}^x\left(J_1\mathbb{1}_{\Omega_\delta}\right) = \sum_{j=1}^{d}\frac{1}{t}\,\mathbb{E}^x\left[(X_t^j - x_j)\mathbb{1}_{\Omega_\delta}\right]\partial_j u(x) \xrightarrow[t\to 0]{(21.4b)} \sum_{j=1}^{d} b_j(x)\,\partial_j u(x).$$

Since $\epsilon > 0$ is arbitrary, the claim follows. □

If $(X_t)_{t\geqslant 0}$ has a transition density, i. e. if $\mathbb{P}^x(X_t \in A) = \int_A p(t,x,y)\,dy$ for all $t > 0$ and $A \in \mathscr{B}(\mathbb{R}^d)$, then we can express the dynamics of the movement as a Cauchy problem. Depending on the point of view, we get Kolmogorov's *first* and *second* differential equation, cf. [128, §§ 12–14].

21.5 Proposition (Backward equation. Kolmogorov 1931). *Let $(X_t)_{t\geqslant 0}$ denote a diffusion with generator $(A, \mathfrak{D}(A))$ such that $A|_{\mathcal{C}_c^\infty(\mathbb{R}^d)}$ is given by (21.1) with bounded drift and diffusion coefficients $b \in \mathcal{C}_b(\mathbb{R}^d, \mathbb{R}^d)$ and $a \in \mathcal{C}_b(\mathbb{R}^d, \mathbb{R}^{d\times d})$. Assume that $(X_t)_{t\geqslant 0}$ has a transition density $p(t,x,y)$, $t > 0$, $x, y \in \mathbb{R}^d$, satisfying*

$$p,\ \frac{\partial p}{\partial t},\ \frac{\partial p}{\partial x_j},\ \frac{\partial^2 p}{\partial x_j \partial x_k} \in \mathcal{C}((0,\infty)\times\mathbb{R}^d\times\mathbb{R}^d),$$

$$p(t,\cdot,\cdot),\ \frac{\partial p(t,\cdot,\cdot)}{\partial t},\ \frac{\partial p(t,\cdot,\cdot)}{\partial x_j},\ \frac{\partial^2 p(t,\cdot,\cdot)}{\partial x_j \partial x_k} \in \mathcal{C}_\infty(\mathbb{R}^d\times\mathbb{R}^d)$$

for all $t > 0$. Then Kolmogorov's first or backward equation holds:

$$\frac{\partial p(t,x,y)}{\partial t} = L(x, D_x)p(t,x,y) \quad \text{for all } t > 0,\ x, y \in \mathbb{R}^d. \tag{21.5}$$

Proof. Denote by $L(x, D_x)$ the second-order differential operator given by (21.1). By assumption, $A = L(x, D)$ on $\mathcal{C}_c^\infty(\mathbb{R}^d)$. Since $p(t,x,y)$ is the transition density, we know that the transition semigroup is given by $T_t u(x) = \int p(t,x,y)u(y)\,dy$.

Ex. 21.2 Using dominated convergence it is not hard to see that the conditions on $p(t,x,y)$ are such that T_t maps $\mathcal{C}_c^\infty(\mathbb{R}^d)$ to $\mathcal{C}_\infty^2(\mathbb{R}^d)$.

Ex. 21.3 Since the coefficients b and a are bounded, $\|Au\|_\infty \leqslant \kappa\|u\|_{(2)}$, where we use the norm $\|u\|_{(2)} = \|u\|_\infty + \sum_j \|\partial_j u\|_\infty + \sum_{j,k}\|\partial_j\partial_k u\|_\infty$ in $\mathcal{C}_\infty^2(\mathbb{R}^d)$, and $\kappa = \kappa(b,a)$ is a constant. Since $\mathcal{C}_c^\infty(\mathbb{R}^d) \subset \mathfrak{D}(A)$ and $(A,\mathfrak{D}(A))$ is closed, cf. Corollary 7.11, it follows that $\mathcal{C}_\infty^2(\mathbb{R}^d) \subset \mathfrak{D}(A)$ and $A|_{\mathcal{C}_\infty^2(\mathbb{R}^d)} = L(x,D)|_{\mathcal{C}_\infty^2(\mathbb{R}^d)}$.

Therefore, Lemma 7.10 shows that $\frac{d}{dt}T_t u = AT_t u = L(\cdot,D)T_t u$, i. e.

$$\begin{aligned}\frac{\partial}{\partial t}\int p(t,x,y)u(y)\,dy &= L(x,D_x)\int p(t,x,y)u(y)\,dy\\ &= \int L(x,D_x)p(t,x,y)u(y)\,dy\end{aligned}$$

for all $u \in \mathcal{C}_c^\infty(\mathbb{R}^d)$. The last equality is a consequence of the continuity and smoothness assumptions on $p(t,x,y)$ which allow us to use the differentiation lemma for parameter-dependent integrals, see e. g. [204, Theorem 11.5]. Applying it to the t-derivative, we get

$$\int\left(\frac{\partial p(t,x,y)}{\partial t} - L(x,D_x)p(t,x,y)\right)u(y)\,dy = 0 \quad \text{for all } u \in \mathcal{C}_c^\infty(\mathbb{R}^d)$$

which implies (21.5). □

Ex. 21.4 If $L(x,D_x)$ is a second-order differential operator of the form (21.1), then the formal adjoint operator is

$$L^*(y,D_y)u(y) = \frac{1}{2}\sum_{i,j=1}^d \frac{\partial^2}{\partial y_i\partial y_j}\big(a_{ij}(y)u(y)\big) - \sum_{j=1}^d \frac{\partial}{\partial y_j}\big(b_j(y)u(y)\big)$$

provided that the coefficients a and b are sufficiently smooth.

Ex. 21.5 **21.6 Proposition** (Forward equation. Kolmogorov 1931). *Let $(X_t)_{t\geqslant 0}$ denote a diffusion with generator $(A,\mathfrak{D}(A))$ such that $A|_{\mathcal{C}_c^\infty(\mathbb{R}^d)}$ is given by (21.1) with drift and diffusion coefficients $b \in \mathcal{C}^1(\mathbb{R}^d,\mathbb{R}^d)$ and $a \in \mathcal{C}^2(\mathbb{R}^d,\mathbb{R}^{d\times d})$. Assume that X_t has a probability density $p(t,x,y)$ satisfying*

$$p,\ \frac{\partial p}{\partial t},\ \frac{\partial p}{\partial y_j},\ \frac{\partial^2 p}{\partial y_j\partial y_k} \in \mathcal{C}((0,\infty)\times\mathbb{R}^d\times\mathbb{R}^d).$$

Then Kolmogorov's second or forward equation holds:

$$\frac{\partial p(t,x,y)}{\partial t} = L^*(y,D_y)p(t,x,y) \qquad \textit{for all } t>0,\ x,y\in\mathbb{R}^d. \tag{21.6}$$

Proof. This proof is similar to the proof of Proposition 21.5, where we use Lemma 7.10 in the form $\frac{d}{dt}T_t u = T_t A u = T_t L(\cdot,D)u$ for $u \in \mathcal{C}_c^\infty(\mathbb{R}^d)$, and then integration by parts in the resulting integral equality. The fact that we do not have to interchange the integral and the operator $L(x,D)$ as in the proof of Proposition 21.5, allows us to relax the boundedness assumptions on p, b and a, but we pay a price in the form of differentiability assumptions for the drift and diffusion coefficients. □

If a and b do not depend on x, the transition probability $p(t, x, y)$ depends only on the difference $x - y$, and the corresponding stochastic process is spatially homogeneous, essentially a Brownian motion with a drift: $aB_t + bt$. Kolmogorov [128, §16] mentions that in this case the forward equation (21.6) was discovered (but not rigorously proved) by Bachelier [6; 7]. Often it is called *Fokker–Planck equation* since it also appears in the context of statistical physics, cf. Fokker [78] and Planck [180].

In order to understand the names 'backward' and 'forward' we have to introduce the concept of a fundamental solution of a parabolic PDE.

21.7 Definition. Let $L(s, x, D) = \sum_{i,j=1}^{d} a_{ij}(s, x)\partial_i\partial_j + \sum_{j=1}^{d} b_j(s, x)\partial_j + c(s, x)$ be a second-order differential operator on $\mathbb{R}^d$ where $(a_{ij}(s, x)) \in \mathbb{R}^{d\times d}$ is symmetric and positive semidefinite. A *fundamental solution* of the parabolic differential operator $\frac{\partial}{\partial s}+L(s, x, D)$ on the set $[0, T] \times \mathbb{R}^d$ is a function $\Gamma(s, x; t, y)$ defined for all $(s, x), (t, y) \in [0, T] \times \mathbb{R}^d$ with $s < t$ such that for every $f \in \mathcal{C}_c(\mathbb{R}^d)$ the function

$$u(s, x) = \int_{\mathbb{R}^d} \Gamma(s, x; t, y) f(y)\, dy$$

satisfies

$$-\frac{\partial u(s, x)}{\partial s} = L(s, x, D)u(s, x) \quad \text{for all } x \in \mathbb{R}^d,\ s < t \leqslant T, \tag{21.7a}$$

$$\lim_{s\uparrow t} u(s, x) = f(x) \quad \text{for all } x \in \mathbb{R}^d. \tag{21.7b}$$

Proposition 21.5 shows that $\Gamma(s, x; t, y) := p(t - s, x, y)$, $s < t$, is the fundamental solution of the parabolic differential operator $L(x, D) + \frac{\partial}{\partial s}$, whereas Proposition 21.6 means that $\Gamma(t, y; s, x) := p(s - t, y, x)$, $t < s$ is the fundamental solution of the (formal) adjoint operator $L^*(y, D) - \frac{\partial}{\partial t}$.

The backward equation has the time derivative at the left end of the time interval, and we solve the PDE in a forward direction with a condition at the right end $\lim_{s\uparrow t} p(t - s, x, y) = \delta_x(dy)$. In the forward equation the time derivative takes place at the right end of the time interval, and we solve the PDE backwards with an initial condition $\lim_{s\downarrow t} p(s - t, y, x) = \delta_y(dx)$. Of course, this does not really show in the time-homogeneous case.... At a technical level, the proofs of Propositions 21.5 and 21.6 reveal that all depends on the identity (7.10): in the backward case we use $\frac{d}{dt}T_t = AT_t$, in the forward case $\frac{d}{dt}T_t = T_t A$.

Using probability theory we can give the following stochastic interpretation. Run the stochastic process in the time interval $[0, t]$. Using the Markov property we can perturb the movement at the beginning, i. e. on $[0, h]$ and then let it run from time h to t, or we could run it up to time $t - h$, and then perturb it at the end in $[t - h, t]$. In the first or backward case, the Markov property gives for all $C \in \mathscr{B}(\mathbb{R}^d)$

$$\begin{aligned}\mathbb{P}^x(X_t \in C) - \mathbb{P}^x(X_{t-h} \in C) &= \mathbb{E}^x\, \mathbb{P}^{X_h}(X_{t-h} \in C) - \mathbb{P}^x(X_{t-h} \in C)\\ &= (T_h - \mathrm{id})\mathbb{P}^{(\cdot)}(X_{t-h} \in C)(x).\end{aligned}$$

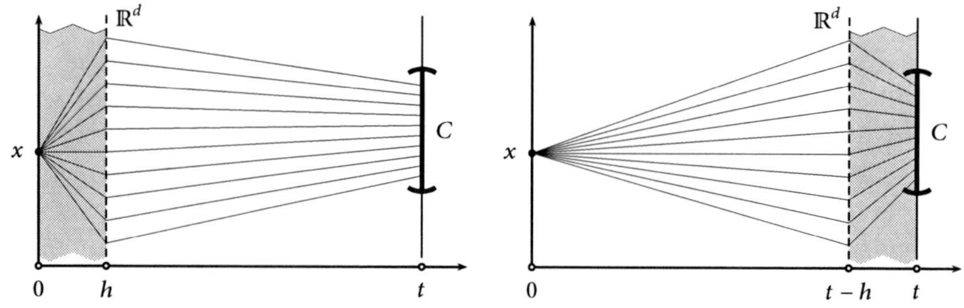

Fig. 21.1. Derivation of the backward and forward Kolmogorov equations.

If $\mathbb{P}^x(X_t \in C) = \int_C p(t,x,y)\,dy$ with a sufficiently smooth density, we get

$$\frac{p(t,x,y) - p(t-h,x,y)}{h} = \frac{T_h - \mathrm{id}}{h}\, p(t-h,\cdot,y)(x)$$

which, as $h \to 0$, yields $\dfrac{\partial}{\partial t} p(t,x,y) = A_x p(t,x,y)$, i. e. the backward equation.

In the second or forward case we get

$$\begin{aligned}\mathbb{P}^x(X_t \in C) - \mathbb{P}^x(X_{t-h} \in C) &= \mathbb{E}^x\, \mathbb{P}^{X_{t-h}}(X_h \in C) - \mathbb{P}^x(X_{t-h} \in C)\\ &= \mathbb{E}^x\left((T_h - \mathrm{id})\mathbb{1}_C(X_{t-h})\right).\end{aligned}$$

Passing to the transition densities and dividing by h, we get

$$\frac{p(t,x,y) - p(t-h,x,y)}{h} = \frac{T_h^* - \mathrm{id}}{h}\, p(t-h,x,\cdot)(y)$$

and $h \to 0$ yields $\dfrac{\partial}{\partial t} p(t,x,y) = A_y^* p(t,x,y)$ (T_t^*, A^* are the formally adjoint operators).

The obvious question is, of course, which conditions on b and a ensure that there is a (unique) diffusion process such that the generator, restricted to $\mathcal{C}_c^\infty(\mathbb{R}^d)$, is a second order differential operator (21.1). If we can show that $L(x,D)$ generates a Feller semigroup or has a nice fundamental solution for the corresponding forward or backward equation for $L(x,D)$, then we could use Kolmogorov's construction, cf. Section 4.2 or Remark 7.6, to construct the corresponding process.

One possibility is to use the Hille–Yosida theorem (Theorem 7.19 and Lemma 7.21); the trouble is then to verify condition c) from Theorem 7.19 which amounts to finding a solution to an elliptic partial differential equation of the type $\lambda u(x) - L(x,D)u(x) = f(x)$ for $f \in \mathcal{C}_\infty(\mathbb{R}^d)$. In a non-Hilbert space setting, this is not an easy task.

Another possibility would be to construct a fundamental solution for the backward equation. Here is a standard result in this direction. For a proof we refer to Friedman [82, Chapter I.6].

21.8 Theorem. *Let* $L(x, D) = \sum_{i,j=1}^d a_{ij}(x)\partial_i\partial_j + \sum_{j=1}^d b_j(x)\partial_j + c(x)$ *be a differential operator on* $\mathbb{R}^d$ *such that*

$$\sum_{i,j=1}^d a_{ij}(x)\xi_i\xi_j \geqslant \kappa\,|\xi|^2 \text{ for some constant } \kappa \text{ and all } x, \xi \in \mathbb{R}^d,$$

$$a_{ij}(\cdot),\ b_j(\cdot),\ c(\cdot) \in \mathcal{C}_b(\mathbb{R}^d) \text{ are locally Hölder continuous.}$$

Then there exists a fundamental solution $\Gamma(s, x; t, y)$ *such that*

- Γ, $\dfrac{\partial\Gamma}{\partial s}$, $\dfrac{\partial\Gamma}{\partial x_j}$ *and* $\dfrac{\partial^2\Gamma}{\partial x_j\partial x_k}$ *are continuous in* $(s, x; t, y)$;
- $\Gamma(s, x; t, y) \leqslant c(t-s)^{-d/2}\, e^{-C|x-y|^2/(t-s)}$,

  $\dfrac{\partial\Gamma(s, x; t, y)}{\partial x_j} \leqslant c(t-s)^{-(d+1)/2}\, e^{-C|x-y|^2/(t-s)}$;
- $-\dfrac{\partial\Gamma(s, x; t, y)}{\partial s} = L(x, D_x)\Gamma(s, x; t, y)$;
- $\lim\limits_{s\uparrow t}\displaystyle\int \Gamma(s, x; t, y) f(x)\, dx = f(y)$ *for all* $f \in \mathcal{C}_b(\mathbb{R}^d)$.

21.2 Itô's theory

It is possible to extend Theorem 21.8 to increasing and degenerate coefficients, and we refer the interested reader to the exposition in Stroock and Varadhan [214, Chapter 3]. Here we will follow Itô's idea to use stochastic differential equations (SDEs) and construct a diffusion directly. In Chapter 19 we have discussed the uniqueness, existence and regularity of SDEs of the form

$$dX_t = b(X_t)\, dt + \sigma(X_t)\, dB_t, \quad X_0 = x, \tag{21.8}$$

for a BM^d $(B_t)_{t\geqslant 0}$ and coefficients $b : \mathbb{R}^d \to \mathbb{R}^d$, $\sigma : \mathbb{R}^d \to \mathbb{R}^{d\times d}$. The key ingredient was the following global Lipschitz condition

$$\sum_{j=1}^d |b_j(x) - b_j(y)|^2 + \sum_{j,k=1}^d |\sigma_{jk}(x) - \sigma_{jk}(y)|^2 \leqslant C^2\,|x-y|^2 \quad \text{for all } x, y \in \mathbb{R}^d. \tag{21.9}$$

If we take $y = 0$, (21.9) entails the linear growth condition.

21.9 Theorem. *Let* $(B_t)_{t\geqslant 0}$ *be a* BM^d *and* $(X_t^x)_{t\geqslant 0}$ *be the solution of the SDE* (21.8). *Assume that the coefficients* b *and* σ *satisfy the global Lipschitz condition* (21.9). *Then* $(X_t^x)_{t\geqslant 0}$ *is a diffusion process in the sense of Definition 21.1 with infinitesimal generator*

$$L(x, D)u(x) = \frac{1}{2}\sum_{i,j=1}^d a_{ij}(x)\frac{\partial^2 u(x)}{\partial x_i\partial x_j} + \sum_{j=1}^d b_j(x)\frac{\partial u(x)}{\partial x_j}$$

where $a(x) = \sigma(x)\sigma^\top(x)$.

Proof. The uniqueness, existence, and Markov property of $(X_t^x)_{t\geqslant 0}$ follow from Corollary 19.12, Theorem 19.13 and Corollary 19.24, respectively. The (stochastic) integral is, as a function of its upper limit, continuous, cf. Theorem 15.13 or Theorem 16.7, and so ! $t \mapsto X_t^x(\omega)$ is continuous. From Corollary 19.31 we know that $\lim_{|x|\to\infty} |X_t^x| = \infty$ for every $t > 0$. Therefore, we conclude with the dominated convergence theorem that

Ex. 19.12
$$x \mapsto T_t u(x) := \mathbb{E}\, u(X_t^x) \text{ is in } \mathcal{C}_\infty(\mathbb{R}^d) \text{ for all } t \geqslant 0 \text{ and } u \in \mathcal{C}_\infty(\mathbb{R}^d).$$

The operators T_t are clearly sub-Markovian and contractive. Let us show that $(T_t)_{t\geqslant 0}$ is strongly continuous, hence a Feller semigroup. Take $u \in \mathcal{C}_c^2(\mathbb{R}^d)$ and apply Itô's formula, Theorem 17.8. Then

$$u(X_t^x) - u(x) = \sum_{j,k=1}^d \int_0^t \sigma_{jk}(X_s^x)\partial_j u(X_s^x)\,dB_s^k + \sum_{j=1}^d \int_0^t b_j(X_s^x)\partial_j u(X_s^x)\,ds + \frac{1}{2}\sum_{i,j=1}^d \int_0^t a_{ij}(X_s^x)\partial_i\partial_j u(X_s^x)\,ds.$$

Since the first term on the right is a martingale, its expectation vanishes, and we get

$$\begin{aligned} |T_t u(x) - u(x)| &= |\,\mathbb{E}\, u(X_t^x) - u(x)| \\ &\leqslant \sum_{j=1}^d \mathbb{E}\left|\int_0^t b_j(X_s^x)\partial_j u(X_s^x)\,ds\right| + \frac{1}{2}\sum_{i,j=1}^d \mathbb{E}\left|\int_0^t a_{ij}(X_s^x)\partial_i\partial_j u(X_s^x)\,ds\right| \\ &\leqslant t\sum_{j=1}^d \|b_j\partial_j u\|_\infty + t\sum_{i,j=1}^d \|a_{ij}\partial_i\partial_j u\|_\infty \xrightarrow[t\to 0]{} 0 \end{aligned}$$

Approximate $u \in \mathcal{C}_\infty(\mathbb{R}^d)$ by a sequence $(u_n)_{n\geqslant 0} \subset \mathcal{C}_c^2(\mathbb{R}^d)$. Then

$$\begin{aligned} \|T_t u - u\|_\infty &\leqslant \|T_t(u - u_n)\|_\infty + \|T_t u_n - u_n\|_\infty + \|u - u_n\|_\infty \\ &\leqslant 2\|u - u_n\|_\infty + \|T_t u_n - u_n\|_\infty. \end{aligned}$$

Letting first $t \to 0$ and then $n \to \infty$ proves strong continuity on $\mathcal{C}_\infty(\mathbb{R}^d)$. We can now determine the generator of the semigroup. Because of Theorem 7.22 we only need to calculate the limit $\lim_{t\to 0}(T_t u(x) - u(x))/t$ for each $x \in \mathbb{R}^d$. Using Itô's formula as before, we see that

$$\begin{aligned} &\frac{\mathbb{E}\, u(X_t^x) - u(x)}{t} \\ &= \sum_{j=1}^d \mathbb{E}\left(\frac{1}{t}\int_0^t b_j(X_s^x)\partial_j u(X_s^x)\,ds\right) + \frac{1}{2}\sum_{i,j=1}^d \mathbb{E}\left(\frac{1}{t}\int_0^t a_{ij}(X_s^x)\partial_i\partial_j u(X_s^x)\,ds\right) \\ &\xrightarrow[t\to 0]{} \sum_{j=1}^d b_j(x)\partial_j u(x) + \frac{1}{2}\sum_{i,j=1}^d a_{ij}(x)\partial_i\partial_j u(x). \end{aligned}$$

For the limit in the last line we use the dominated convergence theorem and the fact that $t^{-1}\int_0^t \phi(s)\,ds \to \phi(0+)$ as $t \to 0$. □

Theorem 21.9 shows what we have to do, if we want to construct a diffusion with a given generator $L(x, D)$ of the form (21.1): the drift coefficient b must be globally Lipschitz, and the positive definite diffusion matrix $a = (a_{ij}) \in \mathbb{R}^{d\times d}$ should be of the form $a = \sigma\sigma^\top$ where $\sigma = (\sigma_{jk}) \in \mathbb{R}^{d\times d}$ is globally Lipschitz continuous.

Let $M \in \mathbb{R}^{d\times d}$ and set $\|M\|^2 := \sup_{|\xi|=1}\langle M^\top M\xi, \xi\rangle$. Then $\|M\|$ is a submultiplicative matrix norm which is compatible with the Euclidean norm on $\mathbb{R}^d$, i. e. $\|MN\| \leqslant \|M\| \cdot \|N\|$ and $|Mx| \leqslant \|M\| \cdot |x|$ for all $x \in \mathbb{R}^d$.

For a symmetric matrix M we write $\lambda_{\min}(M)$ and $\lambda_{\max}(M)$ for the smallest and largest eigenvalue, respectively; we have $\|M\|^2 = \lambda_{\max}(MM^\top)$ and, if M is semidefinite, $\|M\| = \lambda_{\max}(M)$.[1]

21.10 Lemma (van Hemmen–Ando 1980). *Assume that $B, C \in \mathbb{R}^{d\times d}$ are two symmetric, positive semidefinite matrices such that $\lambda_{\min}(B)+\lambda_{\min}(C) > 0$. Then their unique positive semidefinite square roots satisfy*

$$\left\|\sqrt{C} - \sqrt{B}\right\| \leqslant \frac{\|C - B\|}{\max\left\{\lambda_{\min}^{1/2}(B), \lambda_{\min}^{1/2}(C)\right\}}.$$

Proof. A simple change of variables shows that

$$\frac{1}{\sqrt{\lambda}} = \frac{2}{\pi}\int_0^\infty \frac{dr}{r^2 + \lambda}, \quad \text{hence} \quad \sqrt{\lambda} = \frac{2}{\pi}\int_0^\infty \frac{\lambda\, dr}{r^2 + \lambda}.$$

Since B, C are positive semidefinite, we can use this formula to determine the square roots, e. g. for B,

$$\sqrt{B} = \frac{2}{\pi}\int_0^\infty B(s^2\,\mathrm{id} + B)^{-1}\, ds$$

Set $R_s^B := (s^2\,\mathrm{id} + B)^{-1}$ and observe that $BR_s^B - CR_s^C = s^2 R_s^B(B - C)R_s^C$. Thus,

$$\sqrt{C} - \sqrt{B} = \frac{2}{\pi}\int_0^\infty s^2 R_s^C(C - B)R_s^B\, ds.$$

Since the matrix norm is submultiplicative and compatible with the Euclidean norm, we find for all $\xi \in \mathbb{R}^d$ with $|\xi| = 1$ using the Cauchy-Schwarz inequality

$$\left\langle R_s^C(C - B)R_s^B\xi, \xi\right\rangle \leqslant \left|R_s^C(C - B)R_s^B\xi\right| \leqslant \left\|R_s^B\right\| \cdot \left\|R_s^C\right\| \cdot \|C - B\|.$$

Moreover,

$$\left\|R_s^B\right\| = \lambda_{\max}\left(R_s^B\right) = \frac{1}{s^2 + \lambda_{\min}(B)}.$$

1 In fact, if $M = (m_{jk})$, then $\|M\|^2 = \sum_{jk} m_{jk}^2$.

Since C and B play symmetric roles, we may, without loss of generality, assume that $\lambda_{\min}(C) \geqslant \lambda_{\min}(B)$. Thus,

$$\begin{aligned}
\left\|\sqrt{C}-\sqrt{B}\right\| &\leqslant \frac{2}{\pi}\int_0^\infty \left\|R_s^B\right\|\left\|R_s^C\right\| s^2\,ds \cdot \|C-B\| \\
&\leqslant \frac{2}{\pi}\int_0^\infty \frac{s^2}{s^2+\lambda_{\min}(B)}\,\frac{1}{s^2+\lambda_{\min}(C)}\,ds \cdot \|C-B\| \\
&\leqslant \frac{2}{\pi}\int_0^\infty \frac{1}{s^2+\lambda_{\min}(C)}\,ds \cdot \|C-B\| \\
&= \frac{\|C-B\|}{\sqrt{\lambda_{\min}(C)}}.
\end{aligned}$$

□

21.11 Corollary. *Let $L(x,D)$ be the differential operator (21.1) and assume that the diffusion matrix $a(x)$ is globally Lipschitz continuous (Lipschitz constant C) and uniformly positive definite, i. e. $\inf_x \lambda_{\min}(a(x)) = \kappa > 0$. Moreover, assume that the drift vector is globally Lipschitz continuous. Then there exists a diffusion process in the sense of Definition 21.1 such that its generator $(A, \mathfrak{D}(A))$ coincides on the test functions $\mathcal{C}_c^\infty(\mathbb{R}^d)$ with $L(x,D)$.*

Proof. Denote by $\sigma(x)$ the unique positive definite square root of $a(x)$. Using Lemma 21.10 we find

$$\|\sigma(x)-\sigma(y)\| \leqslant \frac{1}{\kappa}\,\|a(x)-a(y)\| \leqslant \frac{C}{\kappa}\,|x-y|, \quad \text{for all } x,y\in\mathbb{R}^d.$$

On a finite dimensional vector space all norms are comparable, i. e. (21.9) holds for σ (possibly with a different Lipschitz constant). Therefore, we can apply Theorem 21.9 and obtain the diffusion process $(X_t)_{t\geqslant 0}$ as the unique solution to the SDE (21.8). □

It is possible to replace the uniform positive definiteness of the matrix $a(x)$ by the weaker assumption that $a_{ij}(x) \in \mathcal{C}_b^2(\mathbb{R}^d,\mathbb{R})$ for all $i,j = 1,\dots,d$, cf. [100, Proposition IV.6.2] or [214, Theorem 5.2.3].

The square root σ is only unique if we require that it is also positive semidefinite. Let us briefly discuss what happens if $\sigma\sigma^\top = a = \rho\rho^\top$ where ρ is some $d{\times}d$ matrix which need not be positive semidefinite. Let $(\sigma, b, (B_t)_{t\geqslant 0})$ be as in Theorem 21.9 and denote by $(X_t^x)_{t\geqslant 0}$ the unique solution of the SDE (21.8). Assume that $(W_t)_{t\geqslant 0}$ is a further Brownian motion which is independent of $(B_t)_{t\geqslant 0}$ and that $(Y_t^x)_{t\geqslant 0}$ is the unique solution of the SDE

$$dY_t = b(Y_t)\,dt + \rho(Y_t)\,dW_t \quad Y_0 = x \in \mathbb{R}^d.$$

The proof of Theorem 21.9 works literally for $(Y_t^x)_{t\geqslant 0}$ since we only need the existence and uniqueness of the solution of the SDE, and the equality $\rho\rho^\top = a$. This means that $(X_t^x)_{t\geqslant 0}$ and $(Y_t^x)_{t\geqslant 0}$ are two Feller processes whose generators, $(A, \mathfrak{D}(A))$ and $(C, \mathfrak{D}(C))$, say, coincide on the set $\mathcal{C}_c^2(\mathbb{R}^d)$. It is tempting to claim that the processes coincide, but for this we need to know that $A|_{\mathcal{C}_c^2} = C|_{\mathcal{C}_c^2}$ implies $(A, \mathfrak{D}(A)) = (C, \mathfrak{D}(C))$. A sufficient criterion would be that $\mathcal{C}_c^2$ is an operator core for A and C, i. e. A and C are the closures of $A|_{\mathcal{C}_c^2}$ and $C|_{\mathcal{C}_c^2}$, respectively. This is far from obvious.

Using the martingale problem technique of Stroock and Varadhan, one can show that – under even weaker conditions than those of Corollary 21.11 – the diffusion process $(X_t^x)_{t\geqslant 0}$ is unique in the following sense: Every process whose generator coincides on $\mathcal{C}_c^2(\mathbb{R}^d)$ with $L(x, D)$, has the same finite dimensional distributions as $(X_t^x)_{t\geqslant 0}$. On the other hand, this is the best we can hope for. For example, we have

$$a(x) = \sigma(x)\sigma^\top(x) = \rho(x)\rho^\top(x) \quad \text{if} \quad \rho(x) = \sigma(x)U$$

for any orthogonal matrix $U \in \mathbb{R}^{d\times d}$. If U is not a function of x, we see that

$$\begin{aligned} dY_t = b(Y_t)\,dt + \rho(Y_t)dB_t &= b(Y_t)\,dt + \sigma(Y_t)U dB_t \\ &= b(Y_t)\,dt + \sigma(Y_t)d(UB)_t. \end{aligned}$$

Of course, $(UB_t)_{t\geqslant 0}$ is again a Brownian motion, but it is different from the original Brownian motion $(B_t)_{t\geqslant 0}$. This shows that the solution $(Y_t^x)_{t\geqslant 0}$ will, even in this simple case, not coincide with $(X_t^x)_{t\geqslant 0}$ in a pathwise sense. Nevertheless, $(X_t^x)_{t\geqslant 0}$ and $(Y_t^x)_{t\geqslant 0}$ have the same finite dimensional distributions. In other words: We cannot expect that there is a strong one-to-one relation between L and the stochastic process.

21.12 Further reading. The interplay of PDEs and SDEs is nicely described in [10] and [80]. The contributions of Stroock and Varadhan, in particular, the characterization of stochastic processes via the martingale problem, opens up a whole new world. Good introductions are [72] and, with the focus on diffusion processes, [214]. The survey [12] gives a brief introduction to the ideas of Malliavin's calculus, see also the *Further reading* recommendations in Chapter 4 and 18.

[10] Bass: *Diffusions and Elliptic Operators.*
[12] Bell: Stochastic differential equations and hypoelliptic operators.
[72] Ethier, Kurtz: *Markov Processes: Characterization and Convergence.*
[80] Freidlin: *Functional Integration and Partial Differential Equations.*
[214] Stroock, Varadhan: *Multidimensional Diffusion Processes.*

Problems

1. Let $(A, \mathfrak{D}(A))$ be the generator of a diffusion process in the sense of Definition 21.1 and denote by a, b the diffusion and drift coefficients. Show that $a \in \mathcal{C}(\mathbb{R}^d, \mathbb{R}^{d\times d})$ and $b \in \mathcal{C}(\mathbb{R}^d, \mathbb{R}^d)$.

2. Show that under the assumptions of Proposition 21.5 we can interchange integration and differentiation: $\frac{\partial^2}{\partial x_j \partial x_k} \int p(t,x,y)u(y)\,dy = \int \frac{\partial^2}{\partial x_j \partial x_k} p(t,x,y)u(y)\,dy$ and that the resulting function is in $\mathcal{C}_\infty(\mathbb{R}^d)$.

3. Let $A : \mathcal{C}_c^\infty(\mathbb{R}^d) \to \mathcal{C}_\infty(\mathbb{R}^d)$ be a closed operator such that $\|Au\|_\infty \leqslant C\,\|u\|_{(2)}$; here $\|u\|_{(2)} = \|u\|_\infty + \sum_j \|\partial_j u\|_\infty + \sum_{jk} \|\partial_j\partial_k u\|_\infty$. Show that $\mathcal{C}_\infty^2(\mathbb{R}^d) = \overline{\mathcal{C}_c^\infty(\mathbb{R}^d)}^{\|\cdot\|_{(2)}} \subset \mathfrak{D}(A)$.

4. Let $L = L(x, D_x) = \sum_{ij} a_{ij}(x)\partial_i\partial_j + \sum_j b_j(x)\partial_j + c(x)$ be a second order differential operator. Determine its formal adjoint $L^*(y, D_y)$, i. e. the linear operator satisfying $\langle Lu, \phi\rangle_{L^2} = \langle u, L^*\phi\rangle_{L^2}$ for all $u, \phi \in \mathcal{C}_c^\infty(\mathbb{R}^d)$. What is the formal adjoint of the operator $L(x, D) + \frac{\partial}{\partial t}$?

5. Complete the proof of Proposition 21.6 (Kolmogorov's forward equation).

6. Let $(B_t)_{t\geqslant 0}$ be a BM^1. Show that $X_t := (B_t, \int_0^t B_s\,ds)$ is a diffusion process and determine its generator.

7. (Carré-du-champ operator) Let $(X_t)_{t\geqslant 0}$ be a diffusion process with infinitesimal generator $L = L(x, D) = A|_{\mathcal{C}_c^\infty}$ as in (21.1). Write $M_t^u = u(X_t) - u(X_0) - \int_0^t Lu(X_r)\,dr$.
   (a) Show that $M^u = (M_t^u)_{t\geqslant 0}$ is an L^2-martingale for all $u \in \mathcal{C}_c^\infty(\mathbb{R}^d)$.
   (b) Show that $M^f = (M_t^f)_{t\geqslant 0}$ is a local martingale for all $f \in \mathcal{C}^2(\mathbb{R}^d)$.
   (c) Denote by $\langle M^u, M^\phi\rangle_t$, $u, \phi \in \mathcal{C}_c^\infty(\mathbb{R}^d)$, the quadratic covariation of the L^2 martingales M^u and M^ϕ, cf. (15.8) and Theorem 15.13 b). Show that

$$\langle M^u, M^\phi\rangle_t = \int_0^t (L(u\phi) - uL\phi - \phi Lu)(X_s)\,ds = \int_0^t \nabla u(X_s)\cdot a(X_s)\nabla\phi(X_s)\,ds.$$

   ($x \cdot y$ denotes the Euclidean scalar product and $\nabla = (\partial_1, \dots, \partial_d)^\top$.)
   Remark: The operator $\Gamma(u, \phi) := L(u\phi) - uL\phi - \phi Lu$ is called *carré-du-champ* or *mean-field operator*.

22 Simulation of Brownian motion

by Björn Böttcher

This chapter presents a concise introduction to simulation, in particular focusing on the normal distribution and Brownian motion.

22.1 Introduction

Recall that for iid random variables $X_1, \dots, X_n$ with a common distribution F for any $\omega \in \Omega$ the tuple $(X_1(\omega), \dots, X_n(\omega))$ is called a *(true) sample* of F. In simulations the following notion of a sample is commonly used.

22.1 Definition. A tuple of values $(x_1, \dots, x_n)$ which is indistinguishable from a true sample of F by statistical inference is called a *sample* of F. An element of such a tuple is called a *sample value* of F, and it is denoted by

$$x_j \overset{s}{\sim} F.$$

Note that the statistical inference has to ensure both the distributional properties and the independence. A naive idea to verify the distributional properties would be, for instance, the use of a Kolmogorov–Smirnov test on the distribution. But this is not a good test for the given problem, since a true sample would fail such a test with confidence level α with probability α. Instead one should do at least repeated Kolmogorov–Smirnov tests and then test the p-values for uniformity. In fact, there is a wide range of possible tests, for example the NIST (National Institute of Standards and Technology)[1] has defined a set of tests which should be applied for reliable results.

In theory, letting n tend to ∞ and allowing all possible statistical tests, only a true sample would be a sample. But in practice it is impossible to get a true sample of a continuous distribution, since a physical experiment as sample mechanism clearly yields measurement and discretization errors. Moreover for the generation of samples with the help of a computer we only have a discrete and finite set of possible values due to the internal number representation. Thus one considers n large, but not arbitrarily large.

Nevertheless, in simulations one treats a sample as if it were a true sample, i. e. any distributional relation is translated to samples. For example, given a distribution function F and the uniform distribution $\mathsf{U}_{(0,1)}$ we have

$$X \sim \mathsf{U}_{(0,1)} \implies F^{-1}(X) \sim F$$

since

$$\mathbb{P}(F^{-1}(X) \leqslant x) = \mathbb{P}(X \leqslant F(x)) = F(x)$$

1 http://csrc.nist.gov/groups/ST/toolkit/rng/stats_tests.html

and thus
$$x \overset{s}{\sim} \mathsf{U}_{(0,1)} \implies F^{-1}(x) \overset{s}{\sim} F \tag{22.1}$$

holds. This yields the following algorithm.

22.2 Algorithm (Inverse transform method). Let F be a distribution function.

1. Generate $u \overset{s}{\sim} \mathsf{U}_{(0,1)}$
2. Set $x := F^{-1}(u)$

Then $x \overset{s}{\sim} F$.

By this algorithm a sample of $\mathsf{U}_{(0,1)}$ can be transformed to a sample of any other distribution. Therefore the generation of samples from the uniform distribution is the basis for all simulations. For this various algorithms are known and they are implemented in most of the common programming languages. Readers interested in the details of the generation of samples of the uniform distribution are referred to [140; 141].

From now on we will assume that we can generate $x \overset{s}{\sim} \mathsf{U}_{(0,1)}$. Thus we can apply Algorithm 22.2 to get samples of a given distribution as the following examples illustrate.

22.3 Example (Scaled and shifted Uniform distribution $\mathsf{U}_{(a,b)}$). For $a < b$ the distribution function of $\mathsf{U}_{(a,b)}$ is given by
$$F(x) = \frac{x-a}{b-a} \quad \text{for } x \in [a,b]$$
and thus
$$u \overset{s}{\sim} \mathsf{U}_{(0,1)} \implies (b-a)u + a \overset{s}{\sim} \mathsf{U}_{(a,b)}.$$

22.4 Example (Bernoulli distribution β_p and $\mathsf{U}_{\{0,1\}}$, $\mathsf{U}_{\{-1,1\}}$). Let $p \in [0,1]$. Given a uniform random variable $U \sim \mathsf{U}_{(0,1)}$, then $\mathbb{1}_{[0,p]}(U) \sim \beta_p$. In particular, we get
$$u \overset{s}{\sim} \mathsf{U}_{(0,1)} \implies \mathbb{1}_{[0,\frac{1}{2}]}(u) \overset{s}{\sim} \mathsf{U}_{\{0,1\}} \quad \text{and} \quad 2 \cdot \mathbb{1}_{[0,\frac{1}{2}]}(u) - 1 \overset{s}{\sim} \mathsf{U}_{\{-1,1\}}.$$

22.5 Example (Exponential distribution Exp_λ). The distribution function of Exp_λ with parameter $\lambda > 0$ is $F(x) = 1 - e^{-\lambda x}$, $x \geqslant 0$, and thus
$$F^{-1}(y) = -\frac{1}{\lambda}\ln(1-y).$$
Now using
$$U \sim \mathsf{U}_{(0,1)} \implies 1 - U \sim \mathsf{U}_{(0,1)}$$
yields with the inverse transform method
$$u \overset{s}{\sim} \mathsf{U}_{(0,1)} \implies -\frac{1}{\lambda}\ln(u) \overset{s}{\sim} \mathsf{Exp}_\lambda.$$

Before focusing on the normal distribution, let us introduce two further algorithms which can be used to transform a sample of a distribution G to a sample of a distribution F.

22.6 Algorithm (Acceptance–rejection method). Let F and G be probability distribution functions with densities f and g, such that there is a constant $c > 0$ with $f(x) \leqslant cg(x)$ for all x.

1. Generate $y \overset{s}{\sim} G$
2. Generate $u \overset{s}{\sim} \mathsf{U}_{(0,1)}$
3. If $ucg(y) \leqslant f(y)$, then $x := y$
   otherwise restart with 1.

Then $x \overset{s}{\sim} F$.

Proof. If $Y \sim G$ and $U \sim \mathsf{U}_{(0,1)}$ are independent, then the algorithm generates a sample value of the distribution μ, given by

$$\mu(A) = \mathbb{P}(Y \in A \,|\, Ucg(Y) \leqslant f(Y)) = \frac{\mathbb{P}(Y \in A, Ucg(Y) \leqslant f(Y))}{\mathbb{P}(Ucg(Y) \leqslant f(Y))} = \int_A f(y)\,dy$$

for $A \in \mathscr{B}(\mathbb{R})$. This follows from the fact that $g(x) = 0$ implies $f(x) = 0$, and

$$\begin{aligned}
\mathbb{P}(Y \in A,\, Ucg(Y) \leqslant f(Y)) &= \int_0^1 \int_A \mathbb{1}_{\{(v,z)\,:\,vcg(z) \leqslant f(z)\}}(u, y)\, g(y)\,dy\,du \\
&= \int_{A\setminus\{g=0\}} \int_0^{\frac{f(y)}{cg(y)}} du\; g(y)\,dy \\
&= \int_{A\setminus\{g=0\}} \frac{f(y)}{cg(y)} g(y)\,dy \\
&= \frac{1}{c} \int_{A\setminus\{g=0\}} f(y)\,dy \\
&= \frac{1}{c} \int_A f(y)\,dy.
\end{aligned}$$

In particular, we see for $A = \mathbb{R}$

$$\mathbb{P}(Ucg(Y) \leqslant f(Y)) = \mathbb{P}(Y \in \mathbb{R},\, Ucg(Y) \leqslant f(Y)) = \frac{1}{c}. \qquad \square$$

22.7 Example (One-sided normal distribution). We can apply the acceptance–rejection method to the one-sided normal distribution

$$f(x) = 2\frac{1}{\sqrt{2\pi}}\, e^{-\frac{1}{2}x^2}\, \mathbb{1}_{[0,\infty)}(x),$$

and the Exp_1 distribution $g(x) = e^{-x}\, \mathbb{1}_{[0,\infty)}(x)$ (for sampling see Example 22.5) with the smallest possible c (such that $f(x) \leqslant cg(x)$ for all x) given by

$$c = \sup_x \frac{f(x)}{g(x)} = \sup_x \sqrt{\frac{2}{\pi}} e^{-\frac{x^2}{2}+x}\, \mathbb{1}_{[0,\infty)}(x) = \sqrt{\frac{2e}{\pi}} \approx 1.315\,.$$

The name of Algorithm 22.6 is due to the fact that in the first step a sample is proposed and in the third step this is either accepted or rejected. In the case of rejection a new sample is proposed until acceptance. The algorithm can be visualized as throwing darts onto the set $B = \{(x,y) \in \mathbb{R}^2 : 0 \leqslant y \leqslant cg(x)\}$ uniformly. Then the x-coordinate of the dart is taken as a sample (accepted) if, and only if, the dart is within $C = \{(x,y) \in \mathbb{R}^2 : 0 \leqslant y \leqslant f(x)\}$; note that $C \subset B$ is satisfied by our assumptions. This also shows that, on average, the algorithm proposes $\lambda^2(B)/\lambda^2(C) = c$ times a sample until acceptance.

In contrast, the following algorithm does not produce *unused* samples.

22.8 Algorithm (Composition method). Let F be a distribution with density f of the form

$$f(x) = \int g_y(x)\,\mu(dy)$$

where μ is a probability measure and g_y is for each y a density.

1. Generate $y \overset{s}{\sim} \mu$
2. Generate $x \overset{s}{\sim} g_y$

Then $x \overset{s}{\sim} F$.

Proof. Let $Y \sim \mu$ and, conditioned on $\{Y = y\}$, let $X \sim g_y$, i. e.

$$P(X \leqslant x) = \int \mathbb{P}(X \leqslant x \mid Y = y)\,\mu(dy) = \int \int_{-\infty}^{x} g_y(z)\,dz\,\mu(dy) = \int_{-\infty}^{x} f(z)dz.$$

Thus we have $X \sim F$. □

22.9 Example (Normal distribution). We can write the density of the standard normal distribution as

$$f(x) = \frac{1}{\sqrt{2\pi}} e^{-\frac{1}{2}x^2} = \underbrace{\frac{1}{2}\,\frac{2}{\sqrt{2\pi}} e^{-\frac{1}{2}x^2}\,\mathbb{1}_{(-\infty,0]}(x)}_{=g_{-1}(x)} + \underbrace{\frac{1}{2}\,\frac{2}{\sqrt{2\pi}} e^{-\frac{1}{2}x^2}\,\mathbb{1}_{(0,\infty)}(x)}_{=g_1(x)}$$

i. e. $f(x) = \int_{-\infty}^{\infty} g_y(x)\,\mu(dy)$ with $\mu(dy) = \frac{1}{2}\delta_{-1}(dy) + \frac{1}{2}\delta_1(dy)$.

Note that $X \sim g_1$ can be generated by Example 22.7 and that

$$X \sim g_1 \implies -X \sim g_{-1}$$

holds. Thus by Example 22.3 and the composition method we obtain the following algorithm.

22.10 Algorithm (Composition of one-sided normal samples). Let f denote the density of the one-sided normal distribution, see Example 22.7.

1. Generate $y \overset{s}{\sim} \mathsf{U}_{\{-1,1\}}$, $z \overset{s}{\sim} f$
2. Set $x := yz$

Then $x \overset{s}{\sim} \mathsf{N}(0,1)$.

In the next section we will see further algorithms to generate samples of the normal distribution.

22.2 Normal distribution

For the one-dimensional standard normal distribution and $\mu \in \mathbb{R}$, $\sigma > 0$ one has

$$X \sim \mathsf{N}(0,1) \implies \sigma X + \mu \sim \mathsf{N}(\mu, \sigma^2).$$

Thus we have the following algorithm to transform samples of the standard normal distribution to samples of a general normal distribution.

22.11 Algorithm (Change of mean and variance). Let $\mu \in \mathbb{R}$ and $\sigma > 0$.

1. `Generate` $y \overset{s}{\sim} \mathsf{N}(0,1)$
2. `Set` $x := \sigma y + \mu$

Then $x \overset{s}{\sim} \mathsf{N}(\mu, \sigma^2)$.

Based on the above algorithm we can focus in the following on the generation of samples of $\mathsf{N}(0,1)$.

The standard approach is the inverse transform method, Algorithm 22.2, with a numerical inversion of the distribution function of the standard normal distribution.

22.12 Algorithm (Numerical inversion).

1. `Generate` $u \overset{s}{\sim} \mathsf{U}_{(0,1)}$
2. `Set` $x := F^{-1}(u)$ (where F is the $\mathsf{N}(0,1)$-distribution function)

Then $x \overset{s}{\sim} \mathsf{N}(0,1)$.

A different approach which generates two samples at once is the following

22.13 Algorithm (Box–Muller-method).

1. `Generate` $u_1, u_2 \overset{s}{\sim} \mathsf{U}_{(0,1)}$
2. `Set` $r := \sqrt{-2\ln(u_1)},\ \ v := 2\pi u_2$
3. `Set` $x_1 := r\cos(v),\ \ x_2 := r\sin(v)$

Then $x_j \overset{s}{\sim} \mathsf{N}(0,1)$.

Proof. Let $X_1, X_2 \sim \mathsf{N}(0,1)$ be independent normal random variables. Their joint density is $f_{X_1,X_2}(x_1,x_2) = \frac{1}{2\pi}\exp\left(-\frac{x_1^2+x_2^2}{2}\right)$. Now we express the vector (X_1, X_2) in polar coordinates, i. e.

$$\begin{pmatrix} X_1 \\ X_2 \end{pmatrix} = \begin{pmatrix} R\cos V \\ R\sin V \end{pmatrix} \tag{22.2}$$

and the probability density becomes, in polar coordinates, $r > 0, v \in [0, 2\pi)$

$$f_{R,V}(r,v) = f_{X_1,X_2}(r\cos v, r\sin v)\det\begin{pmatrix} \cos v & -r\sin v \\ \sin v & r\cos v \end{pmatrix} = e^{-\frac{1}{2}r^2}\frac{r}{2\pi}.$$

Thus R and V are independent, $V \sim \mathsf{U}_{[0,2\pi)}$ and the distribution of R^2 is

$$\mathbb{P}(R^2 \leqslant x) = \int_0^{\sqrt{x}} e^{-\frac{1}{2}r^2} r\,dr = \int_0^x \frac{1}{2}e^{-\frac{1}{2}s}\,ds = 1 - e^{-\frac{1}{2}x},$$

i. e. $R^2 \sim \mathsf{Exp}_{1/2}$.

The second step of the algorithm is just an application of the inverse transform method, Examples 22.3 and 22.5, to generate samples of R and V, in the final step (22.2) is applied. □

A simple application of the central limit theorem can be used to get (approximately) standard normal distributed samples.

22.14 Algorithm (Central limit method).

1. Generate $u_j \overset{s}{\sim} \mathsf{U}_{(0,1)}$, $j = 1, \ldots, 12$
2. Set $x := \sum_{j=1}^{12} u_j - 6$

Then $x \overset{s}{\sim} \mathsf{N}(0, 1)$ approximately.

Proof. Let $\mathsf{U}_j \sim \mathsf{U}_{(0,1)}$, $j = 1, 2, \ldots, 12$, be independent, then $\mathbb{E}(\mathsf{U}_j) = \frac{1}{2}$ and

$$\mathbb{V}\,\mathsf{U}_j = \mathbb{E}\,U_j^2 - (\mathbb{E}\,U_j)^2 = \frac{x^3}{3}\bigg|_0^1 - \frac{1}{4} = \frac{4}{12} - \frac{3}{12} = \frac{1}{12}.$$

By the central limit theorem

$$\frac{\sum_{j=1}^{12} \mathsf{U}_j - 12\,\mathbb{E}\,\mathsf{U}_1}{\sqrt{12\,\mathbb{V}\,\mathsf{U}_1}} = \sum_{j=1}^{12} \mathsf{U}_j - 6 \sim \mathsf{N}(0, 1) \quad \text{(approximately).}$$ □

Clearly one could replace the constant 12 in the first step of the above algorithm by any other integer n, but then the corresponding sum in step 2 would have to be rescaled by $(n/12)^{-\frac{1}{2}}$ and this would require a further calculation. In the above setting it can be shown that the pointwise difference of the standard normal distribution function and the distribution function of $\left(\sum_{j=1}^{12} U_j - 12\,\mathbb{E}\,U_1\right) / \sqrt{12\,\mathbb{V}\,U_1}$ for iid $U_j \sim \mathsf{U}_{(0,1)}$ is always less than 0.0023, see [193, Section 3.1].

Now we have seen several ways to generate samples of the normal distribution, i. e. we know how to simulate increments of Brownian motion.

22.3 Brownian motion

A simulation of a continuous time process is always based on a finite sample, since we can only generate finitely many values in finite time. Brownian motion is usually simulated on a discrete time grid $0 = t_0 < t_1 < \cdots < t_n$, and it is common practice for visualizations to interpolate the simulated values linearly. This is justified by the fact that Brownian Motion has continuous paths.

Throughout this section $(B_t)_{t\geqslant 0}$ denotes a one-dimensional standard Brownian motion. The simplest method to generate a Brownian path uses the property of stationary and independent increments.

22.15 Algorithm (Independent increments). Let $0 = t_0 < t_1 < \cdots < t_n$ be a time grid.
`Initialize` $b_0 := 0$
`For` $j = 1$ `to` n
1. `Generate` $y \overset{s}{\sim} \mathsf{N}(0, t_j - t_{j-1})$
2. `Set` $b_{t_j} := b_{t_{j-1}} + y$

Then $(b_0, b_{t_1}, \dots, b_{t_n}) \overset{s}{\sim} (B_0, B_{t_1}, \dots, B_{t_n})$.

This method works particularly well if the step size $\delta = t_j - t_{j-1}$ is constant for all j, in this case one only has to generate $y \overset{s}{\sim} \mathsf{N}(0, \delta)$ repeatedly. A drawback of the method is, that any refinement of the discretization yields a new simulation and, thus, a different path.

To get a refinement of an already simulated path one can use the idea of Lévy's original argument, see Section 3.4.

22.16 Algorithm (Interpolation. Lévy's argument). Let $0 \leqslant s_0 < s < s_1$ be fixed times and (b_{s_0}, b_{s_1}) be a sample value of (B_{s_0}, B_{s_1}).
1. `Generate` $b_s \overset{s}{\sim} \mathsf{N}\left(\frac{(s_1-s)b_{s_0}+(s-s_0)b_{s_1}}{s_1-s_0}, \frac{(s-s_0)(s_1-s)}{s_1-s_0}\right)$

Then $(b_{s_0}, b_s, b_{s_1}) \overset{s}{\sim} (B_{s_0}, B_s, B_{s_1})$ given that $B_{s_0} = b_{s_0}$ and $B_{s_1} = b_{s_1}$.

Proof. In Section 3.4 we have seen that in the above setting

$$\mathbb{P}(B_s \in \cdot \mid B_{s_0} = b_{s_0}, B_{s_1} = b_{s_1}) = \mathsf{N}\left(m_s, \sigma_s^2\right)$$

with

$$m_s = \frac{(s_1 - s)b_{s_0} + (s - s_0)b_{s_1}}{s_1 - s_0} \quad \text{and} \quad \sigma_s^2 = \frac{(s - s_0)(s_1 - s)}{s_1 - s_0}. \qquad \square$$

Thus, for $0 = t_0 < t_1 < \cdots < t_n$ and $(b_0, b_{t_1}, \dots, b_{t_n}) \overset{s}{\sim} (B_0, B_{t_1}, \dots, B_{t_n})$ we get

$$(b_0, b_{t_1}, \dots, b_{t_j}, b_r, b_{t_{j+1}}, \dots, b_{t_n}) \overset{s}{\sim} (B_0, B_{t_1}, \dots, B_{t_j}, B_r, B_{t_{j+1}}, \dots, B_{t_n})$$

for some $r \in (t_j, t_{j+1})$ by generating the intermediate point b_r given $(b_{t_j}, b_{t_{j+1}})$. Repeated applications yield an arbitrary refinement of a given discrete simulation of a Brownian path. Taking only the dyadic refinement of the unit interval we get the Lévy–Ciesielski representation, cf. Section 3.2.

22.17 Algorithm (Lévy–Ciesielski). Let $J \geqslant 1$ be the order of refinement.
Initialize $b_0 := 0$
Generate $b_1 \overset{s}{\sim} \mathsf{N}(0,1)$
For $j = 0$ to $J-1$
  For $l = 0$ to $2^j - 1$
1. Generate $y \overset{s}{\sim} \mathsf{N}(0,1)$
2. Set $b_{(2l+1)/2^{j+1}} = \frac{1}{2}\left(b_{l/2^j} + b_{(l+1)/2^j}\right) + 2^{-(\frac{j}{2}+1)}y$

Then $(b_0, b_{1/2^J}, b_{2/2^J}, \dots, b_1) \overset{s}{\sim} (B_0, B_{1/2^J}, B_{2/2^J}, \dots, B_1)$.

Proof. As noted at the end of Section 3.4, the expansion of B_t at dyadic times has only finitely many terms and

$$B_{(2l+1)/2^{j+1}} = \frac{1}{2}\left(B_{l/2^j} + B_{(l+1)/2^j}\right) + 2^{-(\frac{j}{2}+1)}Y$$

where $Y \sim \mathsf{N}(0,1)$. □

Analogous to the central limit method for the normal distribution we can also use Donsker's invariance principle, Theorem 3.10, to get an approximation to a Brownian path.

22.18 Algorithm (Donsker's invariance principle). Let n be the number of steps.
Initialize $b_0 := 0$
1. Generate $u_j \overset{s}{\sim} \mathsf{U}_{\{-1,1\}}$, $j = 1, \dots, n$
2. For $k = 1$ to n
   Set $b_{k/n} = \frac{1}{\sqrt{n}}u_k + b_{(k-1)/n}$

Then $(b_0, b_{1/n}, \dots, b_{2/n}, \dots, b_1) \overset{s}{\sim} (B_0, B_{1/n}, B_{2/n}, \dots, B_1)$ approximately.

Proof. The result follows as in Section 3.5. □

Clearly one could also replace the Bernoulli random walk in the above algorithm by a different symmetric random walk with finite variance.

22.4 Multivariate Brownian motion

Since a d-dimensional Brownian motion has independent one-dimensional Brownian motions as components, cf. Corollary 2.9, we can use the algorithms from the last section to generate the components separately.

In order to generate a Q-Brownian motion we need to simulate samples of a multivariate normal distribution with correlated components.

22.19 Algorithm (Two-dimensional Normal distribution). Let an expectation vector $\mu \in \mathbb{R}^2$, variances $\sigma_1^2, \sigma_2^2 > 0$ and correlation $\rho \in [-1, 1]$ be given.

1. Generate $y_1, y_2, y_3 \overset{s}{\sim} \mathsf{N}(0,1)$
2. Set $x_1 := \sigma_1\left(\sqrt{1-|\rho|}\, y_1 + \sqrt{|\rho|}\, y_3\right) + \mu_1$

   $x_2 := \sigma_2\left(\sqrt{1-|\rho|}\, y_2 + \operatorname{sgn}(\rho)\sqrt{|\rho|}\, y_3\right) + \mu_2$

Then $(x_1, x_2)^\top \overset{s}{\sim} \mathsf{N}\left(\mu, \begin{pmatrix} \sigma_1^2 & \rho\sigma_1\sigma_2 \\ \rho\sigma_1\sigma_2 & \sigma_2^2 \end{pmatrix}\right)$.

Proof. Let $Y_1, Y_2, Y_3 \sim \mathsf{N}(0,1)$ be independent random variables. Then

$$\sigma_1\sqrt{1-|\rho|}\begin{pmatrix} Y_1 \\ 0 \end{pmatrix} + \sigma_2\sqrt{1-|\rho|}\begin{pmatrix} 0 \\ Y_2 \end{pmatrix} + \sqrt{|\rho|}\begin{pmatrix} \sigma_1 Y_3 \\ \operatorname{sgn}(\rho)\sigma_2 Y_3 \end{pmatrix}$$

is a normal random variable since it is a sum of independent $\mathsf{N}(0,.)$ distributed random vectors. Observe that the last vector is bivariate normal with completely dependent components. Calculating the covariance matrix yields the statement. □

This algorithm shows nicely how the correlation of the components is introduced into the sample. In the next algorithm this is less obvious, but the algorithm works in any dimension.

22.20 Algorithm (Multivariate Normal distribution). Let a positive semidefinite covariance matrix $Q \in \mathbb{R}^{d\times d}$ and an expectation vector $\mu \in \mathbb{R}^d$ be given.

Calculate the Cholesky decomposition $\Sigma\Sigma^\top = Q$

1. Generate $y_1, \dots, y_d \overset{s}{\sim} \mathsf{N}(0,1)$ and let y be the vector with components $y_1, \dots, y_d$
2. Set $x := \mu + \Sigma y$

Then $x \overset{s}{\sim} \mathsf{N}(\mu, Q)$.

Proof. Compare with the discussion following Definition 2.11. □

Now we can simulate Q-Brownian motion using the independence and stationarity of the increments.

22.21 Algorithm (Q-Brownian motion). Let $0 = t_0 < t_1 < \cdots < t_n$ be a time grid and $Q \in \mathbb{R}^{d\times d}$ be a positive semidefinite covariance matrix.

Initialize $b_k^0 := 0,\ k = 1, \dots, d$

For $j = 1$ to n

1. Generate $y \overset{s}{\sim} \mathsf{N}\left(0, (t_j - t_{j-1})Q\right)$
2. Set $b^{t_j} := b^{t_{j-1}} + y$

Then $(b^0, b^{t_1}, \dots, b^{t_n}) \overset{s}{\sim} (B_0, B_{t_1}, \dots, B_{t_n})$, where $(B_t)_{t\geqslant 0}$ is a Q-Brownian motion.

Fig. 22.1. Simulation of a two-dimensional Q-Brownian motion.

22.22 Example. Figure 22.1 shows the path of a Q-Brownian motion for $t \in [0, 10]$ with step width 0.0001 and correlation $\rho = 0.8$, i. e. $Q = \begin{pmatrix} 1 & \rho \\ \rho & 1 \end{pmatrix}$. Note that it seems as if the correlation were visible, but – especially since it is just a single sample and the scaling of the axes is not given – one should be careful when drawing conclusions from samples.

22.5 Stochastic differential equations

Based on Chapter 19 we consider in this section a process $(X_t)_{t\geqslant 0}$ defined by the stochastic differential equation

$$X_0 = x_0 \quad \text{and} \quad dX_t = b(t, X_t)\,dt + \sigma(t, X_t)\,dB_t. \tag{22.3}$$

Corollary 19.12 and Theorem 19.13 provide conditions which ensure that (22.3) has a unique solution. We present two algorithms that allow us to simulate the solution.

22.23 Algorithm (Euler scheme). Assume that b and σ in (22.3) satisfy for all $x, y \in \mathbb{R}$, $0 \leqslant s < t \leqslant T$ and some $K > 0$

$$|b(t,x) - b(t,y)| + |\sigma(t,x) - \sigma(t,y)| \leqslant K\,|x-y|, \tag{22.4}$$

$$|b(t,x)| + |\sigma(t,x)| \leqslant K(1 + |x|), \tag{22.5}$$

$$|b(t,x) - b(s,x)| + |\sigma(t,x) - \sigma(s,x)| \leqslant K(1+|x|)\sqrt{t-s}. \tag{22.6}$$

Let $n \geqslant 1$, $0 = t_0 < t_1 < \cdots < t_n = T$ be a time grid, and x_0 be the initial position.
For $j = 1$ to n
1. Generate $y \overset{s}{\sim} \mathsf{N}(0, t_j - t_{j-1})$
2. Set $x_{t_j} := x_{t_{j-1}} + b(t_{j-1}, x_{t_{j-1}})(t_j - t_{j-1}) + \sigma(t_{j-1}, x_{t_{j-1}})\,y$

Then $(x_0, x_{t_1}, \ldots, x_{t_n}) \overset{s}{\sim} (X_0, X_{t_1}, \ldots, X_{t_n})$ approximately, where $(X_t)_{t\geqslant 0}$ is the solution to (22.3).

Note that the conditions (22.4) and (22.5) ensure the existence of a unique solution.

The approximation error can be measured in various ways. For this denote, in the setting of the algorithm, the maximal step width by $\delta := \max_{j=1,\dots,n}(t_j - t_{j-1})$ and the sequence of random variables corresponding to step 2 by $(X^\delta_{t_j})_{j=0,\dots,n}$. Given a Brownian motion $(B_t)_{t\geqslant 0}$, we can define $X^\delta_{t_j}$, by $X^\delta_{t_0} := x_0$, and for $j = 1, \dots, n$ by

$$X^\delta_{t_j} := X^\delta_{t_{j-1}} + b(t_{j-1}, X^\delta_{t_{j-1}})(t_j - t_{j-1}) + \sigma(t_{j-1}, X^\delta_{t_{j-1}})(B_{t_j} - B_{t_{j-1}}). \tag{22.7}$$

The process $(X^\delta_{t_j})_{j=0,\dots,n}$ is called the *Euler scheme* approximation of the solution $(X_t)_{t\geqslant 0}$ to the equation (22.3).

We say that an approximation scheme has *strong order* of convergence β if

$$\mathbb{E}\,|X^\delta_T - X_T| \leqslant C\delta^\beta.$$

The following result shows that the algorithm provides an approximate solution to (22.3).

22.24 Theorem. *In the setting of Algorithm 22.23 the strong order of convergence of the Euler scheme is* $1/2$. *Moreover, the convergence is locally uniform, i. e.*

$$\sup_{t\leqslant T} \mathbb{E}\,|X^\delta_t - X_t| \leqslant c_T\sqrt{\delta},$$

where $(X^\delta_t)_{t\leqslant T}$ *is the piecewise constant extension of the Euler scheme from the discrete time set* $t_0, \dots, t_n$ *to* $[0, T]$ *defined by*

$$X^\delta_t := X^\delta_{t_{n(t)}}$$

with $n(t) := \max\{n : t_n \leqslant t\}$.

In the same setting and with some additional technical conditions, cf. [5] and [127], one can show that the *weak order* of convergence of the Euler scheme is 1, i. e.

$$\left|\,\mathbb{E}\, g(X^\delta_T) - \mathbb{E}\, g(X_T)\right| \leqslant c_T\delta$$

for any $g \in \mathcal{C}^4$ whose first four derivatives grow at most polynomially.

Proof of Theorem 22.24.. Let b, σ, T, n, be as in Algorithm 22.23. Furthermore let $(X_t)_{t\geqslant 0}$ be the solution to the corresponding stochastic differential equation (22.3), $(X^\delta_{t_j})_{j=0,\dots,n}$ be the Euler scheme defined by (22.7) and $(X^\delta_t)_{t\leqslant T}$ be its piecewise constant extension onto $[0, T]$.

Note that this allows us to rewrite X^δ_t as

$$X^\delta_t = x_0 + \int_0^{t_{n(t)}} b(t_{n(s)}, X^\delta_{t_{n(s)}})\,ds + \int_0^{t_{n(t)}} \sigma(t_{n(s)}, X^\delta_{t_{n(s)}})\,dB_s.$$

Thus, we can calculate the error by

$$Z(T) := \sup_{t\leqslant T} \mathbb{E}\left[|X_t^\delta - X_t|^2\right] \leqslant \mathbb{E}\left[\sup_{t\leqslant T} |X_t^\delta - X_t|^2\right] \leqslant 3Z_1(T) + 3Z_2(T) + 3Z_3(T), \tag{22.8}$$

where Z_1, Z_2 and Z_3 will be given explicitly below. We begin with the first term and estimate it analogously to the proof of Theorem 19.11, see (19.20), (19.21), and apply the estimate (22.4).

$$\begin{aligned}
Z_1(T) &:= \mathbb{E}\left[\sup_{t\leqslant T}\left|\int_0^{t_{n(t)}} \left(b(t_{n(s)}, X_{t_{n(s)}}^\delta) - b(t_{n(s)}, X_s)\right) ds \right.\right.\\
&\qquad\left.\left. + \int_0^{t_{n(t)}} \left(\sigma(t_{n(s)}, X_{t_{n(s)}}^\delta) - \sigma(t_{n(s)}, X_s)\right) dB_s\right|^2\right]\\
&\leqslant 2T\int_0^T \mathbb{E}\left[|b(t_{n(s)}, X_{t_{n(s)}}^\delta) - b(t_{n(s)}, X_s)|^2\right] ds\\
&\qquad + 8\int_0^T \mathbb{E}\left[|\sigma(t_{n(s)}, X_{t_{n(s)}}^\delta) - \sigma(t_{n(s)}, X_s)|^2\right] ds\\
&\leqslant (2TK^2 + 8K^2)\int_0^T \mathbb{E}\left[|X_{t_{n(s)}}^\delta - X_s|^2\right] ds\\
&\leqslant C\int_0^T Z(s)\, ds.
\end{aligned}$$

The second term in (22.8) can be estimated in a similar way.

$$\begin{aligned}
Z_2(T) &:= \mathbb{E}\left[\sup_{t\leqslant T}\left|\int_0^{t_{n(t)}} \left(b(t_{n(s)}, X_s) - b(s, X_s)\right) ds + \int_0^{t_{n(t)}} \left(\sigma(t_{n(s)}, X_s) - \sigma(s, X_s)\right) dB_s\right|^2\right]\\
&\leqslant 2T\int_0^T \mathbb{E}\left[|b(t_{n(s)}, X_s) - b(s, X_s)|^2\right] ds + 8\int_0^T \mathbb{E}\left[|\sigma(t_{n(s)}, X_s) - \sigma(s, X_s)|^2\right] ds\\
&\leqslant (2TK^2 + 8K^2)\int_0^T (1 + \mathbb{E}\,|X_s|)^2 \left(\sqrt{s - t_{n(s)}}\right)^2 ds\\
&\leqslant C\delta,
\end{aligned}$$

where we used (22.6) for the second inequality, and $|s - t_{n(s)}| < \delta$ and (19.22) for the last inequality. The terms Z_1 and Z_2 only consider the integral up to time $t_{n(t)}$, thus the third term takes care of the remainder up to time t. It can be estimated using (22.5),

$|s - t_{n(s)}| < \delta$ and the estimate (19.22).

$$Z_3(T) := \mathbb{E}\left[\sup_{t\leqslant T}\left|\int_{t_{n(t)}}^{t} b(s, X_s)\, ds + \int_{t_{n(t)}}^{t} \sigma(s, X_s)\, dB_s\right|^2\right]$$
$$\leqslant 2\delta \int_{t_{n(t)}}^{t} \mathbb{E}\left[|b(s, X_s)|^2\right] ds + 8 \int_{t_{n(t)}}^{t} \mathbb{E}\left[|\sigma(s, X_s)|^2\right] ds \leqslant C\delta.$$

Note that all constants appearing in the estimates above may depend on T, but do not depend on δ. Thus we have shown

$$Z(T) \leqslant C_T\delta + C_T \int_0^T Z(s)\, ds,$$

where C_T is independent of δ. By Gronwall's lemma, Theorem A.47, we obtain that $Z(T) \leqslant C_T e^{TC_T}\delta$ and, finally,

$$\sup_{t\leqslant T} \mathbb{E}\left[|X_t^\delta - X_t|\right] \leqslant \sqrt{Z(T)} \leqslant c_T\sqrt{\delta}. \qquad \square$$

In the theory of ordinary differential equations one obtains higher order approximation schemes with the help of Taylor's formula. Similarly for stochastic differential equations an application of a stochastic Taylor–Itô formula yields the following scheme, which converges with strong order 1 to the solution of (22.3), for a proof see [127]. There it is also shown that the weak order is 1, thus it is the same as for the Euler scheme.

22.25 Algorithm (Milstein scheme). Assume that b and σ satisfy the conditions of Algorithm 22.23 and $\sigma(t, \cdot) \in \mathcal{C}^2$. Let $T > 0$, $n \geqslant 1$, $0 = t_0 < t_1 < \cdots < t_n = T$ be a time grid and x_0 be the starting position.

For $j = 1$ to n

1. Generate $y \overset{s}{\sim} \mathsf{N}(0, t_j - t_{j-1})$
2. Set $x_{t_j} := x_{t_{j-1}} + b(t_{j-1}, x_{t_{j-1}})(t_j - t_{j-1}) + \sigma(t_{j-1}, x_{t_{j-1}})\, y$
$$+ \frac{1}{2}\sigma(t_{j-1}, x_{t_{j-1}})\sigma^{(0,1)}(t_{j-1}, x_{t_{j-1}})(y^2 - (t_j - t_{j-1}))$$
(where $\sigma^{(0,1)}(t, x) := \frac{\partial}{\partial x}\sigma(t, x)$)

Then $(x_0, x_{t_1}, \ldots, x_{t_n}) \overset{s}{\sim} (X_0, X_{t_1}, \ldots, X_{t_n})$ approximately.

In the literature one finds a variety of further algorithms for the numerical evaluation of SDEs. Some of them have higher orders of convergence than the Euler or Milstein scheme, see for example [127] and the references given therein. Depending on the actual simulation task one might be more interested in the weak or strong order of convergence. The strong order is important for pathwise approximations whereas the weak order is important if one is only interested in distributional quantities.

We close this chapter by one of the most important applications of samples: The Monte Carlo method.

22.6 Monte Carlo method

The Monte Carlo method is nothing but an application of the strong law of large numbers. Recall that for iid random variables $(Y_j)_{j\geqslant 1}$ with finite expectation

$$\frac{1}{n}\sum_{j=1}^{n} Y_j \xrightarrow{n\to\infty} \mathbb{E}\,Y_1 \quad \text{almost surely.}$$

Thus, if we are interested in the expectation of $g(X)$ where X is a random variable and g a measurable function, we can approximate $\mathbb{E}\,g(X)$ by the following algorithm.

22.26 Algorithm (Monte Carlo method). Let X be a random variable with distribution F and g be a function such that $\mathbb{E}\,|g(X)| < \infty$. Let $n \geqslant 1$.

1. Generate $x_j \overset{s}{\sim} F$, $j = 1, \dots, n$
2. Set $y := \frac{1}{n}\sum_{j=1}^{n} g(x_j)$

Then y is an approximation of $\mathbb{E}\,g(X)$.

The approximation error is, by the central limit theorem and the theorem of Berry–Esseen, proportional to $1/\sqrt{n}$.

22.27 Example. In applications one often defines a model by an SDE of the form (22.3) and one is interested in the value of $\mathbb{E}\,g(X_t)$, where g is some function, and $(X_t)_{t\geqslant 0}$ is the solution of the SDE. We know how to simulate the normal distributed increments of the underlying Brownian motion, e. g. by Algorithm 22.13. Hence we can simulate samples of X_t by the Euler scheme, Algorithm 22.23. Doing this repeatedly corresponds to step 1 in the Monte Carlo method. Therefore, we get by step 2 an approximation of $\mathbb{E}\,g(X_t)$.

---

i **22.28 Further reading.** General monographs on simulation are [5] which focuses, in particular, on steady states, rare events, Gaussian processes and SDEs, and [196] which focuses in particular on Monte Carlo simulation, variance reduction and Markov chain Monte Carlo. More on the numerical treatment of SDEs can be found in [127]. A good starting point for the extension to Lévy processes and applications in finance is [38].

[5] Asmussen, Glynn: *Stochastic Simulation: Algorithms and Analysis.*

[38] Cont, Tankov: *Financial modelling with jump processes.*

[127] Kloeden, Platen: *Numerical Solution of Stochastic Differential Equations.*

[196] Rubinstein, Kroese: *Simulation and the Monte Carlo Method Wiley.*

A Appendix

A.1 Kolmogorov's existence theorem

In this appendix we give a proof of Kolmogorov's theorem, Theorem 4.8, on the existence of stochastic processes with prescribed finite dimensional distributions. In addition to the notation introduced in Chapter 4, we use $\pi_J^L : (\mathbb{R}^d)^L \to (\mathbb{R}^d)^J$ for the projection onto the coordinates from $J \subset L \subset I$. Moreover, $\pi_J := \pi_J^I$ and we write $\mathcal{H}$ for the family of all *finite* subsets of I. Finally

$$\mathscr{Z} := \left\{\pi_J^{-1}(B) \;:\; J \in \mathcal{H},\; B \in \mathscr{B}((\mathbb{R}^d)^J)\right\}$$

denotes the family of *cylinder sets* and we call $Z = \pi_J^{-1}(B)$ a *J-cylinder* with *basis B*. If $J \in \mathcal{H}$ and $\#J = j$, we can identify $(\mathbb{R}^d)^J$ with $\mathbb{R}^{jd}$.

For the proof of Theorem 4.8 we need the following auxiliary result.

A.1 Lemma (Regularity). *Every probability measure μ on* $(\mathbb{R}^n, \mathscr{B}(\mathbb{R}^n))$ *is* outer regular,

$$\mu(B) = \inf\left\{\mu(U) \;:\; U \supset B,\; U \text{ open}\right\} \tag{A.1}$$

and inner (compact) regular,

$$\begin{aligned}\mu(B) &= \sup\left\{\mu(F) \;:\; F \subset B,\; F \text{ closed}\right\} \\ &= \sup\left\{\mu(K) \;:\; K \subset B,\; K \text{ compact}\right\}.\end{aligned} \tag{A.2}$$

Proof. Consider the family

$$\Sigma := \left\{A \subset \mathbb{R}^n : \forall \epsilon > 0\ \exists U \text{ open},\ F \text{ closed},\ F \subset A \subset U,\ \mu(U \setminus F) < \epsilon\right\}.$$

It is easy to see that Σ is a σ-algebra[1] which contains all closed sets. Indeed, if $F \subset \mathbb{R}^n$ is closed, the sets

$$U_n := F + \mathbb{B}(0, 1/n) = \left\{x + y \;:\; x \in F,\; y \in \mathbb{B}(0, 1/n)\right\}$$

---

1 *Indeed:* $\emptyset \in \Sigma$ is obvious. Pick $A \in \Sigma$. For every $\epsilon > 0$ there are closed and open sets $F_\epsilon \subset A \subset U_\epsilon$ such that $\mu(U_\epsilon \setminus F_\epsilon) < \epsilon$. Observe that $U_\epsilon^c \subset A^c \subset F_\epsilon^c$, that U_ϵ^c is closed, that F_ϵ^c is open and that

$$F_\epsilon^c \setminus U_\epsilon^c = F_\epsilon^c \cap (U_\epsilon^c)^c = F_\epsilon^c \cap U_\epsilon = U_\epsilon \setminus F_\epsilon.$$

Thus, $\mu(F_\epsilon^c \setminus U_\epsilon^c) = \mu(U_\epsilon \setminus F_\epsilon) < \epsilon$, and we find that $A^c \in \Sigma$.
Let $A_n \in \Sigma$, $n \geqslant 1$. For $\epsilon > 0$ there exist sets $F_n \subset A_n \subset U_n$ satisfying $\mu(U_n \setminus F_n) < \epsilon/2^n$. Therefore,

$$\overbrace{\Phi_N := F_1 \cup \dots \cup F_N}^{\text{closed}} \subset A_1 \cup \dots \cup A_N \subset \overbrace{U := \bigcup_{j \geqslant 1} U_j}^{\text{open}}.$$

Since

$$\bigcap_{N \geqslant 1} U \setminus \Phi_N = \bigcup_{j \geqslant 1} U_j \setminus \bigcup_{k \geqslant 1} F_k = \bigcup_{j \geqslant 1}\Big(U_j \setminus \bigcup_{k \geqslant 1} F_k\Big) \subset \bigcup_{j \geqslant 1}(U_j \setminus F_j)$$

are open and $\bigcap_{n\geqslant 1} U_n = F$. By the continuity of measures, $\lim_{n\to\infty} \mu(U_n) = \mu(F)$, which shows that $F \subset U_n$ and $\mu(U_n \setminus F) < \epsilon$ for sufficiently large values of $n > N(\epsilon)$. This proves $F \in \Sigma$ and we conclude that

$$\mathscr{B}(\mathbb{R}^n) = \sigma(\text{closed subsets}) \subset \sigma(\Sigma) = \Sigma.$$

In particular, we can construct for every $B \in \mathscr{B}(\mathbb{R}^n)$ sequences of open $(U_n)_n$ and closed $(F_n)_n$ sets such that $F_n \subset B \subset U_n$ and $\lim_{n\to\infty} \mu(U_n \setminus F_n) = 0$. Therefore,

$$\mu(B \setminus F_n) + \mu(U_n \setminus B) \leqslant 2\,\mu(U_n \setminus F_n) \xrightarrow[n\to\infty]{} 0$$

which shows that the infimum in (A.1) and the supremum in the first equality of (A.2) are attained.

Finally, replace in the last step the closed sets F_n with the compact sets

$$K_n := F_n \cap \overline{\mathbb{B}}(0,R), \quad R = R(n).$$

Using the continuity of measures, we get $\lim_{R\to\infty} \mu(F_n \setminus F_n \cap \overline{\mathbb{B}}(0,R)) = 0$, and for sufficiently large $R = R(n)$ with $R(n) \to \infty$,

$$\mu(B \setminus K_n) \leqslant \mu(B \setminus F_n) + \mu(F_n \setminus K_n) \xrightarrow[n\to\infty]{} 0.$$

This proves the second equality in (A.2). □

A.2 Theorem (Theorem 4.8. Kolmogorov 1933). *Let $I \subset [0,\infty)$ and $(p_J)_{J\in\mathcal{H}}$ be a projective family of probability measures p_J defined on $((\mathbb{R}^d)^J, \mathscr{B}((\mathbb{R}^d)^J))$, $J \in \mathcal{H}$. Then there exists a unique probability measure μ on the space $((\mathbb{R}^d)^I, \mathscr{B}^I)$ such that $p_J = \mu \circ \pi_J^{-1}$ where π_J is the canonical projection of $(\mathbb{R}^d)^I$ onto $(\mathbb{R}^d)^J$.*

The measure μ is often called the *projective limit* of the family $(p_J)_{J\in\mathcal{H}}$.

Proof. We use the family $(p_J)_{J\in\mathcal{H}}$ to define a (pre-)measure on the cylinder sets $\mathscr{Z}$,

$$\mu \circ \underbrace{\pi_J^{-1}(B)}_{\in\mathscr{Z}} := p_J(B) \quad \forall B \in \mathscr{B}^J. \tag{A.3}$$

Since $\mathscr{Z}$ is an algebra of sets, i. e. $\emptyset \in \mathscr{Z}$ and $\mathscr{Z}$ is stable under finite unions, intersections and complements, we can use Carathéodory's extension theorem to extend μ onto $\mathscr{B}^I = \sigma(\mathscr{Z})$.

---

we find by the continuity and the σ-subadditivity of the (finite) measure μ

$$\lim_{N\to\infty} \mu(U \setminus \Phi_N) = \mu\Big(\bigcup_{j\geqslant 1} U_j \setminus \bigcup_{k\geqslant 1} F_k\Big) \leqslant \sum_{j\geqslant 1} \mu(U_j \setminus F_j) \leqslant \sum_{j\geqslant 1} \frac{\epsilon}{2^j} = \epsilon.$$

Thus, $\mu(U \setminus \Phi_N) \leqslant 2\epsilon$ for all sufficiently large $N > N(\epsilon)$. This proves $\bigcup_{n\geqslant 1} A_n \in \Sigma$.

Let us, first of all, check that μ is well-defined and that it does not depend on the particular representation of the cylinder set Z. Assume that $Z = \pi_J^{-1}(B) = \pi_K^{-1}(B')$ for $J, K \in \mathcal{H}$ and $B \in \mathscr{B}^J$, $B' \in \mathscr{B}^K$. We have to show that

$$p_J(B) = p_K(B').$$

Set $L := J \cup K \in \mathcal{H}$. Then

$$Z = \pi_J^{-1}(B) = (\pi_J^L \circ \pi_L)^{-1}(B) = \pi_L^{-1} \circ (\pi_J^L)^{-1}(B),$$
$$Z = \pi_K^{-1}(B') = (\pi_K^L \circ \pi_L)^{-1}(B') = \pi_L^{-1} \circ (\pi_K^L)^{-1}(B').$$

Thus, $(\pi_J^L)^{-1}(B) = (\pi_K^L)^{-1}(B')$ and the projectivity (4.6) of the family $(p_J)_{J\in\mathcal{H}}$ gives

$$p_J(B) = \pi_J^L(p_L)(B) = p_L((\pi_J^L)^{-1}(B)) = p_L((\pi_K^L)^{-1}(B')) = p_K(B').$$

Next we verify that μ is an additive set function: $\mu(\emptyset) = 0$ is obvious. Let $Z, Z' \in \mathscr{Z}$ such that $Z = \pi_J^{-1}(B)$ and $Z' = \pi_K^{-1}(B')$. Setting $L := J \cup K$ we can find suitable basis sets $A, A' \in \mathscr{B}^L$ such that

$$Z = \pi_L^{-1}(A) \text{ and } Z' = \pi_L^{-1}(A').$$

If $Z \cap Z' = \emptyset$ we see that $A \cap A' = \emptyset$, and μ inherits its additivity from p_L:

$$\begin{aligned}\mu(Z \uplus Z') &= \mu(\pi_L^{-1}(A) \uplus \pi_L^{-1}(A')) = \mu(\pi_L^{-1}(A \uplus A')) \\ &= p_L(A \uplus A') = p_L(A) + p_L(A') \\ &= \cdots = \mu(Z) + \mu(Z').\end{aligned}$$

Finally, we prove that μ is σ-additive on $\mathscr{Z}$. For this we show that μ is continuous at the empty set, i. e. for every decreasing sequence of cylinder sets $(Z_n)_{n\geqslant 1} \subset \mathscr{Z}$, $\bigcap_{n\geqslant 1} Z_n = \emptyset$ we have $\lim_{n\to\infty} \mu(Z_n) = 0$. Equivalently,

$$Z_n \in \mathscr{Z},\ Z_n \supset Z_{n+1} \supset \cdots,\ \mu(Z_n) \geqslant \alpha > 0 \implies \bigcap_{n\geqslant 1} Z_n \neq \emptyset.$$

Write for such a sequence of cylinder sets $Z_n = \pi_{J_n}^{-1}(B_n)$ with suitable $J_n \in \mathcal{H}$ and $B_n \in \mathscr{B}^{J_n}$. Since for every $J \subset K$ any J-cylinder is also a K-cylinder, we may assume without loss of generality that $J_n \subset J_{n+1}$.

Note that $(\mathbb{R}^d)^{J_n} \cong (\mathbb{R}^d)^{\#J_n}$; therefore we can use Lemma A.1 to see that p_{J_n} is inner compact regular. Thus,

$$\text{there exist compact sets } K_n \subset B_n : \quad p_{J_n}(B_n \setminus K_n) \leqslant 2^{-n}\alpha.$$

Write $Z_n' := \pi_{J_n}^{-1}(K_n)$ and observe that

$$\mu(Z_n \setminus Z_n') = \mu(\pi_{J_n}^{-1}(B_n \setminus K_n)) = p_{J_n}(B_n \setminus K_n) \leqslant 2^{-n}\alpha.$$

Although the Z_n are decreasing, this might not be the case for the Z'_n; therefore we consider $\widehat{Z}_n := Z'_1 \cap \cdots \cap Z'_n \subset Z_n$ and find

$$\begin{aligned}\mu(\widehat{Z}_n) = \mu(Z_n) - \mu(Z_n \setminus \widehat{Z}_n) = \mu(Z_n) - \mu\Bigg(\bigcup_{k=1}^{n} (\underbrace{Z_n \setminus Z'_k}_{\subset Z_k \setminus Z'_k})\Bigg) \\ \geqslant \mu(Z_n) - \sum_{k=1}^{n} \mu\left(Z_k \setminus Z'_k\right) \\ \geqslant \alpha - \sum_{k=1}^{n} 2^{-k}\alpha > 0.\end{aligned}$$

We are done if we can show that $\bigcap_{n\geqslant 1} Z'_n = \bigcap_{n\geqslant 1} \widehat{Z}_n \neq \emptyset$. For this we construct some element $z \in \bigcap_{n\geqslant 1} \widehat{Z}_n$.

1° For every $m \geqslant 1$ we pick some $f_m \in \widehat{Z}_m$. Since the sequence $(\widehat{Z}_n)_{n\geqslant 1}$ is decreasing,

$$f_m \in \widehat{Z}_n \subset Z'_n, \text{ hence } \pi_{J_n}(f_m) \in K_n \text{ for all } m \geqslant n \text{ and } n \geqslant 1.$$

Since projections are continuous, and since continuous maps preserve compactness,

$$\pi_t(f_m) = \pi_t^{J_n} \circ \pi_{J_n}(f_m) \in \underbrace{\pi_t^{J_n}(K_n)}_{\text{compact}} \text{ for all } m \geqslant n,\ n \geqslant 1 \text{ and } t \in J_n.$$

2° The index set $H := \bigcup_{n\geqslant 1} J_n$ is countable; denote by $H = (t_k)_{k\geqslant 1}$ some enumeration. For each $t_k \in H$ there is some $n(k) \geqslant 1$ such that $t_k \in J_{n(k)}$. Moreover, by 1°,

$$(\pi_{t_k}(f_m))_{m\geqslant n(k)} \subset \pi_{t_k}^{J_{n(k)}}(K_{n(k)}).$$

Since the sets $\pi_{t_k}^{J_{n(k)}}(K_{n(k)})$ are compact, we can take repeatedly subsequences,

$$\begin{aligned}&(\pi_{t_1}(f_m))_{m\geqslant n(1)} \subset \text{compact set} \implies \exists (f_m^1)_m \subset (f_m)_m : \ {}^{\exists}\lim_{m\to\infty} \pi_{t_1}(f_m^1)\\ &(\pi_{t_2}(f_m^1))_{m\geqslant n(2)} \subset \text{compact set} \implies \exists (f_m^2)_m \subset (f_m^1)_m : \ {}^{\exists}\lim_{m\to\infty} \pi_{t_2}(f_m^2)\\ &\vdots\\ &(\pi_{t_k}(f_m^{k-1}))_{m\geqslant n(k)} \subset \text{compact set} \implies \exists (f_m^k)_m \subset (f_m^{k-1})_m : \ {}^{\exists}\lim_{m\to\infty} \pi_{t_k}(f_m^k).\end{aligned}$$

The diagonal sequence $(f_m^m)_{m\geqslant 1}$ satisfies

$${}^{\exists}\lim_{m\to\infty} \pi_t(f_m^m) = z_t \text{ for each } t \in H.$$

3° By construction, $\pi_t(f_m) \in \pi_t^{J_n}(K_n)$ for all $t \in J_n$ and $m \geqslant n$. Therefore,

$$\pi_t(f_m^m) \in \pi_t^{J_n}(K_n) \text{ for all } m \geqslant n \text{ and } t \in J_n.$$

Since K_n is compact, $\pi_t^{J_n}(K_n)$ is compact, hence closed, and we see that

$$z_t = \lim_m \pi_t(f_m^m) \in \pi_t^{J_n}(K_n) \text{ for every } t \in J_n.$$

Using the fact that $n \geqslant 1$ is arbitrary, find

$$z(t) := \begin{cases} z_t & \text{if } t \in H; \\ * & \text{if } t \in I \setminus H \end{cases}$$

($*$ stands for any element from $\mathbb{R}^d$) defines an element $z = (z(t))_{t\in I} \in (\mathbb{R}^d)^I$ with the property that for all $n \geqslant 1$

$$\pi_{J_n}(z) \in K_n \text{ if, and only if, } z \in \pi_{J_n}^{-1}(K_n) = Z'_n \supset \widehat{Z}_n.$$

This proves $z \in \bigcap_{n\geqslant 1} Z'_n = \bigcap_{n\geqslant 1} \widehat{Z}_n$ and $\bigcap_{n\geqslant 1} \widehat{Z}_n \neq \emptyset$. □

A.2 A property of conditional expectations

We assume that you have some basic knowledge of conditioning and conditional expectations with respect to a σ-algebra. Several times we will use the following result which is not always contained in elementary expositions. Throughout this section let $(\Omega, \mathscr{A}, \mathbb{P})$ be a probability space.

A.3 Lemma. *Let $X : (\Omega, \mathscr{A}) \to (D, \mathscr{D})$ and $Y : (\Omega, \mathscr{A}) \to (E, \mathscr{E})$ be two random variables. Assume that $\mathscr{X}, \mathscr{Y} \subset \mathscr{A}$ are σ-algebras such that X is $\mathscr{X}/\mathscr{D}$ measurable, Y is $\mathscr{Y}/\mathscr{E}$ measurable and $\mathscr{X} \perp\!\!\!\perp \mathscr{Y}$. Then*

$$\mathbb{E}\left[\Phi(X,Y) \mid \mathscr{X}\right] = \left[\mathbb{E}\,\Phi(x,Y)\right]\big|_{x=X} = \mathbb{E}\left[\Phi(X,Y) \mid X\right] \tag{A.4}$$

holds for all bounded $\mathscr{D} \times \mathscr{E}/\mathscr{B}(\mathbb{R})$ measurable functions $\Phi : D \times E \to \mathbb{R}$.

If $\Psi : D \times \Omega \to \mathbb{R}$ is bounded and $\mathscr{D} \otimes \mathscr{Y}/\mathscr{B}(\mathbb{R})$ measurable, then

$$\mathbb{E}\left[\Psi(X(\cdot),\cdot) \mid \mathscr{X}\right] = \left[\mathbb{E}\,\Psi(x,\cdot)\right]\big|_{x=X} = \mathbb{E}\left[\Psi(X(\cdot),\cdot) \mid X\right]. \tag{A.5}$$

Proof. Assume first that $\Phi(x,y)$ is of the form $\phi(x)\psi(y)$. Then

$$\begin{aligned}\mathbb{E}\big[\underbrace{\phi(X)\psi(Y)}_{=\Phi(X,Y)} \mid \mathscr{X}\big] = \phi(X)\,\mathbb{E}\left[\psi(Y) \mid \mathscr{X}\right] &\overset{Y \perp\!\!\!\perp \mathscr{X}}{=} \phi(X)\,\mathbb{E}\left[\psi(Y)\right] \\ &= \mathbb{E}\big[\underbrace{\phi(x)\psi(Y)}_{=\Phi(x,Y)}\big]\big|_{x=X}.\end{aligned}$$

Now fix some $F \in \mathscr{X}$ and pick $\phi(x) = \mathbb{1}_A(x)$ and $\psi(y) = \mathbb{1}_B(y)$, $A \in \mathscr{D}$, $B \in \mathscr{E}$. Our argument shows that

$$\int_F \mathbb{1}_{A\times B}(X(\omega), Y(\omega))\,\mathbb{P}(d\omega) = \int_F \mathbb{E}\,\mathbb{1}_{A\times B}(x,Y)\big|_{x=X(\omega)}\,\mathbb{P}(d\omega), \quad F \in \mathscr{X}.$$

Both sides of this equality (can be extended from $\mathscr{D} \times \mathscr{E}$ to) define measures on the product σ-algebra $\mathscr{D} \otimes \mathscr{E}$. By linearity, this becomes

$$\int_F \Phi(X(\omega), Y(\omega))\,\mathbb{P}(d\omega) = \int_F \mathbb{E}\,\Phi(x,Y)\big|_{x=X(\omega)}\,\mathbb{P}(d\omega), \quad F \in \mathscr{X},$$

for $\mathscr{D} \otimes \mathscr{E}$ measurable positive step functions Φ; using standard arguments from the theory of measure and integration, we get this equality for positive measurable functions and then for all bounded measurable functions. The latter is, however, equivalent to the first equality in (A.4).

The second equality follows if we take conditional expectations $\mathbb{E}(\cdots \mid X)$ on both sides of the first equality and use the tower property of conditional expectation.

The formulae (A.5) follow in a similar way. □

A simple consequence of Lemma A.3 are the following formulae.

A.4 Corollary. *Let $X : (\Omega, \mathscr{A}) \to (D, \mathscr{D})$ and $Y : (\Omega, \mathscr{A}) \to (E, \mathscr{E})$ be two random variables. Assume that $\mathscr{X}, \mathscr{Y} \subset \mathscr{A}$ are σ-algebras such that X is $\mathscr{X}/\mathscr{D}$ measurable, Y is $\mathscr{Y}/\mathscr{E}$ measurable and $\mathscr{X} \perp\!\!\!\perp \mathscr{Y}$. Then*

$$\mathbb{E}\,\Phi(X,Y) = \int \mathbb{E}\,\Phi(x,Y)\,\mathbb{P}(X \in dx) = \mathbb{E}\left[\int \Phi(x,Y)\,\mathbb{P}(X \in dx)\right] \tag{A.6}$$

holds for all bounded $\mathscr{D} \times \mathscr{E}/\mathscr{B}(\mathbb{R})$ measurable functions $\Phi : D \times E \to \mathbb{R}$.

If $\Psi : D \times \Omega \to \mathbb{R}$ is bounded and $\mathscr{D} \otimes \mathscr{Y}/\mathscr{B}(\mathbb{R})$ measurable, then

$$\mathbb{E}\,\Psi(X(\cdot),\cdot) = \int \mathbb{E}\,\Psi(x,\cdot)\,\mathbb{P}(X \in dx) = \mathbb{E}\left[\int \Psi(x,\cdot)\,\mathbb{P}(X \in dx)\right]. \tag{A.7}$$

A.3 From discrete to continuous time martingales

We assume that you have some basic knowledge of discrete time martingales, e. g. as in [204]. The key to the continuous time setting is the following simple observation: If $(X_t, \mathscr{F}_t)_{t\geqslant 0}$ is a martingale, then

$$(X_{t_j}, \mathscr{F}_{t_j})_{j\geqslant 1} \quad \text{is a martingale for any sequence} \quad 0 \leqslant t_1 < t_2 < t_3 < \cdots$$

We are mainly interested in estimates and convergence results for martingales. Our strategy is to transfer the corresponding results from the discrete time setting to continuous time. The key result is *Doob's upcrossing estimate*. Let $(X_t, \mathscr{F}_t)_{t\geqslant 0}$ be a real-valued martingale, $I \subset [0,\infty)$ be a finite index set with $t_{\min} = \min I$, $t_{\max} = \max I$, and $a < b$. Then

$$U(I;[a,b]) = \max\left\{ m \;\middle|\; \begin{array}{l} \exists\, \sigma_1 < \tau_1 < \sigma_2 < \tau_2 < \cdots < \sigma_m < \tau_m, \\ \sigma_j, \tau_j \in I,\; X(\sigma_j) < a < b < X(\tau_j) \end{array} \right\}$$

is the number of *upcrossings* of $(X_s)_{s\in I}$ across the strip $[a,b]$. Figure A.1 shows two upcrossings. Since $(X_s, \mathscr{F}_s)_{s\in I}$ is a discrete time martingale, we know

$$\mathbb{E}\left[U(I;[a,b])\right] \leqslant \frac{1}{b-a}\,\mathbb{E}\left[(X(t_{\max}) - a)^+\right]. \tag{A.8}$$

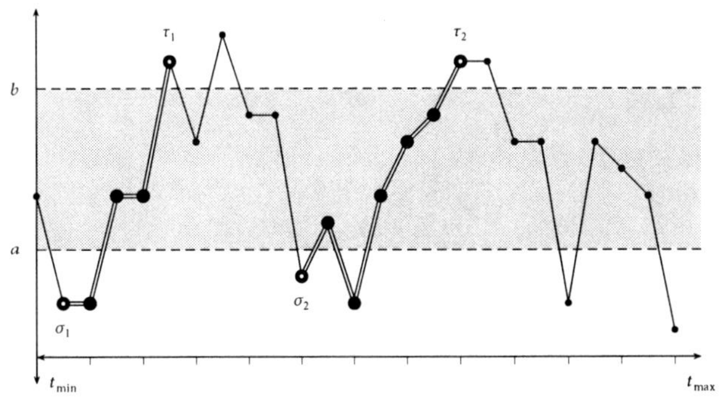

Fig. A.1. Two upcrossings over the interval $[a,b]$.

A.5 Lemma (Upcrossing estimate). *Let $(X_t, \mathscr{F}_t)_{t\geqslant 0}$ be a real-valued submartingale and $D \subset [0,\infty)$ a countable subset. Then*

$$\mathbb{E}\left[U(D \cap [0,t]; [a,b])\right] \leqslant \frac{1}{b-a}\, \mathbb{E}\left[(X(t)-a)^+\right] \tag{A.9}$$

for all $t > 0$ and $-\infty < a < b < \infty$.

Proof. Let $D_N \subset D$ be finite index sets such that $D_N \uparrow D$. If $t_{\max} := \max D_N \cap [0,t]$, we get from (A.8) with $I = D_N \cap [0,t]$

$$\mathbb{E}\left[U(D_N \cap [0,t]; [a,b])\right] \leqslant \frac{1}{b-a}\, \mathbb{E}\left[(X(t_{\max})-a)^+\right] \leqslant \frac{1}{b-a}\, \mathbb{E}\left[(X(t)-a)^+\right].$$

For the last estimate observe that $(X_t - a)^+$ is a submartingale since $x \mapsto x^+$ is convex and increasing. Since $\sup_N U(D_N \cap [0,t]; [a,b]) = U(D \cap [0,t]; [a,b])$, (A.9) follows from monotone convergence. □

A.6 Theorem (Martingale convergence theorem. Doob 1953). *Let $(X_t, \mathscr{F}_t)_{t\geqslant 0}$ be a martingale in $\mathbb{R}^d$ or a submartingale in $\mathbb{R}$ with continuous sample paths. If $\sup_{t\geqslant 0} \mathbb{E}\,|X_t| < \infty$, then the limit $\lim_{t\to\infty} X_t$ exists almost surely and defines a finite $\mathscr{F}_\infty$ measurable random variable X_∞.*

If $(X_t, \mathscr{F}_t)_{t\geqslant 0}$ is a martingale of the form $X_t = \mathbb{E}(Z \mid \mathscr{F}_t)$ for some $Z \in L^1(\mathbb{P})$, then we have $X_\infty = \mathbb{E}(Z \mid \mathscr{F}_\infty)$ and $\lim_{t\to\infty} X_t = X_\infty$ almost surely and in $L^1(\mathbb{P})$.

Proof. Since a d-dimensional function converges if, and only if, all coordinate functions converge, we may assume that X_t is real valued. By Doob's upcrossing estimate

$$\mathbb{E}\left[U(\mathbb{Q}^+; [a,b])\right] \leqslant \frac{\sup_{t\geqslant 0} \mathbb{E}\left[(X_t-a)^+\right]}{b-a} < \infty$$

for all intervals $[a,b]$ and $\mathbb{Q}^+ := \mathbb{Q} \cap [0,\infty)$. Thus,

$$U(\mathbb{Q}^+; [a,b])(\omega) < \infty \quad \text{for all } \omega \notin N_{a,b}$$

where $N_{a,b} \subset \Omega$ with $\mathbb{P}(N_{a,b}) = 0$. Set $N := \bigcup_{a<b,\, a,b\in\mathbb{Q}^+} N_{a,b}$ and $\Omega_0 := \Omega \setminus N$. Then $\mathbb{P}(\Omega_0) = 1$.

We claim that for $\omega \in \Omega_0$ the limit $\lim_{t\to\infty} X_t(\omega)$ exists. Otherwise we could find $\alpha, \beta \in \mathbb{Q}$ such that

$$\varliminf_{s\to\infty} X_s(\omega) < \alpha < \beta < \varlimsup_{s\to\infty} X_s(\omega).$$

By the continuity of the sample paths,

$$\varliminf_{s\to\infty} X_s(\omega) = \varliminf_{\mathbb{Q}^+ \ni r\to\infty} X_r(\omega) \quad \text{and} \quad \varlimsup_{s\to\infty} X_s(\omega) = \varlimsup_{\mathbb{Q}^+ \ni r\to\infty} X_r(\omega),$$

which means that there must be infinitely many upcrossings of $(X_t)_{t\in\mathbb{Q}^+}$ over the interval $[\alpha, \beta]$; this is impossible since $U(\mathbb{Q}^+; [\alpha, \beta])(\omega) < \infty$.

Set $X_\infty := \lim_{t\to\infty} X_t \mathbb{1}_{\Omega_0}$. By Fatou's Lemma,

$$\mathbb{E}\,|X_\infty| = \mathbb{E}\left[\varliminf_{t\to\infty} |X_t|\right] \leqslant \varliminf_{t\to\infty} \mathbb{E}\,|X_t| \leqslant \sup_{t\geqslant 0} \mathbb{E}\,|X_t| < \infty$$

which shows that $X_\infty \in L^1$ and that X_∞ is a. s. finite.

Assume now that $X_t = \mathbb{E}(Z \mid \mathscr{F}_t)$. From

$$|X_t| = |\,\mathbb{E}(Z \mid \mathscr{F}_t)| \leqslant \mathbb{E}(|Z| \mid \mathscr{F}_t)$$

we conclude that $\sup_{t\geqslant 0} \mathbb{E}\,|X_t| \leqslant \mathbb{E}\,|Z| < \infty$ and that $(X_t)_{t\geqslant 0}$ is uniformly integrable. By the L^1-convergence theorem for discrete uniformly integrable martingales we find that for any sequence $t_j \uparrow \infty$ the limit $\lim_{j\to\infty} X_{t_j} = X'_\infty$ exists in L^1 and almost surely. By the uniqueness of a. s. limits we conclude that $X_\infty = X'_\infty$ almost surely.

Fix $t > 0$. For all $t_j \geqslant t$ and $F \in \mathscr{F}_t \subset \mathscr{F}_{t_j}$ we find, because of L^1-convergence,

$$\int_F Z\,d\mathbb{P} \overset{X_{t_j}=\mathbb{E}(Z|\mathscr{F}_{t_j})}{=} \int_F X_{t_j}\,d\mathbb{P} \xrightarrow[j\to\infty]{} \int_F X_\infty\,d\mathbb{P},$$

i. e.

$$\int_F Z\,d\mathbb{P} = \int_F X_\infty\,d\mathbb{P} \quad \text{for all } F \in \bigcup_{t\geqslant 0} \mathscr{F}_t.$$

Since $\bigcup_{t\geqslant 0} \mathscr{F}_t$ is a $\cap$-stable generator of $\mathscr{F}_\infty$, we see $\mathbb{E}(Z \mid \mathscr{F}_\infty) = X_\infty$. □

The next theorem allows us to identify all martingales of the form $X_t = \mathbb{E}(Z \mid \mathscr{F}_t)$.

A.7 Theorem (Closure of a martingale. Doob 1953). *Let $(X_t, \mathscr{F}_t)_{t\geqslant 0}$ be a martingale in $\mathbb{R}^d$ with continuous paths. Then the following assertions are equivalent*

a) $(X_t)_{t\geqslant 0}$ *is uniformly integrable;*

b) $X_\infty = \lim_{t\to\infty} X_t$ *exists a. s. and in L^1 and $(X_t, \mathscr{F}_t)_{t\in[0,\infty]}$ is a martingale.*

Proof. a)⇒b): Since $(X_t)_{t\geqslant 0}$ is uniformly integrable, we have $\sup_{t\geqslant 0} \mathbb{E}\,|X_t| < \infty$. Therefore $\lim_{t\to\infty} X_t = X_\infty$ almost surely, and by Vitali's convergence theorem, see e. g. [204, Theorem 16.6, p. 166], we find that $\lim_{j\to\infty} X_{t_j} = X_\infty$ in L^1 for all increasing sequences $t_j \uparrow \infty$. Thus, the limit $\lim_{t\to\infty} X_t = X_\infty$ exists in L^1, and we find for all $F \in \mathscr{F}_t$ and $T \geqslant t$

$$\int_F X_\infty\,d\mathbb{P} \overset{L^1\text{-conv.}}{=} \lim_{T\to\infty} \int_F X_T\,d\mathbb{P} \overset{F\in\mathscr{F}_t}{=} \int_F X_t\,d\mathbb{P};$$

hence, $\mathbb{E}(X_\infty \mid \mathscr{F}_t) = X_t$ and $(X_t)_{t\in[0,\infty]}$ is a martingale.

b)⇒a): Since $X_t = \mathbb{E}(X_\infty \mid \mathscr{F}_t)$ for $X_\infty \in L^1(\mathbb{P})$, we get $|X_t| \leqslant \mathbb{E}(|X_\infty| \mid \mathscr{F}_t)$ and, therefore, $\mathbb{E}\,|X_t| \leqslant \mathbb{E}\,|X_\infty|$. Hence, we find for all $t \geqslant 0$, $R > 0$ and $K > 0$

$$\begin{aligned}
\int_{\{|X_t|>R\}} |X_t|\,d\mathbb{P} &\leqslant \int_{\{|X_t|>R\}} \mathbb{E}(|X_\infty| \mid \mathscr{F}_t)\,d\mathbb{P} \\
&= \int_{\{|X_t|>R\}} |X_\infty|\,d\mathbb{P} \\
&= \int_{\substack{\{|X_t|>R\}\\ \cap\{|X_\infty|>K\}}} |X_\infty|\,d\mathbb{P} + \int_{\substack{\{|X_t|>R\}\\ \cap\{|X_\infty|\leqslant K\}}} |X_\infty|\,d\mathbb{P} \\
&\leqslant \int_{\{|X_\infty|>K\}} |X_\infty|\,d\mathbb{P} + K\,\mathbb{P}(|X_t| > R) \\
&\leqslant \int_{\{|X_\infty|>K\}} |X_\infty|\,d\mathbb{P} + \frac{K}{R}\,\mathbb{E}\,|X_t| \\
&\leqslant \int_{\{|X_\infty|>K\}} |X_\infty|\,d\mathbb{P} + \frac{K}{R}\,\mathbb{E}\,|X_\infty| \\
&\xrightarrow[R\to\infty]{} \int_{\{|X_\infty|>K\}} |X_\infty|\,d\mathbb{P} \xrightarrow[K\to\infty]{\text{dom. conv.}} 0.
\end{aligned}$$

□

For backwards submartingales, i. e. for submartingales with a decreasing index set, the assumptions of the convergence theorems A.6 and A.7 are always satisfied.

A.8 Corollary (Backwards convergence theorem). *Let $(X_t, \mathscr{F}_t)$ be a submartingale and let $t_1 \geqslant t_2 \geqslant t_3 \geqslant \dots, t_j \downarrow t_\infty$, be a decreasing sequence. Then $(X_{t_j}, \mathscr{F}_{t_j})_{j\geqslant 1}$ is a (backwards) submartingale which is uniformly integrable. In particular, the limit $\lim_{j\to\infty} X_{t_j}$ exists in L^1 and almost surely.*

Proof. That $(X_{t_j})_{j\geqslant 1}$ is a (backwards directed) submartingale is obvious. Now we use that $\mathbb{E}\,X_0 \leqslant \mathbb{E}\,X_t$ and that $(X_t^+)_{t\geqslant 0}$ is a submartingale. Therefore, we get for $t \leqslant t_1$

$$\mathbb{E}\,|X_t| = 2\,\mathbb{E}\,X_t^+ - \mathbb{E}\,X_t \leqslant 2\,\mathbb{E}\,X_t^+ - \mathbb{E}\,X_0 \leqslant 2\,\mathbb{E}\,X_{t_1}^+ - \mathbb{E}\,X_0$$

which shows that $\sup_{j\geqslant 1} \mathbb{E}\,|X_{t_j}| < \infty$. Using Doob's upcrossing estimate we can argue just as in the martingale convergence theorem, Theorem A.6, and conclude that $\lim_{j\to\infty} X_{t_j} = Z$ almost surely for some $\mathscr{F}_{t_\infty}$ measurable random variable $Z \in L^1(\mathbb{P})$.

For a submartingale $\mathbb{E}\,X_{t_j}$ is decreasing, i. e. $\lim_{j\to\infty} \mathbb{E}\,X_{t_j} = \inf_{j\geqslant 1} \mathbb{E}\,X_{t_j}$ exists (and is finite). Thus,

$$\forall \epsilon > 0\ \exists N = N_\epsilon\ \forall j \geqslant N : |\mathbb{E}\,X_{t_j} - \mathbb{E}\,X_{t_N}| < \epsilon.$$

This means that we get for all $j \geqslant N$, i. e. $t_j \leqslant t_N$:

$$\begin{aligned}\int_{\{|X_{t_j}|\geqslant R\}} |X_{t_j}|\,d\mathbb{P} &= 2\int_{\{|X_{t_j}|\geqslant R\}} X_{t_j}^+\,d\mathbb{P} - \int_{\{|X_{t_j}|\geqslant R\}} X_{t_j}\,d\mathbb{P}\\ &= 2\int_{\{|X_{t_j}|\geqslant R\}} X_{t_j}^+\,d\mathbb{P} - \mathbb{E}\,X_{t_j} + \int_{\{|X_{t_j}|< R\}} X_{t_j}\,d\mathbb{P}\end{aligned}$$

and since X_t and X_t^+ are submartingales

$$\begin{aligned}&\leqslant 2\int_{\{|X_{t_j}|\geqslant R\}} X_{t_N}^+\,d\mathbb{P} - \mathbb{E}\,X_{t_j} + \int_{\{|X_{t_j}|< R\}} X_{t_N}\,d\mathbb{P}\\ &\leqslant 2\int_{\{|X_{t_j}|\geqslant R\}} X_{t_N}^+\,d\mathbb{P} + \left(\epsilon - \mathbb{E}\,X_{t_N}\right) + \int_{\{|X_{t_j}|< R\}} X_{t_N}\,d\mathbb{P}\\ &= 2\int_{\{|X_{t_j}|\geqslant R\}} X_{t_N}^+\,d\mathbb{P} + \epsilon - \int_{\{|X_{t_j}|\geqslant R\}} X_{t_N}\,d\mathbb{P}\\ &= \int_{\{|X_{t_j}|\geqslant R\}} |X_{t_N}|\,d\mathbb{P} + \epsilon.\end{aligned}$$

Thus,

$$\int_{\{|X_{t_j}|\geqslant R\}} |X_{t_j}|\,d\mathbb{P} \leqslant \int_{\{|X_{t_j}|\geqslant R\}} |X_{t_N}|\,d\mathbb{P} + \epsilon \leqslant \int_{\substack{\{|X_{t_j}|\geqslant R\}\\ \cap\{|X_{t_N}|\leqslant R/2\}}} |X_{t_N}|\,d\mathbb{P} + \int_{\substack{\{|X_{t_j}|\geqslant R\}\\ \cap\{|X_{t_N}|\geqslant R/2\}}} |X_{t_N}|\,d\mathbb{P} + \epsilon$$

and since we have $|X_{t_N}| \leqslant R/2 \leqslant |X_{t_j}/2|$ on the set $\{|X_{t_j}| \geqslant R\} \cap \{|X_{t_N}| \leqslant R/2\}$, we get

$$\begin{aligned}\int_{\{|X_{t_j}|\geqslant R\}} |X_{t_j}|\,d\mathbb{P} &\leqslant \frac{1}{2}\int_{\substack{\{|X_{t_j}|\geqslant R\}\\ \cap\{|X_{t_N}|\leqslant R/2\}}} |X_{t_j}|\,d\mathbb{P} + \int_{\{|X_{t_N}|\geqslant R/2\}} |X_{t_N}|\,d\mathbb{P} + \epsilon\\ &\leqslant \frac{1}{2}\int_{\{|X_{t_j}|\geqslant R\}} |X_{t_j}|\,d\mathbb{P} + \int_{\{|X_{t_N}|\geqslant R/2\}} |X_{t_N}|\,d\mathbb{P} + \epsilon.\end{aligned}$$

This shows that

$$\int_{\{|X_{t_j}|\geqslant R\}} |X_{t_j}|\,d\mathbb{P} \leqslant 2\int_{\{|X_{t_N}|\geqslant R/2\}} |X_{t_N}|\,d\mathbb{P} + 2\epsilon \xrightarrow[R\to\infty]{\text{dom. conv.}} 2\epsilon \xrightarrow[\epsilon\to 0]{} 0$$

holds uniformly for all $j \geqslant N$. Therefore, $(X_{t_j})_{j\geqslant 1}$ is uniformly integrable. By Vitali's convergence theorem, see e. g. [204, Theorem 16.6, p. 166], we conclude that $L^1\text{-}\lim_{j\to\infty} X_{t_j}$ exists. □

We will also need some maximal inequalities for martingales. Recall that for a real-valued discrete time submartingale $(X(i), \mathscr{F}_i)_{i\in I}$ where $t_{\min} = \min I$, $t_{\max} = \max I$ and $r > 0$ the following estimates hold

$$\mathbb{P}\Big(\max_{s\in I} X(s) \geqslant r\Big) \leqslant \frac{1}{r}\,\mathbb{E}\left[X^+(t_{\max})\right], \tag{A.10}$$

$$\mathbb{P}\Big(\min_{s\in I} X(s) \leqslant -r\Big) \leqslant \frac{1}{r}\Big(\,\mathbb{E}\left[X^+(t_{\max})\right] - \mathbb{E}\left[X(t_{\min})\right]\Big). \tag{A.11}$$

Since for every d-dimensional martingale or positive submartingale $(X(t), \mathscr{F}_t)_{t\geqslant 0}$ such that $X(t) \in L^p(\mathbb{P})$, $p \geqslant 1$, the process $(|X(t)|^p, \mathscr{F}_t)_{t\geqslant 0}$ is a submartingale, we get

$$\mathbb{P}\Big(\max_{s\in I} |X(s)| \geqslant r\Big) \leqslant \frac{1}{r^p}\,\mathbb{E}\left[|X(t_{\max})|^p\right] \quad \text{for all } r > 0. \tag{A.12}$$

A.9 Lemma (Doob 1953). *Let $(X_t, \mathscr{F}_t)_{t\geqslant 0}$ be a positive submartingale or a d-dimensional martingale, $X_t \in L^p(\mathbb{P})$ for some $p \geqslant 1$ and $D \subset [0,\infty)$ a countable subset. Then*

$$\mathbb{P}\Big(\sup_{s\in D\cap[0,t]} |X_s| \geqslant r\Big) \leqslant \frac{1}{r^p}\,\mathbb{E}\left[|X_t|^p\right] \quad \textit{for all } r > 0,\ t \geqslant 0. \tag{A.13}$$

Proof. Let $D_N \subset D$ be finite sets such that $D_N \uparrow D$. We use (A.12) with the index set $I = D_N \cap [0,t]$, $t_{\max} = \max D_N \cap [0,t]$ and observe that for the submartingale $(|X_t|^p, \mathscr{F}_t)_{t\geqslant 0}$ the estimate

$$\mathbb{E}\left[|X(t_{\max})|^p\right] \leqslant \mathbb{E}\left[|X(t)|^p\right]$$

holds. Since

$$\Big\{\max_{s\in D_N\cap[0,t]} |X(s)| \geqslant r\Big\} \uparrow \Big\{\sup_{s\in D\cap[0,t]} |X(s)| \geqslant r\Big\}$$

we obtain (A.13) by using (A.12) and the continuity of measures. □

Finally we can deduce Doob's maximal inequality.

A.10 Theorem (L^p maximal inequality. Doob 1953). *Let $(X_t, \mathscr{F}_t)_{t\geqslant 0}$ be a positive submartingale or a d-dimensional martingale, $X_t \in L^p(\mathbb{P})$ for some $p > 1$ and $D \subset [0,\infty)$ a countable subset. Then for $p^{-1} + q^{-1} = 1$ and all $t > 0$*

$$\mathbb{E}\Big[\sup_{s\in D\cap[0,t]} |X_s|^p\Big] \leqslant q^p \sup_{s\in[0,t]} \mathbb{E}\left[|X_s|^p\right] \leqslant q^p\,\mathbb{E}\left[|X_t|^p\right]. \tag{A.14}$$

If $(X_t)_{t\geqslant 0}$ is continuous, (A.14) *holds for $D = [0,\infty)$.*

Proof. Let $D_N \subset D$ be finite index sets with $D_N \uparrow D$. The first estimate follows immediately from the discrete version of Doob's inequality,

$$\mathbb{E}\Big[\sup_{s\in D_N\cap[0,t]} |X_s|^p\Big] \leqslant q^p \sup_{s\in D_N\cap[0,t]} \mathbb{E}\left[|X_s|^p\right] \leqslant q^p \sup_{s\in[0,t]} \mathbb{E}\left[|X_s|^p\right]$$

and a monotone convergence argument. For the second estimate observe that we have $\mathbb{E}[|X_s|^p] \leqslant \mathbb{E}[|X_t|^p]$ for all $s \leqslant t$.

Finally, if $t \mapsto X_t$ is continuous and D a dense subset of $[0,\infty)$, we know that

$$\sup_{s\in D\cap[0,t]} |X_s|^p = \sup_{s\in[0,t]} |X_s|^p. \qquad \square$$

A.4 Stopping and sampling

Recall that a stochastic process $(X_t)_{t\geqslant 0}$ is said to be *adapted* to the filtration $(\mathscr{F}_t)_{t\geqslant 0}$ if X_t is for each $t \geqslant 0$ an $\mathscr{F}_t$ measurable random variable.

A.4.1 Stopping times

A.11 Definition. Let $(\mathscr{F}_t)_{t\geqslant 0}$ be a filtration. A random time $\tau : \Omega \to [0,\infty]$ is called an *($\mathscr{F}_t$-) stopping time* if

$$\{\tau \leqslant t\} \in \mathscr{F}_t \quad \text{for all } t \geqslant 0. \tag{A.15}$$

Here are a few rules for stopping times which should be familiar from the discrete time setting.

A.12 Lemma. *Let σ, τ, τ_j, $j \geqslant 1$ be stopping times with respect to a filtration $(\mathscr{F}_t)_{t\geqslant 0}$. Then*
a) $\{\tau < t\} \in \mathscr{F}_t$;
b) $\tau + \sigma, \tau \wedge \sigma, \tau \vee \sigma, \sup_j \tau_j$ *are $\mathscr{F}_t$ stopping times;*
c) $\inf_j \tau_j, \underline{\lim}_j \tau_j, \overline{\lim}_j \tau_j$ *are $\mathscr{F}_{t+}$ stopping times.*

Proof. The proofs are similar to the discrete time setting. Let $t \geqslant 0$.
a) $\{\tau < t\} = \bigcup_{k\geqslant 1} \{\tau \leqslant t - \frac{1}{k}\} \in \bigcup_{k\geqslant 1} \mathscr{F}_{t-\frac{1}{k}} \subset \mathscr{F}_t$.
b) We have for all $t \geqslant 0$

$$\begin{aligned}
\{\sigma+\tau > t\} &= \{\sigma = 0,\ \tau > t\} \cup \{\sigma > t,\ \tau = 0\} \cup \{\sigma > 0,\ \tau \geqslant t\} \cup \{0 < \tau < t,\ \tau+\sigma > t\} \\
&= \underbrace{\{\sigma = 0,\ \tau > t\}}_{=\{\sigma\leqslant 0\}\cap\{\tau\leqslant t\}^c\in\mathscr{F}_t} \cup \underbrace{\{\sigma > t,\ \tau = 0\}}_{=\{\sigma\leqslant t\}^c\cap\{\tau\leqslant 0\}\in\mathscr{F}_t} \cup \underbrace{\{\sigma > 0,\ \tau \geqslant t\}}_{=\{\sigma\leqslant 0\}^c\cap\{\tau<t\}^c\in\mathscr{F}_t} \\
&\qquad \cup \bigcup_{r\in(0,t)\cap\mathbb{Q}} \underbrace{\{r < \tau < t,\ \sigma > t - r\}}_{=\{\tau\leqslant r\}^c\cap\{\tau<t\}\cap\{\sigma\leqslant t-r\}^c\in\mathscr{F}_t}.
\end{aligned}$$

This shows that $\{\sigma+\tau \leqslant t\} = \{\sigma+\tau > t\}^c \in \mathscr{F}_t$ for all $t \geqslant 0$. Moreover,

$$\begin{aligned}
\{\sigma \vee \tau \leqslant t\} &= \{\sigma \leqslant t\} \cap \{\tau \leqslant t\} \in \mathscr{F}_t; \\
\{\sigma \wedge \tau \leqslant t\} &= \{\sigma \leqslant t\} \cup \{\tau \leqslant t\} \in \mathscr{F}_t; \\
\left\{\sup\nolimits_j \tau_j \leqslant t\right\} &= \bigcap\nolimits_j \left\{\tau_j \leqslant t\right\} \in \mathscr{F}_t;
\end{aligned}$$

c) Note that $\{\inf_j \tau_j < t\} = \bigcup_j \{\tau_j < t\} = \bigcup_j \bigcup_k \{\tau_j \leqslant t - \frac{1}{k}\} \in \mathscr{F}_t$; thus,

$$\{\inf_j \tau_j \leqslant t\} = \bigcap_{n\geqslant 1} \{\inf_j \tau_j < t + \tfrac{1}{n}\} \in \bigcap_{n\geqslant 1} \mathscr{F}_{t+\frac{1}{n}} = \mathscr{F}_{t+}.$$

Since $\underline{\lim}_j$ and $\overline{\lim}_j$ are combinations of inf and sup, the claim follows from the fact that $\inf_j \tau_j$ and $\sup_j \tau_j$ are $\mathscr{F}_{t+}$ stopping times. □

A.13 Example. The infimum of $\mathscr{F}_t$ stopping times need not be an $\mathscr{F}_t$ stopping time. To see this, let τ be an $\mathscr{F}_{t+}$ stopping time (but not an $\mathscr{F}_t$ stopping time). Then $\tau_j := \tau + \frac{1}{j}$ are $\mathscr{F}_t$ stopping times: **!**

$$\{\tau_j \leqslant t\} = \{\tau \leqslant t - \tfrac{1}{j}\} \in \mathscr{F}_{(t-\frac{1}{j})+} \subset \mathscr{F}_t.$$

But $\tau = \inf_j \tau_j$ is, by construction, not an $\mathscr{F}_t$ stopping time.

We will now associate a σ-algebra with a stopping time.

A.14 Definition. Let τ be an $\mathscr{F}_t$ stopping time. Then

$$\mathscr{F}_\tau := \left\{A \in \mathscr{F}_\infty := \sigma\left(\bigcup_{t\geqslant 0} \mathscr{F}_t\right) \;:\; A \cap \{\tau \leqslant t\} \in \mathscr{F}_t \;\forall t \geqslant 0\right\},$$
$$\mathscr{F}_{\tau+} := \left\{A \in \mathscr{F}_\infty := \sigma\left(\bigcup_{t\geqslant 0} \mathscr{F}_t\right) \;:\; A \cap \{\tau \leqslant t\} \in \mathscr{F}_{t+} \;\forall t \geqslant 0\right\}.$$

It is not hard to see that $\mathscr{F}_\tau$ and $\mathscr{F}_{\tau+}$ are σ-algebras and that for $\tau \equiv t$ we have $\mathscr{F}_\tau = \mathscr{F}_t$. Note that, despite the slightly misleading notation, $\mathscr{F}_\tau$ is a family of sets which does not depend on ω. **!**

A.15 Lemma. *Let $\sigma, \tau, \tau_j, j \geqslant 1$ be $\mathscr{F}_t$ stopping times. Then*

a) *τ is $\mathscr{F}_\tau$ measurable;*

b) *if $\sigma \leqslant \tau$ then $\mathscr{F}_\sigma \subset \mathscr{F}_\tau$;*

c) *if $\tau_1 \geqslant \tau_2 \geqslant \ldots \geqslant \tau_n \downarrow \tau$, then $\mathscr{F}_{\tau+} = \bigcap_{j\geqslant 1} \mathscr{F}_{\tau_j+}$. If $(\mathscr{F}_t)_{t\geqslant 0}$ is right-continuous, then $\mathscr{F}_\tau = \bigcap_{j\geqslant 1} \mathscr{F}_{\tau_j}$.*

Proof. a) We have to show that $\{\tau \leqslant s\} \in \mathscr{F}_\tau$ for all $s \geqslant 0$. But this follows from

$$\{\tau \leqslant s\} \cap \{\tau \leqslant t\} = \{\tau \leqslant s \wedge t\} \in \mathscr{F}_{s\wedge t} \subset \mathscr{F}_t \quad \forall t \geqslant 0.$$

b) Let $F \in \mathscr{F}_\sigma$. Then we have for all $t \geqslant 0$

$$F \cap \{\tau \leqslant t\} = F \cap \{\tau \leqslant t\} \cap \underbrace{\{\sigma \leqslant \tau\}}_{=\Omega} = \underbrace{F \cap \{\sigma \leqslant t\}}_{\in \mathscr{F}_t \text{ as } F \in \mathscr{F}_\sigma} \cap \underbrace{\{\tau \leqslant t\}}_{\in \mathscr{F}_t} \in \mathscr{F}_t.$$

Thus, $F \in \mathscr{F}_\tau$ and $\mathscr{F}_\sigma \subset \mathscr{F}_\tau$.

c) We know from Lemma A.12 that $\tau := \inf_j \tau_j$ is an $\mathscr{F}_{t+}$ stopping time. Applying b) to the filtration $(\mathscr{F}_{t+})_{t\geqslant 0}$ we get

$$\tau_j \geqslant \tau \;\forall j \geqslant 1 \overset{\text{b)}}{\Longrightarrow} \bigcap_{j\geqslant 1} \mathscr{F}_{\tau_j+} \supset \mathscr{F}_{\tau+}.$$

On the other hand, if $F \in \mathscr{F}_{\tau_j+}$ for all $j \geqslant 1$, then

$$F \cap \{\tau < t\} = F \cap \bigcup_{j\geqslant 1}\{\tau_j < t\} = \bigcup_{j\geqslant 1} F \cap \{\tau_j < t\} \in \mathscr{F}_t$$

where we use that

$$F \cap \{\tau_j < t\} = \bigcup_{k\geqslant 1} F \cap \left\{\tau_j \leqslant t - \tfrac{1}{k}\right\} \in \bigcup_{k\geqslant 1} \mathscr{F}_{t-\frac{1}{k}} \subset \mathscr{F}_t.$$

As in (5.9) we see now that $F \cap \{\tau \leqslant t\} \in \mathscr{F}_{t+}$.

If $(\mathscr{F}_t)_{t\geqslant 0}$ is right-continuous, we have $\mathscr{F}_t = \mathscr{F}_{t+}$ and $\mathscr{F}_\tau = \mathscr{F}_{\tau+}$. □

We close this section with a result which allows us to approximate stopping times from above by a decreasing sequence of stopping times with countably many values. This helps us to reduce many assertions to the discrete time setting.

A.16 Lemma. *Let τ be an $\mathscr{F}_t$ stopping time. Then there exists a decreasing sequence of $\mathscr{F}_t$ stopping times with*

$$\tau_1 \geqslant \tau_2 \geqslant \ldots \geqslant \tau_n \downarrow \tau = \inf_j \tau_j$$

and each τ_j takes only countably many values:

$$\tau_j(\omega) = \begin{cases} \tau(\omega), & \text{if } \tau(\omega) = \infty; \\ k2^{-j}, & \text{if } (k-1)2^{-j} \leqslant \tau(\omega) < k2^{-j},\ k \geqslant 1. \end{cases}$$

Moreover, $F \cap \left\{\tau_j = k2^{-j}\right\} \in \mathscr{F}_{k/2^j}$ for all $F \in \mathscr{F}_{\tau+}$.

Proof. Fix $t \geqslant 0$ and $j \geqslant 1$ and pick $k = k(t, j)$ such that $(k-1)2^{-j} \leqslant t < k2^{-j}$. Then

$$\{\tau_j \leqslant t\} = \left\{\tau_j \leqslant (k-1)2^{-j}\right\} = \left\{\tau < (k-1)2^{-j}\right\} \in \mathscr{F}_{(k-1)/2^j} \subset \mathscr{F}_t,$$

i. e. the τ_j's are stopping times.

From the definition it is obvious that $\tau_j \downarrow \tau$: Note that $\tau_j(\omega) - \tau(\omega) \leqslant 2^{-j}$ if $\tau(\omega) < \infty$. Finally, if $F \in \mathscr{F}_{\tau+}$, we have

$$\begin{aligned} F \cap \left\{\tau_j = k2^{-j}\right\} &= F \cap \left\{(k-1)2^{-j} \leqslant \tau < k2^{-j}\right\} \\ &= F \cap \left\{\tau < k2^{-j}\right\} \cap \left\{\tau < (k-1)2^{-j}\right\}^c \\ &= \underbrace{\bigcup_{n\geqslant 1}\left(F \cap \left\{\tau \leqslant k2^{-j} - \tfrac{1}{n}\right\}\right)}_{\in \mathscr{F}_{(k/2^j - 1/n)+} \subset \mathscr{F}_{k/2^j}} \cap \underbrace{\left\{\tau < (k-1)2^{-j}\right\}^c}_{\in \mathscr{F}_{k/2^j}} \in \mathscr{F}_{k/2^j}. \end{aligned}$$

□

A.4.2 Optional sampling

If we evaluate a continuous martingale at an increasing sequence of stopping times, we still have a martingale. The following technical lemma has exactly the same proof as Lemma 6.24 and its Corollary 6.25 – all we have to do is to replace in these proofs Brownian motion B by the continuous process X.

A.17 Lemma. *Let $(X_t, \mathscr{F}_t)_{t\geqslant 0}$ be a d-dimensional adapted stochastic process with continuous sample paths and let τ be an $\mathscr{F}_t$ stopping time. Then $X(\tau)\mathbb{1}_{\{\tau<\infty\}}$ is an $\mathscr{F}_\tau$ measurable random variable.*

Recall the optional sampling theorem for stopping times $\alpha, \beta : \Omega \to \{t_j\}_{j\geqslant 1}$ with values in a discrete set:

$$(X(t_j), \mathscr{F}_{t_j})_{j\geqslant 1} \text{ submartingale, } \alpha \leqslant \beta \leqslant C \implies X(\alpha) \leqslant \mathbb{E}(X(\beta) \mid \mathscr{F}_\alpha).$$

A.18 Theorem (Optional stopping. Doob 1953). *Let $(X_t, \mathscr{F}_t)_{t\geqslant 0}$ denote a submartingale with continuous sample paths and let $\sigma \leqslant \tau$ be $\mathscr{F}_t$ stopping times. Then we have for all $k \geqslant 0$ that $X_{\sigma\wedge k}, X_{\tau\wedge k} \in L^1(\mathbb{P})$ and*

$$X(\sigma \wedge k) \leqslant \mathbb{E}(X(\tau \wedge k) \mid \mathscr{F}_{\sigma\wedge k}). \tag{A.16}$$

If $\sigma \leqslant \tau$ are bounded stopping times, i. e. if $\tau \leqslant K$ a. s. for some constant K, or if we have $\mathbb{P}(\tau < \infty) = 1$, $X(\sigma), X(\tau) \in L^1(\mathbb{P})$ and $\lim_{k\to\infty} \mathbb{E}\left[|X(k)|\,\mathbb{1}_{\{\tau>k\}}\right] = 0$, then

$$X(\sigma) \leqslant \mathbb{E}(X(\tau) \mid \mathscr{F}_\sigma). \tag{A.17}$$

Proof. Using Lemma A.16 we can approximate $\sigma \leqslant \tau$ by discrete-valued stopping times $\sigma_j \downarrow \sigma$ and $\tau_j \downarrow \tau$. The construction in Lemma A.16 shows that we still have $\sigma_j \leqslant \tau_j$.

Using the discrete version of Doob's optional sampling theorem with $\alpha = \sigma_j \wedge k$ and $\beta = \tau_j \wedge k$ reveals that $X(\sigma_j \wedge k), X(\tau_j \wedge k) \in L^1(\mathbb{P})$ and

$$X(\sigma_j \wedge k) \leqslant \mathbb{E}\left[X(\tau_j \wedge k) \,\middle|\, \mathscr{F}_{\sigma_j\wedge k}\right].$$

Since $\mathscr{F}_{\sigma\wedge k} \subset \mathscr{F}_{\sigma_j\wedge k}$, the tower property gives

$$\mathbb{E}\left[X(\sigma_j \wedge k) \mid \mathscr{F}_{\sigma\wedge k}\right] \leqslant \mathbb{E}\left[X(\tau_j \wedge k) \mid \mathscr{F}_{\sigma\wedge k}\right]. \tag{A.18}$$

By discrete optional sampling, $(X(\sigma_j \wedge k), \mathscr{F}_{\sigma_j\wedge k})_{j\geqslant 1}$ and $(X(\tau_j \wedge k), \mathscr{F}_{\tau_j\wedge k})_{j\geqslant 1}$ are backwards submartingales and as such, cf. Corollary A.8, uniformly integrable. Therefore,

$$X(\sigma_j \wedge k) \xrightarrow[j\to\infty]{\text{a. s.}, L^1} X(\sigma \wedge k) \quad \text{and} \quad X(\tau_j \wedge k) \xrightarrow[j\to\infty]{\text{a. s.}, L^1} X(\tau \wedge k).$$

This shows that we have for all $F \in \mathscr{F}_{\sigma\wedge k}$

$$\int_F X(\sigma \wedge k)\, d\mathbb{P} = \lim_{j\to\infty} \int_F X(\sigma_j \wedge k)\, d\mathbb{P} \overset{\text{(A.18)}}{\leqslant} \lim_{j\to\infty} \int_F X(\tau_j \wedge k)\, d\mathbb{P} = \int_F X(\tau \wedge k)\, d\mathbb{P}$$

which proves (A.16).

Assume now that $X(\sigma), X(\tau) \in L^1(\mathbb{P})$. If $F \in \mathscr{F}_\sigma$, then

$$F \cap \{\sigma \leqslant k\} \cap \{\sigma \wedge k \leqslant t\} = F \cap \{\sigma \leqslant k\} \cap \{\sigma \leqslant t\} = F \cap \{\sigma \leqslant k \wedge t\} \in \mathscr{F}_{k\wedge t} \subset \mathscr{F}_t,$$

which means that $F \cap \{\sigma \leqslant k\} \in \mathscr{F}_{\sigma\wedge k}$. Since $\sigma \leqslant \tau < \infty$ we have by assumption

$$\int_{F\cap\{\sigma>k\}} |X(k)|\,d\mathbb{P} \leqslant \int_{\{\sigma>k\}} |X(k)|\,d\mathbb{P} \leqslant \int_{\{\tau>k\}} |X(k)|\,d\mathbb{P} \xrightarrow[k\to\infty]{} 0.$$

Using (A.16) we get by dominated convergence

$$\begin{aligned}\int_F X(\sigma\wedge k)\,d\mathbb{P} &= \overbrace{\int_{F\cap\{\sigma\leqslant k\}} X(\sigma\wedge k)\,d\mathbb{P}}^{\to\int_F X(\sigma)\,d\mathbb{P},\ k\to\infty} + \overbrace{\int_{F\cap\{\sigma>k\}} X(k)\,d\mathbb{P}}^{\to 0,\ k\to\infty}\\ &\leqslant \int_{F\cap\{\sigma\leqslant k\}} X(\tau\wedge k)\,d\mathbb{P} + \int_{F\cap\{\sigma>k\}} X(k)\,d\mathbb{P}\\ &= \underbrace{\int_{F\cap\{\tau\leqslant k\}} X(\tau\wedge k)\,d\mathbb{P}}_{\to\int_F X(\tau)\,d\mathbb{P},\ k\to\infty} + \underbrace{\int_{F\cap\{\tau>k\}} X(k)\,d\mathbb{P}}_{\to 0,\ k\to\infty}\end{aligned}$$

which is exactly (A.17). □

A.19 Corollary. *Let $(X_t, \mathscr{F}_t)_{t\geqslant 0}$, $X_t \in L^1(\mathbb{P})$, be an adapted process with continuous sample paths. The process $(X_t, \mathscr{F}_t)_{t\geqslant 0}$ is a submartingale if, and only if,*

$$\mathbb{E}\,X_\sigma \leqslant \mathbb{E}\,X_\tau \quad \textit{for all bounded } \mathscr{F}_t \textit{ stopping times } \sigma \leqslant \tau. \tag{A.19}$$

Proof. If $(X_t, \mathscr{F}_t)_{t\geqslant 0}$ is a submartingale, then (A.19) follows at once from Doob's optional stopping theorem. Conversely, assume that (A.19) holds true. Let $s < t$, pick any $F \in \mathscr{F}_s$ and set $\rho := t\mathbb{1}_{F^c} + s\mathbb{1}_F$. Obviously, ρ is a bounded stopping time and if we use (A.19) with $\sigma := \rho$ and $\tau := t$ we see

$$\mathbb{E}\,(X_s\mathbb{1}_F) + \mathbb{E}\,(X_t\mathbb{1}_{F^c}) = \mathbb{E}\,X_\rho \leqslant \mathbb{E}\,X_t.$$

Therefore,

$$\mathbb{E}\,(X_s\mathbb{1}_F) \leqslant \mathbb{E}\,(X_t\mathbb{1}_F) \quad \text{for all } s<t \quad \text{and} \quad F\in\mathscr{F}_s,$$

i. e. $X_s \leqslant \mathbb{E}(X_t \mid \mathscr{F}_s)$. □

A.20 Corollary. *Let $(X_t, \mathscr{F}_t)_{t\geqslant 0}$ be a submartingale which is uniformly integrable and has continuous paths. If $\sigma \leqslant \tau$ are stopping times with $\mathbb{P}(\tau < \infty) = 1$, then all assumptions of Theorem A.18 are satisfied.*

Proof. Since $(X_t)_{t\geqslant 0}$ is uniformly integrable, we get $\sup_{t\geqslant 0}\mathbb{E}\,|X(t)| \leqslant c < \infty$. Moreover, since X_t^+ is a submartingale, we have for all $k \geqslant 0$

$$\begin{aligned}\mathbb{E}\,|X(\sigma\wedge k)| = 2\,\mathbb{E}\,X^+(\sigma\wedge k) - \mathbb{E}\,X(\sigma\wedge k) &\overset{\text{(A.16)}}{\leqslant} 2\,\mathbb{E}\,X^+(k) - \mathbb{E}\,X(0)\\ &\leqslant 2\,\mathbb{E}\,|X(k)| + \mathbb{E}\,|X(0)| \leqslant 3c.\end{aligned}$$

By Fatou's lemma we get

$$\mathbb{E}\,|X(\sigma)| \leqslant \varliminf_{k\to\infty} \mathbb{E}\,|X(\sigma\wedge k)| \leqslant 3c,$$

and, similarly, $\mathbb{E}\,|X(\tau)| \leqslant 3c$. Finally,

$$\begin{aligned}\mathbb{E}\left[|X(k)|\,\mathbb{1}_{\{\tau>k\}}\right] &= \mathbb{E}\left[|X(k)|\big(\mathbb{1}_{\{\tau>k\}\cap\{|X(k)|>R\}} + \mathbb{1}_{\{\tau>k\}\cap\{|X(k)|\leqslant R\}}\big)\right]\\ &\leqslant \sup_n \mathbb{E}\left[|X(n)|\,\mathbb{1}_{\{|X(n)|>R\}}\right] + R\,\mathbb{P}(\tau>k)\\ &\xrightarrow[k\to\infty]{} \sup_n \mathbb{E}\left[|X(n)|\,\mathbb{1}_{\{|X(n)|>R\}}\right] \xrightarrow[R\to\infty]{\text{unif. integrable}} 0.\end{aligned}$$

□

A.21 Remark. a) If $(X_t, \mathscr{F}_t)_{t\geqslant 0}$ is a martingale, we have '=' in (A.16), (A.17) and (A.19).

b) Let $(X_t, \mathscr{F}_t)_{t\geqslant 0}$ be a submartingale with continuous paths and ρ an $\mathscr{F}_t$ stopping time. Then $(X_t^\rho, \mathscr{F}_t)$, $X_t^\rho := X_{\rho\wedge t}$, is a submartingale.
Indeed: X^ρ is obviously adapted and continuous. For any two bounded stopping times $\sigma \leqslant \tau$ it is clear that $\rho\wedge\sigma \leqslant \rho\wedge\tau$ are bounded stopping times, too. Therefore,

$$\mathbb{E}\,X_\sigma^\rho \stackrel{\text{def}}{=} \mathbb{E}\,X_{\rho\wedge\sigma} \stackrel{\text{A.19}}{\leqslant} \mathbb{E}\,X_{\rho\wedge\tau} \stackrel{\text{def}}{=} \mathbb{E}\,X_\tau^\rho,$$

and Corollary A.19 shows that X^ρ is a submartingale for $(\mathscr{F}_t)_{t\geqslant 0}$.

c) Note that $(X_t = X_{t+}, \mathscr{F}_{t+})$ is also a submartingale. If $\sigma \leqslant \tau$ are $\mathscr{F}_{t+}$ stopping times, then (A.16) and (A.17) remain valid if we use $\mathscr{F}_{(\sigma\wedge k)+}$, $\mathscr{F}_{\sigma+}$ etc.

We close this section with a uniqueness result for compensators.

A.22 Proposition. *Let $(X_t, \mathscr{F}_t)_{t\geqslant 0}$ be a martingale with paths which are continuous and of bounded variation on compact intervals. Then $X_t = X_0$ a. s. for all $t \geqslant 0$.*

Proof. Without loss of generality we may assume that $X_0 = 0$. Denote by

$$V_t := \mathrm{VAR}_1(X;[0,t]) \quad\text{and}\quad \tau_n = \inf\{t\geqslant 0 : V_t \geqslant n\}$$

the strong variation of $(X_t)_{t\geqslant 0}$ and the stopping time when V_t exceeds n for the first time. Since $t\mapsto X_t$ is continuous, so is $t\mapsto V_t$, cf. Paragraph A.34 f).

By optional stopping, Theorem A.18, we know that $(X_t^{\tau_n}, \mathscr{F}_{t\wedge\tau_n})_{t\geqslant 0}$ is for all $n\geqslant 1$ a martingale. An application of Doob's maximal inequality yields

$$\begin{aligned}\mathbb{E}\left[\sup_{s\leqslant t}(X_s^{\tau_n})^2\right] \stackrel{\text{(A.14)}}{\leqslant} 4\,\mathbb{E}\left[(X_t^{\tau_n})^2\right] &= \mathbb{E}\left[\sum_{j=1}^N \left((X_{\frac{j}{N}t}^{\tau_n})^2 - (X_{\frac{j-1}{N}t}^{\tau_n})^2\right)\right]\\ &= \mathbb{E}\left[\sum_{j=1}^N (X_{\frac{j}{N}t}^{\tau_n} - X_{\frac{j-1}{N}t}^{\tau_n})^2\right]\\ &\leqslant \mathbb{E}\left[V_t \sup_{1\leqslant j\leqslant N}\left|X_{\frac{j}{N}t}^{\tau_n} - X_{\frac{j-1}{N}t}^{\tau_n}\right|\right]\\ &\leqslant n\,\mathbb{E}\left[\sup_{1\leqslant j\leqslant N}\left|X_{\frac{j}{N}t}^{\tau_n} - X_{\frac{j-1}{N}t}^{\tau_n}\right|\right];\end{aligned}$$

in particular, the expectation on the left-hand side is finite. Since the paths are uniformly continuous on compact intervals, we find

$$\sup_{1\leqslant j\leqslant N} \left|X^{\tau_n}_{\frac{j}{N}t} - X^{\tau_n}_{\frac{j-1}{N}t}\right| \xrightarrow[N\to\infty]{} 0 \quad \text{almost surely.}$$

On the other hand, $\sup_{1\leqslant j\leqslant N} |X^{\tau_n}_{\frac{j}{N}t} - X^{\tau_n}_{\frac{j-1}{N}t}| \leqslant \operatorname{VAR}(X^{\tau_n};[0,t]) \leqslant V_{t\wedge\tau_n} \leqslant n$, and dominated convergence gives

$$\mathbb{E}\left[\sup_{s\leqslant t}(X^{\tau_n}_s)^2\right] \leqslant n \lim_{N\to\infty} \mathbb{E}\left[\sup_{1\leqslant j\leqslant N} \left|X^{\tau_n}_{\frac{j}{N}t} - X^{\tau_n}_{\frac{j-1}{N}t}\right|\right] = 0.$$

Therefore, $X_s = 0$ a. s. on $[0, k\wedge\tau_n]$ for all $k \geqslant 1$ and $n \geqslant 1$. The claim follows since $\tau_n \uparrow \infty$ almost surely as $n \to \infty$. □

A.23 Corollary. *Let $(X_t, \mathscr{F}_t)_{t\geqslant 0}$ be an L^2-martingale such that both $(X_t^2 - A_t, \mathscr{F}_t)_{t\geqslant 0}$ and $(X_t^2 - A'_t, \mathscr{F}_t)_{t\geqslant 0}$ are martingales for some adapted processes $(A_t)_{t\geqslant 0}$ and $(A'_t)_{t\geqslant 0}$ having continuous paths of bounded variation on compact sets and $A_0 = A'_0 = 0$. Then A and A' are indistinguishable: $\mathbb{P}\left(A_t = A'_t \;\forall t \geqslant 0\right) = 1$.*

Proof. By assumption $A_t - A'_t = (X_t^2 - A'_t) - (X_t^2 - A_t)$, i. e. $A - A'$ is a martingale with continuous paths which are of bounded variation on compact sets. The claim follows from Proposition A.22. □

A.5 Remarks on Feller processes

A Markov process $(X_t, \mathscr{F}_t)_{t\geqslant 0}$ is said to be a *Feller process* if the transition semigroup (7.1) has the Feller property, i. e. P_t is a strongly continuous Markov semigroup on the space $\mathcal{C}_\infty(\mathbb{R}^d) = \{u \in \mathcal{C}(\mathbb{R}^d) : \lim_{|x|\to\infty} u(x) = 0\}$, cf. Remark 7.4. In Remark 7.6 we have seen how to construct a Feller process from any given Feller semigroup. Throughout this section we assume that the filtration is right-continuous, i. e. $\mathscr{F}_t = \mathscr{F}_{t+}$ for all $t \geqslant 0$, and complete. Let us briefly discuss some properties of Feller processes.

A.24 Theorem. *Let $(X_t, \mathscr{F}_t)_{t\geqslant 0}$ be a Feller process. Then there exists a Feller process $(\tilde{X}_t)_{t\geqslant 0}$ whose sample paths are right-continuous with finite left limits and which has the same finite dimensional distributions as $(X_t)_{t\geqslant 0}$.*

Proof. See Blumenthal and Getoor [18, Theorem I.9.4, p. 46] or Ethier and Kurtz [72, Theorem 4.2.7 p. 169]. □

From now on we will assume that every Feller process has right-continuous paths.

A.25 Theorem. *Every Feller process $(X_t, \mathscr{F}_t)_{t\geqslant 0}$ has the strong Markov property* (6.9).

Proof. The argument is similar to the proof of Theorem 6.5. Let σ be a stopping time and approximate σ by the stopping times $\sigma_j := (\lfloor 2^j\sigma\rfloor + 1)/2^j$, cf. Lemma A.16. Let $t > 0$,

$x \in \mathbb{R}^d$, $u \in \mathcal{C}_\infty(\mathbb{R}^d)$ and $F \in \mathscr{F}_{\sigma+}$. Since $\{\sigma < \infty\} = \{\sigma_n < \infty\}$ we find by dominated convergence and the right-continuity of the sample paths

$$\begin{aligned}\int_{F\cap\{\sigma<\infty\}} u(X_{t+\sigma})\,d\mathbb{P}^x &= \lim_{j\to\infty} \int_{F\cap\{\sigma_j<\infty\}} u(X_{t+\sigma_j})\,d\mathbb{P}^x \\ &= \lim_{j\to\infty} \sum_{k=1}^{\infty} \mathbb{E}^x\left[u(X_{t+\sigma_j})\mathbb{1}_{F\cap\{\sigma_j=k/2^j\}}\right] \\ &= \lim_{j\to\infty} \sum_{k=1}^{\infty} \mathbb{E}^x\left[u(X_{t+k/2^j})\mathbb{1}_{F\cap\{\sigma_j=k/2^j\}}\right].\end{aligned}$$

By Lemma A.16, $F \cap \{\sigma_j = k/2^j\} \in \mathscr{F}_{k/2^j}$; using the tower property and the Markov property, we get

$$\begin{aligned}&\mathbb{E}^x\left[\mathbb{E}^x\left(u(X_{t+k/2^j})\mathbb{1}_{F\cap\{\sigma_j=k/2^j\}} \mid \mathscr{F}_{k/2^j}\right)\right] \\ &= \mathbb{E}^x\left[\mathbb{1}_{F\cap\{\sigma_j=k/2^j\}}\,\mathbb{E}^x\left(u(X_{t+k/2^j}) \mid \mathscr{F}_{k/2^j}\right)\right] \\ &= \mathbb{E}^x\left[\mathbb{1}_{F\cap\{\sigma_j=k/2^j\}}\,\mathbb{E}^{X_{k/2^j}}\, u(X_t)\right] \\ &= \mathbb{E}^x\left[\mathbb{1}_{F\cap\{\sigma_j=k/2^j\}}\,\mathbb{E}^{X_{\sigma_j}}\, u(X_t)\right]\end{aligned}$$

and, again by dominated convergence and the right-continuity

$$\begin{aligned}\int_{F\cap\{\sigma<\infty\}} u(X_{t+\sigma})\,d\mathbb{P}^x &= \lim_{j\to\infty} \sum_{k=1}^{\infty} \mathbb{E}^x\left[\mathbb{1}_{F\cap\{\sigma_j=k/2^j\}}\,\mathbb{E}^{X_{\sigma_j}}\, u(X_t)\right] \\ &= \lim_{j\to\infty} \mathbb{E}^x\left[\mathbb{1}_{F\cap\{\sigma<\infty\}}\,\mathbb{E}^{X_{\sigma_j}}\, u(X_t)\right] \\ &= \mathbb{E}^x\left[\mathbb{1}_{F\cap\{\sigma<\infty\}}\,\mathbb{E}^{X_\sigma}\, u(X_t)\right].\end{aligned}$$

□

A.26 Theorem. *Let $(X_t, \mathscr{F}_t)_{t\geqslant 0}$ be a Feller process and $\tau_x := \inf\{t > 0 : X_t \neq x\}$ the first hitting time of the set $\mathbb{R}^d \setminus \{x\}$. Then there exists some $\lambda(x) \in [0,\infty]$ such that $\mathbb{P}^x(\tau_x \geqslant t) = e^{-\lambda(x)t}$.*

Proof. Under $\mathbb{P}^x$ we have, a. s. $\tau_x > t$ if, and only if, $\sup_{s\leqslant t} |X_s - x| = 0$. Therefore,

$$\mathbb{P}^x(\tau_x > t + s) = \mathbb{P}^x\Big(\tau_x > s, \sup_{r\leqslant t} |X_{r+s} - X_s| = 0\Big)$$

By the Markov property we find

$$\begin{aligned}\mathbb{P}^x\Big(\tau_x > s, \sup_{r\leqslant t} |X_{r+s} - X_s| = 0\Big) &= \mathbb{E}^x\Big(\mathbb{1}_{\{\tau_x>s\}}\mathbb{P}^{X_s}\Big[\sup_{r\leqslant t} |X_r - X_0| = 0\Big]\Big) \\ &\overset{X_s=x}{=} \mathbb{E}^x\Big(\mathbb{1}_{\{\tau_x>s\}}\Big)\,\mathbb{P}^x\Big(\sup_{r\leqslant t} |X_r - x| = 0\Big).\end{aligned}$$

Thus, $\mathbb{P}^x(\tau_x > t+s) = \mathbb{P}^x(\tau_x > s)\mathbb{P}^x(\tau_x > t)$. As

$$\lim_{n\to\infty} \mathbb{P}^x(\tau_x > t - 1/n) = \mathbb{P}^x(\tau_x \geqslant t),$$

we find for all $s, t > 0$ that $\mathbb{P}^x(\tau_x \geqslant t+s) = \mathbb{P}^x(\tau_x \geqslant s)\mathbb{P}^x(\tau_x \geqslant t)$. Since $s \mapsto \mathbb{P}^x(\tau_x \geqslant s)$ is left-continuous and $\mathbb{P}^x(\tau_x \geqslant 0) = 1$, this functional equation has a unique solution of the form $e^{-\lambda(x)t}$ with some $\lambda(x) \in [0,\infty]$. □

A.6 The Doob–Meyer decomposition

We will now show that every martingale $M \in \mathcal{M}_T^{2,c}$ satisfies

$$\begin{gathered}\left(M_t^2 - A_t, \mathscr{F}_t\right)_{t\leqslant T} \quad \text{is a martingale for some} \\ \text{adapted, continuous and increasing process } (A_t)_{t\leqslant T}.\end{gathered} \tag{$\star$}$$

We begin with the uniqueness of the process A. This follows directly from Corollary A.23.

A.27 Lemma. *Assume that $M \in \mathcal{M}_T^{2,c}$ and that there are two continuous and increasing processes $(A_t)_{t\leqslant T}$ and $(A'_t)_{t\leqslant T}$ such that both $(M_t^2 - A_t)_{t\leqslant T}$ and $(M_t^2 - A'_t)_{t\leqslant T}$ are martingales. Then $A_t - A_0 = A'_t - A'_0$ a. s. with the same null set for all $t \leqslant T$.*

Let us now turn to the existence of the process A from ($\star$). The following elementary proof is due to Kunita [134, Chapter 2.2]. We consider first bounded martingales.

A.28 Theorem. *Let $M \in \mathcal{M}_T^{2,c}$ such that $\sup_{t\leqslant T}|M_t| \leqslant \kappa$ for some constant $\kappa < \infty$. Let $(\Pi_n)_{n\geqslant 1}$ be any sequence of finite partitions of $[0,T]$ satisfying $\lim_{n\to\infty}|\Pi_n| = 0$. Then the mean-square limit*

$$L^2(\mathbb{P})\text{-}\lim_{n\to\infty} \sum_{t_{j-1},t_j\in\Pi_n} \left(M_{t\wedge t_j} - M_{t\wedge t_{j-1}}\right)^2$$

exists uniformly for all $t \leqslant T$ and defines an increasing and continuous process $(\langle M\rangle_t)_{t\leqslant T}$ such that $M^2 - \langle M\rangle$ is a martingale.

The proof is based on two lemmas. Let $\Pi = \{0 = t_0 < t_1 < \cdots < t_n = T\}$ be a partition of $[0,T]$ and denote by $M^{\Pi,t} = (M_{t_j}^t, \mathscr{F}_{t_j\wedge t})_{t_j\in\Pi}$, $0 \leqslant t \leqslant T$, where $M_{t_j}^t = M_{t\wedge t_j}$. Consider the martingale transform

$$N_T^{\Pi,t} := M^{\Pi,t} \bullet M_T^{\Pi,t} = \sum_{j=1}^n M_{t\wedge t_{j-1}}\left(M_{t\wedge t_j} - M_{t\wedge t_{j-1}}\right).$$

A.29 Lemma. *$(N_T^{\Pi,t}, \mathscr{F}_t)_{t\leqslant T}$ is a continuous martingale and $\mathbb{E}\left[(N_T^{\Pi,t})^2\right] \leqslant \kappa^4$. Moreover,*

$$M_t^2 - M_0^2 = \sum_{j=1}^n \left(M_{t\wedge t_j} - M_{t\wedge t_{j-1}}\right)^2 + 2N_T^{\Pi,t}, \tag{A.20}$$

and $\mathbb{E}\left[\left(\sum_{j=1}^n (M_{t\wedge t_j} - M_{t\wedge t_{j-1}})^2\right)^2\right] \leqslant 16\kappa^4$.

Proof. The martingale property is just Theorem 15.9 a). For the first estimate we use Theorem 15.4

$$\mathbb{E}\left[(N_T^{\Pi,t})^2\right] = \mathbb{E}\left[(M^{\Pi,t} \bullet M_T^{\Pi,t})^2\right] \overset{(15.6)}{=} \mathbb{E}\left[(M^{\Pi,t})^2 \bullet \langle M^{\Pi,t}\rangle_T\right]$$
$$\leqslant \kappa^2\,\mathbb{E}\left[\langle M^{\Pi,t}\rangle_T\right] = \kappa^2\,\mathbb{E}\left[M_t^2 - M_0^2\right] \leqslant \kappa^4.$$

The identity (A.20) follows from

$$\begin{aligned} M_t^2 - M_0^2 &= \sum_{j=1}^n (M_{t\wedge t_j}^2 - M_{t\wedge t_{j-1}}^2) \\ &= \sum_{j=1}^n (M_{t\wedge t_j} + M_{t\wedge t_{j-1}})(M_{t\wedge t_j} - M_{t\wedge t_{j-1}}) \\ &= \sum_{j=1}^n \left[(M_{t\wedge t_j} - M_{t\wedge t_{j-1}})^2 + 2M_{t\wedge t_{j-1}}(M_{t\wedge t_j} - M_{t\wedge t_{j-1}})\right] \\ &= \sum_{j=1}^n (M_{t\wedge t_j} - M_{t\wedge t_{j-1}})^2 + 2N_T^{\Pi,t}. \end{aligned}$$

From this and the first estimate we get

$$\mathbb{E}\left[\Big(\sum_{j=1}^n (M_{t\wedge t_j} - M_{t\wedge t_{j-1}})^2\Big)^2\right] = \mathbb{E}\left[(\underbrace{M_t^2 - M_0^2}_{\leqslant 2\kappa^2} - 2N_T^{\Pi,t})^2\right] \leqslant 8\kappa^4 + 8\,\mathbb{E}\left[(N_T^{\Pi,t})^2\right] \leqslant 16\kappa^4.$$

In the penultimate inequality we used the fact that $(a+b)^2 \leqslant 2a^2 + 2b^2$. □

A.30 Lemma. *Let $(\Pi_n)_{n\geqslant 1}$ be a sequence of partitions of $[0,T]$ such that the mesh size $|\Pi_n| = \max_{t_{j-1},t_j\in\Pi_n}(t_j - t_{j-1})$ tends to 0 as $n \to \infty$. Then*

$$\lim_{m,n\to\infty} \mathbb{E}\left[\sup_{0\leqslant t\leqslant T} |N_T^{\Pi_n,t} - N_T^{\Pi_m,t}|^2\right] = 0.$$

Proof. For any $\Pi = \{0 = t_0 < t_1 < \cdots < t_n = T\}$ we define $f_u^{\Pi,t} := M_{t\wedge t_{j-1}}$ if $t_{j-1} \leqslant u < t_j$. Note that for any two partitions Π, Π' and all $u \leqslant T$ we have

$$|f_u^{\Pi,t} - f_u^{\Pi',t}| \leqslant \sup_{|r-s|\leqslant|\Pi|\vee|\Pi'|} |M_{t\wedge s} - M_{t\wedge r}|.$$

By the continuity of the sample paths, this quantity tends to zero as $|\Pi|, |\Pi'| \to 0$.

Fix two partitions Π, Π' of the sequence $(\Pi_n)_{n\geqslant 1}$ and denote the elements of Π by t_j and those of $\Pi \cup \Pi'$ by u_k. The process $f_u^{\Pi,t}$ is constructed in such a way that

$$\begin{aligned} N_T^{\Pi,t} = M^{\Pi,t} \bullet M_T^{\Pi,t} &= \sum_{t_{j-1},t_j\in\Pi} M_{t\wedge t_{j-1}}(M_{t\wedge t_j} - M_{t\wedge t_{j-1}}) \\ &= \sum_{u_{k-1},u_k\in\Pi\cup\Pi'} f_{u_{k-1}}^{\Pi,t}(M_{t\wedge u_k} - M_{t\wedge u_{k-1}}) = f^{\Pi,t} \bullet M_T^{\Pi\cup\Pi',t}. \end{aligned}$$

Thus,

$$N_T^{\Pi,t} - N_T^{\Pi',t} = \left(f^{\Pi,t} - f^{\Pi',t}\right) \bullet M_T^{\Pi\cup\Pi',t},$$

and we find by Doob's maximal inequality A.10,

$$\begin{aligned}
&\mathbb{E}\left[\sup_{s\leqslant t} |N_T^{\Pi,s} - N_T^{\Pi',s}|^2\right] \\
&\leqslant 4\,\mathbb{E}\left[|N_T^{\Pi,t} - N_T^{\Pi',t}|^2\right] \\
&= 4\,\mathbb{E}\left[|(f^{\Pi,t} - f^{\Pi',t}) \bullet M_T^{\Pi\cup\Pi',t}|^2\right] \\
&\overset{(15.7)}{=} 4\,\mathbb{E}\left[\sum_{u_{k-1},u_k\in\Pi\cup\Pi'} (f_{u_{k-1}}^{\Pi,t} - f_{u_{k-1}}^{\Pi',t})^2 (M_{t\wedge u_k} - M_{t\wedge u_{k-1}})^2\right] \\
&\leqslant 4\,\mathbb{E}\left[\sup_{u\leqslant T} |f_u^{\Pi,t} - f_u^{\Pi',t}|^2 \sum_{u_{k-1},u_k\in\Pi\cup\Pi'} (M_{t\wedge u_k} - M_{t\wedge u_{k-1}})^2\right] \\
&\leqslant 4\left(\mathbb{E}\left[\sup_{u\leqslant T} |f_u^{\Pi,t} - f_u^{\Pi',t}|^4\right]\right)^{\frac{1}{2}} \left(\mathbb{E}\left[\left(\sum_{u_{k-1},u_k\in\Pi\cup\Pi'} (M_{t\wedge u_k} - M_{t\wedge u_{k-1}})^2\right)^2\right]\right)^{\frac{1}{2}}.
\end{aligned}$$

The first term tends to 0 as $|\Pi|, |\Pi'| \to 0$ while the second factor is bounded by $4\kappa^2$, cf. Lemma A.29. □

Proof of Theorem A.28. Let $(\Pi_n)_{n\geqslant 0}$ be a sequence of refining partitions such that $|\Pi_n| \to 0$ and set $\Pi := \bigcup_{n\geqslant 0} \Pi_n$.

By Lemma A.30, the limit $L^2\text{-}\lim_{n\to\infty} N_T^{\Pi_n,t}$ exists uniformly for all $t \in [0,T]$ and the same is true, because of (A.20), for $L^2\text{-}\lim_{n\to\infty} \sum_{t_{j-1},t_j\in\Pi_n} (M_{t\wedge t_j} - M_{t\wedge t_{j-1}})^2$. Therefore, there is a subsequence such that

$$\langle M\rangle_t := \lim_{k\to\infty} \sum_{t_{j-1},t_j\in\Pi_{n(k)}} (M_{t\wedge t_j} - M_{t\wedge t_{j-1}})^2$$

exists a. s. uniformly for $t \in [0,T]$; due to the uniform convergence, $\langle M\rangle_t$ is a continuous process.

For $s,t \in \Pi$ with $s < t$ there is some $k \geqslant 1$ such that $s,t \in \Pi_{n(k)}$. Obviously,

$$\sum_{t_{j-1},t_j\in\Pi_{n(k)}} (M_{s\wedge t_j} - M_{s\wedge t_{j-1}})^2 \leqslant \sum_{t_{j-1},t_j\in\Pi_{n(k)}} (M_{t\wedge t_j} - M_{t\wedge t_{j-1}})^2$$

and, as $k \to \infty$, this inequality also holds for the limiting process $\langle M\rangle$. Since Π is dense and $t \mapsto \langle M\rangle_t$ continuous, we get $\langle M\rangle_s \leqslant \langle M\rangle_t$ for all $s \leqslant t$.

From (A.20) we see that $M_t^2 - \sum_{t_{j-1},t_j\in\Pi_{n(k)}} (M_{t\wedge t_j} - M_{t\wedge t_{j-1}})^2$ is a martingale. From Lemma 15.10 we know that the martingale property is preserved under L^2 limits, and so $M^2 - \langle M\rangle$ is a martingale. □

Let us, finally, remove the assumption that M is bounded.

A.31 Theorem (Doob–Meyer decomposition. Meyer 1962, 1963). *For all square integrable martingales $M \in \mathcal{M}_T^{2,c}$ there exists a unique adapted, continuous and increasing process $(\langle M\rangle_t)_{t\leqslant T}$ such that $M^2 - \langle M\rangle$ is a martingale and $\langle M\rangle_0 = 0$.*

If $(\Pi_n)_{n\geqslant 1}$ is any refining sequence of partitions of $[0,T]$ with $|\Pi_n| \to 0$, then $\langle M\rangle$ is given by

$$\langle M\rangle_t = \lim_{n\to\infty} \sum_{t_{j-1},t_j\in\Pi_n} (M_{t\wedge t_j} - M_{t\wedge t_{j-1}})^2$$

where the limit is uniform (on compact t-sets) in probability.

Proof. Define a sequence of stopping times $\tau_n := \inf\{t \leqslant T : |M_t| \geqslant n\}$, $n \geqslant 1$, ($\inf\emptyset = \infty$). Clearly, $\lim_{n\to\infty}\tau_n = \infty$, $M^{\tau_n} \in \mathcal{M}_T^{2,c}$ and $|M_t^{\tau_n}| \leqslant n$ for all $t \in [0,T]$. Since for any partition Π of $[0,T]$ and all $m \leqslant n$

$$\sum_{t_{j-1},t_j\in\Pi} \left(M^{\tau_n}_{t\wedge\tau_m\wedge t_j} - M^{\tau_n}_{t\wedge\tau_m\wedge t_{j-1}}\right)^2 = \sum_{t_{j-1},t_j\in\Pi} \left(M^{\tau_m}_{t\wedge t_j} - M^{\tau_m}_{t\wedge t_{j-1}}\right)^2,$$

the following process is well-defined

$$\langle M\rangle_t(\omega) := \langle M^{\tau_n}\rangle_t(\omega) \qquad \text{for all } (\omega,t) : t \leqslant \tau_n(\omega)\wedge T$$

and we find that $\langle M\rangle_t$ is continuous and $\langle M\rangle_t^{\tau_n} = \langle M^{\tau_n}\rangle_t$.

Using Fatou's lemma, the fact that $(M^{\tau_n})^2 - \langle M^{\tau_n}\rangle$ is a martingale, and Doob's maximal inequality A.10, we have

$$\begin{aligned}\mathbb{E}\left(\langle M\rangle_T\right) = \mathbb{E}\left(\varliminf_{n\to\infty}\langle M\rangle_T^{\tau_n}\right) \leqslant \varliminf_{n\to\infty}\mathbb{E}\left(\langle M^{\tau_n}\rangle_T\right) &= \varliminf_{n\to\infty}\mathbb{E}\left((M_T^{\tau_n})^2 - M_0^2\right)\\ &\leqslant \mathbb{E}\left(\sup_{t\leqslant T} M_t^2\right)\\ &\leqslant 4\,\mathbb{E}\left(M_T^2\right).\end{aligned}$$

Therefore, $\langle M\rangle_T \in L^1(\mathbb{P})$. Moreover,

$$\mathbb{E}\left(\sup_{t\leqslant T}\left|\langle M\rangle_t - \langle M^{\tau_n}\rangle_t\right|\right) = \mathbb{E}\left(\sup_{t\leqslant T}\left|\langle M\rangle_t - \langle M\rangle_{\tau_n}\right|\mathbb{1}_{\{t>\tau_n\}}\right) \leqslant \mathbb{E}\left(\langle M\rangle_T\mathbb{1}_{\{T>\tau_n\}}\right)$$

which shows, by the dominated convergence theorem, that $L^1\text{-}\lim_{n\to\infty}\langle M\rangle_t^{\tau_n} = \langle M\rangle_t$ uniformly for $t \in [0,T]$. Similarly,

$$\mathbb{E}\left(\sup_{t\leqslant T}(M_t - M_t^{\tau_n})^2\right) = \mathbb{E}\left(\sup_{t\leqslant T}(M_t - M_{\tau_n})^2\mathbb{1}_{\{t>\tau_n\}}\right) \leqslant 4\,\mathbb{E}\left(\sup_{t\leqslant T} M_t^2\mathbb{1}_{\{T>\tau_n\}}\right).$$

Since, by Doob's maximal inequality, $\sup_{t\leqslant T} M_t^2 \in L^1$, we can use dominated convergence to see that $L^1\text{-}\lim_{n\to\infty}(M_t^{\tau_n})^2 = M_t^2$ uniformly for all $t \in [0,T]$.

Since L^1-convergence preserves the martingale property, $M^2 - \langle M\rangle$ is a martingale. Using that $\langle M\rangle_t = \langle M\rangle_t^{\tau_n} = \langle M^{\tau_n}\rangle_t$ for $t \leqslant \tau_n$, we find for every $\epsilon > 0$

$$\begin{aligned}
&\mathbb{P}\Big(\sup_{t\leqslant T}\Big|\langle M\rangle_t - \sum_{t_{j-1},t_j\in\Pi}(M_{t\wedge t_j} - M_{t\wedge t_{j-1}})^2\Big| > \epsilon\Big)\\
&\quad\leqslant \mathbb{P}\Big(\sup_{t\leqslant T}\Big|\langle M^{\tau_n}\rangle_t - \sum_{t_{j-1},t_j\in\Pi}(M^{\tau_n}_{t\wedge t_j} - M^{\tau_n}_{t\wedge t_{j-1}})^2\Big| > \epsilon,\ T\leqslant\tau_n\Big) + \mathbb{P}(T>\tau_n)\\
&\quad\leqslant \frac{1}{\epsilon^2}\,\mathbb{E}\Big(\sup_{t\leqslant T}\Big|\langle M^{\tau_n}\rangle_t - \sum_{t_{j-1},t_j\in\Pi}(M^{\tau_n}_{t\wedge t_j} - M^{\tau_n}_{t\wedge t_{j-1}})^2\Big|^2\Big) + \mathbb{P}(T>\tau_n)\\
&\quad\xrightarrow[|\Pi|\to 0]{\text{Thm. A.28}} \mathbb{P}(T>\tau_n) \xrightarrow[n\to\infty]{} 0.
\end{aligned}$$

□

A.32 Corollary. *Let $(M_t, \mathscr{F}_t)_{t\geqslant 0} \subset L^2(\mathbb{P})$ be a continuous martingale with respect to a right-continuous filtration. Then $M_\bullet$ and the quadratic variation $\langle M\rangle_\bullet$ are, with probability one, constant on the same t-intervals.*

Proof. Fix $T > 0$. We have to show that there is a set $\Omega^* \subset \Omega$ such that $\mathbb{P}(\Omega^*) = 1$ and for all $[a,b] \subset [0,T]$ and $\omega \in \Omega^*$

$$M_t(\omega) = M_a(\omega) \text{ for all } t \in [a,b] \iff \langle M\rangle_t(\omega) = \langle M\rangle_a(\omega) \text{ for all } t\in[a,b].$$

"⇒": By Theorem A.31 there is a set $\Omega' \subset \Omega$ of full $\mathbb{P}$-measure and a sequence of partitions $(\Pi_n)_{n\geqslant 1}$ of $[0,T]$ such that $\lim_{n\to\infty}|\Pi_n| = 0$ and

$$\langle M\rangle_t(\omega) = \lim_{n\to\infty}\sum_{t_{j-1},t_j\in\Pi_n}(M_{t_j\wedge t}(\omega) - M_{t_{j-1}\wedge t}(\omega))^2, \quad \omega\in\Omega'.$$

If $M_\bullet$ is constant on $[a,b] \subset [0,T]$, we see that for all $t\in[a,b]$

$$\langle M\rangle_t(\omega) - \langle M\rangle_a(\omega) = \lim_{n\to\infty}\sum_{t_{j-1},t_j\in\Pi_n\cap[a,b]}(M_{t_j\wedge t}(\omega) - M_{t_{j-1}\wedge t}(\omega))^2 = 0 \quad \omega\in\Omega'.$$

"⇐": Let Ω' be as above. The idea is to show that for every rational $q \in [0,T]$ there is a set $\Omega_q \subset \Omega'$ with $\mathbb{P}(\Omega_q) = 1$ and such that for every $\omega\in\Omega_q$ a $\langle M\rangle_\bullet(\omega)$-constancy interval of the form $[q,b]$ is also an interval of constancy for $M_\bullet(\omega)$. Then the claim follows for the set $\Omega^* = \bigcap_{q\in\mathbb{Q}\cap[0,T]}\Omega_q$.

The shifted process $N_t := M_{t+q} - M_q$, $0\leqslant t\leqslant T-q$ is a continuous L^2 martingale for the filtration $(\mathscr{F}_{t+q})_{t\in[0,T-q]}$; moreover, $\langle N\rangle_t = \langle M\rangle_{t+q} - \langle M\rangle_q$.

Set $\tau = \inf\{s>0 : \langle N\rangle_s > 0\}$, note that τ is a stopping time (since $\langle N\rangle$ has continuous paths), and consider the stopped processes $\langle N^\tau\rangle_t = \langle N\rangle_{t\wedge\tau}$ and $N_t^\tau = N_{t\wedge\tau}$.

By definition, $\langle N^\tau\rangle_t \equiv 0$ almost surely, hence $\langle M\rangle_{t\wedge\tau+q} \equiv \langle M\rangle_q$ almost surely. Since $\mathbb{E}(N^2_{\tau\wedge t}) = \mathbb{E}\langle N^\tau\rangle_t = 0$, we conclude that $M_{t\wedge\tau+q} - M_q = N_t^\tau = 0$ a. s., i. e. there is a set Ω_q of full measure such that for every $\omega\in\Omega_q$ every interval of constancy $[q,b]$ of $\langle M\rangle_\bullet$ is also an interval of constancy for $M_\bullet$. On the other hand, the direction "⇒" shows that a constancy interval of $M_\bullet$ cannot be longer than that of $\langle M\rangle_\bullet$, hence they coincide. □

A.7 BV functions and Riemann–Stieltjes integrals

Here we collect some material on functions of bounded variation, BV-functions for short, and Riemann–Stieltjes integrals. Our main references are Hewitt & Stromberg [96, Chapters V.8, V.9, V.17], Kestelman [124, Chapter XI], Rudin [197, Chapter 6] and Riesz & Sz.-Nagy [192, Chapters I.4–9, III.49–60].

A.7.1 Functions of bounded variation

Let $f : [a,b] \to \mathbb{R}$ be a function and let $\Pi = \{t_0 = a < t_1 < \cdots < t_n = b\}$ be a finite partition of the interval $[a,b] \subset \mathbb{R}$. By $|\Pi| := \max_{1\leqslant j\leqslant n}(t_j - t_{j-1})$ we denote the *mesh* of Π. We call

$$S_1^{\Pi}(f;[a,b]) := \sum_{t_j,t_{j-1}\in\Pi} |f(t_j) - f(t_{j-1})| = \sum_{j=1}^{n} |f(t_j) - f(t_{j-1})| \tag{A.21}$$

the *variation sum*. The supremum of the variation sums over all finite partitions

$$\mathrm{VAR}_1(f;[a,b]) := \sup\left\{S_1^{\Pi}(f;[a,b]) \;:\; \Pi \text{ finite partition of } [a,b]\right\} \in [0,\infty] \tag{A.22}$$

is the *strong variation* or *total variation*.

A.33 Definition. A function $f : [a,b] \to \mathbb{R}$ is said to be of *bounded variation* (also: *finite variation*) if $\mathrm{VAR}_1(f;[a,b]) < \infty$. We denote by $\mathsf{BV}[a,b]$ the set of all functions $f : [a,b] \to \mathbb{R}$ of bounded variation; if the domain is clear, we just write BV.

Let us collect a few properties of functions of bounded variation.

A.34 Properties of BV-functions. Let $f, g : [a,b] \to \mathbb{R}$ be real functions.

a) If f is monotone, then $f \in \mathsf{BV}$ and $\mathrm{VAR}_1(f;[a,b]) = |f(b) - f(a)|$;

b) If $f \in \mathsf{BV}$, then $\sup_{t\in[a,b]} |f(t)| < \infty$;

c) If $f, g \in \mathsf{BV}$, then $f \pm g$, $f \cdot g$ and $f/(|g| + 1)$ are of bounded variation;

d) Let $a < c < b$. Then $\mathrm{VAR}_1(f;[a,b]) = \mathrm{VAR}_1(f;[a,c]) + \mathrm{VAR}_1(f;[c,b])$ in $[0,\infty]$;

e) $f \in \mathsf{BV}$ if, and only if, there exist two increasing functions $\phi, \psi : [a,b] \to \mathbb{R}$ such that $f = \phi - \psi$.
In particular, $f \in \mathsf{BV}[a,b]$ has for every $t \in (a,b)$ finite left- and right-hand limits $f(t\pm)$, and the number of discontinuities is at most countable.

f) If $f \in \mathsf{BV}$ is continuous at t_0, then $t \mapsto \mathrm{VAR}_1(f;[a,t])$ is continuous at $t = t_0$;

g) Every $f \in \mathsf{BV}$ has a decomposition $f(t) = f_c(t) + f_s(t)$ where $f_c \in \mathsf{BV} \cap \mathcal{C}$ and $f_s(t) = [f(a+) - f(a)] + \sum_{a<s<t}[f(s+) - f(s-)] + [f(t) - f(t-)]$ are uniquely determined;

h) Every $f \in \mathsf{BV}$ has Lebesgue almost everywhere a finite derivative.

A.7.2 The Riemann–Stieltjes Integral

The Riemann–Stieltjes integral is a generalization of the classical Riemann integral. It was introduced by T. J. Stieltjes in his memoir [209, in particular: pp. 69–76] on continued fractions and the general moment problem. While Stieltjes considers only positive integrators ('mass distributions'), it is F. Riesz [190] who remarked that one can consider integrators from BV.

Let $f, \alpha : [a,b] \to \mathbb{R}$ and let $\Pi = \{t_0 = a < t_1 < \cdots < t_n = b\}$ be a partition of the interval $[a,b] \subset \mathbb{R}$ with mesh $|\Pi|$; moreover, ξ_j^Π denotes an arbitrary point from the partition interval $[t_j, t_{j+1}]$, $j = 0, \dots, n-1$.

A.35 Definition. Let $\alpha : [a,b] \to \mathbb{R}$. A function $f : [a,b] \to \mathbb{R}$ is called *(Riemann–Stieltjes) integrable* if the *Stieltjes sums*

$$S^\Pi(f;\alpha) := \sum_{t_j, t_{j-1} \in \Pi} f(\xi_{j-1}^\Pi)(\alpha(t_j) - \alpha(t_{j-1})) \tag{A.23}$$

converge, independently of the choice of the points $\xi_j^\Pi \in [t_j, t_{j+1}]$, $t_j, t_{j+1} \in \Pi$, to a finite limit as $|\Pi| \to 0$. The common value is called the *Riemann–Stieltjes integral* of f, and we write

$$\int_a^b f(t)\, d\alpha(t) := \lim_{|\Pi| \to 0} \sum_{t_j, t_{j-1} \in \Pi} f(\xi_{j-1}^\Pi)(\alpha(t_j) - \alpha(t_{j-1})). \tag{A.24}$$

For $\alpha(t) = t$ we get the Riemann integral as a special case. Let us briefly collect the main features of the Riemann–Stieltjes integral.

A.36 Properties of the Riemann–Stieltjes integral. Let $f, \alpha : [a,b] \to \mathbb{R}$ be real functions and $a \leqslant c \leqslant b$.

a) $(f,\alpha) \mapsto \int_a^b f(t)\, d\alpha(t)$ is bilinear;
b) If $\int_a^b f(t)\, d\alpha(t)$ exists, so does $\int_c^d f(t)\, d\alpha(t)$ for all $a \leqslant c < d \leqslant b$;
c) $\int_a^b f(t)\, d\alpha(t) = \int_a^c f(t)\, d\alpha(t) + \int_c^b f(t)\, d\alpha(t)$;
d) $\left|\int_a^b f(t)\, d\alpha(t)\right| \leqslant \|f\|_\infty \cdot \mathrm{VAR}_1(\alpha; [a,b])$;
e) If f is piecewise continuous, $\alpha \in \mathrm{BV}[a,b]$ and α and f have no common discontinuities, then f is Riemann–Stieltjes integrable;
f) If f is Riemann–Stieltjes integrable, then f and α cannot have common discontinuities;
g) If f is Riemann integrable and if α is absolutely continuous, then f is Riemann–Stieltjes integrable and $\int_a^b f(t)\, d\alpha(t) = \int_a^b f(t)\alpha'(t)\, dt$;
h) If f is Riemann–Stieltjes integrable, then $x \mapsto \int_a^x f(t)\, d\alpha(t)$ is continuous at continuity points of α;
i) If α is right-continuous and increasing, then $\mu_\alpha(s,t] := \alpha(t) - \alpha(s)$ defines a positive measure on $\mathscr{B}[a,b]$ and $\int_a^b f(t)\, d\alpha(t) = \int_{(a,b]} f(t)\, \mu_\alpha(dt)$;

The property A.36 i) implies that we can embed the Riemann–Stieltjes integral into the Lebesgue theory if $\alpha \in \mathsf{BV}$ is right-continuous. Observe that in this case we can make sure that $\alpha = \phi - \psi$ for *right-continuous* increasing functions ϕ, ψ. Then, $\mu_\alpha := \mu_\phi - \mu_\psi$ is a signed measure on $\mathscr{B}[a,b]$ with μ_ϕ, μ_ψ as in A.36 i). The integral $\int_{(a,b]} f(s)\,\mu_\alpha(ds)$ is sometimes called the *Lebesgue–Stieltjes integral*.

The integration by parts formula and the chain rule for Riemann–Stieltjes integrals are particularly simple. The chain rule is the deterministic counterpart of Itô's lemma.

A.37 Theorem (Integration by parts). *Let $\alpha, f : [a,b] \to \mathbb{R}$. If $\int_a^b f(t)\,d\alpha(t)$ exists, so does $\int_a^b \alpha(t)\,df(t)$; in this case*

$$f(b)\alpha(b) - f(a)\alpha(a) = \int_a^b f(t)\,d\alpha(t) + \int_a^b \alpha(t)\,df(t).$$

A.38 Theorem (Chain rule). *Let $\alpha \in \mathcal{C}[0,T]$ be increasing and $f \in \mathcal{C}^1[\alpha(0),\alpha(T)]$. Then*

$$f(\alpha(t)) - f(\alpha(0)) = \int_0^t f'(\alpha(s))\,d\alpha(s) = \int_{\alpha(0)}^{\alpha(t)} f'(x)\,dx \quad \textit{for all } t \leqslant T.$$

Alternatively, if β is strictly increasing and $g \in \mathcal{C}[a,b]$, then

$$\int_a^b g(s)\,d\beta(s) = \int_{\beta(a)}^{\beta(b)} g(\beta^{-1}(x))\,dx.$$

The property A.36 d) implies that $A[f] := \int_a^b f(t)\,d\alpha(t)$, $\alpha \in \mathsf{BV}$, is a continuous linear form on the Banach space $(\mathcal{C}[a,b], \|\cdot\|_\infty)$. In the already mentioned note [190] F. Riesz showed that *all* continuous linear forms are of this type.

A.39 Theorem (Riesz representation theorem). *Let $A : \mathcal{C}[a,b] \to \mathbb{R}$ be a continuous linear form, i. e. a linear map satisfying $|A[f]| \leqslant c_A \|f\|_\infty$. Then there is an $\alpha \in \mathsf{BV}[a,b]$, depending only on A, such that $A[f] = \int_a^b f(t)\,d\alpha(t)$.*

In fact, it is not even necessary to assume that the linear form is *continuous*. As soon as $f \mapsto \int_a^b f(t)\,d\alpha(t)$ is finite for every $f \in \mathcal{C}[a,b]$, then α is necessarily of bounded variation. For this we need

A.40 Theorem (Banach–Steinhaus theorem). *Let $(\mathcal{X}, \|\cdot\|_\mathcal{X})$, $(\mathcal{Y}, \|\cdot\|_\mathcal{Y})$ be two Banach spaces and $S^\Pi : \mathcal{X} \to \mathcal{Y}$ be a family of continuous linear maps indexed by Π. Then*

$$\sup_\Pi \|S^\Pi[x]\|_\mathcal{Y} < \infty,\ \forall x \in \mathcal{X} \quad \textit{implies that} \quad \sup_\Pi \underbrace{\sup_{\|x\|_\mathcal{X} \leqslant 1} \|S^\Pi[x]\|_\mathcal{Y}}_{\text{operator norm } \|S^\Pi\|_{\mathcal{X}\to\mathcal{Y}}} < \infty.$$

A.41 Corollary. *Let $\alpha : [a,b] \to \mathbb{R}$. If $A[f] := \int_a^b f(t)\,d\alpha(t)$ exists for all continuous functions $f \in \mathcal{C}[a,b]$, then $\alpha \in \mathsf{BV}[a,b]$.*

Proof. We apply Theorem A.40 to

$$(\mathcal{X}, \|\cdot\|_{\mathcal{X}}) = (\mathcal{C}[a,b], \|\cdot\|_\infty) \quad \text{and} \quad (\mathcal{Y}, \|\cdot\|_{\mathcal{Y}}) = (\mathbb{R}, |\cdot|)$$

and $S^\Pi[f] := \sum_{t_j, t_{j-1} \in \Pi} f(t_{j-1})(\alpha(t_j) - \alpha(t_{j-1}))$ where $\Pi = \{t_0 = a < t_1 < \cdots < t_n = b\}$ is a partition of $[a, b]$.

For the saw-tooth function

$$f_\Pi(s) := \begin{cases} \operatorname{sgn}(\alpha(t_j) - \alpha(t_{j-1})), & \text{if } s = t_{j-1}, j = 1, \ldots, n, \\ \text{piecewise linear}, & \text{at all other points}, \end{cases}$$

we find

$$S^\Pi[f_\Pi] = \sum_{t_j, t_{j-1} \in \Pi} |\alpha(t_j) - \alpha(t_{j-1})| \leqslant \sup_{\|f\|_\infty \leqslant 1} \left|S^\Pi[f]\right|.$$

By assumption $\lim_{|\Pi| \to 0} S^\Pi[f] = \int_a^b f(t)\, d\alpha(t) < \infty$ for all $f \in \mathcal{C}[a, b]$. In particular, $\sup_\Pi \left|S^\Pi[f]\right| < \infty$ and, by the Banach–Steinhaus theorem,

$$\operatorname{VAR}_1(\alpha; [a, b]) = \sup_\Pi S^\Pi[f_\Pi] \leqslant \sup_\Pi \sup_{\|f\|_\infty \leqslant 1} \left|S^\Pi[f]\right| < \infty. \qquad \square$$

A.42 Remark. Corollary A.41 shows that the stochastic integral $\int_0^t f(s)\, dB_s(\omega)$ cannot be defined for every $\omega \in \Omega$ as a Riemann–Stieltjes integral, if the class of admissible integrands contains all (non-random!) continuous functions f. Since for continuous integrands the Riemann–Stieltjes and the Lebesgue–Stieltjes integrals coincide, it is also not possible to use a Lebesgue–Stieltjes approach.

A.8 Some tools from analysis

A.8.1 Frostman's theorem: Hausdorff measure, capacity and energy

Throughout this section we use the measurable space $(\mathbb{R}^d, \mathscr{B}(\mathbb{R}^d))$ and denote by $\mathcal{M}^+(E)$ the family of positive Borel measures on $(E, \mathscr{B}(E))$ where $E \subset \mathbb{R}^d$. We write $|E| = \sup_{x,y \in E} |x - y|$ for the diameter of the set E, $\mathscr{H}^\alpha(E)$ for the α-dimensional Hausdorff measure and $\dim E$ for the Hausdorff dimension, see Chapter 11. The *energy (integral)* of a measure μ with respect to the kernel $x \mapsto |x|^{-\alpha}$ is

$$I_\alpha(\mu) = \iint_{E \times E} \frac{\mu(dx)\, \mu(dy)}{|x - y|^\alpha} \in [0, \infty],$$

and the *α-capacity* of a set E is

$$\operatorname{Cap}_\alpha(E) = \frac{1}{\inf_{\mu \in \mathcal{M}^+(E)} I_\alpha(\mu)} \in [0, \infty].$$

Standard references for these topics are Carleson [26], Aikawa & Essén [1] or Kahane [115, Chapter 10].

A.43 Theorem (Energy criterion. Frostman 1935). *Let μ be a probability measure on E. Then*

$$\iint\limits_{E\times E} \frac{\mu(dx)\,\mu(dy)}{|x-y|^\alpha} < \infty \implies \mathscr{H}^\alpha(E) > 0 \implies \dim E \geqslant \alpha.$$

Proof. If the double integral is finite, there is some Borel set $B \subset E$ with $\mu(B) > 0$ and

$$\int\limits_E \frac{\mu(dy)}{|x-y|^\alpha} < M \quad \text{for all } x \in B.$$

Let $(B_j)_{j\geqslant 1}$ be any cover of B by Borel sets. Since $|B_j|^\alpha \geqslant |x-y|^\alpha$ for all $x, y \in B_j$ and each $j \geqslant 1$, we get

$$\mu(B_j) \leqslant |B_j|^\alpha \int\limits_{B_j} \frac{\mu(dy)}{|x-y|^\alpha} \quad \text{for all } x \in B_j \cap B.$$

Thus,

$$0 < \frac{1}{M}\,\mu(B) \leqslant \frac{1}{M} \sum_{j\geqslant 1} \mu(B_j) \leqslant \sum_{j\geqslant 1} |B_j|^\alpha,$$

and this shows $\mathscr{H}^\alpha(E) \geqslant \mathscr{H}^\alpha(B) \geqslant M^{-1}\mu(B) > 0$.

The second implication follows directly from the definition of Hausdorff dimension. □

In fact, with a more careful analysis, one can even prove that $\mathscr{H}^\alpha(E) = \infty$, cf. [76, Theorem 4.13 (a)].

A.44 Theorem (Frostman's lemma. Frostman 1935). *Let $K \subset \mathbb{R}^d$ be compact. Then*

$$\alpha < \dim K \implies \mathscr{H}^\alpha(K) > 0 \implies \exists\, \mu \in \mathcal{M}^+(K)\ \forall \text{ balls } B \subset \mathbb{R}^d \ :\ \mu(B) \leqslant c_{\alpha,\mu,d}|B|^\alpha.$$

Proof. The first implication follows directly from the definition of Hausdorff dimension. Since K is compact, we may assume without restriction that $K \subset [0,1]^d$. If $\mathscr{H}^\alpha(K) > 0$, we have $\mathscr{H}^\alpha_\delta(K) > 0$ for some $\delta > 0$, and there is some $\gamma > 0$ such that for all δ-covers $(Q_j)_{j\geqslant 1}$ of K by cubes we have $\sum_{j\geqslant 1} |Q_j|^\alpha \geqslant \gamma > 0$.

Consider the *dyadic cubes*

$$\mathscr{Q}_n := \left\{ \bigtimes_{j=1}^d \left[p_j 2^{-n}, (p_j+1)2^{-n}\right] \ :\ p_j \in \mathbb{Z},\ j = 1, \dots, d \right\}, \qquad \mathscr{Q} := \bigcup_{n=0}^\infty \mathscr{Q}_n.$$

Fix $n \geqslant 1$ and define some measure ν_n on $\mathscr{B}(\mathbb{R}^d)$ satisfying

$$\nu_n(Q) = \begin{cases} |Q|^\alpha, & \text{for all } Q \in \mathscr{Q}_n \text{ such that } Q \cap K \neq \emptyset, \\ 0, & \text{for all } Q \in \mathscr{Q}_n \text{ such that } Q \cap K = \emptyset. \end{cases}$$

For example, $\nu_n(dx) = g_n(x)\,dx$ with the density $g_n(x) = \sum_{Q\in\mathscr{Q}_n, Q\cap K\neq\emptyset} \frac{|Q|^\alpha}{\mathrm{Leb}(Q)} \mathbb{1}_Q(x)$. Looking at the next larger dyadic cubes $Q' \in \mathscr{Q}_{n-1}$, we define a measure ν_{n-1} on $\mathscr{B}(\mathbb{R}^d)$

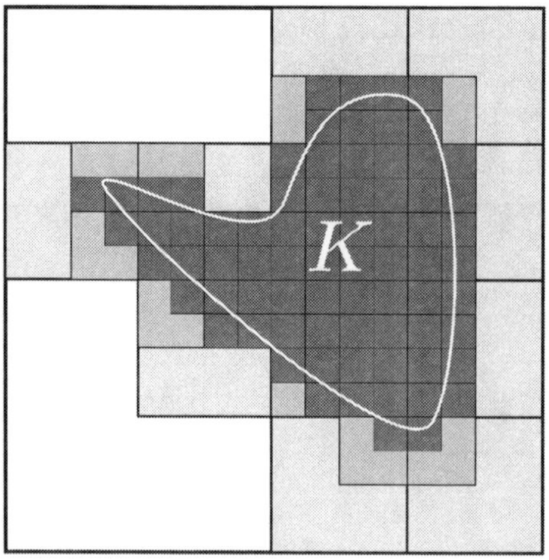

Fig. A.2. The set K is covered by dyadic cubes of different generations.

such that

$$\nu_{n-1}(Q) = \begin{cases} \nu_n(Q), & \text{for all } Q \in \mathscr{Q}_n \text{ with } Q \subset Q' \text{ and } \nu_n(Q') \leqslant |Q'|^\alpha, \\ \dfrac{|Q'|^\alpha}{\nu_n(Q')}\,\nu_n(Q), & \text{for all } Q \in \mathscr{Q}_n \text{ with } Q \subset Q' \text{ and } \nu_n(Q') > |Q'|^\alpha. \end{cases}$$

Similar to ν_n we can represent ν_{n-1} by a density with respect to Lebesgue measure.

Continuing all the way up, we get after n steps a measure ν_0 on $\mathscr{B}(\mathbb{R}^d)$. Since ν_0 depends on the initially fixed n, we write $\nu^{(n)} := \nu_0$. The thus defined measures $(\nu^{(n)})_{n\geqslant 1}$ satisfy:

a) $\nu^{(n)}(Q) \leqslant |Q|^\alpha$ for all $Q \in \mathscr{Q}_0 \cup \ldots \cup \mathscr{Q}_n$.
b) For each $x \in K$ there is some $Q \in \mathscr{Q}$ with $x \in Q$ and $\nu^{(n)}(Q) = |Q|^\alpha$.

Consider for each $x \in K$ the largest cube such that b) holds. This gives a finite covering $(Q_j)_{j\geqslant 1}$ of K by non-overlapping dyadic cubes, and we see

$$\nu^{(n)}(\mathbb{R}^d) = \sum_{j\geqslant 1} \nu^{(n)}(Q_j) = \sum_{j\geqslant 1} |Q_j|^\alpha \geqslant \gamma > 0.$$

Define $\mu^{(n)}(B) := \nu^{(n)}(B)/\nu^{(n)}(\mathbb{R}^d)$. Then

$$\mu^{(n)}(Q) \leqslant \tfrac{1}{\gamma}\,|Q|^\alpha \quad \text{for all } Q \in \mathscr{Q}_0 \cup \ldots \cup \mathscr{Q}_n$$

and $\operatorname{supp}\mu^{(n)} \subset K + \overline{\mathbb{B}}(0,\delta_n)$ where δ_n is the diameter of a cube in $\mathscr{Q}_n$.

Using the Helly–Bray theorem, we can find some weakly convergent subsequence $(\mu^{(n_j)})_{j\geqslant 1} \subset (\mu^{(n)})_{n\geqslant 1}$. Denoting the weak limit by μ we see that

$$\operatorname{supp}\mu \subset K \quad \text{and} \quad \mu(Q) \leqslant \tfrac{1}{\gamma}\,|Q|^\alpha \quad \text{for all } Q \in \mathscr{Q}.$$

Since we may cover any line segment of length h by at most 2 dyadic intervals of length 2^{-k} with $\frac{1}{2}\,2^{-k} \leqslant h < 2^{-k}$, we see that any ball $B \subset \mathbb{R}^d$ of diameter $|B| = h$ is covered by at most 2^d dyadic cubes $Q \in \mathcal{Q}_k$, whence

$$\mu(B) \leqslant 2^d \mu(Q) \leqslant 2^d \gamma^{-1} |Q|^\alpha = 2^d \gamma^{-1} \left(\sqrt{d} 2^{-k}\right)^\alpha \leqslant 2^d \gamma^{-1} \left(2\sqrt{d}\right)^\alpha |B|^\alpha. \qquad \square$$

Theorem A.44 actually holds for all analytic (i. e. Souslin) sets $A \subset \mathbb{R}^d$. For this one needs the fact that any analytic set with $\mathcal{H}^\alpha(A) > 0$ contains a closed set $F \subset A$ with $0 < \mathcal{H}^\alpha(F) < \infty$. This is a non-trivial result due to Davies [44].

A.45 Corollary (Frostman's theorem. Frostman 1935). *Let $K \subset \mathbb{R}^d$ be a compact set. Then*

$$\alpha < \dim K \implies \mathcal{H}^\alpha(K) > 0 \implies \exists \mu \in \mathcal{M}^+(K) \ \forall \beta < \alpha : \iint_{K\times K} \frac{\mu(dx)\,\mu(dy)}{|x-y|^\beta} < \infty.$$

Proof. By Frostman's lemma, Theorem A.44, there is a probability measure μ on K such that $\mu\{x : |x-y| \leqslant \rho\} \leqslant c\,\rho^\alpha$ with an absolute constant $c > 0$. For all $y \in K$ and all $\beta < \alpha$ we find

$$\begin{aligned} \int_K \frac{\mu(dx)}{|x-y|^\beta} = \int_0^\infty \mu\left\{x : |x-y|^{-\beta} \geqslant r\right\} dr &= \int_0^\infty \mu\left\{x : |x-y| \leqslant r^{-1/\beta}\right\} dr \\ &\leqslant \int_0^1 dr + \int_1^\infty \mu\left\{x : |x-y| \leqslant r^{-1/\beta}\right\} dr \\ &\leqslant 1 + c\int_1^\infty r^{-\alpha/\beta}\, dr \leqslant c(\mu,\alpha,\beta) < \infty. \end{aligned}$$

Since μ is a finite measure, we get $I_\beta(\mu) = \iint_{K\times K} |x-y|^{-\beta}\,\mu(dx)\,\mu(dy) < \infty$. $\square$

Similar to the Hausdorff dimension, Definition 11.5, one defines the *capacitary dimension* as the unique number where the capacity changes from strictly positive to zero:

$$\dim_C E := \sup\left\{\alpha \geqslant 0 : \mathrm{Cap}_\alpha E > 0\right\} = \inf\left\{\alpha \geqslant 0 : \mathrm{Cap}_\alpha E = 0\right\}.$$

The following result is an easy consequence of Frostman's theorem

A.46 Corollary. *Let $K \subset \mathbb{R}^d$ be a compact set and $0 < \beta < \alpha < d$. Then*

$$\mathcal{H}^\alpha(K) > 0 \implies \mathrm{Cap}_\beta(K) > 0 \implies \mathcal{H}^\beta(K) > 0.$$

In particular, Hausdorff and capacitary dimension coincide: $\dim K = \dim_C K$.

A.8.2 Gronwall's lemma

Gronwall's lemma is a well-known result from ordinary differential equations. It is often used to get uniqueness of the solutions of certain ODEs.

A.47 Theorem (Gronwall's lemma). *Let $u, a, b : [0, \infty) \to [0, \infty)$ be positive measurable functions satisfying the inequality*

$$u(t) \leqslant a(t) + \int_0^t b(s)u(s)\,ds \quad \text{for all } t \geqslant 0. \tag{A.25}$$

Then

$$u(t) \leqslant a(t) + \int_0^t a(s)b(s)\exp\Big(\int_s^t b(r)\,dr\Big)ds \quad \text{for all } t \geqslant 0. \tag{A.26}$$

If $a(t) = a$ and $b(t) = b$ are constants, the estimate (A.26) simplifies and reads

$$u(t) \leqslant ae^{bt} \quad \text{for all } t \geqslant 0. \tag{A.26'}$$

Proof. Set $y(t) := \int_0^t b(s)u(s)\,ds$. Since this is the primitive of $b(t)u(t)$ we can rewrite (A.25) for Lebesgue almost all $t \geqslant 0$ as

$$y'(t) - b(t)y(t) \leqslant a(t)b(t). \tag{A.27}$$

If we set $z(t) := y(t)\exp\big(-\int_0^t b(s)\,ds\big)$ and substitute this into (A.27), we arrive at

$$z'(t) \leqslant a(t)b(t)\exp\Big(-\int_0^t b(s)\,ds\Big)$$

Lebesgue almost everywhere. On the other hand, $z(0) = y(0) = 0$, hence we get by integration

$$z(t) \leqslant \int_0^t a(s)b(s)\exp\Big(-\int_0^s b(r)\,dr\Big)ds$$

or

$$y(t) \leqslant \int_0^t a(s)b(s)\exp\Big(\int_s^t b(r)\,dr\Big)ds$$

for all $t > 0$. This implies (A.26) since $u(t) \leqslant a(t) + y(t)$. □

A.8.3 Completeness of the Haar functions

Denote by $(H_n)_{n\geqslant 0}$ the Haar functions introduced in Section 3.2. We will show here that they are a complete orthonormal system in $L^2([0,1], ds)$, hence an orthonormal basis. For some background on Hilbert spaces and orthonormal bases see, e. g. [204, Chapters 21, 22].

A.48 Theorem. *The Haar functions are an orthonormal basis of $L^2([0,1], ds)$.*

Proof. We show that any $f \in L^2$ with $\langle f, H_n\rangle_{L^2} = \int_0^1 f(s)H_n(s)\,ds = 0$ for all $n \geqslant 0$ is Lebesgue almost everywhere zero. Indeed, if $\langle f, H_n\rangle_{L^2} = 0$ for all $n \geqslant 0$ we find by induction that

$$\int_{k/2^j}^{(k+1)/2^j} f(s)\,ds = 0 \quad \text{for all } k = 0, 1, \dots, 2^j - 1,\ j \geqslant 0. \tag{A.28}$$

Let us quickly verify the induction step. Assume that (A.28) is true for some j; fix $k \in \{0, 1, \dots 2^j - 1\}$. The interval $I = \left[\frac{k}{2^j}, \frac{k+1}{2^j}\right)$ supports the Haar function H_{2^j+k}, and on the left half of the interval, $I_1 = \left[\frac{k}{2^j}, \frac{2k+1}{2^{j+1}}\right)$, it has the value $2^{j/2}$, on the right half, $I_2 = \left[\frac{2k+1}{2^{j+1}}, \frac{k+1}{2^j}\right)$, it is $-2^{j/2}$. By assumption

$$\langle f, H_{2^j+k}\rangle_{L^2} = 0, \quad \text{thus,} \quad \int_{I_1} f(s)\,ds = \int_{I_2} f(s)\,ds.$$

From the induction assumption (A.28) we know, however, that

$$\int_{I_1 \cup I_2} f(s)\,ds = 0, \quad \text{thus,} \quad \int_{I_1} f(s)\,ds = -\int_{I_2} f(s)\,ds.$$

This is only possible if $\int_{I_1} f(s)\,ds = \int_{I_2} f(s)\,ds = 0$. Since I_1, I_2 are generic intervals of the next refinement, the induction is complete.

Because of the relations (A.28), the primitive $F(t) := \int_0^t f(s)\,ds$ is zero for all dyadic points $t = k2^{-j}$, $j \geqslant 0$, $k = 0, 1, \dots, 2^j$; since the dyadic numbers are dense in $[0, 1]$ and since $F(t)$ is continuous, we see that $F(t) \equiv 0$ and, therefore, $f \equiv 0$ almost everywhere. □

Bibliography

[1] Aikawa, H., Essén, M.: *Potential Theory – Selected Topics*. Springer, Lecture Notes in Mathematics **1633**, Berlin 1996.

[2] Arnold, L.: *Stochastic Differential Equations: Theory and Applications*. Wiley-Interscience, New York 1974.

[3] Arnold, L.: *Random Dynamical Systems*. Springer, Berlin 2009.

[4] Ash, R. B.: *Information Theory*. Interscience Publishers, New York 1965. Reprinted by: Dover, New York 1990.

[5] Asmussen, S., Glynn, P. W.: *Stochastic Simulation: Algorithms and Analysis*. Springer, Berlin 2007.

[6] Bachelier, L.: Théorie de la spéculation. *Ann. Sci. Ec. Norm. Super.* **17** (1900) 21–86. English translations in [39] pp. 17–78 and [8].

[7] Bachelier, L.: *Calcul des Probabilités*. Gauthier-Villars, Paris 1912. Reprinted by Éditions Jacques Gabay, Paris 1992.

[8] Bachelier, L., Davis, M. (ed.), Etheridge, A. (ed.): *Louis Bachelier's Theory of Speculation: The Origins of Modern Finance*. Princeton University Press, Princeton (NJ) 2006.

[9] Bass, R.: *Probabilistic Techniques in Analysis*. Springer, New York 1995.

[10] Bass, R.: *Diffusions and Elliptic Operators*. Springer, New York 1998.

[11] Barlow, M.: One dimensional stochastic differential equations with no strong solution. *J. London Math. Soc.* **26** (1982) 335–347.

[12] Bell, D. R.: Stochastic differential equations and hypoelliptic operators. In: Rao, M. M. (ed.): *Real and Stochastic Analysis. New Perspectives*. Birkhäuser, Boston 2004, 9–42.

[13] Berg, C.: On the potential operators associated with a semigroup. *Studia Math.* **51** (1974) 111–113.

[14] Billingsley, P.: *Convergence of Probability Measures*. Wiley, New York 1968. 2nd edn. Wiley, New York 1999.

[15] Billingsley, P.: *Probability and Measure (3rd edn)*. Wiley, New York 1995.

[16] Blumenthal, R. M.: An extended Markov property. *Trans. Am. Math. Soc.* (1957) **82** 52–72.

[17] Blumenthal, R. M., Getoor, R. K.: Some theorems on stable processes. *Trans. Am. Math. Soc.* **95** (1960) 263–273.

[18] Blumenthal, R. M., Getoor, R. K.: *Markov Processes and Potential Theory*. Academic Press, New York 1968. Reprinted by Dover, Mineola (NY) 2007.

[19] Breiman, L.: On the tail behavior of sums of independent random variables. *Z. Wahrscheinlichkeitstheorie verw. Geb.* **9** (1967) 20–25.

[20] Brown, R.: On the existence of active molecules in organic and inorganic bodies. *The Philosophical Magazine and Annals of Philosophy (new Series)* **4** (1828) 161–173.

[21] Brown, R.: Additional remarks on active molecules. *The Philosophical Magazine and Annals of Philosophy (new Series)* **6** (1828) 161–166.

[22] Bru, B., Yor, M.: Comments on the life and mathematical legacy of Wolfgang Doeblin. *Finance and Stochastics* **6** (2002) 3–47.

[23] Burkholder, D. L.: Martingale Transforms. *Ann. Math. Statist.* (1966) **37** 1494–1504. Reprinted in [25].

[24] Burkholder, D. L., Davis, B. J., Gundy, R. F.: Integral inequalities for convex functions of operators on martingales. In: Le Cam, L. *et al.* (eds.): *Proceedings of the 6th Berkeley Symposium on Mathematical Statistics and Probability, Vol. II*. University of California Press, Berkely (CA) 1972, 223–240. Reprinted in [25].

[25] Burkholder, D. L., Davis, B. (ed.), Song, R. (ed.): *Selected Works of Donald L. Burkholder*. Springer, New York 2011.

[26] Carleson, L.: *Selected Problems on Exceptional Sets*. Van Nostrand, Princeton (NJ) 1967.

[27] Chentsov, N. N.: Weak convergence of stochastic processes whose trajectories have no discontinuities of the second kind and the "heuristic" approach to the Kolmogorov-Smirnov tests, *Theor. Probab. Appl.* **1** (1956) 140–144.

[28] Cherny, A. S., Engelbert, H.-J.: *Singular Stochastic Differential Equations*. Springer, Lecture Notes Math. **1858**, Berlin 2005.

[29] Chung, K. L.: On the maximum partial sums of sequences of independent random variables. *Trans. Am. Math. Soc.* **61** (1948) 205–233.

[30] Chung, K. L.: Excursions in Brownian Motion. *Ark. Mat.* **14** (1976) 155–177.

[31] Chung, K. L.: *Lectures from Markov Processes to Brownian Motion*. Springer, New York 1982. Enlarged and reprinted as [32].

[32] Chung, K. L., Walsh, J. B.: *Markov Processes, Brownian Motion, and Time Symmetry (Second Edition)*. Springer, Berlin 2005. This is the 2nd edn. of [31].

[33] Chung, K. L., Zhao, Z.: *From Brownian Motion to Schrödinger's Equation*. Springer, Berlin 1995.

[34] Ciesielski, Z.: On Haar functions and on the Schauder basis of the space $C_{\langle 0,1\rangle}$. *Bull. Acad. Polon. Sci. Sér. Math. Astronom. Phys.* **7** (1959) 227–232.

[35] Ciesielski, Z.: *Lectures on Brownian Motion, Heat Conduction and Potential Theory*. Aarhus Matematisk Institut, Lecture Notes Ser. **3**, Aarhus 1966.

[36] Ciesielski, Z., Kerkyacharian, G., Roynette, B.: Quelques espaces fonctionnels associés à des processus gaussienes. *Studia Math.* **107** (1993) 171–204.

[37] Cohen, L.: The history of noise. *IEEE Signal Processing Magazine* (November 2005) 20–45.

[38] Cont, R., Tankov, P.: *Financial Modelling With Jump Processes*. Chapman & Hall/CRC, Boca Raton 2004.

[39] Cootner, P. (ed.): *The Random Character of Stock Market Prices*. The MIT Press, Cambridge (MA) 1964.

[40] Csörgő, M., Révész, P.: *Strong Approximations in Probability and Statistics*. Academic Press, New York 1981.

[41] Dambis, K. E.: On the decomposition of continuous submartingales. *Theory Probab. Appl.* **10** (1965) 401–410.

[42] Dautray, R., Lions, J.-L.: *Mathematical Analysis and Numerical Methods for Science and Technology. Volume 1: Physical Origins and Classical Methods*. Springer, Berlin 1990.

[43] DaPrato, G., Zabczyk, J.: *Stochastic Equations in Infinite Dimensions*. Cambridge University Press, Cambridge 1992.

[44] Davies, R. O.: Subsets of finite measure in analytic sets. *Indag. Math.* **14** (1952) 488–489.

[45] Dembo, A., Zeitouni, O.: *Large Deviations Techniques and Applications (2nd ed.)*. Springer, New York 1993.

[46] Deuschel, J.-D., Stroock, D. W.: *Large Deviations*. Academic Press 1989.

[47] Doeblin, W.: Sur l'équation de Kolmogoroff. *C. R. Acad. Sci. Paris* **331** (2000) 1059–1102. (First printing of the *pli cacheté* deposited with the *Académie des Sciences* in 1940)

[48] Doléans-Dade, C.: On the existence and unicity of solutions of stochastic integral equations. *Z. Wahrscheinlichkeitstheorie verw. Geb.* **36** (1976) 93–101.

[49] Donsker, M.: An invariance principle for certain probability limit theorems. *Mem. Am. Math. Soc.* **6** (1951) 12 pp. (paper no. 4)

[50] Doob, J. L.: *Stochastic Processes*. Interscience Publishers, New York 1953. Reprinted by Wiley, New York 1990.

[51] Doob, J. L.: A probability approach to the heat equation. *Trans. Am. Math. Soc.* **80** (1955) 216–280.

[52] Doob, J. L.: *Classical Potential Theory and Its Probabilistic Counterpart*. Springer, Berlin 1984.

[53] Doss, H.: Liens entre équations différentielles stochastiques et ordinaires. *Ann. Inst. H. Poincaré, Sér. B* **13** (1977) 99–125.

[54] Dubins, L. E., Schwarz, G.: On continuous martingales. *Proc. Natl. Acad. Sci. USA* **53** (1965) 913–916.

[55] Dudley, R. M.: Sample functions of the Gaussian process. *Ann. Probab.* **1** (1973) 66–103. Reprinted in [57].

[56] Dudley, R. M.: *Uniform Central Limit Theorems*. Cambridge University Press, Cambridge 1999.

[57] Dudley, R. M., Giné, E. (ed.) et al.: *Selected Works of R. M. Dudley*. Springer, New York 2010.

[58] Dudley, R. M., Norvaiša, R.: *An Introduction to p-variation and Young integrals*. MaPhySto Lecture Notes **1**, Aarhus 1998.

[59] Dudley, R. M., Norvaiša, R.: *Differentiability of Six Operators on Nonsmooth Functions and p-Variation*. Springer, Lecture Notes in Mathematics **1703**, Berlin 1999.

[60] Durrett, R.: *Brownian Motion and Martingales in Analysis*. Wadsworth Advanced Books & Software, Belmont (CA) 1984.

[61] Durrett, R.: *Probability: Theory and Examples (4th edn)*. Cambridge University Press, Cambridge 2010.

[62] Dynkin, E. B.: Criteria of continuity and of absence of discontinuities of the second kind for trajectories of a Markov stochastic process. *Izv. Akad. Nauk SSSR, Ser. Matem.* **16 (6)** (1952) 563–572. (In Russian)

[63] Dynkin, E. B.: Infinitesimal operators of Markov processes. *Theor. Probab. Appl.* **1** (1956) 34–54. Reprinted in [67]

[64] Dynkin, E. B.: *Die Grundlagen der Theorie der Markoffschen Prozesse*. Springer, Berlin 1961. English translation: *Theory of Markov Processes*, Dover, Mineola (NY) 2006.

[65] Dynkin, E. B.: *Markov Processes (2 vols.)*. Springer, Berlin 1965.

[66] Dynkin, E. B., Yushkevich, A. A.: *Markov Processes. Theorems and Problems*. Plenum Press, New York 1969. Russian Original, German Translation: *Sätze und Aufgaben über Markoffsche Prozesse*, Springer 1969.

[67] Dynkin, E. B., Yushkevich, A. A. (ed.) et al.: *Selected Papers of E. B. Dynkin with Commentary*. American Mathematical Society, Providence (RI) 2000.

[68] Einstein, A.: Über die von der molekularkinetischen Theorie der Wärme geforderte Bewegung von in ruhenden Flüssigkeiten suspendierten Teilchen. *Ann. Phys.* **17** (1905) 549–560. Reprinted in [84], English translation in [69].

[69] Einstein, A., Fürth, R. (ed.): *Investigations on the Theory of the Brownian movement*. Dover, New York 1956.

[70] Eliot, G.: *Middlemarch*. William Blackwood and Sons, Edinburgh 1871.

[71] Engelbert, H.-J., Cherny, A. S.: *Singular Stochastic Differential Equations*. Springer, Lecture Notes in Mathematics **1858**, Berlin 2004.

[72] Ethier, S. N., Kurtz, T. G.: *Markov Processes. Characterization and Convergence*. Wiley, New York 1986.

[73] Evans, L. C.: *Partial Differential Equations*. American Mathematical Society, Providence (RI) 1998.

[74] Evans, L. C., Gariepy, R. F.: *Measure Theory and Fine Properties of Functions*. CRC Press, Boca Raton (FL) 1992.

[75] Falconer, K.: *The Geometry of Fractal Sets*. Cambridge University Press, Cambridge 1985.

[76] Falconer, K.: *Fractal Geometry. Mathematical Foundations and Applications. Second Edition*. Wiley, Chichester 2003.

[77] Feyel, D., de La Pradelle, A.: On fractional Brownian processes. *Potential Anal.* **10** (1999) 273–288.

[78] Fokker, A. D.: Die mittlere Energie rotierender elektrischer Dipole im Strahlungsfeld. *Ann. d. Phys.* **43** (1914) 810–820.

[79] Freedman, D.: *Brownian Motion and Diffusion*. Holden-Day, San Francisco (CA) 1971. Reprinted by Springer, New York 1982.

[80] Freidlin, M.: *Functional Integration and Partial Differential Equations*. Princeton University Press, Ann. Math. Studies **109**, Princeton (NJ) 1985.

[81] Freidlin, M. I., Wentzell, A. D.: *Random Perturbations of Dynamical Systems (Second Edition)*. Springer, New York 1998.

[82] Friedman, A.: *Partial Differential Equations of Parabolic Type*. Prentice-Hall, Englewood Cliffs (NJ) 1964. Reprinted by Dover, Mineola (NY) 2008.

[83] Frostman, O.: *Potentiel d'équilibre et capacité des ensembles avec quelques applications a la théorie des fonctions*. Meddelanden från Lunds Universitets Matematiska Seminarium Band **3**, C. W. K. Gleerup, Lund 1953.

[84] Fürth, R. (ed.), Einstein, A., Smoluchowski, M. von: *Untersuchungen über die Theorie der Brownschen Bewegung – Abhandlung über die Brownsche Bewegung und verwandte Erscheinungen*. Harri Deutsch, Ostwalds Klassiker Bd. **199** (reprint of the separate volumes 199 and 207), Frankfurt am Main 1997. English translation [69].

[85] Fukushima, M., Osima, Y., Takeda, M.: *Dirichlet Forms and Symmetric Markov Processes (2nd edn)*. De Gruyter, Berlin 2011.

[86] Garsia, A. M., Rodemich, E., Rumsey, H.: A real variable lemma and the continuity of paths of some Gaussian processes. *Indiana Univ. Math. J.* **20** (1970) 565–578.

[87] Gikhman, I. I., Skorokhod, A. W.: *Stochastic Differential Equations*. Springer, Berlin 1972.

[88] Girsanov, I. V.: On transforming a certain class of stochastic processes by absolutely continuous substitution of measures. *Theor. Probab. Appl.* **5** (1960) 285–301.

[89] Gradshteyn, I. S., Ryzhik, I. M.: *Table of Integrals, Series, and Products (4th corrected and enlarged edn)*. Academic Press, San Diego 1980.

[90] Gunther (Günter), N. M.: *La Théorie du Potentiel et ses Applications aux Problèmes Fondamentaux de la Physique Mathématique*. Gauthier-Villars, Paris 1934.

[91] Günter, N. M.: *Die Potentialtheorie und ihre Anwendung auf Grundaufgaben der mathematischen Physik*. B. G. Teubner, Leipzig 1957. Extended second edition of [90].

[92] Hausdorff, F.: Dimension und äußeres Maß. *Math. Annalen* **79** (1919) 157–179. Reprinted (with English commentary) in [93].

[93] Hausdorff, F., Chatterji, S. D. (ed.), Remmert, R. (ed.), Scharlau, W. (ed.): *Felix Hausdorff – Gesammelte Werke. Band IV: Analysis, Algebra und Zahlentheorie*. Springer, Berlin 2001.

[94] Helms, L. L.: *Introduction to Potential Theory*. Wiley–Interscience, New York 1969.

[95] Helms, L. L.: *Potential Theory*. Springer, London 2009. Extended version of [94].

[96] Hewitt, E., Stromberg, K.: *Real and Abstract Analysis*. Springer, New York 1965.

[97] Hille, E.: *Functional Analysis and Semi-groups*. Am. Math. Society, New York 1948. (2nd ed. 1957 jointly with R. S. Phillips)

[98] Hunt, G. H.: Markoff processes and potentials I, II. *Illinois J. Math.* **1** (1957) 44–93 and 316–369.

[99] Ikeda, N., Nakao, S., Yamato, Y.: A class of approximations of Brownian motion. *Publ. RIMS Kyoto Univ.* **13** (1977) 285–300.

[100] Ikeda, N., Watanabe, S.: *Stochastic Differential Equations and Diffusion Processes (Second Edn)*. North-Holland, Amsterdam 1989.

[101] Itô, K.: Differential equations determining a Markoff process (in Japanese). *J. Pan-Japan Math. Coll.* **1077** (1942). Translation reprinted in [107].

[102] Itô, K.: Semigroups in Probability Theory. In: *Functional Analysis and Related Topics, 1991. Proceedings of the International Conference in Memory of Professor Kôsaku Yosida held at RIMS, Kyoto University, Japan, July 29 - Aug. 2, 1991.* Springer, Lecture Notes in Mathematics **1540**, Berlin 1993, 69–83.

[103] Itô, K.: *Stochastic Processes. Lectures Given at Aarhus University.* Springer, Berlin 2004. Reprint (with commentaries) of Itô's 1969 lectures, Aarhus Lecture Notes Series **16**.

[104] Itô, K.: *Essentials of Stochastic Processes.* Am. Math. Society, Providence (RI) 2006.

[105] Itô, K., McKean, H. P.: *Diffusion Processes and Their Sample Paths*, Springer, Berlin 1965.

[106] Itô, K., Nisio, M.: On the convergence of sums of independent Banach space valued random variables. *Osaka J. Math.* **5** (1968) 35–48.

[107] Itô, K., Stroock, D. W. (ed.), Varadhan, S. R. S. (ed.): *Kiyosi Itô. Selected Papers.* Springer, New York 1987.

[108] Jacod, J., Protter, P.: *Probability Essentials.* Springer, Berlin 2000.

[109] John, F.: *Partial Differential Equations (4th edn).* Springer, Berlin 1982.

[110] Kac, M.: On distributions of certain Wiener functionals. *Trans. Am. Math. Soc.* (1949) **65** 1–13. Reprinted in [112].

[111] Kac, M.: On some connections between probability theory and differential and integral equations. In: Neyman, J. (ed.): *Proceedings of the Second Berkely Symposium on Mathematical Statistics and Probability, August 1950.* University of California Press, Berkeley 1951, 189–225. Reprinted in [112].

[112] Kac, M., Baclawski, K. (ed.), Donsker, M. D. (ed.): *Mark Kac: Probability, Number Theory, and Statistical Physics.* The MIT Press, Cambridge (MA) 1979.

[113] Kahane, J.-P.: Sur l'irrégularité locale du mouvement brownien. *C. R. Acad. Sci, Paris Sér. A* **278** (1974) 331–333.

[114] Kahane, J.-P.: Sur les zéros et les instants de ralentissement du mouvement brownien. *C. R. Acad. Sci, Paris Sér. A* **282** (1976) 431–433.

[115] Kahane, J.-P: *Some random series on functions (2nd edn).* Cambridge University Press, Cambridge 1985.

[116] Kakutani, S.: On Brownian motions in n-space. *Proc. Imperial Acad. Japan* **20** (1944) 648–652. Reprinted in [118].

[117] Kakutani, S.: Two-dimensional Brownian motion and harmonic functions. *Proc. Imperial Acad. Japan* **20** (1944) 706–714. Reprinted in [118].

[118] Kakutani, S., Kallman, R. R. (ed.): *Shizuo Kakutani – Selected Papers (2 vols.).* Birkhäuser, Boston 1986.

[119] Karatzas, I., Shreve, S. E.: *Brownian Motion and Stochastic Calculus.* Springer, New York 1988.

[120] Kaufman, R.: Une propriété métrique du mouvement brownien. *C. R. Acad. Sci, Paris Sér. A* **268** (1969) 727–728.

[121] Kaufman, R.: Temps locaux and dimensions. *C. R. Acad. Sci, Paris Sér. I Math.* **300** (1985) 281–282.

[122] Kaufman, R.: Dimensional properties of one-dimensional Brownian motion. *Ann. Probab.* **17** (1989) 189–193.

[123] Kellogg, O. D.: *Foundations of Modern Potential Theory.* Springer, Berlin 1929. Reprinted by Dover, New York 1956.

[124] Kestelman, H.: *Modern Theories of Integration.* Dover, New York 1960 (2nd edn).

[125] Khintchine, A.: *Asymptotische Gesetze der Wahrscheinlichkeitsrechnung.* Springer, Ergeb. Math. Grenzgeb. Bd. **2**, Heft 4, Berlin 1933.

[126] Kinney, J. R.: Continuity criteria of sample functions of Markov processes. *Trans. Am. Math. Soc.* **74** (1953) 280–302.

[127] Kloeden, P. E., Platen, E.: *Numerical Solution of Stochastic Differential Equations*. Springer, Berlin 1992.

[128] Kolomogoroff, A. N., Über die analytischen Methoden in der Wahrscheinlichkeitsrechnung. *Math. Ann.* **104** (1931) 415–458. English translation in [130].

[129] Kolmogoroff, A. N., *Grundbegriffe der Wahrscheinlichkeitsrechnung*. Springer, Ergeb. Math. Grenzgeb. Bd. **2**, Heft 3, Berlin 1933.

[130] Kolmogorov, A. N., Shiryayev, A. N. (ed.): *Selected Works of A. N. Kolmogorov (3 vols.)*. Kluwer, Dordrecht 1993.

[131] Kowalski, E.: Bernstein polynomials and Brownian Motion. *Am. Math. Monthly* **113** (2006) 865–886.

[132] Krylov, N. V.: *Introduction to the Theory of Diffusion Processes*. American Mathematical Society, Providence (RI) 1995.

[133] Krylov, N. V.: *Introduction to the Theory of Random Processes*. American Mathematical Society, Providence (RI) 2002.

[134] Kunita, H.: *Stochastic Flows and Stochastic Differential Equations*. Cambridge University Press, Cambridge 1990.

[135] Kunita, H.: Stochastic differential equations based on Lévy processes and stochastic flows of diffeomorphisms. In: Rao, M. M. (ed.): *Real and Stochastic Analysis. New Perspectives*. Birkhäuser, Boston 2004, 305–373.

[136] Kunita, H., Watanabe, S.: On square integrable martingales. *Nagoya Math. J.* **30** (1967) 209–245.

[137] Kuo, H.-H.: *Gaussian Measures in Banach Spaces*. Springer, Lecture Notes in Mathematics **463**, Berlin 1975.

[138] Lebesgue, H.: Sur des cas d'impossibilité du problème du Dirichlet ordinaire. *C. R. Séances Soc. Math. France* **41** (1913). Reprinted in [139, Vol. IV, pp. 125–126]

[139] Lebesgue, H.: *Œvres Scientifiques En Cinq Volumes (5 vols.)*. L'enseignement mathématique & Institut de Mathématiques Université de Genève, Geneva 1973.

[140] L'Ecuyer, P.: Uniform random number generation. *Ann. Oper. Res.* **53** (1994) 77–120.

[141] L'Ecuyer, P., Panneton, F.: Fast random number generators based on linear recurrences modulo 2: overview and comparison. In: Kuhl, M. E., Steiger, N. M., Armstrong, F. B., Joines, J. A. (eds.): *Proceedings of the 37th Winter Simulation Conference, Orlando, FL, 2005. ACM 2005*. IEEE, New York 2006, 110–119.

[142] Lévy, P.: *Théorie de l'Addition des Variables Aléatoires*. Gauthier-Villars, Paris 1937.

[143] Lévy, P.: Sur certains processus stochastiques homogènes. *Compositio Math.* **7** (1939) 283–339. Reprinted in [146, vol. 4].

[144] Lévy, P.: Le mouvement brownien plan. *Am. J. Math.* **62** (1940) 487–550. Reprinted in [146, vol. 5].

[145] Lévy, P.: *Processus Stochastiques et Mouvement Brownien*. Gauthier-Villars, Paris 1948.

[146] Lévy, P., Dugué, D. (ed.): *Œvres de Paul Lévy (6 vols.)*. Gauthier-Villars, Paris 1973–1980.

[147] Lifshits, M. A.: *Gaussian Random Functions*. Kluwer, Dordrecht 1995.

[148] Lumer, G., Phillips, R. S.: Dissipative operators in a Banach space. *Pacific J. Math.* **11** (1961) 679–698.

[149] MacDonald, D. K. C.: *Noise and Fluctuations*. Wiley, New York 1962. Reprinted by Dover, Mineola (NY) 2006.

[150] Malliavin, P.: Stochastic calculus of variation and hypoelliptic operators. In: Itô, K. (ed.): *Proceedings Intnl. Symposium on SDEs, Kyoto 1976*, Wiley, New York 1978, 195–263.

[151] Malliavin, P.: *Integration and Probability*. Springer, Berlin 1995.

[152] Malliavin, P.: *Stochastic Analysis*. Springer, Berlin 1997.

[153] Mandelbrot, B. B.: *The Fractal Geometry of Nature*. Freeman, New York 1977.

[154] Marcus, M., Rosen, J.: *Markov Processes, Gaussian Processes, and Local Times*. Cambridge Universtiy Press, Cambridge 2006.

[155] Maruyama, G.: On the transition probability functions of the Markov process. *Natural Sci. Rep., Ochanomizu Univ.* **5** (1954) 10–20. Reprinted in [156].

[156] Maruyama, G., Ikeda, N. (ed.), Tanaka, H. (ed.): *Gisiro Maruyama. Selected Papers*. Kaigai Publications, Tokyo 1988.

[157] Mattila, P.: *Geometry of Sets and Measures in Euclidean Spaces. Fractals and Rectifiability*. Cambridge University Press, Cambridge 1995.

[158] Mazo, R. M.: *Brownian Motion. Fluctuations, Dynamics and Applications*. Oxford University Press, Oxford 2002.

[159] McKean, H. P.: Hausdorff–Besicovitch dimension of Brownian motion paths. *Duke Math. J.* **22** (1955) 229–234.

[160] Meyer, P. A.: A decomposition theorem for supermartingales. *Illinois J. Math.* **6** (1962) 193–205.

[161] Meyer, P. A.: Decompositions of supermartingales. The uniqueness theorem. *Illinois J. Math.* **7** (1963) 1–17.

[162] Mörters, P., Peres, Y.: *Brownian Motion*. Cambridge University Press, Cambridge 2010.

[163] Motoo, M.: Proof of the law of iterated logarithm through diffusion equation. *Ann. Inst. Math. Statist.* **10** (1958) 21–28.

[164] Nelson, E.: *Dynamical Theories of Brownian Motion*. Princeton University Press, Princeton (NJ) 1967.

[165] Novikov, A. A.: On an identity for stochastic integrals. *Theory Probab. Appl.* **16** (1972) 717–720.

[166] Obłój, J.: The Skorokhod embedding problem and its offspring. *Probab. Surveys* **1** (2004) 321–390.

[167] Olver, F. W. J. et al. (eds.): *NIST Handbook of Mathematical Functions*. Cambridge University Press, Cambridge 2010. (Free online access: http://dlmf.nist.gov/)

[168] Orey, S.: Probabilistic methods in partial differential equations. In: Littman, W. (ed.): *Studies in Partial Differential Equations*. The Mathematical Association of America, 1982, 143–205.

[169] Orey, S., Taylor, S. J.: How often on a Brownian path does the law of iterated logarithm fail? *Proc. London Math. Soc.* **28** (1974) 174–192.

[170] Paley, R. E. A. C., Wiener, N., Zygmund, A.: Notes on random functions. *Math. Z.* **37** (1933) 647–668.

[171] Paley, R. E. A. C., Wiener, N.: *Fourier Transforms in the Complex Domain*. American Mathematical Society, Providence (RI) 1934.

[172] Partzsch, L.: *Vorlesungen zum eindimensionalen Wienerschen Prozeß*. Teubner, Texte zur Mathematik Bd. **66**, Leipzig 1984.

[173] Pazy, A.: *Semigroups of Linear Operators and Applications to Partial Differential Equations*. Springer, New York 1983.

[174] Peetre, J.: Rectification à l'article "Une caractérisation abstraite des opérateurs différentiels". *Math. Scand.* **8** (1960) 116–120.

[175] Petrowsky, I.: Zur ersten Randwertaufgabe der Wärmeleitungsgleichung. *Compositio Math.* **1** (1935) 383–419. English translation in [176].

[176] Petrowsky, I. G., Oleinik, O. A. (ed.): *I. G. Petrowsky. Selected Works. Part II: Differential Equations and Probability Theory*. Gordon and Breach, Amsterdam 1996.

[177] Perrin, J.: Mouvement brownien et réalité moléculaire. *Ann. Chimie Physique (8ème sér.)* **18** (1909) 5–114.

[178] Perrin, J.: *Les Atomes*. Librairie Félix Alcan, Paris 1913.

[179] Pinsky, M. A.: *Introduction to Fourier Analysis and Wavelets*. Brooks/Cole, Pacific Grove (CA) 2002.

[180] Planck, M.: Über einen Satz der statistischen Dynamik und seine Erweiterung in der Quantentheorie. *Sitzungsber. Königl. Preuß. Akad. Wiss. phys.-math. Kl.* **10.5** (1917) 324–341.

[181] Poincaré, H.: Sur les équations aux dérivées partielles de la physique mathématique. *Am. J. Math.* **12** (1890) 211–294.

[182] Port, S., Stone, S.: *Brownian Motion and Classical Potential Theory*. Academic Press, New York 1978.

[183] Prevôt, C., Röckner, M.: *A Concise Course on Stochastic Partial Differential Equations*. Springer, Lecture Notes in Mathematics **1905**, Berlin 2007.

[184] Protter, P. E.: *Stochastic Integration and Differential Equations (Second Edn.)*. Springer, Berlin 2004.

[185] Rao, M.: *Brownian Motion and Classical Potential Theory*. Aarhus Universitet Matematisk Institut, Lecture Notes Series **47**, Aarhus 1977.

[186] Rényi, A.: *Probability Theory*. North Holland, Amsterdam 1970. Reprinted by Dover, Mineola (NY) 2007.

[187] Reuter, G. E. H.: Denumerable Markov processes and the associated contraction semigroups on *l*. *Acta Math.* **97** (1957) 1–46.

[188] Révész, P.: *Random Walk in Random and Non-Random Environments (2nd ed.)*. World Scientific, New Jersey 2005.

[189] Revuz, D., Yor, M.: *Continuous Martingales and Brownian Motion*. Springer, Berlin 2005 (3rd edn.).

[190] Riesz, F.: Sur les opérations fonctionnelles linéaires. *C. R. Acad. Sci. Paris* **149** (1909) 974–977. Reprinted in [191, pp. 400–403].

[191] Riesz, F., Császár, Á. (ed.): *Frédéric Riesz – Œuvres complètes*. Gauthier-Villars, Paris 1960.

[192] Riesz, F., Sz.-Nagy, B.: *Leçons d'Analyse Fonctionnnelle*. Akadémiai Kiadó, Budapest 1952. English translation: *Functional Analysis*. Dover, Mineola (NY) 1990.

[193] Ripley, B. D.: *Stochastic Simulation*. Wiley, Series in Probab. Stat., Hoboken (NJ) 1987.

[194] Rogers, C. A.: *Hausdorff Measures*. Cambridge University Press, Cambridge 1970. Unaltered 2nd edition 1998 in the Cambridge Mathematical Library with a new introduction by K. Falconer.

[195] Rogers, L. C. G., Williams, D.: *Diffusions, Markov Processes, and Martingales (2nd edn, 2 vols.)*. Wiley, Chichester 1987 & 1994. Reprinted by Cambridge University Press, Cambridge 2000.

[196] Rubinstein, R. Y., Kroese, D. P.: *Simulation and the Monte Carlo Method*. Wiley, Hoboken (NJ) 2008.

[197] Rudin, W.: *Principles of Mathematical Analysis*. McGraw-Hill, Auckland 1976 (3rd edn).

[198] Rudin, W.: *Real and Complex Analysis*. McGraw-Hill, New York 1986 (3rd edn).

[199] Rudin, W.: *Functional Analysis*. McGraw-Hill, New York 1991 (2nd edn).

[200] Sawyer, S.: The Skorokhod Representation. *Rocky Mountain J. Math.* **4** (1974) 579–596.

[201] Schilder, M.: Some asymptotic formulae for Wiener integrals. *Trans. Am. Math. Soc.* **125** (1966) 63–85.

[202] Slutsky, E. E.: Qualche proposizione relativa alla teoria delle funzioni aleatorie. *Giorn. Ist. Ital. Attuari* **8** (1937) 183–199.

[203] Schilling, R. L.: Sobolev embedding for stochastic processes. *Expo. Math.* **18** (2000) 239–242.

[204] Schilling, R. L.: *Measures, Integrals and Martingales*. Cambridge University Press, Cambridge 2005.